Atmosphere–Ocean Dynamics

ADRIAN E. GILL

*Department of Applied Mathematics
and Theoretical Physics
University of Cambridge
Cambridge, England*

ACADEMIC PRESS, INC.
Harcourt Brace Jovanovich, Publishers
San Diego   New York   Berkeley   Boston
London   Sydney   Tokyo   Toronto

ACADEMIC PRESS, INC.
San Diego, Caliifornia 92101

United Kingdom Edition published by
ACADEMIC PRESS, INC. (LONDON) LTD.
24/28 Oval Road, London NW1 7DX

Library of Congress Cataloging in Publication Data

Gill, Adrian E.
    Atmosphere-ocean dynamics.

    (International geophysics series)
    Includes bibliographical references and index.
    1. Ocean-atmosphere interaction.   I. Title.
II. Series.
GC190.G54 1982   551.47   82-8704
ISBN 0-12-283520-4   AACR2
ISBN 0-12-283522-0 (pbk.)

PRINTED IN THE UNITED STATES OF AMERICA
88 89 90 91   9 8 7 6 5 4 3

Atmosphere–Ocean Dynamics

This is Volume 30 in

INTERNATIONAL GEOPHYSICS SERIES

A series of monographs and textbooks

Edited by WILLIAM L. DONN

A complete list of titles in this series is available from the Publishers upon request.

Contents

Chapter Seven   Effects of Rotation

Chapter Eight   Gravity Waves in a Rotating Fluid

Chapter Nine **Forced Motion**

Chapter Ten **Effects of Side Boundaries**

Chapter Eleven **The Tropics**

Chapter Twelve **Mid-latitudes**

Chapter Thirteen **Instabilities, Fronts, and the General Circulation**

Appendix One **Units and Their SI Equivalents**

Appendix Two **Useful Values**

Appendix Three **Properties of Seawater**

Preface

A systematic, unifying approach to the dynamics of the ocean and atmosphere is given in this book, with emphasis on the larger-scale motions (from a few kilometers to global scale). The foundations of the subject (the equations of state and dynamical equations) are covered in some detail, so that students with training in mathematics should find it a self-contained text. Knowledge of fluid mechanics is helpful but not essential. Simple mathematical models are used to demonstrate the fundamental dynamical principles with plentiful illustrations from field and laboratory. In fact, the search for suitable mathematical models during the eight years of writing stimulated several of my research papers written during that time.

Undergraduates in meteorology and oceanography should find the text a useful introduction to the dynamics of both air and sea. Having grown out of a graduate course, it is equally suitable for more advanced students, and material can be selected from many sections to give a well-structured program. For instance, my graduate course begins with a brief introduction from Chapter 1, skips to Section 5.6 and, by the third lecture, is focusing on rotation effects as covered in Chapter 7. Stratification effects are brought in later with material from Chapters 6 and 8, and then selected sections from the remaining chapters are used. Thus there is scope for considerable flexibility. Elementary courses would use material from the earlier chapters, whereas more advanced courses could be based on in-depth studies of later chapters.

Researchers should find the book attractive, not only for its systematic treatment of the dynamics, but also because of its extensive bibliography and index, the appendixes, and the many useful diagrams and formulas. The treatment of many topics is novel, and considerable historical information is incorporated to make the book more readable and interesting. In fact, I became quite absorbed in pursuing historical aspects during the writing period.

General Description of Contents

The two introductory chapters give an overall picture of how the circulations of both atmosphere and ocean are ultimately driven by the sun's energy. A somewhat novel treatment of basic thermodynamics and hydrostatics follows in Chapter 3, both ocean and atmosphere being discussed together. The fundamental equations for moving fluids are derived in Chapter 4, with particular reference to air containing moisture and water containing dissolved salts. Various types of energy are introduced, and the use of a rotating frame of reference is dealt with.

The fundamental aim is to understand the circulations of the atmosphere and ocean and the observed distributions of physical quantities such as temperature. The temperature distribution can be viewed (following Halley) as the result of a "competition" between the *sun*, which tries to warm the tropics more than the poles (and so create horizontal contrasts), and *gravity*, which tries to remove horizontal contrasts and arrange for warmer fluid to overlie colder fluid. This "competition" is complicated by such effects as the rotation of the earth, the variation of the angle between gravity and the rotation axis (the beta effect), and contrasts between the properties of air and water. Accordingly, we start with as simple a situation as possible and proceed by adding complicating effects one at a time.

The first step is taken in Chapter 5, where we consider adjustment under gravity of a homogeneous layer of fluid in the absence of rotation and external forcing effects. The results are directly applicable to phenomena such as seiches and tides in lakes, estuaries, and narrow seas. This chapter also introduces the important "hydrostatic approximation," which leads to the "shallow-water" equations. Effects of density stratification are then incorporated in Chapter 6, beginning with the two-layer system, like the oil-over-water arrangement that so intrigued Benjamin Franklin in 1762. Several aspects of wave motion are also introduced in Chapters 5 and 6. For instance, group velocity, introduced in Chapter 5 for surface waves, is applied in Chapter 6 to internal gravity waves in a continuously stratified fluid. Waves produced at a horizontal boundary, and possible refraction, reflection, or absorption in the fluid above, are also discussed.

Chapter 7, perhaps the most important in the whole book, introduces effects that are due to the earth's rotation. Although Laplace included these in his tidal equations in 1778, and Kelvin investigated wave motions in a rotating fluid a hundred years later, some of the fundamental ideas were developed relatively recently by Rossby in the 1930s. The "Rossby adjustment problem" brings out many facets of the behavior of rotating fluids, such as the tendency to attain "geostrophic equilibrium," the significance of "potential vorticity," and the importance of the length scale known as the Rossby radius of deformation.

Wave motion in a stratified rotating fluid is examined in Chapter 8 with applications to flow of air over hills and mountains. Propagation in a slowly varying medium, ray-tracing techniques, the internal wave spectrum in the ocean, and effects of waves on the mean flow are also examined. Chapter 9 introduces forcing by effects such as wind action, tide-producing forces, and solar heating. Inertial oscillations in the ocean surface layer are an example of forced motion, and these are dynamically related to the nocturnal jet in the atmosphere. Hurricanes and the ocean's response to storms are also considered.

Phenomena associated with lateral boundaries are treated in Chapter 10. Dynamical studies that stem from Kelvin's work in 1879 can explain the main features of the very destructive North Sea Surge of 1953. Coastal upwelling, which is of great importance to fisheries, can be studied using similar analysis. Other classes of coastally trapped waves are also discussed. Equatorially trapped waves, considered in Chapter 11, have similar dynamics and are used to introduce both the beta effect and the mid-latitude beta-plane approximation for studying quasi-geostrophic flow. The tropical circulations of the atmosphere and ocean are also dealt with in this chapter.

In extratropical latitudes (Chapter 12), slow small-amplitude adjustments take place by means of planetary waves. These can be used, for instance, to describe how the ocean response to the wind has a highly asymmetric character with strong western boundary currents like the Gulf Stream and the Kuroshio. They are also useful for understanding the stationary wave patterns in the atmosphere. The omega equations, which provide a useful diagnostic tool, are also discussed.

The mid-latitude atmosphere is dominated by cyclones and anticyclones, which result from an instability of the basic wind distribution. Models illustrating how the potential energy of the zonal flow is converted into the kinetic energy of the cyclone systems are studied in Chapter 13. Fronts that develop in evolving cyclones and eddies in the ocean are also discussed. The book concludes with a global view of the atmosphere–ocean system.

Acknowledgments

I should like to thank the many colleagues who made comments on the text and helped with the diagrams, especially David Anderson, Brian Hoskins, John Johnson, Peter Killworth, and Michael McIntyre. Support and encouragement from Henry Stommel, Raymond Hide, and the late Jule Charney were much appreciated. Naomi Coyle's help in typing and organizing material was invaluable, and I am grateful to Julian Smith for producing the computer-drawn diagrams. Thanks are due also to Michael Davey, William Hsieh, and Roxana Wajsowicz for their valued assistance at proof stage.

ADRIAN GILL

Chapter One

How the Ocean–Atmosphere System Is Driven

1.1 Introduction

This book is about winds, currents, and the distribution of heat in the atmosphere and ocean. Since these are due to the sun, this first chapter looks at some of the essential processes that determine how the atmosphere and ocean respond to radiation from the sun. Ideally, one would like to be able to deduce this response in all its details from a knowledge of the appropriate properties of the earth and of its ocean and atmosphere, but this is not a simple matter. The nearest approach to a solution of this problem is by means of numerical models, but these still rely to some extent on observations of the real system, e.g., for determining the effects of processes (like those associated with individual clouds) that have a scale small compared with the grid used in the model.

The aim of the numerical models is to include the effects of *all* processes that play a significant part in determining the response of the ocean–atmosphere system. The aim of this chapter, on the other hand, is to consider only the most basic processes and to show how an equilibrium state can be reached. One such basic process is the *absorption of radiation* by certain gases (principally water vapor, carbon dioxide, and ozone), and so the "greenhouse" effect is discussed. The density field that results from radiation processes acting in isolation is not in dynamical equilibrium, because air near the ground is so warm that it is lighter than the air above. Consequently, vertical convection takes place and stirs up the lower atmosphere. Calculations of the equilibrium established when convective and radiative processes are both active is discussed in Section 1.5. These calculations, however, neglect variations in the horizontal,

1

which are, of course, extremely important since they are responsible for the winds and currents that are the main subject of this book. A brief discussion of the effects of horizontal variations is given in Section 1.6. Finally, since radiation is the source of energy for the atmosphere–ocean system, variations in the radiative input are discussed in Section 1.7.

1.2   The Amount of Energy Received by the Earth

Energy from the sun is received in the form of *radiation*, nearly all the energy being at wavelengths between 0.2 and 4 μm. About 40% is in the visible part of the spectrum (0.4–0.67 μm). The average energy flux from the sun at the mean radius of the earth is called the solar constant S and has the value (Willson R. C. 1984)

$$S = 1.368 \quad \text{kW m}^{-2}. \tag{1.2.1}$$

(A great variety of units is used for energy flux. The relation between these is given in Appendix 1.) In other words, a 1-m-diameter dish in space could collect enough energy from the sun to run a 1-kW electric heater! Since the earth's orbit is elliptical rather than circular, the actual energy received varies seasonally by $\pm 3.5\%$ (Kondratyev, 1969, Section 1.1), the maximum amount being received at the beginning of January.

The total energy received from the sun per unit time is

$$\pi R^2 S, \tag{1.2.2}$$

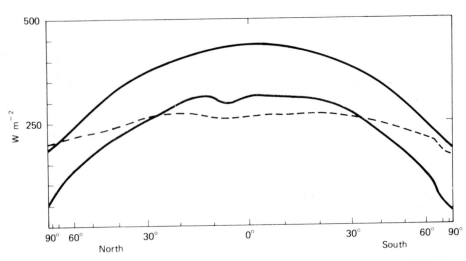

Fig. 1.1. The radiation balance of the earth. The upper solid curve shows the average flux of solar energy reaching the outer atmosphere. The lower solid curve shows the average amount of solar energy absorbed; the dashed line shows the average amount of outgoing radiation. The lower curves are average values from satellite measurements between June 1974 and February 1978, and are taken from Volume 2 of Winston et al. (1979). Values are in watts per square meter. The horizontal scale is such that the spacing between latitudes is proportional to the area of the earth's surface between them, i.e., is linear in the sine of the latitude.

where R is the radius of the earth. Since the area of the earth's surface is $4\pi R^2$, the average amount of energy received per unit area of the earth's surface per unit time is

$$\tfrac{1}{4}S = 344 \quad \text{W m}^{-2}. \tag{1.2.3}$$

If the earth's axis were not tilted, the average flux received would vary from $\pi^{-1}S$ at the equator to zero at the poles. However, the tilt of the earth (23.5°) results in seasonal variations in the distribution of the flux received. When account is taken of these variations, the average flux received in 1 yr is found to vary with latitude as shown in Fig. 1.1.

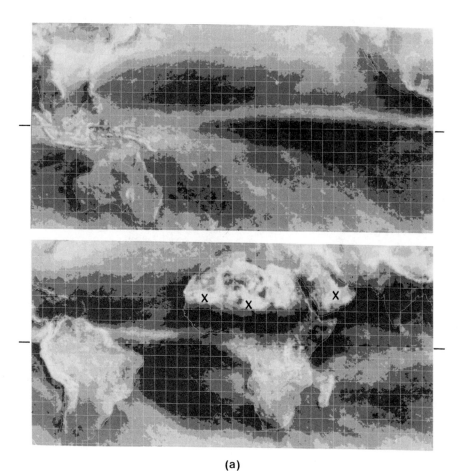

(a)

Fig. 1.2.   The geographical distribution of reflectivity for (a) January 1967–1970 and (b) July 1969–1970, as determined from satellite observations. Most of the bright areas in the figure are characterized by persistent cloudiness and relatively heavy precipitation. However, the following exceptions should be noted: areas indicated by X's denote desert regions where the earth's surface is highly reflective and areas indicated by Y's denote regions of persistent low, nonprecipitating cloud decks. Tick marks along the side denote the position of the equator: the Mercator grid lines are spaced at intervals of 5° of latitude and longitude. [From U.S. Air Force and U.S. Department of Commerce, Global Atlas of Relative Cloud Cover, 1967–1970, Washington, D.C., 1971.]

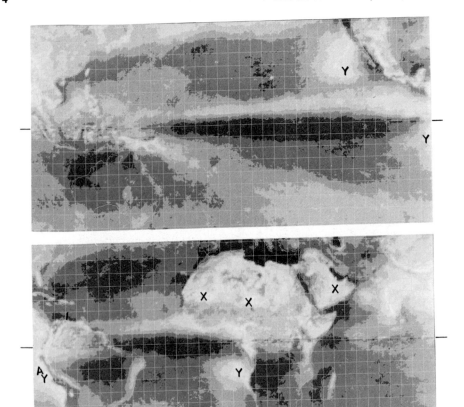

(b)

Fig. 1.2 *(Continued)*

Not all the energy impinging on the earth is absorbed. A fraction $\bar{\alpha}$ is reflected or scattered, so the average flux actually absorbed is

$$\tfrac{1}{4}(1 - \bar{\alpha})S = 240 \quad \text{W m}^{-2}. \tag{1.2.4}$$

The amount reflected or scattered is about 100 W m^{-2} at all latitudes, as shown in Fig. 1.1. (There is no obvious reason that this amount should vary so little with latitude.) The number $\bar{\alpha}$ is called the *albedo* of the earth and has a value (Stephens *et al.*, 1981) of about

$$\bar{\alpha} = 0.3. \tag{1.2.5}$$

Similarly, the albedo α can be defined for a particular place and particular time as the fraction of the impinging radiation that is reflected or scattered. The reflected light is the light by which the earth may be photographed from space, and such photographs (see Fig. 1.2, which is effectively the result of combining many such photographs to give the mean reflectivity) show that the albedo can vary enormously with such factors as the amount of cloud, and whether the ground is covered by ice or

ANNUAL ALBEDO

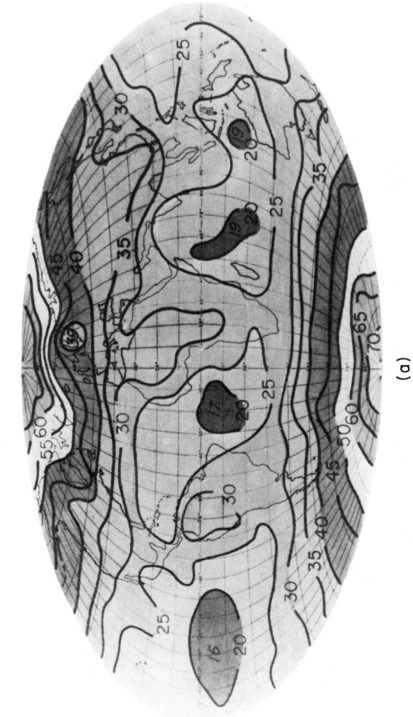

(a)

Fig. 1.3. (a) The average albedo obtained from a composite of 48 months of satellite data obtained between 1964 and 1977. [From Stephens *et al.* (1981, Fig. 6).] (b) The minimum albedo of the earth from Nimbus 3 satellite measurements in 1969–1970. [From Raschke *et al.* (1973, Fig. 23).]

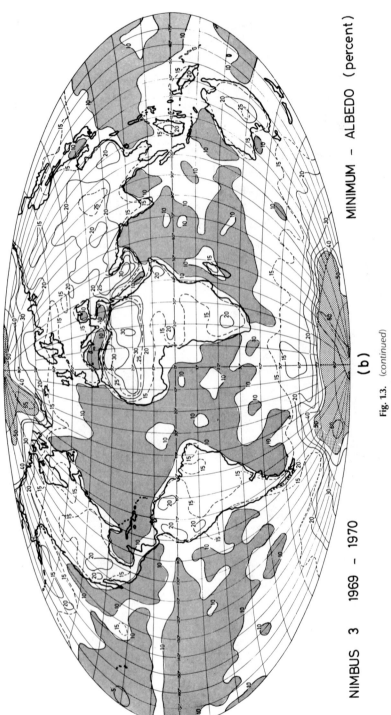

NIMBUS 3   1969 – 1970   MINIMUM – ALBEDO (percent)

(b)   (continued)

Fig. 1.3.   (continued)

snow. Mars, with no cloud cover, has about *half* the albedo of the earth, whereas Venus, with total cloud cover, has about *twice* the albedo of the earth. A quantitative estimate of the degree to which clouds, ice, and snow affect the albedo can be obtained from satellite measurements (Fig. 1.3). The minimum albedo is presumably close to the value in the absence of clouds and of snow-free conditions where these occur. On land, the value is usually about 0.15, with higher values in desert regions (0.2–0.3) and in icy regions, reaching 0.6 in parts of the Antarctic. Comparison of the minimum albedo with the average albedo shows the effect of clouds. For instance, most of the ocean within 40° of the equator has minimum albedo below 0.1, but the average albedo is normally between 0.15 and 0.3. It is clear from these figures that the factors that determine albedo are very important in determining the energy balance of the earth.

1.3 Radiative Equilibrium Models

Since the ocean–atmosphere system is driven by the sun's radiation, it is important to know how radiation is affected by the atmosphere and ocean. Detailed discussion may be found in books such as those of Goody (1964), Kondratyev (1969), and Paltridge and Platt (1976). Only the most basic elements will be discussed here.

To begin with, consider the equilibrium that would be established if the earth had no fluid envelope. The surface would reflect a fraction $\bar{\alpha}$ of the incoming radiation and absorb the remainder. The absorption of energy would cause the surface to warm up until it radiated to space as much energy as it absorbed. When the surface reaches temperature T, the amount of energy E radiated per unit time is given by Stefan's law

$$E = \sigma T^4,$$ (1.3.1)

where

$$\sigma = 5.7 \times 10^{-8} \quad \text{W m}^{-2}\,\text{K}^{-4}.$$ (1.3.2)

For the radiation actually absorbed by the earth (see Fig. 1.1), such an equilibrium would be achieved when the temperature at the equator reached 270 K, the temperature at the South Pole 150 K, and the temperature at the North Pole 170 K. In fact the earth's surface is much warmer, and the contrast in temperature between the equator and the poles is much less. The difference from the observed surface temperature must be due to the existence of the fluid cover of the earth. This can affect the equilibrium reached in two ways. First, radiation can be absorbed within the atmosphere itself. Second, the atmosphere and ocean can carry heat from one area to another, thereby affecting the balance. In this section, the first effect will be considered in isolation from the second. In subsequent sections, the effect of fluid motion on the equilibrium will be discussed. This fluid motion consists of winds, ocean currents, etc., which will be the main concern of this book.

The radiative equilibrium that would be established in the absence of fluid motion has been calculated by Möller and Manabe (1961), and is discussed by Goody (1964, Chapter 8). The average temperature profile thus obtained is shown by the solid line in Fig. 1.4. In some ways the left-hand version of the figure is more appropriate

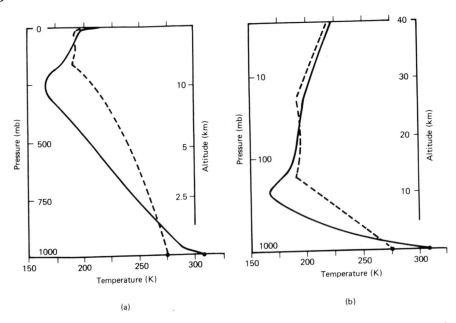

Fig. 1.4. The radiative equilibrium solution (solid line) corresponding to the observed distribution of atmospheric absorbers at 35°N in April, the observed annual average insolation for the whole atmosphere, and no clouds. The dashed line shows the effect of convective adjustment to a constant lapse rate of 6.5 K km^{-1}. In (a) the curves are drawn with a scale linear in pressure, i.e., equal intervals correspond to equal masses of atmosphere. In (b) the scale is linear in altitude. [From Manabe and Strickler (1964, Fig. 4).]

because it gives equal weight to equal masses of air. In the lower 70% (by mass) of the atmosphere, the main physical factor responsible for the equilibrium reached is the absorption of radiation by the water vapor present in the atmosphere. For their calculations, Möller and Manabe used the observed distribution of water vapor with height. At higher levels, other absorbers such as carbon dioxide and ozone become important. Figure 1.4 shows that the presence of the atmosphere results in much higher ground temperatures than would otherwise be achieved. This is due to the "greenhouse" effect, which will be discussed in Section 1.4.

1.4   The Greenhouse Effect

The radiative equilibrium solution shown in Fig. 1.4 has much higher ground temperatures than would exist in the absence of the atmosphere. This is caused by the "greenhouse" effect, the principle of which can be explained as follows. Consider a greenhouse formed by placing a horizontal sheet of glass above the ground as shown in Fig. 1.5. The glass used is transparent to radiation with wavelengths below 4 μm, but partially absorbs radiation of longer wavelengths. Suppose the glass and ground are initially cold, and then a downward flux I of solar radiation is "switched on." This radiation will pass through the glass unattenuated and be absorbed by the ground.

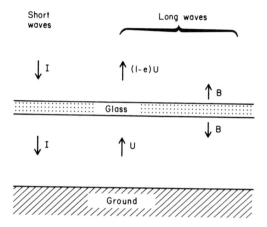

Fig. 1.5.   The greenhouse effect. The glass is transparent to short-wave radiation, the net downward flux of which is I. The balancing upward flux of long-wave radiation from the ground is U, a fraction e of this being absorbed by the glass. This warms the glass, causing it to emit a flux B in both directions.

The ground will warm up to a temperature T_g and emit long-wave radiation with an upward flux U given by Stefan's law:

$$U = \sigma T_g^4. \tag{1.4.1}$$

Practically all the radiation emitted at temperatures typical of the atmosphere has wavelengths *above* 4 μm (the range is 4–100 μm), so a fraction e of this radiation will be absorbed by the glass. Thus the glass will also warm up and emit radiation. Suppose the flux emitted in each direction is B.

Equilibrium will be reached when the upward fluxes balance the downward fluxes, i.e., when

$$I = (1 - e)U + B = U - B. \tag{1.4.2}$$

Solving (1.4.1) and (1.4.2), the result for the ground temperature is

$$\sigma T_g^4 = U = I/(1 - e/2). \tag{1.4.3}$$

Thus T_g is higher (by up to 19%) than it would be in the absence ($e = 0$) of the glass. This is the principle on which a greenhouse operates.

The effect can be most easily understood in the extreme case of glass that absorbs all the long-wave radiation ($e = 1$). Then (Fig. 1.5) $I = B$, which implies that the glass reaches the same temperature that the ground would have in the absence of glass. Since the underside of the glass is at the same temperature, it radiates a downward flux B of long-wave radiation downward, so the ground receives a total flux of $I + B = 2I$. Thus by Stefan's law the ground reaches a temperature that is higher than in the absence of glass by a factor $2^{1/4} = 1.19$. For other nonzero values of e, the ground still receives a back radiation flux B *in addition to* the short-wave flux I, so it reaches a higher temperature than it would otherwise.

In the atmosphere, the absorbing material is distributed continuously in the vertical rather than being confined to a thin sheet. Generalization of the above ideas

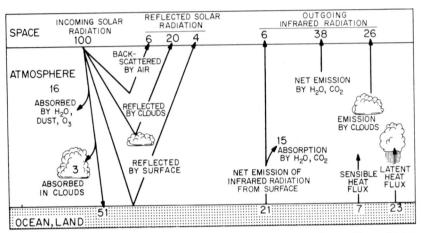

Fig. 1.6.   Radiation balance for the atmosphere. [Adapted from "Understanding Climatic Change," U.S. National Academy of Sciences, Washington, D.C., 1975, p. 14, and used with permission.]

to this case is straightforward however (Goody, 1964, Section 8.4; Chamberlain, 1978, Section 1.2), and gives temperature profiles for the lower atmosphere that are similar to those of Möller and Manabe. More accurate calculations require the radiative energy to be divided up into many wavebands rather than just two (i.e., "long" and "short" waves), and to take account of the absorption in each band separately. Also, reflection and scattering must be allowed for. This depends on the distribution and albedo of clouds and on the albedo of the underlying surface.

An estimate of the radiation balance for the atmosphere is summarized in Fig. 1.6. Setting the incident flux at 100 units, the reflected and scattered flux of short-wave radiation is $100\bar{\alpha} \approx 30$ units. This leaves 70 units of net downward flux of short-wave radiation at the top of the atmosphere, of which 19 units are absorbed in the atmosphere, leaving only 51 units to be absorbed at the surface. There is also a large amount [London and Sasamori (1971) estimate 98 units] of long-wave radiation absorbed at the surface, this representing back radiation from the atmosphere (it is possible for the back radiation to exceed the incident radiation, as a generalization of Fig. 1.5 to several sheets of glass can readily show). The net surface emission (excess of upward over downward radiation) of long-wave radiation is 21 units, the remaining upward flux of 30 units being by convection. The upward flux at the top of the atmosphere is 70 units, as required to balance the short-wave radiation received. The mean surface temperature is that corresponding to the $98 + 51 = 149$ units of radiated energy flux at the ground rather than that corresponding to the 70 units emitted at the top of the atmosphere. The latter flux can be more closely identified with a temperature at "cloud-top" height.

1.5   Effects of Convection

The radiative equilibrium solution was described in Section 1.3 as the solution that would be obtained in the absence of fluid motion. This statement is not strictly true, however, because the radiative equilibrium solution is based on the *observed* distribution of water vapor. This distribution is not predetermined, but is the *result* of a balance that involves fluid motion.

To see how fluid motion can affect the balance, consider an atmosphere that, at some initial time, contained no water vapor, but was in radiative equilibrium. If the atmosphere absorbed no radiation at all, the ground would warm up as in the absence of an atmosphere (see Section 1.3), but the air above would remain cold. Although the system would be in radiative equilibrium, it would not be in *dynamic* equilibrium because the air warmed by contact with the surface could not remain below the cold air above without *convection* occurring, as it does in a kettle full of water that is heated from below. The vigorous motion produced carries not only heat up into the atmosphere, but also *water vapor* produced by evaporation at the surface. The water vapor then affects the radiative balance because of its radiation-absorbing properties, so the final equilibrium depends on a balance between radiative and convective effects and is called *radiative–convective equilibrium*.

Whether or not convection will occur depends on the "lapse" rate, i.e., the rate at which the temperature of the atmosphere decreases with height. Convection will only occur when the lapse rate exceeds a certain value. This value can be calculated by considering the temperature changes of a parcel of air that moves up or down "adiabatically," i.e., without exchanging heat with the air outside the parcel. As such a parcel rises, the pressure falls, the parcel expands, and thus its temperature falls. The rate at which the temperature falls with height, due to expansion, is called the *dry adiabatic lapse rate* and has a value of about 10 K/km. If the temperature of the surroundings fell off more quickly with height, a rising parcel would find itself warmer than its surroundings, and therefore would continue to rise under its own buoyancy. In other words, the situation would not be a stable one, and so convection would occur. Convection carries heat upward and thus will reduce the lapse rate until it falls to the equilibrium value, for then convection can no longer occur. Another way of expressing the same ideas is in terms of potential energy. When the lapse rate exceeds the adiabatic value, the potential energy can be reduced by moving parcels adiabatically to different levels. Thus *energy is released* and is used to drive the convection.

If the atmosphere contained only small amounts of water vapor, convection would only occur if the *dry* adiabatic lapse rate were exceeded. In practice, the situation is complicated by the fact that air at a given temperature and pressure can only hold a certain amount of water vapor. The amount of water vapor relative to this saturation value is called the *relative humidity*. When the relative humidity reaches 100%, water droplets condense out of the air, thereby forming clouds. The condensed water ultimately returns to the earth's surface as precipitation.

This hydrological cycle affects the energy balance of the atmosphere in a number of important ways. First, clouds have an important effect on the *total* amount of energy absorbed by the atmosphere because they reflect and scatter a significant amount of the incoming radiation (see Section 1.2). Second, the radiation-absorbing properties of water vapor are important in determining the temperature of the lower atmosphere, as discussed in Section 1.3. Third, cooling takes place upon evaporation because of the *latent heat* required. This heat is released back into the atmosphere when condensation takes place in clouds. The heat transferred by this means is, on average, about 75% of the convective transport (see Fig. 1.6).

The release of latent heat in clouds also affects the conditions under which convection can take place. The amount of water vapor a parcel of air rising adiabatically can hold decreases with height. Thus if the parcel is already saturated with water vapor, latent heat will be released as the parcel rises, so the rate of decrease of temperature with height will be less than for dry air. The rate of decrease with height is called the *moist adiabatic lapse rate* and has a value that depends on the temperature and pressure. In the lower atmosphere, the value is about 4 deg km^{-1} at 20°C and 5 deg km^{-1} at 10°C [for precise values, see List (1951, Table 79)]. The appropriate lapse rate may also be different if ice is formed instead of liquid water (List, 1951, Table 80). A fuller discussion is given in Section 3.8.

The moist adiabatic lapse rate is appropriate for ascending air, but for descending air the story is different. The amount of water vapor a parcel of air can hold *increases* as the parcel descends, so the parcel is always unsaturated and the dry adiabatic lapse rate is appropriate. Thus in a convecting atmosphere, potential energy may be released where the air is ascending, whereas work is being done against gravity where the air is descending. [For a discussion of convection and models of convection, see Haltiner (1971, Chapter 10) and Holton (1979, Chapter 12).]

Another consequence of the nature of moist convection is the distribution of relative humidity in the atmosphere. The mean value must lie between the 100% of the moist air in rising regions and the lower values of the descending regions. A rough approximation to the observed mean distribution (Manabe and Wetherald, 1967) is a relative humidity that decreases linearly with pressure from 77% at the ground to zero at the top of the atmosphere. The *relative* humidity does not change very much from one season to another, whereas the actual amount of water vapor present varies a great deal.

A problem in modeling the atmosphere is to find a satisfactory way to represent the effects of convection without modeling details of the ascending and descending parcels of air. *Radiative–convective* models represent the effects of convection in a very simple way. First, they ignore horizontal variations, so that the temperature and other quantities are functions only of altitude (or, equivalently, of pressure). Distributions of the radiation-absorbing gases, carbon dioxide and ozone, of clouds, and of either relative humidity or absolute humidity are fixed, as is the downward flux of short-wave radiation at the top of the atmosphere. An initial temperature distribution is allowed to adjust toward equilibrium, taking account not only of radiative fluxes but also of convective fluxes. Convection is assumed to occur only when the radiative fluxes are tending to increase the lapse rate above a certain critical value. Then an opposing convective flux is introduced that redistributes (but does not add or remove) heat in such a way as to keep the lapse rate at the critical value. The difficulty lies in the choice of the critical value. Usually this is simply chosen to be the observed mean lapse rate of the lower atmosphere, namely, 6.5 deg km^{-1}. The result of such a calculation (Manabe and Strickler, 1964) is shown in Fig. 1.4 and gives quite a good approximation to the observed mean temperature profile. As such, it is an improvement over the pure radiative equilibrium model, but its limitations should not be forgotten.

1.6 Effects of Horizontal Gradients

In Section 1.5 it was seen that the large *vertical* temperature gradients that would be produced by radiation acting in isolation result in convection that tends to reduce these gradients. In a similar way, the variations with latitude of the absorbed radiative flux (Fig. 1.1) would lead to large *horizontal* temperature gradients if radiation acted in isolation. Again fluid motion takes place that tends to reduce these gradients. The nature of these motions depends on dynamical processes, which will be the subject of subsequent chapters.

Intuitively, one might expect the nonuniform heating of the atmosphere to cause rising motion in the tropics and descending motion at higher latitudes. Halley (1686) and Hadley (1735) proposed this type of circulation, which is now known as a Hadley cell (see Section 2.3). A similar circulation might be expected to occur in the ocean, so that the excess heat received in the tropics would be transported poleward in both atmosphere and ocean.

The circulation (in the meridional plane) that actually occurs is known quantitatively (but with limited accuracy) for the atmosphere from observations and is shown in Fig. 1.7. By comparison, the meridional circulation in the ocean is very poorly known, but estimates have been made that at least give an order of magnitude. A brief description of the atmospheric part of the circulation is as follows. The Hadley cell is confined to the tropics. Moist air from the trade wind zone, where evaporation exceeds precipitation, is drawn into the areas of rising motion, which, because they are wet and cloudy, show up as regions of high reflectivity in Fig. 1.2. Important regions of rising motion are over Indonesia and the Amazon and Congo basins. Over the Atlantic and Pacific Oceans, the rising motions tends to be concentrated in a fairly narrow band called the Inter-Tropical Convergence Zone (ITCZ), usually found between 5 and 10° to the north of the equator. It can be seen very clearly as a band of high reflectivity in Fig. 1.2. The regions of descending air are dry, and include in particular the desert regions (marked by X's in Fig. 1.2), which are found between latitudes 20° and 30°. These show up as regions of high albedo over land in Fig. 1.3. Where the descent is over cold ocean, low nonprecipitating cloud decks (marked Y in Fig. 1.2) are often found.

In mid-latitudes, the picture is quite different. Because of the rotation of the earth, the motion produced by the horizontal density gradients is mainly east–west, and there is relatively little meridional circulation. (The observed velocity and temperature distribution is shown in Fig. 7.9.) However, the situation is not a stable one, and large transient disturbances (which appear as cyclones and anticyclones on the weather map) develop. These disturbances are very effective at transporting energy poleward.

The effectiveness of fluid motion in reducing horizontal gradients can be judged from a comparison of the two lower curves in Fig. 1.1. The solid curve shows the variation with latitude of the absorbed flux of radiative energy. In a pure radiative equilibrium (or a radiative–convective equilibrium), the outgoing radiation would be equal to the absorbed radiation at all latitudes. In practice, the outgoing flux of

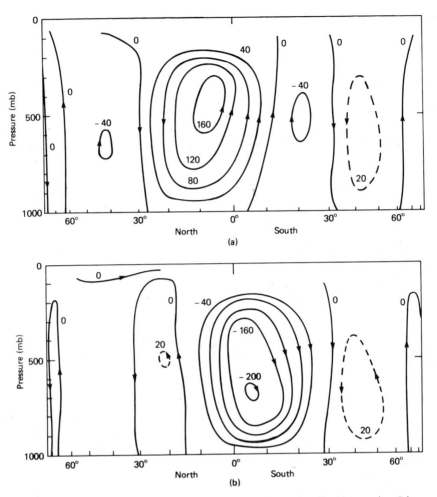

Fig. 1.7. Streamlines of the mean meridional mass flux in the atmosphere for (a) December–February and (b) June–August. Units are megatons per second (Mt s^{-1} = 10^9 kg s^{-1}). The horizontal scale is such that the spacing between latitudes is proportional to the area of the earth's surface between them, i.e., is linear in the sine of the latitude. [Adapted from Newell *et al.* (1972, Vol. 1, p. 45).]

radiative energy, shown by the dashed line in Fig. 1.1, is much more uniform, its departures from the average flux being about one third of those for the absorbed flux. From the difference between the two curves, the amount of energy that must be transported across each circle of latitude by fluid motion can be calculated. The curve so obtained for the northern hemisphere is shown in Fig. 1.8. This curve can be compared with the one for the observed transport of energy by the atmosphere (Oort, 1971; Vonder Haar and Oort, 1973). The difference between the two curves (the shaded region in Fig. 1.8) provides an estimate of the energy transport by the ocean. According to these results, ocean and atmosphere are equally important in transporting energy, the atmosphere being most important at 50°N and the ocean most important at 20°N. There is, however, considerable uncertainty in the measurements,

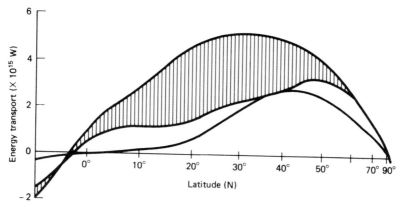

Fig. 1.8.   The northward transport of energy (in units of petawatt = 10^{15} W) as a function of latitude. The outer curve is the net transport deduced from radiation measurements. The white area is the part transported by the atmosphere and the shaded area the part transported by the ocean. The lower curve denotes the part of the atmospheric transport due to transient eddies and is the mean of the monthly values from Oort (1971, Table 3). The horizontal scale is such that the spacing between latitudes is proportional to the area of the earth's surface between them, i.e., is linear in the sine of the latitude. [From Vonder Haar and Oort (1973).]

and probable errors are estimated by Vonder Haar and Oort (1973). For instance, the probable error in the transport of energy by the ocean at 20°N is about 70%.

In calculating the energy transport by the atmosphere from observations, a distinction can be made between the energy transported by the mean (time-averaged) circulation and the energy transported by transient motions. If this calculation is done for each month in turn and the results are averaged, the curve in the unshaded part of Fig. 1.8 is obtained. In latitudes where the transport by the atmosphere is important, the transient motions account for most of the transport. This observation is the basis for simple equilibrium models [e.g., Stone (1972); see also Held and Suarez (1978)] in which the radiative heat flux is balanced not by small-scale convection, as in radiative–convective equilibrium models, but by energy fluxes due to *large-scale* transient motions (such as cyclones and anticyclones). These motions transport heat vertically as well as horizontally [see Palmén and Newton (1969, Chapter 2)], so calculations of both vertical and horizontal gradients can be made.

The method of estimating the transports due to the large-scale transient motions is beyond the scope of the present chapter, but the concept is important. The structure of the atmosphere and ocean depends on the motions driven by radiation and their effectiveness in redistributing heat. If the effect of the dominant energy-transporting mechanism can be estimated in some simple way, one hopes that reasonable estimates of basic features such as the mean horizontal and vertical temperature gradients of the atmosphere can be obtained.

1.7   Variability in Radiative Driving of the Earth

Since the present state of the ocean and atmosphere is a result of their response to the radiation received from the sun, one would like to know what variability there is in this driving. The *total* amount of radiation incident on the earth *in 1 year* depends only on the output of radiation from the sun, which is measured by the solar constant S, whose present value is given by (1.2.1). Measurements since the 1920s (Drummond, 1970) show no variations larger than the probable measurement errors, so S cannot have varied more than 1 or 2% in that time. Thus the hypothesis that S is constant, as suggested by the name "solar constant," is consistent with observations to date, although other possibilities are not ruled out.

The amount of radiation incident at a *particular point* on the earth does, however, vary enormously between day and night and from season to season, and these variations are of obvious importance to life as we know it. Since the emphasis in this book is on periods larger than a day, daily variations will not be discussed explicitly. However, it is important to realize that the *existence* of daily variations can affect the state of the atmosphere over longer periods, the magnitude of the effect depending on the amplitude of the daily variations. An example of such an effect is the mixing of the lower atmosphere. In summer especially, the ground can become very hot during the day, causing strong convection that stirs up a considerable depth of air. The air is not "unmixed" at night, so the net effect is substantially different from that which would be achieved with uniform radiation.

Seasonal variations are due to (i) the tilt of the earth's axis relative to the plane of its orbit (at present 23.5°) and (ii) the ellipticity of the earth's orbit. The ellipticity is such that the total amount of radiation incident on the earth varies by $\pm 3.5\%$, with the maximum in early January. The consequent changes with latitude and time of the incident radiation are given by List (1951, Tables 132 and 134), whereas Stephens *et al.* (1981) give the observed changes in outgoing radiation. These are smaller than the changes in incident radiation, so there is a net gain of energy between October and March when the earth is nearer the sun, and a net loss in the remainder of the year. The variations show a marked asymmetry between the two hemispheres because of the different proportions of land and sea, changes over the latter being relatively small.

The existence of seasonal variations has important effects on the mean state of the atmosphere and ocean, the magnitude of the effect depending on the amplitude of the variations. This fact has been demonstrated by numerical experiments of Wetherald and Manabe (1972). They began with an ocean–atmosphere model driven by the annual mean radiation and then changed to seasonal forcing. The mean state was changed thereby, e.g., surface temperatures in high latitudes were greater and the mean north–south temperature gradient in the atmosphere was reduced. [The *sensitivity*, e.g., to changes in CO_2 content, is also affected—see Wetherald and Manabe (1981).] The most important contributing factor was found to be the melting of snow in high latitudes in summer, thus reducing the net albedo. Another factor was found to be the development of a warm surface layer in the ocean in summer, giving a higher mean sea-surface temperature.

The fact that seasonal variations affect the mean state of the ocean–atmosphere system is the basis of an astronomical theory of climate change due to Milankovich (1930, 1941). Because of perturbations caused by other planets, the tilt of the earth's axis varies between 22 and 24.5°, and the eccentricity of the earth's orbit changes, the time scales of these changes being 10^4–10^5 years. The net radiation incident over a year is altered very little, but the distribution in time and space is changed. The eccentricity varies sufficiently for the amplitude of seasonal variations in the incident radiation to change between 0 and 15% and the time of the maximum also changes. The effects of these changes on the incident radiation are given by Berger (1979), and the theory is discussed by Imbrie and Imbrie (1979) and Monin (1972, Chapter 4). Periods during which the amount of radiation received in summer over the high-latitude continental areas of the northern hemisphere was small appear to coincide with ice ages. Geological evidence in support of the theory is discussed by Hays *et al.* (1976) and Imbrie and Imbrie (1980).

Chapter Two

Transfer
of Properties
between Atmosphere
and Ocean

2.1 Introduction

As stated in the introduction to Chapter 1, one would like to determine the response of the atmosphere–ocean system to the known radiative input from the sun given only the physical properties of air and water, the distribution of land and sea, and other such basic information. Some of that information is given in this chapter, as Section 2.2 discusses the differences between the physical properties of air and water that make their mutual boundary of such importance. The density difference is obviously significant, but contrasts in optical properties are also important since they result in the thermal driving of the ocean being effectively at the surface.

Processes that are responsible for transfer of heat and moisture across the air–sea boundary are briefly discussed in Section 2.4, along with formulas used for calculating the rates of transfer. These can be used to calculate global budgets of heat, moisture, and momentum, which are examined in three different sections. First, the *angular* momentum budget of the atmosphere is discussed in Section 2.3, this having some historical interest in connection with the Hadley circulation. The moisture budget (hydrological cycle) is discussed in Section 2.5, and the heat budget of the ocean is discussed in Section 2.6. Finally, the thermohaline or buoyancy-driven circulation of the ocean is considered in Section 2.7.

2.2   Contrasts in Properties of Ocean and Atmosphere

Water is very much *denser* than air. The density of air varies with temperature, pressure, and humidity (see Chapter 3), typical surface values being 1.2–1.3 kg m^{-3} (0.0013 tonne m^{-3}), whereas the sea is some 800 times more dense (1025 kg m^{-3} = 1.025 tonne m^{-3} at the surface). Thus the interface between air and water is very stable because of the strength of the gravitational restoring force when it is displaced from its equilibrium position. Typical displacements observed in surface waves are of order 1 m. Because of the stability of the interface, the two media do not mix in any significant way (whitecaps and spray are only found close to the interface), so transfers of properties between the two media must take place through a well-defined interface. This contrasts with the atmosphere, for instance, where heat transfer can take place through a plume of hot air rising hundreds of meters and then mixing with the surrounding air. Obviously, such a plume cannot cross the ocean surface, and this is one reason why the air–sea transfer processes need special consideration.

The existence of the interface affects the *radiation* balance because it *reflects* radiation. The fraction α of solar radiation reflected is a function of the angle of incidence and of the surface roughness (Kraus, 1972, Section 3.2). Typical values of α are indicated by the satellite measurements of minimum albedo shown in Fig. 1.3b. It is assumed that the minimum albedo approximates the value that would be obtained in the absence of clouds, and so is close to the surface value. At latitudes below 30°, values less than 0.1 are found. At higher latitudes, the values increase with latitude because of the progressive reduction in the angle between the sun's rays and the surface.

There is not only a discontinuity in density at the ocean surface, but also a discontinuity in *optical* properties that has important consequences for the radiation balance. Consider first the solar radiation impinging on the atmosphere. According to Fig. 1.6, only 19% of this is absorbed within the atmosphere. At the surface, a fraction α is reflected. What happens to the remainder that enters the ocean? This represents about 51% (see Fig. 1.6) of the radiation entering the outer atmosphere. Unlike the atmosphere, the ocean *absorbs* solar radiation very rapidly. The rate of absorption varies with wavelength and with the amount of suspended material (Kraus, 1972, Section 3.2). The *total* energy (in the range of wavelengths appropriate to solar radiation) falls off exponentially with depth. Typical decay rates are such that about 80% (Jerlov 1968, Table 21 and Fig. 50) is absorbed in the top 10 m. In coastal areas where a lot of suspended material is present, the absorption rate can be much greater. A more detailed discussion can be found in Jerlov's (1968) book.

In the atmosphere, long-wave radiation is absorbed much more rapidly than solar radiation, the principal absorber being water vapor. It is hardly surprising, therefore, that long-wave radiation in the ocean is absorbed very rapidly indeed. The result is that the *emission* (and absorption) of long-wave radiation takes place from a very thin layer, less than 1-mm thick (McAlister and McLeish, 1969).

The density contrast between air and water means that the *mass* of the ocean is very much greater than (270 times) that of the atmosphere. The mass per unit area of the atmosphere is approximately 10^4 kg m^{-2} (10 tonne m^{-2}), and since the acceler-

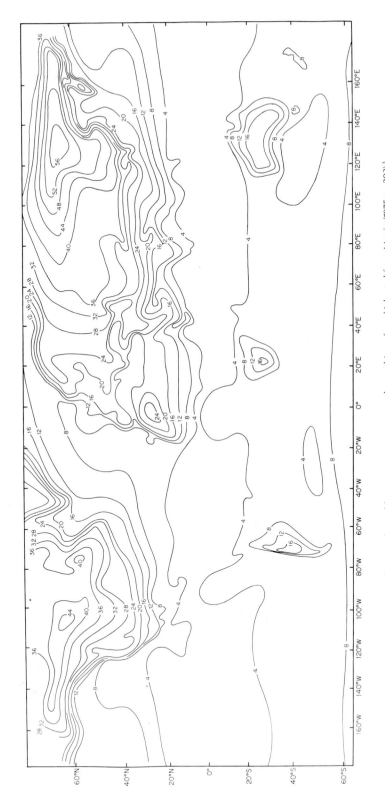

Fig. 2.1. Annual range of monthly mean temperatures at the earth's surface. [Adapted from Monin (1975, p. 203).]

ation due to gravity is about 10 m s^{-2}, the *weight* per unit area, or surface pressure, is about

$$10^5 \text{ Pa} \equiv 10^5 \text{ N m}^{-2} \equiv 1 \text{ bar.}$$

A mere 10-m depth of ocean has the same weight per unit area, so the pressure increases by 1 bar every 10 m. For this reason, oceanographers often express pressures in decibars (dbar) since 1 dbar $\approx$ 1 m in depth (see Section 3.5).

The large difference in mass between air and water also implies a large difference in *heat capacity*. In fact, the specific heat (heat capacity per unit mass) of water is four times that of air, so a mere 2.5-m depth of water has the same heat capacity per unit area (10^7 J m^{-2} K^{-1}) as the whole depth of the atmosphere. In other words, the heat required to raise the temperature of the atmosphere by 1 K can be obtained by changing the temperature of 2.5 m of water by the same amount (or of 25 m by 0.1 K or of 250 m by 0.01 K). Heat can also be stored in latent form, and this same amount of heat can be used to *evaporate* 4 mm of water or to melt 30 mm of ice. (Values of latent heats and specific heats from which these figures are derived are given in Appendixes 3 and 4.) The importance of latent heat can be seen when it is considered that evaporation rates in the tropics are of order 4 mm per day, corresponding to changing the temperature of the atmosphere by 1 K per day. This is consistent with cooling rates by radiation, which are of order 1 K per day (Riehl, 1979).

The large heat capacity of the ocean is of importance for seasonal changes. Although in the long term each *hemisphere* loses by radiation about as much heat as it receives, this is not true of an individual season. The excess heat gained in summer is *not* transported to the winter hemisphere, but is stored in the surface layers (100 m or so) of the ocean and returned to the atmosphere in the winter (Palmén and Newton, 1969, Chapter 2). Because of this ability to store heat, the ocean surface temperature changes by much smaller amounts than the land surface, which cannot store much heat. This contrast between land and sea shows up vividly in Fig. 2.1, which shows the seasonal range in temperature at the earth's surface. Although the outlines of the continents are not drawn in, their position is quite clear. Thermal storage in the ocean is also important at longer time scales, and therefore is of significance for climatic variations.

2.3 Momentum Transfer between Air and Sea, and the Atmosphere's Angular Momentum Balance

How are the winds produced, and what determines their distribution? In offering an explanation for the trade winds found in the tropics, *Halley* (1686, p. 165) pointed out that the driving force is "the Action of the Suns Beams upon the Air and Water." This produces a dynamic effect, namely, "that according to the Laws of *Staticks*, the Air which is less rarified or expanded by heat, and consequently more ponderous, must have a Motion towards those parts thereof, which are more rarified, and less ponderous, to bring it to an *Æquilibrium*." Thus Halley had in mind a steady situation in which there is a balance between the *forcing* effect of radiation, which tends to produce horizontal density differences, and the *dynamic* effects, which tend to reduce the differences.

> But as the cool and dense Air, by reason of its greater Gravity, presses upon the hot and rarified, 'tis demonstrative that this latter must ascend in a continued stream as fast as it Rarifies, and that being ascended, it must disperse it self to preserve the *Æquilibrium*; that is, by a contrary Current, the upper Air must move *from* those parts where the greatest Heat is: So by a kind of Circulation, the North-East Trade Wind below, will be attended with a South Westerly above, and the South Easterly with a North West Wind above (Halley, 1686, p. 167).

Such a circulation in the meridional plane is now known to exist in the tropics (see Fig. 1.7) and Halley's explanation of the circulation is essentially correct. However, this meridional circulation is now called the *Hadley* circulation. This appears to be because Halley's explanation of the easterly component of the trade winds was incorrect, whereas Hadley (1735) gave an explanation that is much closer to the truth. He pointed out that because of the rotation of the earth, the speed of the equator about the earth's axis is greater than that of the tropics (23.5° latitude) by some 2083 miles per day. Thus air at rest relative to the earth at 23.5° would, in the absence of friction, acquire a westward velocity of 2083 miles per day at the equator. Since velocities this large are not observed, "it is to be considered, that before the Air from the Tropicks can arrive at the Equator, it must have gained Some Motion Eastward from the Surface of the Earth or Sea, whereby its relative Motion will be diminished, and in several successive Circulations, may be supposed to be reduced to the Strength it is found to be of. Thus I think the N.E. Winds on this Side of the Equator, and the S.E. on the other Side, are fully accounted for" (Hadley, 1735, p. 61). Halley (1686, facing p. 151) produced the first comprehensive map of these winds over the tropical Atlantic and Indian Oceans based on his own observations and information obtained from "a multitude of Observers."

The principle Hadley appealed to was that of conservation of *angular momentum*, which applies in the absence of friction. Hadley (1735, p. 62) also stated, "The N.E. and S.E. Winds within the Tropicks must be compensated by as much N.W. and S.W. in other Parts, and generally all Winds from any one Quarter must be compensated by a contrary Wind some where or other; otherwise some Change must be produced in the Motion of the Earth round its Axis." This statement is not correct with respect to the northward component of the wind, but the principle Hadley was referring to is clear enough. The net rate of exchange of angular momentum between the atmosphere and the underlying surface must be zero, otherwise the angular momentum of the atmosphere would be continually increasing or decreasing. The quantitative expression of this principle may be derived as follows. Suppose the average eastward *force* (or rate of transfer of eastward momentum) per unit area acting on the earth's surface at latitude φ is

$$\tau^x(\varphi).$$

Then the average *torque* (or rate of transfer of angular momentum) per unit area about the earth's axis is

$$a\tau^x(\varphi)\cos\varphi,$$

where a is the radius of the earth. The area of a zonal strip between latitudes φ and $\varphi + d\varphi$ is $2\pi a^2 \cos\varphi\, d\varphi$, so the torque on this strip is

$$2\pi a^3 \tau^x(\varphi)\cos^2\varphi\, d\varphi.$$

The net torque on the earth's surface (or the net rate of exchange of angular momentum between the atmosphere and the underlying surface) must vanish, so

$$\int_{-\pi/2}^{\pi/2} \tau^x(\varphi)\cos^2\varphi\, d\varphi = 0. \qquad (2.3.1)$$

This is the quantitative expression of the principle that Hadley appealed to.

The force of the atmosphere on the underlying surface may be exerted in two different ways. One is the force exerted on irregularities in the surface associated with pressure differences across the irregularities. The second is by viscous stresses. The irregularities on which forces are exerted may vary in size from mountain ranges like the Andes down to trees, blades of grass, and ocean surface waves. When the

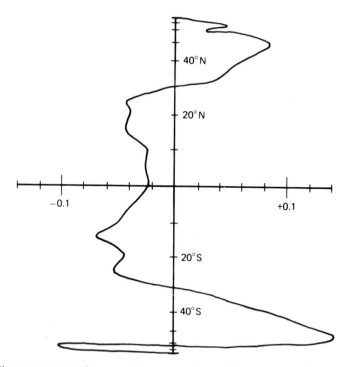

Fig. 2.2. The average eastward stress on the ocean surface as a function of latitude [values are from Eyre (1973)]. The spacing of latitudes is such that the distance between two nearby latitudes is proportional to the square of the cosine of the latitude. With this scale the area under the curve would be zero if the average rate of transfer of momentum from the atmosphere were the same over the land as over the sea at each latitude.

irregularities are small enough (as is the case over the ocean), the associated force per unit area added to the viscous stress is called the *surface stress*, or *wind stress*. Since the earth's surface is mainly ocean, it is not surprising that (2.3.1) is approximately true with $\tau^x(\varphi)$ being the average eastward wind stress over the *ocean* at latitude φ [other contributions to the angular momentum exchange are discussed by Newton (1971)]. Figure 2.2 shows estimated values of the average eastward wind stress $\tau^x(\varphi)$ as a function of latitude. The latitudinal axis is drawn linear in

$$\tfrac{1}{2}\varphi + \tfrac{1}{4}\sin 2\varphi = \int_0^\varphi \cos^2 \varphi' \, d\varphi',$$

so that the area under the curve would be zero if (2.3.1) were exactly correct. Note that there is a westward stress in the trade wind zone (latitudes below 30°) and therefore an eastward stress is required at higher latitudes to give an overall balance. The eastward stress is associated with the prevailing westerly (i.e., eastward) winds at those latitudes. The reason that westerly winds should be found in these latitudes is

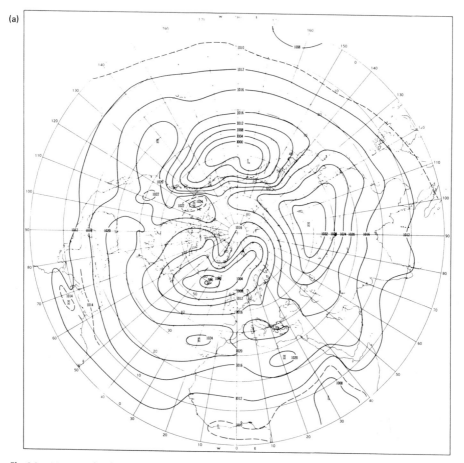

Fig. 2.3. Mean sea-level pressure (mb) for January [(a) and (b)] and July [(c) and (d)]. The northern hemisphere data are from Crutcher and Meserve (1970), and the southern hemisphere data are from Taljaard *et al.* (1969).

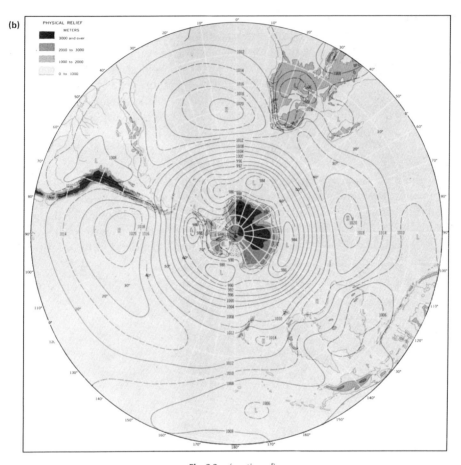

Fig. 2.3.   (continued)

not particularly straightforward and is discussed in relation to the angular momentum balance in Chapter 13 [see also Lorenz (1967, 1969)].

To calculate the ocean currents that are produced by the wind, the detailed distribution of stress with position on the earth's surface is required. The pattern of surface winds away from the equator can be obtained from surface pressure maps (Fig. 2.3), whereas the tropical wind distributions are shown in Figs. 11.24, 11.28, and 11.29. Features like the trade winds, intertropical convergence zone, and westerly wind belts can be clearly seen in these figures. Sources of more detailed information are listed in Appendix 5.

2.4   Dependence of Exchange Rates on Air–Sea Velocity, Temperature, and Humidity Differences

Winds are produced in the atmosphere in response to radiative forcing. These winds transfer momentum to the ocean, producing ocean currents. By what processes

(c)

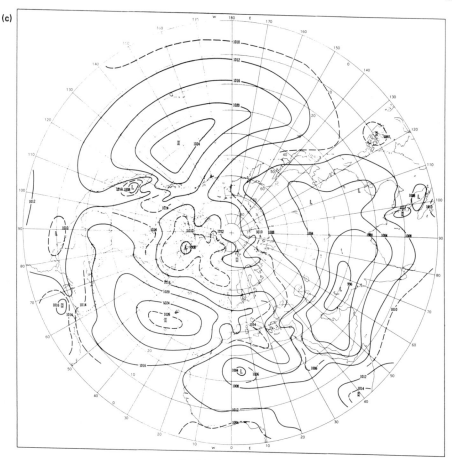

Fig. 2.3. (continued)

is the momentum transferred and on what do the transfer rates depend? This is an important question, about which much has been written (Kraus, 1972, 1977; Garratt, 1977; Liu *et al.*, 1979; Charnock, 1981; Lumley and Panofsky, 1964), and this section is intended to be but a brief introduction.

The radiative forcing of the atmosphere produces pressure gradients that result in wind speeds of order 10 m s^{-1}. If there were no momentum transfer to the lower boundary (i.e., no frictional contact between the atmosphere and surface), such velocities would be expected right down to the surface. However, there is frictional contact at the surface. This means that at *solid* boundaries, the air in immediate contact with the boundary is constrained to have zero velocity. Thus a velocity gradient or *shear* exists near the ground. (An example of the way wind speed varies with height is shown in Fig. 2.4.) The shear flow, however, is not stable because small disturbances can grow to make the flow *turbulent*. The turbulent eddies (which are responsible for the "gusty" nature of the wind) modify the shear, but over a sufficiently long time, a well-defined mean velocity can be determined for each value of z, the distance above the ground. (Typical averaging times required are of the order of

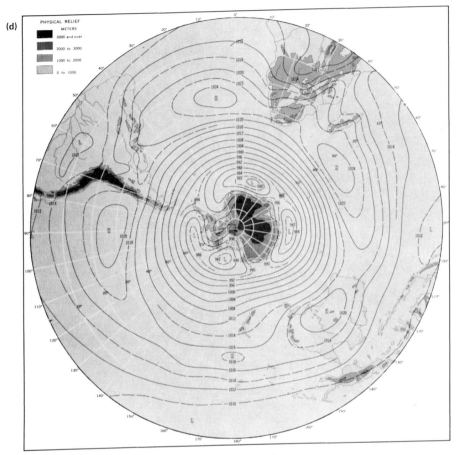

Fig. 2.3. (continued)

minutes for points a few meters above the ground.) In the region of substantial shear, momentum is transferred downward by bodily movement of parcels of air, i.e., by fast-moving parcels moving downward and slow moving parcels moving upward. If u is the horizontal component of velocity, w the vertical component, and ρ the density, then the vertical flux of horizontal momentum per unit area is ρuw, so the mean value of this quantity over a sufficiently large area or sufficiently large time is equal to the mean stress τ.

As the ground is approached, the shear increases in inverse proportion with the distance from the ground. (This law can be deduced on dimensional grounds from the assumption that the shear depends only on τ, ρ, and distance z from the ground. It implies a *logarithmic* mean velocity profile.) The inverse law holds only sufficiently close to the ground where the shear is strong because other effects become important when the shear gets weak. For instance, if the lapse rate (see Section 1.5) is large enough to produce convection, turbulence due to convection will become more important than turbulence due to shear at some level.

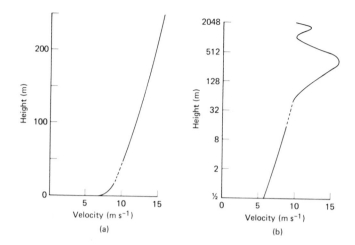

Fig. 2.4. An example of how the wind speed (in m s⁻¹) varies with height (in m). The height scale is (a) linear and (b) logarithmic. Values at 0.5, 1, 2, 4, 8, and 16 m are $\frac{1}{2}$–1 hr averages obtained from anemometers on a mast. Values at 50 m and above were obtained by tracking a pilot baloon released in the same general area. [From Clarke *et al.* (1971, data p. 307, 0900 hr).]

In order to relate the stress τ to the wind speed u, it is necessary to specify the height at which the wind is measured. Once this is done, it follows on dimensional grounds that the relationship between τ and u can be put in the form

$$\tau = c_D \rho u^2, \tag{2.4.1}$$

where c_D is a dimensionless coefficient called the *drag coefficient*. Its value over solid surfaces depends on the roughness of the surface and can also depend on the lapse rate. Values for different types of surface are known from measurement.

The ocean is not a solid surface, but surface velocities are still very much less than those in the atmosphere (typically about 3% of the velocity at 10 m). This is basically due to the *density* difference, for the same momentum can be carried in water with much smaller velocities. Hence the shear over the ocean is just as large as over the land, and turbulence is produced in the same way. However, measurements over the ocean are more difficult than over land, and less is known about how the drag coefficient varies, particularly at high wind speeds. As mentioned in Section 2.3, transfer across the surface can be due to pressure differences across irregularities (in this case waves) or to viscous stresses. The pressure differences across the waves can increase the amplitude of the waves, and waves can carry momentum away without any mean motion of the fluid. However, it seems likely that most of the momentum transferred during a storm is used to drive currents (Manton, 1972).

The drag coefficient c_D for the ocean surface is found to increase with wind speed. Values for low speeds are around 1.1×10^{-3}. For speeds over 6 m s⁻¹ a linear relation between c_D and u is often used, e.g., S. D. Smith (1980) suggests

$$10^3 c_D = 0.61 + 0.063u \qquad \text{for} \quad 6 \text{ m s}^{-1} < u < 22 \text{ m s}^{-1}. \tag{2.4.2}$$

Alternatively, the data can be fitted by a relation obtained on dimensional grounds by Charnock [see Charnock (1981)]. This makes a quantity called the roughness length (see Section 9.5) proportional to the length scale, which can be obtained from τ, ρ, and g. The drag coefficient is then given by

$$c_{\mathrm{D}} = \left[\kappa/\ln(\rho g z/a\tau)\right]^2, \tag{2.4.3}$$

where κ and a are constants (called the von Kármán and Charnock constants, respectively), and z is the anemometer height (normally 10 m). Wu (1980) suggests the values

$$\kappa = 0.4, \qquad a = 0.0185. \tag{2.4.4}$$

An alternative formula for c_{D} is suggested by Liu *et al.* (1979).

Because the eddying motion in the 10 m or so of air near the surface is caused by shear rather than by buoyancy differences (cf. Section 1.5), the rates of transfer of sensible heat and of moisture depend on the wind speed. The heat and moisture are transferred by bodily movement of parcels of air. The direction of transfer usually involves hot and moist fluid being carried upward and relatively cold and dry air being transferred downward. Like the shear, the temperature and humidity gradients increase as the surface is approached, also in inverse proportion with the distance from the surface. Assuming that the upward heat flux Q_{s} depends on (i) the wind speed u, (ii) the difference between the sea temperature T_{s} and the air temperature T_{a} at the standard level, and (iii) the heat capacity $\rho_a c_p$ per unit volume of the air, the relationship

$$Q_{\mathrm{s}}/\rho_a c_p = c_{\mathrm{H}} u (T_{\mathrm{s}} - T_{\mathrm{a}}), \tag{2.4.5}$$

c_{H} being a dimensionless coefficient, is obtained by dimensional arguments. c_{H} is sometimes called the Stanton number. The corresponding rule for the evaporation rate E, defined as the mass of water evaporated per unit area per unit time, is

$$E/\rho_a = c_{\mathrm{E}} u (q_{\mathrm{s}} - q_{\mathrm{a}}), \tag{2.4.6}$$

where q_{a} is the specific humidity (mass of water vapor per unit mass of air) at the standard level, q_{s} is the specific humidity at the sea surface, assumed to be the saturation value of q at the sea-surface temperature, ρ_a is the density of air, and c_{E} is a dimensionless coefficient, sometimes called the Dalton number.

Different formulas for c_{H} and c_{E} have been suggested; e.g., S. D. Smith (1980) finds a good fit to data using

$$10^3 c_{\mathrm{H}} = \begin{cases} 0.83 & \text{for stable conditions,} \\ 1.10 & \text{for unstable conditions,} \end{cases} \tag{2.4.7}$$

whereas a constant value of

$$10^3 c_{\mathrm{E}} = 1.5 \tag{2.4.8}$$

seems to work reasonably well. Alternative formulations are given by Liu *et al.* (1979).

2.5   The Hydrological Cycle

The fundamental importance of water in the atmosphere on the energy balance was pointed out in Chapter 1. If water vapor could only be transported by molecular diffusion, it would presumably diffuse upward until the whole atmosphere was saturated. The atmosphere is *not* saturated, however, because of the *motion* produced by radiation effects. Air is continually moving upward and downward because of convection (caused by radiation tending to heat the bottom of the atmosphere more than the top) and because of the horizontal gradients due to more radiation being received in the tropics than in the polar regions. The upward moving air is carried to levels where the temperature is lower and therefore less moisture can be held. If the air is carried high enough, it becomes saturated, condenses out, and may then fall to the surface as precipitation. The air left behind has less moisture content, so that when it is brought downward again it will be unsaturated. When it gets low enough, some of this relatively dry air will get caught up in the shear-driven eddies and be brought down close to the surface itself. Contact of dry air with the surface leads to evaporation, which moistens the air, and so the cycle continues.

The cooling of the surface due to evaporation represents the main loss of heat to the atmosphere required to balance the radiative gain. The heat removed at the surface is put back into the atmosphere at a higher level when the water vapor condenses, thus providing the upward transfer of heat required by the radiation balance. The mean rate of evaporation over the ocean that provides this transfer is about 1 m yr^{-1} (3 mm day^{-1}). However, the amount of water in the atmosphere at any one time is not large. If precipitated, it would cover the earth's surface to a depth of 23 mm. [This is equivalent to a latent heat content per unit area of the atmosphere of 5.7×10^7 J m^{-2}, which is the value for the northern hemisphere calculated from Oort (1971, Table 1). A change of heat content of this magnitude would change the temperature of the atmosphere by 6°.] Dividing by the mean evaporation rate, a mean residence time for water vapor in the atmosphere of about 1 week is obtained.

In the time during which a particular molecule of water vapor is in the atmosphere, it can be transported considerable distances in the horizontal. From figures for mean winds, the estimated movement of a water molecule in 1 week is of order 10,000 km to the east or west and 1000 km to the north or south. Thus condensation may take place a considerable distance from the evaporation point, and therefore heat is transported horizontally as well as vertically. The water vapor (and hence latent heat) transport can be calculated from measurements of velocity and humidity within the atmosphere (Oort, 1971) or from estimates of differences in the rates of evaporation and precipitation. Figure 2.5 shows estimates based on both techniques. The meridional latent heat flux is an important contributor to the total atmospheric energy flux shown in Fig. 1.8 (Oort, 1971, Table 5); e.g., at 40°N it contributes 1.6×10^{15} W to the total atmospheric flux of 2.9×10^{15} W, whereas at 10°N the latent heat flux is -1.4×10^{15} W with a total atmospheric flux of 1.2×10^{15} W.

Sources of information on the geographical distribution of evaporation and precipitation are given in Appendix 5, and the world water balance is discussed by

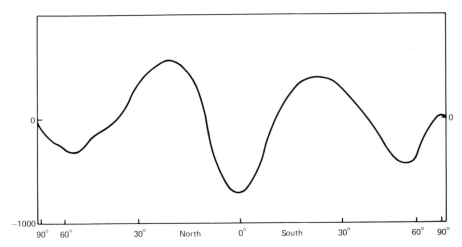

Fig. 2.5. The excess of evaporation over precipitation (in kg m^{-2} yr^{-1}) as a function of latitude. The values between the north pole and 5°S are taken from Oort (1971). The remaining values are from Newell *et al.* (1969).

Baumgartner and Reichel (1975). Figure 2.6 shows a map of the mean annual precipitation. Note the high contrasts in precipitation rates in the tropics. The high-precipitation belt near the equator is the ITCZ (see p. 13), which is a region of ascending air. By contrast, the regions of descending air, which lie on the north and south of this zone in the eastern Atlantic and Pacific, are very dry.

   Although the moisture in the air is very significant for the heat balance of the earth, the amount of water in the atmosphere is only a tiny fraction of the total amount on earth. A useful way of describing the relative amounts of water in various forms is by means of the depth they would occupy in a vertical-sided container with the surface area of the ocean. Then the ocean water would occupy a depth of 3800 m, whereas the moisture in the atmosphere when condensed would only occupy 0.03 m. After the ocean, the next largest repository of water is in the solid form, most of this being in the Antarctic Ice Sheet. If melted, this would occupy 76 m. The water in lakes and rivers would occupy 4 m, whereas that in the ground (mostly very deep) would occupy about 19 m.

   Of the water in solid form, that in the ice sheet can only exchange heat very slowly with the atmosphere because of the poor conductivity of ice and the thickness of the

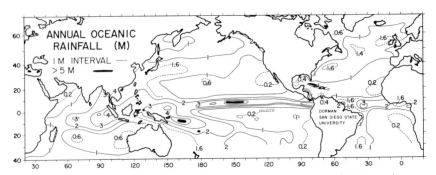

Fig. 2.6. A map showing the annual rainfall over the ocean. [Courtesy of C. Dorman.]

ice sheet. Thus particles in the ice sheet have residence times of order 10^5 years. Of more significance for the heat *balance* of the earth at a given time is the snow cover, which has high albedo, and sea ice, which not only has high albedo but also inhibits heat exchange between the ocean and atmosphere. As far as the heat and water balance of the ocean is concerned, melting and freezing of sea ice are analogous to precipitation and evaporation. During a season, ice may move a distance of order 1000 km, so ice particles will melt in a different place from where they formed. Thus there is a melting–freezing imbalance analogous to the precipitation–evaporation imbalance. Values of this imbalance are not known but could reach several meters per year in special locations of small extent [see Gill (1973)], such as the region in the southern Weddell Sea, where ice is blown off shore by the prevailing winds, thus opening up leads in which rapid freezing can take place.

Where the precipitation rate P exceeds the evaporation rate E (or where the melting rate M exceeds the freezing rate F), the ocean surface would get higher in time if it were not for gravitational restoring forces that tend to keep the surface level constant. At one time [see Stommell (1957)] it was thought that the currents required to keep the level constant might be significant, but it is now known that wind-driven currents are much stronger. [This is basically because the "Ekman suction velocity," to be discussed later, is greater by a factor ~ 30 than typical values of the velocity $(P - E)/\rho$.]

A more important effect of the precipitation–evaporation (or melting–freezing) imbalance is the effect on salinity. If $P - E$ (mass of water per unit area per unit time) is positive, the water is diluted at the same rate it would be if there were a loss of salt at a rate (mass of salt per unit area per unit time)

$$(P - E)s, \tag{2.5.1}$$

where s is the salinity of the seawater, i.e., the mass of salt per unit mass of seawater, which is usually close to 0.035. (*Note:* The symbol s will be used for salinity expressed as a fraction, and S reserved for absolute salinity expressed in parts per thousand (written $^0\!/_{00}$), i.e., $S = 1000s$, or for practical salinity (see Appendix 3) in practical salinity units.) When the ocean is ice-covered, the equivalent upward salt flux is

$$(M - F)(s - s_i), \tag{2.5.2}$$

where s_i is the salinity of sea ice, usually about 0.004 or $4^0\!/_{00}$. The changes in salinity are of dynamic significance because they cause changes of density (see Section 2.7) that produce motion. They also cause differences in water mass properties, and there are considerable differences in the water balances of the different oceans [see Stommel (1980)].

2.6   The Heat Balance of the Ocean

The average amount of solar radiation absorbed at the ocean surface is only about half (see Fig. 1.6) of that incident on the earth and averages about 175 W m^{-2}. In response to this warming, the ocean surface reaches a temperature such that the net losses of heat equal the gain from solar radiation. The upward flux by long-wave radiation is at the rate (1.4.1) determined by the surface temperature, but as discussed in Section 1.4, this radiation is *not* lost to space, most being absorbed and re-emitted in the atmosphere or else reflected downward by clouds. The *net* surface emission (or effective back radiation) is the difference (upward minus downward flux), which varies rather little from a constant value of about 65 W m^{-2}. The remaining non-radiative heat loss from the ocean is principally through cooling by evaporation (the *latent* heat loss) and through direct thermal (or *sensible* heat) transfer.

The calculations of heat flux at a particular locality are usually based on empirical formulas [Kraus (1972) discusses some of them] that involve only quantities that are regularly observed from ships. For instance, the rate Q_1 of absorption of solar radiation is usually calculated as a product of (a) Q_{I_0}, the net downward flux of solar radiation just above the surface in cloudless conditions (this is usually 0.7–0.75 of the flux incident at the top of the atmosphere), (b) $(1 - \alpha_s)$, where α_s is the surface albedo, and (c) a correction factor for cloud effect. A simple example of such a formula is

$$Q_1 = Q_{I_0}(1 - \alpha_s)(1 - 0.7n_c), \qquad (2.6.1)$$

where n_c is the fraction of sky covered by cloud. The correction factor for cloud effect really depends on cloud type and height, and Lumb (1964) has given formulas that take this into account. However, Lumb's formulas cannot be used if n_c is the only cloud parameter available, so global computations are usually based on formulas like (2.6.1).

The net upward flux Q_B of long-wave radiation from the ocean is usually calculated as a product of (a) σT_s^4, which is the flux emitted by a blackbody of temperature T_s, σ being Stefan's constant, which is given by (1.3.2), (b) a correction factor, 0.985 for departure of the ocean surface from blackbody behavior, (c) a correction factor for back radiation in the absence of clouds, and (d) a correction factor for cloud effect. An example of such a formula is

$$Q_B = 0.985\sigma T_s^4(0.39 - 0.05e_a^{1/2})(1 - 0.6n_c^2), \qquad (2.6.2)$$

where e_a is the vapor pressure of water at the standard height (mb). The factor involving e_a is the correction for back radiation, this depending on the vapor content of the atmosphere, and the last factor is obviously the correction for cloud effect.

The total upward flux Q of heat from the ocean is the sum of the fluxes due to the individual processes, namely,

$$Q = Q_B + L_v E + Q_s - Q_1, \qquad (2.6.3)$$

where Q_s is the upward sensible heat flux and $L_v E$ the upward latent heat flux, with E the evaporation rate and L_v the latent heat of vaporization of water, given by

$$L_v = 2.5 \times 10^6 \quad \text{J kg}^{-1}. \qquad (2.6.4)$$

[*Note*: (2.6.3) omits the small effects of condensation on the ocean surface, and heat transferred by precipitation.] Values of Q_s and E are normally calculated from formulas like (2.4.5) and (2.4.6). Figure 2.7 shows the result of such a calculation for the Atlantic Ocean. Note the intense cooling in the western North Atlantic, where cold air from the continent blows over the warm ocean in the winter. Month-by-month

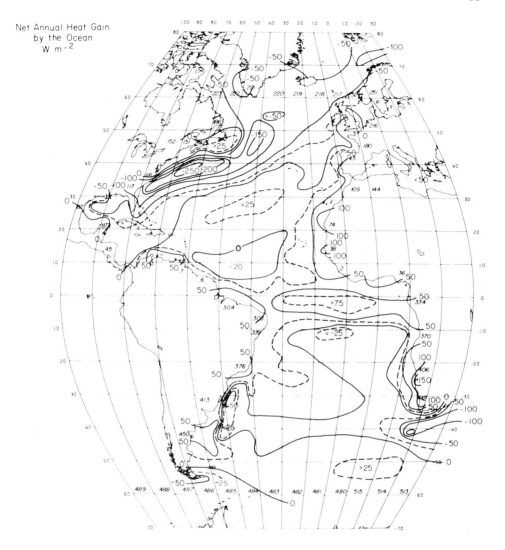

Fig. 2.7.   The annual average net surface heat flux into the ocean (W m^{-2}). [According to Bunker (1980, Fig. 18)].

balances for this and other special regions are given by Bunker (1976). From such a map, heat fluxes across different sections can be computed by integrating southward from the northern extremity of the map. The poleward flux at 24°N is about 1 PW (1 petawatt $= 10^{15}$ watts) and this compares well with calculations from oceanographic data across this section by Bryden and Hall (1980). Integrals over the ocean basins yield some surprises (Hastenrath, 1980; Stommel, 1980), e.g., the heat flux in the Atlantic appears to be northward at all latitudes. (Sources of data on atmosphere–ocean heat fluxes are given in Appendix 5.)

When ice covers the ocean, the flux calculation is more difficult because the ocean exchanges heat with the lower boundary of the ice rather than the atmosphere, and

TABLE 2.1

Values of $\partial Q/\partial T$, $\partial Q/\partial n_c$, and $\partial Q/\partial r$
for the Latitude Ranges Shown[a]

Quantity	Unit	Average value for latitudes	
		0–30°	40–50°
$\partial Q/\partial T$	W m^{-2}	40	32
$\partial Q/\partial n_c$	W m^{-2}	215	135
$\partial Q/\partial r$	W m^{-2}	−555	−290

[a] $\partial Q/\partial T$ is the rate of change of the upward heat flux Q from the ocean with the air–sea temperature difference, $T_s - T_a$, $\partial Q/\partial n_c$ is the rate of change of Q with fractional cloudiness n_c, and $\partial Q/\partial r$ is the rate of change of Q with relative humidity r expressed as a fraction.

the flux depends on the history of the ice layer. Maykut and Untersteiner (1971) have constructed an ice model that takes such effects into account.

The formulas used for calculating fluxes are nonlinear, and it is often useful to replace them with linearized versions, which assume that deviations from some prescribed state are not too large. Haney (1971) has already used the idea for changes in sea–air temperature difference, but it can be extended to cloudiness n_c and relative humidity r as well. The result has the form

$$Q - Q_0 \approx (T_s - T_a)\, \partial Q/\partial T + (n_c - 0.25)\, \partial Q/\partial n_c + (r - 0.75)\, \partial Q/\partial r, \quad (2.6.5)$$

where the derivatives are calculated at $n_c = 0.25$ and $r = 0.75$. Table 2.1 shows values of these derivatives for different latitude ranges calculated from the formulas used by Haney (1971). According to these figures, an increase of 10 W m^{-2} in the tropics will result from either (i) an increase of air–sea temperature difference of 0.25 K, (ii) an increase of cloud of 5% or (iii) a reduction in relative humidity of 1.8%.

2.7 Surface Density Changes and the Thermohaline Circulation of the Ocean

The radiative heating of the atmosphere causes motion because it leads to *density* differences. However, these differences are effective only in producing motion because of gravity, and it is really differences in the *weight* per unit volume $g\rho$ that are important. The quantity $-g\rho$ is called the *buoyancy*, the minus sign being used because a particle is said to be more buoyant when it has less weight. The ocean moves because of buoyancy contrasts, but these are due to *salinity* differences as well as temperature differences. These differences are created by the fluxes of heat and water at the ocean surface, whose combined effect on buoyancy is called the buoyancy flux B, which is given by

$$B = c_w^{-1} g\alpha Q + g\beta(E - P)s, \quad (2.7.1)$$

where Q is the upward heat flux, E the evaporation rate, P the precipitation rate, c_w the specific heat of water, s the surface salinity, $\alpha = -\rho^{-1}\,\partial\rho/\partial T$ the thermal expansion coefficient of seawater at the surface, and $\beta = \rho^{-1}\,\partial\rho/\partial s$ the corresponding coefficient for salinity. An alternative expression may be obtained by using (2.6.3), which gives

$$B = c_w^{-1}g\alpha(Q_B + Q_s - Q_1) - g\beta sP + g(c_w^{-1}\alpha L_v + \beta s)E. \qquad (2.7.2)$$

This expression shows that evaporation decreases buoyancy in two ways—by cooling and by increasing salinity. The former effect is greater by the factor

$$\alpha L_v/c_w\beta s,$$

which is about 4 for typical surface conditions. This relates to the fact that temperature differences generally make greater contributions to density differences than salinity differences in the ocean. There are, however, exceptions such as the polar regions, where α is much smaller than in mid-latitudes (see Appendix 3) or the Baltic, which is a relatively fresh body of water. The circulation that is driven by the buoyancy flux is called the *thermohaline circulation*. (*Note*: This expression has a clear meaning for model oceans in which motion is produced *only* by a buoyancy flux. In practice, however, motion is also driven by the wind and one cannot say that a particular current is, say, 70% wind-driven and 30% buoyancy-driven because the ocean is not a linear system.) Discussions about the nature of the thermohaline circulation followed measurements by Ellis in 1751 of the coldness of subsurface waters in the tropics (Warren, 1981), and Rumford (1800) suggested a model similar to Halley's for the atmosphere.

> But if the water of the ocean, which, on being deprived of a great part of its Heat by cold winds, descends to the bottom of the sea, cannot be warmed *where it descends*, as its specific gravity is greater than that of water at the same depth in warmer latitudes, it will immediately begin to spread on the bottom of the sea, and to flow towards the equator, and this must necessarily produce a current at the surface in an opposite direction.

The element that is different from the atmosphere is bottom topography, which can steer the deep currents from basin to basin.

An interesting feature of a surface-driven thermohaline circulation is the extreme asymmetry between rising and sinking regions. Whenever conditions produce surface water dense enough to sink to the bottom, it does so and spreads over the bottom. If, at a later stage, denser water is produced, it will spread over the bottom in its turn and displace the earlier bottom water upward. When surface water is warmed or freshened, on the other hand, it remains on the surface because it is light. Also, it tends to spread out across the surface due to gravity, covering it with a layer of low density that acts as a barrier against bottom-water formation. Only in extreme conditions can the density of the surface layer be made great enough to cause sinking to the bottom, so sinking tends to be a rare event, found only in localities in which extreme conditions are sufficiently frequent. The main such regions in the ocean appear to be in the Greenland Sea and the Weddell Sea [see Warren (1981)]. The analogous regions in the atmosphere are the regions of rising motion in the tropics [see Section 1.6 and Charney and Flierl (1981)].

The subdivision of the ocean floor into basins gives the circulation a special character since bottom water may form in one basin and then spill over into another. Extreme cases are provided by marginal seas such as the Mediterranean, which is connected to the rest of the ocean by a narrow strait with a shallow sill. The bottom water formed in the Mediterranean, where there is a large excess of evaporation over precipitation, is warm compared with the rest of the ocean and salty. It flows out of the Straits of Gibraltar at a rate of order 1 Mt s^{-1} (megatonne per sec) but is not dense enough to sink to the bottom, and instead spreads out into the Atlantic at an intermediate depth. It is readily identified by its saltiness [see, e.g., Worthington (1976)].

A useful model for discussing the nature of thermohaline circulation is one due to Baines and Turner (1969) [see also Turner (1973)] that is based on laboratory experiments with a source of dense fluid in a container. The descending dense plume entrains fluid from the surroundings in an amount that can be estimated, thus allowing a solution for the circulation to be obtained. When there are competing plumes with different buoyancy fluxes (Peterson, 1982), the weaker one may not penetrate very deeply even if only slightly less strong than its rival. Brass *et al.* (1982) suggest that in the Late Cretaceous, the buoyancy flux due to warm salty sources in marginal seas was greater than that due to cold sources, thus explaining the much warmer bottom waters of the period. (It is of interest to note that the multiple-source solution is also relevant to the distribution of sizes of clouds because these may be regarded as competing convective plumes.)

Details of the thermohaline circulation will in practice depend on dynamical factors to be discussed in later chapters. A review of theories and observations is given by Warren (1981).

Chapter Three

Properties
of a Fluid
at Rest

3.1 The Equation of State

The concept of the state of a fluid comes from comparing different samples of
fluid that are in equilibrium. If the two samples can exist in contact with each other
without a change in properties, the two samples have the same state; if not, their
states are different. There is a choice of variables that may be used to specify the state
of a fluid, but the set normally used to define the state comprises the pressure p,
temperature T, and chemical composition because these are the state variables that
are normally measured. Two samples in contact that have the same state must have
equal pressures, otherwise work will be done by one sample on the other; they must
have equal temperatures, otherwise heat will be transferred from one sample to the
other; and they must have the same concentrations of each of the constituents,
otherwise there will be changes of concentration caused by diffusion. [More detailed
discussions of the concept of state are given in textbooks on thermodynamics such
as that by Morse (1964).]

Equations of state relate properties of state to each other. If pressure, temperature,
and the concentrations of constituents are taken as the set of variables defining the
state, then the equations give other state properties as functions of these variables.
The most important of these equations is the one for the density of the fluid, which is
often simply called "the equation of state."

The two fluids of interest in the context of this book are air and seawater. It
happens that the concentrations of the salts in seawater are very nearly in a constant
proportion (the ions are in the mass ratio: chloride, 55%; sodium, 30%; sulfate, 8%;

magnesium, 4%; potassium, 1%; and calcium, 1%), so that the state may be defined very closely by giving only one concentration. The variable used to describe this concentration is the salinity s, which is equal to the mass of dissolved salts per unit mass of seawater. In practice the most accurate way of determining salinity is by determining the conductivity, and a quantity called practical salinity is now defined as a function of the conductivity (Dauphinee, 1980). For a history of the concept of salinity see Wallace (1974) and Lewis (1980). The equation of state for seawater

$$\rho = \rho(s, T, p) \tag{3.1.1}$$

has been found by experiment correct to better than five parts per million. Appendix 3 gives polynomial approximations that can be used to calculate ρ to the accuracy to which it is known. In addition, a table is included in Appendix 3 that gives ρ correct to 30 parts per million by linear interpolation.

The relative proportions of the gases that are found in air are also very closely constant, with the exception of water vapor. (The dry constituents are in the volume ratio: nitrogen, 78.1%; oxygen, 21.0%; and argon, 0.9%.) Hence the state of air can be very closely approximated by giving only one concentration, the *specific humidity q*, defined as the mass of water vapor per unit mass of air (equivalent ways of defining the humidity are given in Appendix 4). The equation of state of air is approximately given by the ideal gas laws. The ideal gas equation for dry air is

$$p_d = \rho_d R T, \tag{3.1.2}$$

where p_d is the pressure of dry air, ρ_d the mass of dry air per unit volume, T the absolute temperature, and R the gas constant for dry air.

$$R = R_*/m_a = 287.04 \quad \text{J kg}^{-1} \text{ K}^{-1}, \tag{3.1.3}$$

where

$$R_* = 8314.36 \quad \text{J kmol}^{-1} \text{ K}^{-1} \tag{3.1.4}$$

is the universal gas constant and

$$m_a = 28.966 \tag{3.1.5}$$

is the molecular mass of dry air. Similarly, for water vapor,

$$e = \rho_v R_v T, \tag{3.1.6}$$

where e is the water-vapor pressure, ρ_v the mass of water vapor per unit volume, and

$$R_v = R_*/m_w = 461.50 \quad \text{J kg}^{-1} \text{ K}^{-1} \tag{3.1.7}$$

since

$$m_w = 18.016 \tag{3.1.8}$$

is the molecular mass of water.

For a mixture of gases, the pressure p is the sum of the partial pressures of the constituents, i.e.,

$$p = p_d + e. \tag{3.1.9}$$

Also, by definition,

$$\rho_v = q\rho \tag{3.1.10}$$

and

$$\rho = \rho_d + \rho_v, \qquad \text{i.e.} \qquad \rho_d = \rho - \rho_v = (1 - q)\rho. \qquad (3.1.11)$$

Using (3.1.9)–(3.1.11) to substitute for p_d, ρ_v, and ρ_d, two equations may be deduced. The first,

$$e/p = q/(\epsilon + (1 - \epsilon)q), \qquad (3.1.12)$$

which follows by taking the quotient of (3.1.2) and (3.1.6) gives the vapor pressure e in terms of the specific humidity q, assuming that air behaves as a mixture of ideal gases (since specific humidity is easier to measure than vapor pressure, this formula is actually used to *define e* for air, even though it departs slightly from ideal behavior— see Smithsonian Meteorological Tables (List, 1951, Table 93)), where ϵ is defined by

$$\epsilon = m_w/m_a = R/R_v = 0.62197. \qquad (3.1.13)$$

The second equation, which follows by taking the sum of (3.1.2) and (3.1.6), is the *equation of state*. This can be written in the form

$$\rho = p/(RT(1 - q + q/\epsilon)) \equiv p/RT_v, \qquad (3.1.14)$$

where

$$T_v \equiv T(1 - q + q/\epsilon) = T(1 + 0.6078q) \qquad (3.1.15)$$

is called the *virtual temperature*, i.e., it is the temperature that dry air would need to have at the given pressure in order to have the same density as moist air, assuming ideal gas behavior (Smithsonian Meteorogical Tables, Table 72). Departures from ideal behavior (which are given in Table 84 of the Smithsonian Meteorological Tables) amount to 1 part in 1000, so are usually ignored. In fact, the equation for *dry air* ($q = 0$) is accurate enough for most purposes, so humidity effects on density are usually ignored as well. However, in extreme tropical conditions this may not be justified because the amount of water vapor air can hold goes up very rapidly with temperature, and for instance, T_v is 45°C for saturated air at 37°C at 1000 mb.

3.2 Thermodynamic Variables

The laws of thermodynamics lead to the introduction of further state variables whose dependence on p, T, and the concentration variable (q or s) needs to be ascertained. For detailed discussion of these laws, reference should be made to a textbook on thermodynamics such as that by Morse (1964). (*Note*: Textbooks differ considerably in their approach and there are difficulties in obtaining a completely logical treatment.) In this section, it will be assumed that the fluid has *fixed composition*, i.e., that q or s is constant. The state of the fluid therefore depends on two independent variables, usually chosen as p and T. For convenience the *specific volume* v_s will be used as a variable in place of the density $\rho = v_s^{-1}$. The subscript s is used to avoid confusion later with the velocity component v.

The first law of thermodynamics leads to the introduction of a quantity E, called the *internal energy* per unit mass, which is a property only of the state of the fluid.

The second law leads to the introduction of yet another state variable, η, the *specific entropy* (or entropy per unit mass), and to the relation

$$dE = T\,d\eta - p\,dv_s \tag{3.2.1}$$

among the five state variables E, T, η, p, and v_s. This is the fundamental equation from which thermodynamic relationships are derived, and it should be emphasized that the variables involved depend only on the *state* of the fluid. [Strictly speaking, this is only true when the fluid is in equilibrium, so that (3.2.1) is valid only for changes that are slow enough for the fluid to be almost in equilibrium—see Batchelor (1967, Section 3.4). In practice, the circumstances in which changes are not "slow enough" are rather rare and are not of importance for the topics discussed in this book.]

In Eq. (3.2.1), the term $T\,d\eta$ represents the increase in the *heat content* per unit mass of the fluid (and this is the only physical meaning that need be attached to η for the present). The rate of change of heat content with temperature is called the *specific heat* of the fluid. Since η is a function of *two* state variables (p, T), the change in η for a fixed change in temperature will depend on how p (or some other state variable) changes at the same time. If, for instance, the volume v_s is kept fixed, the value is given by

$$c_v \equiv T(\partial\eta/\partial T)_v = (\partial E/\partial T)_v \tag{3.2.2}$$

and is called the specific heat at constant volume. The right-hand side is derived from (3.2.1) and the subscript v denotes that the partial derivative is taken at constant v_s. Similarly, the specific heat c_p at constant pressure is given by

$$c_p \equiv T(\partial\eta/\partial T)_p = (\partial E/\partial T)_p + p(\partial v_s/\partial T)_p. \tag{3.2.3}$$

For the ocean and atmosphere, the two independent state variables used to describe the state are (for fixed composition) the pressure and temperature. Therefore it is desirable to have expressions for the rate of change of entropy with pressure as well as with temperature. It follows from (3.2.1) that

$$T(\partial\eta/\partial p)_T = (\partial E/\partial p)_T + p(\partial v_s/\partial p)_T \tag{3.2.4}$$

and by subtracting the p derivative of (3.2.3) from the T derivative of (3.2.4) that

$$(\partial\eta/\partial p)_T = -(\partial v_s/\partial T)_p. \tag{3.2.5}$$

Therefore

$$T\,d\eta = T(\partial\eta/\partial T)_p\,dT + T(\partial\eta/\partial p)_T\,dp,$$

i.e.,

$$T\,d\eta = c_p\,dT - T(\partial v_s/\partial T)_p\,dp, \tag{3.2.6}$$

which is the desired expression for the change of entropy (and hence of heat content per unit mass) with temperature and pressure. This is the form of the thermodynamic equation that will usually be employed rather than that of (3.2.1).

In the case of an ideal gas (and therefore a good approximation for air), there are some simplifications. First, the internal energy per unit mass is a function only of temperature and is zero when the absolute temperature is zero, so (3.2.2) becomes

$$E = c_v T. \tag{3.2.7}$$

Also, the derivative in (3.2.6) can be calculated from the equation of state (3.1.14), and (3.2.6) simplifies to give

$$T \, d\eta = c_p \, dT - v_s \, dp. \tag{3.2.8}$$

This is the expression normally used by meteorologists. It is slightly inaccurate because of departures from ideal gas behavior. Equation (3.1.14) can also be used in (3.2.3) to give [after use of (3.2.7)]

$$c_p = c_v + R(1 - q + q/\epsilon). \tag{3.2.9}$$

Another thermodynamic quantity that is often used is the *enthalpy* h, defined by

$$h = E + pv_s. \tag{3.2.10}$$

From (3.2.1) it follows that the differential dh satisfies

$$dh = T \, d\eta + v_s \, dp. \tag{3.2.11}$$

In the case of an ideal gas, (3.2.8) and (3.2.11) give on integration

$$h = c_p T. \tag{3.2.12}$$

3.3   Values of Thermodynamic Quantities for the Ocean and Atmosphere

For seawater, values of c_p at atmospheric pressure are found by experiment. Values at higher pressures can be found from the thermodynamic equation and the equation of state since (3.2.3) and (3.2.5) give

$$(\partial c_p/\partial p)_T = -T(\partial^2 v_s/\partial T^2)_p. \tag{3.3.1}$$

Direct measurements of c_p at high pressure are not available, but the right-hand side can be estimated from experimental data. A formula for c_p obtained in this way is used in Appendix 3. In addition, a table is included and values obtained by linear interpolation from the table are correct to 1 part in 1000.

Air follows the ideal gas laws quite closely, and these are sufficiently accurate for most purposes. Corrections for small departures from ideal gas behavior are given in the Smithsonian Meteorological Tables. The specific heat of an ideal gas is proportional to the gas constant R, the constant of proportionality depending on the number of atoms in a molecule [$\frac{5}{2}$ for monatomic, $\frac{7}{2}$ for diatomic, and 4 for polyatomic; see Morse (1964, Chapter 22)]. *Dry air* is composed of 99% diatomic molecules for which

$$c_p = \tfrac{7}{2}R. \tag{3.3.2}$$

For water vapor, which is triatomic, $c_p = 4R_v$. Thus for a mixture of two such ideal gases, the specific heat is given by

$$c_p = (1 - q)\tfrac{7}{2}R + q4R_v = \tfrac{7}{2}R(1 - q + 8q/7\epsilon),$$

i.e.,

$$c_p = 1004.6(1 + 0.8375q) \quad \text{J kg}^{-1} \text{ K}^{-1}. \tag{3.3.3}$$

For temperatures, pressures, and humidities normally found in the atmosphere, this formula is correct to 0.3%. Corrections need to be made (a) for departures from ideal gas behavior and (b) for the presence of monatomic gases like argon. These corrections are given in Table 88 of the Smithsonian Meteorological Tables. (*Note*: These tables give values per unit mass of *dry* air, whereas the above formula gives values per unit mass of *moist* air.)

The specific heat at constant volume c_v can be calculated from (3.2.9), which gives, when combined with (3.3.3),

$$c_v = \tfrac{5}{2}R(1 - q + 6q/5\epsilon), \tag{3.3.4}$$

so the ratio γ of specific heats is

$$\gamma \equiv c_p/c_v = [7(1 - q) + 8q/\epsilon]/[5(1 - q) + 6q/\epsilon]. \tag{3.3.5}$$

3.4   Phase Changes

In the atmosphere, changes from vapor to liquid phases of water are of great importance, as was seen in Chapter 1. At a given temperature T, vapor and liquid phases can coexist in equilibrium if the vapor pressure has the value $e_w(T)$, called the *saturation vapor pressure* (e_w is the value over a plane water surface; the subscript w is used to distinguish this from the value over a plane ice surface). If a change of phase occurs, an amount of heat $L_v(T)$ per unit mass is required to vaporize the liquid, L_v being called the *latent heat of vaporization*. It is only necessary to determine the values of e_w and L_v at one temperature, since the values at other temperatures can then be deduced from the laws of thermodynamics, as follows.

Let the subscript w refer to the liquid phase (water) and v to the vapor phase. Then by definition,

$$L_v(T) = T(\eta_v - \eta_w) = E_v - E_w + e_w(T)(v_v - v_w), \tag{3.4.1}$$

the right-hand side being obtained from (3.2.1), where v_v and v_w are the specific volumes of vapor and liquid phases, respectively. Differentiating with respect to T, the pressure being assumed kept at the saturation value $e_w(T)$, gives

$$dL_v/dT = \eta_v - \eta_w + T(d\eta_v/dT - d\eta_w/dT)$$
$$= dE_v/dT - dE_w/dT + e_w(dv_v/dT - dv_w/dT) + de_w/dT\,(v_v - v_w). \tag{3.4.2}$$

(Ordinary derivatives are used because the values at saturation depend only on temperature.) Using (3.2.1) for each phase in turn, the right-hand side of the equation gives

$$\eta_v - \eta_w = (v_v - v_w)\,de_w/dT. \tag{3.4.3}$$

Hence by the first half of Eq. (3.4.1),

$$de_w/dT = L_v/T(v_v - v_w). \tag{3.4.4}$$

This is known as the *Clausius–Clapeyron equation*.

To calculate how L_v changes with temperature, use (3.4.3) to substitute for $\eta_v - \eta_w$ in the first line of (3.4.2), and use (3.2.6) to calculate the derivatives of η. This gives

$$dL_v/dT = (v_v - v_w)\, de_w/dT + c_{pv} - c_{pw} - T(dv_v/dT - dv_w/dT)\, de_w/dT$$

$$= (c_{pv} - c_{pw}) + L_v(v_v - T\, dv_v/dT - v_w + T\, dv_w/dT)/T(v_v - v_w). \quad (3.4.5)$$

The term involving L_v is only one-half percent of the first term, and so can be neglected. The integrated form is

$$L_v(T) = L_v(T_0) + (c_{pv} - c_{pw})(T - T_0),$$

i.e., $\qquad\qquad\qquad\qquad\qquad\qquad\qquad\qquad\qquad\qquad\qquad\qquad (3.4.6)$

$$L_v(T) \simeq 2.5008 \times 10^6 - 2.3 \times 10^3\, t \quad \text{J kg}^{-1},$$

where T_0 is the zero point of the Celsius scale and t is the temperature in degrees Celsius.

The change of vapor pressure can now be found by integrating (3.4.4). Using the approximation $v_w \ll v_v$ and the ideal gas equation (3.1.6) for v_v, Eq. (3.4.4) becomes

$$e_w^{-1}\, de_w/dT \approx L_v/R_v T^2, \quad (3.4.7)$$

which, using (3.4.6), gives the approximate result

$$\ln(e_w(T)/e_w(T_0)) \approx R_v^{-1}(L_v(T_0) + \tfrac{1}{2}(c_{pv} - c_{pw})t)(T_0^{-1} - T^{-1}). \quad (3.4.8)$$

A similar formula, but one based on curve fitting, is given in Appendix 4, and this gives the saturation vapor pressure correct to 1 part in 500 for temperatures between -40 and $+40°$C. Similar results are also given for the saturation vapor pressure over ice. Considerations akin to these apply to the depression of the freezing point with pressure. Effects of pressure and salinity on the freezing point of seawater are given in Appendix 3.

3.5   Balance of Forces in a Fluid at Rest

Consider the balance of forces in a fluid at rest on the earth. A small volume of the fluid is subject to two types of forces: (i) the pressure due to the surrounding fluid and (ii) body forces due to gravity and to the rotation of the earth. If the pressure on the boundaries of the volume of fluid being considered were uniform, there would be no net force on the volume since the pressure forces would balance. The net force per unit volume is rather the pressure *gradient* ∇p, so the force per unit mass is $\nabla p/\rho$, where ρ is the density (see Fig. 3.1).

The body force per unit mass can be expressed as the gradient of a potential Φ, called the "geopotential." It is the sum of the earth's gravitational potential and the centrifugal potential associated with the earth's rotation [see Phillips (1973) and Chapter 4]. The direction of $\nabla\Phi$ is called "vertical" and the magnitude of $\nabla\Phi$ is called the acceleration g due to gravity. For most purposes it is sufficiently accurate to take g *as a constant, given approximately by*

$$g \approx g_c = 9.8 \quad \text{m s}^{-2}. \quad (3.5.1)$$

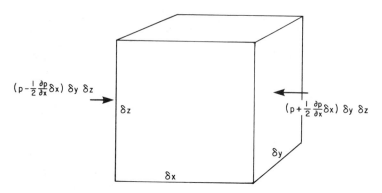

Fig. 3.1. The pressure forces acting on a small volume element of a fluid. The net force in the direction of the x axis is approximately $-\delta x\, \delta y\, \delta z\, \partial p/\partial x$, so in the limit, as the size of the element shrinks to zero, the force *per unit volume* on the element is $(-\partial p/\partial x, -\partial p/\partial y, -\partial p/\partial z) = -\nabla p$. The force *per unit mass* is therefore $-\rho^{-1}\nabla p$.

In practice, the shape of the earth is such that the value of g at sea level varies with latitude by $\pm 0.3\%$ and the inverse square law for gravitation results in a change in g of 0.3% for a change in height of 10 km (see Appendix 2). If the sea were at rest, its surface would coincide with the geopotential surface. This geopotential surface is called sea level and is defined as $\Phi = 0$. To a good approximation, so the vertical coordinate z measures distance upward from this reference level, so

$$\Phi \approx gz \approx g_c z. \tag{3.5.2}$$

Geopotential is sometimes given in units of the geopotential meter (gpm) defined by

$$1 \text{ gpm} = 9.8 \text{ m}^2 \text{ s}^{-2} \equiv 9.8 \text{ J kg}^{-1}, \tag{3.5.3}$$

so that the value of the geopotential in geopotential meters is close to the height in meters. Alternatively, the *geopotential height* Z is defined by

$$Z = \Phi/g_c, \tag{3.5.4}$$

so that the geopotential height in meters is numerically the same as the geopotential in geopotential meters. Equation (3.5.2) states that the geopotential height is approximately equal to the geometric height. Differences are less than 1% for heights within 22 km of the surface. [*Note*: The geopotential meter replaced an earlier unit called the dynamic or geodynamic meter that used a factor of 10 instead of 9.8 in (3.5.3)—see List (1951, Section IV). Also, other values of g_c are sometimes used, e.g., 9.80665 m s^{-2} is used for the U.S. standard atmosphere tables (NOAA/NASA/USAF, 1976).]

When a fluid is at rest and in equilibrium, the body force is balanced by the pressure force, and so

$$\rho^{-1}\,\nabla p + \nabla\Phi = 0. \tag{3.5.5}$$

This equation can be satisfied throughout the fluid only when p and ρ are constant on geopotential surfaces, i.e., p and ρ are functions of Φ only and satisfy

$$dp/d\Phi = -\rho. \tag{3.5.6}$$

Since surfaces of constant height z are defined in such a way that they are almost coincident with geopotential surfaces, the equilibrium condition can be expressed in the form

$$p = p(z), \qquad \rho = \rho(z), \tag{3.5.7}$$

where by (3.5.6) and the above definition of g,

$$dp/dz = -g\rho. \tag{3.5.8}$$

This is known as the *hydrostatic equation*. [For a more detailed discussion see Batchelor (1967, Section 1.4).]

In order to apply (3.5.8) to find the equilibrium pressure distribution in the ocean and atmosphere, it is necessary to know how the density varies. Consider first the ocean, for which the density varies (apart from a few exceptional places) by less than 2% from a constant value ρ_c given by

$$\rho_c = 1035 \quad \text{kg m}^{-3}. \tag{3.5.9}$$

Thus (3.5.8) integrates to give approximately

$$p = p_a - g\rho_c z, \tag{3.5.10}$$

where p_a is the atmospheric pressure at the surface. The value of p_a is close to 1 bar and the density is such that the pressure increases by close to 1 bar every 10 m. Because of this, pressure in the ocean is often quoted in decibars (db) because the pressure in decibars corresponds very closely to the depth in meters.

A more accurate description can be obtained by taking depth variations of density into account. Figure 3.2 shows the range of temperature, salinity, and density encountered for 98% of the ocean at each depth. Variations of potential density (see

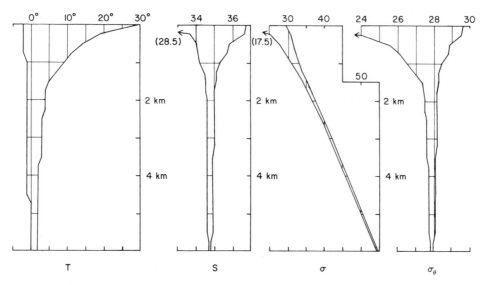

Fig. 3.2. The ranges of temperature T (in °C) and salinity S for 98% of the ocean as a function of depth [From Bryan and Cox (1972)], and the corresponding ranges of density σ and potential density σ_θ (see Appendix 3).

Section 3.7) are usually due mainly to temperature rather than salinity variations. Temperature gradients are usually small below 1500 m (1500 db or 150 bars). The region of large gradient at smaller depths is called the *thermocline*. The corresponding region of large gradient of potential density is called the *pycnocline*. Sometimes there is a density gradient due to salinity rather than temperature variations, the corresponding region being called a *halocline*. Often there is a region of large temperature gradient near the surface that appears only in summer and autumn. This is called the *seasonal thermocline*.

The *in situ* density increases with depth, as shown in Fig. 3.2, mainly because of the pressure effect. Taking into account the mean value of gravity at the appropriate depth (see Appendix 2), the midpoint density values make 1 bar equivalent to 9.95 m at the surface and to 9.69 m at a depth of 5000 m.

To calculate pressure variations with height in the atmosphere, the hydrostatic equation (3.5.8) is usually combined with the equation of state (3.1.2) for *dry* air since moisture effects on density are only significant in the tropics at near-surface levels, and even then the maximum departure from (3.1.2) is only 2%. The combination of the two equations gives

$$p^{-1}\,dp/dz = -g/RT, \qquad (3.5.11)$$

so that if T is known as a function of z, the pressure variations can be calculated by integrating (3.5.11) with respect to height. Altimetric tables are based on such an integration and a standard T-profile representing mean mid-latitude conditions. Figure 3.3 shows the "1976 U.S. Standard Atmosphere" (Minzner, 1977) up to a height of 86 km, in which range it is defined as a continuous function with piecewise constant gradients. Also shown is the range of temperatures encountered at each height and the pressures and densities corresponding to the standard atmosphere at intervals of 10 km. These quantities are tabulated in "U.S. Standard Atmosphere 1976" (NOAA/NASA/USAF, 1976) with other tables such as acceleration due to gravity, viscosity, and thermal conductivity as functions of height.

The atmosphere is divided into different segments, as shown in Fig. 3.3, because of its different characteristics in different height ranges. The *troposphere* is characterized by rather strong vertical mixing (it is the "convective layer" in the radiative–convective models discussed in Chapter 1), largely associated with latent heat effects and clouds. This layer contains 80% of the mass and nearly all of the water vapor and clouds. It is capped by the *tropopause*, which is at 11 km in the model atmosphere. The *stratosphere* is poorly mixed as shown by the persistence of thin layers of aerosol and by the long residence time of debris from past nuclear explosions. The strong stability is associated with the increase in temperature with height that results from radiative balances, as described in Chapter 1. This increase stops at the *stratopause* and only one-thousandth of the atmospheric mass lies above this level. The *mesosphere* is a region in which temperature decreases again, the *mesopause* that marks the top being at about 86 km. The region from the tropopause up to levels of about 100 km is also called the *middle atmosphere*.

Above the mesopause, the proportions of the different atmospheric constituents do not remain the same because of diffusive separation. The temperature increases rapidly with height in the so-called *thermosphere*, reaching values that may be 600 K

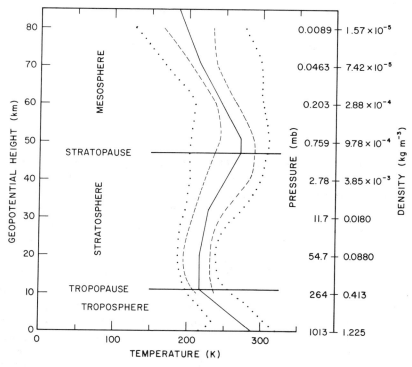

Fig. 3.3. Temperature variation with geopotential height for the U.S. Standard Atmosphere (solid line). This consists of straight-line segments with breaks at 11, 20, 32, 47, 51, and 71 km. The surface temperature is 15°C and the gradients, starting from the surface, are −6.5, 0. 1.0, 2.8, 0, −2.8, and −2.0 K km⁻¹. The dashed line shows the lowest and highest monthly mean temperatures obtained for any location between equator and pole, whereas the dotted line shows estimates of the 1% maximum and minimum temperatures that occur during the warmest and coldest months, respectively, in the most extreme locations. The scale at the right gives pressures and densities at 10-km intervals for the standard profile. [From NOAA/NASA/USAF, 1976.]

in periods of quiet sun but more like 2000 K in active sun periods. Besides the subdivisions based on temperature structure, there are others based on other properties. For instance, important electromagnetic effects are found in the region between 80 and 500 km, called the *ionosphere*, and the region above, where complete magnetic control is encountered, is called the *magnetosphere*. At these levels, it is also possible for energetic particles to emerge from the earth's gravity field, so this region is also called the *exosphere*.

Between the surface and the 70 km level, the absolute temperature for the standard atmosphere is within 15% of a constant value of $T_c = 250$ K, so a rough approximation to the integral of (3.5.11) is provided by the solution for an isothermal atmosphere, namely,

$$p = p_a \exp(-gz/RT_c) = p_a \exp(-z/H_s). \qquad (3.5.12)$$

The quantity H_s is the height at which the pressure has fallen to e^{-1} of its surface value (i.e., to about 370 mb) and is called the *scale height* of the atmosphere, given by

$$H_s = RT_c/g. \qquad (3.5.13)$$

For the median value of $T_c = 250$ K, H_s is 7.4 km. It is also possible to define H_s as a continuously varying quantity such that H_s^{-1} is defined by the left-hand side of (3.5.11). With this definition, the standard atmosphere gives 8.4 km for H_s at the ground, 6.4 km at 11 km, and values lie between these limits up to 71 km.

3.6   Static Stability

A fluid can be at rest and in equilibrium provided that the density ρ is a function of z only. But is the equilibrium a stable one? The stability can be tested by considering the exchange of two parcels of fluid at different levels. If the parcel moved to the higher level finds itself to be heavier than the surrounding fluid, gravitational forces will cause the parcel to descend back toward its original level, and in this case the equilibrium is stable. If, however, the parcel is lighter than its surroundings, the equilibrium is unstable.

In order to make this calculation, it is necessary to know the changes in properties of the parcel that is displaced. These changes can be calculated if it is assumed that the time scale of the motion is too short for any change in *composition* or of *heat content* of the parcel to take place. Two alternative words are used to describe changes in which the heat content is not changed: *adiabatic* (meaning no exchange of heat with surroundings) or *isentropic* (meaning no change of entropy). For an isentropic change, (3.2.6) states that the temperature change dT is related to the pressure change dp by the formula

$$dT = \frac{T}{c_p}\left(\frac{\partial v_s}{\partial T}\right)_{p,s} dp = -\frac{T}{\rho^2 c_p}\left(\frac{\partial \rho}{\partial T}\right)_{p,s} dp = \frac{\alpha T}{\rho c_p} dp, \qquad (3.6.1)$$

where

$$\alpha = -\rho^{-1}(\partial \rho/\partial T)_{p,s} \qquad (3.6.2)$$

is called the *thermal expansion coefficient*. The subscript s is used to emphasize the fact that the composition is kept fixed, i.e., the salinity s is constant for the case of seawater. (In Section 3.2 the same assumption was made, but the subscript s was not used.) The same formulas apply in the atmosphere with humidity kept fixed, but in that case they are simplified because for an ideal atmosphere, (3.1.14) implies that

$$\alpha = T^{-1}. \qquad (3.6.3)$$

(Departures from ideal gas behavior alter this by less than 0.3% in normal atmospheric conditions.) Now using the hydrostatic equation (3.5.8), Eq. (3.6.1) gives

$$dT = -\Gamma\, dz, \qquad (3.6.4)$$

where

$$\Gamma = g\alpha T/c_p \qquad (3.6.5)$$

is called the *adiabatic lapse rate*. In the special case of an ideal atmosphere,

$$\Gamma = g/c_p, \qquad (3.6.6)$$

which is approximately 10 K km^{-1}. More accurate values can be calculated, using the expression for g in Appendix 2 and formula (3.3.3) for the specific heat. For seawater, Γ can be calculated from (3.6.5), using values given in Appendix 3.

The density change $d\rho$ for a parcel of fixed composition that is moved isentropically is given by

$$d\rho = (\partial\rho/\partial p)_{T,s}\, dp + (\partial\rho/\partial T)_{p,s}\, dT = \left[(\partial\rho/\partial p)_{T,s} - \alpha^2 T/c_p\right] dp$$

$$= \rho\left[-g(\partial\rho/\partial p)_{T,s} + \alpha\Gamma\right] dz, \tag{3.6.7}$$

where (3.6.1), (3.6.2), (3.6.5), and the hydrostatic equation (3.5.8) have been used. For stability, $d\rho$ must exceed the density change of the surroundings, namely,

$$\frac{d\rho}{dz}\, dz \equiv \left[\left(\frac{\partial\rho}{\partial p}\right)_{T,s}\frac{dp}{dz} + \left(\frac{\partial\rho}{\partial T}\right)_{p,s}\frac{dT}{dz} + \left(\frac{\partial\rho}{\partial s}\right)_{p,T}\frac{ds}{dz}\right] dz. \tag{3.6.8}$$

In other words, the condition for stability is

$$\alpha(dT/dz + \Gamma) - \beta\, ds/dz > 0, \tag{3.6.9}$$

or

$$\alpha\, dT/dz + c_p^{-1} g\alpha^2 T - \beta\, ds/dz > 0, \tag{3.6.10}$$

where

$$\beta = \rho^{-1}(\partial\rho/\partial s)_{p,T} \tag{3.6.11}$$

is the expansion coefficient for salinity. The formulas or table in Appendix 3 give the coefficients that appear in (3.6.10) for the case of seawater. Sometimes the ocean is only marginally stable, and accurate values are required.

For the atmosphere, (3.6.9) applies with specific humidity q replacing s. For ideal gas behavior, the expression simplifies to

$$dT_v/dz + \Gamma > 0, \tag{3.6.12}$$

where T_v is the virtual temperature. Except in some tropical situations, the gradient of T is not very different from the gradient of T_v, and so (3.6.12) is used with T replacing T_v.

3.7 Quantities Associated with Stability

The stability considerations of the last section lead to the definition of some useful quantities that will be introduced in this section.

3.7.1 Buoyancy Frequency (or Brunt–Väisälä Frequency)

It follows from (3.6.9) that the quantity N^2, defined by

$$N^2 = g\alpha(dT/dz + \Gamma) - g\beta\, ds/dz = g\alpha\, dT/dz + c_p^{-1} g^2\alpha^2 T - g\beta\, ds/dz, \tag{3.7.1}$$

gives a measure of the degree of stability; for when N^2 is positive, the medium is stable and when N^2 is negative, the medium is unstable. In a stable medium, N is

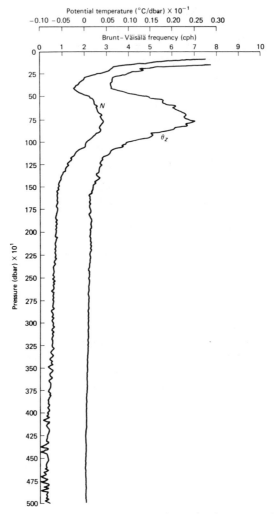

Fig. 3.4.   A profile of buoyancy frequency *N* in the ocean. [From the North Atlantic near 28°N, 70°W, courtesy of Dr. R. C. Millard.]

real and has the dimensions of frequency. It is, in fact, the frequency of vibration of a parcel in purely vertical motion. It has also been called the stability frequency, the Brunt frequency, and the Väisälä frequency. A sample profile for N for the ocean is shown in Fig. 3.4. Values in the large gradient regions of the upper ocean typically reach 0.01 s^{-1} corresponding to a period $2\pi/N$ of 10 min. The value of N in the troposphere is also 0.01 s^{-1} (for the standard atmosphere) with 70% larger values in the lower stratosphere. In contrast, values for the deep ocean are 0.001 s^{-1} or smaller.

3.7.2   Potential Temperature

This is defined as the temperature θ that a parcel of fixed composition would acquire if moved adiabatically to a given pressure level p_r, usually taken as 1 bar.

(This is called the reference pressure.) Its value can be obtained by integrating (3.6.1) with respect to pressure from the observed value to the reference pressure. Values for seawater are given in Appendix 3. For *air*, an explicit formula can be obtained if ideal gas behavior is assumed. Then substitution of the equation of state (3.1.14) and the expression (3.3.3) for the specific heat in (3.6.1) give

$$dT/T = \kappa \, dp/p, \tag{3.7.2}$$

where

$$\kappa \equiv \frac{\gamma - 1}{\gamma} = \frac{2}{7}\left(\frac{1 - q + q/\epsilon}{1 - q + 8q/7\epsilon}\right) \approx \frac{2}{7}(1 - 0.23q) \tag{3.7.3}$$

and γ is the ratio of specific heats given by (3.3.5). Integrating from pressure p and temperature T to the pressure p_r and temperature θ gives

$$\theta/T = (p_r/p)^\kappa. \tag{3.7.4}$$

Even in extreme conditions, the factor involving q in (3.7.3) is only 1% less than unity, and so is usually ignored. In that case

$$\theta/T = (p_r/p)^{2/7} = (1000/p)^{2/7}, \tag{3.7.5}$$

the latter expression being valid if p is in millibars. Table 75 of the Smithsonian Meteorological Tables gives θ as calculated by this formula.

3.7.3 Relation between Potential Temperature and Entropy

By definition, θ is constant when the entropy is constant, and so for a fluid of fixed composition, η is a function only of θ. To find this function in the general case, use is made of (3.2.6) at a fixed pressure p_r equal to the reference pressure. At this pressure, the temperature is equal to θ, so (3.2.6) gives

$$d\eta/d\theta = c_p(p_r, \theta)/\theta. \tag{3.7.6}$$

For an ideal gas [cf. (3.3.3)], c_p is independent of pressure and temperature, so (3.7.6) integrates to give

$$\eta = c_p \ln \theta + \text{const.} \tag{3.7.7}$$

3.7.4 Use of Potential Temperature as a State Variable

Sometimes it is convenient to use θ and p as state variables in place of T and p. This is equivalent to using entropy and pressure as the state variables. The change of density for a parcel of fixed composition moved isentropically [cf. (3.6.7)] is

$$d\rho = (\partial\rho/\partial p)_{\theta,s} \, dp. \tag{3.7.8}$$

The density change of the surroundings is given by (3.6.8) with θ replacing T, so the condition for stability is

$$g^{-1}N^2 = \alpha' \, d\theta/dz - \beta' \, ds/dz > 0, \tag{3.7.9}$$

where

$$\alpha' = -\rho^{-1}(\partial\rho/\partial\theta)_{p,s} = \alpha(\partial T/\partial\theta)_{p,s} \tag{3.7.10}$$

and

$$\beta' = \rho^{-1}(\partial\rho/\partial s)_{p,\theta} = \beta - \alpha(\partial T/\partial s)_{p,\theta} = \beta + \alpha'(\partial\theta/\partial s)_{p,T}. \tag{3.7.11}$$

Values of $(\partial\theta/\partial s)_{p,T}$ for seawater are given in Appendix 3, and show that β' differs from β by less than 0.3%. To find α', use (3.2.6), which implies that

$$\left(\frac{\partial T}{\partial\theta}\right)_{p,s} = \frac{T}{c_p}\frac{d\eta}{d\theta} = \frac{Tc_p(p_r, \theta)}{\theta c_p(p, T)}, \tag{3.7.12}$$

where (3.7.6) is used for $d\eta/d\theta$. Hence

$$\alpha' = \alpha(p, T)Tc_p(p_r, \theta)/\theta c_p(p, T). \tag{3.7.13}$$

In the case of an ideal atmosphere, c_p is independent of p and T and $\alpha T = 1$; therefore

$$\alpha' = \theta^{-1}. \tag{3.7.14}$$

It follows from (3.7.9) that if humidity effects are ignored,

$$N^2 = g\theta^{-1}\,d\theta/dz, \tag{3.7.15}$$

and so stability in the atmosphere is measured by the potential temperature gradient.
The quantity

$$c_s^2 = (\partial p/\partial\rho)_{\theta,s}, \tag{3.7.16}$$

whose inverse appears in (3.7.8), is useful in a number of contexts. c_s is the speed of sound in the medium. For an ideal atmosphere, (3.7.4) and the equation of state (3.1.14) can be used to give p as a function of ρ, θ, and q. The ρ-derivative for fixed θ and q (in the general treatment, q and s are interchangeable) gives

$$c_s^2 = \frac{\gamma p}{\rho} = \frac{7}{5}RT\frac{(1 - q + q/\epsilon)(1 - q + 8q/7\epsilon)}{(1 - q + 6q/5\epsilon)}. \tag{3.7.17}$$

For small q, the factor involving q is approximately $1 + 0.51q$ and can usually be ignored, in which case

$$c_s^2 \approx \tfrac{7}{5}RT. \tag{3.7.18}$$

In the case of the ocean, c_s can be measured in situ, and thus (3.7.16) can be used to find information about the equation of state at high pressures.

3.7.5 Potential Density

This is defined as the density ρ_{pot} that a parcel of fixed composition would acquire if moved adiabatically to a given pressure level p_r (called the reference pressure), usually taken as 1 bar. For a fluid of fixed composition, ρ_{pot} is a function of η only, and hence a function of θ only. The relation between ρ_{pot} and θ is obtained by applying the equation of state at the reference pressure p_r where the temperature is equal to θ, i.e.,

$$\rho_{pot} = \rho(p_r, s, \theta). \tag{3.7.19}$$

For dry air, (3.1.14) gives

$$\rho_{\text{pot}} = p_r/R\theta, \qquad (3.7.20)$$

i.e., potential density is inversely proportional to θ. Because of this relationship, potential density is not used much in meteorology. In oceanography, on the other hand, potential density is useful, particularly near the reference pressure, where its vertical gradient determines the stability. Since its value is always close to 1000 kg m^{-3}, the potential density is usually expressed in terms of a quantity called σ_θ (see Appendix 3). Figure 3.2 shows the range of values of σ_θ found in the ocean.

The stability condition can be found by the same method used at the beginning of Section 3.7.4, namely, by comparison of the density change of a parcel moving isentropically with that of the surroundings. The condition for stability obtained is

$$g^{-1}N^2 = -C\rho_{\text{pot}}^{-1}\, d\rho_{\text{pot}}/dz - B\, ds/dz > 0, \qquad (3.7.21)$$

where

$$C = \frac{\rho_{\text{pot}}}{\rho}\left(\frac{\partial\rho}{\partial\rho_{\text{pot}}}\right)_{p,s} = \frac{\rho_{\text{pot}}}{\rho}\left(\frac{\partial\rho}{\partial\theta}\right)_{p,s}\left(\frac{\partial\rho_{\text{pot}}}{\partial\theta}\right)_{s}^{-1} \qquad (3.7.22)$$

and

$$B = \frac{1}{\rho}\left(\frac{\partial\rho}{\partial s}\right)_{p,\rho_{\text{pot}}} = \frac{1}{\rho}\left(\frac{\partial\rho}{\partial s}\right)_{p,\theta} - \frac{1}{\rho}\left(\frac{\partial\rho}{\partial\rho_{\text{pot}}}\right)_{p,s}\left(\frac{\partial\rho_{\text{pot}}}{\partial s}\right)_{\theta}. \qquad (3.7.23)$$

Expressions for B and C in terms of quantities tabulated in Appendix 3 can be obtained, using the above formula together with Eqs. (3.7.19), (3.7.10)–(3.7.13), and the definitions (3.6.2) and (3.6.11) of the expansion coefficients. The result is

$$C = \alpha(p, T)c_p(p_r, \theta)T/\alpha(p_r, \theta)c_p(p, T)\theta, \qquad (3.7.24)$$

$$B = \beta' - C\beta(p_r, \theta) \simeq \beta(p, T) - C\beta(p_r, \theta), \qquad (3.7.25)$$

where all quantities are evaluated at the given salinity s. The approximation in (3.7.25) is based on the fact that β' is within 0.3% of β. There has been a tendency to assume that it is a reasonable approximation to take $C = 1$ and $B = 0$, i.e., that the stability depends only on the potential density gradient. However, this is not true (Lynn and Reid, 1968), largely because of the variations of α with pressure (Gill, 1973). In fact, for the temperature range found at depths over 3000 m, C is more than twice the surface value $C = 1$, and $-B$ is greater than β.

3.8 Stability of a Saturated Atmosphere

When the atmosphere is saturated with water vapor, the preceding stability arguments are no longer applicable. The calculations of buoyancy changes for *descending* air are still valid because the amount of moisture a descending parcel can hold generally increases. For *ascending* air, however, the amount of moisture that can be retained decreases. Therefore condensation takes place, releasing latent heat and so making the parcel more buoyant than would otherwise be the case. A lapse rate Γ_s can be calculated on the assumption that the air remains saturated and that all the liquid water formed by condensation is removed as precipitation (without affecting the buoyancy of the parcel). The process envisaged is not truly adiabatic since material is continually being removed, so Γ_s is called the *pseudoadiabatic lapse rate* (or sometimes simply the *moist adiabatic lapse rate*). Γ_s is smaller than Γ (the dry adiabatic lapse rate). Thus it is possible that in a saturated atmosphere one of the following is valid: (a) The lapse rate is less than Γ_s, in which case the equilibrium is stable. (*Note*: Layers in which the sign of the temperature gradient is reversed are called *inversions*. In the atmosphere, this implies greater than usual stability.) (b) The lapse rate lies between Γ and Γ_s. In this case parcels displaced downward will tend to be restored, whereas saturated parcels displaced upward will continue to move upward. The atmosphere is said to be *conditionally unstable* when the lapse rate is between Γ and Γ_s, whatever the moisture content. (c) The lapse rate exceeds Γ, in which case the situation is clearly unstable.

To calculate the lapse rate Γ_s one needs to consider changes in which the mass of dry air remains constant, even though the mass of water vapor changes through condensation. If the humidity is expressed in terms of the *mixing ratio r* (see Appendix 4), defined as the mass of vapor divided by the mass of dry air, then the change in vapor content per unit mass of *dry* air is dr. The change per unit mass of *moist* air is therefore $dr/(1 + r)$, i.e., by (A4.2) the change is

$$dr/(1 + r) \equiv dq/(1 - q). \tag{3.8.1}$$

In the pseudoadiabatic process this is equal to the mass of condensed water per unit mass of moist air, and q (or r) always has its saturation value q_w (or r_w). The change in heat content per unit mass of moist air is thus L_v times (3.8.1), and so the entropy equation is [cf. (3.2.6)]

$$L_v \, dq_w/(1 - q_w) + c_p \, dT - T(\partial v_s/\partial T)_p \, dp = 0. \tag{3.8.2}$$

This differs from Eq. (3.6.1) for a dry adiabatic process only by the addition of the term involving the latent heat L_v.

Now the saturation specific humidity q_w (see Appendix 4) is a known function of temperature and pressure, so dq_w can be written as

$$dq_w = (\partial q_w/\partial T)_p \, dT + (\partial q_w/\partial p)_T \, dp.$$

Substituting in (3.8.2) and using (3.6.2) and (3.6.4) give

$$\Gamma_s \left(1 + \frac{L_v}{c_p(1 - q_w)} \left(\frac{\partial q_w}{\partial T} \right)_p \right) = \Gamma \left(1 - \frac{\rho L_v}{\alpha T(1 - q_w)} \left(\frac{\partial q_w}{\partial p} \right)_T \right). \tag{3.8.3}$$

Values of Γ_s are given in Table 79 of the Smithsonian Meteorological Tables, and Appendix 4 gives an approximate formula.

An alternative form of (3.8.2) can be obtained by using potential temperature as the variable in place of the temperature. If, moreover, the ideal gas approximation is made, use of (3.2.6) and (3.7.6) in (3.8.2) gives

$$c_p^{-1} T^{-1} L_v \, dq_w/(1 - q_w) + d\theta/\theta = 0. \tag{3.8.4}$$

Now the curve on which the temperature varies with pressure in accordance with (3.8.2) is called a *saturation pseudoadiabat* or moist adiabat (Table 78 of Smithsonian

Meteorological Tables). The quantity that is constant on this curve is called the *equivalent potential temperature,* θ_e^* for saturated air. It is defined as the potential temperature that a parcel would have if all its moisture were condensed out and the latent heat thus released were used to warm the parcel. The value of θ_e^* can be obtained by integrating (3.8.4). Since the temperature change in the process is not very large and q_w is small, an approximate integral of (3.8.4) is

$$L_v q_w / c_p T + \ln(\theta/\theta_e^*) = 0,$$

i.e.,

$$\theta_e^* = \theta \exp(L_v q_w / c_p T). \tag{3.8.5}$$

An alternative label for moist adiabats is the *wet-bulb potential temperature* θ_w^* for saturated air, defined as the temperature at which the moist adiabat crosses the line $p = p_r = 1000$ mbar. For instance $\theta_w^* = 0°C$ refers to the same curve as $\theta_e^* = 10.2°C$, whereas $\theta_w^* = 10, 20,$ and $30°$ correspond to $\theta_e^* = 31.2, 62.3,$ and $113.0°C$, respectively (see Fig. 3.6).

The quantity θ_e^* defined above is a function of p and T only since it refers solely to saturation conditions. However, an equivalent potential temperature θ_e (which depends on r, p, and T) can be defined for any parcel, whether saturated or not, as the value of θ_e^* acquired after adiabatic expansion to the saturation level. In other words, θ_e is the temperature acquired if a parcel is expanded (adiabatically until it reaches saturation, and pseudoadiabatically thereafter) until all its moisture is removed, and then is compressed adiabatically to the reference pressure p_r (normally 1 bar). It follows that θ_e is constant for a parcel *whether or not it is saturated*, provided that the changes are adiabatic when unsaturated and pseudoadiabatic when saturated.

When air is unsaturated, there are two terms used to express the possibility of instability due to moisture effects during upward motion. The first term, *conditional instability*, meaning the lapse rate is between Γ_s and Γ, takes no account of the relative

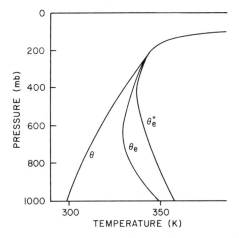

Fig. 3.5. Profiles of potential temperature θ and equivalent potential temperature θ_e for the tropical atmosphere. These profiles are means for the West Indies rainy season compiled by Jordan (1958). The third profile is of θ_e^*, the equivalent potential temperature of a hypothetically saturated atmosphere with the same temperature at each level.

moisture levels of neighboring parcels. The potential for instability can, however, exist even when the profile is stable everywhere in the sense that the lapse rate is everywhere less than Γ_s. This can occur when the upper of two parcels is relatively dry, so that if both parcels are forced upward beyond their lifting condensation levels, its potential temperature becomes less than that of the lower parcel. The condition for this so-called *convective* (or *potential*) *instability* is that

$$d\theta_e/dz < 0. \tag{3.8.6}$$

Figure 3.5 shows typical profiles for potential temperature θ and equivalent potential temperature θ_e for the *tropics*. $d\theta/dz$ is everywhere positive, indicating that the tropical atmosphere in these conditions is stable to dry adiabatic processes. $d\theta_e/dz$, on the other hand, is negative below 700 mb, showing that the atmosphere is convectively unstable in this region, whereas negative values of $d\theta_e^*/dz$ (indicating conditional instability) occur up to even higher levels. Despite this instability, which is typical throughout the tropics, deep convection takes place only in a small fraction of the total area (Riehl, 1979).

3.9   Graphical Representation of Vertical Soundings

The basic properties of moist air can be given graphically by diagrams (sometimes called pseudoadiabatic charts) such as those shown in Fig. 3.6, where potential temperature θ, saturation mixing ratio r_w, and equivalent potential temperature θ_e^* for saturated air are shown as functions of pressure p and temperature T. Consider first Fig. 3.6a (sometimes called a Stüve diagram), for which the ordinate is p^κ and the abscissa is T. This choice makes the θ contours (i.e., the dry adiabats shown by the sloping solid lines) straight by virtue of formula (3.7.4) for potential temperature. The slope of the lines is inversely proportional to θ. The curves for the saturation mixing ratio, given by the formulas in Appendix 4, turn out to be close to straight lines, and thus are indicated by the short sloping lines at the upper and lower edges of the diagram. Finally, the saturation pseudoadiabats, or contours of θ_e^*, are given approximately by (3.8.5) and are shown by the dashed curves. As their definition requires, they asymptotically approach the dry adiabats with the corresponding value of θ as $p \to 0$.

Figure 3.6b, which is not shown in the same detail, is known as a tephigram ($T-\phi$ gram) because the axes (which are tilted through 45° and shown by solid lines) are $\log \theta$ (which is proportional to entropy, a quantity often denoted by ϕ rather than by η as in this book) and T. The pressure contours in this representation are slightly curved but near horizontal. An advantage of this diagram is that area is proportional to energy, and any diagram with this property is called a thermodynamic diagram. Other examples are discussed by Hess (1959).

Radiosonde soundings can be represented on such diagrams by lines indicating how temperature T and, say, dew-point temperature T_d vary with pressure. The diagram (which may then be called an aerological diagram) can be used to make deductions about stability, the effect of lifting of samples of air, etc. An example

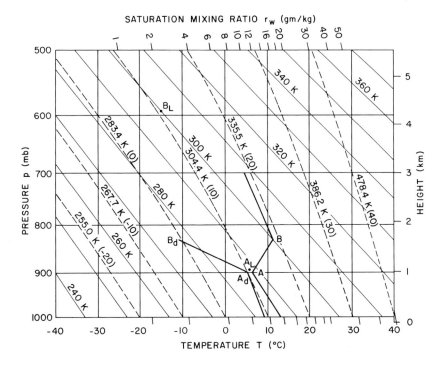

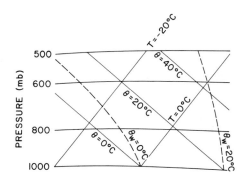

Fig. 3.6.   Examples of two forms of pseudoadiabatic chart that display the properties of moist air in graphic form. (a) Axes are p^κ and T (Stüve diagram). The sloping solid straight lines are dry adiabats, i.e., contours of potential temperature θ, with a contour interval of 10 K. The r_w contours are very nearly straight, and are indicated by the sloping line segments at the top and bottom. The saturation pseudoadiabats are marked by dashed lines and are labeled by the value of equivalent potential temperature θ_e (in degrees Kelvin) with the corresponding value of the wet-bulb potential temperature θ_w (in degrees Celsius) in brackets. The contour interval for the latter quantity is 10°C. The thick lines show a sample sounding of temperature T on the right and dew point T_d on the left. B_L is the lifting condensation level corresponding to the parcel at 830 mb. with temperature point B and dew point B_d. It is the intersection of the θ contour ($\theta = 300$ K) through B and the r_w contour ($r_w = 2$ gm kg^{-1}) through B_d. A_L is the corresponding intersection for the point $A - A_d$. (b) Another form of pseudoadiabatic chart, the tephigram. The axes, which are for $\log\theta$ and T, are rotated through 45°, so that the pressure contours are approximately horizontal. The dashed lines are moist pseudoadiabats.

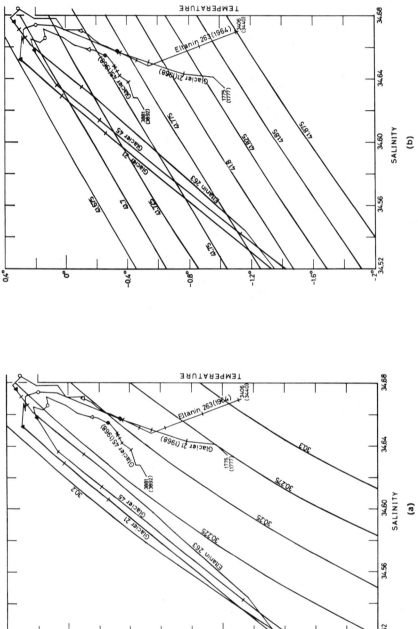

Fig. 3.7. Temperature–salinity curves for three stations in the Antarctic (Weddell Sea) shown against the isopycnals (constant-density lines) for two different depths: (a) 500 m and (b) 3000 m. Depths (in meters) are 500 (▲), 1000 (○), 2000 (△), 3000 (●), and 1500 or 3500 (×). The short line segments are at intervals of 100 m. The diagram illustrates that stability can be judged only in relation to the isopycnals for the depth concerned. If the isopycnals shown in (b) were appropriate at all depths, one would conclude that, in the upper levels, the density distribution was unstable, whereas diagram (a) shows this not to be the case. Similarly, if the isopycnals shown in (a) were supposed to be appropriate for all depths, the deep water at Glacier station 45 would appear to be unstable, but diagram (b) shows that this is not so. [From Gill (1973).]

(from Capetown, South Africa) is shown by the thick lines in Fig. 3.6a. It is clear at a glance that the air in the lower kilometer is much closer to saturation than is the air above because the difference between T and T_d is relatively small there. Stability properties can be assessed by comparing the slope of the temperature sounding with the slopes of the dry and moist adiabats. In the example, the air near the ground is conditionally stable since the observed slope is between that of the two adiabats. The bottom kilometer of relatively moist air is capped by an inversion, where T increases with height (up to the 830 mb level), and dry air is found at higher levels. It is so dry that only the slope relative to the dry adiabat is relevant, and this indicates stability (θ increases with height).

If a parcel of air such as that (marked B) at the top of the inversion layer is lifted adiabatically, the temperature changes experienced are those obtained by following the dry adiabat (in this case $\theta = 300$ K) until the lifting condensation level is reached. This level can be found by using the fact that the mixing ratio r for the parcel is equal to r_w at the dew-point temperature T_d (marked B_d in the diagram), which in this case is 2 gm kg^{-1}. The $r_w = 2$ and $\theta = 300$ lines intersect at the point B_L, so the lifting condensation level is the value of p at this point, namely, 590 mb. The corresponding moist adiabat is given by $\theta_e^* = 306$ K, so the equivalent potential temperature of the air at the top of the inversion is $\theta_e = 306$ K. It is hardly likely that a parcel from the top of the inversion layer would be lifted so far, but a similar construction for a parcel $(A - A_d)$ at the base of the inversion gives an intersection at A_L, showing that the lifting condensation level is only about 10 mb above the actual level, and the equivalent potential temperature is again 306 K. It follows that the inversion layer is neutral (i.e., $d\theta_e/dz = 0$) as far as the convective instability criterion (see Section 3.8) is concerned, and similar calculations for the parcel at the surface show that the same holds true for the moist layer below the inversion. Further constructions, using thermodynamic diagrams, are discussed, e.g., by Godske *et al.* (1957, Chapter 3).

The state of seawater depends on *three* variables: temperature, salinity, and pressure, so a three-dimensional diagram would be needed to assess stability properties. Two-dimensional representations of soundings can be in the form of temperature and salinity versus pressure (or depth) as in Fig. 3.2 or in terms of a *temperature–salinity* diagram (or potential-temperature–salinity diagram) as shown in Fig. 3.7. A useful feature of such a diagram is due to the fact that if two parcels of water with different values of T and S mix, the values of (T, S) for the mixture lie on the straight line joining the two original points. Often the T, S values for a given station lie on a straight line over a significant range of depths, and these points can be interpreted as mixtures of suitably defined water masses in proportions that vary along the straight line. Stability properties can be assessed by comparing the slope of such a line with that of the isopycnals, but this is only valid if the isopycnals are those appropriate to the depth range concerned, as Fig. 3.7 shows.

Chapter Four

Equations
Satisfied
by a Moving Fluid

4.1 Properties of a Material Element

When a fluid is in motion, its properties are functions both of spatial position

$$\mathbf{x} \equiv (x, y, z) \tag{4.1.1}$$

and time t. In other words, for any property γ,

$$\gamma = \gamma(x, y, z, t) \equiv \gamma(\mathbf{x}, t). \tag{4.1.2}$$

(The symbol $\equiv$, meaning "is identical to," is used here to relate different ways of writing the same expression, e.g., in scalar notation on one side and vector notation on the other side.) Now the concepts of the state of a fluid apply to a particular sample (or "parcel") that will move around when the fluid is in motion. Since nearby particles of fluid may move apart in time, it is necessary to think of an infinitesimally small sample that will retain its identity. This will be called a material element of fluid (Batchelor, 1967, Chapter 2).

Suppose now that this material element has position $\mathbf{x}$ at time t given by

$$\mathbf{x} = \mathbf{x}(t). \tag{4.1.3}$$

Then the property γ for this material element will vary with time according to

$$\gamma = \gamma(x(t), y(t), z(t), t) \equiv \gamma(\mathbf{x}(t), t). \tag{4.1.4}$$

It follows that the rate of change of γ for the material element is given by

$$\frac{d\gamma}{dt} = \frac{\partial\gamma}{\partial t} + \frac{\partial\gamma}{\partial x}\frac{dx}{dt} + \frac{\partial\gamma}{\partial y}\frac{dy}{dt} + \frac{\partial\gamma}{\partial z}\frac{dz}{dt} \equiv \frac{\partial\gamma}{\partial t} + \frac{d\mathbf{x}}{dt}\cdot\nabla\gamma. \tag{4.1.5}$$

Now $d\mathbf{x}/dt$ is the rate of change of position of a material element, i.e., the fluid velocity

$$d\mathbf{x}/dt = \mathbf{u} \equiv (u, v, w). \tag{4.1.6}$$

Therefore, for a material element, $d\gamma/dt$ is equal to $D\gamma/Dt$, where $D\gamma/Dt$ is defined by

$$\frac{D\gamma}{Dt} \equiv \frac{\partial\gamma}{\partial t} + u\frac{\partial\gamma}{\partial x} + v\frac{\partial\gamma}{\partial y} + w\frac{\partial\gamma}{\partial z} \equiv \frac{\partial\gamma}{\partial t} + \mathbf{u}\cdot\mathbf{V}\gamma. \tag{4.1.7}$$

Note that the symbol D/Dt is *defined* by (4.1.7) and so has this meaning irrespective of the context. The symbol d/dt, on the other hand, means the time derivative of a quantity that is a function of time only. (Despite this fact, the symbol d/dt is used in some textbooks to have the same meaning as that given to D/Dt here. This does not often lead to confusion, but it is better to have different symbols for operators that have different meanings.)

The usefulness of the operator D/Dt can be illustrated immediately by considering the "concentration" equations for air and seawater. If molecular diffusion can be ignored, the material element will always consist of the same particles and so the *mass* of each constituent will remain constant. Since the salinity s is the mass of dissolved salt per unit mass of fluid, s will also remain constant, and so

$$Ds/Dt = 0. \tag{4.1.8}$$

Similarly, for the atmosphere, the specific humidity q is the mass of water vapor per unit mass of air. Thus if no phase changes are taking place,

$$Dq/Dt = 0. \tag{4.1.9}$$

A similar equation holds for any quantity that is conserved by material elements.

4.2 Mass Conservation Equation

As a material element moves, its mass remains constant but its volume may alter. Therefore, its density may change, but in a way that is dependent on the field of motion. The equation relating the rate of change of density to the field of motion is called the mass conservation equation. There are two equivalent ways of writing this equation, corresponding to different methods of derivation. The first method, which was used by Euler (1755) in his paper on the equations of motion, considers the changes following a material element. The second method considers the changes for a fixed volume element. These two different approaches may be applied to the other equations of motion as well, and both will be considered here.

The first method requires calculation of the fractional rate of change $v_s^{-1} Dv_s/Dt$ of the specific volume v_s of a material element. Here this will be calculated for an infinitesimal Cartesian element as shown in Fig. 4.1 [for a more general discussion, see Batchelor (1967, Sections 2.2 and 3.1)]. Consider an element that (Fig. 4.1a) is initially rectangular, with sides δx, δy, and δz. A short time later (Fig. 4.1b), the element will be slightly distorted. To first order in δx, δy, and δz, the volume changes only

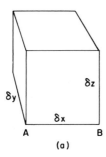 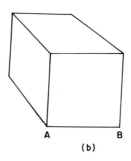

$$(a) \qquad\qquad\qquad\qquad (b)$$

Fig. 4.1. A material volume element that is initially rectangular (a) with sides δx δy, δz will a short time later (b) be displaced and slightly distorted. To first order in δx, δy, and δz, the volume change is due only to changes in lengths of the sides. Changes in angles between the sides do not alter the volume to this order.

because of small changes in the lengths of the sides, slight rotations of the edges not being significant to this order. Therefore the fractional rate of change of volume is

$$\frac{1}{\delta x\, \delta y\, \delta z}\frac{D}{Dt}(\delta x\, \delta y\, \delta z) = \frac{1}{\delta x}\frac{D}{Dt}(\delta x) + \frac{1}{\delta y}\frac{D}{Dt}(\delta y) + \frac{1}{\delta z}\frac{D}{Dt}(\delta z). \qquad (4.2.1)$$

But the first term

$$\frac{1}{\delta x}\frac{D}{Dt}(\delta x) = \frac{1}{\delta x}\frac{D}{Dt}(x(B) - x(A)) = \frac{1}{\delta x}(u(B) - u(A)) \to \frac{\partial u}{\partial x} \quad \text{as} \quad \delta x \to 0,$$

and similarly for the other terms. It follows that the fractional rate of change of specific volume is equal to the divergence $\mathbf{V}\cdot\mathbf{u}$ of the velocity, i.e.,

$$\frac{1}{v_s}\frac{Dv_s}{Dt} = \mathbf{V}\cdot\mathbf{u} \equiv \frac{\partial u}{\partial x} + \frac{\partial v}{\partial y} + \frac{\partial w}{\partial z}. \qquad (4.2.2)$$

For fluid mechanics problems it is usually more convenient to use density ρ as a variable rather than its reciprocal v_s. Then (4.2.2) takes the form

$$\rho^{-1}\, D\rho/Dt + \mathbf{V}\cdot\mathbf{u} = 0. \qquad (4.2.3)$$

The mass conservation equation (4.2.3) is fundamental in all problems involving fluid motion. An alternative form of it, which will be derived from first principles, can be obtained from (4.2.3) using (4.1.7), the definition of D/Dt, and (4.2.2), the definition of the divergence operator. First, (4.2.3) gives

$$\frac{\partial \rho}{\partial t} + u\frac{\partial \rho}{\partial x} + v\frac{\partial \rho}{\partial y} + w\frac{\partial \rho}{\partial z} + \rho\left(\frac{\partial u}{\partial x} + \frac{\partial v}{\partial y} + \frac{\partial w}{\partial z}\right) = 0,$$

i.e.,

$$\frac{\partial \rho}{\partial t} + \frac{\partial}{\partial x}(\rho u) + \frac{\partial}{\partial y}(\rho v) + \frac{\partial}{\partial z}(\rho w) = 0,$$

or

$$\partial \rho/\partial t + \mathbf{V}\cdot(\rho\mathbf{u}) = 0. \qquad (4.2.4)$$

The second method of derivation considers the mass balance for a small volume element *fixed in space* (Fig. 4.2) and leads directly to the form (4.2.4). For such an

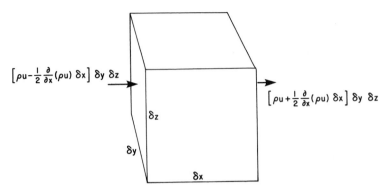

Fig. 4.2.   The mass balance for a fixed rectangular volume element with sides δx, δy, and δz. The mass fluxes through the left- and right-hand faces are shown, where u is the x-component of velocity and ρ the density at the center of the element. The errors in these expressions are small compared with $\delta x \, \delta y \, \delta z$, for small δx, δy, and δz. For the pair of faces there is a net efflux of mass of $\partial(\rho u)/\partial x \cdot \delta x \, \delta y \, \delta z$. Similar expressions can be obtained for the contributions to the mass balance from the other two pairs of faces.

element, mass is continually being carried, or "advected" through the sides by fluid motion. (A property transferred bodily by the flow is said to be "advected" by the flow. The term "advection" will be used in preference to "convection," which, as in Chapter 1, will be used to refer to the process that occurs when heavy fluid overlies light fluid.) Take, for instance, the side of area $\delta y \, \delta z$ on the left-hand side of Fig. 4.2. The mass crossing this area per unit time is approximately

$$(\rho u - \tfrac{1}{2}\,\delta x\,\partial(\rho u)/\partial x)\,\delta y\,\delta z,$$

where ρ, u are values at the center of the element. That crossing the opposite side is approximately

$$(\rho u + \tfrac{1}{2}\,\delta x\,\partial(\rho u)/\partial x)\,\delta y\,\delta z,$$

and so adding contributions from all six sides, one obtains for the net rate of increase of mass the expression

$$-(\partial(\rho u)/\partial x + \partial(\rho v)/\partial y + \partial(\rho w)/\partial z)\,\delta x\,\delta y\,\delta z.$$

In the limit as the volume shrinks to zero, the rate of increase of mass *per unit volume* is therefore

$$-\partial(\rho u)/\partial x - \partial(\rho v)/\partial y - \partial(\rho w)/\partial z \equiv -\mathbf{\nabla}\cdot(\rho\mathbf{u}). \tag{4.2.5}$$

Since by definition ρ is the mass per unit volume, the rate of increase of mass per unit volume is $\partial\rho/\partial t$ and is therefore equal to the above expression. Equation (4.2.4) follows.

4.3   Balances for a Scalar Quantity like Salinity

The ideas applied above to the mass balance of a fixed volume can also be applied to other scalar quantities. The basic requirement is an estimate of the rate at which the scalar quantity is transported across the sides of the volume element. In general a direction can be assigned to the transport of a scalar quantity, so one can define a vector

$$\mathbf{F} \equiv (F_x, F_y, F_z)$$

that has this direction and whose magnitude gives the rate of transport of the scalar quantity across the unit area normal to the direction of $\mathbf{F}$. $\mathbf{F}$ is called the flux per unit area or *flux density* of the scalar quantity. For example, the flux due to fluid motion is in the direction of motion, and the amount crossing the unit area normal to the motion is Q_v times the velocity, where Q_v is the amount of the scalar quantity contained in the unit volume. In this case,

$$\mathbf{F} = Q_v\mathbf{u}.$$

The flux (or rate of transport) through an element of area δA, which is not at right angles to $\mathbf{F}$, can be calculated by simple geometry. Figure 4.3 shows a cut through the element of area, the plane of the drawing being the one that contains both $\mathbf{F}$ and the normal to δA. $\delta S = \delta A \cos \alpha$ is the projection of δA on the plane normal to $\mathbf{F}$ and α is the angle between the planes of δA and δS. The flux is equal to

$$F \, \delta S = F \, \delta A \cos \alpha = F \cos \alpha \, \delta A,$$

where F is the magnitude of $\mathbf{F}$, and so the flux per unit area is $F \cos \alpha$, i.e., the component of $\mathbf{F}$ normal to the area concerned.

For the volume element depicted in Fig. 4.2 the flux across the side of area $\delta y \, \delta z$ is $F_x \, \delta y \, \delta z$, where F_x is the x-component of the flux. Following the same argument as for mass, the difference in flux between the two sides of area $\delta y \, \delta z$ is

$$\partial F_x / \partial x \cdot \delta x \, \delta y \, \delta z$$

and the rate of loss of the scalar quantity per unit volume is

$$\mathbf{V} \cdot \mathbf{F} \equiv \partial F_x / \partial x + \partial F_y / \partial y + \partial F_z / \partial z.$$

In other words, the equation satisfied by Q_v, the quantity per unit volume, is

$$\partial Q_v / \partial t + \mathbf{V} \cdot \mathbf{F} = 0. \tag{4.3.1}$$

The mass conservation equation (4.2.4) is the special case in which $Q_v = \rho$ and $\mathbf{F} = \rho\mathbf{u}$. The equation for salinity or humidity is another special case, in which $Q_v = \rho s$ is the mass of salt (or water vapor) per unit volume. The advective flux (i.e., the flux due to fluid motion) is $\rho s\mathbf{u}$, so if there is no other means of transporting salt

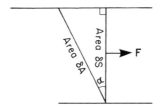

Fig. 4.3.   A diagram for calculating the flux across an area element δA when the flux density is $\mathbf{F}$. The sketch shows a section through the area element in the plane that contains $\mathbf{F}$ and the normal to the area element. Lines are drawn through the boundary of δA parallel to $\mathbf{F}$, and δS is the projection of δA on a plane normal to $\mathbf{F}$.

(or water vapor), (4.3.1) becomes

$$\partial(\rho s)/\partial t + \mathbf{V} \cdot (\rho s \mathbf{u}) = 0. \tag{4.3.2}$$

This equation could also be derived from (4.1.8) and (4.2.3), so it is merely an alternative way of expressing the salinity (or water vapor) balance.

However, there is another means of transporting salt (or water vapor), molecular diffusion, which occurs when there are salinity (or humidity) gradients. This is a very slow process and is therefore neglected in most problems considered in this book. The diffusive flux is in the opposite direction to $\mathbf{V}s$, the gradient of s, i.e., it carries salt from regions of high concentration to regions of low concentration, and is equal to (Batchelor, 1967, Section 1.6)

$$-\rho \kappa_D \mathbf{V}s,$$

where κ_D, the *diffusivity* of salt in water, is a coefficient that determines the rate of diffusion and can depend on the state of the fluid, i.e., on the temperature, pressure, and salinity. Values of κ_D for common substances can be found in Weast (1971–1972), e.g., the value for salt in water is 1.5×10^{-9} m^2 s^{-1} at 25°C, and that for water vapor in air is 2.4×10^{-5} m^2 s^{-1} at 8°C [see also List (1951, Table 113)]. When the diffusive flux is included,

$$\mathbf{F} = \rho s \mathbf{u} - \rho \kappa_D \mathbf{V}s \tag{4.3.3}$$

and (4.3.1) becomes

$$\partial(\rho s)/\partial t + \mathbf{V} \cdot (\rho s \mathbf{u} - \rho \kappa_D \mathbf{V}s) = 0. \tag{4.3.4}$$

[*Note*: Salt diffusion can also be caused by temperature and pressure gradients (Fofonoff, 1962), but the effects are minor on the very small scales for which diffusion is important.]

4.3.1   Finite-Difference Formulation for Numerical Models

The concepts of fluxes across the sides of volume elements are also used in *numerical models* of the atmosphere and ocean [see, e.g., Haltiner (1971), Bryan (1969), Mesinger and Arakawa (1976)]. Such models may be divided into a set of volume elements like that shown in Fig. 4.2, except that now the dimensions are finite. Each volume element is identified by integers (i, j, k) that determine its position in a grid (see Fig. 4.4) and the value $Q_v(i, j, k; t)$ denotes the average value of Q_v over the volume element (i, j, k) at time t. The change in Q_v in time δt can be calculated from the sum of the fluxes over the sides of the volume element, giving as the finite-difference analog of (4.3.1),

$$\frac{Q_v(i, j, k; t + \delta t) - Q_v(i, j, k; t)}{\delta t} + \frac{F_x(i + \frac{1}{2}, j, k; \bar{t}) - F_x(i - \frac{1}{2}, j, k; \bar{t})}{\delta x}$$

$$+ \frac{F_y(i, j + \frac{1}{2}, k; \bar{t}) - F_y(i, j - \frac{1}{2}, k; \bar{t})}{\delta y}$$

$$+ \frac{F_z(i, j, k + \frac{1}{2}; \bar{t}) - F_z(i, j, k - \frac{1}{2}; \bar{t})}{\delta z} = 0, \tag{4.3.5}$$

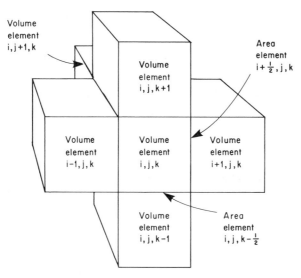

Fig. 4.4.   A typical arrangement of volume elements (of finite size) in a numerical model. Volume elements, and quantities associated with them, are identified by integers as shown. To identify area elements, one of the integers is replaced by a value halfway between the integer values associated with the two volume elements that it separates.

where δx, δy, and δz are the sides of the volume element and $F_x(i + \frac{1}{2}, j, k; \bar{t})$ is the average x-component of the flux per unit area across face $(i + \frac{1}{2}, j, k)$ during the time interval from t to $t + \delta t$. Face $(i + \frac{1}{2}, j, k)$ is the one common to volume elements (i, j, k) and $(i + 1, j, k)$.

   With the interpretation given above, (4.3.5) is exact. The approximation comes in when the components of **F** are calculated in terms of other quantities. For instance, in the *mass conservation* equation, where $Q_v = \rho$, $F_x(i + \frac{1}{2}, j, k; \bar{t})$ is the average value of ρu across the *face* $(i + \frac{1}{2}, j, k)$. However, changes in ρu are calculated from the *momentum* equations, where ρu is interpreted as the average value of the x-component of momentum per unit volume over a *volume* element. Some form of approximation is required to relate the average value over a face to average values over appropriate volume elements. This approximation must have the property that if the size of the volume elements tends toward zero, the approximation becomes more and more accurate. This should be true of all finite-difference schemes, whether or not the formulas are readily interpreted in the way described above. Unfortunately, it is not always possible to choose elements small enough for the finite-difference solutions to be close to the exact solutions. In that case, the numerical model is best interpreted as a system distinct from the exact one, but having, it is hoped, closely analogous behavior.

4.3.2   Changes Following a Material Element

   Often it is desirable to modify an equation like (4.3.4) to gives changes following a material element rather than changes at a fixed position. This can be done by using, for any variable γ (usually a quantity per unit mass), an expression for $\rho \, D\gamma/Dt$ that is modified from that given by (4.1.7) by adding γ times the quantity equated to zero in (4.2.4). Thus

$$\rho \, D\gamma/Dt = \rho \, \partial\gamma/\partial t + \rho \mathbf{u} \cdot \nabla\gamma + \gamma \, \partial\rho/\partial t + \gamma \, \nabla \cdot (\rho \mathbf{u})$$

i.e.,

$$\rho \, D\gamma/Dt = \partial(\rho\gamma)/\partial t + \nabla \cdot (\rho\gamma\mathbf{u}). \tag{4.3.6}$$

This can be regarded as an identity because the continuity equation (4.2.4) is exact. Applying this to (4.3.4), for example, gives

$$\rho \, Ds/Dt = \nabla \cdot (\rho\kappa_D \, \nabla s). \tag{4.3.7}$$

The coefficient $\rho\kappa_D$ is in general a function of the state of the fluid, but the variations are sufficiently small in most cases to take $\rho\kappa_D$ as a constant. Then (4.3.7) becomes

$$Ds/Dt = \kappa_D \, \nabla^2 s, \tag{4.3.8}$$

where

$$\nabla^2\gamma \equiv \nabla \cdot (\nabla\gamma) \equiv \frac{\partial^2\gamma}{\partial x^2} + \frac{\partial^2\gamma}{\partial y^2} + \frac{\partial^2\gamma}{\partial z^2}. \tag{4.3.9}$$

4.4   The Internal Energy (or Heat) Equation

This equation has a simple form for fluid elements that do not exchange heat with their surroundings and that retain a fixed composition. In such cases, the motion is said to be *isentropic*, i.e., the entropy of a material element is fixed, and the state of such an element does not change during the motion. Therefore the relationships among state variables given in Chapter 3 apply to the element at all times, and the form of equation depends on which variables are used to describe the state of the fluid. In terms of specific entropy η or potential temperature θ,

$$D\eta/Dt \equiv c_p(p_r, \theta)\theta^{-1} \, D\theta/Dt = 0 \tag{4.4.1}$$

by (3.7.6), where c_p is the specific heat and p_r is the reference pressure. Alternative forms follow from (3.2.1), (3.2.6), and (3.6.1), namely,

$$T\frac{D\eta}{Dt} \equiv \frac{DE}{Dt} + p\frac{Dv_s}{Dt} \equiv c_p\frac{DT}{Dt} - \frac{\alpha T}{\rho}\frac{Dp}{Dt} = 0, \tag{4.4.2}$$

where T is temperature, E internal energy per unit mass, v_s specific volume, and α the thermal expansion coefficient. All the above equations can be expressed as balances for a fixed volume element by using the general relationship (4.3.6), and diagrams like Fig. 4.2 can be drawn to visualize the balances that occur. For instance, the equation in terms of the internal energy E may be written

$$\rho \, DE/Dt \equiv \partial(\rho E)/\partial t + \nabla \cdot (\rho E\mathbf{u}) \equiv -pv_s^{-1} \, Dv_s/Dt = -p \, \nabla \cdot \mathbf{u}, \tag{4.4.3}$$

where (4.2.2) has been used to give Dv_s/Dt. The physical interpretation of (4.4.3) is that the internal energy in a fixed volume can change by advection across the sides (the term $\rho E\mathbf{u}$) or by compression or expansion of the fluid in the volume (the right-hand side is equal to the rate per unit volume at which work is done on the fluid when it is compressed).

When the motion is *not* isentropic, additional terms must be added to (4.4.3) to include the additional effects. A thorough discussion is given by Batchelor (1967, Section 3.4). The additional terms are of three types:

(i) Radiative exchange with the surroundings. This requires a knowledge of the radiative flux density $\mathbf{F}^{\mathrm{rad}}$ of energy that can be calculated if the distribution and state of absorbing, emitting, reflecting, and scattering agents are known.

(ii) Heat exchange by molecular conduction. The flux of heat by this means is proportional to the temperature gradient, and is given by

$$-k\,\nabla T,$$

where k is called the thermal conductivity. [Values may be found in Weast (1971–1972), e.g., 0.6 W m^{-1} K^{-1} for water and 0.023 W m^{-1} K^{-1} for air.]

(iii) Heating due to change of phase (latent heat release), to chemical reaction, or to viscous dissipation. The effect of these processes can be represented by a term Q_{H}, which gives the rate of heating per unit volume. The modified form of (4.4.3) is then

$$\partial(\rho E)/\partial t + \nabla \cdot (\rho E \mathbf{u} + \mathbf{F}^{\mathrm{rad}} - k\,\nabla T) = Q_{\mathrm{H}} - p\,\nabla \cdot \mathbf{u}, \qquad (4.4.4)$$

where the quantity

$$\mathbf{F} = \rho E \mathbf{u} + \mathbf{F}^{\mathrm{rad}} - k\,\nabla T \qquad (4.4.5)$$

that appears on the left-hand side may be called the heat flux density.

Alternative forms of this equation can be obtained by using the relations among state variables [summarized in (4.4.1) and (4.4.2)] and the expression (4.3.6) relating point derivatives to derivatives following the motion. With temperature as state variable, the equation becomes

$$\rho c_p\, DT/Dt - \alpha T\, Dp/Dt = \nabla \cdot (k\,\nabla T - \mathbf{F}^{\mathrm{rad}}) + Q_{\mathrm{H}}. \qquad (4.4.6)$$

Alternatively, in terms of potential temperature

$$\rho T c_p(p_r, \theta)\theta^{-1}\, D\theta/Dt = \nabla \cdot (k\,\nabla T - \mathbf{F}^{\mathrm{rad}}) + Q_{\mathrm{H}}. \qquad (4.4.7)$$

Note that if k is a constant, it may be taken outside the bracket in the above expressions, giving rise to the combination

$$\kappa = k/\rho c_p, \qquad (4.4.8)$$

where κ is called the thermal diffusivity. Typical values are 1.4×10^{-7} m^2 s^{-1} for water and 2×10^{-5} m^2 s^{-1} for air. These values are so small that thermal conduction is not of direct importance for the scales mainly considered in this book, and hence is usually neglected. The radiative term may be quite important in the atmosphere, but not in the ocean except for the top 30 m or so. The internal heating term Q_{H} is rarely important except in those parts of the atmosphere where latent heat release is taking place due to condensation. Assuming the latent heat release to be pseudoadiabatic (see Section 3.8), Q_{H} is nonzero only when (a) q has attained the saturation value q_{w}

and (b) the rates of change of pressure and temperature for a material particle are such that the saturation humidity $q_w(p, T)$ is falling. The latter condition is usually equivalent to requiring upward motion ($w > 0$). Thus when the internal heating term Q_H is due to latent heat released by a pseudoadiabatic process,

$$Q_H = \begin{cases} 0 & \text{if } q < q_w \text{ or } Dq_w/Dt \le 0, \\ -\dfrac{\rho L_v}{1 - q_w} \dfrac{Dq_w}{Dt} & \text{otherwise.} \end{cases} \tag{4.4.9}$$

An alternative way to include condensational heating when it occurs is to put $Q_H = 0$ in (4.4.7) but replace θ by the equivalent potential temperature θ_e. This follows from the definition of θ_e given in Section 3.8. If ice forms instead of water q_w should be replaced by q_i, the saturation value with respect to ice. Note that in practice q_w is small, so that the factor $(1 - q_w)$ in (4.4.9) can be approximated by unity. Then (4.4.9) can be replaced by

$$Q_H \approx -L_v \rho \, Dq/Dt, \tag{4.4.10}$$

assuming that diffusion of water vapor can be ignored; for if the conditions of the first line of the right-hand side are satisfied, (4.1.9) makes the right-hand side of (4.4.10) zero as required. Otherwise $q = q_w$, so (4.4.10) is again consistent with (4.4.9).

4.5 The Equation of Motion

The equation of motion is the expression of Newton's second law of motion for a material volume element, namely, that the rate of change of momentum of the element is equal to the net force acting on the element. For the scales considered in this book, the main forces are those considered in Section 3.5, i.e., the pressure force and the gravitational force, which is the gradient of a potential Φ_v. (This is *not* the same as the geopotential Φ introduced in Section 3.5 because of rotation effects, to be discussed in Section 4.5.1.) The resultant of these two forces per unit mass is, from Section 3.5,

$$-\rho^{-1} \, \nabla p - \nabla \Phi_v$$

and so must be equal to the rate $D\mathbf{u}_f/Dt$ of change of momentum per unit mass of a material volume element (Batchelor, 1967, Section 3.2). In other words, the equation is

$$D\mathbf{u}_f/Dt = -\rho^{-1} \, \nabla p - \nabla \Phi_v. \tag{4.5.1}$$

The subscript f is used to denote the fact that the velocity $\mathbf{u}_f$ is measured relative to a fixed frame. In geophysics, position $\mathbf{x}_r$ and velocity $\mathbf{u}_r$ are measured relative to a rotating frame, the earth, so an expression for $D\mathbf{u}_f/Dt$ in terms of these quantities is required. The derivation of such an expression is purely a matter of geometry and calculus.

4.5.1   Motion Relative to a Rotating Frame

Let the subscript f refer to quantities measured relative to the fixed frame and subscript r refer to measurements relative to the rotating frame. Let the angular velocity of the rotating frame be $\mathbf{\Omega}$ (i.e., $\mathbf{\Omega}$ is the vector along the axis of rotation with magnitude equal to the magnitude of the angular velocity about this axis, where the sense of rotation is clockwise when looking down the axis in the direction of $\mathbf{\Omega}$). Then a point with fixed position $\mathbf{x}_r$ in the rotating frame has (Fig. 4.5) velocity $\mathbf{\Omega} \times \mathbf{x}_r$. When the point $\mathbf{x}_r$ is moving relative to the rotating frame, its velocity relative to the fixed frame is therefore given by

$$d\mathbf{x}_f/dt = d\mathbf{x}_r/dt + \mathbf{\Omega} \times \mathbf{x}_r. \tag{4.5.2}$$

A repetition of this operation gives the acceleration

$$\frac{d^2\mathbf{x}_f}{dt^2} = \frac{d}{dt}\left(\frac{d\mathbf{x}_r}{dt} + \mathbf{\Omega} \times \mathbf{x}_r\right) + \mathbf{\Omega} \times \left(\frac{d\mathbf{x}_r}{dt} + \mathbf{\Omega} \times \mathbf{x}_r\right),$$

i.e.,

$$\frac{d^2\mathbf{x}_f}{dt^2} = \frac{d^2\mathbf{x}_r}{dt^2} + 2\mathbf{\Omega} \times \frac{d\mathbf{x}_r}{dt} + \mathbf{\Omega} \times (\mathbf{\Omega} \times \mathbf{x}_r). \tag{4.5.3}$$

Thus in addition to the acceleration measured relative to the rotating frame, there are two other contributions to the acceleration relative to the fixed frame. The one given by the second term on the right-hand side of (4.5.3) is called the *Coriolis acceleration* [after Coriolis (1835) who discussed it, although the term appeared earlier in the tidal equations of Laplace (1778, 1779)]. The last term can be written as the gradient of a scalar

$$\mathbf{\Omega} \times (\mathbf{\Omega} \times x_r) = -\mathbf{V}(\tfrac{1}{2}\Omega^2 x_r^2),$$

and so (4.5.3) may be written

$$du_f/dt = du_r/dt + 2\mathbf{\Omega} \times u_r - \mathbf{V}(\tfrac{1}{2}\Omega^2 x_r^2). \tag{4.5.4}$$

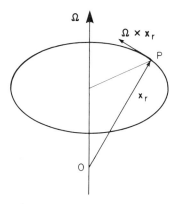

Fig. 4.5.   A point P with fixed position $\mathbf{x}_r$ in a frame of reference rotating with angular velocity $\mathbf{\Omega}$ about an axis through O moves in the circular path shown with velocity $\mathbf{\Omega} \times \mathbf{x}_r$.

When the point $\mathbf{x}_r$ is the position of a material volume element, the derivative d/dt is the same as D/Dt (see Section 4.1), so (4.5.1) becomes

$$D\mathbf{u}/Dt + 2\mathbf{\Omega} \times \mathbf{u} = -\rho^{-1}\,\nabla p - \nabla\Phi, \tag{4.5.5}$$

where the subscript r will be implied rather than written explicitly from this point onward. Φ is the geopotential defined by

$$\Phi = \Phi_v - \tfrac{1}{2}\Omega^2 x_r^2, \tag{4.5.6}$$

i.e., is the sum of the gravitational potential Φ_v and centrifugal potential $-\tfrac{1}{2}\Omega^2 x_r^2$ as defined in Section 3.5. Note also that (4.5.5) is not altered by a change in origin of the axes, so it is not necessary to have the origin on the axis of rotation when using (4.5.5).

4.5.2   Momentum Balance for a Fixed Volume Element

Multiplication of (4.5.5) by ρ and use of the identity (4.3.6) for each component in turn give an alternative form of the equation of motion, namely,

$$\partial(\rho u)/\partial t + \nabla\cdot(\rho u\mathbf{u}) + 2\Omega_y \rho w - 2\Omega_z \rho v = -\partial p/\partial x, \tag{4.5.7}$$

$$\partial(\rho v)/\partial t + \nabla\cdot(\rho v\mathbf{u}) + 2\Omega_z \rho u - 2\Omega_x \rho w = -\partial p/\partial y, \tag{4.5.8}$$

$$\partial(\rho w)/\partial t + \nabla\cdot(\rho w\mathbf{u}) + 2\Omega_x \rho v - 2\Omega_y \rho u = -\partial p/\partial z - \rho g. \tag{4.5.9}$$

The axes (at the element concerned) have been chosen so that the z axis points vertically upward, i.e., in the direction of

$$\mathbf{g} = \nabla\Phi, \tag{4.5.10}$$

where $-\mathbf{g}$ is the acceleration due to gravity of magnitude g [cf. (3.5.2)]. The angular velocity $\mathbf{\Omega}$ has been written in terms of its components $(\Omega_x, \Omega_y, \Omega_z)$. These equations can be interpreted in terms of the momentum balance (per unit volume) for a fixed volume element (such as that shown in Fig. 4.2). The rate of change of momentum (first term) is determined by the flux of momentum across the sides of the element (second term), the Coriolis force acting on the element (last two terms on the left-hand side), the net force resulting from the pressure on the sides (first term on the right-hand side) and the gravitational force [last term in (4.5.9)].

4.5.3   Effects of Viscosity

Although viscosity is not of direct importance for the scales of motion considered in this book, it is of indirect importance as a means of removing mechanical energy from the system. Viscosity gives rise to stresses on the surface of a material volume element that may be related to the rate of strain. A detailed discussion is given by Batchelor (1967, Section 3.3). It turns out that if, on the scales for which viscosity is important, viscosity changes and compressibility effects can be ignored, then the effect on momentum is the same as that of a diffusion process (see Section 4.3) and can be taken into account by adding diffusive fluxes to (4.5.7)–(4.5.9), i.e.,

$$\partial(\rho u)/\partial t + \mathbf{V} \cdot (\rho u \mathbf{u} - \mu \, \mathbf{V} u) + 2\Omega_y \rho w - 2\Omega_z \rho v = -\partial p/\partial x, \qquad (4.5.11)$$

$$\partial(\rho v)/\partial t + \mathbf{V} \cdot (\rho v \mathbf{u} - \mu \, \mathbf{V} v) + 2\Omega_z \rho u - 2\Omega_x \rho w = -\partial p/\partial y, \qquad (4.5.12)$$

$$\partial(\rho w)/\partial t + \mathbf{V} \cdot (\rho w \mathbf{u} - \mu \, \mathbf{V} w) + 2\Omega_x \rho v - 2\Omega_y \rho u = -\partial p/\partial z - \rho g, \qquad (4.5.13)$$

where μ is called the *viscosity* of the fluid. [Values of μ are tabulated by Weast (1971–1972), e.g., 10^{-3} kg m^{-1} s^{-1} for water and 1.7×10^{-5} kg m^{-1} s^{-1} for air.] The value of μ depends on the state of the fluid, but the variations are sufficiently slow in most cases to take μ as a constant. Then the modified form of (4.5.5) is

$$D\mathbf{u}/Dt + 2\mathbf{\Omega} \times \mathbf{u} = -\rho^{-1} \, \mathbf{V}p - \mathbf{g} + v \, \nabla^2 \mathbf{u}, \qquad (4.5.14)$$

where

$$v = \mu/\rho \qquad (4.5.15)$$

is called the *kinematic viscosity* [which has a value of 10^{-6} m^2 s^{-1} for water and 1.4×10^{-5} m^2 s^{-1} for air at 1000 mbar—see List (1951, Table 113)], and $\nabla^2 \mathbf{u}$ is the vector with components

$$\nabla^2 \mathbf{u} = (\nabla^2 u, \nabla^2 v, \nabla^2 w). \qquad (4.5.16)$$

4.5.4   Perturbation Pressure and Perturbation Density

For large-scale motions in the ocean and atmosphere, the dominant terms by far in the equation of motion (4.5.14) are the gravitational acceleration $\mathbf{g}$ and the vertical component of the pressure gradient, which approximately balances it. In other words, none of the other acceleration terms in (4.5.14) approaches the gravitational acceleration. In the atmosphere, for instance, winds are of order 10 m s^{-1}, so the Coriolis acceleration is about 10^{-3} m s^{-2}, i.e., less than the gravitational acceleration by a factor of 10,000!

Hence it is desirable to define a perturbation pressure and a perturbation density as departures from an equilibrium solution

$$p = p_0(z), \qquad \rho = \rho_0(z), \qquad (4.5.17)$$

of the type considered in Section 3.5, i.e., which satisfied the hydrostatic equation

$$dp_0/dz = -g\rho_0. \qquad (4.5.18)$$

The perturbation pressure p' and perturbation density ρ' are defined by

$$p = p_0(z) + p', \qquad \rho = \rho_0(z) + \rho', \qquad (4.5.19)$$

in which case (4.5.14) becomes

$$\rho(D\mathbf{u}/Dt + 2\mathbf{\Omega} \times \mathbf{u}) = -\mathbf{V}p' - \rho'\mathbf{g} + \mathbf{V} \cdot (\mu \, \mathbf{V}\mathbf{u}). \qquad (4.5.20)$$

In the special case of a homogeneous fluid, i.e., one of uniform density, ρ' is zero. Otherwise the term $-\rho'\mathbf{g}$ represents a force per unit volume called the *buoyancy* force since an element with negative ρ' is relatively buoyant and therefore experiences an upward force due to the action of gravity.

4.6 Mechanical Energy Equation

The set of equations governing the behavior of the ocean and atmosphere has now been derived. They are summarized in Section 4.10 and consist of (a) the mass conservation equation, (b) the equation of motion, (c) the internal energy or heat equation, (d) the equation of state, and (e) the equations for the concentrations of constituents such as salt and water vapor. From this set, other useful equations can be derived, using elementary calculus. In this section, the equation for *mechanical* or *kinetic* energy is considered.

The kinetic energy per unit mass is *defined* as $\frac{1}{2}\mathbf{u}^2$.* An equation for the rate of change of this quantity following a material volume element is obtained by taking the scalar product of (4.5.20) with $\mathbf{u}$, which gives

$$\rho \, D(\tfrac{1}{2}\mathbf{u}^2)/Dt = -wg\rho' + \nabla\cdot(-p'\mathbf{u} + \mu \, \nabla(\tfrac{1}{2}\mathbf{u}^2)) - \rho\epsilon + p' \, \nabla\cdot\mathbf{u}, \qquad (4.6.1)$$

where

$$\epsilon = v\left(\left(\frac{\partial \mathbf{u}}{\partial x}\right)^2 + \left(\frac{\partial \mathbf{u}}{\partial y}\right)^2 + \left(\frac{\partial \mathbf{u}}{\partial z}\right)^2\right) \qquad (4.6.2)$$

is always positive and is called the *dissipation* rate (see below). Another version of (4.6.1) has the primes removed and is obtained from (4.5.14) by the same procedure. Note that the scalar product of $\mathbf{u}$ with the Coriolis acceleration in (4.5.20) is identically zero, so there is no Coriolis term in (4.6.1).

Equation (4.6.1) can be converted to one for a fixed volume element by applying (4.3.6), which gives

$$\partial(\tfrac{1}{2}\rho\mathbf{u}^2)/\partial t + \nabla\cdot\mathbf{F}' = -wg\rho' - \rho\epsilon + p' \, \nabla\cdot\mathbf{u}, \qquad (4.6.3)$$

where

$$\mathbf{F}' = (p' + \tfrac{1}{2}\rho\mathbf{u}^2)\mathbf{u} - \mu \, \nabla(\tfrac{1}{2}\mathbf{u}^2) \qquad (4.6.4)$$

will be called the *energy flux density vector* because it gives a rate of flow of energy per unit area. It is not uniquely defined, however. For instance, any vector with zero divergence could be added to $\mathbf{F}'$ without altering (4.6.3). As before, there is an alternative version with the primes removed, i.e., with pressure in place of perturbation pressure and density in place of perturbation density. In this case $\mathbf{F}$ is used in place of $\mathbf{F}'$ in (4.6.3) and (4.6.4).

In the special case of a fluid of uniform density, $\rho' = 0$ and $\nabla\cdot\mathbf{u} = 0$ by (4.2.3), so (4.6.3) simplifies to

$$\partial(\tfrac{1}{2}\rho\mathbf{u}^2)/\partial t + \nabla\cdot\mathbf{F}' = -\rho\epsilon. \qquad (4.6.5)$$

---

* In order to simplify notation, $\mathbf{u}^2$ has been written in place of $\mathbf{u}\cdot\mathbf{u}$ throughout.

The physical interpretation of this equation is illustrated in Fig. 4.6 (cf. Fig. 4.2). The mass of the fluid in the volume element shown is $\rho\, \delta x\, \delta y\, \delta z$, so its kinetic energy is $\frac{1}{2}\rho\mathbf{u}^2\, \delta x\, \delta y\, \delta z$ by definition. This can change (a) by energy transfer across the sides of the element or (b) by energy loss within the element. The rate of transfer of energy due to a flux $\mathbf{F}'$ is shown for two of the sides in the figure, F'_x being the component of $\mathbf{F}'$ in the direction of the x axis. Adding the contributions from all sides gives a net gain of energy of

$$-\left(\frac{\partial}{\partial x}F'_x + \frac{\partial}{\partial y}F'_y + \frac{\partial}{\partial z}F'_z\right)\delta x\, \delta y\, \delta z \equiv -\mathbf{V}\cdot\mathbf{F}'\, \delta x\, \delta y\, \delta z$$

to the appropriate order of approximation. This gives rise to the term $\mathbf{V}\cdot\mathbf{F}'$ in (4.6.5) after dividing by the volume $\delta x\, \delta y\, \delta z$ and taking the limit as the volume shrinks to zero.

By (4.6.4), the individual contributions to the rate $F'_x\, \delta y\, \delta z$ of energy transfer across a face of area $\delta y\, \delta z$ are

(i) $p'u\, \delta y\, \delta z,$
(ii) $\frac{1}{2}\rho\mathbf{u}^2 u\, \delta y\, \delta z,$
(iii) $-\mu\, \partial(\frac{1}{2}\mathbf{u}^2)/\partial x\cdot \delta y\, \delta z.$

The first contribution is the product of $p'\, \delta y\, \delta z$, the normal force on the face of the element due to the pressure perturbation p', and u, the rate of movement in the direction of the force. It is therefore the rate of working by the pressure force on that side. The second contribution is the rate of advection of kinetic energy across the face. The third contribution can be interpreted as the rate of diffusion of kinetic energy across the face due to viscous processes.

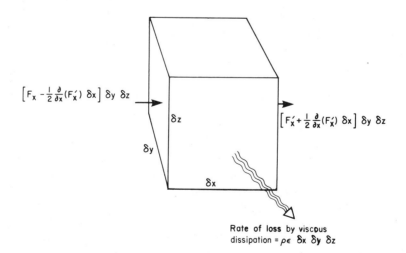

Rate of loss by viscous
dissipation $= \rho\epsilon\ \delta x\ \delta y\ \delta z$

Fig. 4.6. The mechanical energy balance for a fixed rectangular volume element in a homogeneous fluid of density ρ. Fluxes across one pair of faces are shown, where F'_x is the x-component of the mechanical energy flux density $\mathbf{F}'$. These contribute a net rate of loss of energy *per unit volume* of $\partial F'_x/\partial x$ and the two other pairs of faces contribute $\partial F'_y/\partial y$ and $\partial F'_z/\partial z$, where F'_y and F'_z are the y- and z-components of $\mathbf{F}'$. The balance of energy for the element cannot be described completely in terms of fluxes across the sides. There is an additional loss of energy per unit volume of $\rho\epsilon$, where ϵ is a positive quantity called the dissipation rate.

Returning to (4.6.5), the remaining contribution to the rate of change of kinetic energy of the element is

$$- \rho \epsilon \, \delta x \, \delta y \, \delta z,$$

which is interpreted as the rate of loss of energy *within* the element due to viscous processes. Thus ϵ is called the rate of *dissipation* of mechanical energy per unit mass, or simply the dissipation rate.

The same ideas can be applied to a large volume of fluid that can be subdivided mentally into small volume elements like that shown in Fig. 4.6. The transfers of energy across the faces of the elements merely represent a flow of energy from one part of the fluid to another, and therefore make no contribution to the energy balance of the large volume except for the contributions from the outer surface of that volume. On the other hand, the dissipation in each volume element contributes to the total energy loss in the large volume. In other words, integration of (4.6.5) over a volume gives an equation for the rate of change of

$$K = \int \int \int \frac{1}{2} \rho \mathbf{u}^2 \, dx \, dy \, dz, \tag{4.6.6}$$

the kinetic energy of the volume of fluid, the equation being

$$dK/dt + \int \int F'_n \, dS = - \int \int \int \rho \epsilon \, dx \, dy \, dz, \tag{4.6.7}$$

where F'_n denotes the outward normal component of the flux across the surface of the volume and dS an element of area, so that the integral is the total rate of transfer of energy across the surface. The integral on the right-hand side is over the volume concerned. For instance, the volume concerned could be that part of an ocean below some fixed level surface. Then (4.6.7) says that changes in kinetic energy result from transfer across this surface, transfer across the bottom, and dissipation within.

It is also possible to derive an equation for a large *material* volume of fluid, i.e., a volume with mobile boundaries but always consisting of the same fluid particles. An example is the ocean, which is bounded above not by a fixed surface but by its (constantly moving) free surface. As before, this volume could be subdivided into material volume elements and the balances for each element added. This is equivalent to integrating (4.6.1) over the material volume, which for a homogeneous fluid again leads to (4.6.7), with the integrations now being over the material volume and its surface. The advective contribution to the transfer across the surface is zero since by definition no advection takes place across the surface of a material volume.

In the case of a homogeneous ocean or lake, energy is transferred across the free surface from the atmosphere through the working of the pressure force and through "diffusion" of energy, which represents the action of viscous stresses. Since the normal velocity is zero at the bottom, the pressure force cannot do work there, so the only means of losing energy involve viscosity, namely, viscous stress acting on the bottom and viscous dissipation within the ocean or lake. Since the kinetic energy of such bodies of water does not continually increase, the energy losses through viscous effects must balance the energy inputs over a long period of time. At first sight, this seems to contradict the statement that viscous effects are *not* important on the larger scales such as those characteristic of the energy inputs. The implication is that energy is transferred from one scale to another [which is possible because of the nonlinear terms in (4.5.14)] and significant dissipation takes place only at scales where velocity gradients are large enough for (4.6.2) to give values such that dissipation balances inputs. These scales are very small in practice, and can be estimated if it is assumed that the scale depends only on ϵ and ν. The only combination of these parameters with the dimensions of length is

$$(\nu^3/\epsilon)^{1/4},$$

and typical values for the ocean and atmosphere are of the order of a millimeter. [For a discussion of dissipation processes in the ocean, see Woods (1982).] This fact creates a problem for numerical models that cannot hope to cope with scales ranging from the size of the earth to the dissipation scale (ten factors of ten!). A common technique is to make the viscosity artificially large (in which case it is called an *eddy viscosity*) so that sufficient energy dissipation can occur on scales resolvable by the numerical scheme. Since the vertical resolution is usually much better than the horizontal resolution, smaller values can be used for vertical "diffusion" of momentum and energy than are used for horizontal diffusion. [Vertical eddy viscosities may be 10^2 or 10^3 times the molecular value, i.e., typical of oils rather than air or water. Horizontal eddy viscosities used are often 10^{10} or even 10^{11} times molecular, i.e., akin to very viscous substances like glucose (Weast, 1971–1972).] There is, however, no guarantee that this procedure will remove energy in a realistic way, and a major problem of numerical modeling is to find schemes that will remove energy realistically.

Energy principles are such that the loss of mechanical energy by dissipation represents a rate of *conversion* of energy into a different form, namely, heat. Thus there is a contribution $\rho\epsilon$ to the term in (4.4.4) that represents a rate of gain of internal energy per unit volume. This contribution is, however, so tiny that it is nearly always neglected. The terms $-wg\rho'$ and $p' \, \mathbf{V} \cdot \mathbf{u}$ that appear in (4.6.3) also represent rates of conversions of energy from one form to another. The former term is the product of the upward buoyancy force per unit volume $-g\rho'$ and the rate w of movement in the direction of that force, and therefore represents the rate of working per unit volume by the buoyancy force. The latter term is the product of a pressure p' and the fractional rate of change of volume [see (4.2.2)] of a material element, and therefore represents the rate of release of energy per unit volume of the fluid by expansion. The energy conversions associated with these terms are dealt with in the next section.

4.7 Total Energy Equation

Thus far, equations have been derived for energy in two forms: internal energy [Eq. (4.4.4)] and kinetic energy [Eq. (4.6.3)]. If the version of the latter equation with primes is used and the two equations are added, the term $p' \, \mathbf{V} \cdot \mathbf{u}$ disappears because it represents a rate of conversion of energy from the internal form to the kinetic form. The same is true of the term $\rho\epsilon$ (which cancels a contribution to Q_H). Another term that requires interpretation as a rate of conversion of energy from one form to another is the buoyancy term $-wg\rho'$ in (4.6.3). This represents work done by gravitational forces when fluid crosses geopotential surfaces, and the appropriate form of energy per unit mass is the geopotential Φ defined by (4.5.6). With this interpretation, Φ is called the *potential* energy per unit mass associated with gravitational and centrifugal forces. Φ depends only on z, and so by definition (4.1.7) its rate of change, following a material element, is

$$D\Phi/Dt \equiv \mathbf{u}\cdot\nabla\Phi \equiv \mathbf{u}\cdot\mathbf{g} = wg \tag{4.7.1}$$

by (4.5.10). This can be converted into a rate of change for a fixed volume element by using the standard formula (4.3.6), giving

$$\partial(\rho\Phi)/\partial t + \nabla\cdot(\rho\Phi\mathbf{u}) = wg\rho. \tag{4.7.2}$$

The total energy equation is now obtained by adding (4.7.2), the internal energy equation (4.4.4), and the version of the kinetic energy equation (4.6.3) without primes. The result is

$$\partial(\rho(E + \Phi + \tfrac{1}{2}\mathbf{u}^2))/\partial t + \nabla\cdot\mathbf{F}^{\text{tot}} = Q_{\text{H}}, \tag{4.7.3}$$

where $\mathbf{F}^{\text{tot}}$ is the total energy flux vector given by

$$\mathbf{F}^{\text{tot}} = \rho\mathbf{u}(E + \Phi + \tfrac{1}{2}\mathbf{u}^2) + p\mathbf{u} + \mathbf{F}^{\text{rad}} - k\,\nabla T - \mu\,\nabla(\tfrac{1}{2}\mathbf{u}^2). \tag{4.7.4}$$

The terms contributing to $\mathbf{F}^{\text{tot}}$ in order of appearance are the advective flux, the rate of working by pressure forces per unit area, the radiative flux, the flux by diffusion of heat, and the flux by diffusion of kinetic energy. Equation (4.7.3) gives the changes for a fixed volume element. Using (4.3.6), the equation for changes of a material volume element is

$$\rho\,D(E + \Phi + \tfrac{1}{2}\mathbf{u}^2)/Dt + \nabla\cdot(p\mathbf{u} + \mathbf{F}^{\text{rad}} - k\,\nabla T - \mu\,\nabla(\tfrac{1}{2}\mathbf{u}^2)) = Q_{\text{H}}. \tag{4.7.5}$$

As with the kinetic energy equation, (4.7.3) can be interpreted in terms of balances for a fixed volume element like those illustrated in Fig. 4.6. Also, by adding contributions of many such elements, an equation for the rate of change of total energy of a large volume can be obtained. The internal energy I of the volume is defined by

$$I = \iiint \rho E\, dx\, dy\, dz, \tag{4.7.6}$$

and the potential energy P by

$$P = \iiint \rho\Phi\, dx\, dy\, dz \approx \iiint \rho gz\, dx\, dy\, dz \tag{4.7.7}$$

by (3.5.2). The total energy [see (4.6.6)] is thus

$$K + I + P,$$

whose rate of change is given by

$$d(K + I + P)/dt + \iint F_{\text{n}}^{\text{tot}}\, dS = \iiint Q_{\text{H}}\, dx\, dy\, dz. \tag{4.7.8}$$

F_n^{tot} denotes the outward component of the flux across the surface of the volume and dS an element of area, so that the integral is the total rate of transfer of energy across the surface. As found in the previous section, the equation applies both to a fixed volume, across which fluid may flow, or to a material volume, which may have a moving surface but always consists of the same particles.

The above discussion of energy is satisfactory and consistent when no phase changes (or chemical changes, etc.) occur, for then Q_H is zero. If Q_H is nonzero, it must represent a conversion of energy from one form to another. For instance, the effect of latent heat release by means of the pseudoadiabatic process can be included by using expression (4.4.10) for Q_H. Substituting in (4.7.8) and using the standard relation (4.3.6) then give

$$d(K + I + P + L)/dt + \int \int (F_n^{tot} + \rho q L_v u_n)\, dS = 0, \qquad (4.7.9)$$

where

$$L = \int \int \int L_v \rho q\, dx\, dy\, dz \qquad (4.7.10)$$

is the energy that could be released by moving each parcel adiabatically upward to the saturation level and then pseudoadiabatically upward until all moisture is removed.

Estimates of the terms K, I, P, and L for the atmosphere have been made by Oort (1971) for each month of the year, and fluxes of energy across circles of latitude have also been made. The largest contributions to the mean total energy are I (73%), defined by (4.7.6) and (3.2.7), and P (25%), defined by (4.7.7). However, with these definitions, $I + P$ represents the energy that could be obtained by lowering the temperature of the atmosphere to absolute zero and bringing the mass of the atmosphere down to sea level. Since little of this energy could be obtained by a process that could readily occur, Lorenz (1955) has introduced the concept of *available potential energy* as the energy that could be obtained by some well-defined process. Usually the process considered is an adiabatic redistribution of mass without phase changes to a statically stable state of rest (see Sections 3.5 and 3.6). With this definition, the available potential energy of the atmosphere has been estimated (Price, 1975) to be about 23×10^{20} J, giving a mean over the whole earth of about 4.5×10^6 J m^{-2}. This may be compared with the mean available potential energy in a typical mid-latitude ocean gyre, estimated by Gill *et al.* (1974) to be of the order of 10^5 J m^{-2}. An alternative definition for the atmosphere [a discussion of the concept of available potential energy is given by Dutton and Johnson (1967)] would also include the energy L that could be released by condensing all the moisture. This amounts to 64×10^6 J m^{-2} (Peixoto *et al.*, 1981).

For numerical models of the atmosphere and ocean, finite-difference approximations of the balances of mass, momentum, internal energy, etc., for finite volume elements are used, as discussed in Section 4.3. It does not follow automatically that a finite-difference equivalent of the total energy equation will exist. All that can be said is that the total energy equation will be satisfied, correct to a certain order, as the size of the element shrinks to zero. However, it is always possible to write the finite-difference equations in such a way that a finite-difference form of the total energy equation is *exactly* satisfied. In other words, there are no energy sources or sinks within the body of fluid, only conversions of energy from one form to another. Such formulations have been found to eliminate problems that may be encountered otherwise, namely, an artificial increase or decrease of energy over a long period of time (Arakawa, 1966).

4.8   Bernoulli's Equation

A variant of (4.7.5) can be deduced from the identity

$$\mathbf{V}\cdot(p\mathbf{u}) = \mathbf{u}\cdot\mathbf{V}p + p\,\mathbf{V}\cdot\mathbf{u} = \frac{Dp}{Dt} - \frac{\partial p}{\partial t} - \frac{p}{\rho}\frac{D\rho}{Dt},$$

which follows from the definition (4.1.7) and from the continuity equation (4.2.3). The result may also be written

$$\mathbf{V}\cdot(p\mathbf{u}) = \rho\,\frac{D}{Dt}\left(\frac{p}{\rho}\right) - \frac{\partial p}{\partial t}, \tag{4.8.1}$$

and so (4.7.5) becomes

$$\rho\,D(E + p/\rho + \Phi + \tfrac{1}{2}\mathbf{u}^2)/Dt + \mathbf{V}\cdot(\mathbf{F}^{\mathrm{rad}} - k\,\mathbf{V}T - \mu\,\mathbf{V}(\tfrac{1}{2}\mathbf{u}^2)) = Q_{\mathrm{H}} + \partial p/\partial t. \tag{4.8.2}$$

For applications not connected with acoustic waves, the term $\partial p/\partial t$ is often relatively small, changes in pressure due to a change in level of a fluid element being large compared with changes at a fixed point. Also, viscous and diffusive effects can be ignored except on the smallest scales. Thus in situations for which radiative heating and latent heat release can also be ignored, Eq. (4.8.2) becomes

$$D(E + p/\rho + \Phi + \tfrac{1}{2}\mathbf{u}^2)/Dt = 0. \tag{4.8.3}$$

This is known as Bernoulli's equation since both Daniel and John Bernoulli contributed to special forms of it. [A historical discussion is given by Truesdell (1954b).] A discussion of circumstances in which it is valid may be found in Batchelor (1967; Section 3.5).

The quantity $E + p/\rho$, which appears in (4.8.3), often occurs in thermodynamics and is called the *enthalpy* per unit mass. For a perfect gas, it follows from (3.2.10) and (3.2.12) that

$$E + p/\rho = c_p T, \tag{4.8.4}$$

and this approximation is used for applications to the atmosphere. Corrections to (4.8.4) for moist air can be found in Table 85 of the Smithsonian Meteorological Tables (List, 1951). In applications to the atmosphere, the quantity $E + p/\rho + \Phi$ is sometimes called the *dry static energy* per unit mass. An approximate expression for this quantity for air is

$$E + p/\rho + \Phi \simeq c_p T + gz. \tag{4.8.5}$$

The Bernoulli equation (4.8.3) can be modified to include effects of latent heat release in a pseudoadiabatic process by using the approximation for Q_H in (4.8.2). The modified version is

$$D(E + p/\rho + \Phi + L_v q + \tfrac{1}{2}\mathbf{u}^2)/Dt = 0. \tag{4.8.6}$$

The quantity

$$E + p/\rho + \Phi + L_v q \simeq c_p T + gz + L_v q, \tag{4.8.7}$$

which appears in the equation, is sometimes called the *moist static energy* per unit mass.

4.9   Systematic Effects of Diffusion

The rates of diffusion of salt in water ($\kappa_D = 1.5 \times 10^{-9}$ m^2 s^{-1}) and of water vapor in air ($\kappa_D = 2.4 \times 10^{-5}$ m^2 s^{-1}) are so small that diffusion plays no direct role on larger-scale motions. However, diffusion is systematic in that it always acts to *reduce* gradients. An equation that shows this effect can be derived from (4.3.8) in a similar way to that by which the mechanical energy equation (4.6.5) was derived from the momentum equations. Multiplication of (4.3.8) by s gives

$$\rho\, D(\tfrac{1}{2}s^2)/Dt = \mathbf{V}\cdot(\rho\kappa_D\, \mathbf{V}(\tfrac{1}{2}s^2)) - \rho\kappa_D(\mathbf{V}s)^2, \tag{4.9.1}$$

or, using the identity (4.3.6), this can be written

$$\partial(\tfrac{1}{2}\rho s^2)/\partial t + \mathbf{V}\cdot(\tfrac{1}{2}\rho s^2\mathbf{u} - \rho\kappa_D\, \mathbf{V}(\tfrac{1}{2}s^2)) = -\rho\kappa_D(\mathbf{V}s)^2. \tag{4.9.2}$$

An alternative version has s in (4.9.1) and (4.9.2) replaced by the salinity perturbation s', defined by

$$s' = s - s_0. \tag{4.9.3}$$

s_0 could be any constant since a constant value satisfies (4.3.8), (4.9.1), and (4.9.2) identically, but a natural choice would be the mean salinity for the volume of fluid being considered (e.g., the value for the ocean is about 0.0348 or 34.8‰; see Fig. 3.2).

Equation (4.9.2) can be given a physical interpretation similar to that illustrated in Fig. 4.6 for the energy equation. The squared salinity in a small volume is changed not only by fluxes across the surface of the volume, but also there is a systematic *loss* represented by the negative definite term on the right-hand side of (4.9.2).

If (4.9.2), or the equivalent version with primes, is integrated over the whole of the ocean, the same apparent contradiction is obtained as for that with the mechanical energy equation. There is an input across the ocean surface at large scale [because surface salinity tends to be high where there is a salt flux into the ocean—see, e.g., Defant (1961, Volume 1, Fig. 68)], but losses by diffusion are insignificant at the large scales. As with energy, transfers from one scale to another take place because of the nonlinear advection terms in (4.3.8), the significant contributions to the right-hand side of (4.9.2) coming at very small scales. On this basis, Stern (1968) estimated the root-mean-square salinity gradient in the upper ocean to be about 1000 times the mean value.

In order to represent diffusive losses in numerical models, the device of introducing artificially large coefficients is often used, the artificial values being called *eddy diffusivities*. Values used are of the same order as those used for eddy viscosities, and such values are also used for diffusion of heat in model calculations.

4.10   Summary List of the Governing Equations

The equations that govern the changes of properties of the atmosphere and ocean have now been derived and are summarized below. The equations are the mass conservation equation (4.2.3)

$$\rho^{-1}\, D\rho/Dt + \mathbf{V}\cdot\mathbf{u} = 0; \tag{4.10.1}$$

the momentum equation (4.5.14)

$$D\mathbf{u}/Dt + 2\mathbf{\Omega} \times \mathbf{u} = -\rho^{-1}\,\mathbf{V}p - \mathbf{g} + v\,\nabla^2\mathbf{u}; \tag{4.10.2}$$

the heat, or energy equation (4.4.7)

$$\rho T c_p(p_{\mathrm{r}}, \theta)\theta^{-1}\, D\theta/Dt = \mathbf{V}\cdot(k\,\mathbf{V}T - \mathbf{F}^{\mathrm{rad}}) + Q_{\mathrm{H}} \tag{4.10.3}$$

[which requires knowledge of the heating function Q_{H}, e.g., from Equation (4.4.9), of the radiative flux $\mathbf{F}^{\mathrm{rad}}$, and of T as a function of p, s and θ]; the concentration equation (4.3.7) for salt or humidity

$$\rho\, Ds/Dt = \mathbf{V}\cdot(\rho\kappa_{\mathrm{D}}\,\mathbf{V}s); \tag{4.10.4}$$

and the equation of state (3.1.1)

$$\rho = \rho(p, s, \theta). \tag{4.10.5}$$

The last of these may be combined with the mass conservation equation in a way that depends on which state variables are being used to describe the system. If p, s, and θ are used, as anticipated by the form in which (4.10.5) is written, then the density derivative in (4.10.1) can be expanded as

$$\frac{1}{\rho}\frac{D\rho}{Dt} = \frac{1}{\rho}\frac{\partial\rho}{\partial p}\frac{Dp}{Dt} + \frac{1}{\rho}\frac{\partial\rho}{\partial s}\frac{Ds}{Dt} + \frac{1}{\rho}\frac{\partial\rho}{\partial\theta}\frac{D\theta}{Dt}, \tag{4.10.6}$$

where for each partial derivative the other two variables of the set p, s, θ are kept constant. Thus in the notation of Section 3.7.4,

$$\frac{1}{\rho}\frac{D\rho}{Dt} = \frac{1}{\rho c_{\mathrm{s}}^2}\frac{Dp}{Dt} + \beta'\frac{Ds}{Dt} - \alpha'\frac{D\theta}{Dt}, \tag{4.10.7}$$

where c_{s} is the speed of sound and α' and β' are the expansion coefficients with respect to potential temperature and salinity. Now the set of equations gives expressions for the rates of change of p, θ, s, and $\mathbf{u}$, and thus changes in all the properties of the system can be calculated. It is not necessary, however, to have *explicit* expressions for the rates of change of all these variables—the requirement is to have the same number of equations as unknowns. In fact, the equations are rarely used in their complete form because approximate forms usually have some advantage in applications—e.g., for analytic simplification or because larger time steps can be used with suitable approximate forms when numerical methods are being employed.

A special case of particular interest is that of isentropic motion in the absence of viscous or diffusive effects. Then (4.10.3) and (4.10.4) become (4.4.1) and (4.1.8), respectively, i.e.,

$$D\theta/Dt = 0, \tag{4.10.8}$$

$$Ds/Dt = 0. \tag{4.10.9}$$

The mass conservation equation (4.10.1) is unchanged, but when combined with (4.10.7)–(4.10.9), it becomes

$$(\rho c_s^2)^{-1}\, Dp/Dt + \mathbf{V} \cdot \mathbf{u} = 0. \tag{4.10.10}$$

The momentum equation (4.10.2) becomes

$$D\mathbf{u}/Dt + 2\mathbf{\Omega} \times \mathbf{u} = -\rho^{-1}\, \mathbf{V}p - \mathbf{g}. \tag{4.10.11}$$

In many circumstances, the pressure derivative term in (4.10.10) can be neglected, this being equivalent to assuming that changes in density with pressure are negligible, i.e., that the fluid is *incompressible*. Then (4.10.10) becomes

$$\mathbf{V} \cdot \mathbf{u} \equiv \partial u/\partial x + \partial v/\partial y + \partial w/\partial z = 0, \tag{4.10.12}$$

i.e., the velocity field is "nondivergent" or "solenoidal." Batchelor (1967, Section 3.6) has discussed circumstances for which this is a good approximation. The requirements are as follows:

(i)   The particle velocity should be small compared with the sound speed.

(ii)   The phase speed (or wavelength divided by period) of disturbances should be small compared with c_s.

(iii)   The vertical scale of the motion should be small compared with the *scale height* H_s [defined as a median value of $\rho/|d\rho/dz|$].

The last condition is automatically satisfied in the ocean because H_s is about 40 times the depth, but is not true of some atmospheric motions.

4.11 Boundary Conditions

In order to compute changes in the structure of the atmosphere and ocean, it is necessary to know not only the governing equations but also the appropriate conditions to apply at boundaries. The appropriate conditions depend on the properties the fluid is assumed to have, and are simplest for an inviscid nondiffusive fluid. The conditions for this case are given below, and the modifications necessary when diffusion and viscous effects are included will be discussed later in the section.

4.11.1   An Inviscid Nondiffusive Fluid

For this case, the conditions applying at three types of boundary will be considered.

(a) *Fixed Solid Boundary.*   No fluid flows across such a boundary (see Fig. 4.3 and Section 4.3), i.e., the component of the mass flux density $\mathbf{F} = \rho\mathbf{u}$ normal to the boundary must vanish. In other words, the condition is

$$u_n = 0, \tag{4.11.1}$$

where u_n is the normal component of velocity. This condition automatically ensures that fluxes of conserved quantities like θ, q, or s across the boundary are zero.

(b) *A Material Boundary* (such as a free surface or an interface between two fluids).   By definition, no fluid particles cross this boundary, and thus a particle on the boundary will remain on the boundary. Thus if

$$G(x, y, z, t) = 0 \tag{4.11.2}$$

is the equation of the boundary surface, G will always be zero for a material particle on this surface, and therefore

$$DG/Dt = 0 \tag{4.11.3}$$

on the boundary. [This condition is due to Lagrange (1781).] Equation (4.11.1) is, in fact, a special case for which G is independent of t. For then (4.11.3) becomes

$$\mathbf{u} \cdot \nabla G = 0,$$

which is equivalent to (4.11.1) since ∇G is perpendicular to the boundary.

Another condition is required at any nonsolid boundary, whether a material surface or otherwise, to ensure a balance of forces at the boundary. Since the only force exerted by an inviscid fluid on a boundary is due to the pressure, which acts at right angles to the boundary, the condition to apply is that the pressure be the same on both sides: in other words, the pressure is continuous across the boundary. If subscript 1 refers to the value on one side and subscript 2 to the value on the other, then

$$p_1 = p_2. \tag{4.11.4}$$

Actually, Eq. (4.11.4) is also true at a solid boundary, but is not needed if the boundary is rigid, for then it can sustain any pressure applied.

(c) *An Internal Boundary.*   Sometimes a fluid is divided into separate regions for convenience of calculation, the boundary between the regions being completely contained within the fluid. This may be done, for instance, to give each subregion a shape for which solutions can conveniently be obtained. In such cases, appropriate conditions must be applied at the internal boundary to connect the solutions on the two sides. One condition is (4.11.4), i.e., continuity of pressure, which is required for a balance of forces on the boundary. The remaining conditions are continuity of the flux densities

$$\rho u_n, \quad \rho q u_n, \quad \rho s u_n, \quad \rho \theta u_n$$

normal to the boundary of conserved quantities like mass, water vapor, salt, and potential temperature, where u_n is the component of velocity normal to the boundary.

4.11.2 A Diffusive Fluid

When diffusive effects are included, an additional boundary condition is required for each diffusive substance, and the ideal fluid conditions may need to be modified. For instance, at an *internal boundary* (see Section 4.11.1c), the normal flux density must be continuous as before, but now the full expression (4.3.3) or (4.4.5), including the diffusive part, must be used. The extra condition is *continuity of temperature* (or concentration), i.e., using subscript 1 for one side of the boundary and subscript 2 for the other,

$$T_1 = T_2. \tag{4.11.5}$$

For an *external boundary*, the new condition that is added is usually a specification of either (i) the normal flux density, (ii) the temperature (or concentration), or (iii) a relationship between (i) and (ii).

A *phase boundary*, e.g., between air and water or between ice and water, requires special consideration. Such a boundary is *not* a material boundary since molecules can cross it, such as when evaporation of water takes place at the ocean surface. The interface between two fluids is not merely a boundary, but is observed to have a finite thickness over which the properties differ from those of either fluid. Furthermore, soluble material may be preferentially adsorbed into the surface "film," or insoluble material may spread over the surface. Such contaminants may have considerable effect on the properties of the surface [measurements of contaminants found on the ocean surface are reported by W. D. Garrett (1967)]. Thus a proper description of the surface film requires a consideration of its thermodynamics (with quantities per unit area appearing where quantities per unit volume were used in Chapter 3 for the bulk of the fluid), equations of state, as well as a description of motion within the surface, etc. Such matters are discussed in books such as that by Davies and Rideal (1963), and in series, such as that by Matijević (1969ff).

Consider first the boundary conditions which involve diffusive processes that apply to the *air–water* interface. The notation F_{na} will be used (see Fig. 4.7) for the normal flux density (in the upward direction) in the air and F_{nw} for the normal flux density (upward) in the water. A superscript will be used to indicate the quantity whose flux is being considered. Let E be the mass of water being evaporated per unit area per unit time. Then the normal flux $F_{na}^{(q)}$ of water vapor in the air is given by (4.3.3), with q replacing s, and its value at the surface is equal to E. Thus

$$F_{na}^{(q)} = E \tag{4.11.6}$$

is one condition on humidity. This flux must also be equal to the flux of vapor out of the surface film, which is related to the humidity q at the surface by

$$E = k_I \rho_a (q_w - q), \tag{4.11.7}$$

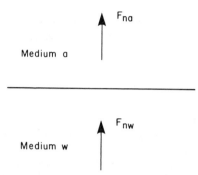

Fig. 4.7. A phase boundary between air and water. F_{na} and F_{nw} are the components of the flux normal to the boundary in the sense shown.

where q_w is the saturation value of the specific humidity corresponding to the surface conditions, and k_l is a property of the surface called the *permeability coefficient*. This coefficient measures how easily water molecules can cross the surface film. For clear water (Davies and Rideal, 1963, Chapter 7), k_l is equal to 5 m s^{-1}, but it can be very much less (down to 10^{-3} m s^{-1}) for substances like cetyl alcohol that are used to reduce evaporation from reservoirs. In normal conditions, k_l is considered large enough for (4.11.7) to be approximately equivalent to the condition

$$q = q_w, \qquad (4.11.8)$$

or, equivalently, that the vapor pressure e of water vapor is equal to its saturation value e_w (see Appendix 4). Note, however, that the values of q_w and e_w quoted in meteorological tables are normally for equilibrium over a plane surface of pure water. In practice [see Kraus (1972)], the value depends on the salinity (e_w is 2% smaller for salinities of $35^0/_{00}$ than for pure water) and on the surface curvature (only significant for drops of spray).

The thermal boundary conditions that are used at the air–sea boundary are continuity of temperature, i.e., (4.11.5), and the flux condition

$$F_{na}^{(T)} - F_{nw}^{(T)} = -L_v E, \qquad (4.11.9)$$

where $F_{na}^{(T)}$ is the upward normal flux of heat in the air at the surface, $F_{nw}^{(T)}$ is the corresponding flux in the water, and L_v is the latent heat of vaporization of water. An expression for the heat flux is given by (4.4.5). Equation (4.11.9) expresses the fact that when evaporation takes place, latent heat is removed at a rate $L_v E$ per unit area per unit time.

The boundary condition on salinity is simply one of no flux into the atmosphere. However, when evaporation is taking place, water is being removed and therefore the salinity of the fluid left behind tends to increase. The surface moves downward relative to fluid particles at a rate E/ρ_w, where ρ_w is the water density, and thus there is an effective flux density of salt of

$$(E/\rho_w)s$$

into the surface film, where s is the salinity at the base of the film. If this is not balanced by a diffusive flux downward, then the salinity of the surface film will increase. When precipitation is occurring, drops of fresh water are introduced at the surface, after which their motion and properties will be determined by the usual equations.

For the phase boundary between *ice* and *water*, the conditions are similar but differ in a few details. For instance, the heat flux condition is equivalent to (4.11.9) and may be written

$$F_{ni}^{(T)} - F_{nw}^{(T)} = -L_m M, \qquad (4.11.10)$$

where $F_{ni}^{(T)}$ is the normal heat flux density into the ice, M is the mass of ice melting per unit area per unit time (negative when freezing takes place), and L_m is the latent heat of melting. The condition (4.11.5) of continuity of temperature is the same, but there is an extra condition, namely, that this temperature be equal to the freezing point. The conditions on salinity again merely express conservation of salt. If melting is occurring, then the condition is that the melted water have the salinity s_i (mass of salt per unit mass of salt and water mixture) of the ice from which it was formed. If freezing is occurring, the salinity of the ice being formed is a function of the conditions occurring at the interface [see Maykut and Untersteiner (1971)]. This salinity is usually a few parts per thousand and much less than the salinity of the water, so the salinity of the remaining liquid water is increased. Considerations that are similar to those for the salinity near the air–water boundary when evaporation takes place also apply.

4.11.3 A Viscous Fluid

For real fluids, additional conditions on velocity are required at boundaries, and there are additional requirements for a balance of forces. The conditions on velocity are that the velocity be continuous, whatever the type of boundary, i.e., (4.11.1) is replaced by

$$\mathbf{u} = \mathbf{u}_b, \qquad (4.11.11)$$

where $\mathbf{u}_b$ is the velocity of the boundary. Conditions expressing a balance of forces acting on the boundary are required only when the boundary is nonrigid. These conditions involve the components of the stress exerted by the fluid on elements of the boundary. Expressions for the components of the stress may be found in textbooks on fluid mechanics such as that by Batchelor (1967). For instance, the normal component of the stress (or force per unit area) on the boundary is given by

$$p - 2\mu \, \partial u_n/\partial n,$$

where p is the pressure, μ the viscosity, u_n the normal component of velocity, and $\partial/\partial n$ denotes the derivative in the direction normal to the boundary. At internal boundaries, the fluid stress is continuous, but discontinuities can exist at the air–sea phase boundary because of forces within the surface film. The normal stress in the water exceeds that in the air by an amount

$$\gamma(r_1^{-1} + r_2^{-1}),$$

where r_1, r_2 are the principal radii of curvature of the surface, and γ is the surface tension. The value of γ depends on temperature, salinity, and the concentration of contaminants. For uncontaminated salt water at normal temperature, γ has values near 0.08 N m^{-1}. Contaminants can cause considerable reductions in the value of γ [see Kraus (1972)].

For contaminated surfaces, there can also be discontinuities in tangential stress. Convergence in surface motion increases contaminant concentration, lowering surface tension at these points, so the high surface tension elsewhere tends to pull the local concentration out again. This resistance to surface convergence is responsible for the rapid damping of ripples on contaminated surfaces ("oil on troubled waters"). The phenomenon and associated boundary conditions are discussed by Lucassen-Reynders and Lucassen (1969).

4.11.4   Representation of Viscous and Diffusive Effects without Detailed Resolution

Because the viscosity and diffusivities of air and water are small, one might think that their effects could be ignored altogether. However, their importance for large-scale motion has already been discussed, and their effects near boundaries are particularly important. For instance, condition (4.11.11) requires the tangential component of velocity of the atmosphere and ocean to be continuous at the interface, where an inviscid model would give a large discontinuity in tangential velocity. In practice, this gives rise to a large *shear*, or velocity gradient, near the boundary. The thickness of the region of large shear (called the *boundary layer*) is determined by the viscosity if the shear is small enough, as in some laboratory situations. In the atmosphere and ocean, however, the shear (see Section 2.4) is nearly always so large that small disturbances grow spontaneously, taking energy from the shear flow and thereby forming a turbulent boundary layer. The transfer of momentum, heat, humidity, salt, etc., in such layers is achieved by the eddy motion, except for a very thin layer near the boundary, at which molecular transfer processes dominate. The nature of the eddy motion, and hence the rates of transfer, is not entirely dependent on the shear. Convection due to heavy fluid overlying light fluid can also produce eddies or modify shear-produced eddies. The rates of transfer can also be influenced by surface properties, either by some direct effect (such as the effect of low permeability or evaporation) or indirectly by influencing the shape of the surface (contaminants modify wave properties and their rates of transfer of momentum to the waves).

For modeling large-scale motions of the atmosphere and ocean, the detailed structure of the boundary layer cannot be resolved. Instead, transfer rates across the boundary are related to properties of the boundary and of the air or ocean some distance from the boundary. In particular, those representing the effect of a turbulent shear flow take the forms given in Section 2.4. For instance, the tangential stress at the bottom of the ocean or atmosphere can be calculated from (2.4.1). The existence of this stress implies that energy is being removed from the ocean or atmosphere, so this effect is sometimes referred to as "*bottom friction*." The fluxes of heat and water between ocean and atmosphere are dealt with in a similar way, using empirically derived boundary conditions of the type considered in Chapter 2.

4.12   A Coordinate System for Planetary Scale Motions

Because the gravitational force is so dominant in the equations of motion, great care is required [see N. A. Phillips (1973)] in defining a suitable coordinate system. If, for instance, spherical polar coordinates were used, it would be found that an important term in the equations for large-scale motion tangential to the spherical surfaces would be the component of gravity along those surfaces! It is therefore preferable to use geopotential surfaces rather than spheres in defining the coordinate system.

The shape of geopotential surfaces is known quite well from satellite data. As a first approximation, the sea-level geopotential surface is an oblate spheroid with eccentricity ϵ given by $1/\epsilon = 298.257$ and a semimajor axis of 6378.139 km. In other words, a section through the earth which includes the earth's axis is an ellipse with the polar radius (or semiminor axis) smaller than the equatorial radius (or semimajor axis) by a fraction ϵ, i.e., by 21.385 km. Departures from this ellipsoid of best fit are relatively small and can be shown on a map [e.g., Fig. 3 in the work by Lerch *et al.* (1979)]. The largest departure is a depression of "depth" 100 m just to the south of India, i.e., departures are of order 10^{-5} earth radii.

If the geopotential surfaces were exactly oblate spheroids, the natural coordinate system to use would be oblate spheroidal coordinates (λ, φ, r), which are related (Morse and Feshbach, 1953, p. 662) to spherical polar coordinates $(\lambda, \varphi_s, r_s)$, where r_s is radial distance, φ_s latitude, and λ longitude, by

$$r_s^2 = r^2 + \tfrac{1}{2}d^2 - d^2 \sin^2 \varphi, \tag{4.12.1}$$

$$r_s^2 \cos^2 \varphi_s = (r^2 + \tfrac{1}{2}d^2) \cos^2 \varphi, \tag{4.12.2}$$

where d is a constant equal to half the distance between the foci of the ellipsoid ($d = 521.854$ km for the earth). With this definition, the ellipsoid of best fit to the sea-level geopotential surface is given by

$$r_0 = 6367.456 \quad \text{km}. \tag{4.12.3}$$

The equatorial radius (semimajor axis) is

$$(r_0^2 + \tfrac{1}{2}d^2)^{1/2} = 6378.139 \quad \text{km}, \tag{4.12.4}$$

and the polar radius (semiminor axis) is

$$(r_0^2 - \tfrac{1}{2}d^2)^{1/2} = 6356.754 \quad \text{km}. \tag{4.12.5}$$

The coordinate system to be used is a slight distortion from the oblate spheroidal set in that the surfaces $r = $ const are defined to be geopotential surfaces. The value of r assigned to each geopotential surface is the one for the spheroid such that the area average of the distance between the two surfaces is zero. The difference between the expressions for the various terms in the equations is so small that it can be ignored, with the exception of the fact that in the system to be used gravity is exactly perpendicular to the surfaces $r = $ const. The sea-level geopotential surface is now given by (4.12.3).

The expressions in terms of λ, φ, r for the operators that occur in the equations depend on the coefficients h_λ, h_φ, h_r that appear in the metric

$$h_\lambda^2 \, d\lambda^2 + h_\varphi^2 \, d\varphi^2 + h_r^2 \, dr^2,$$

i.e., the expression for the square of an infinitesimal displacement. For oblate spheroidal coordinates, these coefficients are given by (Morse and Feshbach, 1953)

$$
\begin{aligned}
h_\lambda^2 &= (r^2 + \tfrac{1}{2}d^2) \cos^2 \varphi, \\
h_\varphi^2 &= r^2 - \tfrac{1}{2}d^2 + d^2 \sin^2 \varphi, \\
h_r^2 &= r^2(r^2 - \tfrac{1}{2}d^2 + d^2 \sin^2 \varphi)(r^4 - \tfrac{1}{4}d^4)^{-1}.
\end{aligned}
\tag{4.12.6}
$$

However, the maximum error in using the approximations, valid for small d, namely,

$$h_\lambda = r \cos \varphi, \qquad h_\varphi = r, \qquad h_r = 1, \tag{4.12.7}$$

is only $d^2/4r^2$, which is less than 0.17% in the neighborhood of the earth's surface. If this approximation is used, the *equations are the same* as those written in spherical polar coordinates. However, the *meaning* given to the variables *is different*, and it is exactly true to say that gravity is perpendicular to the surfaces $r = \text{const}$. It is important to remember that vertical displacements (e.g., of the sea surface) are expressed relative to *geopotential* (and *not* spherical) surfaces.

Now the velocity components u, v, w that are associated with the coordinates λ, φ, r are defined, i.e., u is in the direction of increasing λ (which will be referred to as "eastward"), v is in the direction of increasing φ (which will be referred to as "northward"), and w is in the direction of increasing r, i.e., "upward" or in the direction opposite to gravity. The coordinate z is defined by

$$z = r - r_0, \tag{4.12.8}$$

i.e., the distance measured upward from the sea-level geopotential surface. The equations in these coordinates [using the approximation $d \ll r_0$, which gives (4.12.7)] are listed below, with the derivative following the motion (see Section 4.1) being given by

$$\frac{D\gamma}{Dt} \equiv \frac{\partial \gamma}{\partial t} + \frac{u}{r \cos \varphi} \frac{\partial \gamma}{\partial \lambda} + \frac{v}{r} \frac{\partial \gamma}{\partial \varphi} + w \frac{\partial \gamma}{\partial z}, \tag{4.12.9}$$

and the divergence operator (see Section 4.2) being given by

$$\mathbf{V} \cdot \mathbf{F} = \frac{1}{\cos \varphi} \left\{ \frac{\partial}{\partial \lambda} \left(\frac{F_\lambda}{r} \right) + \frac{\partial}{\partial \varphi} \left(\frac{F_\varphi \cos \varphi}{r} \right) + \frac{1}{r^2} \frac{\partial}{\partial z} (r^2 F_r \cos \varphi) \right\}, \tag{4.12.10}$$

where F_λ, F_φ, F_r are the components of $\mathbf{F}$ in the directions of increasing λ, φ, and r.

The mass conservation equation (4.10.1) can be written as

$$\frac{1}{\rho} \frac{D\rho}{Dt} + \frac{1}{r} \frac{\partial u}{\partial \lambda} + \frac{1}{r \cos \varphi} \frac{\partial}{\partial \varphi} (v \cos \varphi) + \frac{\partial w}{\partial z} = 0 \tag{4.12.11}$$

or in the alternative form (4.2.4), namely,

$$\frac{\partial}{\partial t}(\rho \cos \varphi) + \frac{\partial}{\partial \lambda}\left(\frac{\rho u \cos \varphi}{r}\right) + \frac{\partial}{\partial \varphi}\left(\frac{\rho v \cos \varphi}{r}\right) + \frac{\partial}{\partial z}(\rho w \cos \varphi) = 0. \quad (4.12.12)$$

Combining this with (4.12.9), the identity (4.3.6) can be written

$$\rho \cos \varphi \frac{D\gamma}{Dt} \equiv \frac{\partial}{\partial t}(\rho \gamma \cos \varphi) + \frac{\partial}{\partial \lambda}\left(\frac{\rho u \gamma \cos \varphi}{r}\right)$$

$$+ \frac{\partial}{\partial \varphi}\left(\frac{\rho v \gamma \cos \varphi}{r}\right) + \frac{\partial}{\partial z}(\rho w \gamma \cos \varphi). \quad (4.12.13)$$

The momentum equations have a rather complicated form in terms of the coordinates λ, φ, r. They are

$$\frac{Du}{Dt} - \left(2\Omega + \frac{u}{r \cos \varphi}\right)(v \sin \varphi - w \cos \varphi) = -\frac{1}{\rho r \cos \varphi}\frac{\partial p}{\partial \lambda} + \mathscr{V}_\lambda, \quad (4.12.14)$$

$$\frac{Dv}{Dt} + \frac{wv}{r} + \left(2\Omega + \frac{u}{r \cos \varphi}\right)u \sin \varphi = -\frac{1}{\rho r}\frac{\partial p}{\partial \varphi} + \mathscr{V}_\varphi, \quad (4.12.15)$$

$$\frac{Dw}{Dt} - \frac{v^2}{r} - \left(2\Omega + \frac{u}{r \cos \varphi}\right)u \cos \varphi = -\frac{1}{\rho}\frac{\partial p}{\partial z} - g + \mathscr{V}_r, \quad (4.12.16)$$

where $\mathscr{V}_\lambda$, $\mathscr{V}_\varphi$, $\mathscr{V}_r$ are the components of the viscous term. For constant μ, they are given by

$$\mathscr{V}_\lambda = \nu\left\{\Delta u - \frac{1}{r^2 \cos^2 \varphi}\left(u + 2\frac{\partial}{\partial \lambda}(v \sin \varphi - w \cos \varphi)\right)\right\}, \quad (4.12.17)$$

$$\mathscr{V}_\varphi = \nu\left(\Delta v - \frac{v}{r^2 \cos^2 \varphi} + \frac{2 \sin \varphi}{r^2 \cos^2 \varphi}\frac{\partial u}{\partial \lambda} + \frac{2}{r^2}\frac{\partial w}{\partial \varphi}\right), \quad (4.12.18)$$

$$\mathscr{V}_r = \nu\left(\Delta w - \frac{2w}{r^2} - \frac{2}{r^2 \cos \varphi}\frac{\partial u}{\partial \lambda} - \frac{2}{r^2 \cos \varphi}\frac{\partial}{\partial \varphi}(v \cos \varphi)\right), \quad (4.12.19)$$

where

$$\Delta\gamma \equiv \frac{1}{r^2 \cos^2 \varphi}\frac{\partial^2 \gamma}{\partial \lambda^2} + \frac{1}{r^2 \cos \varphi}\frac{\partial}{\partial \varphi}\left(\cos \varphi \frac{\partial \gamma}{\partial \varphi}\right) + \frac{1}{r^2}\frac{\partial}{\partial r}\left(r^2 \frac{\partial \gamma}{\partial r}\right). \quad (4.12.20)$$

Equation (4.12.14) can be rearranged to become an equation for the angular momentum per unit mass, $ur \cos \varphi + \Omega r^2 \cos^2 \varphi$, namely,

$$\rho D(ur \cos \varphi + \Omega r^2 \cos^2 \varphi)/Dt = -\partial p/\partial \lambda + \rho r \cos \varphi \, \mathscr{V}_\lambda. \quad (4.12.21)$$

Angular momentum balances over the whole earth were referred to in Chapter 2.

An additional approximation that is often made is to treat r as a constant in (4.12.11) and (4.12.14)–(4.12.16). Since the ocean depth rarely exceeds 6 km, the maximum error in applications to the ocean is $\pm0.05\%$, i.e., smaller than the error involved in (4.12.7). For applications to the stratosphere, however, the error is about 1%.

The heat and salinity equations (4.10.3) and (4.10.4) can be expressed in terms of λ, φ, r coordinates by using the expressions (4.12.9) and (4.12.10) and the fact that the components of the gradient vector $\nabla\gamma$ are

$$\nabla\gamma \equiv \left(\frac{1}{r\cos\varphi}\frac{\partial\gamma}{\partial\lambda},\ \frac{1}{r}\frac{\partial\gamma}{\partial\varphi},\ \frac{\partial\gamma}{\partial z} \right). \qquad (4.12.22)$$

Chapter Five

Adjustment
under Gravity
in a Nonrotating System

5.1 Introduction: Adjustment to Equilibrium

The first two chapters sought to give some insight into the way energy from the sun is absorbed by the atmosphere–ocean system, how fluid motion results, and how this motion affects the mean distribution of temperature. As Halley (1686, p. 165) stated (see Chapter 2), the system is driven by "the Action of the Suns Beams upon the Air and Water," and so "according to the Laws of *Staticks*, the Air which is less rarified or expanded by heat, and consequently more ponderous, must have a Motion towards those parts thereof, which are more rarified, and less ponderous, to bring it to an *Æquilibrium*." This chapter marks the beginning of a more detailed study of the way the atmosphere–ocean system *tends to adjust* to equilibrium. The adjustment processes are most easily understood in the absence of driving forces. Suppose, for instance, that the sun is "switched off," leaving the atmosphere and ocean with some nonequilibrium distribution of properties. How will they respond to the gravitational restoring force? Presumably there will be an adjustment to some sort of equilibrium. If so, what is the nature of the equilibrium? How long does the adjustment take? In what way is the adjustment process most readily described and understood?

The problem will be studied in stages, roughly following the historical development. In this chapter, for instance, complications due to the rotation and shape of the earth will be ignored and only small departures from the hydrostatic equilibrium of Section 3.5 will be considered. The nature of the adjustment processes will be found by deduction from the equations of motion developed in Chapters 3 and 4.

95

This method was not available in the seventeenth century, but it was possible instead to study simpler systems in the laboratory and thereby gain an improved understanding of Nature. A remarkable example is found in the work of Marsigli (1681). It seems (Deacon, 1971, pp. 147–149) that when Marsigli went to Constantinople in 1679 he was told about an undercurrent in the Bosphorus that was well-known to local fishermen. The undercurrent was in fact referred to in a sixth century discussion of flows through straits by Procopius of Caesarea (History of the Wars VIII, vi. 27) "... for the fishermen of the towns on the Bosphorus say that the whole stream does not flow in the direction of Byzantium, but while the upper current which we can see plainly does flow in this direction, the deep water of the abyss, as it is called, moves in a direction exactly opposite to that of the upper current and so flows continuously against the current which is seen." [That is, the undercurrent flows toward the Black Sea from the Mediterranean. Defant (1961, Chapter 16) gives a modern description.] By observing the distortions and feel of a rope lowered into the water, Marsigli found that the current reversal occurred at depths varying between 8 and 12 Turkish feet. He reasoned that the effect was due to density differences, and so made measurements of these differences using a hydrostatic balance. He found that water from the Black Sea is lighter than water from the Mediterranean, giving readings on his instrument up to $29\frac{1}{4}$ grains lower. He attributed the low density of the Black Sea to lower salinity resulting from river runoff. Marsigli then measured the density of samples taken from the surface of the Bosphorus and from the undercurrent. Here the difference was 10 grains, values being consistent with a Mediterranean origin for the undercurrent and a Black Sea origin for the surface water. To clinch the point, he performed a laboratory experiment, which he illustrated as shown in Fig. 5.1. A tank

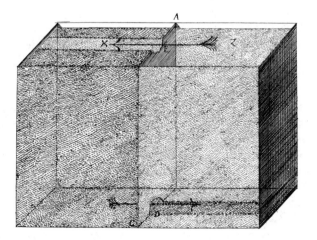

Fig. 5.1. A figure from Marsigli (1681) illustrating adjustment under gravity of two fluids of different density. Initially the container was divided in two by a partition. Side X contained water taken from the undercurrent in the Bosphorus. Side Z contained dyed water having the density of surface water in the Black Sea. The experiment was to put holes in the partition at D and E and to observe the resulting flow. The flow through the lower hole was in the direction of the undercurrent in the Bosphorus, while the flow through the upper hole was in the direction of the surface flow.

was divided in two by a partition. Side X was filled with water taken from the under-current and side Z with dyed water having density equal to that of the Black Sea. Holes in the partition at D and E were then opened and water from X was observed to pass through hole D to side Z, whereas the movement through hole E was in the opposite direction. Marsigli noted that this internal adjustment could give surface currents in the observed direction without requiring a difference in sea level between the Black Sea and the Mediterranean. He had already attempted to measure such a difference using a mercury barometer.

The force that produces the motion when the holes in Marsigli's apparatus are opened is gravity, which produces pressure differences between the two sides. An arrangement that will receive further study in this chapter is the one shown in Fig. 5.2a. A fluid of uniform density ρ_c is at rest on two sides of a partition, the surface levels on the two sides differeing by an amount h. By (3.5.8), the pressure at A exceeds the pressure at B by an amount $\rho_c gh$, so that if the partition is suddenly removed, the fluid near the partition will start toward the right. A situation closer to that of Marsigli's experiment is illustrated in Fig. 5.2b, where now there are two fluids, one of smaller density ρ_1 and the other of larger density ρ_2. A partition divides the lower fluid into two parts as shown. The upper fluid is in equilibrium and is assumed *not* to be completely divided by the partition. It follows that the pressures at any level above the top of the partition are equal, and the same is true at the level of C and D by the hydrostatic equation (3.5.8). However, the same equation shows that the pressure difference between C and A exceeds the difference between D and B by an amount $(\rho_2 - \rho_1)gh$, so the pressure difference between A and B is $(\rho_2 - \rho_1)gh$. When the partition is removed, motion toward the right will take place as in case (a), but the adjustment is slower since the pressure difference is reduced by a factor $(\rho_2 - \rho_1)/\rho_2$.

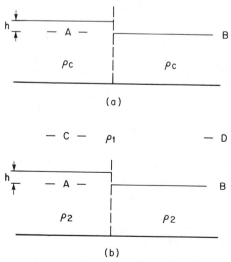

(a)

(b)

Fig. 5.2. Initial states considered for problems of adjustment under gravity. The dashed line marks a partition (of finite height) which is removed at the initial instant. (a) The fluid of density ρ_c has different depths on the two sides and is bounded above by a free surface. The difference h in depths is supposed to be very small, so that the problem to be solved is linear. (b) The only difference is that a fluid of density $\rho_1 < \rho_2$ lies above the lower fluid.

The adjustment processes are, in fact, exactly the same as they would be in case (a) if the gravitational acceleration were reduced to a value g' given by

$$g' = g(\rho_2 - \rho_1)/\rho_2. \tag{5.1.1}$$

g' is called *reduced gravity* for this reason. This example also shows that the driving force is proportional to

$$(\rho_2 - \rho_1)g,$$

the density difference times g. This product is called the *buoyancy* force per unit volume (see Section 4.5).

In the quantitative treatment of this problem, which will be developed in this chapter, the difference h in initial levels will be assumed to be small. This simplifies the mathematics because it leads to a linear problem, and so solutions can be superposed. This chapter is concerned with adjustment of a homogeneous fluid with a free surface (including the case depicted in Fig. 5.2a). Chapter 6 deals with internal adjustment of a density-stratified fluid, including problems such as that illustrated in Fig. 5.2b.

Although the qualitative ideas in the above argument (relating to Fig. 5.2) can be found in the work of Archimedes (287–212 BC) ["On Floating Bodies," English translation in Hutchins (1952); see also discussion in Dugas (1957)], quantitative treatment of the problem required first the development of the laws of motion and of the calculus needed to apply the laws. Both of these developments are among the achievements of Newton (whose *Principia* was printed in 1687 with the help and encouragement of Halley). Also required was a proper understanding of hydrostatics and the nature of pressure forces. The work of Stevin (1548–1620) and Pascal on this subject is translated by Spiers and Spiers (1937). The main credit for developing the equations of motion goes to Euler (1755), who commented:

> But here we see well enough how far distant we yet are from the complete knowledge of the motion of fluids, and that which I have just explained contains only a feeble beginning. Nevertheless, all that the theory of fluids includes is contained in the two equations presented above, so that it is not the principles of mechanics which we lack in the pursuit of these researches, but solely analysis, which is not yet sufficiently cultivated for this purpose. And thus we see clearly what discoveries remain for us to make in this science before we can arrive at a perfect theory of the motion of fluids.

(This translation appears in the commentary by Truesdell (1954b, p. LXXXIX) in Euler's *Opera Omnia*. The two equations referred to are the continuity equation, derived by the first method used in Section 4.2, and the inviscid momentum equations, derived in Section 4.5.)

Among the first problems treated using the equations of motion were problems of the response of the ocean and atmosphere to gravitational forces. Laplace (1778–1779) developed the equations for motion on a rotating sphere under the action of tide-generating forces and found solutions for the "equilibrium" tide in a constant-depth worldwide ocean. He also encountered the problem of treating thermal forcing of the atmosphere.

Our atmosphere consists of an elastic fluid whose density is a function of pressure and temperature. These are not constant at a given point in the atmosphere because the rotating earth presents a different point to the sun at each instant of the day, and, because of the inclination of the ecliptic, each day has a different length and the elevation of the sun increases or decreases. It is readily seen that the variations in heating due to these different causes must excite oscillations which it seems impossible to submit to calculation because the law of these variations ... has not been sufficiently well determined.

[My paraphrasing; see "On oscillations of the atmosphere," Laplace "Œuvres" (1893, pp. 283–301).] However, by making some assumptions Laplace reduced the problem to one of forced oscillations of an isothermal atmosphere, which he found to obey the tidal equations already deduced for the ocean, but with the ocean depth replaced by the scale height of the atmosphere, which he calculated to be 27,000 ft (the surface pressure in feet of water multiplied by the density ratio between air and water, i.e., 32×850).

The final part of Laplace's paper ["On waves" (Laplace, "Œuvres" (1893, pp. 301–310))] treats the problem to be dealt with in the next two sections, namely, the adjustment of a homogeneous fluid of constant depth that initially has a small displacement of its free surface. He found, in particular, the relation (5.3.8), which showed that disturbances propagate away from the disturbed region at a speed that depends on the curvature of the surface.

5.2 Perturbations from the Rest State for a Homogeneous Inviscid Fluid

The equilibrium state considered here is one of a fluid of uniform density ρ_e that is at rest and that has uniform depth H. A good example would be an artificial pond with a flat bottom. To be able to give a precise description of the motion that occurs when the system is perturbed (e.g., by throwing in a stone), a coordinate system is required. A convenient one consists of the Cartesian coordinates (x, y, z) chosen so that the z axis points vertically upward, the free surface being at $z = 0$ and the bottom at $z = -H$. The equilibrium solution has zero velocity, and the pressure is determined by the hydrostatic equation (3.5.8) or (4.5.18). The equilibrium pressure $p_0(z)$ in this case is given by

$$p_0(z) = -g\rho z, \tag{5.2.1}$$

where ρ is the in-situ density, i.e., ρ_e in the fluid and zero above, and g is the acceleration due to gravity. (If there is any fluid in the region $z > 0$, it is assumed to have negligible density.)

Suppose now that the equilibrium is slightly disturbed. The perturbations are assumed to be small enough for products of perturbation quantities to be neglected in comparison with the perturbation quantities themselves. Suppose that (u, v, w) are the velocity components corresponding to the coordinates (x, y, z) and that the disturbed position of the free surface (see Fig. 5.3) is given by

$$z = \eta(x, y, t). \tag{5.2.2}$$

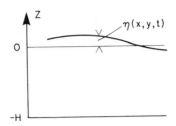

Fig. 5.3. The geometry of the disturbed surface. The displacement from the rest position is η and the undisturbed depth is H.

For this problem, it is convenient to define the perturbation pressure by

$$p = -g\rho z + p', \tag{5.2.3}$$

where ρ is the insitu density, i.e., ρ_c in the fluid and zero above. (This differs from the definition of Section 4.5 only in the infinitesimal region between the disturbed and undisturbed positions of the free surface.)

The equations of motion consist of the continuity equation (4.2.3) and the momentum equations (4.5.7)–(4.5.9) for an inviscid fluid. Since the density is constant within the fluid, the rotation rate Ω is 0, and products of perturbation quantities can be neglected, the continuity equation is in this case

$$\partial u/\partial x + \partial v/\partial y + \partial w/\partial z = 0, \tag{5.2.4}$$

and the momentum equation is

$$\rho \, \partial u/\partial t = -\partial p'/\partial x, \qquad \rho \, \partial v/\partial t = -\partial p'/\partial y, \tag{5.2.5}$$

$$\rho \, \partial w/\partial t = -\partial p'/\partial z. \tag{5.2.6}$$

Adding the x, y, and z derivatives of the above three components of the momentum equation, and using the continuity equation (5.2.4), there results an equation for p', namely, Laplace's equation

$$\nabla^2 p' \equiv \partial^2 p'/\partial x^2 + \partial^2 p'/\partial y^2 + \partial^2 p'/\partial z^2 = 0. \tag{5.2.7}$$

(In connection with the present problem, Laplace "Œuvres" (1893, pp. 301–310) found this as an equation for the vertical displacement of a material particle). The condition (see Section 4.11) that must be satisfied at the bottom, where $Z = -H$, is one of no normal flow, i.e.,

$$w = 0 \qquad \text{at} \quad z = -H. \tag{5.2.8}$$

The condition that a particle in the free surface $z = \eta$ will remain in it (see Section 4.11.1) is in this case

$$D(z - \eta)/Dt = 0,$$

i.e.,

$$w = \partial\eta/\partial t + u \, \partial\eta/\partial x + v \, \partial\eta/\partial y, \tag{5.2.9}$$

which, for small perturbations, reduces to

$$w = \partial\eta/\partial t \qquad \text{at} \quad z = \eta. \tag{5.2.10}$$

In addition, the pressure must vanish at the free surface, i.e.,

$$p = p_0 + p' = 0 \qquad \text{or} \qquad p' = \rho g \eta \quad \text{at} \quad z = \eta \qquad (5.2.11)$$

by (5.2.3). Also, since the differences in the solutions for w and p' between $z = \eta$ and $z = 0$ are small, Eqs. (5.2.10) and (5.2.11) can both be applied at $z = 0$ and will be correct to the same order of approximation.

The problem is to solve Laplace's equation (5.2.7) subject to the boundary conditions (5.2.8) at the bottom and (5.2.10) and (5.2.11) at $z = 0$. There are in fact a great variety of solutions depending on the initial condition, i.e., the nature of the perturbation at the beginning. In the next section, solutions will be considered where p' varies *sinusoidally* with horizontal position. This is not a real restriction because an arbitrary disturbance can be described as a superposition of such waves by Fourier's theorem.

5.3   Surface Gravity Waves

A disturbance that is sinusoidal in the horizontal can take the form of a traveling wave or of a standing wave. In particular, a "long-crested" traveling wave has the form

$$\eta = \eta_0 \cos(kx + ly - \omega t), \qquad (5.3.1)$$

where η_0 is the *amplitude*, the vector

$$\mathbf{k} = (k, l)$$

is the *wavenumber* (proportional to the number of waves per unit distance), ω is the *frequency*, and the quantity

$$\hat{\Phi} = kx + ly - \omega t = \mathbf{k} \cdot \mathbf{x} - \omega t \qquad (5.3.2)$$

is called the *phase* of the wave. Such a wave consists of a sinusoidal corrugation of the surface that moves at uniform speed. A sketch of the wave is shown in Fig. 5.4 with sections cut normal to the wave crests and along the x axis. In the section normal to the wave crests, one sees a series of waves with wavelength $2\pi\kappa^{-1}$ where κ, given by

$$\kappa^2 = k^2 + l^2, \qquad (5.3.3)$$

is the magnitude of the wavenumber. In this plane, the crests move at a speed

$$c = \omega/\kappa, \qquad (5.3.4)$$

called the *phase speed* (i.e., the speed of lines of constant phase $\hat{\Phi}$). Their rate of movement in any other plane *appears* to be faster by a factor equal to the secant of the angle between that plane and the plane normal to the crests. For instance, in Fig. 5.4 the cut along the x axis shows a greater apparent wavelength $2\pi k^{-1}$ than that for the plane normal to the crests, and the apparent propagation speed is proportionately higher. This should be borne in mind when propagation is observed along only one plane.

Although one is free to choose the initial disturbance to be sinusoidal in the horizontal, it does not follow that the traveling wave (5.3.1) is a possible form of

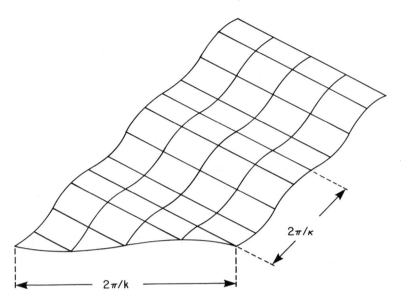

Fig. 5.4. A plane sinusoidal wave train moving at an angle to the axis. The wavenumber (k, l) has magnitude κ. Note that the wavelength $2\pi/k$ observed in a section along the x axis is larger than the actual wavelength $2\pi/\kappa$.

motion. This can be deduced only from the equations. If, however, it is assumed that p' is proportional to η, as given by (5.3.1), Laplace's equation (5.2.7) gives

$$\partial^2 p'/\partial z^2 - \kappa^2 p' = 0, \tag{5.3.5}$$

so that, at a given horizontal position and time, the vertical variation of p' must be a sum of exponentials or hyperbolic functions. The boundary condition (5.2.8), together with (5.2.6), shows that $\partial p'/\partial z$ must vanish at $z = -H$. Since p' is also given by (5.2.11) at $z = 0$, the solution must be

$$p' = \frac{\rho g \eta_0 \cos(kx + ly - \omega t) \cosh \kappa(z + H)}{\cosh \kappa H}, \tag{5.3.6}$$

with the vertical velocity component $\left[\text{see } (5.2.6)\right]$ given by

$$w = \frac{\kappa g \eta_0 \sin(kx + ly - \omega t) \sinh \kappa(z + H)}{\omega \cosh \kappa H}. \tag{5.3.7}$$

It remains to satisfy condition (5.2.10) at $z = 0$. Substitution shows that this *is* consistent with the assumed form (5.3.1), provided that

$$\omega^2 = g\kappa \tanh \kappa H. \tag{5.3.8}$$

This important equation determines the frequency and hence the phase speed of waves of a given wavenumber, such an equation being called a *dispersion relation*. The above dispersion relation was obtained by Laplace "Œuvres" (1893, pp. 301–310). Figure 5.5 shows graphs of ω and $c = \omega/\kappa$ as functions of κ.

One important property is that the frequency does not depend on the *direction* of the wave, but only on the *magnitude* of the wavenumber. Thus waves of a given

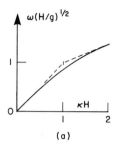
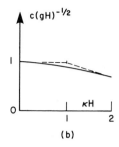

Fig. 5.5. The dispersion relation for surface gravity waves on water of depth H. (a) Frequency ω and (b) phase speed c as functions of wavenumber κ. The dashed line shows the long-wave approximation for $\kappa H < 1$ and the short-wave approximation for $\kappa H > 1$. The maximum error in these approximations is 13% at $\kappa H = 1$.

wavenumber that move in different directions all do so at the same speed. Consider the sort of perturbation that may be obtained by the superposition of such waves. For instance, the wave given by

$$\eta = \eta_0\left[\cos(kx + ly - \omega t) + \cos(kx - ly - \omega t)\right] = 2\eta_0 \cos ly \cos(kx - \omega t) \quad (5.3.9)$$

represents a wave with crests parallel to the y axis that moves in the x direction with speed ω/k (faster than ω/κ). The height varies along the crest with wavelength $2\pi/l$. Another example is the standing wave

$$\eta = \eta_0\left[\cos(kx + ly - \omega t) + \cos(kx + ly + \omega t)\right] = 2\eta_0 \cos(kx + ly)\cos \omega t \quad (5.3.10)$$

for which the wave crests remain stationary, but the surface moves up and down with frequency ω. For each wave form, the velocity field can be calculated from (5.2.5) and (5.2.6). Figure 5.6 shows how velocities relate to the free surface for (a) a traveling wave and (b) a standing wave.

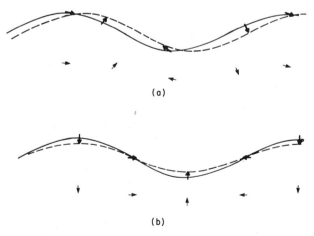

Fig. 5.6. The motion, shown by arrows, of fluid particles associated with a traveling wave (a) and a standing wave (b). The solid line shows the free surface at some initial time and the dotted line shows the position of the surface a short time later. The arrows mark particle displacements in this time. For the standing wave, the particle paths are straight-line segments whose orientation depends on position relative to the crests. For the traveling wave the particle paths are ellipses that become circular for large κH and straight line segments for small κH. In each case, the perturbation pressure is highest below crests and lowest below troughs.

5.4   Dispersion

Another important consequence of (5.3.8) is that the phase speed $c = \omega/\kappa$ varies with κ (see Fig. 5.5). Thus waves of different wavelengths, starting at the same place, will move away at different speeds and thus will *disperse* or spread out. The phenomenon is called *dispersion*—hence the name dispersion relation for equation (5.3.8). The concept is quite a general one, and any waves whose speed varies with wavenumber are called dispersive waves. The effect of dispersion is particularly noticeable with ocean waves that are generated by a distant storm (Barber and Ursell, 1948). Since long waves (small κ) travel fastest, these arrive first and may precede shorter waves from the same storm by one or two days. The fact that waves of different length become separated and arrive at different times explains why swell is so regular compared with waves produced by local winds.

The dispersion effect has been used to identify the point of origin of waves that have traveled extraordinarily large distances (Snodgrass *et al.*, 1966). One set of waves observed in the North Pacific was estimated to have traveled halfway around the world from the Indian Ocean, the great circle route passing south of Australia. The direction of travel is determined from the orientation (in deep water) of the wave crests, and the distance is calculated from the difference in arrival time of waves of different length, and hence of different frequency. The dominant frequency increases progressively with time as the progressively shorter waves arrive, and the rate of change of this frequency gives the distance of travel.

Despite the effects of dispersion, in practice waves are never purely sinusoidal, but are instead a mixture of waves of different wavenumber. As a wave train travels away from its source region, the waves at a particular point become more "pure" in the sense that the wavenumbers that give significant contributions to the wave become confined to a narrower band. Hence there is particular interest in waves made up of components with nearly equal wavenumber. The simplest example (Stokes, 1876) consists of a superposition of two plane waves with equal amplitude:

$$\eta = \cos[(k + \delta k)x - (\omega + \delta\omega)t] + \cos[(k - \delta k)x - (\omega - \delta\omega)t], \quad (5.4.1)$$

i.e.,

$$\eta = 2\cos(\delta k\, x - \delta\omega\, t)\cos(kx - \omega t), \quad (5.4.2)$$

and this example is shown in Fig. 5.7. Equation (5.4.2) shows that this can be interpreted as a wave that is approximately sinusoidal with phase $\hat{\Phi} = kx - \omega t$ but with amplitude

$$2\cos(\delta k\, x - \delta\omega\, t) \approx 2\cos[\delta k(x - t\, d\omega/dk)], \quad (5.4.3)$$

which varies from place to place and from time to time. Because δk is small, however, the amplitude change from one wave crest to the next is only slight, and so (5.4.2) is an example of what is called a "slowly varying wave train." Individual wave crests move with the phase speed ω/k, but the region in which waves have large amplitudes moves with speed

$$c_{\mathrm{g}} = d\omega/dk, \quad (5.4.4)$$

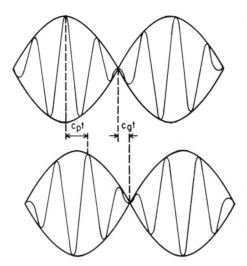

Fig. 5.7. A superposition of two sinusoidal traveling waves, illustrating the difference between the speed c_p of the wave crests and the speed c_g of the envelope of the waves, i.e., of the regions of large amplitude. The group velocity c_g equals $d\omega/dk$, which in this case is equal to $\tfrac{1}{2}c_p$ as for deep-water waves.

as given by (5.4.3). This speed is called the *group velocity*. It depends on the *derivative* of ω because the region of large amplitudes occurs where the phase *difference* between these two component waves has a certain value.

A more general example (Landau and Lifshitz, 1959, Section 66) consists of a superposition of many waves with wavenumber close to k, but with a range of different values of δk. This combination also may be considered as a superposition of solutions of the form (5.4.2) with a common factor $\cos(kx - \omega t)$. It follows that if the initial superposition has the form

$$\eta = f(x)\cos kx, \tag{5.4.5}$$

then, to the same level of approximation as (5.4.3), the solution at time t will have the form

$$\eta = f(x - t\, d\omega/dk)\cos(kx - \omega t). \tag{5.4.6}$$

In other words, since each *contribution* to the amplitude moves with the group velocity, the amplitude function f moves with the group velocity as well. In cases in which f is significant only in a finite region, the waves in this region are called a *wave group*. Hence the name "group velocity" for the velocity at which the group moves.

A description of the phenomenon (which preceded the explanation) was given by Scott Russell (1844) [see Lamb (1932, Section 236)]:

> It has often been noticed that, when an isolated group of waves, of sensibly the same length, is advancing over relatively deep water, the velocity of the group as a whole is less than that of the individual waves composing it. If attention is fixed on a particular wave, it is seen to advance through the group, gradually dying out as it approaches the front, whilst its former place in the group is occupied in succession by other waves which have come forward from the rear.

The above discussion of wave groups was restricted to a rather special case in which all of the components have wave crests that are exactly parallel. In order to remove this restriction, the whole argument can be repeated, but beginning with two waves with y-dependence to replace (5.4.1). The expression replacing (5.4.2) for the superposition of two waves will then be

$$\eta = 2\cos(\delta k\, x + \delta l\, y - \delta\omega\, t)\cos(kx + ly - \omega t). \tag{5.4.7}$$

As before, this can be interpreted as a wave of slowly varying amplitude, and the amplitude factor is now [instead of (5.4.3)]

$$2\cos(\delta k\, x + \delta l\, y - \delta\omega\, t) \approx 2\cos\left[\delta k(x - t\, \partial\omega/\partial k) + \delta l(y - t\, \partial\omega/\partial l)\right]. \tag{5.4.8}$$

If a set of such waves with different values of δk and δl is superposed and the initial form of the waves [cf. (5.4.5)] is

$$\eta = f(x, y)\cos(kx + ly), \tag{5.4.9}$$

then, to the level of approximation of (5.4.8), the form at time t will be

$$\eta = f(x - t\, \partial\omega/\partial k,\, y - t\, \partial\omega/\partial l)\cos(kx + ly - \omega t). \tag{5.4.10}$$

The velocity $\mathbf{c}_g$ of translation of the group is therefore the *vector quantity*

$$\mathbf{c}_g = (\partial\omega/\partial k,\, \partial\omega/\partial l), \tag{5.4.11}$$

i.e., $\mathbf{c}_g$ is the *gradient* of the frequency ω in the wavenumber plane. The concept of group velocity is quite general since the above argument makes no reference to any particular type of wave, and it will be useful again and again in future chapters. Further discussion is given by Lighthill (1965, 1978).

5.5   Short-Wave and Long-Wave Approximations

The length scale that appears in the dispersion relation (5.3.8) and hence determines the character of the waves is the fluid depth H. Different approximations apply, depending on how κ^{-1} relates to H. For the case of *short waves*, i.e., for $\kappa^{-1} \ll H$, (5.3.8) is approximated by (see dashed line in Fig. 5.5)

$$\omega^2 = g\kappa \tag{5.5.1}$$

and (5.3.6) by

$$p' = \rho g\eta_0 \cos(kx + ly - \omega t)\exp(\kappa z). \tag{5.5.2}$$

These are also called *deep-water* waves because $H \gg \kappa^{-1}$. The pressure perturbation and the motion are confined to a distance of order κ^{-1} from the surface, so propagation is unaffected by the bottom. For instance, the dominant waves that one sees in the ocean have periods $2\pi\omega^{-1}$ of order 10 s. By (5.5.1) a deep-water wave of period 10 s has a wavelength $2\pi\kappa^{-1}$ of about 150 m, its amplitude has an e-folding depth of 25 m, and the phase speed is 15 m s^{-1}. [Such a phase speed is typical because it matches the wind speeds found near the water surface in the generation regions, and the period follows by (5.5.1).]

The deep-water approximation is reasonable for such waves when the depth is greater than 25 m. Since the ocean is about 5-km deep, these waves move over large distances as deep-water waves, and only feel the effects of the bottom when they come near the shore. The frequency remains constant as they move into shallow water, so by (5.3.8) the waves become shorter and the phase speed decreases. For instance, when the depth is reduced to 1 m, the wavelength of a 10-s wave is 30 m and the phase speed is only 3 m s^{-1}. Thus deductions about wavelength and phase speed that are made by observing waves on a beach can lead to erroneous conclusions about their properties in deep water.

The dividing line between deep-water and shallow-water waves depends on the depth. For the deep ocean, which has depth 5 km, deep-water waves must have wavelength $2\pi\kappa^{-1}$ less than $2\pi H \approx 30$ km and periods $2\pi\omega^{-1}$ less than $2\pi(g^{-1}H)^{1/2} \approx 2$ min. Phase speeds must be less than 200 m s^{-1}. For a continental shelf of depth 50 m, on the other hand, deep-water waves must have wavelength less than 300 m, period less than 15 s, and phase speed less than 20 m s^{-1}. Since this book is primarily about large-scale motions, such short waves will not be studied further, and the reader is referred for additional information to Kinsman (1965), Lamb (1932), O. M. Phillips (1977), and Stoker (1957).

The approximation to (5.3.8) for *long waves*, i.e., for $\kappa^{-1} \gg H$, is (see dashed line in Fig. 5.5)

$$\omega^2 = g\kappa^2 H, \tag{5.5.3}$$

i.e.,

$$c^2 = gH. \tag{5.5.4}$$

These are also called *shallow-water* waves because $H \ll \kappa^{-1}$, and they are *nondispersive* because the phase speed c does not depend on wavenumber. This speed is about 200 m s^{-1} for deep water, i.e., such waves could cross the Atlantic Ocean in 7 hr. The speed on a continental shelf of depth 50 m is less by a factor of ten, i.e., about 20 m s^{-1}. The corresponding approximation to (5.3.6) is

$$p' = \rho g \eta_0 \cos(kx + ly - \omega t), \tag{5.5.5}$$

i.e., the pressure perturbation is independent of depth. Since the density perturbation is zero, this is precisely the result that would be obtained if the pressure were calculated from the hydrostatic equation (3.5.5). It will be shown in the next section that if it is assumed that the pressure is approximately equal to that given by the hydrostatic equation (called the hydrostatic approximation), (5.5.3) is obtained as well as (5.5.5). In other words, in this case at least, the *hydrostatic approximation and the long-wave (or shallow-water) approximation are equivalent*. Note also that in the small κH limit, Eq. (5.3.7) for w shows that the vertical velocity increases linearly with z from zero at the bottom to a maximum of $\partial\eta/\partial t$ at the surface.

5.6   Shallow-Water Equations Derived Using the Hydrostatic Approximation

The emphasis of this book is on motions with horizontal scale large enough compared with the vertical scale for the hydrostatic approximation to be valid. In this section, the pressure is assumed at the outset to satisfy the hydrostatic equation

$$\partial p/\partial z = -\rho g. \tag{5.6.1}$$

This leads to simplifications in the treatment of the equations, and the result is found to be the same as that obtained by applying the limit $\kappa H \to 0$ to the more general solution.

For a homogeneous fluid, (5.6.1) implies that the perturbation pressure p' satisfies

$$\partial p'/\partial z = 0, \tag{5.6.2}$$

and so the boundary condition (5.2.11) at the surface implies

$$p' = \rho g \eta \tag{5.6.3}$$

at all points within the fluid [in agreement with (5.5.5)]. The momentum equations (5.2.5) therefore become

$$\partial u/\partial t = -g\, \partial \eta/\partial x, \tag{5.6.4}$$

$$\partial v/\partial t = -g\, \partial \eta/\partial y, \tag{5.6.5}$$

showing that time-varying currents are independent of depth. This simplifies the continuity equation (5.2.4), which can now be integrated with respect to depth, using as boundary conditions (5.2.8) and (5.2.10). The result is

$$\partial \eta/\partial t + H(\partial u/\partial x + \partial v/\partial y) = 0. \tag{5.6.6}$$

The quantity $(\partial u/\partial x + \partial v/\partial y)$ is called the *horizontal divergence*, being the divergence of the horizontal component of the velocity.

The continuity equation can also be derived from first principles by considering the fluid column above a fixed element of area, as shown in Fig. 5.8. Suppose (u, v) is the velocity at the center of the element, and η the surface elevation there. Since (u, v) is independent of depth, the rate of mass flux across the central section normal to the x axis is ρu times the area $(H + \eta)\, \delta y$ of the section. The difference between the

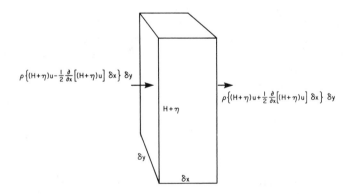

Fig. 5.8. The mass balance for a fluid column of area $\delta x\, \delta y$ when the horizontal velocity components u, v are independent of depth. The mass fluxes across two of the planes are shown.

outward flux from the right-hand face and the inward flux across the left-hand face is therefore, to the appropriate level of approximation,

$$\delta x \, \delta y \, \partial(\rho u(H + \eta))/\partial x.$$

Taking account of the other two sides and equating the net rate of inflow to the rate of change of the total mass $\rho(H + \eta) \, \delta x \, \delta y$ then give

$$\frac{\partial \eta}{\partial t} + \frac{\partial}{\partial x}\left[(H + \eta)u\right] + \frac{\partial}{\partial y}\left[(H + \eta)v\right] = 0. \tag{5.6.7}$$

This is valid even for large perturbations, provided that the horizontal velocity components u and v are independent of depth. Equation (5.6.7) can in fact be derived by integrating (5.2.4), using (5.2.8) and (5.2.9) as boundary conditions. If the perturbation is small, (5.6.7) reduces to the linear equation

$$\partial \eta/\partial t + \partial(Hu)/\partial x + \partial(Hv)/\partial y = 0, \tag{5.6.8}$$

which in turn reduces to (5.6.6) when H is constant.

An equation with only one dependent variable η can be obtained by eliminating u, v from (5.6.4), (5.6.5), and (5.6.8). The result is (Lagrange, 1781)

$$\frac{\partial^2 \eta}{\partial t^2} = \frac{\partial}{\partial x}\left(gH \frac{\partial \eta}{\partial x}\right) + \frac{\partial}{\partial y}\left(gH \frac{\partial \eta}{\partial y}\right). \tag{5.6.9}$$

In the particular case of constant depth, this may be written in the form

$$\partial^2 \eta/\partial t^2 = c^2(\partial^2 \eta/\partial x^2 + \partial^2 \eta/\partial y^2) \equiv c^2 \, \nabla^2 \eta \tag{5.6.10}$$

where c^2 is given by (5.5.4). This is the wave equation which has solutions of the form (5.3.1), showing that the hydrostatic approximation leads to the same results as the long-wave approximation. Also, as Lagrange (1781) pointed out, (5.6.10) is the same as the equation for sound propagation, so there is a complete analog between small-amplitude shallow-water waves and small-amplitude sound waves in two dimensions.

The wave equation (5.6.10) has very simple solutions when there is no dependence on y. In particular, if the fluid is initially at rest and has surface displacement

$$\eta = G(x),$$

then the solution of (5.6.10) is

$$\eta = \tfrac{1}{2}\left[G(x + ct) + G(x - ct)\right]. \tag{5.6.11}$$

The corresponding fluid velocity distribution obtained from (5.6.4) is

$$u = -\tfrac{1}{2}c^{-1}g\left[G(x + ct) - G(x - ct)\right]. \tag{5.6.12}$$

Figure 5.9 shows two special cases. (These will be contrasted in Chapter 7 with the corresponding solutions in a rotating system.) Case (a) has initial displacement the same as that in Fig. 5.2a, namely,

$$\eta = -\eta_0 \, \text{sgn}(x), \tag{5.6.13}$$

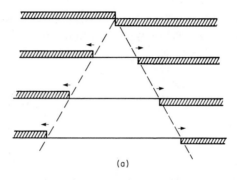

(a)

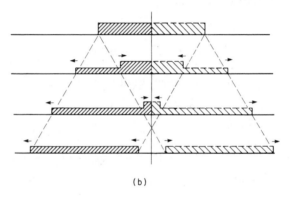

(b)

Fig. 5.9. Solutions of the shallow-water wave equation for two different initial surface displacements. In case (a) waves move out from the initial discontinuity with speed c, leaving behind zero displacement but a steady motion from left to right with velocity $c^{-1}g\eta_0$, where η_0 is the magnitude of the initial surface elevation. In case (b) there are two pairs of wave fronts. The velocity is zero everywhere except where the surface elevation is $\frac{1}{2}\eta_0$, where η_0 is the initial displacement at the center. In these places, the velocity is $\frac{1}{2}c^{-1}g\eta_0$ and directed away from the axis of symmetry. Since no motion occurs at the axis of symmetry, a wall could be placed there without altering the solution.

where sgn(x) is the sign function (sign of x) defined by

$$\operatorname{sgn}(x) = \begin{cases} 1 & \text{for} \quad x > 0, \\ -1 & \text{for} \quad x < 0. \end{cases} \tag{5.6.14}$$

"Wave fronts," consisting of discontinuities in both surface elevation and fluid velocity, propagate out from the initial discontinuity as shown. The fluid at any point remains at rest until a wave front passes, after which the surface elevation is zero and there is a current directed toward the region of low surface elevation.

The case shown in Fig. 5.9b has the initial perturbation confined to a finite region, the initial surface elevation being given by

$$\eta = \begin{cases} \eta_0, & |x| < L, \\ 0, & |x| > L. \end{cases} \tag{5.6.15}$$

In this case there is symmetry about the center line, which could therefore be replaced by a solid boundary, with no motion taking place across this line.

5.7 Energetics of Shallow-Water Motion

The energy equations for shallow-water motion can be derived directly from the momentum equations (5.6.4) and (5.6.5) and the continuity equation (5.6.8). The *mechanical* energy equation [cf. (4.6.3)] is obtained by multiplying (5.6.4) by ρHu, (5.6.5) by ρHv, and adding. This gives

$$\frac{\partial}{\partial t}\left[\tfrac{1}{2}\rho H(u^2 + v^2)\right] = -\rho g\left[Hu\frac{\partial \eta}{\partial x} + Hv\frac{\partial \eta}{\partial y}\right], \tag{5.7.1}$$

the quantity $\tfrac{1}{2}\rho H(u^2 + v^2)$ being the *kinetic* energy per unit area. Now by the definition in Section 4.7, the *potential* energy per unit area is

$$\int_{-H}^{\eta} \rho \Phi\, dz = \int_{-H}^{\eta} \rho g z\, dz = \tfrac{1}{2}\rho g(\eta^2 - H^2), \tag{5.7.2}$$

and so the perturbation potential energy per unit area is $\tfrac{1}{2}\rho g\eta^2$. The equation corresponding to the potential energy equation (4.7.2) is obtained by multiplying (5.6.8) by $\rho g\eta$ to give

$$\frac{\partial}{\partial t}(\tfrac{1}{2}\rho g\eta^2) = -\rho g\left[\eta\frac{\partial}{\partial x}(Hu) + \eta\frac{\partial}{\partial y}(Hv)\right]. \tag{5.7.3}$$

The equation for perturbation *total* energy is obtained by adding (5.7.1) and (5.7.3), namely,

$$\frac{\partial}{\partial t}\left[\tfrac{1}{2}\rho H(u^2 + v^2) + \tfrac{1}{2}\rho g\eta^2\right] + \frac{\partial}{\partial x}(\rho g Hu\eta) + \frac{\partial}{\partial y}(\rho g Hv\eta) = 0. \tag{5.7.4}$$

For the special case in which there is no variation with y, the integral of (5.7.4) with respect to x over the region $|x| < X$ gives

$$\partial E/\partial t + F(X, t) - F(-X, t) = 0, \tag{5.7.5}$$

where

$$E = \int_{-X}^{X}\left[\tfrac{1}{2}\rho H(u^2 + v^2) + \tfrac{1}{2}\rho g\eta^2\right] dx \tag{5.7.6}$$

is the total perturbation energy per unit length in the y direction in the region $|x| < X$ and

$$F(x, t) = \rho g H u\eta \tag{5.7.7}$$

is the rate per unit length in the y direction of transfer of energy in the x direction at point x.

For the case shown in Fig. 5.9a, the perturbation potential energy per unit area is $\tfrac{1}{2}\rho g\eta_0^2$ in the undisturbed region in which the kinetic energy is zero. After the wave has passed, the perturbation potential energy has dropped to zero, but the kinetic energy per unit area has attained the value $\frac{1}{2}\rho H u^2 = \frac{1}{2}\rho H(g\eta_0/c)^2 = \frac{1}{2}\rho g\eta_0^2$. Thus there is no change in total energy, but there is a conversion from potential energy to kinetic energy that takes place at the instant when the wave front passes. After a sufficient length of time, therefore, all the perturbation potential energy in a fixed region will be converted into the kinetic energy associated with the steady current that remains after the wave front has passed.

In the case shown in Fig. 5.9b, in which the initial perturbation is given by (5.6.15), the initial energy per unit length is finite and is given by

$$E(0) = \rho g \eta_0^2 L. \tag{5.7.8}$$

After time L/c, the energy is all contained in two blocks, each of which has length $2L$, these blocks moving outward at speed c. Since the surface elevation of each block is only $\frac{1}{2}\eta_0$, the potential energy associated with each is $\frac{1}{4}E(0)$. The kinetic energy of each block is also $\frac{1}{4}E(0)$, so the total energy is still $E(0)$, but is now partitioned equally between the potential and kinetic forms. If attention is restricted to a *finite* region, the blocks containing the energy will eventually pass out of the region. The fluid within the finite region has now adjusted to equilibrium, the surface elevation and fluid velocity both being zero. The energy loss from the region takes place as the blocks containing the energy cross the boundaries, and is said to be lost by *radiation*. By (5.7.7), the outward energy flux at such times is equal to

$$\rho g H \cdot \tfrac{1}{2} c^{-1} g \eta_0 \cdot \tfrac{1}{2}\eta_0 = \tfrac{1}{4}\rho g c \eta_0^2.$$

The block of energy takes time $2L/c$ to pass, giving an energy loss of $\frac{1}{2}E(0)$ for each block as required to conserve the total energy.

The last example is taken to typify adjustment under gravity in the absence of rotation effects. The final state is one of rest with a horizontal free surface, and all the energy initially present has been lost by radiation. The time for the energy contained in a finite region to be radiated out of the region can, at most, be the time taken for a gravity wave to cross the region. In Chapter 7 the effect of rotation on this adjustment process will be examined.

5.8 Seiches and Tides in Channels and Gulfs

The shallow-water equations of Section 5.7 are used a great deal for the calculation of tides, seiches, storm surges, etc., on the continental shelf and in marginal seas. For the shelf and for "broad" seas like the North Sea, it is essential to include the effect of the rotation of the earth, as will be seen later. However, for sufficiently *narrow* gulfs and channels like the Bay of Fundy, the Bristol Channel, the Gulf of California, and the Adriatic Sea, rotation effects can be neglected, at least to a first approximation.

Let x measure distance along the channel, and consider motions for which the surface elevation does not vary across the channel, but depends only on x and time t. It then follows from (5.6.4) and (5.6.5) that the motion is in the x direction (i.e., parallel to the axis of the channel) and independent of distance across the channel. In this circumstance, the continuity equation simplifies, and it may be derived either by integrating (5.6.7) with respect to y or from first principles as follows. If A is the area of water in any cross section, then the rate of mass transfer across this section is $\rho A u$, and the rate of transfer into the volume between two sections a distance δx apart is $-\delta x\, \partial(\rho A u)/\partial x$ to first order. This must equal the time rate of change of the mass $\rho A\, \delta x$ of water between the two sections, so the continuity equation obtained in the limit as δx tends to zero is

$$\partial A/\partial t + \partial(Au)/\partial x = 0. \tag{5.8.1}$$

For small perturbations, A changes by only a small amount $W\eta$ from its equilibrium value, where W is the width of the channel at the surface and η the surface elevation, so (5.8.1) becomes

$$W\, \partial \eta/\partial t + \partial(Au)/\partial x = 0, \tag{5.8.2}$$

where W and A are now given functions of x. For applications to particular channels, this is the form that must be used [examples are given by Defant (1961) and Proudman (1953)].

However, the phenomena that occur can be illustrated by considering a channel of constant width W and constant depth H. Then (5.8.2) reduces to a special case of (5.6.6), namely,

$$\partial \eta/\partial t + H\, \partial u/\partial x = 0. \tag{5.8.3}$$

The other equation required to complete the problem is the momentum equation (5.6.4). If this is substituted in the time derivative of the continuity equation, there results (Green, 1838)

$$W\, \partial^2 \eta/\partial t^2 = g\, \partial(A\, \partial \eta/\partial x)/\partial x \tag{5.8.4}$$

in the general case and

$$\partial^2 \eta/\partial t^2 = c^2\, \partial^2 \eta/\partial x^2 \tag{5.8.5}$$

in the case of a channel of uniform width and uniform depth, where c^2 is equal to gH [Eq. (5.5.4)].

The solution that satisfies the condition of no flow across the closed end of the channel has the form of a standing wave [cf. (5.3.10) and Fig. 5.6b] or of a superposition of them. Choosing the origin $x = 0$ to be at the closed end, the solution is

$$\eta = \eta_0 \cos kx \cos \omega t, \qquad u = H^{-1} c \eta_0 \sin kx \sin \omega t, \tag{5.8.6}$$

where the wavenumber k and frequency ω are related by

$$\omega = kc. \tag{5.8.7}$$

What conditions must now be applied at the open end $x = L$? First, the pressure $\rho g \eta$ in the channel must equal the pressure outside the channel (otherwise an infinite acceleration would result). Second, the mass flux $\rho A u = \rho H W u$ out of the channel must equal the flux into the open sea. Hence the *ratio Z* of these two quantities, given by

$$Z = g\eta/Au = W^{-1}(g/H)^{1/2} \cot kL \cot \omega t, \tag{5.8.8}$$

must also equal the value just outside the channel. The reason for considering this ratio is that it is independent of the amplitude of the wave. Z is called the *impedance*, and hence the boundary condition to apply at $x = L$ is that the impedance of the channel, given by (5.8.8), match the impedance of the open sea.

Now (5.8.8) shows that the impedance tends to zero as the width W or depth H of a channel approaches infinity, so the impedance of the open sea is taken to be zero, at least to a first approximation. Thus the impedance of the channel is required to be zero, so (5.8.8) gives

$$kL = (n + \tfrac{1}{2})\pi, \qquad n = 0, 1, 2, \ldots, \tag{5.8.9}$$

for possible wavenumbers, or using (5.8.7), the possible frequencies of oscillation are given by

$$\omega L = (n + \tfrac{1}{2})\pi c, \qquad n = 0, 1, 2, \ldots. \tag{5.8.10}$$

These are the frequencies for the natural modes of oscillation of the channel, and are known as *seiches*.

The problem just discussed of seiches in a channel is completely analogous to that of sound waves in a pipe that is stopped at one end but open at the other. The case treated corresponds to a pipe of uniform bore (like a flute), whereas a constant-depth channel whose area varies as the square of x is the analog of a conical pipe (like a clarinet). For musical instruments, account must be taken of the junctions where the finger holes are inserted. These can be treated as small pipes branching off the main one. At a junction, the pressure is continuous and the sum of the mass fluxes into the junction is zero. Hence the sum of the ratios of mass flux to pressure is zero, i.e., the sum of the *admittances* (inverse of impedance) is zero. The same technique can be applied to deal with side channels from a gulf or estuary (Defant, 1961). The problem of seiches in lakes (which are closed at both ends) can be treated in the same way as the problem for an open channel.

Oscillations in gulfs, estuaries, and lakes may be stimulated by changes in the wind stress at the surface, by changes in atmospheric pressure at the surface, or by changes in the gravitational attraction of the moon and the sun (the so-called tide-generating forces). Often the largest oscillations are *not* produced by local forcing, but are a response to oscillations in the open sea that are produced by these forces acting there. If, for instance, the action of these forces results in an oscillation of frequency ω whose amplitude at the *mouth* of the channel is η_L, then (5.8.6) shows that the amplitude η_0 at the *head* of the channel is given by

$$\eta_0 = \eta_L \sec kL. \tag{5.8.11}$$

The increase in amplitude is very large (i.e., resonance occurs) when the frequency of forcing is close to one of the natural frequencies of oscillation given by (5.8.10). This gives rise to the spectacular tides found, for instance, in the Bay of Fundy/Gulf of Maine system and in the Bristol Channel. The interpretation of (5.8.10), with $n = 0$, in this case is that a long gravity wave needs to take about a quarter of a period (about 3 hr for the semidiurnal tide) to travel the length of the estuary or gulf. For a depth of 20 m, the required length for resonance is about 150 km (the value increases in proportion with the square root of depth). Note that Eq. (5.8.11), is a relation only between η_0 and η_L and does not show how these quantities depend on open sea conditions. That question is discussed by Garrett and Greenberg (1977).

The subject of seiches and tides in lakes and channels is dealt with in much greater detail by Proudman (1953) and Defant (1961, Volume 2). Both authors give examples in which the theory has successfully been applied. The prediction of seiches and tides is important for shipping, flood warnings, etc., and numerical models of some estuaries have been set up to help with this prediction. Another application is for studying the effect that barriers across the estuary may have on the tides where tidal power generation is contemplated. The advantage of numerical models is that they can include effects additional to those considered above, such as friction, cross-channel variations, and large-amplitude effects.

Chapter Six

Adjustment under Gravity
of a Density-Stratified Fluid

6.1 Introduction

Chapter 5 provided an introduction to the study of adjustment to equilibrium under gravitational forces in the absence of rotation. Attention was restricted, however, to the case of a fluid of uniform density with gravitational restoring forces coming into play when the free surface was perturbed from the horizontal. In this chapter, the study is extended to fluids of variable density.

As a first introduction to the effects of stratification, the case of two superposed shallow layers, each of uniform density, is considered in Sections 6.2 and 6.3. This serves to introduce the concepts of barotropic and baroclinic modes and two widely used approximations: the rigid lid approximation and the Boussinesq approximation. "Shallow" in this case means that the depth of each layer is small compared with the horizontal scale of the perturbation, i.e., the horizontal scale is large compared with the vertical scale.

In reality, of course, the atmosphere and ocean are continuously stratified, although the two-layer model can be quite useful and appropriate for many situations. The study of continuously stratified fluids begins in Section 6.4 with the case of an *incompressible* fluid, i.e., one whose density depends on temperature and composition but not on pressure. No restriction on scale is made at first, but toward the end of the chapter (from Section 6.11 on) special consideration is given to the case in which the horizontal scale is large compared with the vertical scale. This is partly to prepare the way for the introduction of rotation effects in Chapter 7, for with the exception of some rather special situations, rotation is important only for motions with this property. An additional reason is that most energy in the atmosphere and ocean is in components with this property.

No scale restriction applies to the motions studied in Sections 6.4–6.10. In Section 6.4, the equations for the general case are obtained and the buoyancy frequency N, which is of fundamental importance to the subject, is introduced. The perturbations have a wavelike behavior, the waves concerned being called internal gravity waves. Their most basic properties are most readily studied in the case for which N is constant and the Boussinesq approximation is made. This case is treated in Sections 6.5–6.7. In particular, the "polarization relations" for a plane wave are found in Section 6.5, dispersion properties are discussed in Section 6.6, and energetics in Section 6.7.

It is then but a small step to consider internal waves generated at a (slightly perturbed) horizontal boundary, and this is done in Section 6.8. A particular case of interest is that of waves generated by flow of uniform velocity U over a gently undulating surface. For small-wavelength undulations (wavenumber k greater than N/U), the fluid is only significantly perturbed within a certain distance of the ground, but for large-wavelength undulations ($kU < N$), waves are produced that propagate energy and momentum to large distances from the generating surface.

In practice, the buoyancy frequency N is not uniform, so some important effects of variations in N are considered in Section 6.9, with special reference to the simple case in which N is piecewise uniform. In particular, waves can be refracted and reflected from a discontinuity in N. This can lead to wave energy being confined or "trapped" in a particular structure that is called a duct or waveguide. This concept is applied to free wave propagation in Section 6.10, where methods of treating the general case (N is any function of vertical coordinate z) are developed.

In Section 6.11, there is a return to discussion of cases for which the "hydrostatic approximation" applies, i.e., for which the horizontal scale is large compared with that of the vertical. This serves as a prelude to consideration of adjustment problems rather like that considered in Section 5.6 for the homogeneous case, except that the initial perturbation is now a function of z. However, there is some dependence on the nature of the boundaries. In Section 6.12, we consider adjustment in a semi-infinite region, i.e., a case such as the atmosphere, in which there is a solid boundary below but no definite boundary above. In Section 6.13, we deal with a region of finite depth such as the ocean. The structure of these solutions is of special interest because the way in which they are modified by rotation effects will be considered in Chapter 7.

In Section 6.14, effects of compressibility are considered. For problems of adjustment under gravity, the main new effect of interest is connected with waves that carry energy most rapidly in the horizontal in the atmosphere. These are called Lamb waves, whose vertical scale is the "scale height" and which propagate horizontally at the speed of sound. They were responsible for carrying the pressure pulses observed all around the world following the eruption of Krakatoa, and more recently from nuclear explosions. An example of adjustment in a compressible atmosphere is considered in Section 6.15, and weak dispersion effects, which characterize pressure pulses from distant sources, are considered in Section 6.16. This section, and many others in the book, can be regarded as fairly general discussions of wave properties. Wave dispersion, for instance, is first treated in Section 6.6, and the behavior of waves in a medium of variable properties is discussed in Section 6.9. Further aspects of wave behavior are to be found in Chapter 8.

When the hydrostatic equation is valid, i.e., when horizontal scales are large compared with vertical scales, it is often advantageous to use pressure as an independent variable in place of height. In particular, the continuity equation has a simple form even when the fluid is compressible. The equations in isobaric coordinates are derived in Section 6.17 and are used later in the book, particularly in connection with slow adjustment processes in the atmosphere (Chapter 12). The energy equations in these coordinates are discussed in Section 6.18.

6.2   The Case of Two Superposed Fluids of Different Density

As a first example of stratification effects, consider the case of two fluids of different density that are immiscible or for which the effect of mixing can be ignored. This system is easily set up in the laboratory, and Marsigli's experiment is an early example. The hydrostatic approximation will be made from the outset, so the results should be applied only to cases in which the horizontal scale is large compared with the depth. The problem was first treated by Stokes (1847).

The means of describing the situation is shown in Fig. 6.1. Subscript 1 is used for the upper layer whose density is ρ_1, whose equilibrium depth is H_1, and whose horizontal velocity components are u_1 and v_1. The free surface, whose equilibrium position is $z = 0$, has perturbed position $z = \eta$. The interface displacement (upward) is h. It follows from the hydrostatic equation

$$\partial p / \partial z = -\rho g \tag{6.2.1}$$

and the surface condition $p = 0$, that the pressure p_1 in the upper layer is given by

$$p_1 = \rho_1 g(\eta - z), \qquad -H_1 + h < z < \eta. \tag{6.2.2}$$

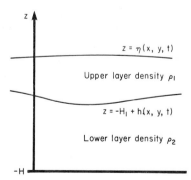

Fig. 6.1.   The notation used to describe the motion of two superposed shallow homogeneous layers of fluid. H_1, H_2 are the depths of the layers when at rest and $H = H_1 + H_2$ is the total depth. The z axis points vertically upward, $z = \eta(x, y, t)$ is the surface elevation, and $z = -H + h(x, y, t)$ gives the disturbed position of the interface between the two fluids.

Therefore the momentum equations (4.5.7) and (4.5.8) for small disturbances become

$$\partial u_1/\partial t = -g\,\partial\eta/\partial x, \qquad \partial v_1/\partial t = -g\,\partial\eta/\partial y, \tag{6.2.3}$$

whereas the continuity equation obtained by the same method as that for (5.6.7) is

$$\partial(\eta + H_1 - h)/\partial t + H_1(\partial u_1/\partial x + \partial v_1/\partial y) = 0. \tag{6.2.4}$$

Taking the time derivative and substituting from (6.2.3) for $\partial u_1/\partial t$ and $\partial v_1/\partial t$ eliminate u_1 and v_1 to give

$$\frac{\partial^2}{\partial t^2}(\eta - h) = H_1\left(\frac{\partial^2}{\partial x^2} + \frac{\partial^2}{\partial y^2}\right)g\eta \equiv gH_1\,\nabla^2\eta, \tag{6.2.5}$$

where ∇^2 is defined by (4.3.9).

Similarly, for the lower layer, denoted by subscript 2, the pressure p_2 obtained by integrating (6.2.1) and using continuity of pressure at the interface is

$$p_2 = \rho_1 g(\eta + H_1 - h) + \rho_2 g(-H_1 + h - z), \qquad z < -H_1 + h. \tag{6.2.6}$$

Thus the momentum equations are

$$\frac{\partial u_2}{\partial t} = -\frac{\rho_1}{\rho_2}g\frac{\partial\eta}{\partial x} - g'\frac{\partial h}{\partial x}, \qquad \frac{\partial v_2}{\partial t} = -\frac{\rho_1}{\rho_2}g\frac{\partial\eta}{\partial y} - g'\frac{\partial h}{\partial y}, \tag{6.2.7}$$

where g' is the *reduced gravity* [see (5.1.1)], defined by

$$g' = g(\rho_2 - \rho_1)/\rho_2. \tag{6.2.8}$$

The continuity equation for the layer in this case is

$$\partial h/\partial t + H_2(\partial u_2/\partial x + \partial v_2/\partial y) = 0. \tag{6.2.9}$$

As before, the velocity components are eliminated from these equations to give

$$\frac{\partial^2 h}{\partial t^2} = H_2\left(\frac{\partial^2}{\partial x^2} + \frac{\partial^2}{\partial y^2}\right)\left(\frac{\rho_1}{\rho_2}g\eta + g'h\right) = H_2\,\nabla^2(g\eta - g'\eta + g'h), \tag{6.2.10}$$

use being made of (6.2.8).

The adjustments of the two-fluid system are thus governed by Eqs. (6.2.5) and (6.2.10). If, say, η were eliminated from these, a fourth-order partial differential equation for h would be obtained. However, the problem can be greatly simplified by looking for solutions with a special structure, namely, those for which η and h are proportional, i.e.,

$$h(x, y, t) = \mu\eta(x, y, t), \tag{6.2.11}$$

where μ is independent of x, y, and t. Then (6.2.5) and (6.2.10) both reduce to the second-order equation

$$\partial^2\eta/\partial t^2 = c_e^2\,\nabla^2\eta, \tag{6.2.12}$$

provided that μ and c_e^2 satisfy

$$gH_1/(1 - \mu) = \mu^{-1}(g - g'(1 - \mu))H_2 = c_e^2. \tag{6.2.13}$$

This simplification is an example of a method that can be used for a wide class of mechanical problems involving small oscillations. In fact, Lamb (1932), in his treatise *Hydrodynamics*, spends the first section of his chapter on tidal waves discussing the general theory because of its wide applicability. For the present problem, there are two values of μ, and hence two values of c_e that satisfy the above equation, and the motions corresponding to these particular values are called normal modes of oscillation. In a system consisting of n layers of different density, there are n such modes corresponding to the n degrees of freedom. A continuously stratified fluid corresponds to an infinite number of layers, and so there is an infinite set of modes. The fact that each mode behaves independently is of great utility, and application of the concept will be made repeatedly. The independence of each mode can be seen from the fact that if h and η satisfy (6.2.11) at some initial time, they will then satisfy (6.2.11) for all subsequent times, so only one mode will be in oscillation. If, on the other hand, any given initial state can be represented as a sum of modes the change of each in time and space can be followed independently. The fluid state can then be found by adding together the contributions of each mode.

The structure of the modes is obtained by solving the quadratic (6.2.13) [cf. Stokes (1847, Section 17)], which can be written in the alternative form

$$c_e^4 - gHc_e^2 + gg'H_1H_2 = 0, \tag{6.2.14}$$

where

$$H = H_1 + H_2 \tag{6.2.15}$$

is the total depth of fluid in the equilibrium state. To each of the two solutions (or *eigenvalues*) c_e^2 of (6.2.14) there corresponds a particular normal-mode structure represented by (6.2.11) and the appropriate value of μ. For the case of several layers, there is a value μ_i corresponding to the displacement h_i of each interface, so μ represents an *eigenvector* of the problem.

Another way of expressing the equivalence between the normal mode of the two-layer system and the motion of the one-layer system is to define an *equivalent depth* H_e by

$$c_e^2 = gH_e. \tag{6.2.16}$$

Then η_e satisfies the same equation as that for the surface elevation in a homogeneous fluid of depth H_e. By (6.2.14), the eigenvalue equation for H_e is

$$gH_e^2 - gHH_e + g'H_1H_2 = 0. \tag{6.2.17}$$

For applications to the ocean, approximations can be made because the fractional changes in density are small: of the order of $3^0/_{00}$, i.e., $g'/g = 1 - \rho_1/\rho_2 \approx 0.003$. This results in two widely separated roots c_e^2 of (6.2.14). The larger one, c_0^2, is given approximately by

$$c_0^2 = gH(1 - g'H_1H_2/gH^2 \cdots), \tag{6.2.18}$$

and the ratios η/h and u_2/u_1 are approximately

$$\eta/h \approx H/H_2, \qquad u_2/u_1 = 1 - g'H_1/gH \cdots. \tag{6.2.19}$$

In the limit as $g'/g \to 0$, this becomes the surface gravity wave obtained for a fluid of uniform density. It is often called the *barotropic* mode. The strict meaning of the term "barotropic" is that the pressure is constant on surfaces of constant density, and hence is constant on the interface. This is only approximately true, but it is conventional to call this mode barotropic nevertheless.

The smaller root c_1^2 of (6.2.14) is given for small g'/g by

$$c_1^2 = (g'H_1H_2/H)(1 + g'H_1H_2/gH^2 \cdots), \tag{6.2.20}$$

and the corresponding values of the ratios η/h and u_2/u_1 are approximately

$$\eta/h \approx -g'H_2/gH, \qquad u_2/u_1 \approx -H_1/H_2. \tag{6.2.21}$$

This mode is called the *baroclinic* mode, the word "baroclinic" meaning that pressure is not constant on surfaces of constant density. Typical values of c_1 for the ocean are 2 or 3 m s^{-1}, corresponding to an equivalent depth of 0.5–1 m. Use of the two-layer model for the atmosphere is not so common, but on occasions when it is used, c_1 typically has values of 10–20 m s^{-1} and the equivalent depth of 10–50 m. Often one layer is deep compared with the other, e.g., $H_2 \gg H_1$, and then (6.2.20) is approximated by

$$c_1^2 \approx g'H_1. \tag{6.2.22}$$

Then the internal wave is just the same as a surface gravity wave would be if the acceleration due to gravity were g' instead of g. This is because it is g' that determines pressure differences rather than g (see Section 5.1).

Because $g' \ll g$, the wave speed of internal waves is very much less than that of surface waves, so that the internal waves look like surface waves in slow motion. This difference accounts for a phenomenon noted by Benjamin Franklin (1762, p. 438) in a letter dated December 1, 1762.

> At Madeira we got oil to burn, and with a common glass tumbler or beaker, slung in wire, and suspended to the ceiling of the cabbin .... I made an Italian lamp.... The glass at bottom contained water to about one third of its height; another third was taken up with oil .... At supper, looking on the lamp, I remarked that tho' the surface of the oil was perfectly tranquil, and duly preserved its position with regard to the brim of the glass, the water under the oil was in great commotion, rising and falling in irregular waves ....

Franklin's experiment can be set up in the kitchen and completed within a minute or two, and readers are urged to try it. The instructions are given in the next paragraph (p. 439) of Franklin's letter.

> Since my arrival in America, I have repeated the experiment frequently thus. I have put a pack-thread round a tumbler, with strings of the same, from each side, meeting above it in a knot at about a foot distance from the top of the tumbler. Then putting in as much water as would fill about one third part of the tumbler, I lifted it up by the knot, and swung it to and fro in the air; when the water appeared to keep its place in the tumbler as steadily as if it had been ice. But pouring gently in upon the water about as much oil, and then again swinging it in the air as before, the tranquility before possessed by the water, was trans-

ferred to the surface of the oil, and the water under it was agitated with the same commotions as at sea.

Perhaps the first explanation of an oceanic phenomenon in terms of internal waves was V. Bjerknes' explanation of "dead water," a hitherto mysterious effect in which ships in certain coastal localities would be unable to maintain their normal speed. Ekman (1904) cites a large number of examples of the phenomenon going back as far as Pliny the Naturalist, who reported that the effect was attributed either to a mollusk or a certain type of fish that attached itself to the keel. In a preface to Ekman's paper (p. III), Bjerknes says:

> The present investigation of "Dead-Water" was occasioned by a letter in Novem-
> ber 1898 from Prof. NANSEN asking my opinion on the subject. In my reply
> to Prof. NANSEN I remarked that in the case of a layer of fresh water resting
> on the top of salt water, a ship will not only produce the ordinary visible waves
> at the boundary between the water and the air, but will also generate invisible
> waves in the salt-water fresh-water boundary below; I suggested that the great
> resistance experienced by the ship was due to the work done in generating these
> invisible waves.

Ekman substantiated this view with extensive laboratory experiments, and includes photographs of his experiments (an example is shown in Fig. 6.2b) and slick patterns behind ships. Figure 6.2a shows such a pattern observed off British Columbia, where freshwater from river outflow forms a relatively light upper layer over the heavier salt water. The internal waves on the interface are associated with horizontal move-ments at the surface that affect the ripple pattern and thus become visible. The motion of the interface that corresponds to this sort of situation can be seen in Ekman's laboratory experiment (Fig. 6.2b). A nice demonstration is also given in a cine film by Long [see National Committee for Fluid Mechanics Films (1972, pp. 136–142)].

The main result of this section is that the motion can be represented in terms of two normal modes, and for each mode η satisfies the wave equation (6.2.12), i.e., the same equation as that for a homogeneous fluid (but with a different time scale). Thus the results of Chapter 5 for shallow-water motion can be applied equally well to the two-layer system. For instance, Fig. 6.3 shows the structure of the *progressive waves* associated with the two different modes in a particular case. The value of g'/g has been chosen to be small but still much larger than for the ocean. This has been done so that certain features peculiar to internal motion would be visible in the diagram, namely, the slight differences in velocity between the two layers in the "barotropic" mode of motion and the free surface movement associated with the baroclinic motion. In the ocean, the free surface movement associated with the baroclinic mode is only about $1/400$ of the interface movement, but this is still sufficient for baroclinic motions to be detectable by sea-level changes (Wunsch and Gill, 1976).

Figure 6.3 shows the variation in space of a progressive internal wave at a fixed time, but the variation in time at a fixed point has the same character. For comparison, Fig. 6.4 shows simultaneous observations [from Lee (1961)] of the motion of an isotherm at three points about 170 m apart, thus giving some information about variations in space as well as time. Since particles conserve their temperature over

Fig. 6.2. (a) Surface "slicks" showing the presence of internal waves in the wake of a ship in Bute Inlet, British Columbia. The vessel was traveling at 0.5 m s^{-1} in a surface layer of almost fresh water only slightly deeper than its 3.4 m draft. The internal waves caused horizontal motion at the surface that affects the ripple pattern and so renders the internal wave pattern visible at the surface during calm conditions. [Photo courtesy of Defence Research Establishment Pacific, Victoria, British Columbia.] (b) A laboratory experiment [from Ekman (1904)], showing internal waves being generated by a model ship. The tank is filled with two fluids of different density, the heavier one being dyed to make the interface clearly visible. The model ship (the superstructure of the "Fram" has been drawn in subsequently) is towed from right to left, causing a wake of waves on the interface.

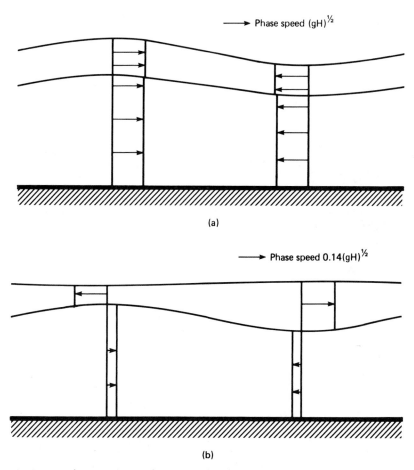

Fig. 6.3. Layer configuration in a two-layer system for a barotropic wave (a) and a baroclinic wave (b) propagating from left to right. For the case shown, the lower layer is three times deeper than the upper layer and has density 10% greater. Also shown are the directions of flow at troughs and crests, and the relative velocities of the two layers at these points.

the short period of the record, these contours mark the vertical excursion of fluid particles. Although there is in reality a continuous change of temperature with depth, the dominant features of the motion are close to those displayed by a system with two layers of different density. La Fond (1962) summarizes the main features observed at the site of these observations, where the water is 20 m deep. The period of the waves is mainly in the range 4–10 min. Since $g' \approx 0.01$ m s^{-2} and H_1 is 3–10 m, (6.2.20) gives c_1 as 0.15–0.22 m s^{-1}. This agrees well with the speed of the waves as observed by the movement of slicks and by taking simultaneous measurements at three locations. In 80% of the cases, the surface slick was found to be located between a crest and the following trough, where Fig. 6.3 shows the surface velocity field is convergent. This convergence is presumed to be responsible for the slicks.

Another extension of the results of Chapter 5 is to internal seiches and tides in channels, gulfs, and lakes. These have the form of a standing wave as given by (5.8.6)

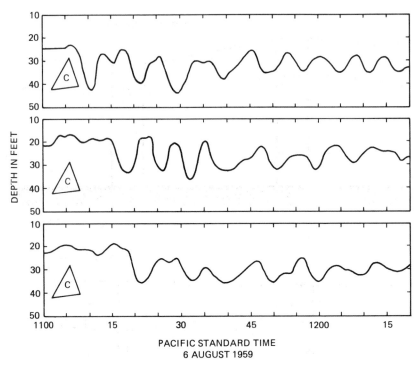

Fig. 6.4.   Observations of progressive internal waves in 20 m of water about 1.3 km offshore at San Diego. The records are from three points on the triangle C as shown, the sides of the triangle being approximately 170 m. The waves moved to the right (onshore) at 0.2 m s^{-1}. The continuous line is the depth of an isotherm (64°F) located in the thermocline. [From Lee (1961, Fig. 3).]

and (5.8.7), c now being the internal wave speed. Quite detailed observations of this phenomenon were made in Loch Ness by Watson (1904, pp. 435–436).

> These observations revealed a pendulous swinging of the ends of the isotherms, the amplitude of the swing being greatest for the isotherm in the region 200 feet below the surface, and dying off both above and below this region. A few other observations taken simultaneously with these at other parts of the loch show that the isotherms are swinging as a whole about a transverse central axis.
>
> To what can this swinging be due?
>
> If we take a long rectangular trough with glass sides, and put into it a layer of water, and above the water a layer of lighter oil, and then disturb the arrangement, one of the movements observed will be a swinging of the interface between the oil and water…. The time, of swing can be calculated from the formula….

The formula given is the standing-wave period, namely, twice the time to travel the length of the lake at speed c_1, given by the first term of (6.2.20). Using values of $g' \approx 2.6 \times 10^{-3}$ m s^{-2}, $H_1 \approx 60$ m, $H_2 \approx 120$ m, and a length of 40 km, Watson obtained a period of 68 hr, "which is of the same order as the period observed." He also explained how seiches could be generated by wind action, and continued even during a prolonged period of calm. [See also Watson (1903). Internal seiches were reported earlier by Thoulet (1894), who reports an experiment with three superposed layers, but did not make observations of the period.]

6.3   The Baroclinic Mode and the Rigid Lid Approximation

The disparity in the values of g' and g means that approximations can be made to the equations and boundary conditions, depending on the mode being studied. For the barotropic mode, the approximation is simply to ignore density differences altogether and treat the fluid as one of uniform density as in Chapter 5. There are two approximations used to obtain the *baroclinic* mode. The first uses the fact that for this mode surface displacements are *small* compared with interface displacements (as can be seen in Fig. 6.2, for instance). Thus the continuity equation (6.2.4) for the upper layer is approximated by

$$-\partial h/\partial t + H_1(\partial u_1/\partial x + \partial v_1/\partial y) = 0. \tag{6.3.1}$$

The momentum equations for the upper layer are given by (6.2.3) as before. This is called the *rigid lid approximation*, although the name is somewhat misleading because free surface displacements are required to give pressure gradients in the upper layer [i.e., (6.2.3) involves $\nabla\eta$]. The justification for the name lies in the fact that if there *were* a rigid lid at $z = 0$, the identical pressure gradients would be achieved because the rigid lid would provide the necessary pressure.

The second approximation is simply to replace the ratio ρ_1/ρ_2 by unity in (6.2.7) [and hence in (6.2.10)], giving

$$\partial u_2/\partial t = -g\,\partial\eta/\partial x - g'\,\partial h/\partial x, \qquad \partial v_2/\partial t = -g\,\partial\eta/\partial y - g'\,\partial h/\partial y. \tag{6.3.2}$$

This is usually referred to as the *Boussinesq approximation*, which will be discussed in a more general context later.

Since the two continuity equations (6.3.1) and (6.2.9) do not involve η, it is desirable to obtain a combination of the momentum equations that does not involve η. This combination is obtained by subtracting (6.3.2) from (6.2.3), giving

$$\partial\hat{u}/\partial t = g'\,\partial h/\partial x, \qquad \partial\hat{v}/\partial t = g'\,\partial h/\partial y, \tag{6.3.3}$$

where $(\hat{u}, \hat{v})$ is given by

$$\hat{u} = u_1 - u_2, \qquad \hat{v} = v_1 - v_2, \tag{6.3.4}$$

and so represents the *difference in velocity* between the two layers. $(\hat{u}, \hat{v})$ can also be thought of as the amplitude of the baroclinic mode.

Now a combination of the continuity equations that involves only $(\hat{u}, \hat{v})$ is required. This is obtained by subtracting $1/H_2$ times (6.2.9) from $1/H_1$ times (6.3.1), giving

$$-\left(\frac{1}{H_1} + \frac{1}{H_2}\right)\frac{\partial h}{\partial t} + \frac{\partial\hat{u}}{\partial x} + \frac{\partial\hat{v}}{\partial y} = 0. \tag{6.3.5}$$

Equations (6.3.3) and (6.3.5) are, apart from multiplying constants, the same as Eqs.

(5.6.4)–(5.6.6) for the homogeneous fluid. Thus when the velocity components $\hat{u}$ and $\hat{v}$ are eliminated, the result is the wave equation

$$\frac{\partial^2 h}{\partial t^2} = c_1^2 \left(\frac{\partial^2 h}{\partial x^2} + \frac{\partial^2 h}{\partial y^2} \right) \equiv c_1^2 \, \nabla^2 h, \qquad (6.3.6)$$

where

$$c_1^2 = g' H_1 H_2 / (H_1 + H_2) \qquad (6.3.7)$$

is the square of the speed of propagation of the baroclinic mode. This is the same value as that given by (6.2.20) in the limit as $g'/g \to 0$. An alternative form of (6.3.7) is the equation

$$\frac{1}{gH_e} = \frac{1}{g'H_1} + \frac{1}{g'H_2} \qquad (6.3.8)$$

for the equivalent depth $H_e = c_1^2/g$. For typical oceanic values of $g' = 0.03$ m s^{-2}, $H_1 = 400$ m, $H_2 = 4000$ m, one finds H_e is ≈ 1 m.

6.4 Adjustments within a Continuously Stratified Incompressible Fluid

So far the study of adjustment under gravity has been restricted to a fluid that has uniform density, or to a system consisting of two immiscible fluids, each of uniform density. Particular emphasis has been placed on motions with horizontal scale large compared with that of the vertical scale. In the remainder of this chapter, the study of adjustment processes will be extended to continuously stratified fluids, i.e., fluids with continuously varying density. The scale restriction will not be made at first, although the emphasis in subsequent chapters will be on motions with relatively large horizontal scale since these contain by far the most energy.

To begin, the fluids to be considered will be restricted to a class such that the density depends only on entropy and on composition, i.e., ρ depends only on the potential temperature θ and on the concentrations of the constituents, e.g., the salinity s or humidity q. Then for fixed θ and q (or s), ρ is *independent of pressure*:

$$\rho = \rho(\theta, q). \qquad (6.4.1)$$

The motion that takes place is assumed to be isentropic and without change of phase, so that θ and q are constant for a material element. Therefore

$$\frac{D\rho}{Dt} \equiv \frac{\partial \rho}{\partial \theta} \frac{D\theta}{Dt} + \frac{\partial \rho}{\partial q} \frac{Dq}{Dt} = 0. \qquad (6.4.2)$$

In other words, ρ is constant for a material element *because* θ and q are, and ρ depends only on θ and q. Such a fluid is said to be *incompressible*, and because of (6.4.2), the continuity equation (4.2.3) becomes (4.10.12), i.e.,

$$\partial u/\partial x + \partial v/\partial y + \partial w/\partial z = 0. \qquad (6.4.3)$$

The equilibrium state to be perturbed is the state of rest, so the distribution of density and pressure is the hydrostatic equilibrium distribution given by (4.5.17) and (4.5.18). In the absence of rotation and of friction, the momentum equations (4.10.11) for *small* perturbations p' in pressure and ρ' in density become

$$\rho_0 \, \partial u/\partial t = -\partial p'/\partial x, \qquad \rho_0 \, \partial v/\partial t = -\partial p'/\partial y, \qquad (6.4.4)$$

$$\rho_0 \, \partial w/\partial t = -\partial p'/\partial z - \rho' g, \qquad (6.4.5)$$

where $\rho_0(z)$ is the unperturbed density and g the acceleration due to gravity. For the moment, *no restriction on horizontal scale* is being made, so there are no approximations other than for that of the smallness of the perturbation. The governing equations are (6.4.3)–(6.4.5) and the linearized form of (6.4.2) appropriate to small perturbations, namely,

$$\partial \rho'/\partial t + w \, d\rho_0/dz = 0. \qquad (6.4.6)$$

Calculations based on these equations were made by Rayleigh (1883, p. 170) "in order to illustrate the theory of cirrous clouds propounded by the late Prof. Jevons."

The initial step in dealing with the equations is the same as that used for the one- and two-layer system, namely, to eliminate u, v from the horizontal part of the momentum equations and the continuity equation. This is done by using (6.4.4) to substitute expressions for the acceleration components in the time derivative of (6.4.3). The result is

$$\rho_0 \frac{\partial^2 w}{\partial z \, \partial t} = \left(\frac{\partial^2}{\partial x^2} + \frac{\partial^2}{\partial y^2} \right) p' \equiv \nabla_H^2 p'. \qquad (6.4.7)$$

This equation may be thought of as a relation between the *horizontal divergence* $\partial u/\partial x + \partial v/\partial y = -\partial w/\partial z$ and the perturbation pressure p'.

For the stratified system, another relation between w and p' is required. This is obtained by eliminating ρ' from (6.4.5) and (6.4.6) to give

$$\partial^2 w/\partial t^2 + N^2 w = -\rho_0^{-1} \, \partial^2 p'/\partial z \, \partial t, \qquad (6.4.8)$$

where $N(z)$ is a quantity of fundamental importance to this problem (see Section 3.7.1 for expressions for N), defined by

$$N^2 = -g\rho_0^{-1} \, d\rho_0/dz. \qquad (6.4.9)$$

N has the dimensions of frequency, and is variously known as the Brunt–Väisälä frequency, the Brunt frequency [after Brunt (1927)], the Väisälä frequency [after Väisälä (1925)], and the *buoyancy frequency* [see, e.g., Turner (1973)]. Other names (such as stability frequency and intrinsic frequency) have also been used, but Brunt–Väisälä seems to be the most common appellation. However, Rayleigh (1883) drew attention to this frequency (as the maximum possible in a stratified layer) well before Brunt and Väisälä, and buoyancy frequency is the most appropriate name physically. This is because of the solution for purely vertical motion for which p' vanishes, and hence (6.4.8) shows that the frequency of oscillation is N. The restoring force that produces the oscillation is the buoyancy force [see (6.4.5)].

There are now two equations to be satisfied, namely, (6.4.7) and (6.4.8). It is useful to think of (6.4.7) as being associated with the *horizontal* part of the motion since it is derived from the horizontal part of the momentum equations, and to think of (6.4.8) as being associated with the *vertical* part of the motion since it comes from the vertical component (6.4.5) of the momentum equation. When p' is eliminated, a single equation for w results, namely,

$$\frac{\partial^2}{\partial t^2}\left[\frac{\partial^2}{\partial x^2} + \frac{\partial^2}{\partial y^2} + \frac{1}{\rho_0}\frac{\partial}{\partial z}\left(\rho_0\frac{\partial}{\partial z}\right)\right]w + N^2\left(\frac{\partial^2}{\partial x^2} + \frac{\partial^2}{\partial y^2}\right)w = 0. \qquad (6.4.10)$$

It is this equation that determines how small amplitude adjustments within a continuously stratified incompressible fluid take place.

Exact solutions can be found in special cases such as those in which the density varies exponentially with height. However, there is a simplifying approximation that is always a good one in the ocean and that is valid for many applications to the atmosphere. This is based on the observation that if w varies with z much more rapidly than ρ_0, then

$$\frac{1}{\rho_0}\frac{\partial}{\partial z}\left(\rho_0\frac{\partial}{\partial z}\right)w \approx \frac{\partial^2 w}{\partial z^2}, \qquad (6.4.11)$$

and so (6.4.10) can be approximated by

$$\frac{\partial^2}{\partial t^2}\left[\frac{\partial^2}{\partial x^2} + \frac{\partial^2}{\partial y^2} + \frac{\partial^2}{\partial z^2}\right]w + N^2\left(\frac{\partial^2}{\partial x^2} + \frac{\partial^2}{\partial y^2}\right)w = 0. \qquad (6.4.12)$$

In the ocean, ρ_0 never departs by more than 2% from its mean value, so it is a very good approximation to treat ρ_0 as a constant as implied by (6.4.11).

Another way of stating the condition for (6.4.11) to be valid is that the vertical scale for variations of w be small compared with the vertical scale for variations of ρ_0, i.e., be small compared with the *scale height* H_s (see Section 3.5). If this condition is satisfied, it turns out, as will be shown later, that (6.4.12) is a good approximation even when the fluid is compressible (conversely, if the condition is not satisfied, compressibility should *not* be ignored). Since vertically propagating internal waves in the atmosphere are usually found to satisfy this condition, (6.4.12) can be used in applications to the atmosphere as well.

The approximation that applies when the motion has vertical scale small compared with the scale height is called the *Boussinesq approximation* [see, e.g., Spiegel and Veronis (1960)] and is attributed to Boussinesq (1903). Basically, it consists of taking the density to be constant in computing rates of change of momentum from accelerations, but taking full account of density variations when they give rise to *buoyancy* forces, i.e., when there is a multiplying factor g in the vertical component of the momentum equations. For the case considered in this chapter, this means taking ρ_0 as a constant in (6.4.4) and (6.4.5) and hence in (6.4.7) and (6.4.8). Buoyancy effects come in through the term $\rho'g$ in (6.4.5), which gives rise to the term N^2w in (6.4.8).

6.5 Internal Gravity Waves

Consider the case in which the buoyancy (Brunt–Väisälä) frequency N is constant throughout the fluid. Traveling wave solutions of (6.4.12) can be found of the form

$$w = w_0 \cos(kx + ly + mz - \omega t), \tag{6.5.1}$$

where w_0 is the *amplitude* of vertical velocity fluctuations, the vector

$$\mathbf{k} = (k, l, m) \tag{6.5.2}$$

is the *wavenumber* of the disturbance, and ω is the frequency. In order for (6.5.1) to satisfy Eq. (6.4.12), ω and $\mathbf{k}$ must be related by the *dispersion relation*

$$\omega^2 = (k^2 + l^2)N^2/(k^2 + l^2 + m^2). \tag{6.5.3}$$

Thus internal waves can have any frequency between zero and a *maximum* value of N. As Rayleigh (1883, p. 174) put it, "Contrary to what is met with in most vibrating systems, there is a limit on the side of rapidity of vibration, but none on the side of slowness."

The dispersion relation for internal waves is of quite a different character compared to that for surface waves. In particular, the frequency of surface waves depends only on the *magnitude* κ of the wavenumber, whereas the frequency of internal waves is *independent* of the magnitude of the wavenumber and depends only on the *angle* φ' that the wavenumber vector makes with the horizontal. To bring this out, it is useful to specify the wavenumber in spherical polar coordinates $(\lambda', \varphi', \kappa)$ in wavenumber space (see Fig. 6.5), namely,

$$k = \kappa \cos \varphi' \cos \lambda', \qquad l = \kappa \cos \varphi' \sin \lambda', \qquad m = \kappa \sin \varphi'. \tag{6.5.4}$$

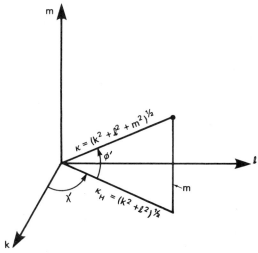

Fig. 6.5. The system of spherical polar coordinates in wavenumber space used to express the dispersion relation for internal waves. For these waves, the frequency ω does not depend on the magnitude κ of the wavenumber, but only on the direction φ' between the wavenumber and the horizontal plane. The dispersion relation is $\omega = N \cos \varphi'$.

The prime is used to denote wavenumber angles as opposed to angles in physical space. Then the dispersion relation (6.5.3) becomes simply

$$\omega = N \cos \varphi'. \tag{6.5.5}$$

The way p', ρ', u, and v vary for the plane wave (6.5.1) can be deduced from the appropriate equations. The relationships among these variables are sometimes called the *polarization relations*. The perturbation pressure p' is, from (6.4.7), given by

$$p' = -(k^2 + l^2)^{-1}\omega m \rho_0 w_0 \cos(kx + ly + mz - \omega t), \tag{6.5.6}$$

whereas (6.4.6) gives for the perturbation density

$$\rho' = -(N^2/\omega g)\rho_0 w_0 \sin(kx + ly + mz - \omega t). \tag{6.5.7}$$

Note that the last two equations, together with (6.5.3), imply that for a plane progressive wave,

$$\partial p'/\partial z = -(m^2/(k^2 + l^2 + m^2))g\rho'. \tag{6.5.8}$$

The horizontal velocity components can be found from (6.4.4), which gives

$$(u, v) = -(k, l)(k^2 + l^2)^{-1} m w_0 \cos(kx + ly + mz - \omega t)$$

$$= (k, l)(\omega \rho_0)^{-1} p'. \tag{6.5.9}$$

The above relations between pressure and velocity fluctuations can be useful for deducing wave properties from observations at a fixed point. For instance, if the horizontal velocity components and perturbation pressure of a progressive wave are measured, the horizontal component of the wavenumber vector can be deduced from (6.5.9). This device was used, for instance, by Gossard and Munk (1954).

A sketch showing the properties of a plane progressive internal wave in the vertical plane that contains the wavenumber vector is presented in Fig. 6.6. The particle motion is along wave crests, and there is no pressure gradient in this direction. The restoring force on a particle is therefore due solely to the component $g \cos \varphi'$ of gravity in the direction of motion. The restoring force is also proportional to the component of the density change in this direction, which is $\cos \varphi' \, d\rho/dz$ per unit displacement. It follows from Newton's second law that the square of the frequency of vibration is

$$\rho^{-1}g \cos \varphi' \cos \varphi' \, d\rho/dz = (N \cos \varphi')^2,$$

thus giving a physical interpretation of the dispersion relation (6.5.5).

Consider now the succession of solutions as φ' progressively increases from zero to a right angle. When $\varphi' = 0$, a vertical line of particles moves together like a rigid wire undergoing longitudinal vibrations. When the line of particles is displaced from its equilibrium position, buoyancy restoring forces come into play just as if the line of particles were on a spring, resulting in oscillations of frequency N. The solutions for other values of φ' correspond to lines of particles moving together at angle φ' to the vertical. The restoring force per unit displacement is less than it is for $\varphi' = 0$, and so the frequency of vibration is less. As φ' tends to $\pi/2$, the frequency of vibration tends to zero.

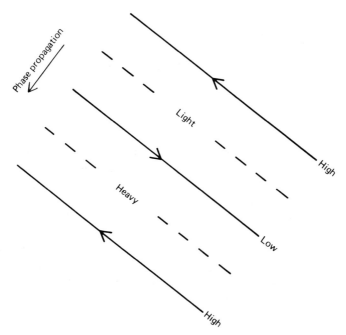

Fig. 6.6.   Sketch showing in a vertical plane the phase relationships for a progressive internal wave with downward phase velocity (this implies upward group velocity). The solid lines mark lines of maximum (High) and minimum (Low) pressure, which are also lines of maximum and minimum velocity, the direction of motion being as shown. The dashed lines mark the positions of maximum (Heavy) and minimum (Light) density perturbations. If the direction of phase propagation is reversed, the only change in the diagram is a reversal of the direction of motion.

The extreme case of purely horizontal motion requires special consideration because this is a singular limit for which the solution of (6.4.12) is the trivial one, $w = 0$. In this case, the general solution of Eqs. (6.4.3)–(6.4.6) has $p' = \rho' = 0$, with u and v any functions of x, y, and z alone that satisfy

$$\partial u/\partial x + \partial v/\partial y = 0. \tag{6.5.10}$$

In other words, each horizontal plane of particles can move independently of any other plane, but the motion within each plane must be nondivergent. An alternative form of this solution is

$$w = p' = \rho' = 0, \qquad u = -\partial\psi/\partial y, \qquad v = \partial\psi/\partial x, \tag{6.5.11}$$

where ψ is an arbitrary function of x, y, and z. This solution is *not* an internal wave, or even a limiting form of one, but it represents an important form of motion that is often observed. For instance, it is quite common on airplane journeys to see thick layers of cloud that are remarkably flat and extensive. Each cloud layer is moving in its own horizontal plane, but different layers are moving relative to each other as described by (6.5.11). If a mountain pierces such a layer, it is possible to have motion of the form (6.5.11) with two-dimensional flow around the mountain in each horizontal layer. Spectacular consequences are the vortex streets observed behind islands

(Gjevik, 1980), and related experimental work is discussed by Brighton (1978). However, it is not possible to have a solution of the form (6.5.11) representing uniform flow normal to a ridge at levels below the crest of the ridge. Ridges are sometimes found to block flow in this way, and the general phenomenon is called *blocking*.

6.6   Dispersion Effects

In practice, internal gravity waves never have the pure form (6.5.1), so it is necessary to consider superpositions of such waves. Then dispersion effects become evident when different waves have different phase velocitites, as discussed in Section 5.4. The dispersion of internal waves is quite different from surface waves, one reason being that the frequency of internal waves is independent of the wavenumber *magnitude*, whereas the frequency of surface waves is independent of wave *direction*.

For internal waves, the surfaces of constant frequency in wavenumber space are the cones $\varphi' = $ const shown in Fig. 6.7. The phase velocity is directed along the wavenumber vector and therefore lies on the cone, its magnitude being

$$\omega/\kappa = (N/\kappa) \cos \varphi'.$$

The *group velocity* $\mathbf{c}_g$ is by (5.4.11) the gradient of ω in wavenumber space and therefore is normal to the surface of constant ω. It follows that (as for any waves whose frequency is independent of wavenumber magnitude) the group velocity is at right angles to the wavenumber vector. When the group velocity has an upward component, therefore, the phase velocity has a downward component, and vice versa. By (5.4.11)

$$\mathbf{c}_g = (N/\kappa) \sin \varphi' (\sin \varphi' \cos \lambda', \sin \varphi' \sin \lambda', -\cos \varphi'). \qquad (6.6.1)$$

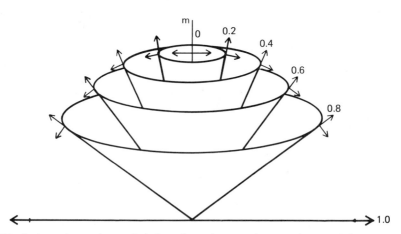

Fig. 6.7.   For internal waves (no rotation), the surfaces of constant frequency in wavenumber space are cones as shown, contours being values of ω/N, where ω is the frequency and N the buoyancy frequency. The group velocity is in a direction perpendicular to the cone in the direction of increasing frequency as shown by one set of arrows, whereas the phase velocity is in a direction along the cone away from the origin as shown by the other set of arrows.

Therefore the magnitude of the group velocity is $(N/\kappa) \sin \varphi'$ and its direction is at an angle φ' to the *vertical*.

Figure 6.8 is designed to illustrate how the dispersion properties of internal waves differ from those of surface gravity waves. In each case the wave field shown is a simple combination of four waves of equal amplitude with wavenumbers $\mathbf{k} \pm \delta\mathbf{k} \pm \delta\mathbf{k}'$, where $\delta\mathbf{k}$ and $\delta\mathbf{k}'$ are small compared with $\mathbf{k}$, and $\delta\mathbf{k}'$ is in a direction for which no change in ω occurs. This combination has the form [cf. (5.4.1)]

$$\cos[(\mathbf{k} + \delta\mathbf{k} + \delta\mathbf{k}')\cdot\mathbf{x} - (\omega + \delta\omega)t] + \cos[(\mathbf{k} + \delta\mathbf{k} - \delta\mathbf{k}')\cdot\mathbf{x} - (\omega + \delta\omega)t]$$

$$+ \cos[(\mathbf{k} - \delta\mathbf{k} + \delta\mathbf{k}')\cdot\mathbf{x} - (\omega - \delta\omega)t] + \cos[(\mathbf{k} - \delta\mathbf{k} - \delta\mathbf{k}')\cdot\mathbf{x} - (\omega - \delta\omega)t]$$

$$= 4 \cos(\delta\mathbf{k}' \cdot \mathbf{x}) \cos(\delta\mathbf{k} \cdot \mathbf{x} - \delta\omega t) \cos(\mathbf{k} \cdot \mathbf{x} - \omega t)$$

$$\approx 4 \cos(\delta\mathbf{k}' \cdot \mathbf{x}) \cos[\delta\mathbf{k} \cdot (\mathbf{x} - \mathbf{c}_{\mathrm{g}}t)] \cos(\mathbf{k} \cdot \mathbf{x} - \omega t). \tag{6.6.2}$$

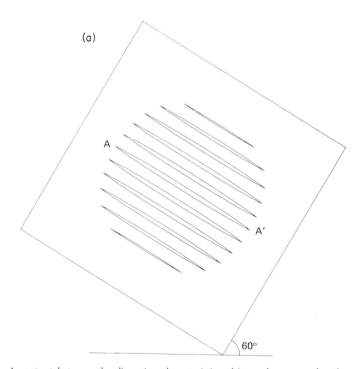

Fig. 6.8. A contrast between the dispersion characteristics of internal waves and surface gravity waves, illustrated by the behavior of a suitable combination of four progressive waves. (a) The initial configuration of a group of internal waves with wave crests at 60° to the vertical. Contours are of pressure perturbation, where this is equal to $+0.5$ times the maximum value. (b) The configuration four periods later, the group having moved parallel to the crests and upward, while the individual crest AA' has moved four wavelengths downward and to the left. To compare this with the way surface waves behave, suppose (a) now shows a plan view of a similar combination of waves, contours now being where the surface elevation is $+0.5$ times the maximum value. Then (c) shows the configuration four periods later. The individual wave crest AA' has again moved four wavelengths, but now the group has moved two wavelengths in the same direction.

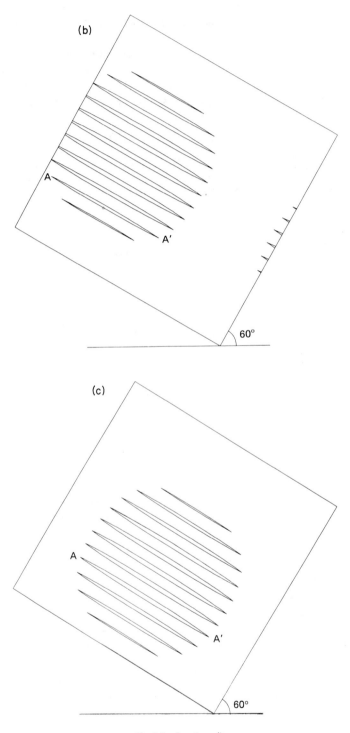

Fig. 6.8.   (*continued*)

Figure 6.8a represents such a combination of internal gravity waves. It is a vertical section in the plane of propagation, showing contours only where the pressure perturbation is $+0.5$ times the maximum value for the whole wave field. The wavenumber vector is at angle $\varphi' = 60°$ to the vertical and pointing downward, $\delta\mathbf{k}'$ is chosen to be equal to $0.03\mathbf{k}$, and $\delta\mathbf{k}$ has the same magnitude but is at right angles to $\delta\mathbf{k}'$. Figure 6.8b shows the same waves four periods later. Wave crest AA' has moved four wavelengths downward to the left, but the wavegroup has moved upward parallel to the crests, i.e., at right angles to the direction of phase propagation, For comparison, Fig. 6.8c shows the behavior of a similar combination of surface gravity waves. In this case, Fig. 6.8a is interpreted as a plan view, showing contours of surface elevation, with $\delta\mathbf{k} = 0.03\mathbf{k}$ and $\delta\mathbf{k}'$ of the same magnitude but at right angles. Figure 6.8c is the view four periods later, crest AA' having moved four wavelengths. The group as a whole has moved in the same direction but at half the speed.

The difference in the directions of phase and group propagation for internal waves is nicely illustrated in laboratory experiments (Mowbray and Rarity, 1967), in which density perturbations can be made visible by using shadowgraph or Schlieren techniques. In their experiment, results from which are shown in Fig. 6.9, energy propagates outward from a vibrating cylinder that can be regarded as a point source of waves with a fixed frequency ω. Consequently, energy propagates radially outward in the direction of group propagation, i.e., it travels in beams whose angle φ' with the *vertical* is given by (6.5.5). Lines of constant phase are observed to cross the beams transversely, their motion being directed toward the horizontal plane through the source region. These and other laboratory experiments on internal waves are discussed by Turner (1973).

In Chapter 5 reference was made to the effect of dispersion in separating out waves of different length that come from a distant source of *short duration*. With surface waves, this effect has been used to calculate the position and time of the source of swell arriving at some distant location. The way the properties of the waves arriving at the distant point change with time is easily calculated for *any* type of dispersive wave. Choose the origin of the coordinate system to be at the source (see Fig. 6.9c), and let $t = 0$ be the time of generation. Since waves move outward with the group velocity $\mathbf{c}_\mathrm{g}$, the waves found at point $\mathbf{x}$ at time t will satisfy

$$\mathbf{x} = \mathbf{c}_\mathrm{g} t$$

(provided that the properties of the medium are uniform), and so the wavenumber $\mathbf{k}$ can be found by solving

$$\mathbf{c}_\mathrm{g}(\mathbf{k}) = \mathbf{x}/t. \tag{6.6.3}$$

Consider the special case of internal waves in which $\mathbf{c}_\mathrm{g}$ is given by (6.6.1) and hence $O\mathbf{x}$ makes angle φ' with the vertical. Since this angle is fixed for a given point $\mathbf{x}$ of observation, the frequency ω of the waves passing this point will have the fixed value given by (6.5.5). In other words, wave crests will pass at fixed intervals of time. However, the spacing between crests will decrease with time; for (6.6.3) gives (for magnitudes)

$$c_\mathrm{g} = r/t,$$

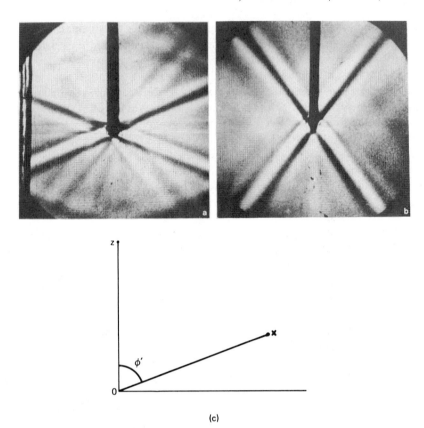

(c)

Fig. 6.9. Patterns, obtained by Schlieren techniques, of internal waves propagating away from a cylinder that is vibrating at a frequency given by (a) $\omega/N = 0.366$, (b) $\omega/N = 0.699$. Waves propagate outward with group velocity along the dark lines, which indicate an extreme of refractive index and hence a wave crest. The orientation of the crest is the one expected for the frequency of vibration. The dark lines continually move toward the horizontal plane containing the cylinder, new lines appearing at the top edge of the beam and old ones disappearing at the lower edge of the beam, thus showing that the phase propagation is at right angles to the group velocity. [From D. E. Mowbray and B. S. H. Rarity (1967). A theoretical and experimental investigation of the phase configuration of internal waves of small amplitude in a density stratified fluid. *J. Fluid Mech.* **28,** 1 (Plate 1). Cambridge University Press.] (c) The geometry of the situation. For a point source at O, waves received at **x** travel in the direction of the group velocity, and hence $\mathbf{c_g}$ is in the direction of $O\mathbf{x}$. The angle marked is ϕ' since $\mathbf{c_g}$ makes this angle to the vertical. ϕ' is also the angle the wavenumber vector makes with the horizontal since the phase velocity is at right angles to the group velocity. For a source of fixed frequency ω as in the experiments shown in (a) and (b), waves are observed only along the beam for which ϕ' is given by (6.5.5). For an impulsive source applied at $t = 0$, all frequencies will be present, but only those with frequency ω, given by (6.5.5), will be observed at **x**. The wavenumber of the dominant wave will, however, increase with time as calculated in the text.

and substitution for c_g from (6.6.1) then gives

$$\kappa = r^{-1}Nt \sin \varphi'. \qquad (6.6.4)$$

Thus κ increases in proportion with time, i.e., the spacing between wave crests decreases inversely in proportion with time. More detailed discussion of the solution for an impulsive source of internal waves is given by Bretherton (1967), and further discussion of the general problem is given by Whitham (1974) and Lighthill (1978).

Sometimes geometric factors cause the wavenumber to be constrained to lie on some surface or line in wavenumber space, in which case the dispersion properties depend on how the frequency varies on that surface or line. For example, the vertical component m of the wavenumber may be fixed because the waves are contained in a region of finite vertical extent. The dispersion properties then depend on how ω, given by (6.5.3), varies with k and l. This shows that the long waves, which have low frequency, have the largest group velocity, equal to N/m. As the horizontal component of wavenumber increases, the frequency increases toward a maximum value of N and the group velocity decreases toward zero. (Further discussion of this case can be found in Section 6.10.)

Another example corresponds to the case in which the horizontal component (k, l) of the wavenumber is fixed, so that (6.5.3) is regarded as a relation between ω and m. In this case, the frequency is a maximum and the group velocity c_{gz} is zero when m = 0. The group velocity also tends to zero as $m \to \infty$, and has a maximum at a value of m, corresponding to propagation at $35°$ to the vertical ($\cot \varphi' = 2^{1/2}$). The associated frequency is $(2/3)^{1/2}N$ and the group velocity is $2N/3^{3/2}(k^2 + l^2)^{1/2}$.

6.7   Energetics of Internal Waves

The energy equation for internal waves can be obtained by multiplying (6.4.4) by (u, v), (6.4.5) by w, and (6.4.6) by $g^2\rho'/\rho_0N^2$, then adding the results. With the use of (6.4.3) and (6.4.9), this gives

$$\frac{\partial}{\partial t}\left[\frac{1}{2}\rho_0(u^2 + v^2 + w^2) + \frac{1}{2}\frac{g^2\rho'^2}{\rho_0N^2}\right] + \frac{\partial}{\partial x}(p'u) + \frac{\partial}{\partial y}(p'v) + \frac{\partial}{\partial z}(p'w) = 0. \quad (6.7.1)$$

This is a special case of the total energy equation discussed in Section 4.7. As found in Chapter 4, the energy equation can be integrated over a large volume, thereby giving useful results about overall balances.

The identification of the perturbation kinetic energy density term $\frac{1}{2}\rho_0(u^2 + v^2 + w^2)$ in (6.7.1) with the corresponding term in (4.7.3) is obvious, as is the correspondence between the perturbation energy flux term $(p'u, p'v, p'w)$ in (6.7.1) and the full expression (4.7.4) for the flux when account is taken of the perturbations being infinitesimal, incompressible, inviscid, and nondiffusive. The identification of the perturbation potential energy term in (6.7.1) is less obvious, and it is helpful first to consider the case of the two-layer fluid of Section 6.2. There the potential energy (see Section 5.7) is equal to

$$\iiint \rho\Phi \, dx \, dy \, dz$$

$$= \iiint \rho g z \, dz \, dx \, dy$$

$$= \iint \left\{\frac{1}{2}\rho_1 g[\eta^2 - (H_1 - h)^2] + \frac{1}{2}\rho_2 g[(H_1 - h)^2 - H^2]\right\} dx \, dy,$$

and so the perturbation potential energy is equal to

$$\iint \left[\frac{1}{2} \rho_1 g \eta^2 + \frac{1}{2} (\rho_2 - \rho_1) g h^2 \right] dx\, dy. \tag{6.7.2}$$

It follows that in a many-layered system, each interface will contribute a term like

$$\iint \frac{1}{2} (\rho_2 - \rho_1) g h^2 \, dx\, dy,$$

and in the limit of a continuously stratified fluid this becomes

$$-\iiint \frac{1}{2} \left(\frac{d\rho_0}{dz} \right) g h^2 \, dz\, dx\, dy = \iiint \frac{1}{2} \rho_0 N^2 h^2 \, dx\, dy\, dz, \tag{6.7.3}$$

where h is the displacement of a fluid element from its equilibrium position and the second expression is obtained by using (6.4.9). Since the density of a fluid element at its perturbed level $z + h$ is equal to the density $\rho_0(z)$ at its equilibrium position, the perturbation density ρ' is given by

$$\rho' = \rho_0(z) - \rho_0(z + h) \approx -h \, d\rho_0/dz, \tag{6.7.4}$$

and so an alternative form of (6.7.3) is

$$\iiint \frac{1}{2} \left(\frac{g^2 \rho'^2}{\rho_0 N^2} \right) dx\, dy\, dz. \tag{6.7.5}$$

The connection with (6.7.1) is now clear.

In the case of periodic waves in a medium of uniform properties, the integral over each wavelength is the same, and so the mean over any large volume becomes equal to the mean over one wavelength in the limit as the volume tends to infinity. Hence it is useful to consider *mean* quantities rather than integrated quantities, the mean being defined as the mean over a wavelength, and denoted by an overbar. The energy density E of an internal wave is defined as the mean perturbation energy per unit volume, i.e., by

$$E = \frac{1}{2} \rho_0 \overline{(u^2 + v^2 + w^2)} + \frac{1}{2} g^2 \overline{\rho'^2} / \rho_0 N^2. \tag{6.7.6}$$

(*Note*: This should not be confused with the internal energy per unit mass, which is also denoted by the symbol E in Chapters 3 and 4.) As with all conservative non-rotating dynamic systems undergoing small oscillations, the mean energy is equally divided between the kinetic and potential forms. For the plane wave given by (6.5.1),

$$E = \frac{1}{2} \rho_0 (w_0/\cos \varphi')^2, \tag{6.7.7}$$

where $w_0/\cos \varphi'$ is the amplitude of the velocity fluctuations [see (6.5.9)] and w_0 the amplitude of the vertical component.

When integrated over a large volume, (6.7.1) shows that the rate of change of energy over that volume is equal to the flux of energy across the sides. Since this flux is also periodic, the average over a large plane area is approximately the same as the average over one wavelength, so it is convenient to consider the spatial mean for fluxes as well. Thus the energy flux density vector $\mathbf{F}'$ is defined by (see Chapter 4)

$$\mathbf{F}' = \overline{p'\mathbf{u}}. \tag{6.7.8}$$

$\mathbf{F}'$ has the property that its outward normal component, when integrated over the surface of a large volume, gives the rate of energy flow out of the volume concerned.

For the plane wave given by (6.5.1), the pressure perturbation p' and disturbance velocity $\mathbf{u}$ are in phase (see Fig. 6.6), so $\mathbf{F}'$ has magnitude [see (6.5.6) and (6.5.9)]

$$F' = \frac{1}{2} \frac{\omega \sin \varphi'}{\kappa \cos^2 \varphi'} \rho_0 w_0 \frac{w_0}{\cos \varphi'}$$

and is in the direction of the particle velocity on the high pressure crests, i.e., is in the direction of the group velocity. It follows from (6.7.7), (6.6.1), and (6.5.5) that

$$\mathbf{F}' = E\mathbf{c}_g. \tag{6.7.9}$$

This result is in fact true for a large class of waves [see Whitham (1974, Chapter 11)].

For problems in which vertical propagation is being considered, e.g., of energy produced at the ground, one is interested in the vertical component F_z' of $\mathbf{F}'$, which for the plane wave solution (6.5.1) is given by

$$F_z' = -\tfrac{1}{2}\omega m \rho_0 w_0^2 / (k^2 + l^2). \tag{6.7.10}$$

Thus if a situation were produced with no motion at some lower boundary but with a traveling wave of the form (6.5.1) crossing the upper boundary, then the energy would be increasing when the phase propagation was upward ($\omega/m > 0$) and decreasing when the phase propagation was downward ($\omega/m < 0$). This is indicative of the property that energy is transferred upward when phase propagation is downward, and vice versa.

Another form of the energy equation (6.7.1) is obtained by integrating vertically, e.g., from the bottom $z = -H$ to the surface $z = \eta$ of a continuously stratified ocean. In that case, (6.7.1) becomes

$$\frac{\partial}{\partial t} \int \frac{1}{2} \rho_0 (u^2 + v^2 + w^2)\, dz + \frac{\partial}{\partial t} \int \frac{1}{2} \frac{g^2 \rho'^2}{\rho_0 N^2}\, dz + \frac{\partial}{\partial x} \int p'u\, dz$$

$$+ \frac{\partial}{\partial y} \int p'v\, dz + [p'w] = 0, \tag{6.7.11}$$

where the square brackets denote the difference between the surface value and the value at the bottom. If (6.7.4) is used to substitute for ρ' and the boundary conditions (5.2.8), (5.2.10), and (5.2.11) are used to evaluate $[p'w]$, the result is

$$\frac{\partial}{\partial t} \int \frac{1}{2} \rho_0 (u^2 + v^2 + w^2)\, dz + \frac{\partial}{\partial t} \left\{ \frac{1}{2} \rho_0(0)g\eta^2 + \int \frac{1}{2} \rho_0 N^2 h^2\, dz \right\}$$

$$+ \frac{\partial}{\partial x} \int p'u\, dz + \frac{\partial}{\partial y} \int p'v\, dz = 0. \tag{6.7.12}$$

If this is further integrated over a horizontal area, it is seen that the perturbation potential energy P' per unit area is given by

$$P' = \int \int \left\{ \frac{1}{2} \rho_0(0) g \eta^2 + \int \frac{1}{2} \rho_0 N^2 h^2 \, dz \right\} dx \, dy. \tag{6.7.13}$$

Thus P' has contributions from two terms. The first term, as in (6.7.2), is the contribution from surface displacements, and the second [also given by (6.7.3) and (6.7.5)] is the contribution from vertical displacements of isopycnals within the body of the fluid.

6.8 Internal Waves Generated at a Horizontal Boundary

Internal waves in the atmosphere and ocean can be generated by a variety of mechanisms. Often the source region is approximately horizontal, so the vertical velocity component can effectively be specified on some horizontal surface, and the motion away from the source region can be calculated from the equations of motion. As an example, take the case in which air or water is moving with uniform horizontal velocity over a succession of hills and valleys whose elevation h above a plane horizontal surface $z = 0$ can be regarded as small. Such a topography can be represented as a superposition of sine waves. In this section the response to a single wave component will be calculated. From this result more general cases can be solved by superposition. (*Note*: The discussion until now has been about *free* waves produced by some initial perturbation, and the discussion returns to free waves in Section 6.10. In contrast, *forced* waves are considered in this section and the next, i.e., there is a continuous source of energy at the boundary.)

The x axis is chosen to be perpendicular to the crests of the sinusoidal range of hills, and the axes are chosen to be fixed relative to the mean motion of the air. Thus if U is the velocity component normal to the crests of the air relative to the ground, the topography in the chosen coordinate frame has phase velocity $-U$, i.e., is given by

$$h = h_0 \sin[k(x + Ut)]. \tag{6.8.1}$$

It follows that the frequency ω of the motion produced is given by

$$\omega = -Uk, \tag{6.8.2}$$

so $\omega/2\pi$ is the frequency with which air particles at the surface encounter crests. The vertical component of velocity at the surface is that experienced by particles that follow the undulations of the boundary while maintaining the prescribed horizontal motion, i.e.,

$$w = U \, \partial h / \partial x = w_0 \exp(i(kx - \omega t)) \qquad \text{on} \quad z = 0, \tag{6.8.3}$$

where

$$w_0 = Ukh_0, \tag{6.8.4}$$

and the convention is adopted that when a complex expression is used, the physical quantity of interest is understood to be the real part of the expression [i.e., $w_0 \cos(kx - \omega t)$ in the above case].

When the stratification is uniform, i.e., when N is constant, the appropriate solution of (6.4.12) is

$$w = w_0 \exp(i(kx + mz - \omega t)), \qquad (6.8.5)$$

where m is the positive root of (6.5.3), which in this case may be written as

$$m^2 = k^2(N^2 - \omega^2)/\omega^2 = (N/U)^2 - k^2. \qquad (6.8.6)$$

The sign of m is determined by the condition that energy propagates *away* from the source region, i.e., the group velocity is upward and the phase velocity downward. There is a continual upward propagation of energy at a rate given by (6.7.10), namely,

$$F_z' = -\tfrac{1}{2}\omega m \rho_0 w_0^2/k^2 = \tfrac{1}{2}k\rho_0 h_0^2 U^2(N^2 - U^2 k^2)^{1/2}. \qquad (6.8.7)$$

This is therefore the rate at which energy is being supplied to the atmosphere at the ground.

The solution (6.8.5), however, is only valid when the frequency of encounter of air particles with crests is *less* than the buoyancy frequency N. If $\omega^2 > N^2$, (6.8.6) does not have real solutions, so the solution of (6.4.12) is

$$w = w_0 \exp(-\gamma z + i(kx - \omega t)), \qquad (6.8.8)$$

where

$$\gamma^2 = k^2(\omega^2 - N^2)/\omega^2 = k^2 - (N/U)^2. \qquad (6.8.9)$$

Because of the exponential fall-off with height, the disturbance energy in this case is said to be *trapped* near the ground. Also the word "evanescent," meaning "vanishing" or "quickly fading," is associated with waves like these whose amplitude decays exponentially in one direction. By (6.4.7) the perturbation pressure p' is

$$p' = -i\omega\gamma k^{-2}\rho_0 w_0 \exp(-\gamma z + i(kx - \omega t)). \qquad (6.8.10)$$

This is *out of phase* with the vertical velocity, i.e., is zero when w is a maximum or minimum, and is a maximum or minimum when w is zero. Thus the rate F_z' of doing work by the boundary is *zero* by (6.7.8).

The distinction between the above two types of solution is important and is illustrated in Fig. 6.10. For the smaller wavelengths ($kU > N$), for which the disturbance energy is trapped, the air particles at all levels undergo vertical displacements of the same character as the terrain below but with amplitudes that fall off with height. The solution is an equilibrium one, with no inputs or outputs of energy occurring. The perturbation pressure is low on the hills and high in the valleys, so there is no net horizontal force between the atmosphere and the underlying surface. For very small wavelengths ($kU \gg N$), stratification has little effect and the solution is that for irrotational flow in a homogeneous fluid.

For the larger wavelengths ($kU < N$) for which the waves are not trapped, the energy density is the same at all levels. Energy is continually being supplied at the lower boundary at the rate, per unit area, given by (6.8.7) and is propagated upward. This implies that if, for some reason, energy were *absorbed* at some upper level, there would be an effective transfer of energy from the ground to some remote region by

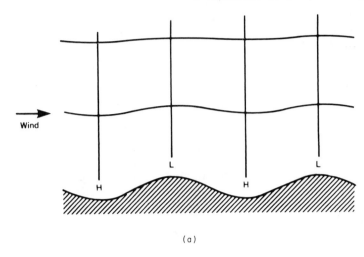

(a)

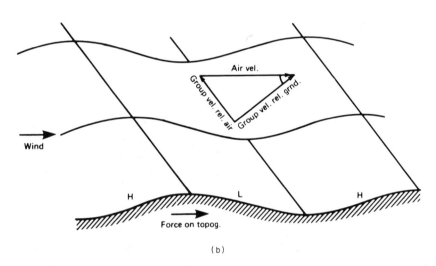

(b)

Fig. 6.10. The motion produced by uniform flow of a uniformly stratified fluid over sinusoidal topography of small amplitude. The sinuous lines indicate the displacement of isopycnal surfaces whose equilibrium configurations are horizontal, and the straight lines join crests and troughs. (a) For *small-wavelength* topography, i.e., wavenumber $k > N/U$, where N is buoyancy frequency and U is a fluid velocity relative to the ground (a typical value of U/N for the atmosphere is 1 km). The drawing is for $kU = 1.25N$. Note the decay of amplitude with height, showing that energy is trapped near the ground. H and L indicate positions of maximum and minimum pressure perturbation, respectively, i.e., there is suction over crests. When the lower half plane is fluid, this can lead to instability (Kelvin–Helmholtz) when the relative velocity between fluids is great enough for the suction to overcome gravity. (b) The response to *large-wavelength* topography, i.e., $k < N/U$ (the drawing is for $kU = 0.8N$). Now the displacement of isopycnals is uniform with height, but wave crests move upstream with height, i.e., phase lines are tilted as shown. The group velocity *relative to the air* is along these phase lines, but the group velocity *relative to the ground* is at right angles, i.e., upward and in the downstream direction. High and low pressures are now at the nodes, so there is a net force on the topography in the direction of flow.

radiation without any energy changes being necessary in the intervening medium, whose function would merely be to act as a *carrier* for the waves. As can be seen from Fig. 6.6, high pressure is found where the particle velocity is upward and hence on the windward side, and so there is a net force exerted by wind on the ground. This force is exerted normal to the crests in the direction of the corresponding wind component. From the point of view of the atmosphere, there is a *drag* exerted by the topography due to the generation of internal waves. The magnitude of the drag force per unit area is equal to the rate τ per unit area at which horizontal momentum is transferred vertically by the waves. This is given by

$$\tau = -\rho_0 \overline{uw} = F_z'/U = \tfrac{1}{2}k\rho_0 h_0^2((N/U)^2 - k^2)^{1/2}U^2. \qquad (6.8.11)$$

As Bretherton (1969) pointed out, the momentum removed from the atmosphere at the ground in this way will be returned to the atmosphere at the levels where the wave energy is absorbed. He estimated that on 29 December 1966, a day favorable for wave generation, the wave drag in the hills of North Wales had an average value per unit area of 0.4 N m^{-2} and that absorption levels for this momentum would be above 20 km. This calculation took account of the effects of variation of N and U with height, which have been ignored in the above discussion. The value of U at the lower levels was 15 m s^{-1}, which by (6.8.11) gives a vertical energy flux of 6 W m^{-2}. These figures show how strong the effect can be under favorable conditions. Lilly (1971) has since reported actual measurements in the American Rockies that give vertical momentum fluxes of 0.5–1 N m^{-2} over a region of width 100–200 km. He further showed (Lilly, 1972) that if this transfer is incorporated into a general circulation model of the atmosphere, the effect is quite large and improves the agreement between the model and observation.

The cutoff wavenumber

$$k_c = N/U, \qquad (6.8.12)$$

which divides the two types of solution, corresponds to a wavelength $2\pi/k_c$ equal to the horizontal distance traveled by a particle in one buoyancy period. A typical value of this wavelength for the atmosphere (Bretherton, 1969) is about 10 km (corresponding to k_c^{-1} of order 1 km), and a typical value for the deep-ocean floor (Bell, 1975) is about 300 m $(k_c^{-1} \sim 50$ m). As the horizontal distance between wave crests increases (i.e., k decreases), (6.8.6) shows that the distance between wave crests along the perpendicular remains the same, i.e., the total wavenumber $(k^2 + m^2)^{1/2} = k_c$. Thus the angle φ' between wave crests and the vertical changes according to the formula

$$\cos \varphi' = k/k_c = Uk/N. \qquad (6.8.13)$$

The group velocity (relative to the air) can be calculated from (6.6.1). It is directed upward *along* wave crests and has magnitude $U \sin \varphi'$. It follows by vector addition that the velocity of group propagation *relative to the ground* is directed upward normal to the wave crests with magnitude $U \cos \varphi'$ as shown in Fig. 6.10. The vertical component of group velocity is equal to $U \sin \varphi' \cos \varphi'$ and has a maximum value of $\tfrac{1}{2}U$ when $\varphi' = 45°$.

Another quantity of interest is the energy density E, which can be calculated from

(6.7.7), (6.8.4), and (6.8.13). The result

$$E = \tfrac{1}{2}\rho_0 N^2 h_0^2 \qquad\qquad (6.8.14)$$

has a simple form because h_0 is the amplitude of vertical displacement, so $\tfrac{1}{4}\rho_0 N^2 h_0^2$ is the disturbance potential energy density. E is twice this value because of the equipartition property. As (6.7.9) requires, the vertical energy flux density F_z' is E times the vertical component $U \sin \varphi' \cos \varphi'$ of the group velocity. This is consistent with (6.8.7).

6.9   Effects on Boundary-Generated Waves of Variations of Buoyancy Frequency with Height

Section 6.8 was about disturbances produced at a horizontal boundary, but was restricted to the rather special case of a medium of uniform properties. In practice, the ocean and atmosphere do not have uniform properties, and this leads to a great variety of phenomena, many of which are beyond the scope of this book. The main nonuniformity, however, is due to changes with *height*, and some basic effects of variation of properties with height will be considered in this section. The main effects can be illustrated by the simple case of a medium with *piecewise-uniform* properties [cf. Rayleigh (1883)]. In particular, the special case is taken in which the velocity U is uniform but the buoyancy frequency N is given by

$$N = \begin{cases} N_1, & 0 < z < H \\ N_2, & z > H, \end{cases} \qquad\qquad (6.9.1)$$

since this is not only a simple and convenient case for illustrative purposes, but is, on occasions, a useful approximation to the structure of the atmosphere [see, e.g., Stilke (1973)]. The same principles may be applied to a much more general class of waves for which the parameter m^2 is piecewise-uniform.

6.9.1   Refraction and Reflection of Waves at a Discontinuity in N

There are four possible forms of the solution since the solution may be wavelike or exponential in each of the two regions. This subsection and the next are restricted to the case in which the frequency $|\omega| = Uk$ is less than *both* values of N, so that waves can propagate through both media. Attention will be given first to the properties of the waves near the interface. Since the horizontal component k of the wavenumber is the same in both media, (6.8.6) shows that the vertical component m, and hence the angle φ', is larger in the region of larger N. In other words, waves are *refracted* at the boundary. The law of refraction

$$N_1 \cos \varphi_1' = N_2 \cos \varphi_2' \qquad\qquad (6.9.2)$$

follows from the fact that the frequency, given by (6.5.5), is the same in each layer. This is reminiscent of Snell's law in optics, but in the present case φ' is the angle the wavenumber vector makes with the *horizontal* irrespective of the orientation of the boundary (if the boundary were sloping φ' would still be the angle with the horizontal).

Not only is there refraction of the upgoing wave at $z = H$, but part of the energy is *reflected*. The proportion involved can be found by applying the appropriate conditions at $z = H$, namely that the perturbation pressure p' and vertical velocity w be continuous. Alternatively, this condition can be expressed in terms of the ratio

$$Z = p'/\rho_0 w, \tag{6.9.3}$$

which must be the same on both sides of the boundary. It is convenient to refer to Z as the "impedance," following a similar usage in Section 5.8. The upper medium, because it is unbounded, has as solution a plane wave with upward group velocity. From (6.5.1) and (6.5.6), the impedance of a plane wave is given by

$$Z = -\omega m/(k^2 + l^2), \tag{6.9.4}$$

where in this case $l = 0$ and m has the value m_2 that (6.8.6) gives when $N = N_2$.

Below $z = H$, the solution contains both upward and downward propagating waves, and thus can be expressed in the form

$$
\begin{aligned}
w &= w_1 \big[\exp(im_1(z - H)) + r \exp(-im_1(z - H)) \big] \exp(i(kx - \omega t)), \\
k^2 p' &= m_1 \omega \rho_0 w_1 \big[-\exp(im_1(z - H)) + r \exp(-im_1(z - H)) \big] \exp(i(kx - \omega t)), \\
Z &= \frac{\omega m_1}{k^2} \frac{-\exp(im_1(z - H)) + r \exp(-im_1(z - H))}{\exp(im_1(z - H)) + r \exp(-im_1(z - H))},
\end{aligned}
\tag{6.9.5}
$$

where the *reflection coefficient* r is a number whose magnitude gives the ratio between the amplitudes of the upward and downward moving waves. Its value is obtained by equating the two expressions (6.9.4) and (6.9.5) for Z. This gives

$$r = (m_1 - m_2)/(m_1 + m_2). \tag{6.9.6}$$

If $r = 0$, the solution below $z = H$ is a pure traveling wave, whereas if $|r| = 1$, it is a standing wave.

The behavior of the velocity amplitude below $z = H$ is of interest because it oscillates between extreme values spaced at intervals of a quarter of a wavelength, $\pi/2m_1$, as (6.9.5) shows. The amplitude at even multiples of $\pi/2m_1$ below $z = H$ is the same as at $z = H$. At odd multiples of $\pi/2m_1$, the amplitude is

$$(1 - r)/(1 + r) = m_2/m_1 \tag{6.9.7}$$

times the value at $z = H$. It follows that the extremum at $z = H$ is a *minimum* when $N_1 < N_2$ (so $m_1 < m_2$) and a *maximum* when $N_1 > N_2$ (so $m_1 > m_2$). The latter case therefore offers the possibility of the most efficient generation of waves in the upper medium, depending on how many quarter wavelengths the ground is below the discontinuity.

6.9.2 Effect of the Change in N on Vertical Fluxes

To calculate how the wave amplitude and other quantities like energy flux depend on conditions at the ground, the condition that

$$w = w_0 \exp(i(kx - \omega t)) \qquad \text{at} \quad z = 0$$

must be used. Comparison with (6.9.5) shows that w_0 and w_1 are related by

$$w_0 = w_1(\exp(-im_1 H) + r \exp(im_1 H)). \tag{6.9.8}$$

Particular attention will be given to the vertical energy flux density F'_z since this quantity is independent of height, as is necessary for the energy equation to be satisfied. (The vertical flux $\tau = \rho \overline{uw}$ of momentum is also constant but is related to F'_z by the simple expression $\tau = F'_z/U$.) The definition (6.7.8) gives

$$F'_z = \tfrac{1}{2}\mathscr{R}(p'w^*) \tag{6.9.9}$$

when p' and w are given in complex form, where $\mathscr{R}$ denotes "the real part of" and the asterisk denotes a complex conjugate. Substitution from (6.9.5), using (6.8.2), (6.9.6), and (6.9.8), then gives

$$F'_z = (Um_2\rho_0 w_0^2/2k)/[1 + ((m_2/m_1)^2 - 1)\sin^2 m_1 H]. \tag{6.9.10}$$

The numerator is the value F'_z would have if there were no discontinuity [see (6.7.10)], so the denominator shows the effect of the discontinuity on the wave flux.

As found in Section 6.9.1, there is a dependence on how many quarter wavelengths occur between the ground and the discontinuity. Extreme values occur when there are an integral number. If the number is even, then $\sin(m_1 H) = 0$, so (6.9.10) gives

$$F'_z = Um_2\rho_0 w_0^2/2k, \tag{6.9.11}$$

i.e., the same result as if the lower layer were not present. If the number of quarter waves is an odd integer, then $\sin^2(m_1 H) = 1$, so

$$F'_z = Um_1^2\rho_0 w_0^2/2m_2 k. \tag{6.9.12}$$

The flux is enhanced when $m_2 < m_1$ ($N_2 < N_1$) and becomes infinite when $m_2 = 0$.

An important observation to make about (6.9.10) is the fact that the flux depends on m_2, i.e., on the properties of the atmosphere well away from the boundary. Of course, if the wind were suddenly switched on at some initial time, it would take some time for the steady-state solution studied here to be established: the presence of different values of N aloft can only affect conditions at the ground after sufficient time has elapsed for waves to propagate to those regions and be reflected back to the ground. The dependence on $\sin(m_1 H)$ in (6.9.10) shows that the steady-state flux depends on the phase of the wave that is reflected back to the ground from the discontinuity relative to those being newly produced. For instance, when $m_2 < m_1$ and $\sin^2(m_1 H) = 1$, (6.9.12) shows that the phase is such as to reinforce the wave and cause the flux to be greater than it would be in the absence of a discontinuity. The extreme case $m_2 = 0$ corresponds to a resonance, where the wave continues to be reinforced without limit, so there is no finite steady-state solution.

6.9.3   The Case in Which the Weakly Stratified Region Is near the Ground

This subsection is a summary of the results when $N_1 < N_2$, and Subsection 6.9.4 concerns the case in which $N_1 > N_2$. When $N_1 < N_2$, the solution in both regions is wavelike for sufficiently long waves, i.e., when $k < N_1/U$. The reflection coefficient r is negative and (6.9.10) shows that the vertical flux is generally *reduced* by the presence of the weakly stratified layer near the ground. If N_1/N_2 is small, the reduction is substantial *unless* $\sin(m_1 H)$ is small, i.e., the layer is very thin or H is close to a multiple of π/m_1. The smallest value of H for the latter possibility to occur is $\pi U/N_1$, which follows from (6.8.6).

When the wavelength of the topography is small enough ($k > N_1/U$), internal waves cannot exist in the lower medium, so one might expect vertical propagation of energy to cease. This is not so, however, because the exponentially decaying solution (6.8.8) always gives some movement, however weak, at $z = H$, and this movement can generate waves in the upper region where vertical propagation is possible. The solution in the lower region in such a case can be found by replacing m_1 in (6.9.5) and (6.9.6) by $i\gamma$, where γ is given by (6.8.9) with $N = N_1$. The expression for F_z' then includes a factor $\exp(-2\gamma H)$, showing that little energy "tunnels through" the non-wave region unless γH is small.

6.9.4   The Case in Which the Strongly Stratified Layer Is near the Ground

When the topographic wavelength is large enough ($k < N_2/U$), waves can be supported in both layers. The reflection coefficient in this case is positive and (6.9.10) shows that the vertical flux is generally *enhanced* by the presence of the strongly stratified layer near the ground. The enhancement is greatest when $\sin^2(m_1 H) = 1$, and then (6.9.12) applies. As k tends to N_2/U, $m_2 \to 0$, and then (6.9.12) shows that $F_z' \to \infty$ because of the resonant effect due to reflected waves interfering with those newly generated in such a way as to produce large amplitudes aloft.

When the wavenumber k of the topography lies between N_2/U and N_1/U, waves can exist in the lower medium, but the solution in the upper layer is evanescent in character, i.e., has the form given by (6.8.8)–(6.8.10) with $N = N_2$. It follows that the impedance Z is given by

$$Z = -i\omega\gamma/k^2. \tag{6.9.13}$$

Comparing this with (6.9.4), it follows that the solution *below* $z = H$ is given by (6.9.5) but with m_2 replaced by $i\gamma$. In particular, the reflection coefficient

$$r = (m_1 - i\gamma)/(m_1 + i\gamma) \tag{6.9.14}$$

has magnitude unity, i.e., total reflection occurs at the discontinuity, and this is due to the upper medium being unable to propagate energy. As a consequence, the vertical energy flux density F_z' must be zero. Another consequence is that the reflected waves can interfere with the waves produced at the ground to produce a resonance if the phases are right, i.e., if the distance between the discontinuity and the ground is the right number of wavelengths. From (6.9.8), the condition for resonance ($w_1 = \infty$) is

$$r = -\exp(-2im_1 H), \tag{6.9.15}$$

which by (6.9.14) gives the condition

$$\cot m_1 H = -\gamma/m_1. \tag{6.9.16}$$

To show how the response of the atmosphere depends on the parameters, some measure A of wave amplitude is needed. The wave flux is not a suitable measure since it is identically zero. A suitable quantity is the mean wave energy in the lower layer, normalized by the value this quantity would have if N_2 were equal to N_1. Calculations lead to the following formulas, with $\epsilon = N_2/N_1$ (which is less than unity):

$$A = \frac{m_1^2 + m_2^2 + (\epsilon^2 m_1^2 - m_2^2)(\sin 2m_1 H/2m_1 H)}{2(m_1^2 \cos^2 m_1 H + m_2^2 \sin^2 m_1 H)} \quad \text{for} \quad k < N_2/U,$$

$$A = \frac{(m_1^2 + \gamma^2)^2 + (\epsilon^2 m_1^2 + \gamma^2)[4\gamma m_1 \sin^2 m_1 H + (m_1^2 - \gamma^2)\sin 2m_1 H]/2m_1 H}{2(m_1^2 + \gamma^2)[m_1 \cos m_1 H + \gamma \sin m_1 H]^2} \tag{6.9.17}$$

$$\text{for} \quad N_2/U < k < N_1/U.$$

When $k > N_1/U$, the solution is not wavelike in either region, so the response is not of great interest. Figure 6.11 shows for the case $\epsilon = 0.3$ (a reasonable value for the atmosphere) how A varies with the nondimensional wavenumber kH and with the nondimensional flow parameter $N_1 H/U$. The inverse

$$F = U/N_1 H \tag{6.9.18}$$

of this quantity, which is the ratio of the fluid speed U to a measure $N_1 H$ of the internal wave speed, is called the *Froude number*. The figure clearly shows the conditions under which a large response should be anticipated. In particular, it shows that resonance can occur only when the Froude number is less than $2/\pi$, and the number of wavelengths at which resonance can occur goes up as the Froude number decreases.

For topography with a fixed wavenumber k, the resonance condition (6.9.16) will be satisfied only in rather special circumstances. Real topographies, however, have contributions from all wavenumbers, so the effect of resonance is to emphasize the response at those values of the wavenumber for which resonance occurs. In particular, "lee" waves that are generated in the lee of isolated hills have been studied a great deal (Alaka, 1960; Nicholls, 1973; Queney, 1973, 1977; Gossard and Hooke, 1975; R. B. Smith, 1979; Turner, 1973; Yih, 1980), and many features of these waves can be deduced from Fig. 6.11, e.g., the condition that the Froude number must be below a certain value before a resonant wave will be found and that the number of resonant waves goes up as the Froude number decreases. Figure 6.11 also suggests that when $1/F$ is near an odd multiple of $\pi/2$, a considerable amount of long-wave energy will be generated, and this will involve considerable upward radiation of energy, as discussed earlier.

Lee waves in the atmosphere have usually been noticed because of their clearly defined wavelength, i.e., in conditions favorable for resonance. Typical values of the

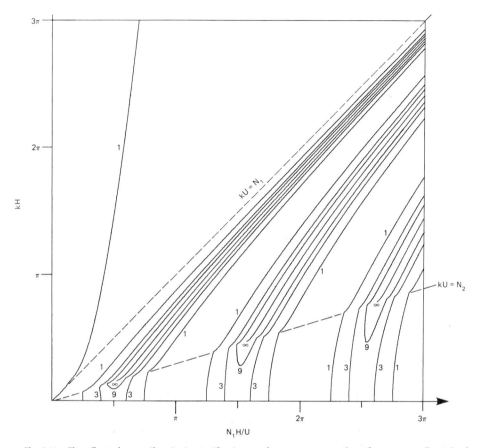

Fig. 6.11.   The effect of nonuniformity in stratification on the response to uniform flow over small-amplitude topography. In this case the buoyancy frequency has one value N_1 up to a height H and then another value $N_2 = 0.3N_1$ above that. The graph shows contours of the wave energy below $z = H$ divided by the value it would have had the buoyancy frequency been N_1 at all levels. The contours shown are for the values 1, 3, 9, and ∞. The vertical axis is kH, a nondimensional wavenumber, and the horizontal axis is the inverse of the Froude number U/N_1H. There are three regimes, depending on the frequency kU with which air particles meet crests. (a) If $kU > N_1$, the energy falls off with height and the variation in buoyancy frequency has no major effect. (b) If $N_2 < kU < N_1$, internal waves are generated in the region of higher N, but they are confined to this region because the solution decays with height above $z = H$. In these circumstances resonance is possible, and wave energy can build up indefinitely, hence the lines of infinite wave energy and the nearby regions of very large response. The lower atmosphere is then acting as a waveguide or duct for the wave energy and the resonance occurs when the wavelength of the topography matches the wavelength of the ducted mode. (c) If $kU < N_2$, internal waves occur in both regions so that a perfect waveguide no longer exists. However, there are still regions of large response as the figure shows.

wavelengths observed are 10–20 km. Figure 6.12 shows an example of lee waves that have been observed. In practice, the detailed behavior of such waves involves a number of effects not yet considered, so further discussion of these waves will be left for a later section.

The condition (6.9.16) for resonance is also the condition [see (6.9.8)] for a non-trivial solution to exist that has $w = 0$ at the surface. In other words, the resonant

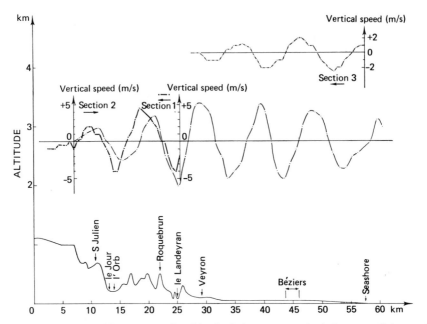

Fig. 6.12. Observations of lee waves produced by the Espinouse mountains in France on 16 January 1970. The wind is from left to right and the observations are from three traverses made by an aircraft, two at 2800 m and one at 4200 m in the directions shown. The topography along the track of the aircraft is shown. The observations are of vertical velocity. The wavelength of the lee waves is about 10 km. [From Cruette (1976, Fig. 22).]

modes are identical with the modes that can exist in the absence of forcing. If such a mode is stimulated, the discontinuity in buoyancy frequency prevents energy from escaping vertically, and so energy is carried horizontally without loss. In these circumstances, the region in which the energy is trapped is called a *waveguide* or a *duct*. The modes that propagate along these waveguides play an important part in the description of motion in the ocean and the atmosphere, and so will receive a great deal of attention in subsequent sections.

6.9.5 More General Profiles of N(z)

The discussion so far has been about a very simple model with two layers of different N. However, this model contains most of the important features that occur when N is a more general function of z. A model with several layers of different N can obviously be treated by the same methods, and such models can be used to approximate solutions for continuously varying N [see, e.g., Tolstoy (1973)]. Also, the effect of a density discontinuity can be obtained [cf. Rayleigh (1883)] by taking a thin layer of large N and letting its thickness tend to zero whilst keeping the density jump across it constant. The results of the two-layer model of Section 6.2 could, for instance, be obtained in this way.

For continuously varying N, the equation satisfied by w can be obtained by looking for a solution of (6.4.12) that is proportional to $\exp[i(kx - \omega t)] = \exp[ik(x + Ut)]$.

Then (6.4.12) gives

$$d^2w/dz^2 + m^2w = 0, \tag{6.9.19}$$

where $m^2(z)$ is given in terms of $N^2(z)$ by (6.8.6) for the case of uniform velocity U. When U is nonuniform (stratified shear flow), it can be shown (see Section 8.9) that (6.9.19) still applies, but $m^2(z)$ is now given by

$$m^2 = (N/U)^2 - k^2 - U^{-1}\, d^2U/dz^2. \tag{6.9.20}$$

The condition at the surface is (6.8.3) and the condition at infinity is one of upward radiation if solutions are wavelike there, but tend to zero otherwise. One approximation that is often very useful is called the Liouville–Green or WKB or WKBJ approximation (see Section 8.12). This applies when the properties of the medium vary slowly enough, i.e., when $m(z)$ changes by only a small amount over a distance of $1/m$. In this approximation, wave reflection *does not occur*, so the solution is given by (6.8.5) at all heights, with mz replaced by $\int m\, dz$. The amplitude w_0 varies in such a way that the upward energy flux density F'_z, given by (6.8.7), remains independent of height, and this defines the approximate solution completely. The way in which different wave quantities vary with height follows automatically. For instance, comparison of (6.8.11) and (6.8.14) gives the expression

$$E = k^{-1}(N^2 - k^2U^2)^{-1/2}(N/U)^2F'_z \tag{6.9.21}$$

for energy density E. This shows, for instance, that for small k

$$E \propto N, \tag{6.9.22}$$

i.e., the energy density is largest where the buoyancy frequency is largest, and this relationship is frequently observed.

The study of waves that are continuously forced at a boundary will not be pursued further in this chapter. Instead, freely propagating waves will be considered, but in the presence of boundaries such as the ground, the sea surface, and the sea floor. Only *horizontal* boundaries, however, will be considered, so solutions can be considered as a Fourier superposition of waves that vary sinusoidally in the horizontal. Then the equation has the same form as that of (6.9.19), but the boundary conditions will be those appropriate for free waves rather than for forced waves.

6.10   Free Waves in the Presence of Boundaries

The study of stratified fluids in this chapter began with the simple example of two superposed layers of different density, which gives a quite good approximation to the behavior of the ocean and of stratified lakes. Now the case of a continuously stratified ocean or lake will be studied. Attention is restricted to the case in which the bottom is flat, but neither the hydrostatic approximation nor long-wave approximation will be made to begin with. (The long-wave limit is considered in Section 6.11.) The equilibrium state that is being perturbed is one at rest, so density, and hence buoyancy frequency, is a function only of the vertical coordinate z. The atmosphere is somewhat different from the ocean in that it has no definite upper boundary, so a study of waves in this situation is made later in the section.

6.10.1   The Oceanic Waveguide

Because, in the undisturbed state, fluid properties are constant on horizontal surfaces and, furthermore, the boundaries are horizontal, solutions of the perturbation equations can be found in the form

$$w = \hat{w}(z) \exp[i(kx + ly - \omega t)]. \tag{6.10.1}$$

The equation for $\hat{w}$ can be found by substitution in (6.4.10). If the Boussinesq approximation is made, as it is in this section, Eq. (6.4.12) is used instead, and this gives

$$d^2\hat{w}/dz^2 + ((N^2 - \omega^2)/\omega^2)(k^2 + l^2)\hat{w} = 0. \tag{6.10.2}$$

This is in fact the same equation as (6.9.19), which corresponds to the case $l = 0$, $\omega = -Uk$. For an ocean or lake of depth H, the condition at the bottom is

$$\hat{w} = 0 \qquad \text{at} \quad z = -H. \tag{6.10.3}$$

At the free surface, a combination of (5.2.10) and (5.2.11) gives

$$\partial p'/\partial t = \rho_0 g w \qquad \text{at} \quad z = 0, \tag{6.10.4}$$

i.e., using (6.10.1) and (6.4.7),

$$\omega^2 \, d\hat{w}/dz = (k^2 + l^2)g\hat{w} \qquad \text{at} \quad z = 0. \tag{6.10.5}$$

The two boundaries have the effect of confining wave energy to a region of finite vertical extent, so the ocean can be considered as a *waveguide* that causes energy to propagate horizontally.

A useful piece of imagery is to picture internal waves propagating obliquely through the ocean, reflections at the upper and lower boundaries ensuring no loss of energy from the wave guide, whereas horizontal propagation is uninhibited.

A characteristic feature of a waveguide such as the ocean is that solutions of (6.10.2) that satisfy the boundary conditions exist only for particular values (eigenvalues) ω, for each of which there is a corresponding waveguide mode or eigenfunction $\hat{w}(z)$. In general, these have a different structure for each different wavenumber, but in the long-wave limit the structure becomes independent of wavenumber. Then the eigenfunctions are called normal modes, and these will be studied in Section 6.11.

Before calculating solutions for a special case, the approximations that follow from the fact that density differences within the fluid are small compared with the difference across the free surface will be made. These were introduced in Sections 6.2 and 6.3. As for the two-layer case, the largest eigenvalue ω_0 corresponds to a *surface* wave, and stratification has very little effect on its properties. ω_0 is very large compared with N, so Eq. (6.10.2) for $\hat{w}$ is approximately of the same form as Eq. (5.3.5) for p', and the dispersion relation (5.3.8) for ω_0 follows on application of the boundary condition (6.10.5). The other *internal* modes have ω of the same order (or smaller) as N, and for these the rigid lid approximation (see Section 6.3) can be made, i.e., the left-hand side of (6.10.5) is small compared with the right, so the approximate condition is

$$\hat{w} = 0 \qquad \text{at} \quad z = 0. \tag{6.10.6}$$

Now the internal modes will be found, using this approximation, for the particularly simple case of constant N that was considered by Rayleigh (1883). This case is often applicable to laboratory experiments. Equation (6.10.2) has the sinusoidal solution

$$\hat{w} = \sin m(z + H), \tag{6.10.7}$$

with m given by the internal wave dispersion relation [cf. (6.8.6)]

$$m^2 = (k^2 + l^2)(N^2 - \omega^2)/\omega^2. \tag{6.10.8}$$

The values that m can take are restricted to a discrete set by the need to satisfy the surface condition. If the approximate condition (6.10.6) is used, mH is required to be a multiple of π, so the eigenfrequencies ω_n for this waveguide are, from (6.10.8), given by

$$\omega_n^2 = \frac{(k^2 + l^2)N^2 H^2}{n^2 \pi^2 + (k^2 + l^2)H^2}, \qquad n = 1, 2, 3 \ldots. \tag{6.10.9}$$

Rayleigh found the equivalent dispersion relation for the case in which the Boussinesq approximation is not made, and it was for this equation that he made the remark previously quoted about ω_n that there is a "limit on the side of rapidity of variation, but none on the side of slowness." The dispersion diagram is shown in Fig. 6.13a and corresponds to the variation shown in Fig. 6.7 along the appropriate planes $m = $ const. Long waves have the maximum group velocity of $NH/n\pi$ for each mode, whereas short waves have frequency approaching N from below. If the rigid lid approximation is not made, m is slightly larger for each mode, but the effect on Fig. 6.13a would hardly be noticeable.

For application to the ocean, a constant-N model is rarely suitable because the variation of N with z is more like the sample profile shown in Fig. 3.4. There are analytic models that have profiles of N with characteristics like these [see Eckart (1960), Krauss (1966), Phillips (1977), or Roberts (1975)], but in practice it is not difficult to compute solutions of (6.10.2) for actually observed profiles $N(z)$. The computation is just as simple if the rigid lid and Boussinesq approximations are *not* made, but these approximations make little difference to the solutions. As an example of computed solutions, Fig. 6.14 shows eigenfunctions corresponding to the $N(z)$ profile of Fig. 3.4.

If the ocean is perturbed with the spatial structure of one of the eigensolutions, then the subsequent behavior in time is described by (6.10.1), i.e., there is an oscillation with a particular frequency. Such a situation, however, is unlikely, so it is necessary to represent the initial structure in space as a *superposition* of eigensolutions. Then each of these will behave in time as found above, and so the solution can be constructed at all times by taking the appropriate superposition of modes.

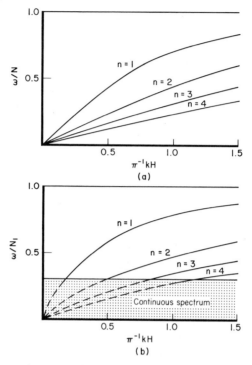

Fig. 6.13. (a) The dispersion diagram for ducted modes in a uniformly stratified fluid of depth H. These correspond to the vertical wavenumber m, having values given by $mH = n\pi$, where n is an integer that gives the number of the mode. (b) The dispersion diagram for a semi-infinite region for which the buoyancy frequency has the uniform value N_1 below $z = H$ and another uniform value $N_2 = 0.3N_1$ above $z = H$. For $N_2 < \omega < N_1$, there is a discrete set of ducted modes. For $\omega < N_2$, there is a continuous spectrum, i.e., solutions exist for all ω, k in this region. However, there tends to be an enhanced response in the neighborhood of the dashed lines, which corresponds to an odd number of quarter wavelengths that are below $z = H$.

6.10.2   Free Waves in a Semi-infinite Region

The atmosphere does not have a definite upper boundary as does the ocean, so solutions of (6.10.2) will now be considered for the case of a semi-infinite domain $z > 0$. In this case there are two types of solution, the first being typified by the case $N = $ const. Then the only solutions of (6.10.2) that satisfy the condition at the ground $z = 0$ and remain bounded at infinity are sinusoidal, i.e.,

$$\hat{w} = \sin mz, \tag{6.10.10}$$

with m given by (6.10.8). There is now no restriction on m, so by (6.10.8) the frequency ω can have any value in the range $0 \leq \omega < N$, i.e., there is a continuous spectrum of solutions. Superpositions of such solutions can be used to solve initial-value problems, and have the form of Fourier integrals.

When N varies with z, there is another type of solution possible, namely, one that satisfies the condition at the ground yet decays as $z \to \infty$. These are the waveguide modes, and there are, in general, only a finite number possible. A simple example is

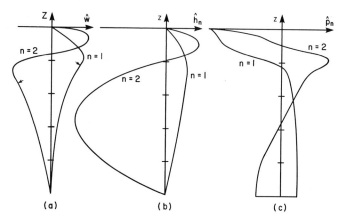

Fig. 6.14. Eigenfunctions for internal waves when buoyancy frequency N varies with z as shown in Fig. 3.4 (i.e., as observed in the North Atlantic near 28°N 70°W). The depth in the region is a little over 5000 m, so the markings on the vertical axis are about 1000 m apart. (a) The first two eigenfunctions for vertical velocity $\hat{w}$ when the inverse wavenumber k^{-1} is 1 km. The eigenfrequencies are $\omega_1 = 2.1 \times 10^{-3}$ s^{-1} and $\omega_2 = 1.0 \times 10^{-3}$ s^{-1}. The arrows mark the position of inflection points at which the eigenfrequency is equal to the local buoyancy frequency. (b) Vertical velocity (or vertical displacement) eigenfunctions $\hat{h}_n$. (c) Horizontal velocity (or pressure perturbation) eigenfunctions $\hat{p}_n$ for very small wavenumber. The eigenfunctions in this case are called baroclinic normal modes and the figure shows the first two of them. In the limit as $k \to 0$, the eigenfrequency $\omega_n \to 0$, but the wave speed $c_n = \omega_n/k$ tends to a finite limit. In the case shown, $c_1 = 3.0$ m s^{-1} and $c_2 = 1.2$ m s^{-1}.

provided by the case studied in Section 6.9, in which a region of depth H of uniform large buoyancy frequency N_1 underlies a semi-infinite region of uniform small buoyancy frequency N_2. Similar models for electromagnetic and sound waves are treated by Budden (1961, Chapters 9 and 10). The waveguide modes have frequencies in the range $N_2 < \omega < N_1$, for then $\hat{w}$ varies sinusoidally in the lower layer [as given by (6.10.10) and (6.10.8)] but decays exponentially in the upper layer with decay factor γ [cf. (6.8.9)], where

$$\gamma^2 = (k^2 + l^2)(\omega^2 - N_2^2)/\omega^2. \tag{6.10.11}$$

The solutions are, in fact, the "resonant" modes found in Section 6.9.4. The equation that determines the eigenfrequencies is (6.9.16), where by (6.10.8) and (6.10.11)

$$(\gamma/m_1)^2 = (1 - \epsilon^2)(k^2 + l^2)m_1^{-2} - \epsilon^2, \tag{6.10.12}$$

and ϵ is the ratio N_2/N_1, as in Section 6.9. Figure 6.13b depicts the eigenvalues for the case $\epsilon = 0.3$. These correspond to the contours $A = \infty$ in Fig. 6.11. Mode n has fractionally over $2n - 1$ quarter wavelengths between the discontinuity and the ground.

In addition to the discrete set of waveguide modes, there is a continuous infinity of solutions of the first type, i.e., solutions for which $\hat{w}$ varies sinusoidally in both regions. These solutions have $\omega < N_2$. Thus to deduce how the perturbation will change with time from some initial state, it is necessary to represent this state as a superposition both of discrete waveguide modes and of the continuous spectrum of sinusoidal modes. The relative amplitudes of the different modes depends on the initial state. If, however, the scale of the initial perturbation is large compared with

H and ϵ is small, enhanced contributions are found to come from the neighborhood of the broken lines shown in Fig. 6.13b, i.e., from the "continuation" of the eigenvalue curves into the region where the spectrum is continuous. This implies that, in time, the perturbation will develop a quite regular structure. Such regular structures are indeed seen on occasions, and Fig. 6.15 shows an example (which is discussed below).

Sometimes a large proportion of energy appears to be concentrated in a single waveguide mode, and then comparison of observations with theory is fairly straightforward. Profiles of $N(z)$ can be obtained from radiosonde ascents, and waveguide modes can be computed. Often a model that has analytical solutions gives a sufficiently accurate approximation to observed conditions, and many such models are described by Gossard and Hooke (1975, Chapter 5). For example, Gossard and Munk (1954) found that a model with three layers of constant N was quite adequate for studies of waveguide modes observed near San Diego, California. Long-crested waves, coherent over several tens of kilometers, were observed to propagate at a speed consistent with theory. A discussion of such observations was presented by Gossard and Hooke (1975, Chapter 10).

Figure 6.15 shows oscillations of surface pressure and of the height of isentropic surfaces at Hamburg on an occasion [July 8, 1967, reported by Stilke (1973)] when radiosonde ascents showed the two-layer model (6.9.1) to be quite a good one, with $N_1 = 37 \times 10^{-3}$ s^{-1}, $N_2 = 8.5 \times 10^{-3}$ s^{-1}, and $H = 150$ m. The longest possible waveguide mode (see Fig. 6.13b) has a wavelength of 2.5 km and a period of $2\pi/N_2 = 12$ min. Particle displacements for this mode increase from zero at the ground like a sine function with maximum at the top of the inversion layer. Pressure fluctuations [see (6.4.7)], on the other hand, increase from zero at the top of the layer to a maximum at the ground. The observed fluctuations have a similar behavior.

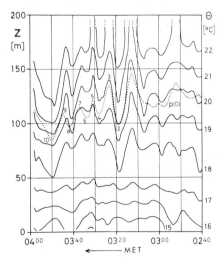

Fig. 6.15. Height (solid lines) of surfaces of constant potential temperature θ observed at Hamburg between 2:40 and 4:00 A.M. on 8 July 1967. Also shown (dotted line) is the pressure $p(0)$ observed at ground level. At this time, radiosonde ascents showed nighttime inversion conditions with a stable layer of depth around 150 m near the ground and with buoyancy frequency 4–5 times greater than that of the air above. The observations are interpreted as internal waves ducted in the lower layer of high buoyancy frequency. [From Stilke (1973, Fig. 7a).]

Stilke reports that such waves have an important effect on microwave transmission, the strength of the signal received at a given station having variations with internal wave periods. A similar effect on acoustic wave propagation is found in the ocean.

6.11   Waves of Large Horizontal Scale: Normal Modes

The ocean and atmosphere are thin sheets of fluid in the sense that their horizontal extent is very much larger than their vertical extent. It comes as no surprise, therefore, to find that most of the energy associated with motion lies in components with *horizontal scale much larger than the vertical*. For such components, there are certain simplifications that can be made, and these have been exploited since the time of Laplace (1778-1779). The simplification is that a *separation of variables* technique can be used, i.e., the solution can be expressed as a sum of *normal modes*, each of which has a fixed vertical structure and behaves in the horizontal dimension and in time in the same way as does a homogeneous fluid with a free surface. This is true even when rotation effects are introduced, and thus provides a useful simplification in the study of these effects in later chapters.

The idea of normal modes was introduced near the beginning of this chapter for the special case of two superposed layers of fixed density. This already provides a good description of much of the behavior of an ocean or lake, and the theory will now be extended to the case of *continuously stratified* oceans or lakes. The main difference here is that there is now an infinite set of baroclinic modes instead of just one, but each mode behaves in a similar fashion. The long-wave approximation can also be used in the atmosphere where there is no fixed upper boundary. Then there is no longer a discrete set of modes, but a continuous spectrum.

Consider first the normal modes for a continuously stratified flat-bottomed ocean or lake. These are the limiting forms of the waveguide modes of the last section in the limit as $k^2 + l^2 \to 0$. The case of constant N provides a good example. Here the dispersion relation for the baroclinic modes is (6.10.9) and this is shown in Fig. 6.13a. For small $(k^2 + l^2)H^2/n^2\pi^2$, the eigenfrequencies ω_n are given approximately by

$$\omega_n^2/(k^2 + l^2) \equiv c_n^2 \approx N^2 H^2/n^2\pi^2. \tag{6.11.1}$$

Thus the horizontal phase speed $c_n \equiv \omega_n/(k^2 + l^2)^{1/2}$ has a fixed value in the limit, and this is true for other density structures as well. It follows from this that the long waves have *low frequency* since $\omega_n \to 0$ as $k^2 + l^2 \to 0$. Another feature follows from (6.5.8), which shows that when $k^2 + l^2 \ll m^2$, the *hydrostatic approximation* can be made. Conversely, if the hydrostatic approximation is made at the outset, the equations obtained are the same as those obtained in the long-wave limit. Thus the hydrostatic and long-wave approximations are identical, just as they were for a homogeneous fluid (see Section 5.5).

Now consider how the equations of Section 6.4 are modified when the hydrostatic approximation is made. Of the three basic equations, (6.4.4) and (6.4.6) are unaltered, but (6.4.5) is replaced by the hydrostatic equation

$$\partial p'/\partial z = -\rho'g. \tag{6.11.2}$$

Of the two derived equations, (6.4.7) and (6.4.8), the former is unaltered, i.e., the equation for the horizontal divergence is still

$$\rho_0\, \partial^2 w/\partial z\, \partial t = (\partial^2/\partial x^2 + \partial^2/\partial y^2)p', \tag{6.11.3}$$

but (6.4.8), which pertains to vertical motion, is now approximated by

$$N^2 w = -\rho_0^{-1}\, \partial^2 p'/\partial z\, \partial t. \tag{6.11.4}$$

The two equations (6.11.3) and (6.11.4) clearly possess separable solutions of the form

$$w = \hat{h}(z)\tilde{w}(x, y, t), \qquad p' = \hat{p}(z)\tilde{\eta}(x, y, t), \tag{6.11.5}$$

where $\hat{p}$ and $\hat{h}$ satisfy

$$\rho_0(z)^{-1}\hat{p} = c_e^2\, d\hat{h}/dz, \tag{6.11.6}$$

$$\rho_0(z)^{-1}\, d\hat{p}/dz = -N^2(z)\hat{h}, \tag{6.11.7}$$

and c_e is a separation constant with the dimensions of velocity. The notation $\hat{h}$ is used since $\hat{h}(z)$ also gives the variation with z of the vertical displacement h of a fluid particle, and because (6.11.6) and (6.11.7) then have dimensional consistency if $\hat{h}$ has the dimensions of displacement and $\hat{p}$ the dimensions of pressure. The corresponding equations for variations with x, y, and t are

$$\tilde{w} = \partial\tilde{\eta}/\partial t, \tag{6.11.8}$$

$$\partial\tilde{w}/\partial t = c_e^2(\partial^2\tilde{\eta}/\partial x^2 + \partial^2\tilde{\eta}/\partial y^2). \tag{6.11.9}$$

These have the same form as do the equations for a homogeneous fluid with free surface displacement $\tilde{\eta}$, and yield the wave equation (5.6.10). The notation $\tilde{\eta}$ also seems appropriate when it is considered that the vertical displacement h of a fluid particle is, from (6.11.5) and (6.11.8), given by

$$h = \hat{h}(z)\tilde{\eta}(x, y, t). \tag{6.11.10}$$

The analogy can be taken even further to include the equations for horizontal velocity. The components (u, v) for the separable solution have the form

$$u = \tilde{u}(x, y, t)\hat{p}(z)/g\rho_0(z), \qquad v = \tilde{v}(x, y, t)\hat{p}(z)/g\rho_0(z), \tag{6.11.11}$$

and substitution in (6.4.4) yields equations of the same form as do (5.6.4) and (5.6.5) for a one-layer system, namely,

$$\partial\tilde{u}/\partial t = -g\, \partial\tilde{\eta}/\partial x, \qquad \partial\tilde{v}/\partial t = -g\, \partial\tilde{\eta}/\partial y, \tag{6.11.12}$$

whereas the incompressibility condition (6.4.3) yields the one layer form (5.6.6) of the continuity equation, namely,

$$\partial\tilde{\eta}/\partial t + H_e(\partial\tilde{u}/\partial x + \partial\tilde{v}/\partial y) = 0. \tag{6.11.13}$$

The depth H_e appearing in this equation is the *equivalent depth*, i.e., the depth for an equivalent homogeneous system, and is related to the separation constant c_e by

$$c_e^2 = gH_e. \tag{6.11.14}$$

In fact, the homogeneous fluid is the special case $N = 0$ for which (6.11.7) gives $\hat{p}$ independent of depth. The case of two superposed homogeneous layers that was considered at the beginning of this chapter is another special case, and the notion of equivalent depth has already been introduced in that connection.

The values that the separation constant c_e may take are determined by the two boundary conditions on $\hat{p}$ and $\hat{h}$. For an ocean of constant depth H, the condition (5.2.11) at the free surface becomes

$$\hat{p} = \rho_0 g \hat{h} \qquad \text{at} \quad z = 0, \tag{6.11.15}$$

and that at the ocean floor is

$$\hat{h} = 0 \qquad \text{at} \quad z = -H. \tag{6.11.16}$$

The equations to be satisfied are (6.11.6) and (6.11.7), which can be reduced to a single equation for either $\hat{h}$ or $\hat{p}$. The one for $\hat{h}$, for example, is

$$\frac{1}{\rho_0} \frac{d}{dz}\left(\rho_0 \frac{d\hat{h}}{dz}\right) + \frac{N^2}{c_e^2}\hat{h} = 0. \tag{6.11.17}$$

This is of the Sturm–Liouville form, and analogies can be made with the normal modes of oscillation for other systems, e.g., the stretched string [see appendix to Lighthill (1969)]. In circumstances in which the Boussinesq approximation applies (i.e., ρ_0 varies slowly compared with $\hat{h}$), (6.11.17) can be approximated by

$$d^2\hat{h}/dz^2 + (N/c_e)^2\hat{h} = 0. \tag{6.11.18}$$

For a continuously stratified ocean or lake, there is an infinite sequence

$$c_n, \qquad n = 0, 1, 2, 3, \ldots,$$

of possible values (eigenvalues) of c_e, arranged in descending order, and the corresponding eigenfunctions, the normal modes, are denoted by

$$\hat{h}_n(z), \qquad \hat{p}_n(z), \qquad n = 0, 1, 2, 3, \ldots.$$

Mode zero is the barotropic mode for which the approximate solution of (6.11.18) is the one obtained when $N = 0$, namely,

$$\hat{p}_0 = \rho_0 g H, \qquad \hat{h}_0 = z + H, \qquad c_0^2 = gH. \tag{6.11.19}$$

Since H is between 4 and 6 km over most of the ocean, c_0 is usually between 200 and 250 m s^{-1}, but in shallow seas and on continental shelves, where H may be between 40 and 160 m, c_0 lies between 20 and 40 m s^{-1}.

The values of c_n for the baroclinic modes can readily be found by computing solutions of (6.11.17), using observed density profiles. Examples of computed eigenfunctions are shown in Fig. 6.14b,c. Often profiles for which analytic solutions can be obtained [see, e.g., Krauss (1966)] give good approximations. For the ocean, values of c_1 are usually about 2 or 3 m s^{-1}, whereas for large n, c_n is inversely proportional to n as suggested by the constant-N solution (6.11.1). For all baroclinic modes, the rigid lid approximation can be made, i.e., the displacement $\hat{h}$ of the surface is small compared with displacements in the interior, so good approximations to the solutions are obtained by replacing the surface condition (6.11.15) by the rigid lid condition

$$\hat{h} = 0 \quad \text{at} \quad z = 0. \tag{6.11.20}$$

It should be borne in mind, however, that the pressure is not zero at $z = 0$, and surface displacements, although small compared with interior displacements, can be measured by tide gauges (Wunsch and Gill, 1976). The magnitude of the surface displacement can still be found from the rigid lid solution because $\hat{p}$ does not vanish at $z = 0$ and thus (6.11.15) gives an estimate of $\hat{h}$!

The separation of variables technique can also be used for a semi-infinite region $z > 0$ that may model the atmosphere. The conditions are (6.11.20) at the ground $z = 0$ and the condition that solutions that grow exponentially as $z \to \infty$ must be excluded. These conditions allow c_e to have any value in a continuous range. Consider two special cases that have been studied before. First, the medium of constant buoyancy frequency N: the allowed solutions of (6.11.18) are

$$\hat{h} = m^{-1} \sin mz, \quad \rho_0^{-1}\hat{p} = (N/m)^2 \cos mz, \quad c_e = N/m, \quad 0 < m < \infty. \tag{6.11.21}$$

The second case is that studied in Section 6.9, namely, that for which the buoyancy frequency is a constant N up to height H and has the value $N_2 = \epsilon N$ above that level. In this case, the solution of (6.11.18) is

$$\hat{h} = \begin{cases} m^{-1} \sin mz & \text{for} \quad 0 < z \leq H \\ m^{-1} \sin mH \cos\left[\epsilon m(z - H)\right] + (\epsilon m)^{-1} \cos mH \sin\left[\epsilon m(z - H)\right] \\ \qquad \text{for} \quad z > H, \end{cases} \tag{6.11.22}$$

where

$$m = m_1 = N/c_e, \quad 0 < m < \infty, \tag{6.11.23}$$

is the vertical wavenumber in the layer near the ground and $m_2 = N_2/c_e = \epsilon m$ the vertical wavenumber in the region above $z = H$.

If the initial structure happens to be that of one of the normal modes, then it will retain that structure at subsequent times. The adjustment to equilibrium will then be according to the wave equation (6.11.8) and (6.11.9). For general initial conditions, adjustment problems can be solved by superposition of normal modes, and examples are given in the next two sections.

6.12 An Example of Adjustment to Equilibrium in a Stratified Fluid

In Chapter 5, examples were considered in which the free surface of a homogeneous fluid was initially displaced from its equilibrium position and adjustment by gravitational forces took place. This adjustment was illustrated in Fig. 5.9. Now a similar example for a continuously stratified fluid will be considered, and it will be found that many features are the same.

Mathematically, the problem is straightforward in principle: one merely requires an appropriate superposition of the simple solutions studied in earlier sections. This requires first that the initial perturbation be analyzed into components of the appropriate form, and orthogonality properties of the eigensolutions allow this to be done in a straightforward manner. The behavior with time of each component of the solution is known from the previous sections, so the solution at any given time is known as a synthesis of these component functions. Evaluation of the details, however, can be difficult. The methods are common to all sorts of waves (electromagnetic, acoustic, electric, etc.), and general textbooks on such waves can be consulted [e.g., Brekhovskikh (1980), Tolstoy (1973), Whitham (1974), Lighthill (1978)]. The example chosen for study in this section is a particularly simple case for which the exact solution can be found analytically, but it illustrates the principles involved and the behavior of a stratified fluid quite well.

As with the problem considered in Chapter 5, the hydrostatic approximation is made. This means that the solution can be expressed in terms of the response solutions considered in Section 6.11. Also, the Boussinesq approximation is made (the change in density over the vertical scale D of the disturbance is considered to be small). The fluid has uniform buoyancy frequency N and occupies the semi-infinite region $z > 0$. It is initially at rest, but isopycnal surfaces are displaced from their equilibrium position. How will these surfaces adjust with time to their equilibrium position? To illustrate the adjustment process, an example will be given in which each isopycnal surface has initially a step function profile as in the example of Section 5.6, i.e., the vertical displacement field h has the form [cf. (5.6.13)]

$$h = -h_{in}(z)\,\text{sgn}(x) \qquad \text{for} \quad t = 0. \tag{6.12.1}$$

The function $h_{in}(z)$ gives the way in which the initial displacement changes with height and $\text{sgn}(x)$ is the sign of x, i.e., the function defined by (5.6.14). Although convenient for illustration, the step function distribution is not strictly compatible with the hydrostatic approximation because of the infinite horizontal gradient. It is best thought of therefore as representing a more gradual change in the horizontal, whose details are not resolved in the approximation.

The solution for h for a semi-infinite region of constant N can be represented as a superposition of solutions of the form (6.11.10) with vertical structure h given by (6.11.21). The superposition has the form of a Fourier integral, which it is convenient to express in terms of the vertical wavenumber m rather than c_e. Then it may be written

$$h = \int_0^\infty \tilde{\eta}(x, t; m)\,\sin mz\,dm, \tag{6.12.2}$$

where $\tilde{\eta}$ satisfies the wave equation with wave speed $c_e = N/m$ by (6.11.21). Solutions are given in Section 5.6, and for the particular case considered here they have the form [c.f. (5.6.11)]

$$\tilde{\eta} = -\tfrac{1}{2}\eta_0(m)\{\text{sgn}(x + Nt/m) + \text{sgn}(x - Nt/m)\}. \tag{6.12.3}$$

Comparison of (6.12.1) and (6.12.2) for this case shows that $\eta_0(m)$ is a function such that

$$h_{in}(z) = \int_0^\infty \eta_0(m)\,\sin mz\,dm, \tag{6.12.4}$$

i.e., $h_{in}(z)$ and $\eta_0(m)$ are Fourier sine transforms of each other.

For illustrative purposes, it is convenient to choose a simple analytic function for $\eta_0(m)$ in order that the solution have a simple form. The choice

$$\eta_0(m) = 2h_0 De^{-mD} \tag{6.12.5}$$

corresponds to the initial variation with height

$$h_{in}(z) = 2Dh_0 z/(D^2 + z^2). \tag{6.12.6}$$

The maximum value is h_0, this being obtained at $z = D$.

The exact solution can be found for this case by substituting (6.12.5) and (6.12.3) in (6.12.2) to give

$$h = -2D(D^2 + z^2)^{-1}h_0 e^{-sD}\{z \cos sz + D \sin sz\} \operatorname{sgn}(x), \tag{6.12.7}$$

where

$$s = Nt/|x| \tag{6.12.8}$$

is a similarity variable in the sense that x and t appear only in the combination (6.12.8). This combination arises because it is equal to the value of m for the wave that arrives at point x at time t. This solution is shown in Fig. 6.16. Figure 6.16a shows the initial

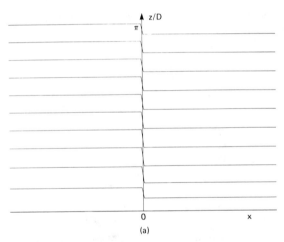

(a)

Fig. 6.16.   Adjustment to equilibrium in a fluid of constant buoyancy frequency N. The results shown are based on calculations that assume infinitesimal disturbances that satisfy the hydrostatic equation. (a) The initial displacement from equilibrium of selected isopycnals, as given by (6.12.1) and (6.12.6), the maximum displacement being at height z = D. (b) The displacements of these same isopycnals at time t if the medium is *incompressible* and of *semi-infinite* extent. The results apply to an atmosphere of constant N when D is small compared with the scale height. The horizontal lines at the two sides of the picture show the initial positions of the isopycnals. The broken line joins points of maximum displacement, i.e., is a line of constant phase. (c) The displacements of the isopycnals at time t if the medium is an *incompressible* fluid of *finite* depth πD with a free surface. The wave front associated with the barotropic waves is well out of the picture since it moves so fast relative to the baroclinic modes. Adjustment takes place in a series of steps since there is a discrete set of modes. The first step, corresponding to the first mode, is of the same sign at all levels and varies sinusoidally with half a wave between top and bottom. The second jump shows the second mode structure, etc. If lines are drawn to smooth out the jump, a fair approximation to solution (b) is obtained except near the top where boundary effects are significant. (d) The displacements of the isopycnals at time t if the medium is a *compressible* isothermal atmosphere of *semi-infinite* extent and D = 2H$_s$, i.e., is comparable with the scale height. In this case the maximum disturbance speed is 2NH$_s$, i.e., 90% of the speed of sound, and the associated front is clearly visible.

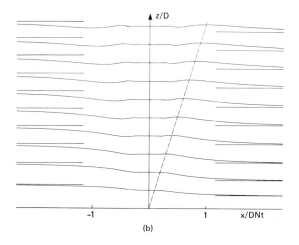

(b)

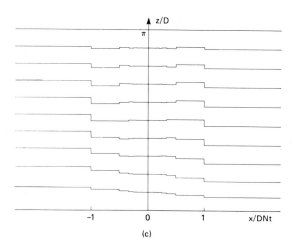

(c)

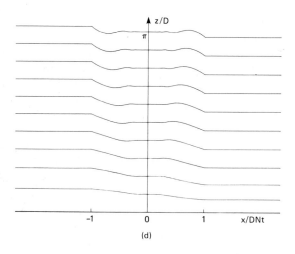

(d)

displacement of various lines of particles, and Fig. 6.16b shows the configuration at subsequent times. In this approximation, the pattern is similar at all times and merely stretches out horizontally as if it were drawn on a sheet of rubber that is pulled out sideways at a uniform rate. The behavior is also characteristic of the simpler case illustrated in Fig. 5.9a.

Because the initial perturbation has a definite vertical scale D, the dominant changes in structure propagate horizontally at speeds of order ND as appropriate to this scale. There is no sharp wave front as in the homogeneous case because different components travel at different speeds, e.g., all the components with speeds between $\frac{1}{2}ND$ and ND will make significant contributions to the solution. Otherwise the solution behaves like the homogeneous case. Consider, for instance, a point x to the right of the initial discontinuity. After a time of order x/ND, the particle at this point begins to rise noticeably and to move toward the right. When t becomes large compared with x/ND, the particle becomes close to its equilibrium level. The potential energy in the initial field has been converted into kinetic energy, and the particle has a uniform velocity to the right.

Details of the velocity field and of the energetics can be found either by adding the contributions from individual modes, or by direct calculation of exact solutions. For instance, u can be found from (6.12.7) by integrating the continuity equation

$$\partial u/\partial x + \partial w/\partial z \equiv \partial u/\partial x + \partial^2 h/\partial z\, \partial t = 0,$$

i.e.,

$$s\, \partial u/\partial s = N\, \partial^2 h/\partial s\, \partial z. \tag{6.12.9}$$

(*Note*: Other exact solutions can be found by writing the equations in terms of the variables s and z.) This gives

$$u = 2ND(D^2 + z^2)^{-1}h_0\{D - e^{-sD}(D \cos sz - z \sin sz)\}. \tag{6.12.10}$$

After a long time has passed, the velocity is a function of z only, namely,

$$u = 2ND^2 h_0/(D^2 + z^2) \tag{6.12.11}$$

and it may be confirmed that the kinetic energy associated with this motion is equal to the potential energy associated with the initial displacement, given by (6.12.1) and (6.12.6), of fluid particles.

A feature of the solution (6.12.7) that is not present in the shallow water case is the upward propagation of energy. Evidence for this can be seen from phase lines. If these are defined as linking points where h is a maximum or minimum in time, they correspond to the quantity $Ntz/|x|$, being equal to an integral multiple of π. Thus phase lines are straight lines from the origin that *descend* toward the horizontal plane as time goes on. This corresponds to *upward* propagation of energy. One such phase line is shown in Fig. 6.16b.

By combining two solutions of the form (6.12.7), one can obtain the solution for the case in which the initial disturbance has finite horizontal extent, i.e., fluid particle displacements at any level have the form (5.6.15). As in the homogeneous shallow-water case, the energy initially present in the bounded region is lost by radiation, but now there can be upward components of energy propagation as well as horizontal components.

It is possible to construct solutions for cases in which N varies with z by similar means, although the simplifications in the analysis, such as those found in the above example, are unlikely to be possible. For instance, when N has one value up to a height H and then a different constant value, the superposition of solutions can be expressed in the form

$$h = \int_0^\infty \tilde{\eta}(x, t; m)\hat{h}(z; m)\, dm,$$

where $\hat{h}$ is now the function given by (6.11.22) and c_e is related to m by (6.11.23). The case in which ϵ is small is interesting, since then most of the energy is found in components for which m is close to an odd multiple of π. (This property is, in fact, closely related to the occurrence of large values of A [given by (6.9.17)] at such intervals along the horizontal axis in Fig. 6.11.) A consequence is that most of the changes at a fixed point are associated with passage of these energetic components and so tend to take place in a series of steps rather than by a smooth transition over a period of time.

6.13 Resolution into Normal Modes for the Ocean

The solutions obtained by separation of variables are widely used in studies of transient ocean currents. In contrast to the case of an unbounded medium, there is only a discrete set of such solutions, so the superposition has the form of a sum rather than an integral, namely [see (6.11.5) and (6.11.10)],

$$w = \sum_{n=0}^\infty \hat{h}_n(z)\tilde{w}_n(x, y, t),$$

$$p' = \sum_{n=0}^\infty \hat{p}_n(z)\tilde{\eta}_n(x, y, t), \qquad (6.13.1)$$

$$h = \sum_{n=0}^\infty \hat{h}_n(z)\tilde{\eta}_n(x, y, t),$$

and similarly for u, v [see (6.11.11)]. $\hat{p}_n$ and $\hat{h}_n$ satisfy (6.11.6) and (6.11.7), with c_e taking the allowed values c_n ($n = 0, 1, 2, \ldots$) arranged in descending order. $\tilde{w}_n$ and $\tilde{\eta}_n$ satisfy the same equations [(6.11.8) and (6.11.9)] as those for a homogeneous shallow layer of depth c_n^2/g, so solving for $\tilde{\eta}_n$ is straightforward.

To find how adjustment to equilibrium takes place for some given initial configuration, it is necessary to express this configuration in terms of normal modes. Fortunately, this is straightforward because normal modes are *orthogonal* to each other in a sense that is defined in what follows. The orthogonality conditions are found by standard procedures for Sturm–Liouville systems, of which (6.11.6) and (6.11.7), or equivalent forms such as (6.11.17), form an example. Such procedures are outlined in textbooks on differential equations such as that by Birkhoff and Rota (1962). For example, if Eq. (6.11.17) for $\hat{h}_m$ is itself multiplied by $\rho_0\hat{h}_n$ and is then subtracted from $\rho_0 \hat{h}_m$ times the corresponding equation for $\hat{h}_n$, there results

$$\frac{d}{dz}\left(\rho_0 \hat{h}_m \frac{d\hat{h}_n}{dz} - \rho_0 \hat{h}_n \frac{d\hat{h}_m}{dz}\right) = \left(\frac{1}{c_m^2} - \frac{1}{c_n^2}\right)\rho_0 N^2 \hat{h}_m \hat{h}_n. \tag{6.13.2}$$

Integration over the ocean depth and use of the boundary conditions (6.11.15) and (6.11.16) give

$$\int_{-H}^{0} \rho_0 N^2 \hat{h}_m(z)\hat{h}_n(z)\, dz + \rho_0 g \hat{h}_m(0)\hat{h}_n(0) = 0 \qquad \text{if} \quad m \neq n, \tag{6.13.3}$$

which is the orthogonality condition for $\hat{h}$. The equivalent procedure for the equation for $\hat{p}$ yields

$$\int_{-H}^{0} \rho_0^{-1} \hat{p}_m(z)\hat{p}_n(z)\, dz = 0 \qquad \text{if} \quad m \neq n, \tag{6.13.4}$$

which can also be deduced from (6.13.3) and the equations and boundary conditions. This is the orthogonality condition for $\hat{p}$. If the Boussinesq approximation is made, ρ_0 in (6.13.3) and (6.13.4) can be regarded as constant rather than as a function of z. If the rigid lid approximation is also made, the second term of (6.13.3) can be neglected, and so

$$\int_{-H}^{0} N^2 \hat{h}_m(z)\hat{h}_n(z)\, dz = 0 \qquad \text{if} \quad m \neq n. \tag{6.13.5}$$

To resolve w, p', or h into normal modes, i.e., to express them in the form (6.13.1), requires that the coefficient $\tilde{w}_n$ and $\tilde{\eta}_n$ be determined. This is achieved by applying the operators that appear in (6.13.3) and (6.13.4) to (6.13.1). For instance, if the expression (6.13.1) for p' is multiplied by $\rho_0^{-1}\hat{p}_m$ and integrated over the ocean depth, all terms except the mth on the right-hand side become zero by (6.13.4). Therefore

$$\int_{-H}^{0} \rho_0^{-1} p' \hat{p}_m(z)\, dz = \tilde{\eta}_m \int_{-H}^{0} \rho_0^{-1} \hat{p}_m^2\, dz, \tag{6.13.6}$$

and this gives the required expression for $\tilde{\eta}_m$. In practice, the normal modes $\hat{p}_m$ that correspond to a particular situation can be obtained numerically, and the coefficients $\tilde{\eta}_m$ can be evaluated by numerical evaluation of the integrals in (6.13.6). Similar procedures can be followed when an expansion in terms of $\hat{h}$ is required, e.g., (6.13.1) and (6.13.3) yield

$$\int_{-H}^{0} \rho_0 N^2 h \hat{h}_m\, dz + \rho_0 g h(0)\hat{h}_m(0) = \tilde{\eta}_m \left\{ \int_{-H}^{0} \rho_0 N^2 \hat{h}_m^2\, dz + \rho_0 g (\hat{h}_m(0))^2 \right\}. \tag{6.13.7}$$

To illustrate, consider an ocean of constant buoyancy frequency N that is initially at rest but has isopycnals displaced from equilibrium with the step function distribution (6.12.1). Suppose also that the variation of h_{in} with height is the same as in Section 6.12, namely,

$$h_{\text{in}}(z) = 2Dh_0(z + H)/(D^2 + (z + H)^2). \tag{6.13.8}$$

This differs from (6.12.6) only in that the floor is now at $z = -H$ instead of $z = 0$.

Initial values $\tilde{\eta}_m(x, 0)$ of the coefficients in (6.13.1) can be found by application of (6.13.7) to the initial distribution. The solution at subsequent times has the form (5.6.11) appropriate to the wave equation, namely [cf. (6.12.3)],

$$\tilde{\eta}_m(x, t) = \tfrac{1}{2}\tilde{\eta}_m(x, 0)\{\operatorname{sgn}(x + c_m t) + \operatorname{sgn}(x - c_m t)\}, \tag{6.13.9}$$

where c_0 is given by (6.11.19) and other values of c_m are given by (6.11.1) if the rigid lid and Boussinesq approximations are appropriate for the baroclinic modes.

Each mode behaves with time as illustrated in Fig. 5.9a, i.e., a step change in isopycnal height moves at the speed c_m appropriate to that mode. Thus the changes with time at a given point z take place in a series of step changes. No effect is felt at all until time x/c_0 when the barotropic mode arrives. Much later, the first baroclinic mode arrives at time x/c_1, and thereafter the successive modes arrive at equal intervals of time (for the case $N = $ const). Figure 6.16c shows the solution for the case $H = \pi D$ for comparison with the unbounded case shown in Fig. 6.16b. In the earlier stages, the vertical structure is very simple because only one or two modes contribute; later on, the structure is rich in modes and can be quite complicated, involving several changes of sign with height. Such effects are observed in laboratory experiments in which disturbances spread out from a localized source.

6.14   Adjustment to Equilibrium in a Stratified Compressible Fluid

Small perturbations from the rest state for a *compressible* fluid can be treated by the same methods as used hitherto, but the equations are increased in complexity because of the additional effects introduced. Viscous effects, diffusion, and changes of state (e.g., condensation) will again be neglected, so the potential temperature and humidity (or salinity) are conserved following a fluid particle. However, the equation of state is different because the density of a compressible fluid changes with pressure. Thus Eqs. (6.4.2) and (6.4.3) will be altered. The former, which comes from the equation of state, is replaced by the corresponding version of (4.10.7), namely,

$$D\rho/Dt = c_s^{-2}\, Dp/Dt, \tag{6.14.1}$$

where c_s, the speed of sound, is defined by (3.7.16). Consequently, the latter equation, which expresses conservation of mass, takes the form (4.10.10), namely,

$$c_s^{-2}\, Dp/Dt + \rho\, \nabla \cdot \mathbf{u} = 0. \tag{6.14.2}$$

For small perturbations from the rest state, the linearized versions of these equations may be used. Equation (6.14.1) is replaced by

$$\partial\rho'/\partial t + w\, d\rho_0/dz = c_s^{-2}(\partial p'/\partial t - g\rho_0 w),$$

i.e.,

$$\partial\rho'/\partial t - g^{-1}\rho_0 N^2 w = c_s^{-2}\, \partial p'/\partial t, \tag{6.14.3}$$

where the buoyancy (or Brunt–Väisälä) frequency N for a compressible fluid is defined by

$$N^2 = -g(\rho_0^{-1}\, d\rho_0/dz + g/c_s^2). \tag{6.14.4}$$

The equivalence of this definition to the one given in Chapter 3 $\left[\text{Eq. (3.7.9)}\right]$ follows from the expression

$$\frac{1}{\rho}\frac{d\rho}{dz} = \frac{1}{\rho}\frac{\partial\rho}{\partial p}\frac{dp}{dz} + \frac{1}{\rho}\frac{\partial\rho}{\partial\theta}\frac{d\theta}{dz} + \frac{1}{\rho}\frac{\partial\rho}{\partial s}\frac{ds}{dz};$$

therefore,

$$\frac{1}{\rho_0}\frac{d\rho_0}{dz} = -\frac{g}{c_s^2} - \alpha'\frac{d\theta_0}{dz} + \beta'\frac{ds_0}{dz}, \tag{6.14.5}$$

where use has been made of the definitions (3.7.10), (3.7.11), (3.7.16), and the hydrostatic equation.

The linearized form of the mass conservation equation (6.14.2) is

$$\frac{\partial u}{\partial x} + \frac{\partial v}{\partial y} + \frac{\partial w}{\partial z} + \frac{1}{\rho_0 c_s^2}\left(\frac{\partial p'}{\partial t} - \rho_0 g w\right) = 0. \tag{6.14.6}$$

As before, p' and ρ' are the perturbation pressure and density, i.e., the deviations from the hydrostatic equilibrium values, and use has been made of the hydrostatic balance (4.5.18) for the undisturbed flow. The full set of equations governing the motion is now made up of (6.14.3), (6.14.6), and the momentum equations (6.4.4) and (6.4.5).

The perturbation energy equation is obtained by multiplying (6.4.4) by (u, v), (6.4.5) by w, (6.14.3) by $g^2(\rho' - p'/c_s^2)/\rho_0 N^2$, (6.14.6) by p', and adding the results. This gives

$$\frac{\partial}{\partial t}\left\{\frac{1}{2}\rho_0(u^2 + v^2 + w^2) + \frac{1}{2}\frac{g^2}{\rho_0 N^2}\left(\rho' - \frac{p'}{c_s^2}\right)^2 + \frac{1}{2}\frac{p'^2}{\rho_0 c_s^2}\right\}$$

$$+ \frac{\partial}{\partial x}(p'u) + \frac{\partial}{\partial y}(p'v) + \frac{\partial}{\partial z}(p'w) = 0. \tag{6.14.7}$$

Compared with the corresponding equation (6.7.1) for the incompressible case, the change is in the expression for perturbation potential plus internal energy. The part of this quantity associated with internal motion is now given by

$$A' = \iiint\left\{\frac{1}{2}\frac{g^2}{\rho_0 N^2}\left(\rho' - \frac{p'}{c_s^2}\right)^2 + \frac{1}{2}\frac{p'^2}{\rho_0 c_s^2}\right\}\,dx\,dy\,dz. \tag{6.14.8}$$

This is also called the available potential energy for an infinitesimal disturbance (see Section 7.8). An alternative form is that given in terms of the fluid particle displacement h. Since $Dh/Dt = w$, (6.14.3) gives

$$h = g(\rho' - p'/c_s^2)/\rho_0 N^2, \tag{6.14.9}$$

and so

$$A' = \iiint\left(\frac{1}{2}\rho_0 N^2 h^2 + \frac{1}{2}\rho_0^{-1}(p'/c_s)^2\right)\,dx\,dy\,dz. \tag{6.14.10}$$

The first term inside the brackets is associated with vertical displacements. The second term can be nonzero when there is no vertical displacement because pressure perturbations in a compressible medium can be produced by purely horizontal motion if this is convergent or divergent. Yet another form of (6.14.10) is in terms of potential temperature θ and concentration s (which can denote salinity or humidity). The conservation equations (4.1.8) and (4.1.9) reduce for small perturbations to

$$\theta' + h \, d\theta_0/dz = 0, \qquad s' + h \, ds_0/dz = 0, \tag{6.14.11}$$

or by (3.7.9), to

$$g(\alpha'\theta' - \beta's') + N^2h = 0, \tag{6.14.12}$$

which could also be deduced from (6.14.9). Thus (6.14.10) can also be written

$$A' = \int\int\int \left\{ \frac{1}{2} \rho_0(g/N)^2(\alpha'\theta' - \beta's')^2 + \frac{1}{2}\rho_0^{-1}(p'/c_s)^2 \right\} dx \, dy \, dz. \tag{6.14.13}$$

Two derived equations involving only w and p' can be obtained as in Section 6.4. The equation for the horizontal divergence obtained from (6.4.4) is altered only by the expression for the divergence in terms of w and p' that comes from (6.14.6). The result is

$$\rho_0 \frac{\partial}{\partial t}\left(\frac{\partial w}{\partial z} - \frac{g}{c_s^2} w \right) = \left(\frac{\partial^2}{\partial x^2} + \frac{\partial^2}{\partial y^2} - \frac{1}{c_s^2}\frac{\partial^2}{\partial t^2} \right) p', \tag{6.14.14}$$

and this replaces (6.4.7). The equation for the vertical motion that replaces (6.4.8) is obtained by eliminating ρ' between (6.4.5) and (6.14.3). The result is

$$\frac{\partial^2 w}{\partial t^2} + N^2w = -\frac{1}{\rho_0}\frac{\partial}{\partial t}\left(\frac{\partial p'}{\partial z} + \frac{g}{c_s^2} p' \right). \tag{6.14.15}$$

Equations (6.14.14) and (6.14.15) should be used rather than (6.4.7) and (6.4.8) when the motion covers a large range of heights. In such circumstances, however, a change of dependent variables is advantageous. The reason can be seen for motions for which the energy density terms in (6.14.7) do not change with height. Then the velocity increases with height in proportion with $\rho_0^{-1/2}$, whereas the pressure perturbation decreases in proportion with $\rho_0^{1/2}$. To remove this dependence, new variables (distinguished by capital letters) are defined by

$$\begin{aligned} (U, V, W) &= \rho_0^{1/2}(u, v, w), \\ P &= \rho_0^{-1/2}p'. \end{aligned} \tag{6.14.16}$$

In terms of the new variables, the equations become

$$\frac{\partial}{\partial t}\left(\frac{\partial W}{\partial z} - \Gamma W \right) = \left(\frac{\partial^2}{\partial x^2} + \frac{\partial^2}{\partial y^2} - \frac{1}{c_s^2}\frac{\partial^2}{\partial t^2} \right) P, \tag{6.14.17}$$

$$\frac{\partial^2 W}{\partial t^2} + N^2W = -\frac{\partial}{\partial t}\left(\frac{\partial P}{\partial z} + \Gamma P \right), \tag{6.14.18}$$

where

$$\Gamma = \frac{1}{2\rho_0}\frac{d\rho_0}{dz} + \frac{g}{c_s^2} = \frac{1}{2}\left(\frac{g}{c_s^2} - \frac{N^2}{g} \right). \tag{6.14.19}$$

(This Γ should not be confused with the adiabatic lapse rate, which was also denoted by Γ in Chapter 3).

The character of the solutions of these equations is most conveniently illustrated by the case of an isothermal atmosphere, i.e., one in which the temperature T_c is independent of height. The rest state in this case has the properties discussed in Section 3.5, i.e., pressure and density fall off exponentially with height [see (3.5.12) and (3.5.8)]:

$$p \propto \exp(-z/H_s), \qquad \rho \propto \exp(-z/H_s), \tag{6.14.20}$$

where the scale height H_s is defined by (3.5.13), i.e.,

$$H_s = RT_c/g, \tag{6.14.21}$$

where R is the gas constant. The sound speed c_s is given by (3.7.17), or the approximate expression (3.7.18), which would be exact for a perfect diatomic gas, namely,

$$c_s^2 = \tfrac{7}{5}RT_c = \tfrac{7}{5}gH_s. \tag{6.14.22a}$$

Consequently,

$$N^2 = \tfrac{2}{7}gH_s^{-1}, \qquad \Gamma = \tfrac{3}{14}H_s^{-1}, \tag{6.14.22b}$$

so (6.14.17) and (6.14.18) have constant coefficients and yield the dispersion relation

$$c_s^{-2}\omega^4 - \omega^2(k^2 + l^2 + m^2 + (N/c_s)^2 + \Gamma^2) + (k^2 + l^2)N^2 = 0. \tag{6.14.23}$$

This may be compared with (6.5.3), which is obtained in the limit $c_s \to \infty$, $\Gamma \to 0$.

Figure 6.17 shows contours of ω in the wavenumber plane. The curves for $\omega < N$ are hyperbolic, representing internal gravity waves. As $mH_s \to \infty$, these become the dispersion curves for an incompressible fluid as shown in Fig. 6.7. An additional set of curves is obtained for $\omega > N_A$, where

$$N_A = (N^2 + c_s^2\Gamma^2)^{1/2} \tag{6.14.24}$$

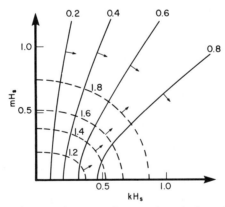

Fig. 6.17.   Curves of constant frequency in wavenumber space for an isothermal atmosphere (perfect diatomic gas). The wavenumber (k, l, m) is scaled by the scale height H_s and contours are of ω/N. Since the surfaces of constant frequency are surfaces of revolution about the m axis, only the cut through the k, m plane is shown. The group velocity, since it is the gradient of frequency in wavenumber space, is perpendicular to the surfaces of constant frequency and in the direction shown by the arrows. The surfaces with $\omega < N$ (solid lines) are hyperboloids representing internal gravity waves, and far from the origin approximate the cones of Fig. 6.7, which correspond to the incompressible limit. The surface with $\omega > N_A \approx 1.11N$ (broken lines) are ellipsoids and represent acoustic gravity or infrasonic waves.

is called the acoustic cutoff frequency. For a perfect diatomic gas, $N_A = 1.11N$. These are acoustic or infrasonic waves and will not be treated in detail here, the emphasis being on lower frequencies. Further information on these waves can be found in the works by Gossard and Hooke (1975), Tolstoy (1973), or Brekhovskikh (1980).

The dispersion characteristics of the internal gravity waves can be inferred from Fig. 6.17. As for the incompressible limit, the group velocity, which is normal to the curves of constant ω, always has a component toward the plane $m = 0$, i.e., waves with upward phase velocity have downward group velocity and *vice versa*. Also, the group velocity never exceeds the speed of sound, whereas in the incompressible approximation it can be infinite.

There is an important *additional* solution to (6.14.17) and (6.14.18) that was, in fact, the first to be discovered. This is the solution referred to in Section 5.1 that was discovered by Laplace (1778-1779) in discussing thermal oscillations in the atmosphere. He effectively stated that there is a mode with equivalent depth equal to the scale height of the atmosphere, and this will satisfy the tidal equations. This mode is now usually called a Lamb wave, following a more complete discussion by Lamb (1910). A feature of this mode is that the velocity is everywhere parallel to the earth's surface, i.e., $W = 0$. Thus (6.14.17) shows that P satisfies the wave equation, i.e., the same equation as for a homogeneous fluid of constant depth (5.6.10). The propagation speed is the speed of sound and has a value of about 300 m s^{-1}, i.e.,

$$\omega^2/(k^2 + l^2) = c_s^2. \tag{6.14.25}$$

By (6.14.22a) the *equivalent depth* for a diatomic perfect gas is 1.4 times the scale height, i.e., 40% greater than that Laplace assumed. (Laplace supposed the particles to retain a fixed temperature when displaced, rather than to retain a fixed entropy.)

The vertical structure of the perturbation pressure for a Lamb wave is determined by (6.14.18), which gives

$$P \propto \exp(-\Gamma z), \qquad p' \propto \exp(-(\Gamma + \tfrac{1}{2}H_s^{-1})z) \approx \exp(-\tfrac{5}{7}z/H_s), \tag{6.14.26}$$

i.e., the e-folding depth for p' is equal to the equivalent depth. This structure is typical of many waves, further examples of which will be encountered later in this book, in that the energy density falls off with distance from a boundary (in this case the earth's surface). The disturbance could therefore be said to be confined to the neighborhood of the boundary, and thus these waves are called "edge waves" or sometimes "surface waves." Surface gravity waves over deep water are another example of such waves.

In the study of motions of large horizontal scale and low frequency, the Lamb wave is an important feature of atmospheric motion and the most important new feature introduced by the inclusion of compressibility effects. The internal waves are modified, it is true, but their essential character is not changed. Figure 6.18 illustrates this by showing ω as a function of k. The dispersion curves for the internal waves are much the same as those for the incompressible case (cf. Fig. 6.7). However, the Lamb wave is an entirely new feature since the restoring forces that act are due to compression rather than buoyancy and the energy exchange is between kinetic and internal forms rather than between kinetic and potential forms.

Now consider what simplifications are introduced when the *hydrostatic approximation* is made. This has the effect of removing the second time derivative from the

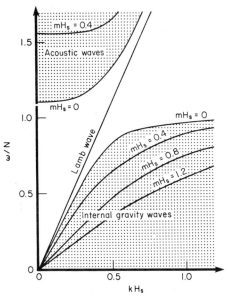

Fig. 6.18. Dispersion diagram for an isothermal atmosphere. Waves are possible only in the stippled regions and on the line for the Lamb wave. The curves in the stippled regions give the same information as that shown in Fig. 6.17 but in a different way. Acoustic waves (modified by gravity) have frequency greater than the acoustic cutoff frequency $N_A \approx 1.11N$, whereas internal waves have frequency less than the buoyancy frequency N. k is the horizontal wavenumber, m the vertical wavenumber, and H_s the scale height. The horizontal component of the group velocity is given by the slope of the curves of constant m. The maximum value for internal waves is $\frac{2}{9}(10)^{1/2}c_s \approx 0.9c_s$, this value being obtained when $m = 0$ and $k = 0$, i.e., for long waves.

left-hand side of (6.14.18), i.e., corresponds to the frequency being small compared with N. For an isothermal atmosphere, the dispersion relation (6.14.23) reduces to

$$\omega^2/(k^2 + l^2) = N^2/(m^2 + \Gamma^2 + (N/c_s)^2) \qquad (6.14.27)$$

and now includes only long internal waves that, as in the incompressible case, have a fixed horizontal propagation velocity. The Lamb wave satisfies the hydrostatic approximation exactly and has the separate dispersion relation (6.14.25).

For general density profiles, the phase speed and perturbation structure can be found by using separation of variables, as pointed out by Taylor (1936). Solutions have a form similar to that of (6.11.5), namely,

$$W = \hat{h}_v(z)\tilde{w}(x, y, t), \qquad P = \hat{p}_v(z)\tilde{\eta}(x, y, t). \qquad (6.14.28)$$

[Note that by (6.11.5) and (6.14.16), $\hat{h}_v = \rho_0^{1/2}\hat{h}$ and $\hat{p}_v = \rho_0^{-1/2}\hat{p}$.] The functions $\hat{p}_v, \hat{h}_v$ that define the vertical structure now satisfy

$$(c_e^{-2} - c_s^{-2})\hat{p}_v = d\hat{h}_v/dz - \Gamma\hat{h}_v, \qquad (6.14.29)$$

$$d\hat{p}_v/dz + \Gamma\hat{p}_v = -N^2\hat{h}_v, \qquad (6.14.30)$$

where c_e is a separation constant with dimensions of velocity as before. The equations for $\tilde{w}$ and $\tilde{\eta}$ are again the wave equations, i.e., (6.11.8) and (6.11.9), and the equivalent depth H_e [see Taylor (1936)] is again defined by (6.11.14). For the special case of an isothermal atmosphere, for which c_s is a constant, (6.14.29) and (6.14.30) have a special solution that corresponds to a Lamb wave for which $c_e = c_s$ and $\hat{h} = 0$. For a non-isothermal atmosphere, the Lamb wave is modified in structure and no longer has the property of involving no vertical motion. Details for a realistic atmosphere are discussed by Lindzen and Blake (1972).

As was found for the incompressible case discussed in Section 6.11, other variables can also be expressed in separable form, the full set being summarized by

$$u = \rho_0^{-1/2} \hat{p}_v(z) \tilde{u}(x, y, t)/g,$$

$$v = \rho_0^{-1/2} \hat{p}_v(z) \tilde{v}(x, y, t)/g,$$

$$p' = \rho_0^{1/2} \hat{p}_v(z) \tilde{\eta}(x, y, t), \qquad (6.14.31)$$

$$w = \rho_0^{-1/2} \hat{h}_v(z) \tilde{w}(x, y, t),$$

$$h = g(\rho' - p'/c_s^2)/\rho_0 N^2 = \rho_0^{-1/2} \hat{h}_v(z) \tilde{\eta}(x, y, t).$$

The expressions for p' and w come from (6.14.16) and (6.14.28), those for u and v are the analog of (6.11.11), whereas that for h follows from (6.14.3) and the expression for w. When these expressions are substituted in the equations, they all reduce to their counterparts for a shallow homogeneous layer of fluid. It can also be shown that the vertical integral of the long-wave approximation to the energy equation (6.14.7) for a single mode leads to the shallow-water form (5.7.4), provided that an equivalent density ρ_e is used, where ρ_e is given by

$$g\rho_e = \int (\hat{p}_v/c_e)^2 \, dz = \int (N^2 \hat{h}_v^2 + (\hat{p}_v/c_s)^2) \, dz. \qquad (6.14.32)$$

6.15   Examples of Adjustment in a Compressible Atmosphere

To illustrate the type of behavior to be expected, consider the case of an isothermal atmosphere. Then the internal wave solutions of (6.14.29) and (6.14.30) have the form [cf. (6.11.21)]

$$\hat{h}_v = m^{-1} \sin mz, \qquad (6.15.1)$$

$$\hat{p}_v = N^2(\cos mz - (\Gamma/m) \sin mz)/(m^2 + \Gamma^2), \qquad (6.15.2)$$

with

$$c_e^2 = N^2/(m^2 + \Gamma^2 + (N/c_s)^2), \qquad (6.15.3)$$

and the Lamb wave solution is

$$\hat{h}_v = 0, \qquad \hat{p}_v \propto e^{-\Gamma z}. \qquad (6.15.4)$$

Two examples will be considered. The first is the one discussed in Section 6.12 in which the fluid is initially at rest, but isopycnal surfaces are displaced from their equilibrium positions according to (6.12.1) and (6.12.6). The eigenfunction $\hat{h}_v$ is the same as that for the incompressible case, so the only difference in the solution for h arises from differences in the propagation speed c_e. Since c_e is now bounded, h remains equal to its initial value until the first wave arrives. Thereafter it is given by

(6.12.7) as before, but s, which is equal to the value of m for the wave arriving at point x at time t, is now given by

$$s^2 = (Nt/x)^2 - ((N/c_s)^2 + \Gamma^2) = (Nt/x)^2 - (1/2H_s)^2. \qquad (6.15.5)$$

The first expression comes from (6.15.3) with $x^2 = c_e^2 t^2$, and the second expression makes use of formulas (6.14.22) for a perfect diatomic gas. Equation (6.12.7) for h applies wherever s^2 is positive, and h has its initial value elsewhere. If D/H_s is small, the solution shown in Fig. 6.16b will not be altered noticeably, showing that the Boussinesq and incompressibility approximations are applicable in this case. If D is comparable with H_s, however, the predominant internal waves travel at speeds comparable with that of sound, and so the solution is significantly different. Figure 6.16d shows the solution for $D = 2H_s$, and the wave front associated with the fastest wave is quite apparent. Other differences from the incompressible case are in the expressions for u, p', etc. For the incompressible case, fixing h determines the pressure field completely, apart from an arbitrary constant. For the compressible case, an arbitrary Lamb wave may be added to the solution.

The second example is again one in which the fluid is initially at rest, but now the initial pressure perturbation is specified rather than particle displacements. Particular attention is given to the case in which the perturbation is confined to a small region near the ground. The expression for P obtained by superposition of the separable solutions (6.14.28) is by (6.15.2) and (6.15.4) of the form

$$P(x, z, t) = \int_0^\infty \tilde{\eta}(x, t; m)(m \cos mz - \Gamma \sin mz)\, dm + \tilde{\eta}_L(x, t)e^{-\Gamma z}, \quad (6.15.6)$$

where the subscript L denotes the part associated with the Lamb wave. To invert such an expression to find $\tilde{\eta}$ and $\tilde{\eta}_L$ in terms of P, use can be made of the orthogonality relationships, which can be deduced from (6.14.29) and (6.14.30). The result (which may be compared with expressions for inverting a Fourier transform) is

$$\tilde{\eta} = 2\pi^{-1}(m^2 + \Gamma^2)^{-1} \int_0^\infty (m \cos mz - \Gamma \sin mz)P\, dz,$$

$$\tilde{\eta}_L = 2\Gamma \int_0^\infty e^{-\Gamma z}P\, dz. \qquad (6.15.7)$$

Now consider the case in which the initial perturbation is confined to a thin region close to the ground and therefore can be represented in terms of a delta function by

$$P = \delta(z)G(x) \qquad \text{at} \quad t = 0, \qquad (6.15.8)$$

where G is equal to the integral of P over the whole depth of the atmosphere. In this case, (6.15.7) gives

$$\tilde{\eta}(x, 0; m) = 2mG(x)/\pi(m^2 + \Gamma^2), \qquad \tilde{\eta}_L(x, 0) = 2\Gamma G(x). \qquad (6.15.9)$$

Since $\tilde{\eta}$ satisfies the wave equation, the solutions for $\tilde{\eta}$ and $\tilde{\eta}_L$ at subsequent times are given by (5.6.11) with $c = c_e$ for the internal waves and $c = c_s$ for the Lamb waves, i.e.,

$$P = \Gamma e^{-\Gamma z}\{G(x + c_s t) + G(x - c_s t)\}$$

$$+ \pi^{-1} \int_0^\infty m(m^2 + \Gamma^2)^{-1}(m \cos mz - \Gamma \sin mz)$$

$$\times \{G(x + c_e t) + G(x - c_e t)\}\, dm. \tag{6.15.10}$$

The Lamb wave part is nondispersive, so the pulse created by the initial disturbances moves at constant speed c_s without change of shape. The part associated with the internal waves spreads out because different components with different m travel at different speeds. This spreading effect can be demonstrated easily for the case in which the initial disturbance is confined in the horizontal as well as the vertical, so that G is also a delta function. Then for positive x, the contribution to P from the integral in (6.15.10) is

$$\pi^{-1} m(m^2 + \Gamma^2)^{-1}(m \cos mz - \Gamma \sin mz)/(-t\, dc_e/dm),$$

evaluated at the point where $x = c_e t$, i.e., where $m = s$. In particular, the value of P at the ground is given, after use of (6.14.22) and (6.15.5), by

$$\frac{14}{3} x H_s P = \begin{cases} \delta(t' - 1), & t'^2 < \dfrac{49}{40} \\[2ex] -\dfrac{7 t'^2}{3\pi(t'^2 - 1)}\left(\dfrac{40 t'^2}{49} - 1\right)^{1/2}, & t'^2 > \dfrac{49}{40}, \end{cases} \tag{6.15.11}$$

where

$$t' = c_s t/x.$$

The pulse arriving at $t' = 1$ is the Lamb wave. The fastest internal wave arrives at $t' = 1.1$, causing a rapid buildup in pressure and then a much slower increase. However, the internal waves are quite weak compared with the Lamb wave and are therefore less likely to be seen.

6.16 Weak Dispersion of a Pulse

The discussion in Section 6.15 refers to an atmosphere that in the undisturbed state is isothermal and at rest. Although it gives a good first approximation to the behavior of disturbances in the real atmosphere, there are weak effects that can make large differences, given sufficient time, i.e., a sufficiently large number of wave periods. For instance, pressure pulses resulting from localized high-energy atmospheric events have been picked up all over the world and therefore have received considerable study. The events responsible for the observed pulses include the eruption of Krakatoa in 1883 (Symons, 1888), the impact of the great Siberian meteor in 1908 (Whipple, 1930), nuclear explosions (Donn and Shaw, 1967), and other less spectacular events

(Klostermeyer, 1977). The pulses travel at approximately the predicted speed (i.e., the speed of sound, or a little faster than most commercial jet aircraft), but there are small variations from place to place, as observed with the Krakatoa pulse (Symons, 1888), whose progress was traced for four transits around the globe. Another effect is due to variations of thermal structure and wind velocity with height. This causes weak dispersion of the waves, i.e., the phase speed ω/k is not constant but varies a little from the speed of sound c_s. Salby (1979) has shown that disturbance energy profiles corresponding to the 1976 U.S. Standard Atmosphere are not very different from those for the isothermal case, but there is a slow leak of energy due to upward radiation at large heights. Lindzen and Blake (1972) have incorporated dissipative effects and found decay times of 10–15 days.

The weak dispersion effect (Garrett, 1969) can be represented by an approximate expression for phase speed that is valid for small wavenumbers, namely,

$$\omega/k = c_s(1 - L^2k^2 \cdots), \tag{6.16.1}$$

where k is the horizontal wavenumber. Such an expression will be valid when k^{-1} is small compared with the length scale L, which depends on the way the density and velocity vary with height. Garrett estimated a value of 0.8 km for one particular case. The effect is quite like the weak dispersion of tsunamis due to the finite depth of the ocean. Figure 6.19a shows how the dispersion is observed in microbarograph records of pressure fluctuations produced by a nuclear explosion. Instead of a single pulse, a dispersed group of waves is observed, and Fig. 6.19 illustrates very well how the wave group spreads out or disperses with distance from the source.

Methods for calculating effects of dispersion are given in books on waves such as that by Whitham (1974, Chapter 11). If the initial pulse can be represented as a delta function, an expression for the weakly dispersed form of the pulse is not difficult to obtain. The expression for a delta function in terms of its Fourier components is [see, e.g., Lighthill (1958)]

$$\delta(x) = 2\pi^{-1} \int_0^\infty \cos(kx)\, dk. \tag{6.16.2}$$

After weak dispersion according to (6.16.1), the pulse will have the form

$$\tilde{\eta} = 2\pi^{-1} \int_0^\infty \cos(kx - kc_st + k^3c_sL^2t)\, dk. \tag{6.16.3}$$

The expression on the right-hand side is a definition of the Airy function [see, e.g., Abramowitz and Stegun (1964, Section 10.4)], i.e., the Airy function can be thought of as the shape of a weakly dispersed pulse. Thus (6.16.3) can be rewritten

$$\tilde{\eta} = 2(3c_sL^2t)^{-1/3}\, \text{Ai}\{(x - c_st)(3c_sL^2t)^{-1/3}\}. \tag{6.16.4}$$

The pulse arrives at a given point x at approximately $t = x/c_s$. When t has about this value, (6.16.4) is given approximately by

$$\tilde{\eta} = 2(3L^2x)^{-1/3}\, \text{Ai}\{(x - c_st)(3L^2x)^{-1/3}\}. \tag{6.16.5}$$

Thus weak dispersion causes the delta function on the right-hand side of (6.15.11) to

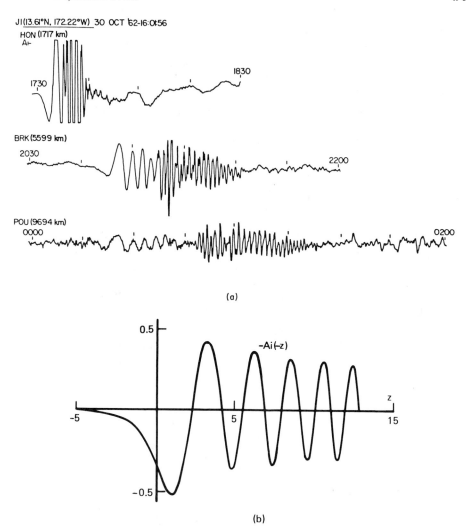

Fig. 6.19. (a) Microbarograph records following a nuclear test near Johnston Island in the Pacific Ocean. Honolulu is 1717 km from the source, Berkeley 5599 km, and Poughkeepsie 9694 km. The dispersive effects on the pulse are evident. [From Donn and Shaw (1967, Fig. 1).] (b) The theoretical solution for a weakly dispersed pulse, namely, the Airy function. Signs have been chosen for easy comparison with the observed pressure records.

be replaced by

$$2(x^2/3L^2)^{1/3}\,\mathrm{Ai}\{(1 - t')(x^2/3L^2)^{1/3}\}, \tag{6.16.6}$$

where t' is given by (6.15.11). The factor in front is the one required by the fact that the integral with respect to t' should be unity. It shows, incidentally, that the internal wave contribution to (6.15.11) is of order $(L/x)^{2/3}$ relative to the Lamb wave part, and therefore will usually be much too weak to be observed.

Figure 6.19b shows a graph of the Airy function that gives a fairly good approximation to the first part of the pulse as seen in the records of Fig. 6.19a. The fact that the first displacement is downward indicates a negative pulse. The waves spread out as (6.16.5) predicts and the observations agree with the predicted rule that the time scale should vanish like the one-third power of distance from the source. A value of L of 0.93 km fits the observations quite well.

Solutions for initial pulse shapes other than a delta function can be obtained by integration, e.g., an initial step function will give rise to a pulse with the shape of an integral of an Airy function. This would describe, for instance, the effect of weak dispersion for the problem considered in Section 5.6.

6.17   Isobaric Coordinates

When horizontal scales are large compared with vertical scales, i.e., in circumstances for which the hydrostatic approximation can be made, it is sometimes advantageous to replace the z-coordinate with another variable to use with x and y as independent coordinates. Sometimes potential temperature θ is used, in which case the variables (x, y, θ) are known as *isentropic* coordinates. More commonly, pressure p is used and then the variables (x, y, p) are known as *isobaric coordinates* (Sutcliffe and Godart, 1942; Eliassen, 1949). This set of coordinates is in fact widely used in meteorology, and the equations in these coordinates are derived below.

First consider the *hydrostatic equation*. This is used in the form (3.5.5), i.e., at fixed values of x and y,

$$dp = -\rho \, d\Phi = -\rho g \, dz.$$

Thus

$$\partial\Phi/\partial p = -1/\rho, \tag{6.17.1}$$

where Φ is the *geopotential* defined in Section 3.5, and this is the usual way of writing the hydrostatic equation in isobaric coordinates.

The fluid velocity (u, v, ϖ) in isobaric coordinates is defined as the rate of change of the isobaric coordinates of a fluid particle, i.e.,

$$(u, v, \varpi) = d(x, y, p)/dt, \tag{6.17.2}$$

and therefore the equations of motion need to be given in terms of (u, v, ϖ). These involve the derivative D/Dt, following a fluid particle that (cf. Section 4.1) is given by

$$\frac{D\gamma}{Dt} \equiv \frac{\partial\gamma}{\partial t} + \frac{\partial\gamma}{\partial x}\frac{Dx}{Dt} + \frac{\partial\gamma}{\partial y}\frac{Dy}{Dt} + \frac{\partial\gamma}{\partial p}\frac{Dp}{Dt} \equiv \frac{\partial\gamma}{\partial t} + u\frac{\partial\gamma}{\partial x} + v\frac{\partial\gamma}{\partial y} + \varpi\frac{\partial\gamma}{\partial p}. \tag{6.17.3}$$

Also, an expression is needed for the horizontal pressure gradient. Since

$$0 = \left(\frac{\partial p}{\partial x}\right)_p = \left(\frac{\partial p}{\partial x}\right)_z + \frac{\partial p}{\partial z}\left(\frac{\partial z}{\partial x}\right)_p,$$

it follows from the hydrostatic equation that

$$\left(\frac{\partial p}{\partial x}\right)_z = g\rho\left(\frac{\partial z}{\partial x}\right)_p = \rho\left(\frac{\partial\Phi}{\partial x}\right)_p. \tag{6.17.4}$$

The relationship is illustrated in Fig. 6.20, which also shows that a "high" in pressure on a level surface appears as a "high" in geopotential on an isobaric surface. For an inviscid nonrotating fluid, the horizontal components of the momentum equation (4.10.11) take the form

$$Du/Dt = -\partial\Phi/\partial x, \qquad Dv/Dt = -\partial\Phi/\partial y. \qquad (6.17.5)$$

The continuity equation is best derived from first principles by the methods used in Section 4.2. Consider a fluid volume that initially has rectangular projection in the horizontal with sides δx and δy and is bounded above and below by isobaric surfaces separated by a pressure difference δp. This corresponds to a vertical separation $\delta z = -\delta p/g\rho$ by the hydrostatic equation. The *mass* of the element is therefore

$$\rho\,\delta x\,\delta y\,\delta z = -\delta x\,\delta y\,\delta p/g.$$

The fractional rate of change of mass must be zero, and taking the limit as $\delta x\,\delta y\,\delta p \to 0$ therefore gives (cf. Section 4.2)

$$\partial u/\partial x + \partial v/\partial y + \partial\varpi/\partial p = 0. \qquad (6.17.6)$$

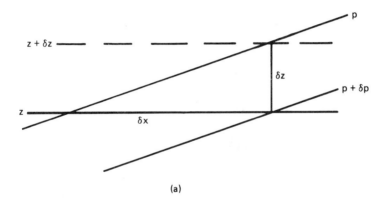

(a)

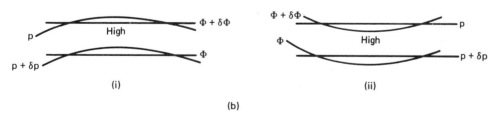

(b)

Fig. 6.20.   The relation between horizontal variations of pressure at a fixed level z and horizontal variations of level (or geopotential $\Phi = gz$) at a fixed pressure. (a) Two neighboring pressure surfaces in the x, z plane such that the change in pressure in a horizontal distance δx is δp, i.e., the horizontal pressure gradient at a fixed level is $\delta p/\delta x$. If δz is the vertical distance between these surfaces, it can be seen that the gradient of z on an isobaric surface is $\delta z/\delta x$. Since by the hydrostatic equation $\delta p = \rho g\,\delta z$, it follows that $\delta p/\delta x = \rho g\,\delta z/\delta x$, i.e., $\rho^{-1}(\partial p/\partial x)_z = g(\partial z/\partial x)_p = (\partial\Phi/\partial x)_p$. In the case shown in the figure, $(\partial p/\partial x)_z$ and $(\partial z/\partial x)_p$ are both positive. (b) Representation of a high-pressure region in the x, z or x, Φ plane (i). When the same situation is redrawn in the x, p plane, as shown in (ii), the feature appears as a high in geopotential on an isobaric surface.

This is the continuity equation in isobaric coordinates, and its simple form is one of the main reasons for using this coordinate system.

If the motion is isentropic and free of diffusive effects, Eqs. (4.10.8) and (4.10.9) are valid and the equation of state leads to the following form of (4.10.7):

$$D\rho/Dt = c_s^{-2} \, Dp/Dt = \varpi/c_s^2. \tag{6.17.7}$$

The motion is therefore governed by the five equations (6.17.1), (6.17.5)–(6.17.7) for the five variables u, v, ϖ, Φ, and ρ. The condition (4.11.3) satisfied at a material boundary is unaltered.

A variant of the above scheme is to use the logarithm of the pressure as the third coordinate since this is not too different from height in the atmosphere and leads to more convenient forms of solutions for small disturbances. The heightlike coordinate z_* is defined by

$$p = p_r \exp(-z_*/H_s), \tag{6.17.8}$$

where p_r is a constant reference pressure [other functions of pressure are sometimes used—see, e.g., Hoskins and Bretherton (1972)] and H_s is a fixed scale height whose value can be chosen at will (see below). It follows that w_*, the rate of change of the z_* coordinate of a fluid particle, is given by

$$w_* = -H_s\varpi/p \tag{6.17.9}$$

and that the derivative following a material element is

$$\frac{D}{Dt} \equiv \frac{\partial}{\partial t} + u\frac{\partial}{\partial x} + v\frac{\partial}{\partial y} + w_*\frac{\partial}{\partial z_*}. \tag{6.17.10}$$

The momentum equations (6.17.5) have the same form as before, and the continuity equation, obtained by substituting the expression (6.17.9) for ϖ in (6.17.6), is

$$\partial u/\partial x + \partial v/\partial y + \partial w_*/\partial z_* - w_*/H_s = 0. \tag{6.17.11}$$

It remains to consider the new forms of the hydrostatic and buoyancy equations. These forms depend on whether ρ or another quantity, such as temperature or potential temperature, is used as the state variable. For instance, the hydrostatic equation can be written

$$H_s \, \partial\Phi/\partial z_* = \rho^{-1}p \equiv RT_v. \tag{6.17.12}$$

The last expression, which comes from (3.1.14), *defines* the virtual temperature T_v (where R is the gas constant for dry air) and does not depend on the assumption of ideal gas behavior. For most applications to the atmosphere, however, the distinction between virtual temperature and actual temperature is unimportant, so T is often used in place of T_v. The equation for changes in T_v can be obtained by substituting the expression (6.17.12) for ρ in terms of T_v in (6.17.7). The result is

$$DT_v/Dt + \kappa w_* T_v/H_s = 0, \tag{6.17.13}$$

where

$$\kappa = 1 - \gamma^{-1} \tag{6.17.14}$$

and

$$\gamma = c_s^2/RT_v. \tag{6.17.15}$$

In the case of dry air with perfect gas behavior, (3.7.17) shows that γ is the quantity defined by (3.3.5), namely, the ratio c_p/c_v of specific heats, and so κ is the quantity defined by (3.7.3) with a value that is close to $\frac{2}{7}$ for air. Further treatment of the equation when expressed in terms of T_v can be found in the work by Holton (1975).

An alternative is to use potential temperature as the working thermodynamic variable because it satisfies the simple conservation equation (4.10.8), and this can be used in place of (6.17.13). However, the hydrostatic equation (6.17.12) becomes more complicated because it is necessary to use (3.1.15) to express T_v in terms of T and (3.7.4) to express T in terms of θ. For *dry* air, the result of the substitutions is

$$H_s \, \partial\Phi/\partial z_* = RT = R\theta(p/p_r)^\kappa. \tag{6.17.16}$$

The behavior of *small perturbations* from the rest state (which will be denoted by subscript 0) can be studied by the same methods as used in Section 6.14. However, a perturbation is now the deviation from the equilibrium value for the same pressure and will be denoted by a double prime. It is generally *different* from the deviation from the equilibrium value for the same *level*, which was denoted by a prime in Section 6.14. For instance, p' is the pressure perturbation from equilibrium at a fixed level, but p'' is zero by definition. The equivalent quantity in isobaric coordinates is the perturbation geopotential Φ'', which is related to p' through the hydrostatic equation by

$$p' = \rho_0 \Phi''. \tag{6.17.17}$$

Also, $\partial/\partial t$ in this section denotes the time derivative at a fixed pressure, whereas in Section 6.14, $\partial/\partial t$ was used to denote a different operator, namely, the time derivative at a fixed geopotential level.

From (6.17.5), the perturbation momentum equations are

$$\partial u/\partial t = -\partial\Phi''/\partial x, \qquad \partial v/\partial t = -\partial\Phi''/\partial y. \tag{6.17.18}$$

By taking the divergence and using the mass conservation equation (6.17.11), there results an equation that involves only w_* and Φ'', namely,

$$\frac{\partial}{\partial t}\left(\frac{\partial w_*}{\partial z_*} - \frac{w_*}{H_s}\right) = \left(\frac{\partial^2}{\partial x^2} + \frac{\partial^2}{\partial y^2}\right)\Phi''. \tag{6.17.19}$$

Another relation between w_* and Φ'' can be obtained from the hydrostatic and buoyancy equations. The former, which takes the form (6.17.12) or (6.17.16), yields

$$\partial\Phi''/\partial z_* = -p\rho''/H_s\rho_0^2 = RT_v''/H_s = \alpha_* g\theta'', \tag{6.17.20}$$

where α_* plays the role of an expansion coefficient and is given for dry air by

$$\alpha_* = (R/gH_s)(p/p_r)^\kappa = (R/gH_s)T_0/\theta_0 = T_0/T_r\theta_0. \tag{6.17.21}$$

At this point, and in the remainder of the section, the subscript r (indicating a reference value) will be used to denote the equilibrium value at $z_* = 0$, and the scale height H_s is now *defined* by

$$gH_s = p_r/\rho_r = RT_r = H_s \, \partial\Phi_0(0)/\partial z_*. \tag{6.17.22}$$

The subscript * is used to distinguish quantities defined in terms of log-pressure coordinates from the equivalent quantity when height is used as the coordinate.

The perturbation form of the thermodynamic equation (6.17.7), (6.17.13), or (4.10.8) is, with use of (6.17.9) where necessary,

$$N_*^2 w_* = (p/\rho_0^2 H_s)\, \partial\rho''/\partial t = -(R/H_s)\, \partial T_v''/\partial t$$

$$= -\alpha_* g\, \partial\theta''/\partial t = -\partial^2\Phi''/\partial t\, \partial z_*, \tag{6.17.23}$$

where the last expression comes from the hydrostatic equation (6.17.20) and gives the required relation between w_* and Φ''. The quantity N_*^2 is defined by

$$N_*^2 = (p/\rho_0 H_s)^2(d\rho_0/dp - c_s^{-2}) = (R/H_s)(dT_0/dz_* + \kappa T_0/H_s) = \alpha_* g\, d\theta_0/dz_*, \tag{6.17.24}$$

where T_0 refers to the equilibrium virtual temperature, and equality between the different expressions follows from (6.17.12) and (6.17.16). N_* may be related to the buoyancy (or Brunt–Väisälä) frequency N defined by (6.14.4) if use is made of the hydrostatic equation (3.5.8) that relates z and p. This shows that

$$N_*/N = p/\rho_0 g H_s = T_0/T_r = \partial z/\partial z_*, \tag{6.17.25}$$

use having been made of (6.17.22), (6.17.12) and (3.5.2). The quantity called the buoyancy frequency by Holton (1975) is N_*, not N as is normally the case. In the case of an isothermal atmosphere, in which (3.5.13) defines the scale height, N_* and N are the same.

It is useful to define the analog h_* of a displacement by

$$w_* = \partial h_*/\partial t, \tag{6.17.26}$$

in which case (6.17.23) yields

$$N_*^2 h_* = -(R/H_s)T_v'' = (p/\rho_0^2 H_s)\rho'' = -\partial\Phi''/\partial z_*. \tag{6.17.27}$$

The relation between h_* and h follows from the definition $\varpi = Dp/Dt$. One can use (6.17.9) and (6.17.26) to replace ϖ in this equation by $-pH_s^{-1}\,\partial h_*/\partial t$, whereas Dp/Dt can be equated, in the limit of infinitesimal perturbations, to $\partial p'/\partial t - \rho_0 g\, \partial h/\partial t$, use having been made of the hydrostatic equation (6.17.1). Consequently, the required relation is

$$-ph_*/H_s = p' - \rho_0 gh. \tag{6.17.28}$$

The perturbation equations are now (6.17.11), (6.17.18), (6.17.20), and (6.17.23), which can be reduced to the pair of equations (6.17.19) and (6.17.23), which involve w_* and Φ'' only. It is also possible to obtain the perturbation energy equations by adding $\rho_*(u, v)$ times (6.17.18), $\rho_*\Phi''$ times (6.17.11), and $N_*^{-2}\rho_*\,\partial\Phi''/\partial z_*$ times (6.17.23), where ρ_*, which plays a role like that of density, is defined by

$$\rho_* = \rho_r \exp(-z_*/H_s) = \rho_0 \partial z/\partial z_*, \tag{6.17.29}$$

the last equality following from (6.17.8), (6.17.12), (6.17.22), and (6.17.25). The result is

$$\frac{\partial}{\partial t}\left\{\frac{1}{2}\rho_*(u^2 + v^2 + (N_*^{-1}\,\partial\Phi''/\partial z_*)^2)\right\}$$

$$+ \frac{\partial}{\partial x}(\rho_*\Phi''u) + \frac{\partial}{\partial y}(\rho_*\Phi''v) + \frac{\partial}{\partial z_*}(\rho_*\Phi''w_*) = 0. \tag{6.17.30}$$

Now consider the boundary condition at the surface. If this is flat, the condition is (4.11.3) with $G = \Phi$, i.e., $D\Phi/Dt = 0$. The linearized form is

$$\partial\Phi''/\partial t + w_* \, d\Phi_0/dz_* = 0, \tag{6.17.31}$$

which should be applied at the surface, i.e., at a value of the pressure only slightly different from the equilibrium surface value. This slight difference can be ignored in the linearized theory. An alternative form follows from (6.17.22) and (6.17.26), namely,

$$\Phi'' + gh_* = 0 \qquad \text{at} \quad z_* = 0. \tag{6.17.32}$$

Strangely enough, this condition is of the same form as that which applies at the free surface of the ocean [see (6.11.15) and (5.2.11)]! It is this condition that gives rise to the Lamb wave solution, which therefore can be regarded as being analogous to the barotropic wave in the ocean. For changes that are slow compared with the Lamb wave speed, i.e., the speed c_s of sound, an equivalent of the rigid lid approximation (see Section 6.3) can be made, and (6.17.31) and (6.17.32) reduce to

$$w_* = h_* = 0 \qquad \text{at} \quad z_* = 0. \tag{6.17.33}$$

When this approximation applies, there is no real distinction at the level of linear theory between use of pressure coordinates and height coordinates apart from effects of the stretching factor $\partial z/\partial z_*$ given by (6.17.25) and the change of dependent variable defined by (6.17.17). Thus when oceanographers use as the dependent variable the pressure perturbation divided by the density, they are following an equivalent procedure to that of meteorologists who use perturbation geopotential as the dependent variable.

In Section 6.14, a scaling factor was inserted in (6.14.16) and a separation of variables was defined by (6.14.28), the combined effect being given by (6.14.31). The analog for isobaric coordinates is achieved by seeking solutions of the form

$$w_* = \rho_*^{-1/2}\hat{h}_*(z_*)\tilde{w}(x, y, t),$$
$$\Phi'' = \rho_*^{-1/2}\hat{\Phi}(z_*)\tilde{\eta}(x, y, t),$$
$$u = \rho_*^{-1/2}\hat{\Phi}(z_*)\tilde{u}(x, y, t)/g, \tag{6.17.34}$$
$$v = \rho_*^{-1/2}\hat{\Phi}(z_*)\tilde{v}(x, y. t)/g,$$
$$h_* = \rho_*^{-1/2}\hat{h}_*(z_*)\tilde{\eta}(x, y, t).$$

The equations for $\tilde{u}$, $\tilde{v}$, and $\tilde{\eta}$ are the shallow-water equations, i.e., (5.6.4)–(5.6.6), or equivalently, the equations for $\tilde{w}$ and $\tilde{\eta}$ are the wave equations given by (6.11.8) and (6.11.9), provided that $\hat{h}_*$ and $\hat{\Phi}$ satisfy [see (6.17.19), (6.17.26) and (6.17.27)]

$$\hat{\Phi}/c_e^2 = d\hat{h}_*/dz_* - \hat{h}_*/2H_s, \tag{6.17.35}$$
$$d\hat{\Phi}/dz_* + \hat{\Phi}/2H_s = -N_*^2\hat{h}_*, \tag{6.17.36}$$

where c_e is a separation constant with the dimensions of velocity. The depth that appears in the equivalent shallow-water equations is the equivalent depth $H_e = c_e^2/g$, not the scale height H_s. If $\hat{\Phi}$ is eliminated from the above pair of equations, the result is a simple second-order equation for $\hat{h}_*$, namely,

$$d^2\hat{h}_*/dz_*^2 + ((N_*/c_e)^2 - (1/2H_s)^2)\hat{h}_* = 0. \tag{6.17.37}$$

The boundary condition (6.17.32) at the ground requires that

$$\hat{\Phi} + g\hat{h}_* = 0 \qquad \text{at} \quad z_* = 0. \tag{6.17.38}$$

The solutions for an isothermal atmosphere are easily calculated since the solutions then have constant coefficients and N_* is identical to N. For the case of a perfect diatomic gas, for which (6.14.22b) applies, the internal wave solution has the form

$$g^{-1}\hat{\Phi} = \tfrac{3}{14}(mH_s)^{-1} \sin mz_* - \cos mz_*,$$
$$\hat{h}_* = \cos mz_* - (\tfrac{7}{2}mH_s + \tfrac{3}{8}(mH_s)^{-1}) \sin mz_*, \tag{6.17.39}$$

where

$$c_e^2 = N^2/(m^2 + (1/2H_s)^2), \tag{6.17.40}$$

which is the same result as that of (6.15.3). The Lamb wave solution corresponds to no vertical motion, so (6.17.38) is satisfied at all levels. This is consistent with (6.17.35) and (6.17.36), provided that

$$\hat{\Phi} = -gH_s \exp(-\Gamma z_*), \qquad \hat{h}_* = H_s \exp(-\Gamma z_*), \qquad c_e = c_s. \tag{6.17.41}$$

The above solutions are, of course, the same as those found in Section 6.14, but with a different form in the new coordinate system.

6.18   The Vertically Integrated Perturbation Energy Equation in Isobaric Coordinates

The perturbation energy equation (6.17.30) can be integrated vertically as was done in Sections 6.7 and 6.14. When this is done, the last term in (6.17.30) gives

$$[\rho_*\Phi''w_*],$$

where the brackets denote the difference between the values at the top and bottom of the region of integration. For the case of an atmosphere overlying a flat surface and extending to zero pressure ($z_* = \infty$), conditions (6.17.32) and (6.17.26) show that this term is equal to

$$\frac{\partial}{\partial t}\left(\frac{1}{2}g\rho_r h_*^2(0)\right)$$

Hence the integrated form of (6.17.30) is

$$\frac{\partial}{\partial t}\int \frac{1}{2}\rho_*(u^2 + v^2)\,dz_* + \frac{\partial}{\partial t}\left\{\frac{1}{2}\rho_r g h_*^2(0) + \int \frac{1}{2}\rho_* N_*^2 h_*^2\,dz_*\right\}$$
$$+ \frac{\partial}{\partial x}\int \rho_*\Phi''u\,dz_* + \frac{\partial}{\partial y}\int \rho_*\Phi''v\,dz_* = 0. \tag{6.18.1}$$

Since (see Eq. (6.17.29)) a mass element dM is given by

$$dM = \rho_0 \, dx \, dy \, dz = \rho_* \, dx \, dy \, dz_*, \tag{6.18.2}$$

the perturbation kinetic energy K' is [cf. (4.6.6)] given by

$$K' = \frac{1}{2} \int \int \int \rho_* (u^2 + v^2) \, dx \, dy \, dz_*, \tag{6.18.3}$$

with the contribution from vertical motion missing because of the hydrostatic approximation. For a particular mode, use of (6.17.34) shows that this reduces to the shallow-water form

$$K' = \frac{1}{2} \rho_e H_e \int \int (\tilde{u}^2 + \tilde{v}^2) \, dx \, dy, \tag{6.18.4}$$

provided that the equivalent density ρ_e is defined by

$$g\rho_e = \int \left(\frac{\hat{\Phi}}{c_e} \right)^2 dz_*. \tag{6.18.5}$$

The perturbation internal plus potential energy (or perturbation available potential energy) A' is given by (6.14.10). The corresponding form in isobaric coordinates can be found by use of (6.17.17), (6.17.28), and (6.18.2), which give

$$A' = \frac{1}{2} \int \int \int \left\{ \frac{N^2}{g^2} \left(\Phi'' + \frac{ph_*}{\rho_0 H_s} \right)^2 + \frac{\Phi''^2}{c_s^2} \right\} \rho_0 \, dx \, dy \, dz.$$

Expanding the quadratic, putting $N^2/g + g/c_s^2 = -(1/\rho_0) \, d\rho_0/dz$ by (6.14.4), using (6.17.12) and (6.17.25), and putting $N_*^2 h_* = -\partial \Phi''/\partial z_*$ by (6.17.27), there results

$$A' = \frac{1}{2} \int \int \int \left\{ -\frac{d(\rho_0 \Phi''^2/g)}{dz_*} + \rho_* N_*^2 h_*^2 \right\} dx \, dy \, dz_*.$$

Integrating the derivative and using the boundary condition (6.17.38), there results

$$A' = \int \int \left\{ \frac{1}{2} \rho_r g h_*^2(0) + \int \frac{1}{2} \rho_* N_*^2 h_*^2 \, dz_* \right\} dx \, dy, \tag{6.18.6}$$

as would be expected from the form of (6.18.1). For a particular mode, substitution from (6.17.34) shows that this reduces to the shallow-water from $A' = \frac{1}{2} g \rho_e \int \int \tilde{\eta}^2 \, dx \, dy$, provided that ρ_e satisfies

$$g\rho_e = g\hat{h}_*^2(0) + \int N_*^2 \hat{h}_*^2 \, dz_*. \tag{6.18.7}$$

This can be shown to be equivalent to (6.18.5) by substracting $\hat{\Phi}$ times (6.17.35) from $\hat{h}_*$ times (6.17.36), integrating with respect to z_*, and using the boundary condition (6.17.38). Substitution from (6.17.34) also shows that the flux terms reduce to the shallow-water form, e.g.,

$$\int \rho_* \Phi'' u \, dz_* = \rho_e g H_e \tilde{\eta} \tilde{u}. \tag{6.18.8}$$

As in Section 6.14, alternative expressions for A' can be found in terms of potential temperature θ and concentration s (salinity or humidity). The conservation equations (4.1.8) and (4.1.9) now reduce for small perturbations to

$$\theta'' + h_* \, d\theta_0/dz_* = 0, \qquad s'' + h_* \, ds_0/dz_* = 0, \qquad (6.18.9)$$

which have form similar to (6.14.11). However, θ'' is the perturbation from the equilibrium value at the same *pressure* and is not the same as θ'. In the case of a Lamb wave, for instance, for which the motion is purely horizontal, there is no change in potential temperature at any given level, so θ' vanishes; but since the pressure changes, θ'' is nonzero.

An alternative expression, obtained from (6.18.9) after use of (3.7.9) and (6.17.25), is

$$g(\alpha'\theta'' - \beta's'')/N + N_* h_* = 0. \qquad (6.18.10)$$

Substituting in (6.18.6) and using (6.17.28) and (6.17.22) at the surface give

$$A' = \int\int \left\{ \frac{1}{2} \frac{p'(0)^2}{g\rho_r} + \int \frac{1}{2} \rho_* \left(\frac{g}{N} \right)^2 (\alpha'\theta'' - \beta's'')^2 \, dz_* \right\} dx \, dy. \qquad (6.18.11)$$

In the case of dry air ($s'' \equiv q'' = 0$) that is treated as a perfect gas, $\alpha' = 1/\theta_0$ by (3.7.14) and $N^2 = (g/\theta_0) \, d\theta_0/dz$ by (3.7.15). Using also the hydrostatic equation and the definition (6.17.8) of z_* to convert to pressure derivatives, and (6.17.22) for the definition of H_s, (6.18.11) becomes

$$A' = \int\int \left\{ \frac{1}{2} \frac{p'(0)^2}{g\rho_r} + \int_0^{p_r} \frac{1}{2} \frac{\theta''^2}{(-g\rho\theta_0 \, d\theta_0/dp)} \, dp \right\} dx \, dy. \qquad (6.18.12)$$

This is a form that is useful for calculating A' for the atmosphere, along with the perfect gas laws (3.1.2) and (3.7.4), which gives density ρ in terms of pressure and potential temperature, namely,

$$1/\rho = RT/p = R\theta p^{\kappa-1}/p_r^\kappa \qquad (6.18.13)$$

with κ given by (6.17.14). An expression in terms of temperature rather than potential temperature can be derived from this, or simply by substitution from (6.17.27) in (6.18.6) and using (6.18.2). Alternative forms follow from (6.17.25), the ideal gas equation (6.18.13), and expressions for N_* [see (6.17.24)] or N [see (6.14.4)]. Examples are

$$A' = \int\int \frac{1}{2} \frac{p'(0)^2}{g\rho_r} \, dx \, dy + \int\int\int \frac{1}{2} \frac{g^2 T_v''^2}{N^2 T_0^2} \, dM$$

$$= \int\int \frac{1}{2} \frac{p'(0)^2}{g\rho_r} \, dx \, dy + \int\int\int \frac{R\theta_0 T_v''^2 \, dM}{-2T_0 p \, d\theta_0/dp} . \qquad (6.18.14)$$

Expressions like this have been used to estimate the available potential energy of the atmosphere.

Chapter Seven

Effects of Rotation

7.1 Introduction

In the seventeenth century, a picture of how the atmosphere is set in motion began to emerge with Halley's (1686) work. However, arguments that neglected the rotation of the earth failed to explain the easterly component of the trade winds. Hadley (1735) showed how rotation (see Section 2.3) could explain this, using the concept of conservation of angular momentum. Laplace (1778-1779) recognized the importance of rotation in his theory of the tides and developed the necessary equations for studying rotation effects. Despite these equations being available for so many years, much of the work (based on these equations) that gives a proper foundation for the understanding of rotation effects is quite recent. One reason for the delay is the difficulty in setting up experiments like that of Marsigli in a rotating system [see, e.g., Saunders (1973)].

The problem that tells us a great deal about this question is the one discussed in Section 5.6, i.e., the one associated with Marsigli's experiment of adjustment of a fluid under gravity, but now with rotation effects included. The question of how a fluid, not initially in equilibrium, adjusts in a uniformly rotating system was not completely discussed until the time of Rossby (1938a), although transient wave solutions had been considered much earlier by Kelvin (Thomson, 1879). In a series of papers, Rossby (1936, 1937, 1938a,b, 1940; Rossby et al., 1939) was concerned with how the mass and resulting pressure distributions in the ocean and atmosphere are established. In particular, he studied a problem in which momentum was supposed to be put into the ocean to give a nonequilibrium velocity distribution. Rossby (1938a) then considered the process of adjustment to equilibrium. A similar problem is introduced in the next section, and the rest of the chapter is devoted to the repercussions of this.

A key feature of the adjustment process in a rotating fluid is that the fluid adjusts rapidly (in a time of the order of the rotation period) to an equilibrium that is *not* a state of rest and contains more potential energy than does the rest state. In fact, very little of the potential energy initially present may be converted into kinetic energy. Also, the equilibrium state achieved (called a geostrophic equilibrium) cannot be found by solving the steady-state equations because these are degenerate in that any solution of the momentum equations satisfies the continuity equation exactly. It is this degeneracy, exemplified by the fact that the equilibrium fields of mass and momentum are related to each other, that causes the difficulties with which Rossby was concerned.

The equilibrium state achieved thus depends on the initial state, and Rossby showed the connection between the two states through conservation of a quantity he called potential vorticity. Using this property, the final state can be found, and this is shown in Section 7.2. Details of the transient motions require further analysis, and this is done in Section 7.3.

The analysis presented in this chapter is for a fluid that is rotating with uniform angular velocity about a vertical axis. However, application of the results to the atmosphere and ocean is possible in an approximate sense, and this point is discussed in Section 7.4. Section 7.5 is about the fundamental horizontal length scale that appears in problems dealing with adjustment under gravity of a rotating fluid. This is called the Rossby radius of deformation. Since the analysis can be applied to any of the normal modes of a stratified fluid, there is an infinite set of Rossby radii, one Rossby radius being associated with each of the modes.

The equilibrium solution is discussed in Sections 7.6 and 7.7. The large-scale motion in the ocean and atmosphere is nearly always close to such an equilibrium, and the implied connection between mass and velocity fields is of great importance in practice. In fact, much of our knowledge of the circulation of the ocean and atmosphere was deduced from the mass distribution before direct measurements were made. The relationship is used a great deal both as a means of estimating the velocity field and as an approximation in theoretical studies.

The discussion of energetics is taken up again in Section 7.8, which is concerned with the concept of available potential energy. This is the difference between the internal plus potential energy at any time and the minimum value to which it could be reduced by an inviscid isentropic rearrangement of fluid particles. The quantity is therefore a valuable measure of how much kinetic energy is potentially obtainable, and is widely used in studies of the circulation of the atmosphere and ocean.

Sections 7.9–7.12 are about the concept of vorticity and the results concerning circulation and potential vorticity that are of such great utility for rotating fluids. It is perhaps unfortunate that the name "potential vorticity" is given to more than one quantity, but in any given context it is usually quite clear which quantity is referred to! One form is that used for homogeneous shallow layers of fluid and the appropriate conservation equation is derived in Section 7.10. The potential vorticity in this case is defined as the total vorticity (assumed to be close to vertical) divided by the depth. A different form is appropriate in the continuously stratified case, and this is derived in Section 7.11. The conservation relation in this case requires no assumptions about the direction of the vorticity or about the ratio of horizontal to vertical scales, and the conserved quantity is called Ertel's potential vorticity. Section 7.12 discusses the perturbation forms of the conservation equations for both types of potential vorticity. Finally, in Section 7.13 there is a discussion of an important practical problem for numerical weather prediction, namely, the initialization of fields, because this problem has much to do with the ideas developed in the remainder of the chapter.

7.2   **The Rossby Adjustment Problem**

In Section 5.6, the adjustment under gravity of a homogeneous shallow layer of fluid was considered, the particular case being one in which the fluid was initially at rest but had a discontinuity (or discontinuities) in surface level. Now the same problem will be considered for a rotating fluid, i.e., one that is initially at rest *relative to a frame of reference rotating with uniform angular velocity* $\frac{1}{2}f$ about a vertical axis. The motion is considered relative to this frame and is supposed to be a small perturbation from the state of relative rest at all times. The z axis is vertical; the bottom $z = -H$ is horizontal (i.e., a geopotential surface); and the surface elevation $z = \eta$ relative to a geopotential surface is assumed to be small. The horizontal scale is assumed to be large compared with the depth, so that the hydrostatic approximation can be made.

The equations are the same as those in Section 5.6 apart from the addition of the Coriolis acceleration $(-fv, fu)$, which produces the effects of rotation (see Section 4.5.1). Thus the momentum equation (4.10.11), after use of the hydrostatic equation [see (5.6.3)], gives

$$\partial u/\partial t - fv = -g\,\partial\eta/\partial x, \qquad (7.2.1)$$

$$\partial v/\partial t + fu = -g\,\partial\eta/\partial y. \qquad (7.2.2)$$

Since η is independent of z, the velocity (u, v) is independent of depth as in the non-rotating case. The letter f is used for *twice* the rotation rate to avoid having factors of 2 appear in these equations. The continuity equation is (5.6.6), namely,

$$\partial\eta/\partial t + H(\partial u/\partial x + \partial v/\partial y) = 0. \qquad (7.2.3)$$

The method of dealing with these equations in the nonrotating case was to take the divergence of the momentum equations $[\partial/\partial x$ of (7.2.1) plus $\partial/\partial y$ of (7.2.2)] and substitute from (7.2.3) for the horizontal divergence $\partial u/\partial x + \partial v/\partial y$. In the rotating case this gives

$$\partial^2\eta/\partial t^2 - c^2(\partial^2\eta/\partial x^2 + \partial^2\eta/\partial y^2) + fH\zeta = 0, \qquad (7.2.4)$$

where c^2 is given by (5.5.4), namely,

$$c^2 = gH, \qquad (7.2.5)$$

and

$$\zeta = \partial v/\partial x - \partial u/\partial y \qquad (7.2.6)$$

is the *relative vorticity* of the fluid, i.e., the vertical component of the vorticity relative to the rotating frame (the horizontal components are identically zero). When $f = 0$, this gives an equation in one variable η only, namely, the wave equation (5.6.10). In the rotating case, this equation points to the necessity of considering how the relative vorticity changes.

7.2.1   Conservation of Potential Vorticity

Equations (7.2.1)–(7.2.3) were given by Kelvin (Thomson, 1879) in his paper "On gravitational oscillations of rotating water," in which he sought to simplify Laplace's tidal theory by considering "an area of water so small that the equilibrium figure of its surface is not sensibly curved." From these equations he derived an equation that is of fundamental importance in the theory of rotating fluids. This is obtained in two steps. First, the curl of the momentum equations $[\partial/\partial y$ of (7.2.1) minus $\partial/\partial x$ of (7.2.2)] eliminates η and gives the vorticity equation

$$\partial \zeta / \partial t + f(\partial u / \partial x + \partial v / \partial y) = 0, \tag{7.2.7}$$

i.e., the rate of change of ζ/f is equal to minus the horizontal divergence. Second, the continuity equation (7.2.3) is used to eliminate the horizontal divergence, to give

$$\frac{\partial}{\partial t} \left(\frac{\zeta}{f} - \frac{\eta}{H} \right) = 0. \tag{7.2.8}$$

The fact that this equation is easily integrated with respect to time is a very powerful result. Equation (7.2.8) is in fact a linearized form of the equation (to be considered later) expressing the conservation of *potential vorticity* for a homogeneous rotating fluid. The quantity Q', defined by

$$Q' = \zeta / H - f\eta / H^2, \tag{7.2.9}$$

may be called the perturbation potential vorticity, and (7.2.8) expresses the fact that Q' retains its initial value at each point for all time, i.e.,

$$Q'(x, y, t) = Q'(x, y, 0). \tag{7.2.10}$$

This infinite memory of an inviscid rotating fluid can be exploited, as will be shown, to find the final equilibrium solution for a particular initial state *without considering* details of the transient motion at finite times. Kelvin (Thomson, 1879), being interested only in oscillations, took Q' to be zero, whereas the most interesting cases of adjustment to equilibrium are those for which Q' is nonzero. Solutions with nonzero Q' were apparently not considered until the time of Rossby (1938a).

The particular initial condition to be considered here is the same as that in Section 5.6, namely, $u = v = 0$ and surface elevation given by (5.6.13), i.e.,

$$\eta = -\eta_0 \, \text{sgn}(x). \tag{7.2.11}$$

(This is an initial condition different from that considered by Rossby, but the analysis is similar.) The integral of (7.2.8) in this case is

$$\zeta / f - \eta / H = (\eta_0 / H) \, \text{sgn}(x), \tag{7.2.12}$$

and substitution in (7.2.4) gives an equation for η alone, namely,

$$\partial^2\eta/\partial t^2 - c^2(\partial^2\eta/\partial x^2 + \partial^2\eta/\partial y^2) + f^2\eta = -fH^2Q'(x, y, 0)$$
$$= -f^2\eta_0\, \text{sgn}(x). \qquad (7.2.13)$$

7.2.2 The Steady Solution: Geostrophic Flow

If the gravitational adjustment processes lead ultimately to a steady state, that state will be given by the time-independent solution of (7.2.13). Since the initial condition is independent of y, the solution at all subsequent times can be assumed to be independent of y, and thus the vorticity ζ is equal to $\partial v/\partial x$. Furthermore, a steady solution of (7.2.1) and (7.2.2) must entail a balance between the Coriolis acceleration $(-fv, fu)$ and the pressure gradient. This is known as a *geostrophic balance*

$$fu = -g\,\partial\eta/\partial y, \qquad fv = g\,\partial\eta/\partial x, \qquad (7.2.14)$$

and has the property that the flow is along contours of constant pressure (i.e., along isobars, as is familiar from weather maps).

The steady-state solution has a very special property in that *any* solution satisfying the geostrophic balance happens to satisfy the time-independent version of the continuity equation (7.2.3) *exactly*, i.e., is nondivergent with

$$\partial u/\partial x + \partial v/\partial y = 0. \qquad (7.2.15)$$

An alternative way of viewing this result is to use (7.2.15) to introduce a stream function ψ such that

$$u = -\partial\psi/\partial y, \qquad v = \partial\psi/\partial x. \qquad (7.2.16)$$

Then the geostrophic balance may be written

$$f\,\partial\psi/\partial y = g\,\partial\eta/\partial y, \qquad f\,\partial\psi/\partial x = g\,\partial\eta/\partial x. \qquad (7.2.17)$$

If η is eliminated from this pair of equations in order to obtain an equation for ψ, i.e., if the y derivative of the second equation is subtracted from the x derivative of the first, all that emerges is the trivial statement that zero equals zero. In fact (7.2.17) shows that the stream function (with suitable choice of reference value) is related to pressure perturbation by

$$f\psi = g\eta = p'/\rho. \qquad (7.2.18)$$

Any distribution $\eta(x, y)$ of surface elevation gives a stream function ψ by (7.2.18) that satisfies all the steady-state equations.

In this sense, the steady-state equations are *degenerate* and cannot yield the ultimate steady solution by themselves. An added piece of information is required, and this is the fact that each element of fluid retains its initial potential vorticity, i.e., (7.2.10) is satisfied, which, for the special case being considered, takes the form (7.2.12). For a geostrophically balanced flow, substitution of (7.2.14) in (7.2.6) shows that the vorticity is given by

$$\zeta = f^{-1}g(\partial^2\eta/\partial x^2 + \partial^2\eta/\partial y^2), \qquad (7.2.19)$$

and so (7.2.9) and (7.2.10) yield the steady-state version of (7.2.13),

$$-c^2(\partial^2\eta/\partial x^2 + \partial^2\eta/\partial y^2) + f^2\eta = -fH^2Q'(x, y, 0). \tag{7.2.20}$$

For the present case, this gives

$$-c^2\, d^2\eta/dx^2 + f^2\eta = -f^2\eta_0\, \mathrm{sgn}(x). \tag{7.2.21}$$

The solution η that is continuous and antisymmetric about $x = 0$ is given by

$$\frac{\eta}{\eta_0} = \begin{cases} -1 + e^{-x/a} & \text{for} \quad x > 0 \\ 1 - e^{x/a} & \text{for} \quad x < 0, \end{cases} \tag{7.2.22}$$

where

$$a = c/|f| = (gH)^{1/2}/|f| \tag{7.2.23}$$

is a length scale of fundamental importance for the behavior of rotating fluids subject to gravitational restoring forces. It is called the Rossby *radius of deformation*, following the name given by Rossby (1938a, p. 242), or simply the "Rossby radius" or the "radius of deformation." The modulus sign is used in (7.2.23) to ensure that a be a positive quantity since f can have either sign.

The velocity field associated with the solution (7.2.22) follows from the geostrophic equation (7.2.14), which gives $u = 0$ and

$$v = -(g\eta_0/fa)\exp(-|x|/a). \tag{7.2.24}$$

The flow is *not* in the direction of the pressure gradient, but at right angles, i.e., along contours of surface elevation that are parallel to the line of the initial discontinuity. The solution is depicted in Fig. 7.1.

7.2.3 Energy Considerations

The energy equations for rotating shallow-water motion may be found by the same methods as those in Section 5.7. In particular, the mechanical energy equation is

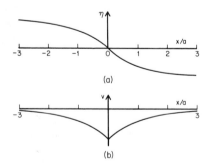

(a)

(b)

Fig. 7.1. The geostrophic equilibrium solution corresponding to adjustment from an initial state that is one of rest but has uniform infinitesimal surface elevation $-\eta_0$ for $x > 0$ and elevation η_0 for $x < 0$. (a) The equilibrium surface level η, which tends toward the initial level as $x \to \pm\infty$. The unit of distance in the figure is the Rossby radius $a = (gH)^{1/2}/f$, where g is the acceleration due to gravity, H the depth of fluid, and f twice the rate of rotation of the system about a vertical axis. (b) The corresponding equilibrium velocity distribution, there being a "jet" directed *along* the initial discontinuity in level with maximum velocity equal to $(g/H)^{1/2}$ times η_0.

obtained by adding ρHu times (7.2.1) to ρHv times (7.2.2). This operation *eliminates* the terms that come from the Coriolis acceleration, so that rotation terms do not appear *explicitly* in the energy equations, which therefore have exactly the same form as those in Section 5.7. However, the solution of the adjustment problem is drastically affected by rotation, and thus the energy changes in the rotating case are quite different from those of the nonrotating case discussed in Section 5.7.

Consider first the perturbation potential energy. This is infinite at the initial moment but, unlike the nonrotating case, it is *still* infinite when the steady equilibrium solution is established (assuming such an equilibrium does occur). However, the *change* in potential energy per unit length is finite and is given by

$$\text{P.E. released per unit length} = 2 \cdot \tfrac{1}{2}\rho g \eta_0^2 \int_0^\infty \{1 - (1 - e^{-x/a})^2\} \, dx$$

$$= \tfrac{3}{2}\rho g \eta_0^2 a. \tag{7.2.25}$$

In the nonrotating case, *all* the potential energy available in the initial perturbation is converted into kinetic energy. For the rotating case, only a finite amount of potential energy is released. The amount of kinetic energy per unit length found in the equilibrium solution is given by

$$\text{K.E. per unit length} = 2 \cdot \tfrac{1}{2}\rho H g^2 \eta_0^2 (fa)^{-2} \int_0^\infty e^{-2x/a} \, dx$$

$$= \tfrac{1}{2}\rho g \eta_0^2 a. \tag{7.2.26}$$

This is only one-third of the potential energy released! What happens to the other two-thirds? Rossby (1938a, p. 244) suggested that a fluid particle must "continue its displacement beyond the equilibrium point until an excessive pressure gradient develops which forces it back. An *inertia* oscillation around the equilibrium position results." These speculative comments are fairly close to the truth, but give the mistaken impression that an equilibrium solution is never reached in any finite domain. What really happens will be found in Section 7.3, where details of the transients will be calculated.

7.2.4 Summary

The problem considered above, even though only partially completed, gives a great many insights into the behavior of rotating fluids responding to gravitational forces. Five notable features are listed below, and various concepts arising from these results are discussed in more detail in later sections of this chapter.

(a)   The energy analysis indicates that *energy is hard to extract* from a rotating fluid. In the problem studied, there was an infinite amount of potential energy available for conversion into kinetic energy, but only a finite amount of this available energy was released. The reason was that a geostrophic equilibrium was established, and such an equilibrium retains potential energy—an infinite amount in the case studied!

(b)   The steady equilibrium solution is *not* one of rest, but is a *geostrophic balance,* i.e., a balance between the Coriolis acceleration and the pressure gradient divided by density.

(c)   The steady solution is *degenerate* in the sense that any velocity field in geostrophic balance satisfies the continuity equation exactly. Therefore the steady solution cannot be found by looking for a solution of the steady-state equations—some other item of information is required.

(d)   This information is supplied by the conservation of potential vorticity principle, i.e., the potential vorticity of each fluid element is the same as that at the initial instant. With this knowledge, a steady solution can be found.

(e)   The equation determining this steady solution contains a length scale a, called the Rossby radius of deformation, which is equal to $c/|f|$, where c is the wave speed in the absence of rotation effects, i.e., $(gH)^{1/2}$. If f tends to zero, a tends to infinity, indicating that for length scales small compared with a, rotation effects are small, whereas for scales comparable with or large compared with a, rotation effects are important.

7.3   The Transients

To complete the solution of the adjustment problem, i.e., of (7.2.13), it is necessary to add a solution of the homogeneous equation

$$\partial^2\eta/\partial t^2 - c^2(\partial^2\eta/\partial x^2 + \partial^2\eta/\partial y^2) + f^2\eta = 0 \tag{7.3.1}$$

to a particular solution, which can be taken as the steady solution $\eta_{\text{steady}}(x)$, given by (7.2.22). The solution of (7.3.1) must satisfy the initial condition

$$\eta = -\eta_0 \operatorname{sgn}(x) - \eta_{\text{steady}},$$

i.e.,

$$\eta = -\eta_0 e^{-|x|/a} \operatorname{sgn}(x) \qquad \text{at} \quad t = 0. \tag{7.3.2}$$

Equation (7.3.1) is sometimes known as the Klein–Gordon equation (Morse and Feshbach, 1953) because of applications in physics, and Morse and Feshbach discuss an analog provided by a stretched string embedded in a rubberized medium. The equation is also discussed by Whitham (1974). The transient solution for Rossby's original problem was found by Cahn (1945) and is discussed by Blumen (1972).

Equation (7.3.1) has wavelike solutions of the form

$$\eta \propto \exp i(kx + ly - \omega t), \tag{7.3.3}$$

which on substitution in (7.3.1) gives the dispersion relation

$$\omega^2 = f^2 + \kappa_H^2 c^2, \tag{7.3.4}$$

where

$$\kappa_H^2 = k^2 + l^2 \tag{7.3.5}$$

is the square of the horizontal wavenumber. Waves with this dispersion relation (k, l real) will be referred to as "Poincaré waves" here, although this name is sometimes reserved for the subset that satisfies the boundary conditions for a channel (see

Chapter 10). Despite the name, such waves were first discussed by Kelvin (Thomson, 1879). In meteorology, they are usually referred to simply as gravity waves, with rotation effects being understood. A graph of the dispersion relation is shown in Fig. 7.2. It can be seen from the dispersion relation that the properties of these waves depend on how the wavelength compares with the Rossby radius. The limiting cases are as follows:

(i)   Short waves ($\kappa_H a \gg 1$), i.e., waves short compared with the Rossby radius, for which (7.3.4) becomes approximately

$$\omega \sim \kappa_H c; \tag{7.3.6}$$

hence "short" waves are ordinary nondispersive shallow-water waves. It will be recalled, however, that shallow-water theory requires that the waves have horizontal scale large compared with the depth, so these waves have the above form only when the Rossby radius is large compared with the depth. This condition is satisfied in the atmosphere and ocean (see Section 7.4).

(ii)   Long waves ($\kappa_H a \ll 1$), i.e., waves long compared with the Rossby radius, for which (7.3.4) gives approximately

$$\omega \sim f, \tag{7.3.7}$$

i.e., the frequency is approximately constant and equal to f or twice the rotation rate. In this limit, gravity has no effect, so fluid particles are moving under their own inertia. For this reason f is often called the "*inertial*" *frequency.*

The group velocity $\mathbf{c}_g$ of Poincaré waves is equal to the slope of the dispersion curve in Fig. 7.2 and thus has a maximum value of c obtained in the shortwave limit, whereas it tends to zero as the wavelength tends to infinity. The consequences of

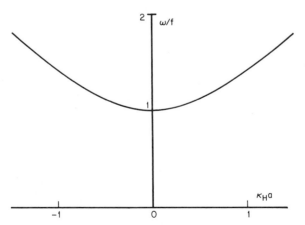

Fig. 7.2.   The dispersion relation for Poincaré waves. For small wavenumber κ_H (waves long compared with the Rossby radius a), the frequency ω is only slightly above the "inertial" frequency f. For large wavenumber (waves short compared with the Rossby radius), the waves are little affected by rotation and so approximate the nondispersive shallow-water waves found in a nonrotating system. Note that the group velocity, which is the gradient of the curve shown in the figure, is zero for zero wavenumber (infinitely long waves) and increases monotonically in magnitude with κ_H to a maximum of $(gH)^{1/2}$ for very short waves.

these variations will be seen in the transient solution because short waves move off rapidly from the initial discontinuity, whereas long waves move off only slowly—the longer the wave, the smaller the group velocity, which velocity is given by

$$\mathbf{c}_g = c^2 \mathbf{k}/\omega \approx c^2 \mathbf{k}/f \qquad \text{for small} \quad \mathbf{k}. \tag{7.3.8}$$

Other properties of Poincaré waves are discussed later.

The solution of the transient problem can now be found by finding a suitable superposition of wave solutions (7.3.3) (Gill, 1976). The appropriate combination of waves is the one that gives the initial distribution (7.3.2), and this can be found from tables of integral transforms [see, e.g., Erdélyi *et al.* (1954, Vol. I, p. 72)]. Thus (7.3.2) is equivalent to

$$\eta = -\frac{2\eta_0}{\pi} \int_0^\infty \frac{k \sin kx}{k^2 + a^{-2}} \, dk. \tag{7.3.9}$$

At later times, η will consist of the same superposition of Poincaré waves, but allowance must be made for their propagation. Thus $2 \sin kx$ in (7.3.9) will be replaced by the combination of Poincaré waves that preserves antisymmetry, namely,

$$\sin(kx + \omega t) + \sin(kx - \omega t) = 2 \sin kx \cos \omega t. \tag{7.3.10}$$

In other words, the solution at time t is given by

$$\eta = -\frac{2\eta_0}{\pi} \int_0^\infty \frac{k \sin kx \cos \omega t}{k^2 + a^{-2}} \, dk, \tag{7.3.11}$$

where ω is given by (7.3.4) with $l = 0$.

The solutions for u and v can also be obtained by reference to the standing-wave solutions that follow directly from Eqs. (7.2.2) and (7.2.3) with $\partial/\partial y = 0$, namely,

$$\eta = \sin kx \cos \omega t,$$
$$u = -(\omega/kH) \cos kx \sin \omega t, \tag{7.3.12}$$
$$v = -(f/kH) \cos kx \cos \omega t.$$

Thus u and v are obtained by replacing $\sin kx \cos \omega t$ in (7.3.11) by the appropriate expression from (7.3.12). In particular, the expression for u, after use of (7.3.4), is

$$u = 2(g\eta_0/\pi c) \int_0^\infty (k^2 + a^{-2})^{-1/2} \sin \omega t \cos kx \, dk, \tag{7.3.13}$$

where ω is given by (7.3.4). It happens that the transform on the right-hand side can be evaluated exactly (Erdélyi *et al.*, 1954, Vol. I, p. 26), giving

$$u = \begin{cases} (g\eta_0/c)J_0(f(t^2 - x^2/c^2)^{1/2}) & \text{for} \quad |x| < ct \\ 0 & \text{for} \quad |x| > ct, \end{cases} \tag{7.3.14}$$

where J_0 is a Bessel function of order zero. This is a special solution of the Klein–Gordon equation, corresponding to a point impulse at $x = 0$ and $t = 0$ (Morse and Feshbach, 1953, p. 139), i.e., the acceleration $\partial u/\partial t$ has the form of a delta function as a result of the infinite pressure gradient that exists at the initial instant. The solution

(7.3.14) is useful for computing solutions, and integrals for η, v in terms of the Bessel function can be obtained from (7.2.2) and (7.2.3). Cahn (1945) used expressions of this type for Rossby's initial values.

The solutions for η, u, v are displayed in Fig. 7.3 and can be compared with the solutions for the nonrotating case discussed in Section 5.6 (see Fig. 5.9a). Instead of the wave front transmitting just the initial step, as in the nonrotating case, the step is now followed by a "wake" of waves that trail behind because of dispersion. The short waves that make up the step still travel at speed c, but longer waves travel more slowly (i.e., their group velocity is smaller), so they lag behind the front. At a fixed point, this is made evident by the fact that the frequency appears to decrease with time after the wave front has passed (i.e., the time between wave crests increases) and soon approaches the inertial frequency f, as can be seen from (7.3.14). Figure 7.4b shows how u changes with time at $x = a$. Another property that can be seen from

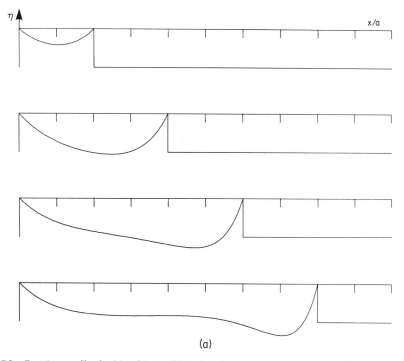

(a)

Fig. 7.3.   Transient profiles for (a) η, (b) u, and (c) v for adjustment under gravity of a fluid with an initial infinitesimal discontinuity in level of $2\eta_0$ at $x = 0$. The solution is shown in the region $x > 0$, where the surface was initially depressed, at time intervals of $2f^{-1}$, where f is twice the rate of rotation of the system about a vertical axis. The marks on the x axis are at intervals of a Rossby radius, i.e., $(gH)^{1/2}/f$, where g is the acceleration due to gravity and H is the depth of fluid. The solutions retain their initial values until the arrival of a wave front that travels out from the position of the initial discontinuity at speed $(gH)^{1/2}$. When the front arrives, the surface elevation rises by η_0 and the u component of velocity rises by $(g/H)^{1/2}\eta_0$ just as in the nonrotating case depicted in Fig. 5.9a. This is because the first waves to arrive are the very short waves, which are unaffected by rotation. Behind the front, however, is a "wake" of waves produced by dispersion, which in the case of u, have the slope given by the Bessel function (7.3.14). This is the point impulse solution to the Klein–Gordon equation. The "width" of the front narrows in inverse proportion with time. Well behind the front, the solution adjusts to the geostrophic equilibrium solution depicted in Fig. 7.1.

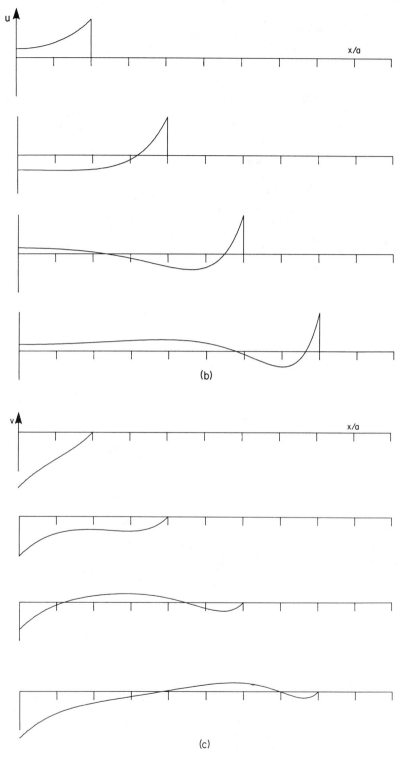

Fig. 7.3. (continued)

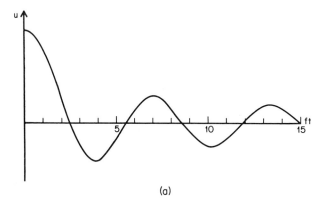

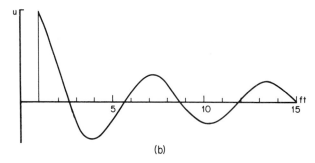

Fig. 7.4.   The u velocity as a function of time t (a) at the position of the initial discontinuity in level and (b) one Rossby radius away. The time axis is marked at intervals of f^{-1}, where f is the inertial frequency. The solutions show oscillations with frequency near f, and these oscillations decay with time like $t^{-1/2}$ at large times.

Fig. 7.4b is the shortening of the length scale just behind the wave front at $x = ct$. This is because the expression in (7.3.14) is approximated by

$$t^2 - x^2/c^2 \equiv (t + x/c)(t - x/c) \simeq 2t(t - x/c),$$

so the length scale diminishes in inverse proportion with time.

It can also be seen that the solution approaches the steady solution of the previous paragraph as time goes on. Details can be calculated from the asymptotic behavior of the Bessel function for large times. It is also clear where the potential energy that was not converted into kinetic energy of the equilibrium solution has gone. The wave fronts moving away from the initial discontinuity carry energy with them, so for any finite region energy is lost through the sides by "radiation" of Poincaré waves until the only energy left is that associated with the steady geostrophic equilibrium.

The other new information provided by the transient solution is the *time scale* of the adjustment process. Near the origin, i.e., within a distance of the order of the Rossby radius, the time scale is f^{-1}, i.e., the rotation time scale or "inertial" time scale ($2\pi/f$ is also half the period of a Foucault pendulum). However, as Rossby suggested, the solution does not adjust monotonically to the equilibrium solution, but overshoots

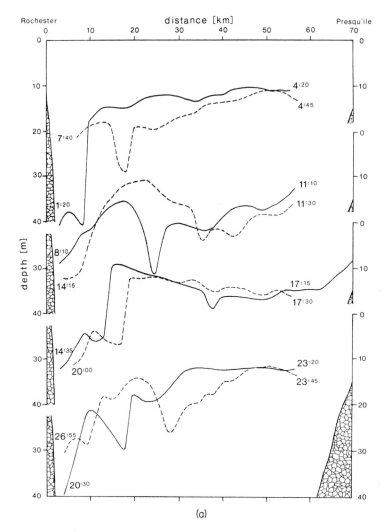

Fig. 7.5. (a) An (internal) Poincaré wave front observed in Lake Ontario following a storm on 9 August 1972. Lines show the thermocline depth as measured by the 10° isotherm. Times of the beginning and end of each transect are shown. The first transect shows the large downwelling produced by the passage of the storm, and subsequent sections show the geostrophic adjustment process involving radiation of Poincaré waves. (b) Results of a (nonlinear two-layer) model simulation of this event by Simons (1978). The diagrams are from Simons (1978, 1980) and may be compared with the solution shown in Fig. 7.3 for a very simple initial condition.

and tends to oscillate about it. The behavior is typified by the value of u at $x = 0$, namely,

$$u = (g\eta_0/c)J_0(ft), \tag{7.3.15}$$

shown in Fig. 7.4a. For large ft, this solution is approximated asymptotically by

$$u \sim (g\eta_0/c)(2/\pi ft)^{1/2} \sin(ft + \pi/4). \tag{7.3.16}$$

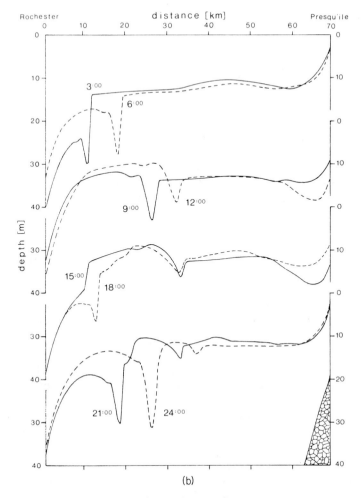

Fig. 7.5. (continued)

Thus oscillations of frequency f are found because these are associated with the long waves of zero group velocity that are left behind. However, the group velocity is not exactly zero for any nonzero wavenumber, so energy disperses slowly and this causes the algebraic decay in the oscillations as given by (7.3.16).

Some waves with characteristics very similar to those depicted in Figs. 7.3 and 7.4 were observed in Lake Ontario in August 1972 and are shown in Fig. 7.5. The initial condition, produced by storm-induced downwelling near the boundary, has a steplike structure as assumed for the solution shown in Figs. 7.3 and 7.4. However, because of the boundary at the coast, there are also reflected waves. A linear calculation could easily deal with them by the method of images used to construct Fig. 5.9b in the nonrotating case. The model solution shown in Fig. 7.5b includes nonlinear and other effects as well.

7.4 Applicability to the Rotating Earth

The Rossby adjustment problem teaches us a lot about the behavior of rotating fluids, but the analysis is for a fluid rotating about a vertical axis. The rotation axis of the earth is not vertical, except at the poles, and furthermore, its angle with the vertical changes from place to place. Does this mean that the analysis is not applicable to the rotating earth, or is it applicable in some approximate sense? Kelvin (Thomson, 1879) stated that his wave solutions (the so-called Poincaré waves) are applicable

> in any narrow lake or portion of the sea covering not more than a few degrees of the earth's surface, if for $\frac{1}{2}f$ we take the component of the earth's angular velocity round a vertical through the locality—that is to say
>
> $$\frac{1}{2}f = \Omega \sin \varphi, \tag{7.4.1}$$
>
> where Ω denotes the earth's angular velocity, and φ the latitude.

(The notation has been changed in the quotation to comply with that used here.)

Kelvin's statement can be justified by examining the linearized version of the momentum equations (4.12.14)–(4.12.16), which are the ones appropriate to the earth. These reduce to (7.2.1) and (7.2.2) if three conditions are satisfied. The first condition is that referred to by Kelvin, namely, that the range of latitude be small enough for f to be regarded as constant and to allow use of local rectangular coordinates. The second is that the additional term $2\Omega u \cos \varphi$ in the vertical component (4.12.16) of the momentum equations should not upset the hydrostatic balance. This is easily checked by calculating the additional pressure at the bottom due to this term when a Poincaré wave given by (7.3.12) is present. This causes negligible change in the pressure gradient if

$$g\kappa_H \gg (2\Omega)^2 \sin \varphi \cos \varphi, \tag{7.4.2}$$

which is well satisfied since κ_H^{-1} cannot be greater than the radius of the earth, which happens to be much smaller than

$$g/(2\Omega)^2 \simeq 460,000 \quad \text{km}.$$

The third condition is that it be possible to neglect the additional term $2\Omega w \cos \varphi$ in the horizontal momentum equation. The largest value of w is its value $\partial \eta / \partial t$ at the surface. Using the Poincaré wave solution (7.3.12), it is found that the condition for its neglect is that

$$H\kappa_H \ll \tan \varphi, \tag{7.4.3}$$

which is implied by the condition $H\kappa_H \ll 1$, used already to justify the hydrostatic approximation, provided φ is not too small, i.e., the area considered is not too close to the equator.

Further discussion of the approximation will be made later. For the moment, the main point is that f is interpreted as the quantity defined by (7.4.1), which is called the *Coriolis parameter*. This parameter is positive in the northern hemisphere and negative in the southern hemisphere. The sign of f is very important in many applications, so special terminology is used, namely, when a rotation is in the same sense as f, it is called *cyclonic* and when it is in the reverse sense, it is said to be *anticyclonic*.

The Rossby adjustment problem explains why the atmosphere and ocean are nearly always close to geostrophic equilibrium, for if any force tries to upset such an equilibrium, the gravitational restoring force acts in the way described in Sections 7.2 and 7.3 to quickly restore a near-geostrophic equilibrium. However, there is much more to the story than that, because the geostrophic equilibrium solution (7.2.14) does *not* satisfy the equations exactly when account is taken of the fact that f is not constant. Because the constant-f solution is degenerate, the processes that actually take place are rather subtle, and much of what happens in the ocean and atmosphere can be described as "quasi-geostrophic" motion that has this subtle character.

The use of a constant-f approximation to describe motion on the earth is sometimes called an f-plane approximation. It is adequate to handle the rapid or "gross" adjustment processes of the sort already considered, and these are characterized by time scales of order f^{-1} or smaller. The more subtle adjustment processes, which are characterized by time scales large compared with f^{-1}, will not be considered until **Chapter 11**.

7.5   The Rossby Radius of Deformation

The Rossby radius of deformation a is a length scale of fundamental importance in atmosphere–ocean dynamics. Basically, it is the *horizontal scale* at which *rotation* effects (of the "gross" variety) become as important as *buoyancy* effects. More specifically, it is the scale for which the middle and last terms on the left-hand side of (7.2.13) are of the same order.

Consider first its significance in transient problems. In the early stages of adjustment from an initial discontinuity, the change of level is confined to a small distance, the pressure gradient is consequently very large, and gravity dominates the behavior. In other words, at scales small compared with the Rossby radius, the adjustment is approximately the same as in a nonrotating system. Later, however, when the change in level is spread over a distance comparable with the Rossby radius, the Coriolis acceleration becomes just as important as the pressure gradient term and thus rotation causes a response that is markedly different from the nonrotating case.

The same considerations apply to Poincaré waves, so the short waves ($\kappa_H^{-1} \ll a$) are very much like gravity waves in a nonrotating system, as discussed in Section 7.3. For waves with scales comparable with the radius of deformation, the buoyancy term $\kappa_H^2 c^2$ in the dispersion relation (7.3.4) is of the same order as the rotation term f^2. Long waves ($\kappa_H^{-1} \gg a$), on the other hand are *dominated* by rotation effects and have frequency close to the inertial frequency f, which for applications to the ocean and atmosphere is also the Coriolis parameter given by (7.4.1). The inertial period $2\pi/f$ is also half the period of a Foucault pendulum, and therefore is sometimes called a half pendulum day. This varies with latitude and is 12 hr at the poles, 17 hr at 45° latitude, 1 day at 30°, and nearly 3 days at 10°. At the equator, it becomes infinite, but by that stage the f-plane approximation breaks down.

The Rossby radius of deformation is not only significant for the behavior of transients, but is also an important scale for the geostrophic equilibrium solution as well. That was seen in the adjustment from the initial discontinuity, because the discontinuity did not spread out indefinitely, but only over a distance of the order of the Rossby radius.

For geostrophic flow, the Rossby radius is the scale for which the two contributing terms in (7.2.9) to the perturbation potential vorticity Q' are of the same order. For a sinusoidal variation of surface elevation with wavenumber κ_H, the contribution to Q' from the vorticity ζ is in the ratio

$$-\frac{\zeta}{f} : \frac{\eta}{H} = \kappa_H^2 a^2 : 1 \qquad (7.5.1)$$

to the contribution from the surface elevation, by (7.2.19). For short waves ($\kappa_H^{-1} \ll a$) therefore, the vorticity term dominates, whereas the surface elevation term dominates for long waves ($\kappa_H^{-1} \gg a$).

The ratio (7.5.1) gives not only the partition of perturbation potential vorticity, but also the *partition of energy*. This may be seen by multiplying the terms on the left-hand side of (7.5.1) by $\frac{1}{2}\rho g H \eta$ and integrating over a wavelength. The second term

$$\frac{1}{2} \rho g \int \int \eta^2 \, dx \, dy$$

is the potential energy, whereas the first is, by (7.2.19),

$$-\frac{1}{2} \rho \left(\frac{g}{f}\right)^2 H \int \int \eta \left(\frac{\partial^2 \eta}{\partial x^2} + \frac{\partial^2 \eta}{\partial y^2}\right) dx \, dy$$

$$= \frac{1}{2} \rho \left(\frac{g}{f}\right)^2 H \int \int \left\{ \left(\frac{\partial \eta}{\partial x}\right)^2 + \left(\frac{\partial \eta}{\partial y}\right)^2 - \frac{\partial}{\partial x}\left(\eta \frac{\partial \eta}{\partial x}\right) - \frac{\partial}{\partial y}\left(\eta \frac{\partial \eta}{\partial y}\right) \right\} dx \, dy$$

$$= \frac{1}{2} \rho H \int \int (u^2 + v^2) \, dx \, dy,$$

the last equality making use of (7.2.14). Thus the first term is the perturbation kinetic energy, and so

$$\text{K.E.} : \text{P.E.} = \kappa_H^2 a^2 : 1, \qquad (7.5.2)$$

i.e., short-wavelength geostrophic flow contains mainly kinetic energy, whereas long-wavelength geostrophic flow has most of its energy in the potential form.

Now *typical values* of the Rossby radius will be calculated. These vary somewhat with latitude because of the variation of f that is given by (7.4.1), i.e., by

$$f = 1.47 \times 10^{-4} \sin \varphi \quad \text{s}^{-1}. \qquad (7.5.3)$$

Estimates will be based on the value $f = 1.0 \times 10^{-4} \text{ s}^{-1}$ appropriate to 45° latitude, but it should be remembered that the Rossby radius is considerably *larger near the equator*, e.g., four times bigger at $10°$, where $f = 0.25 \times 10^{-4}$ s^{-1}.

For deep water in the ocean, where H is 4 or 5 km, c is about 200 m s^{-1} and therefore the Rossby radius $a = c/f \approx 2000$ km. This is large compared with the depth, so the hydrostatic approximation is valid at this scale, although the scale is rather large for f to be taken as constant. For applications on the continental shelves and in shallow seas like the North Sea much smaller values apply because the depth is much smaller. For $H = 40$ m, for instance, $c = 20$ m s^{-1} and $a = c/f = 200$ km. Since the North Sea has larger dimensions than this, rotation has a strong effect on transient motions such as tides and surges.

The above values are calculated for a homogeneous shallow layer of fluid. However, the adjustment problem can also be done for a stratified fluid using the separation of variables technique discussed in Chapter 6. The Coriolis acceleration has the same structure in the vertical as has the acceleration relative to the rotating frame, so the separation technique works in the same way and the analysis of Sections 7.2 and 7.3 applies to each of the normal modes, the only difference (see Sections 6.11 and 6.14) being that H is replaced by the equivalent depth H_e, which is related to the separation constant c_e by (6.11.14).

Thus there is a Rossby radius associated with each of the normal modes. The values calculated above are for the barotropic mode and are therefore called values of the *barotropic Rossby radius*. Each of the baroclinic modes has an associated Rossby radius

$$a_n = c_n/|f|, \qquad n = 1, 2, \ldots, \tag{7.5.4}$$

which can be called the *nth baroclinic Rossby radius*, c_n being the nth value of the separation constant c_e (see Section 6.11), which is equal to the wave speed of the nth mode in a nonrotating system. If a value of n is not given, the first baroclinic mode is understood. For the ocean, the value of c_1 is usually 1–3 m s^{-1}, so typical values of the baroclinic Rossby radius are 10–30 km, with larger values in low latitudes. This is large compared with the vertical scale (which may be taken as the thermocline depth of about 1 km), so the hydrostatic approximation is valid at this scale. The baroclinic Rossby radius is a natural scale in the ocean that is often associated with boundary phenomena, such as boundary currents and fronts, and with eddies.

For the atmosphere, the Lamb wave is the fastest mode with $c \approx 300$ m s^{-1}. The associated Rossby radius of 3000 km is too large for the f-plane approximation to be valid. For internal modes, there is a continuous set of modes and therefore a continuous set of Rossby radii. In the isothermal case, $c \approx N/m$, where N is the buoyancy frequency and m the vertical wavenumber, so

$$a \approx N/mf. \tag{7.5.5}$$

The ratio N/f is typically of order 100, so the Rossby radius is about 100 times the vertical scale m^{-1}. For a vertical scale associated with the height of the tropopause, a is about 1000 km. This is the predominant scale seen on weather charts as the scale of cyclones and **anticyclones**, and is often called the "synoptic scale."

For both ocean and atmosphere it happens that

$$N \gg f \tag{7.5.6}$$

except for rather limited regions; the horizontal scale a is therefore large compared with the vertical scale m^{-1}, and therefore the hydrostatic approximation is justified for motions with these scales. The fact that (7.5.6) is generally true has strongly influenced the way rotation effects have been introduced in this chapter, in particular, the restriction to motions for which the hydrostatic approximation is valid. For a planet for which (7.5.6) were not true, a rather different approach would be needed.

For baroclinic modes, the results (7.5.1) and (7.5.2) are still valid, but the term involving η is then associated with vertical displacements of isopycnals and in the case of a compressible medium, with compression and expansion of fluid elements, i.e., with changes of internal and potential energy. Thus it can be said that the term represented by η corresponds to changes in the mass field, whereas that represented by ζ corresponds to changes in the velocity field. Thus for large scales ($\kappa_H a \ll 1$) (7.5.1) and (7.5.2) show that the potential vorticity perturbation is mainly associated with perturbations in the mass field, and that the energy changes are in the potential and internal forms. On the other hand, for small scales ($\kappa_H a \gg 1$) potential vorticity perturbations are associated with the velocity field, and the energy perturbation is mainly kinetic. It follows that a distinction can be made between the adjustment processes at different scales. At large scales ($\kappa_H^{-1} \gg a$), it is the mass field that is determined [through (7.2.10)] by the initial potential vorticity, and the velocity field is merely that which is in geostrophic equilibrium with the mass field. It is said, therefore, that the large-scale velocity field adjusts to be in equilibrium with the large-scale mass field. On the other hand, at small scales ($\kappa_H^{-1} \ll a$) it is the velocity field that is determined by the initial potential vorticity, and the mass field is merely that which is in geostrophic equilibrium with the velocity field. In this case it can be said that the mass field adjusts to be in equilibrium with the velocity field.

7.6　The Geostrophic Balance

An important feature of the response of a rotating fluid to gravity is that it does not adjust to a state of rest, but rather to a geostrophic equilibrium [the name geostrophic is due to Shaw (1916)]. Consequently, the ocean and atmosphere tend to be close to a state of geostrophic equilibrium all the time [see Phillips (1963) for a review].

The development of an awareness of this fact has been very slow. The barometer was invented by Torricelli in 1643, and its potential for weather prediction was soon realized. Barometer readings, along with temperature, wind direction, and state of the sky, were taken daily in the first network of stations set up by Anitoni, secretary to the Grand Duke Ferdinand II of Tuscany in 1654 [see, e.g., Khrgian (1970, Chapter 6)]. This network included stations as far apart as Florence, Warsaw, and Paris and operated until 1667. Various other attempts were made in the eighteenth century, the most notable being an international effort with standardized instruments organized by the Mannheim (or Palatine) Meteorological Society, beginning in 1781.

However, a clear idea of the relationship between wind direction and pressure gradient does not appear to have emerged until 200 years after the invention of the barometer. Many writers in the mid-nineteenth century showed some awareness of the relationship, so one cannot easily associate the idea with any particular person. Hildebrandsson and Teisserenc de Bort (1898, Chapter 3) give examples, the earliest being Brandes (1820), who studied data collected by the Mannheim Society for the year 1783. He did not publish diagrams, but Fig. 7.6 shows a chart constructed from Brandes' figures by Hildebrandsson and Teisserenc de Bort. Contours are pressure deviations in lines ($\frac{1}{12}$ Parisian inch of mercury or about 3 mb) from mean values at each locality. Brandes noted that the wind direction was closely related to the pressure distribution and attributed the turning to the right (from the direction opposite to the pressure gradient) to the rotation of the earth. Another interesting example is shown in Fig. 7.7 from Birt (1847). In the first report of the British Association in 1832, Forbes expressed his hopes for future networks of meteorological stations as a means for detecting "great atmospheric tidal waves" like those treated by Laplace. As a result, a committee was set up under Herschell, and Birt gave five reports on work he did for the committee. Birt was influenced by Scott Russell's report on waves (1844), referred to in Section 5.4, and proposed that the wave description shown in Fig. 7.7 could explain much of the available observations.

Let the strata *aaa'a'*, *b'b'bb*, fig. 2, represent two parallel aerial currents, *aaa'a'* being from S.W. and *b'b'bb* from N.E., and conceive them both to advance from

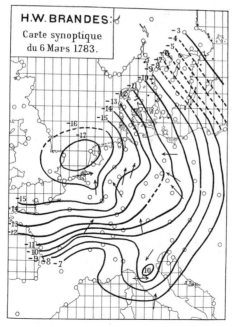

Fig. 7.6. A reconstruction by Hidebrandsson and Teisserenc de Bort (1898) of the early synoptic maps of Brandes (1820), based on data collected by the Mannheim Society. The contours are of pressure deviation from the mean value in lines ($\frac{1}{12}$ of a Parisian inch or about 3 mb), and the arrows show wind direction.

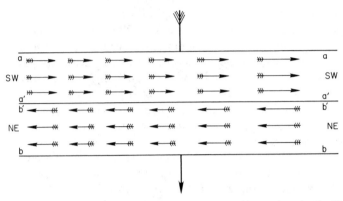

Fig. 7.7.   A wave description of wind and pressure changes proposed by Birt (1847 fig. 2), which included the concept of wind being along isobars. The lines aa and bb are lines of low pressure, whereas the line a′a′ or b′b′ is one of high pressure. Winds in between are in the direction shown, and the whole system propagates in the direction shown by the large arrow.

the N.W. in the direction of the large arrow, that is the strata themselves will advance with a *lateral* motion. Now conceive the barometer to commence rising just as the edge *bb* passes any line of country, and continue rising until the edge *b′b′* arrives at that line, when the maximum is attained. The wind now changes and the barometer immediately begins to fall and continues to fall until the edge *aa* coincides with the line of the country on which *bb* first impinged (Birt, 1847 p. 135).

Birt's description is not only of interest in connection with the relation between wind direction and pressure gradient, but also in connection with waves, which will be studied in later chapters.

Despite these insights, the rules that "the wind is in general perpendicular to the barometric slope" and that "if you turn your back to the wind, the lower pressure will be on your left and the higher pressure your right" are sometimes referred to as Buys-Ballot's law (for the northern hemisphere) since he expressed them thus in his yearbooks of 1857 and 1860 (Khrgian, 1970).

On the theoretical side, interest in the effects of the rotation of the earth was stimulated by the experiments of Foucault (1851), which were followed 8 years later by the "bathtub" experiment of Perrot (1859). In this, a small hole in the center of the base of a large cylindrical container was opened after the water in the container had been left a whole day to settle down. As he expected from theory, Perrot found that fluid particles were deflected to the right, thereby acquiring, in modern parlance, a cyclonic rotation. A repetition of this experiment can be seen in the film "Vorticity" by Shapiro [see National Committee for Fluid Mechanics Films (1972, pp. 63–74)].

Perrot's experiment prompted Babinet (1859) to attribute preferential erosion on the right banks of Siberian rivers, among other things, to the rotation of the earth. He was immediately "jumped on" by his colleagues for this. In particular, Delaunay (1859), after showing that the horizontal force per unit mass due to rotation is f times the velocity, put it this way: Consider a straight canal in the northern hemisphere. If the fluid is at rest it will exert equal pressures on its two banks. If it moves, "the pressure will diminish a little on the bank left of the current and increase a little on the right bank (p. 692)." He also said that the changes would be quite small. Combes (1859) went further and showed that the surface would slope up to the right in the northern hemisphere (and to the left in the southern hemisphere) with an inclination given by

$$\text{surface slope} = 2\Omega \sin \varphi \, v/g, \tag{7.6.1}$$

which is another way of expressing the geostrophic balance (7.2.14). He calculates that for a river 4 km wide flowing at 3 m s^{-1} at 45° latitude, the difference in level between the two sides would be 12 cm.

These discussions at the Paris Academy were not directed toward meteorological questions, but had some "spinoff" in European meteorology later [see Abbe (1877, 1893, 1910)]. In the United States at this time, however, Ferrel was concerned with applying the equations of fluid motion on a rotating sphere to meteorological problems, and in particular to the global circulation. He appears to be the first person (Ferrel, 1859, pp. 397–398) to deduce that large-scale motions of the atmosphere are approximately hydrostatic and geostrophic. His approach was to first integrate the hydrostatic equation, neglecting changes of temperature with height to obtain the exponential falloff (3.5.12) with height. This was substituted in the north–south component (4.12.15) of the momentum equation and then approximations were made that amount to approximating (4.12.15) by

$$2\Omega u \sin \varphi = -(\rho r)^{-1} \, \partial p/\partial \varphi + \text{friction term}. \tag{7.6.2}$$

He then argued that the friction term would be relatively small and used surface pressure measurements to calculate from his formula the zonal winds at the surface and at a height of 3 miles (5 km), using observed pressures and a reasonable approximation for dependence of temperature on latitude. At 5 km he obtained westerly (i.e., eastward) winds at all latitudes with maxima of 13 m s^{-1} at 55°N and 23 m s^{-1} at 40°S. He did not find easterlies near the equator as observed (see Fig. 7.9), but as he stated: "Very near the equator the formula ... fails practically, since, on account of the small value of sin φ there, the effect of" friction and inertia "may be very great" (p. 401).

The geostrophic relationship found in (7.2.14) applies to any one mode. The more general result (which could be obtained by adding contributions from modes) comes from balancing the pressure gradient and Coriolis terms in (4.10.11) to give

$$-fv = -\rho^{-1} \, \partial p/\partial x, \tag{7.6.3}$$

$$fu = -\rho^{-1} \, \partial p/\partial y. \tag{7.6.4}$$

The nonlinear terms and friction terms, which were automatically excluded in the linear inviscid analysis, tend to be important only in regions of strong gradient such as fronts and boundary currents, and friction effects are significant, though not dominant, near the surface. Thus surface winds tend to be along isobars, in the direction given by Buys-Ballot, and pressure is related to a stream function by (7.2.18), i.e., winds are strongest when isobars are closest together. Another way of remembering direction is in terms of the direction of rotation around a "high" (anticyclone) or "low" (cyclone). The air has a cyclonic rotation around a cyclone, as the term

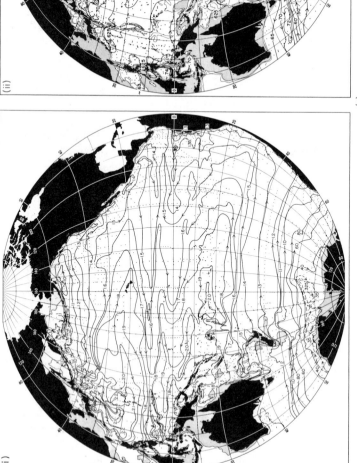

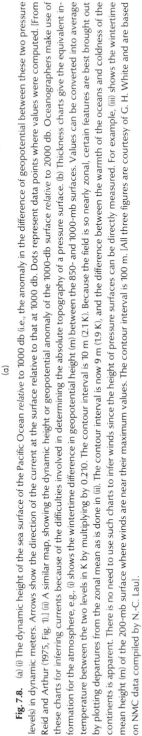

Fig. 7.8. (a) (i) The dynamic height of the sea surface of the Pacific Ocean *relative* to 1000 db (i.e., the anomaly in the difference of geopotential between these two pressure levels) in dynamic meters. Arrows show the direction of the current at the surface relative to that at 1000 db. Dots represent data points where values were computed. [From Reid and Arthur (1975, Fig. 1).] (ii) A similar map, showing the dynamic height or geopotential anomaly of the 1000-db surface relative to 2000 db. Oceanographers make use of these charts for inferring currents because of the difficulties involved in determining the absolute topography of a pressure surface. (b) Thickness charts give the equivalent information for the atmosphere, e.g., (i) shows the wintertime difference in geopotential height (m) between the 850- and 1000-mb surfaces. Values can be converted into average temperature between the two levels in K by multiplying by 0.210. The contour interval is 10 m (2.1 K). Because the field is so nearly zonal, certain features are best brought out by plotting departures from the zonal mean as is done in (ii). The contour interval is now 9 m (1.9 K), and the difference between the warmth of the oceans and coldness of the continents is apparent. There is no need to use such charts to infer winds since the height of pressure surfaces can be directly measured. For example, (iii) shows the wintertime mean height (m) of the 200-mb surface where winds are near their maximum values. The contour interval is 100 m. [All three figures are courtesy of G. H. White and are based on NMC data compiled by N.-C. Lau.

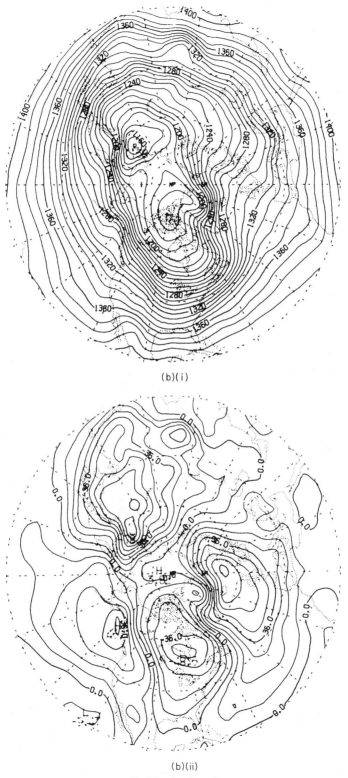

(b)(i)

(b)(ii)

Fig. 7.8. (*continued*)

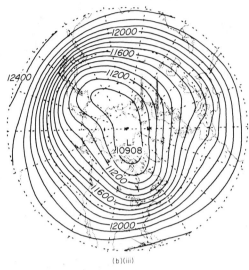

(b)(iii)

Fig. 7.8.   (*continued*)

implies (i.e., anticlockwise, looking downward, in the northern hemisphere and clockwise in the southern hemisphere) and anticyclonic rotation around an anticyclone.

There is a significant correction to geostrophy for surface winds. One way of expressing this is to define the *geostrophic wind* (u_g, v_g) by

$$fu_g = -\rho^{-1}\,\partial p/\partial y, \qquad fv_g = \rho^{-1}\,\partial p/\partial x, \qquad (7.6.5)$$

and express the surface wind in terms of the geostrophic wind. This amounts to a reduction in magnitude (which increases as the distance from the ground decreases) and a change in direction toward the low pressure, typical angles being around 20°. In practice, the correction depends on stability of the air and on whether or not equilibrium conditions have been established. (Diurnal variations in heating rate and variations in terrain work against this over land.) The deviation from geostrophy decreases with height and is usually quite small above 1 km. A verification of the closeness of geostrophic balance aloft was obtained by Gold (1908).

Although (7.6.3) and (7.6.4) are a convenient way of expressing the geostrophic relationship at the surface, a more convenient form for other levels is in terms of isobaric coordinates (see Section 6.17), i.e., the velocity on a constant-pressure surface is given by

$$-fv = -\partial\Phi/\partial x, \qquad (7.6.6)$$

$$fu = -\partial\Phi/\partial y, \qquad (7.6.7)$$

where Φ is the geopotential on that surface. This is the form used in practice in both meteorology and oceanography. The advantage in meteorology is clear because density is eliminated and thus charts of Φ at different pressure levels have the same scale for converting into velocities. As an example, Fig. 7.8b includes charts of the geopotential height (see Section 3.5) of the 200 mbar surface. Winds at this pressure can be calculated from (7.6.6), (7.6.7), and (3.5.2).

7.7   Relative Geostrophic Currents: The Thermal Wind

Ferrel not only found that the atmosphere was approximately in hydrostatic and geostrophic equilibrium, but also showed that this fact could be exploited to calculate upper air winds, using surface pressure measurements and temperature observations. For lack of other information, Ferrel's winds at an elevation of 3 miles were calculated on the assumption that temperatures up to this level were not too different from surface values.

Nowadays, radiosonde ascents of the atmosphere (and lowerings of salinity–temperature–depth recorders in the ocean) are routine, so accurate information about variations of temperature and humidity (or temperature and salinity) with pressure can be obtained. The equation of state gives the density as a function of pressure, so the geopotential Φ can be calculated from the hydrostatic equation (3.5.6), i.e.,

$$d\Phi = -\frac{dp}{\rho} = -v_s \, dp. \tag{7.7.1}$$

The information from radiosondes is often recorded in terms of values at "significant points," i.e., places at which there is a significant change in temperature gradient. A good approximation to the profile is obtained by joining these points by straight lines on thermodynamic diagrams (see Section 3.9). Graphical methods of calculating geopotential changes from these diagrams are discussed, e.g., by Godske *et al.* (1957, Chapter 3).

In the atmosphere, the dynamic height of any pressure surface can be calculated because the surface pressure is known. The same is not true in the ocean because the elevation of the free surface relative to a geopotential is not usually known. However, differences in the dynamic height of given pressure surfaces can still be calculated, so the geostrophic velocity at one level can be calculated relative to that at another.

Temperature and salinity values in the ocean, if obtained by STD (salinity–temperature–depth) or CTD (conductivity–temperature–depth) recorders, are usually listed in a cruise report and sent to a data center as values at certain standard depths, with some additional values where changes of gradient occur. If obtained by Nansen bottles, which record temperatures and collect samples of water for analysis at prearranged depths, values at those depths are given. There are standard computer routines to calculate values of density and of dynamic height. For calculations of the latter quantity, the density is calculated in terms of the *specific volume anomaly* δ, defined as the specific volume $v_s = \rho^{-1}$ related to the value at the same pressure for a temperature of $0°C$ and a practical salinity of 35, i.e.,

$$\delta = v_s(S, T, p) - v_s(35, 0, p). \tag{7.7.2}$$

δ can be calculated using (A3.3), and is usually given in units of $10^{-8} \, m^3 \, kg^{-1}$. The geopotential anomaly Φ' (the usual notation is $-\Delta D$) is then defined by

$$\Delta D = -\Phi' = \int_0^p \delta \, dp, \tag{7.7.3}$$

i.e., is obtained by integration from zero pressure (the surface) to the pressure concerned. Since the anomalies are expressed relative to a function of pressure only, the horizontal gradients of Φ' along isobaric surfaces are the same as horizontal gradients of Φ. Hence

$$-f\{v_g(p) - v_g(0)\} = -\frac{\partial \Phi'(p)}{\partial x},$$

$$f\{u_g(p) - u_g(0)\} = -\frac{\partial \Phi'(p)}{\partial y}, \qquad (7.7.4)$$

give geostrophic currents relative to the surface. Similarly, geostrophic velocities at pressure p_1 can be calculated relative to another pressure p_2 by

$$-f\{v_g(p_1) - v_g(p_2)\} = -\frac{\partial}{\partial x}\{\Phi'(p_1) - \Phi'(p_2)\},$$

$$f\{u_g(p_1) - u_g(p_2)\} = -\frac{\partial}{\partial y}\{\Phi'(p_1) - \Phi'(p_2)\}. \qquad (7.7.5)$$

Usually the reference level (subscript 2) is chosen to be the *lower* of the two levels, and hence at the higher pressure ($p_2 > p_1$).

As an example, Fig. 7.8a shows the dynamic topography of (i) the surface of the Pacific Ocean relative to the 1000-decibar level and (ii) the 1000-db relative to the 2000-db level. [In oceanography the *decibar* (dbar) (see Section 3.5) is often used as a unit of pressure since a pressure change of 1 db corresponds to a change of depth of $1/\rho g$ times this value, which is very close to 1 m: typically 1 db = 0.995 m near the surface and 0.969 m at the 5000-db level. Often the distinction between a decibar and a meter is of little importance and is ignored.] Usually, currents at the deeper levels are small compared with surface values, so the surface currents relative to 1000 or 2000 db are assumed to be a good approximation to actual surface currents. There is always the question, however, of which reference level gives the best approximation to surface currents. The ideal reference level would be a "level of no motion," but such a level does not necessarily exist in practice because *both* components of velocity need to vanish at the same level. Methods of deducing a reference level from temperature and salinity observations in a neighborhood are discussed by Stommel and Schott (1977) and Killworth (1980b).

In meteorology, the dynamic height of one pressure surface relative to another is called the *thickness*. If the perfect gas law (3.1.2) is satisfied, (7.7.1) gives

$$\Phi_1 - \Phi_2 = \int_{p_1}^{p_2} p^{-1} RT \, dp = R\bar{T} \ln(p_2/p_1), \qquad (7.7.6)$$

where $\bar{T}$ is the temperature averaged with respect to the logarithm of the pressure between the two levels. Hence the thickness, and therefore relative winds, is associated with a mean temperature $\bar{T}$. Another interpretation of $\bar{T}$ is obtained by integrating the hydrostatic equation in the form (3.5.11), which gives

$$\ln(p_2/p_1) = \int_{p_2}^{p_1} (g/RT) \, dz = (g/R\bar{T}) \int_{p_2}^{p_1} dz, \qquad (7.7.7)$$

i.e., $1/\bar{T}$ is the reciprocal of the temperature averaged with respect to distance z

between the two pressure surfaces. Figure 7.8b shows an example of a thickness chart. In meteorology these are more often used as a measure of mean temperature than of relative winds.

In the above discussion, the information in the hydrostatic and geostrophic equations has been combined, following Ferrel, by first integrating the hydrostatic equation and then using the geostrophic relation. Alternatively, the geostrophic equations (7.6.6) and (7.6.7) can be differentiated with respect to pressure and then the hydrostatic relation used to substitute for $\partial\Phi/\partial p = -\rho^{-1}$. The result is

$$f\ \partial v/\partial p = \rho^{-2}\ \partial\rho/\partial x, \qquad f\ \partial u/\partial p = -\rho^{-2}\ \partial\rho/\partial y. \tag{7.7.8}$$

Alternatively, using the hydrostatic equation again to put $dp = -\rho g\ dz$, this may be written

$$f\ \partial v/\partial z = -g\rho^{-1}(\partial\rho/\partial x)_p, \qquad f\ \partial u/\partial z = g\rho^{-1}(\partial\rho/\partial y)_p, \tag{7.7.9}$$

where the derivatives on the right-hand side are taken on constant-pressure surfaces. (The difference between the gradients on constant-pressure and constant-level surfaces is usually so small that the distinction is unimportant for practical purposes.)

For a perfect gas $\rho = p/RT$, so the derivatives on the right can be reexpressed in terms of temperature to give

$$f\ \partial v/\partial z = gT^{-1}(\partial T/\partial x)_p, \qquad f\ \partial u/\partial z = -gT^{-1}(\partial T/\partial y)_p. \tag{7.7.10}$$

This form of the equation is called the *thermal wind* equation, and gives a relation between temperature gradient (on an isobaric surface) and wind shear. It follows that, as Ferrel found, when temperature decreases toward the poles, winds become more westerly (i.e., stronger toward the east) with height. Figure 7.9 shows observed distributions of temperature and wind with latitude and height, and the relationship between the two fields, as expressed by (7.7.10), is apparent.

It is useful to think of the thermal wind as the wind at one level (denoted by subscript 1, say) *relative to* the wind at a *lower* level (denoted by subscript 2). Then the thermal wind blows along isotherms (or, more precisely, along contours of constant thickness) with, in the northern hemisphere, *cold* air on the *left* and warm air on the right. There are various consequences that are useful to remember.

Suppose, for instance, the geostrophic wind at the reference level has a component from cold to warm. Then the thermal wind will be directed to the left in the northern hemisphere (see Fig. 7.10), so the wind will *back* with height (i.e., the wind vector will rotate **anticlockwise, or cyclonically, with height**). Conversely, the wind will *veer* with height (i.e., rotate anticyclonically) if the wind has a component from warm to cold. Thus, *backing* of the wind with height is associated with cold air advection and veering with height is associated with warm air advection.

On a chart showing isotherms on a constant-pressure surface, the shear vector is directed cyclonically around low temperatures (or low thickness) and anticyclonically around high temperatures (or high thickness). Thus if a *low*-pressure disturbance has a *cold* core, the cyclonic flow around the core will *increase* with height, and vice versa. Similarly, if a high-pressure disturbance has a warm core, the anticyclonic flow will increase with height, and vice versa. If temperature and pressure centers do not

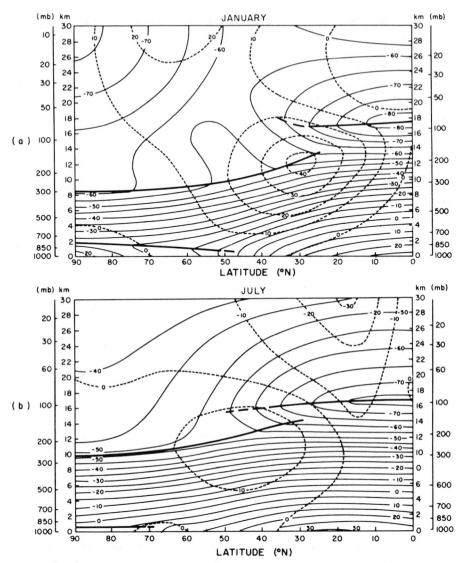

Fig. 7.9. Mean meridional cross sections of wind and temperature for (a) January and (b) July. Thin solid temperature lines are in degrees Celsius and dashed wind speed lines are in meters per second. Heavy solid lines represent tropopause and inversion discontinuities. [After *Arctic Forecast Guide*, Navy Weather Research Facility, April 1962.]

coincide, the *cyclones* will *shift* with height *toward cold air*, and anticyclones will shift toward warm air.

For the ocean, the equation of state can be used to give the density gradients in (7.7.9) in terms of temperature and salinity gradients. This gives (see Section 3.6)

$$f\, \partial v/\partial z = g\alpha\, \partial T/\partial x - g\beta\, \partial s/\partial x, \qquad f\, \partial u/\partial z = -\, g\alpha\, \partial T/\partial y + g\beta\, \partial s/\partial y. \quad (7.7.11)$$

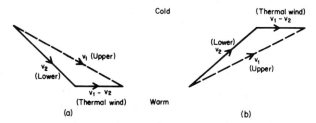

Fig. 7.10. The association between the direction of rotation of the wind vector with height and the direction of heat advection. Subscript 2 denotes the lower level and subscript 1 the upper level. In the northern hemisphere, cold air is on the left of an observer traveling with the thermal wind $v_1 - v_2$. Consequently, cold air advection [case (a)] corresponds to a wind vector that backs (i.e., rotates cyclonically) with height, whereas warm air advection [case (b)] is associated with a wind vector that veers (i.e., rotates anticyclonically) with height in the northern hemisphere.

Alternatively (see Section 3.7), T can be replaced by potential temperature θ and α, β by α', β'. When salinity gradients are small and temperature decreases toward the poles, currents become more strongly westward with depth. A manifestation of this result is in the poleward displacement with depth of the centers of the subtropical gyres, which are anticyclonic. This can be seen in Fig. 7.8a.

A limiting form of the thermal wind is obtained in the case of a "front" or sloping surface of discontinuity of density. If this surface is given by $z = h(x, y)$, subscript 2 denotes values *below* the surface, and subscript 1 values above, then the additional pressure created below the surface by the discontinuity will be

$$p_2 - p_1 = (\rho_2 - \rho_1)g(h - z) \tag{7.7.12}$$

by the hydrostatic equation. From the geostrophic relation, the additional velocity *below* the surface will be given by

$$f(v_2 - v_1) = g'\, \partial h/\partial x, \qquad f(u_2 - u_1) = -g'\, \partial h/\partial y, \tag{7.7.13}$$

where

$$g' = (\rho_2 - \rho_1)g/\rho_2. \tag{7.7.14}$$

Thus the additional velocity is directed along contours of the surface of discontinuity and has magnitude g'/f times the slope of the surface. A special case is the discontinuity at a free surface, in which (7.7.13) reduces to the result (7.6.1) obtained by Combes (1859). The relation (7.7.13) for a general surface of discontinuity is often referred to as Margules' (1906) relation. A more general form of the relation for zonal flows was obtained by Helmholtz (1888).

7.8 Available Potential Energy

In Section 5.7, the energy changes associated with the adjustment under gravity of a homogeneous fluid were considered for the case of small perturbations from a state of rest. The perturbation potential energy associated with the surface elevation

η was found to be the positive definite quantity

$$A' = \int\int \frac{1}{2}\rho g\eta^2 \, dx \, dy, \tag{7.8.1}$$

whose minimum value, zero, is achieved in the rest state. In any frictionless motion, the sum of kinetic and potential energies remains constant, so the maximum amount of kinetic energy that can be obtained by conversion of potential into kinetic energy is equal to A'. Thus A' is called *available potential energy* since it represents energy that is available for conversion into the kinetic form if the constraints imposed by the equation of motion will allow this to happen. In the nonrotating case studied in Section 5.7, it was found that complete conversion occurred, i.e., *all* the available potential energy was released and appeared in the fluid state as kinetic energy. *Rotation* can, however, *prevent conversion* from taking place. In fact, in the problem studied in Section 7.2, very little of the available potential energy was released because the geostrophic equilibrium obtained had nearly as much available potential energy as the initial state. Hence the idea was put forward that potential energy is "hard to extract" in a rotating system.

There is more to the story, however, because the adjustment of a stratified fluid is often sensitive to the initial condition. For instance, an atmosphere that is constrained to have no variations with longitude may adjust with very little change in potential energy. Slight irregularities on such a flow, however, can often grow and draw on the potential energy available in the zonal flow, and therefore much more potential energy is released. In practice, the energy released is manifested in the form of cyclones and anticyclones, which not only dominate our weather but also play a vital role in the general circulation of the atmosphere. Thus the concept of available energy, first put forward by Margules (1903) for discussing storms, is a valuable tool for analyzing the general circulation of the atmosphere and was developed with this in mind by Lorenz (1955).

The discussion of the first paragraph of this section was in terms of a homogeneous fluid, but the concept of available potential energy applies equally well to stratified compressible fluids (see Section 4.7) and is usually based on the behavior of an ideal fluid in the absence of changes of state (i.e., of latent heat release). Then, as found in Section 4.7, the sum of kinetic, internal, and potential energies remains constant, so available potential energy is defined as the *difference* between the *internal plus potential* energy observed and the minimum value of this quantity that could be achieved by a redistribution of mass in which the entropy and composition of each fluid element is conserved. Defined this way the following hold:

(a)   The available potential energy is a positive definite quantity.

(b)   The available potential energy depends only on the distribution of potential temperature (or temperature), of pressure, and of constituents throughout the fluid.

(c)   The sum of available potential and kinetic energies remains constant in any isentropic motion without diffusion, dissipation, latent heat release, or radiative heat exchanges.

As mentioned in Section 4.7, the mean value (per unit area) of the available potential energy of the atmosphere has been estimated as 4.5×10^6 J m^{-2}, which is about ten times the value of the mean kinetic energy. If the available potential energy were released to give a uniform speed over the whole of the atmosphere, that speed would be about 30 m s^{-1}.

The expression obtained in Section 6.14 for perturbation internal plus potential energy A', namely, (6.14.10) or

$$A' = \frac{1}{2} \int \int \int \left(\rho_0 N^2 h^2 + \rho_0^{-1}(p'/c_s)^2 \right) dx\, dy\, dz, \tag{7.8.2}$$

where h is the vertical particle displacement, N the buoyancy frequency, c_s the speed of sound, p' the perturbation pressure, and ρ_0 the undisturbed density, shows that the rest state is a minimum for internal plus potential energy, but it is conceivable that other minima could exist for which the energy was smaller. This is not the case, however, which can be proved by calculating the change δA in internal plus potential energy during a displacement $\delta \mathbf{x}$ of fluid particles that conserves mass and entropy [cf. Van Mieghem (1956, 1957)].

From the definitions (4.7.6) and (4.7.7),

$$\delta A = \delta(I + P) = \int \int \int \rho \delta(E + \Phi)\, dx\, dy\, dz, \tag{7.8.3}$$

but by the first law of thermodynamics (3.2.1), the change of the internal energy per unit mass for an isentropic displacement is given by

$$\delta E = -p\, \delta v_s = p\, \delta \rho / \rho^2, \tag{7.8.4}$$

where $v_s = \rho^{-1}$ is the specific volume. But the mass conservation equation (4.2.2), when written in terms of the material displacement $\delta \mathbf{x}$, has the form

$$\delta v_s / v_s = -\delta \rho / \rho = \mathbf{V} \cdot (\delta \mathbf{x}). \tag{7.8.5}$$

Also, the change $\delta \Phi$ in a quantity Φ associated with a displacement $\delta \mathbf{x}$ is given by

$$\delta \Phi = \delta x\, \partial \Phi / \partial x + \delta y\, \partial \Phi / \partial y + \delta z\, \partial \Phi / \partial z = \mathbf{V}\Phi \cdot \delta \mathbf{x}. \tag{7.8.6}$$

It follows by substituting (7.8.6) for $\delta \Phi$ and the result of (7.8.5) and (7.8.4) for δE in (7.8.3) that

$$\delta A = \int \int \int (-p\, \mathbf{V} \cdot \delta \mathbf{x} + \rho\, \mathbf{V}\Phi \cdot \delta \mathbf{x})\, dx\, dy\, dz$$

$$= \int \int \int (-\mathbf{V} \cdot (p\, \delta \mathbf{x}) + \mathbf{V}p \cdot \delta \mathbf{x} + \rho\, \mathbf{V}\Phi \cdot \delta \mathbf{x})\, dx\, dy\, dz,$$

i.e.,

$$\delta A = -\int \int p(\delta x)_n\, dS + \int \int \int (\mathbf{V}p + \rho\, \mathbf{V}\Phi) \cdot \delta \mathbf{x}\, dx\, dy\, dz, \tag{7.8.7}$$

where (cf. Section 4.6) the first term on the right-hand side is obtained by use of the divergence theorem, $(\delta x)_n$ denotes the displacement along the outward normal to the volume in question, and dS denotes an element of surface area. This term vanishes if there is a rigid boundary where $(\delta x)_n = 0$ as at the ocean floor and if $p = 0$ as at the "top" of the atmosphere. Therefore the condition for δA to vanish for the atmosphere–ocean system is

$$\nabla p + \rho \, \nabla \Phi = 0, \qquad (7.8.8)$$

i.e., the *full* hydrostatic equation must be satisfied, including the condition of no horizontal pressure gradient. From the equation of motion (4.5.5), the only state of motion consistent with this and the boundary conditions is one of rest.

However, there are many possible rest states since the density could be any function of pressure. To find which one is the state of minimum internal plus potential energy, the change in A can be calculated to second order in the displacement [cf. Van Mieghem (1956, 1957)], and this gives the result (7.8.2). This shows that the minimum corresponds to a state in which mass is distributed in such a way that the buoyancy frequency is real, i.e., $N^2 \geq 0$.

Equation (7.8.7) is closely related to the energy equation discussed in Section 4.7; for when written in terms of rates of change of energy associated with a certain rate of change of displacement, it becomes

$$dA/dt = -\iint pu_n \, dS + \iiint (\nabla p + \rho \, \nabla \Phi) \cdot \mathbf{u} \, dx \, dy \, dz \qquad (7.8.9)$$

since $\mathbf{u} = D(\delta\mathbf{x})/Dt$, the integral being over a material volume. When the equation of motion is used, $\nabla p/\rho + \nabla\Phi$ can be replaced by minus the acceleration, so the last term becomes minus the rate of change of kinetic energy and (7.8.9) becomes a special case of (4.7.8) for the conservation of total energy.

To illustrate the concept of available potential energy, consider Marsigli's experiment, discussed in Section 5.1. The initial state consists of two equal volumes of homogeneous fluid in a rectangular container separated by a partition as illustrated in Fig. 7.11a(i). Using the base of the container as a reference level, the average potential energy P per unit area is given by

$$P = L^{-1} \int_{-L/2}^{L/2} \left\{ \int_0^H \rho\Phi \, dz \right\} dx = L^{-1} \int_{-L/2}^{L/2} \left\{ \int_0^H \rho g z \, dz \right\} dx = \frac{1}{4}(\rho_1 + \rho_2)gH^2.$$

The origin $x = 0$, $z = 0$ has been placed at the center of the base, which has width L. The equilibrium state is shown in Fig. 7.11a(iii), and calculating the same integral for this gives

$$P = \tfrac{1}{4}(\tfrac{3}{2}\rho_1 + \tfrac{1}{2}\rho_2)gH^2.$$

Subtracting the two expressions gives the available energy A per unit area as

$$A = \tfrac{1}{8}(\rho_2 - \rho_1)gH^2. \qquad (7.8.10)$$

If all the available energy were converted into kinetic energy, the mean square velocity of the motion would be equal to

$$U^2 = \tfrac{1}{2}(\rho_2 - \rho_1)gH/(\rho_2 + \rho_1). \qquad (7.8.11)$$

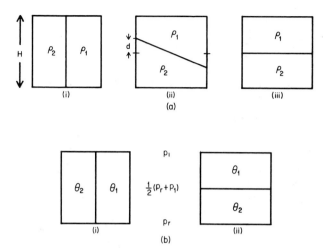

Fig. 7.11. (a) Three different arrangements of two incompressible fluids, each of uniform density ρ_i, with $\rho_2 > \rho_1$. Case (i) is the initial configuration of Marsigli's experiment (see Fig. 5.1). Case (iii) shows the arrangement at the end of the experiment when a condition of minimum potential energy is obtained. Case (ii) shows an alternative initial condition. The available potential energy is defined as the potential energy relative to that in case (iii). The average values per unit area are $\frac{1}{8}(\rho_2 - \rho_1)gH^2$ in case (i), $\frac{1}{6}(\rho_2 - \rho_1)gd^2$ in case (ii), and zero in case (iii). (b) Different arrangements of two equal masses of dry air, each with uniform potential temperature θ_i, with $\theta_1 > \theta_2$. Margules (1903) calculated the potential plus internal energy in each case. Case (i) is an arrangement similar to Marsigli's experiment, but the upper surfaces are at the same pressure, not the same level. Case (ii) shows the arrangement with minimum potential plus internal energy. Since the mass above any level is proportional to pressure, the interface is at pressure $\frac{1}{2}(p_r + p_1)$, p_r being the pressure at the ground and p_1 the pressure at the upper surface. The available potential energy is defined as the potential energy relative to that in case (ii) and its value for case (i) is given by (7.8.21).

Similar calculations can be made for other configurations, such as the one with a sloping interface in Fig. 7.11a(ii), for which

$$A = \tfrac{1}{6}(\rho_2 - \rho_1)gd^2. \tag{7.8.12}$$

Margules (1903) considered a similar example for a compressible fluid, which is illustrated in Fig. 7.11b(i). Equal masses of a perfect gas are on the two sides of the partition, one with potential temperature θ_1, and the other with lower potential temperature θ_2. The pressure at the ground is taken as the reference pressure p_r used in the definition of potential temperature, and the pressure at the upper surface is p_1. In the rest state, the fluid with potential temperature θ_1 lies on top as shown in Fig. 7.11b(ii), with the interface at pressure $\frac{1}{2}(p_r + p_1)$ by continuity of mass.

To calculate the available potential energy, the state variables need to be expressed in terms of potential temperature and pressure. By (3.7.4) the temperature T is given by

$$T = \theta(p/p_r)^{\kappa}, \tag{7.8.13}$$

where

$$\kappa = 1 - \gamma^{-1} = R/c_p \tag{7.8.14}$$

by (3.2.9) and (3.3.5) with $q = 0$. The ideal gas equation (3.1.2) gives for the density

$$\frac{1}{\rho} = \frac{RT}{p} = \frac{R\theta}{p_r}\left(\frac{p}{p_r}\right)^{\kappa-1}, \tag{7.8.15}$$

and so the hydrostatic equation (3.5.6) or (3.5.8) gives for an isentropic region

$$[gz] = [\Phi] = -\int \rho^{-1}\, dp = -c_p\theta[(p/p_r)^\kappa]. \tag{7.8.16}$$

This shows that pressure varies as some power of the height in a homentropic fluid. The square brackets here and in other formulas in this section denote differences in values between two pressure levels.

Now the internal energy I per unit area for a perfect gas is given by (3.2.7) and (4.7.6), i.e., by

$$I = \int \rho c_v T\, dz = R^{-1} c_v \int p\, dz. \tag{7.8.17}$$

The potential energy P per unit area is, from the definition (4.7.7) and the hydrostatic equation (3.5.8), given by

$$P = \int \rho\Phi\, dz = -g^{-1}\int \Phi\, dp = -g^{-1}[p\Phi] + \int p\, dz. \tag{7.8.18}$$

Since $[p\Phi]$ vanishes over the whole depth of atmosphere, these formulas show that internal energy and potential energy for a whole column of atmosphere are in the ratio $c_v : R$. In general, the sum of (7.8.17) and (7.8.18) gives, after use of (3.2.9) with $q = 0$ and (7.8.14),

$$P + I = -g^{-1}[p\Phi] + \kappa^{-1}\int p\, dz. \tag{7.8.19}$$

Now the values for the situations shown in Fig. 7.11b can be calculated, using the result that

$$\int p\, dz = -\int \frac{p\, dp}{\rho g} = -\frac{R}{g}\int T\, dp = -\frac{R}{g}\int \theta\left(\frac{p}{p_r}\right)^\kappa dp = -\frac{R\theta p_r}{(1+\kappa)g}\left[\left(\frac{p}{p_r}\right)^{1+\kappa}\right], \tag{7.8.20}$$

use having been made of (7.8.15) and the hydrostatic equation (3.5.8). The last formula applies only in the case of an isentropic layer. Subtracting the result for the case shown in Fig. 7.11b(ii) from that for the case shown in Fig. 7.11b(i) gives, after some calculation, the available potential energy A per unit area as

$$A = \frac{c_p p_r}{2g}(\theta_1 - \theta_2)\left\{\frac{1}{1+\kappa} - r - \frac{\kappa}{1+\kappa}r^{1+\kappa} - \frac{2}{1+\kappa}\left(\frac{1+r}{2}\right)^{1+\kappa} + 2r\left(\frac{1+r}{2}\right)^\kappa\right\}, \tag{7.8.21}$$

where

$$r = p_1/p_r, \tag{7.8.22}$$

In particular, if the columns extend over the whole depth of atmosphere, so that $p_1 = r = 0$, (7.8.21) gives, using (3.7.3) with $q = 0$,

$$A = \frac{c_p p_r(\theta_1 - \theta_2)}{2g(1 + \kappa)}\left(1 - \left(\frac{1}{2}\right)^\kappa\right) \approx 0.07\frac{c_p p_r(\theta_1 - \theta_2)}{g}. \tag{7.8.23}$$

In the opposite extreme, when the layer concerned is very shallow, so that $p_1 - p_r$ is small,

$$A \approx \frac{c_p(\theta_1 - \theta_2)\kappa(1 - \frac{1}{2}\kappa)}{4g}\frac{(p_1 - p_r)^2}{p_r} \approx 0.06\frac{c_p(p_r - p_1)^2(\theta_1 - \theta_2)}{p_r g}. \tag{7.8.24}$$

Since the mass per unit area is $(p_r - p_1)/g$, the second formula gives the mean square velocity U^2 that would be obtained as a result of conversion of the available energy by

$$U^2 \approx 120(\theta_1 - \theta_2)(p_r - p_1)/p_r \tag{7.8.25}$$

in SI units, use being made of (3.3.3) with $q = 0$. A formula for calculating U was obtained by Margules in a similar form. As an example, if p_1 is 700 mbar and $\theta_1 - \theta_2$ is 5 K, U is 14 m s^{-1}. Comparison of the approximation (7.8.24) when $p_1 = 0$ with the exact result (7.8.23) for this extreme case suggests that the approximate formula will usually give an underestimate. This is consistent with Dutton and Johnson's (1967) experience that the small-perturbation formula for A applied to real situations gave answers that were generally 5% below the exact value.

Lorenz (1955, 1960) generalized this result by considering the atmosphere as made up of stratified columns each in equilibrium. If the whole depth of the atmosphere is involved, (7.8.19) and (7.8.20) give

$$P + I = c_p \iiint (p/p_r)^\kappa \theta \, dM, \tag{7.8.26}$$

where

$$dM = -g^{-1} dp \, dx \, dy \tag{7.8.27}$$

is interpreted as the mass of one of the layers that make up the column. When moved to its state of minimum potential plus internal energy, each isentropic layer will become flat. Its mass dM will be conserved, and so will the mean pressure $\bar{p}$ since this is proportional to the mass of fluid above the isentropic layer. Thus when the surface is flat, the pressure on it will be $\bar{p}$ everywhere, so (7.8.26) gives the available potential energy A as

$$A = c_p \iiint \{(p/p_r)^\kappa - (\bar{p}/p_r)^\kappa\}\theta \, dM. \tag{7.8.28}$$

Lorenz also adopted the convention that the pressure on any isentrope that descends below the surface is equal to the surface pressure at the same horizontal position. He also showed that for small perturbations this reduces to the formulas found in Section 6.18, using isobaric coordinates, because the hydrostatic approximation is basic to that description as well. He did not, however, discuss the surface contribution in detail. Such a discussion may be found in the work by Dutton and Johnson (1967).

7.9 Circulation and Vorticity

The study of adjustment to equilibrium in Section 7.2 showed the importance of the concept of "potential vorticity," which is based on properties possessed by the vorticity

$$\zeta = (\xi, \eta, \zeta) \tag{7.9.1}$$

of a fluid [the modern name for this quantity was introduced by Lamb (1916) in the fourth edition of his "Hydrodynamics"], which is defined as the curl of the velocity, i.e.,

$$\zeta = \mathbf{V} \times \mathbf{u} = \left(\frac{\partial w}{\partial y} - \frac{\partial v}{\partial z}, \frac{\partial u}{\partial z} - \frac{\partial w}{\partial x}, \frac{\partial v}{\partial x} - \frac{\partial u}{\partial y} \right). \tag{7.9.2}$$

This section reviews the concept of vorticity, and the related one of circulation.

Vorticity may be identified in various ways with the local rate of rotation of a fluid element [Truesdell (1954a, Chapter 3) gives details with historical references]. Such an identification requires an analysis of the relative motion of neighboring fluid particles. If these are denoted by A and B and their relative displacement by $\delta \mathbf{x} = \mathbf{x}_A - \mathbf{x}_B$, the rate of change of this displacement is

$$D(\delta\mathbf{x})/Dt = D\mathbf{x}_A/Dt - D\mathbf{x}_B/Dt = \mathbf{u}_A - \mathbf{u}_B = \delta\mathbf{u}. \tag{7.9.3}$$

Now the change $\delta\gamma$ in any quantity γ for a displacement $\delta\mathbf{x}$ is given by

$$\delta\gamma = \delta x \,\partial\gamma/\partial x + \delta y \,\partial\gamma/\partial y + \delta z \,\partial\gamma/\partial z = (\delta\mathbf{x} \cdot \mathbf{V})\gamma, \tag{7.9.4}$$

and since this applies to each component of $\delta\mathbf{x}$,

$$\delta\mathbf{u} = (\delta\mathbf{x} \cdot \mathbf{V})\mathbf{u}. \tag{7.9.5}$$

where $\delta\mathbf{x} \cdot \mathbf{V}$ is the operator defined by (7.9.4).

Combining (7.9.3) and (7.9.5) now gives

$$D(\delta\mathbf{x})/Dt = (\delta\mathbf{x} \cdot \mathbf{V})\mathbf{u}. \tag{7.9.6}$$

A useful alternative expression for the right-hand side is

$$D(\delta\mathbf{x})/Dt = \tfrac{1}{2}\mathbf{V}_\delta G + \tfrac{1}{2}\zeta \times \delta\mathbf{x}, \tag{7.9.7}$$

where

$$G = \frac{\partial u}{\partial x}\delta x^2 + \frac{\partial v}{\partial y}\delta y^2 + \frac{\partial w}{\partial z}\delta z^2 + \left(\frac{\partial u}{\partial y} + \frac{\partial v}{\partial x} \right)\delta x\,\delta y$$

$$+ \left(\frac{\partial u}{\partial z} + \frac{\partial w}{\partial x} \right)\delta x\,\delta z + \left(\frac{\partial v}{\partial z} + \frac{\partial w}{\partial y} \right)\delta y\,\delta z, \tag{7.9.8}$$

and $\mathbf{V}_\delta$ denotes the gradient with respect to $\delta\mathbf{x}$, velocity derivatives being regarded as constant. The equality of the right-hand sides of (7.9.6) and (7.9.7) is easily confirmed by calculating components such as that in the direction of the x axis, namely,

$$\frac{\partial u}{\partial x}\delta x + \frac{\partial u}{\partial y}\delta y + \frac{\partial u}{\partial z}\delta z = \frac{1}{2}\frac{\partial G}{\partial(\delta x)} + \frac{1}{2}\left(\frac{\partial u}{\partial z} - \frac{\partial w}{\partial x} \right)\delta z - \frac{1}{2}\left(\frac{\partial v}{\partial x} - \frac{\partial u}{\partial y} \right)\delta y.$$

Thus the motion in the neighborhood of a fluid particle can always be resolved into two parts corresponding to the terms on the right-hand side of (7.9.7). The second part [cf. (4.5.2)] is a pure rotation with angular velocity $\frac{1}{2}\zeta$, which explains why $\frac{1}{2}\zeta$ is identified with the local rate of rotation of fluid elements. Early writers used names like "angular velocity" or "rotation velocity" for this quantity. The motion due to the $\frac{1}{2}\nabla_\delta G$ term follows from the fact that contours of G are ellipsoidal surfaces, so the motion due to this term is normal to these surfaces. In particular, the motion along the axes of the ellipsoid is linear, and a surface that is initially spherical will distort into an ellipsoid with these axes. This is called a pure straining motion, so (7.9.7) shows that motion in the neighborhood of a fluid particle can always be regarded as the resultant of a pure straining motion and a pure rotation. More detailed discussion can be found, e.g., in the work of Batchelor (1967, Section 2.3).

Now the vorticity ζ appears in an important formula for the acceleration of a fluid element, which was derived by Lagrange (1781). The formula is

$$D\mathbf{u}/Dt \equiv \partial\mathbf{u}/\partial t + \zeta \times \mathbf{u} + \nabla(\tfrac{1}{2}\mathbf{u}^2) \tag{7.9.9}$$

and is easily verified by considering individual components, e.g.,

$$\frac{Du}{Dt} \equiv \frac{\partial u}{\partial t} + u\frac{\partial u}{\partial x} + v\frac{\partial u}{\partial y} + w\frac{\partial u}{\partial z}$$

$$\equiv \frac{\partial u}{\partial t} + u\frac{\partial u}{\partial x} + v\frac{\partial v}{\partial x} + w\frac{\partial w}{\partial x} - v\left(\frac{\partial v}{\partial x} - \frac{\partial u}{\partial y}\right) + w\left(\frac{\partial u}{\partial z} - \frac{\partial w}{\partial x}\right).$$

It can be seen (cf. Section 4.5) that this formula anticipates the result of Coriolis (1835) for the components of acceleration relative to a frame rotating with the fluid.

Although the quantity ζ appears in the early papers in which the equations of fluid mechanics were first developed, some hundred years elapsed before Helmholtz (1858) deduced from these equations some of the important properties possessed by vorticity, which established the fundamental role it plays in fluid mechanics. The vorticity equation itself follows from the identity

$$\nabla \times D\mathbf{u}/Dt \equiv D\zeta/Dt - (\zeta\cdot\nabla)\mathbf{u} + \zeta(\nabla\cdot\mathbf{u}), \tag{7.9.10}$$

which can be derived from (7.9.9), e.g., by calculating components such as

$$\frac{\partial}{\partial x}\left(\frac{\partial v}{\partial t} + \zeta u - \xi w\right) - \frac{\partial}{\partial y}\left(\frac{\partial u}{\partial t} + \eta w - \zeta v\right) \equiv \frac{\partial\zeta}{\partial t} + u\frac{\partial\zeta}{\partial x} + v\frac{\partial\zeta}{\partial y} - w\left(\frac{\partial\xi}{\partial x} + \frac{\partial\eta}{\partial y}\right)$$

$$- \xi\frac{\partial w}{\partial x} - \eta\frac{\partial w}{\partial y} + \zeta\left(\frac{\partial u}{\partial x} + \frac{\partial v}{\partial y}\right)$$

and using the identity

$$\nabla\cdot\zeta \equiv \partial\xi/\partial x + \partial\eta/\partial y + \partial\zeta/\partial z \equiv 0, \tag{7.9.11}$$

which follows from the definition (7.9.2). An alternative form of (7.9.10) is obtained by combining it with the mass conservation equation (4.2.3) to give

$$\frac{1}{\rho} \mathbf{V} \times \frac{D\mathbf{u}}{Dt} \equiv \frac{D}{Dt}\left(\frac{\zeta}{\rho}\right) - \left(\frac{\zeta}{\rho} \cdot \mathbf{V}\right)\mathbf{u}. \tag{7.9.12}$$

The dynamical equations are now utilized by substituting for $D\mathbf{u}/Dt$ in (7.9.12), and the result depends on what assumptions are made about the fluid. Helmholtz' (1858) results were, in particular, for the case of a *homogeneous inviscid* fluid, considering motion relative to a *fixed frame of reference*. Then $D\mathbf{u}/Dt$ is the gradient of a scalar, $-\Phi - p/\rho$, and so the curl vanishes. Hence (7.9.12) reduces to

$$\frac{D}{Dt}\left(\frac{\zeta}{\rho}\right) = \left(\frac{\zeta}{\rho} \cdot \mathbf{V}\right)\mathbf{u}. \tag{7.9.13}$$

Since ρ is constant, the factor ρ may be removed from both sides. Helmholtz' results can also be applied to a nonhomogeneous fluid when the pressure is a function of density only; for then $D\mathbf{u}/Dt$ is still the gradient of a scalar, $-\Phi - \int dp/\rho$, and the left-hand side of (7.9.12) still vanishes, yielding (7.9.13) once again. Such a fluid automatically has pressure surfaces and density surfaces that coincide, and is sometimes called (cf. Section 6.2) an *autobarotropic* fluid. An example is an ideal homentropic fluid (i.e., a fluid with uniform potential temperature θ) with uniform composition s; for then (4.10.8) and (4.10.9) ensure that these uniform values will be maintained, and hence the equation of state (4.10.5) reduces to the statement that density is a function of pressure only.

Helmholtz' deductions from (7.9.13) follow because it has the same form as (7.9.6) for what can be called a material line element $\delta\mathbf{x}$. It follows that a material line element initially aligned with the vorticity vector ζ will remain so aligned. Helmholtz expressed this idea by introducing the notion of a *vortex filament*, i.e., a material line of particles such that the tangent to the line is everywhere in the direction of the vorticity. Since each segment of the filament remains a material line element, Helmholtz' result that a vortex filament retains its identity follows.

A further result follows by introducing a quantity χ that varies along the filament in such a way that the displacement between neighboring points on the filament is given by

$$\delta\mathbf{x} = \rho^{-1}\zeta\,\delta\chi. \tag{7.9.14}$$

In other words, χ is the integral of $\rho/|\zeta|$ along the vortex filament. Substituting in (7.9.6) gives

$$\frac{D}{Dt}\left(\frac{\zeta}{\rho}\,\delta\chi\right) \equiv \delta\chi\,\frac{D}{Dt}\left(\frac{\zeta}{\rho}\right) + \frac{\zeta}{\rho}\,\frac{D}{Dt}(\delta\chi) = \delta\chi\left(\frac{\zeta}{\rho} \cdot \mathbf{V}\right)\mathbf{u},$$

and applying (7.9.13) reduces this to

$$D(\delta\chi)/Dt = 0. \tag{7.9.15}$$

In other words $\delta\chi$ (and hence χ), if measured from a fixed particle, is constant. This is a way of expressing Helmholtz' result that ζ/ρ varies in proportion with the length of the local segment of the vortex filament.

A closely related way of expressing these properties of a fluid is in terms of the circulation C introduced by Kelvin (Thomson, 1869). This is the property of a closed curve in space, and the circulation is defined as the line integral around the closed curve denoted by

$$C = \oint \mathbf{u} \cdot d\mathbf{s} = \oint (u\,dx + v\,dy + w\,dz). \tag{7.9.16}$$

Figure 7.12a shows the meaning of this expression, and Fig. 7.12b shows the contributions for a particular rectangular circuit for which the connection between circulation and vorticity is apparent. In general, the relationship is given by Stokes' theorem

$$C = \oint \mathbf{u} \cdot d\mathbf{s} = \int \int \nabla \times \mathbf{u} \cdot d\mathbf{S} = \int \int \zeta \cdot d\mathbf{S}, \tag{7.9.17}$$

where the double integrals can be evaluated over any surface whose perimeter is the circuit with which the circulation is associated and $d\mathbf{S}$ denotes an element of area, i.e., has magnitude equal to the area of the element and direction normal to the surface. Equation (7.9.17) also allows vorticity to be interpreted as circulation per unit area.

Kelvin's result is for a *material circuit*, i.e., a closed curve that always consists of the same fluid particles. (One can find results for other circuits, but they are not so useful.) Then the rate of change of the contribution $\mathbf{u} \cdot \delta\mathbf{s}$ of a small part of the circuit is given by

$$\frac{D}{Dt}(u\,\delta x + v\,\delta y + w\,\delta z) = \frac{Du}{Dt}\delta x + \frac{Dv}{Dt}\delta y + \frac{Dw}{Dt}\delta z + u\,\delta u + v\,\delta v + w\,\delta w.$$

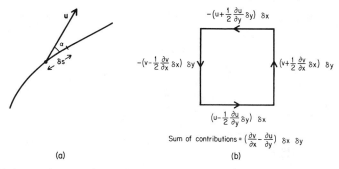

(a)   (b)

Fig. 7.12.   (a) The contribution to the circulation from a small length δs of a material circuit is defined as $\mathbf{u} \cdot \delta\mathbf{s} = u\,\delta s \cos\alpha$, where $\mathbf{u}$ is the fluid velocity at the section in question, $\delta\mathbf{s}$ is a vector of length δs and directed along the circuit, and α is the angle between the two vectors $\mathbf{u}$ and $\delta\mathbf{s}$. In terms of Cartesian components, $\mathbf{u} = (u, v, w)$, $\delta\mathbf{s} = (\delta x, \delta y, \delta z)$, and $\mathbf{u} \cdot \delta\mathbf{s} = u\,\delta x + v\,\delta y + w\,\delta z$. (b) Calculation of the circulation of a small rectangular circuit. The arrows show the direction in which the circuit is traversed, and hence the direction of $\delta\mathbf{s}$. In this case, the circulation is equal to the vorticity component $\zeta = \partial v/\partial x - \partial u/\partial y$ normal to the circuit times the area $\delta x\,\delta y$ of the circuit. The generalization of this result to an arbitrary circuit can be obtained by dividing up the circuit into a large number of infinitesimal sections for each of which the above result can be applied. Thus, in general, the circulation is equal to $\int\int \zeta \cdot d\mathbf{S}$, where $d\mathbf{S}$ is a vector with magnitude equal to the area of an infinitesimal section and with direction normal to that area. The integral can be evaluated over any surface whose perimeter is the circuit involved.

It follows that for any segment of the circuit

$$\frac{D}{Dt} \int \mathbf{u} \cdot d\mathbf{s} = \int \frac{D\mathbf{u}}{Dt} \cdot d\mathbf{s} + [\tfrac{1}{2}\mathbf{u}^2],$$

where the last term is the difference in the values between the end points. This vanishes for a closed curve, giving Kelvin's identity

$$\frac{D}{Dt} \oint \mathbf{u} \cdot d\mathbf{s} \equiv \oint \frac{D\mathbf{u}}{Dt} \cdot d\mathbf{s}. \tag{7.9.18}$$

Now the dynamical equations can be used to substitute for $D\mathbf{u}/Dt$ and the results will depend on the assumptions made about the fluid. Kelvin considered the particular case of a homogeneous inviscid fluid for which $D\mathbf{u}/Dt$ is the gradient of a scalar, $-\Phi - p/\rho$. For any segment of the circuit

$$\int \mathbf{\nabla}\left(\frac{p}{\rho} + \Phi\right) \cdot d\mathbf{s} = \int \left\{ \frac{\partial}{\partial x}\left(\frac{p}{\rho} + \Phi\right) dx + \frac{\partial}{\partial y}\left(\frac{p}{\rho} + \Phi\right) dy + \frac{\partial}{\partial z}\left(\frac{p}{\rho} + \Phi\right) dz \right\}$$

$$= \left[\frac{p}{\rho} + \Phi\right],$$

where the right-hand side is the difference between the values of $\Phi + p/\rho$ at the two ends of the segment. This vanishes for a closed circuit, so (7.9.18) gives *Kelvin's circulation theorem*, namely, that the circulation of a material circuit is constant for a homogeneous inviscid fluid. As found above, $D\mathbf{u}/Dt$ is also the gradient of a scalar for an autobarotropic fluid, so Kelvin's result applies in that case as well.

Another useful concept is that of a *vortex tube*, i.e., the tube made up of all vortex filaments passing through a particular material circuit. Figure 7.13 shows a segment of such a tube. Since by (7.9.11) the divergence of ζ is zero, the flux $\iint \zeta \cdot d\mathbf{S}$ out of any

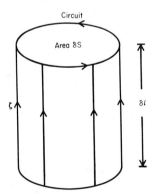

Circuit

Area δS

ζ

δl

Fig. 7.13. A vortex tube is made up of all the vortex filaments passing through a particular material circuit. The diagram shows a small length δl of such a tube, which in this case has small cross-sectional area δS. The flux $\iint \zeta \cdot d\mathbf{S}$ through any cross section of the tube is constant since the divergence of ζ is zero by definition, and hence the circulation C for any circuit around the tube has a fixed value. In the case shown, for which the cross-sectional area is small, $C = \zeta \delta S$ to first order. In a homogeneous fluid, vortex filaments are material lines, and so a segment of vortex tube moves as a material volume, preserving its mass $\delta M = \rho \delta S \delta l$. Hence $C/\delta M = \zeta/\rho \delta l$ is constant, so during the motion ζ/ρ varies in proportion to the length δl of the segment of vortex tube.

volume is zero (cf. Section 4.6). Now the flux over the sides of a vortex tube is zero, so the fluxes across the two ends of the vortex tube segment must be equal, i.e., by (7.9.17) the circulation around any circuit of the tube is equal to that around any other, so a unique value can be assigned to the circulation of a vortex tube.

Now, since the segment of the vortex tube is a material segment, its mass $dM = \rho \, dS \, dl$ will remain constant, dl being its length and dS its cross-sectional area. But the circulation $C = \zeta \, dS$ is also constant, so the ratio

$$C/dM = \zeta/\rho \, dl = 1/d\chi \tag{7.9.19}$$

is also constant, reproducing Helmholtz' result. It also shows that the quantity χ, defined by (7.9.14), is the ratio of the mass of a piece of vortex tube to its circulation.

7.10   Conservation of Potential Vorticity
for a Shallow Homogeneous Layer

A key factor in Rossby's solution of the adjustment problem was the manipulation of the equations to give one equation, namely, (7.2.8), that could be integrated immediately and thus give a relationship between the initial state and later ones of the system. The conserved quantity was called perturbation potential vorticity. The conservation law is, in fact, much more general than the special one for small perturbations, and it will now be investigated. Although the result can be obtained directly from those of Section 7.9, there is some advantage in repeating the manipulations involved for the simpler case of shallow-water motion of a homogeneous fluid.

As found in Section 5.6, the hydrostatic equation (which applies to shallow water motion, i.e., motion with horizontal scale large compared with the depth) gives perturbation pressure independent of depth, so the velocity will also have this property if it does so initially. Assuming this, the shallow-water approximation to the momentum equations (4.10.2) is

$$\frac{\partial u}{\partial t} + u\frac{\partial u}{\partial x} + v\frac{\partial u}{\partial y} - fv = -g\frac{\partial \eta}{\partial x},$$
$$\frac{\partial v}{\partial t} + u\frac{\partial v}{\partial x} + v\frac{\partial v}{\partial y} + fu = -g\frac{\partial \eta}{\partial y}. \tag{7.10.1}$$

When he derived the equations of motion, Euler (1755) saw the advantage of eliminating the pressure (i.e., η) from these equations by taking the curl. However, to take full advantage of this requires some manipulation of the acceleration terms, namely, that derived by Lagrange (1781) and given by (7.9.9). Applied to (7.10.1), (7.9.9) gives

$$\partial u/\partial t - (f + \zeta)v = -\partial B/\partial x,$$
$$\partial v/\partial t + (f + \zeta)u = -\partial B/\partial y, \tag{7.10.2}$$

where ζ is the vertical component of relative vorticity defined by (7.2.6) and B is the

Bernoulli function (see Section 4.8) defined by

$$B = g\eta + \tfrac{1}{2}(u^2 + v^2).$$ (7.10.3)

The term $(f + \zeta)$ in (7.10.2) is the *absolute vorticity*, i.e., the vorticity relative to a fixed frame of reference. This follows from (4.5.2), which gives

$$\mathbf{u}_f = \mathbf{u} + \mathbf{\Omega} \times \mathbf{x},$$ (7.10.4)

where $\mathbf{\Omega}$, the angular velocity, is a vertical vector of magnitude $\tfrac{1}{2}f$, $\mathbf{x}$ is position relative to the rotation axis, and subscript f denotes values relative to a fixed frame. It can be shown that (7.10.4) implies that

$$\partial v_f/\partial x_f - \partial u_f/\partial y_f = f + \zeta.$$ (7.10.5)

The result also follows from the identification of $\tfrac{1}{2}\zeta$ as the local rate of rotation of fluid elements. The value relative to a fixed frame will be the sum of the rotation vector for the frame and the rotation vector for the fluid element relative to the rotating frame.

The vorticity equation can now be obtained by eliminating B from (7.10.2) to give

$$D\zeta/Dt + (f + \zeta)(\partial u/\partial x + \partial v/\partial y) = 0,$$ (7.10.6)

or alternatively,

$$\frac{1}{f + \zeta} \frac{D}{Dt}(f + \zeta) + \left(\frac{\partial u}{\partial x} + \frac{\partial v}{\partial y}\right) = 0.$$ (7.10.7)

This is the shallow water version of (7.9.13) (but expressed relative to the rotating frame rather than the fixed frame), and follows from that equation and the fact that the horizontal velocity components are depth-independent and hence the vorticity vector is vertical. Equation (7.10.7) shows that if the motion is divergent, i.e., if $\partial u/\partial x + \partial v/\partial y > 0$, then the magnitude of the absolute vorticity will decrease. In particular, if there is no initial relative vorticity, divergent motion will cause the fluid to acquire anticyclonic vorticity. Conversely, convergent motion would cause the fluid to acquire cyclonic vorticity. These relationships are important and have been used, for instance, to explain the acquisition of cyclonic vorticity in a storm resulting from convergence of fluid toward the storm center. [An early qualitative argument can be found, for instance, in the paper by Tracy (1843).]

The next step is based on the fact that convergence leads to accumulation of mass and hence to raising of the free surface, as expressed by the continuity equation (5.6.7), which can be written as

$$\frac{1}{H + \eta} \frac{D}{Dt}(H + \eta) + \left(\frac{\partial u}{\partial x} + \frac{\partial v}{\partial y}\right) = 0.$$ (7.10.8)

Subtraction of (7.10.8) from (7.10.7) eliminates the divergence and gives

$$DQ/Dt = 0,$$ (7.10.9)

where

$$Q = (f + \zeta)/(H + \eta)$$ (7.10.10)

is called the *potential vorticity* for homogeneous shallow-water motion. The perturbation potential vorticity Q' for infinitesimal disturbances is readily confirmed to be given by (7.2.9). Equation (7.10.9) was obtained in a form similar to this by Rossby (1936), who later introduced the name potential vorticity (Rossby, 1940).

The result (7.10.9) can also be derived directly from Helmholtz' result as obtained in Section 7.9. Since the absolute vorticity $f + \zeta$ is purely vertical and independent of depth, it varies in proportion with the length $H + \eta$ of a vortex filament, which in this case is a vertical line of fluid confined between the upper and lower boundaries. Q is in fact the inverse of the quantity χ introduced in Section 7.9 and therefore [see (7.9.19)] can also be identified as the volume of a vortex tube (in this case a vertical cylinder of fluid extending from top to bottom) divided by its circulation. The important result can be expressed very simply: stretching of a vortex line causes acquisition of *cyclonic* vorticity; shrinking produces anticyclonic vorticity. This result will be applied over and over again in later chapters.

The concept of potential vorticity is of fundamental importance, because of the conservation property (7.10.9), in the same way that the concept of vorticity is important in the study of strictly two-dimensional (i.e., z-independent) flows. In fact, shallow water flow in a region of constant depth becomes strictly two dimensional in the limit in which $g \to \infty$, but $g\eta$ remains finite in order to preserve pressure gradients because then the free surface cannot move (the rigid lid approximation—see Section 6.3) and $H + \eta$ remains constant. The relationship is worth remembering because many useful results derived for two-dimensional flows can readily be generalized to shallow-water motions.

Consider, in particular, the case of steady flow. Then (7.10.8) allows a stream function ψ to be defined by

$$(H + \eta)u = -\partial\psi/\partial y, \qquad (H + \eta)v = \partial\psi/\partial x, \qquad (7.10.11)$$

and (7.10.9) shows that Q is constant along a streamline, i.e.,

$$Q = Q(\psi). \qquad (7.10.12)$$

Furthermore, when (7.10.11) is used to substitute for u and v in (7.10.2), the two equations show that the Bernoulli function B is constant along streamlines as well, and

$$Q(\psi) = dB(\psi)/d\psi. \qquad (7.10.13)$$

For the case of constant H in the limit as $g \to \infty$, this reduces to the result obtained by d'Alembert (1761) and Lagrange (1781), that vorticity is constant on streamlines in strictly two-dimensional flow.

Another useful way of looking at (7.10.1) in the case of steady motion is in terms of local (or "natural") coordinates (s, n) corresponding to axes parallel and perpendicular to the local direction of motion (see Fig. 7.14). Thus **u** is written

$$\mathbf{u} = U\hat{t}, \qquad (7.10.14)$$

where U is the speed and $\hat{t}$ is a vector of unit magnitude in the direction of the velocity vector **u**. The coordinate s measures distance along the path of a fluid particle, and

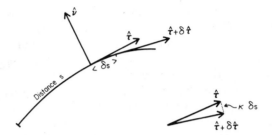

Fig. 7.14. Local coordinates (s, n) for a fluid particle moving along the horizontal path shown by the curved line, where s is distance along the path and n is distance normal to the path. $\hat{t}$ represents the unit vector tangential to the path and $\hat{v} = \mathbf{k} \times \hat{t}$ is the unit vector normal to the path, where $\mathbf{k}$ is the unit vector pointing vertically upward. n increases in the direction of $\hat{v}$. In a small distance δs, the unit tangent rotates. Since the length of the tangent vector remains constant, $\delta\hat{t}$ must be normal to $\hat{t}$. The curvature κ is defined by (7.10.17), and so the distance between the ends of the two unit vectors is $\kappa\,\delta s$ as shown. The inverse of κ is called the radius of curvature R.

thus its speed U is given by

$$U = Ds/Dt. \tag{7.10.15}$$

The acceleration of a fluid particle can be obtained by differentiating (7.10.14). Since s is the only coordinate that varies with time along a streamline, the derivative may be written

$$\frac{D\mathbf{u}}{Dt} = \left(\frac{\partial U}{\partial s} \hat{t} + U \frac{\partial \hat{t}}{\partial s} \right) \frac{Ds}{Dt}. \tag{7.10.16}$$

Now since $\hat{t}$, is a unit vector, it cannot change in length, and therefore can change only in direction. Hence the derivative $\partial \hat{t}/\partial s$ is at right angles to $\hat{t}$. The direction of this vector is defined by $\hat{v}$, called the unit normal to the curve, and the magnitude of the vector is defined as the *curvature* κ of the curve (not to be confused with wavenumber magnitude, which is given the same symbol elsewhere), i.e.,

$$\partial \hat{t}/\partial s = \kappa \hat{v}. \tag{7.10.17}$$

This definition allows an arbitrariness of sign, which is removed if $\hat{v}$ is defined by

$$\hat{v} = \mathbf{k} \times \hat{t}, \tag{7.10.18}$$

where $\mathbf{k}$ is the unit vector pointing vertically upward. The reciprocal R of κ is called the *radius of curvature*. The definition (7.10.18) implies that the curvature is positive when the particle is curving toward the left and negative when it is curving toward the right.

Now use of (7.10.15)–(7.10.18) in (7.10.1) gives

$$U \frac{\partial U}{\partial s} \hat{t} + (\kappa U^2 + fU)\hat{v} = -g\,\nabla\eta. \tag{7.10.19}$$

The component of $\nabla\eta$ along the particle path is $\partial\eta/\partial s$ and if the coordinate at right angles to the path is called n, the component in this direction is $\partial\eta/\partial n$. Thus (7.10.19) yields, in component form,

$$U \, \partial U/\partial s = -g \, \partial \eta/\partial s, \tag{7.10.20}$$

$$\kappa U^2 + f U = -g \, \partial \eta/\partial n. \tag{7.10.21}$$

The first equation says that the rate of change of particle speed is due only to the component of pressure gradient along the path and may be integrated to show that the Bernoulli function

$$B = g\eta + \tfrac{1}{2} U^2 \tag{7.10.22}$$

is constant along the path, as already found. The second equation shows that the pressure gradient across streamlines is balanced by a Coriolis acceleration term $f U$ and a centrifugal acceleration term $\kappa U^2 = U^2/R$. For the flow to be in approximate geostrophic balance, the centrifugal term must be small compared with the Coriolis term, i.e.,

$$\kappa U/f \equiv U/fR \ll 1. \tag{7.10.23}$$

An alternative form of (7.10.20) and (7.10.21) is (7.10.2), which gives in local coordinates

$$\partial B/\partial s = 0, \quad (f + \zeta)U = -\partial B/\partial n = -g \, \partial \eta/\partial n - U \, \partial U/\partial n \tag{7.10.24}$$

by (7.10.22). Comparison with (7.10.21) shows that

$$\zeta = -\partial U/\partial n + \kappa U = -\partial U/\partial n + U/R, \tag{7.10.25}$$

which could also, of course, be deduced from first principles. This form gives the vorticity as the sum of two parts—one associated with the horizontal shear $\partial U/\partial n$ and the other with the curvature of the path.

Now all of the above analysis for a single shallow homogeneous layer applies equally well for any homogeneous layer in a multilayer system, provided that the gradient of $g\eta$ be replaced by ρ^{-1} times the gradient of perturbation pressure for the layer in question and provided that the horizontal scale of motion in the layer be large compared with the depth of the layer. Since a continuously stratified fluid can be obtained as the limit of a multilayer system as the number of layers goes to infinity, the result applies to a continuously stratified fluid as well, provided that the vertical scale be small compared with the horizontal scale. Thus (7.10.21) becomes

$$\kappa U^2 + f U = -\rho^{-1} \, \partial p/\partial n \quad \text{or} \quad U^2/R + f U = -\rho^{-1} \, \partial p/\partial n. \tag{7.10.26}$$

A wind satisfying this balance is called the *gradient wind*, and this equation needs to replace the geostrophic wind only when (7.10.23) is not satisfied, i.e., in strong flows of large curvature.

The extension of the idea of potential vorticity to multilayer systems, and hence, by taking the limit, to continuously stratified fluids, is due to Rossby (1940). However, there is a more general result, which does not require any scale assumption, and this will be examined in the next section.

It is interesting that some of the ideas that come from the study of Rossby's adjustment problem were anticipated by Shaw (1908) many years earlier. He was considering under what conditions a particle could move over the earth's surface at constant speed and come to the conclusion that there are "three items which may arrange themselves" for this to happen, "viz. pressure gradient, the Earth's rotation and the curvature of the path," i.e., the items that produce the gradient wind. Persistence of such motion, Shaw (1955, p. 141) stated,

> becomes a question of adjusting the combined effect of the two accelerations to balance the effect of pressure. The question at once arises whether it is possible, under practical meteorological conditions, that such an adjustment should be automatic; whether, indeed, by natural process the velocity and the curvature tend to adjust themselves so as to balance the pressure distribution. If that question be answered in the affirmative, a great advance is made in dynamical meteorology.
>
> The adoption of this idea changes the point of view of the meteorologist very materially. If persistent motion with constant speed, the "gradient velocity," becomes the stable and in a sense the "natural" condition of moving air, we must seek first to explain, not the actual motion, but the divergence of the actual motion from the gradient velocity. The gradient velocity thus becomes the starting point of meteorological study.

Shaw's ideas are expounded in the preface to a paper by Gold, in which he, at Shaw's suggestion, compared measured velocities aloft with gradient wind estimated from isobaric charts and found good agreement. Shaw (1955, p. 142) concluded: "The general result of the investigation is, in my opinion, to confirm the suggestion that the adjustment of wind velocity to the gradient is an automatic process which may be looked upon as a primary meteorological law."

Shaw (1955, p. 143) also refers to the fact that Gold

> gives a calculation of the length of time that would be required for air starting from a state of rest to take up the gradient velocity and adjust its motion to the direction of the isobars. The computed times are short compared with such an interval as a day, an ordinary period in meteorological discussion, so that we need not be surprised that, under the slow changes of condition due to the progress of diurnal or of non-periodic changes, the wind appears to adapt itself continuously to these changes.

(Gold's calculation was *not* of the *mutual* adjustment of wind and pressure fields as was Rossby's, but of the motion of a particle in the horizontal due to a fixed force.)

Shaw (1955, p. 145) concludes his discussion by saying

> The whole question of the cause and meaning of the discrepancies between the gradient wind and actual wind is, of course, bound up with the origin of pressure differences. To put the point in a crude form, I do not know whether, in practice, the winds have to adjust themselves to the pressure conditions, or the pressure distribution is the *result* of the motion of the air. I presume, however, that as a matter of fact, it is a case of action and reaction in which each modifies the other.

This conjecture was shown to be true by Rossby's study, discussed in Section 7.2 and 7.3.

7.11   Circulation in a Stratified Fluid and Ertel's Potential Vorticity

The extension of the ideas introduced in Section 7.9 to nonbarotropic fluids first followed Kelvin's concept of circulation rather than Helmholtz' concept of vorticity. The circulation theorem for a stratified fluid is associated with the name of Bjerknes (1898), who not only derived the result but showed its usefulness in applications. Bjerknes was apparently unaware of an earlier derivation of the result by Silberstein (1896). (Bjerknes also derived a result for the quantity $\oint \rho \mathbf{u} \cdot d\mathbf{s}$. This is called his second circulation theorem, but is not as useful as the one discussed here.)

The rate of change of circulation for a nonhomogeneous fluid is easily calculated from Kelvin's result (7.9.18) upon substitution for $D\mathbf{u}/Dt$ from the momentum equation (4.5.1), namely,

$$D\mathbf{u}/Dt = -\rho^{-1} \nabla p - \nabla \Phi. \tag{7.11.1}$$

The result of the substitution is (see 7.9.16)

$$\frac{DC}{Dt} \equiv \frac{D}{Dt} \oint \mathbf{u} \cdot d\mathbf{s} = -\oint \frac{1}{\rho} \nabla p \cdot d\mathbf{s} = -\oint \frac{dp}{\rho}. \tag{7.11.2}$$

The term on the far right vanishes for a homogeneous (or an autobarotropic) fluid, but not, in general, for a stratified fluid. However, in special circumstances, one can choose *special circuits* for which the term does vanish.

Consider an inviscid fluid of uniform composition, i.e., fixed salinity or fixed humidity. Then the equation of state gives ρ to be a function only of pressure and potential temperature, i.e.,

$$\rho = \rho(p, \theta). \tag{7.11.3}$$

Any fluid with this property is called *potentiotropic* (Eliassen and Kleinschmidt, 1957, Section 5). For such a fluid, one can choose a circuit that lies entirely on an equipotential (i.e., isentropic) surface $\theta = $ const. Since θ is conserved in the motion, θ will remain constant on such a circuit, on which density will be a function of pressure only, so the term on the far right in (7.11.2) vanishes, and hence

$$\frac{D}{Dt} \oint \mathbf{u} \cdot d\mathbf{s} = 0 \tag{7.11.4}$$

for isentropic material circuits. This is the circulation theorem for a potentiotropic fluid. The result does *not* hold for a fluid of variable composition *unless* the composition is uniform on isentropic surfaces, i.e.,

$$q = q(\theta) \quad \text{or} \quad s = s(\theta). \tag{7.11.5}$$

This condition is sufficient to make (7.11.3) true, i.e., to ensure that the fluid is potenti-otropic. [Note that the essential point is to be able to choose a circuit on which the term on the far right of (7.11.2) always vanishes. Another possibility, for instance, would be to choose, for a homentropic fluid, a circuit of fixed composition.]

Now consider the vorticity equation for a stratified fluid. This is obtained by substituting for $D\mathbf{u}/Dt$ from (7.11.1) in the identity (7.9.12), the result being

$$\frac{D}{Dt}\left(\frac{\boldsymbol{\zeta}}{\rho}\right) = \left(\frac{\boldsymbol{\zeta}}{\rho} \cdot \nabla\right)\mathbf{u} + \frac{\mathbf{B}}{\rho}, \tag{7.11.6}$$

where

$$\mathbf{B} = \frac{\nabla\rho \times \nabla p}{\rho^2} = -\nabla\left(\frac{1}{\rho}\right) \times \nabla p = -\nabla \times \left(\nabla\frac{p}{\rho}\right) \tag{7.11.7}$$

is called the *baroclinicity vector*. (The vector $\mathbf{B}$ should not be confused with the scalar B denoting the Bernoulli function in Section 7.10.) Thus there is a tendency to generate vorticity [represented by the term $\mathbf{B}/\rho$ in (7.11.6)] whenever density surfaces are inclined to pressure surfaces. An example is provided by the initial condition of Marsigli's experiment (see Section 5.1) but with the density jump smoothed out into a uniform horizontal gradient G. This is illustrated in Fig. 7.15a. The density is given by $\rho = \rho_r - Gx$, and the fluid is assumed to be at rest, so the pressure p is given by $p = -g(\rho_r - Gx)z$. The gradient vectors for ρ and p have the directions shown, so the baroclinicity vector is directed out of the page. Thus when the partition is removed,

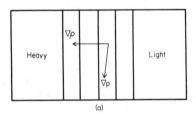

(a)

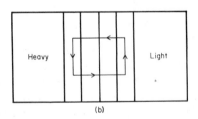

(b)

Fig. 7.15. The initial condition of Marsigli's experiment, but with the horizontal density step smoothed out into a uniform gradient G so that density ρ is given by $\rho = \rho_r - Gx$. The fluid is assumed at rest, so the hydrostatic equation gives the pressure by $p = -g(\rho_r - Gx)z$. (a) The directions of $\nabla\rho = (-G, 0)$ and $\nabla p = g(Gz, -\rho_r + Gx)$ are shown, and therefore the baroclinicity vector $\mathbf{B} = (\nabla\rho \times \nabla p)/\rho^2$ points out of the page toward the reader. Hence vorticity with this direction is generated. (b) The circulation generating term $-\oint dp/\rho$ can be calculated for the rectangular circuits shown. Contributions $-\int \rho^{-1} dz \, \partial p/\partial z$ from the two vertical sides cancel, but the contributions $-\int \rho^{-1} dx \, \partial p/\partial x = -\int gGz(\rho_r - Gx)^{-1} dx$ from the two horizontal sides do not because of the different values of z. If the integration is in the direction of the arrows, the value is positive and thus circulation in this direction is generated.

vorticity is generated with direction the same as that of **B**. The same is true of any situation in which there is a tendency to create horizontal gradients of density, e.g., by heating or cooling, for then $\nabla \rho$ will not be vertical and therefore will not coincide with the direction that ∇p has when the fluid is at rest.

The same situation may be analyzed in terms of circulation by calculating $-\oint dp/\rho$ for a small circuit as shown in Fig. 7.15b. The contributions from the two horizontal sides of the circuit do not cancel because the horizontal pressure changes are different. Thus circulation is generated, corresponding to rotation in the same sense as the vorticity—which must be so because of the identity (7.9.17), i.e., since vorticity is circulation per unit area. In fact (7.11.2) can be written in the form

$$\frac{D}{Dt} \oint \mathbf{u} \cdot d\mathbf{s} = \int\int \mathbf{B} \cdot d\mathbf{S} \tag{7.11.8}$$

since

$$-\oint \frac{1}{\rho} \nabla p \cdot d\mathbf{s} = -\int\int \nabla \times \left(\frac{\nabla p}{\rho}\right) \cdot d\mathbf{S} = \int\int \mathbf{B} \cdot d\mathbf{S}, \tag{7.11.9}$$

the equality between the second and first expressions resulting from Stokes' theorem. In general $\nabla \rho$ can be expanded in terms of ∇p, $\nabla\Theta$, and ∇s as in Section 3.7, the result for **B** being

$$\rho \mathbf{B} = (-\alpha' \nabla\theta + \beta' \nabla s) \times \nabla p. \tag{7.11.10}$$

Now consider Helmholtz' interpretation of the vorticity equation (Helmholtz, 1858). It is no longer true that a material line element in the direction of the vorticity vector remains aligned with the vorticity vector and changes its length in proportion with changes in the magnitude of the vorticity. This vector relationship is altered by the new term **B**, but this only affects the component in the direction of **B**. The components perpendicular to **B** are not affected.

Consider in particular the case of a potentiotropic fluid. Then $\nabla\theta$ is perpendicular to **B**, so the component of vorticity in the direction of $\nabla\theta$ is not affected. The ratio between the component of ζ/ρ in the direction of $\nabla\theta$ and the distance along this direction is (see Fig. 7.16)

$$\frac{\zeta \cdot \nabla\theta}{\rho \, \delta\theta} = \frac{1}{\delta\chi}, \tag{7.11.11}$$

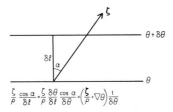

$$\frac{\zeta}{\rho}\frac{\cos\alpha}{\delta l} = \frac{\zeta}{\rho}\frac{\delta\theta}{\delta l}\frac{\cos\alpha}{\delta\theta} = \left(\frac{\zeta}{\rho}\cdot\nabla\theta\right)\frac{1}{\delta\theta}$$

Fig. 7.16. Ertel's potential vorticity $Q = \rho^{-1}\zeta \cdot \nabla\theta$ for a potentiotropic fluid. Helmholtz' result is still true for the component of vorticity in the direction of $\nabla\theta$, i.e., the component $\zeta \cos\alpha$ in this direction changes in proportion with $\rho \, \delta l$ for a material element. Since θ is conserved by a material element as well, it follows from the equation under the diagram that Q is conserved by a material element.

where $\delta\theta$ is the difference in potential temperature between two neighboring isentropic surfaces. The term $\delta\chi$ has the same meaning as in Section 7.9, but here it applies only to one component of vorticity and not to all three. $\delta\chi$ is conserved following the motion, i.e., satisfies (7.9.15), but $\delta\theta$ is conserved as well, so the quantity

$$Q = \rho^{-1}\boldsymbol{\zeta}\cdot\nabla\theta = d\theta/d\chi \tag{7.11.12}$$

is also preserved following the motion, i.e.,

$$DQ/Dt = 0. \tag{7.11.13}$$

This result is due to Ertel (1942a), and Q is called Ertel's potential vorticity or just the potential vorticity. It is *not* the same as the potential vorticity for a homogeneous fluid (although χ is), but has in common the property of being conserved and being proportional to a vorticity over a length. Use of the same name for different quantities does not lead to confusion because one version can be applied only to a homogeneous fluid and the other version only to a stratified fluid. It should be noted that in the above θ could be replaced by any function of ρ and p that is conserved following the motion, as Ertel demonstrated.

The relationship between potential vorticity and circulation follows (Ertel, 1942b) by integrating the potential vorticity over a material mass of fluid, for this integral will be conserved:

$$\int\int\int \rho Q\, dx\, dy\, dz = \int\int\int \boldsymbol{\zeta}\cdot\nabla\theta\, dx\, dy\, dz = \int\int\int \nabla\cdot(\boldsymbol{\zeta}\theta)\, dx\, dy\, dz = \int\int \theta\zeta_n\, dS. \tag{7.11.14}$$

The equality between the second and third expressions follows from the identity (7.9.11) and the last expression from the divergence theorem, ζ_n being the component of $\boldsymbol{\zeta}$ along the outward normal to the material volume. If now the volume is chosen as shown in Fig. 7.17, the outward normal will be oppositely directed on the upper and lower isentropic surfaces and contributions from the sides will be small, so that the

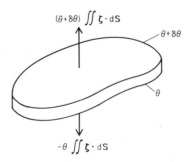

Fig. 7.17. Relation between Ertel's potential vorticity Q and Bjerknes' circulation theorem for a potentiotropic fluid. Since Q is conserved, it follows from (7.11.14) that $\int\int \theta\boldsymbol{\zeta}\cdot d\mathbf{S}$ is conserved for a material volume. If this volume is chosen as the sandwich between two isentropic surfaces, the contributions to the surface integral from these surfaces are as shown, and the contributions from the sides are negligible. Since $\int\int \boldsymbol{\zeta}\cdot d\mathbf{S} = C$ is the circulation on an isentropic surface, the surface integral is equal to $\delta\theta\cdot C$, so C must be conserved for a material circuit.

last expression equals

$$\delta\theta \int\int \boldsymbol{\zeta} \cdot d\mathbf{S} = \delta\theta \cdot C, \tag{7.11.15}$$

where C is the circulation. Since $\delta\theta$ is conserved, C will be also. It can also be seen from (7.11.12) that χ can again be equated with mass divided by circulation.

The results thus far have not taken into account *rotation*, i.e., they apply to motion relative to a fixed frame. To interpret the results in a rotating system, all that is required is to use the relationship (4.5.2) between the velocity relative to a fixed frame (denoted by subscript f) and that relative to a rotating frame (no subscript). Consider first the vorticity $\boldsymbol{\zeta}$. From the interpretation of $\boldsymbol{\zeta}$ as half the local rate of rotation of a material element, it is clear that the absolute vorticity $\boldsymbol{\zeta}_f$ must be given by

$$\boldsymbol{\zeta}_f = \boldsymbol{\zeta} + 2\boldsymbol{\Omega}, \tag{7.11.16}$$

where $\boldsymbol{\Omega}$ is the angular velocity of the rotating frame, and this is readily confirmed from (4.5.2) and the definition (7.9.2) of vorticity. Ertel's potential vorticity is thus given by

$$Q = \rho^{-1}\boldsymbol{\zeta}_f \cdot \nabla\theta = \rho^{-1}(\boldsymbol{\zeta} + 2\boldsymbol{\Omega}) \cdot \nabla\theta. \tag{7.11.17}$$

Similarly, the *absolute circulation* C_f is, by (7.9.17), given by

$$C_f = \int\int \boldsymbol{\zeta}_f \cdot d\mathbf{S} = \int\int \boldsymbol{\zeta} \cdot d\mathbf{S} + \int\int 2\boldsymbol{\Omega} \cdot d\mathbf{S},$$

i.e.,

$$C_f = C + 2\Omega S_{eq}, \tag{7.11.18}$$

where S_{eq} is the projection of the area of the circuit on the equatorial plane (or any other plane perpendicular to the axis of rotation). Thus the circulation theorem (7.11.4) becomes

$$DC_f/Dt \equiv D(C + 2\Omega S_{eq})/Dt = 0 \tag{7.11.19}$$

for isentropic material circuits. This form for a rotating fluid was obtained by Bjerknes (1901).

7.12   Perturbation Forms of the Vorticity Equations in a Uniformly Rotating Fluid

In an inviscid fluid that is rotating with uniform angular velocity about a vertical axis, the horizontal components of the perturbation momentum equations are, from (4.10.2),

$$\partial u/\partial t - fv = -\rho_0^{-1} \, \partial p'/\partial x, \tag{7.12.1}$$

$$\partial v/\partial t + fu = -\rho_0^{-1} \, \partial p'/\partial y. \tag{7.12.2}$$

In the shallow-water case, these reduce to (7.2.1) and (7.2.2). Elimination of the pressure perturbation from (7.12.1) and (7.12.2) yields exactly the same equation (7.2.7) for the vertical component of vorticity ζ as in the shallow-water case, namely,

$$\partial\zeta/\partial t + f(\partial u/\partial x + \partial v/\partial y) = 0. \qquad (7.12.3)$$

A perturbation form of the potential vorticity equation can now be obtained by substituting for the horizontal divergence from the continuity equation, putting $w = \partial h/\partial t$, where h is the vertical displacement of a fluid particle from its equilibrium level. The precise form of the resulting equation depends on whether the fluid is compressible or not.

For an *incompressible* fluid, the continuity equation is (6.4.3), and thus (7.12.3) gives

$$\partial\zeta/\partial t - f\, \partial w/\partial z = 0 \qquad (7.12.4)$$

or

$$\partial(\zeta - f\, \partial h/\partial z)/\partial t = 0. \qquad (7.12.5)$$

If the separable forms for a normal mode are substituted from (6.11.5) and (6.11.10), this reduces to the shallow-water form (7.2.8) of the perturbation potential vorticity equation.

For a *compressible* fluid, the linearized form of the mass conservation equation is (6.14.6), so (7.12.3) in this case yields

$$\frac{\partial}{\partial t}\left(\zeta - \frac{fp'}{\rho_0 c_s^2}\right) - f\left(\frac{\partial w}{\partial z} - \frac{gw}{c_s^2}\right) = 0$$

or

$$\frac{\partial}{\partial t}\left\{\zeta - \frac{fp'}{\rho_0 c_s^2} - f\left(\frac{\partial h}{\partial z} - \frac{gh}{c_s^2}\right)\right\} = 0. \qquad (7.12.6)$$

Alternative forms in terms of density perturbation rather than pressure perturbation are, by (6.14.3),

$$\frac{\partial}{\partial t}(\zeta - f\rho'/\rho_0) - \rho_0^{-1}f\frac{\partial}{\partial z}(\rho_0 w) = 0 \qquad (7.12.7)$$

or

$$\frac{\partial}{\partial t}\left\{\zeta - f\rho'/\rho_0 - \rho_0^{-1}f\frac{\partial(\rho_0 h)}{\partial z}\right\} = 0. \qquad (7.12.8)$$

For the particular case of motion in the form of a normal mode, the variables have the separable form (6.14.31), and substitution in (7.12.8) yields the shallow-water form (7.2.8) of the potential vorticity equation once again, but with H replaced by the equivalent depth H_e.

Equation (7.12.8) may also be related to the perturbation form of Ertel's potential vorticity equation. Using, as usual, a prime to denote the perturbation and subscript zero to denote the value in the rest state, Ertel's equation (7.11.13) has linearized form

$$\partial(Q' + h\, dQ_0/dz)/\partial t = 0, \qquad (7.12.9)$$

where by the definition (7.11.12) of Ertel's potential vorticity,

$$Q_0 = \rho_0^{-1}f\, d\theta_0/dz \qquad (7.12.10)$$

and

$$Q' = \frac{f}{\rho_0} \frac{\partial \theta'}{\partial z} + \frac{1}{\rho_0} \frac{d\theta_0}{dz} \left(\zeta - f \frac{\rho'}{\rho_0} \right). \tag{7.12.11}$$

If (6.14.11) is used to give θ' in terms of h, then (7.12.9) when divided by $\rho_0^{-1} d\theta_0/dz$ gives (7.12.8). Note, however, that the expression inside the curly brackets in (7.12.8) is *not* proportional to perturbation potential vorticity Q' but is proportional to $Q' + h \, dQ_0/dz$.

When the hydrostatic approximation may be made and thus isobaric coordinates can be used, the mass conservation equation is (6.17.6) or (6.17.11) and so (7.12.3) becomes

$$\partial \zeta / \partial t - f \, \partial \varpi / \partial p = 0, \tag{7.12.12}$$

$$\partial \zeta / \partial t - f(\partial w_* / \partial z_* - w_* / H_s) = 0, \tag{7.12.13}$$

or

$$\partial \{ \zeta - f(\partial h_* / \partial z_* - h_* / H_s) \} / \partial t = 0. \tag{7.12.14}$$

When the perturbation is in the form of a normal mode, substitution from (6.17.34) shows that this reduces to the shallow-water form (7.2.8) of the potential vorticity equation. Equation (7.12.14) can also be related to Ertel's equation. Use of the hydrostatic equation (3.5.8) shows that the part of Ertel's potential vorticity (7.11.17) associated with the component normal to isobaric surfaces is

$$-g(\zeta + f) \, \partial \theta / \partial p,$$

and hence

$$Q_0 = -gf \, d\theta_0/dp = \rho_*^{-1} f \, d\theta_0/dz_*, \tag{7.12.15}$$

$$Q'' = -gf \frac{\partial \theta''}{\partial p} - g\zeta \frac{d\theta_0}{dp} = \frac{1}{\rho_*} \left(f \frac{\partial \theta''}{\partial z_*} + \zeta \frac{d\theta_0}{dz_*} \right), \tag{7.12.16}$$

where z_* is related to p and ρ_* by (6.17.8), (6.17.22) and (6.17.29). The perturbation form of Ertel's equation

$$\partial (Q'' + h_* \, dQ_0/dz_*) / \partial t = 0 \tag{7.12.17}$$

reduces to (7.12.14) when the relation (6.18.9) between θ'' and h_* is used.

7.13 Initialization of Fields for Numerical Prediction Schemes

The idea of using numerical methods for weather forecasting was developed by L. F. Richardson before the First World War, and details of the proposed method were eventually published (Richardson, 1922). In the summary Richardson states:

The fundamental idea is that atmospheric pressures, velocities, etc. should be expressed as numbers, and should be tabulated at certain latitudes, longitudes and heights, so as to give a general account of the state of the atmosphere at any instant, over an extended region, up to a height of say 20 kilometres. The numbers in this table are supposed to be given, at a certain initial instant, by means of observations. It is shown that there is an arithmetical method of operating upon these tabulated numbers, so as to obtain a new table representing approximately the subsequent state of the atmosphere after a brief interval of time, δt say. The process can be repeated so as to yield the state of the atmosphere after $2\,\delta t$, $3\,\delta t$ and so on (Richardson, 1922, p. 1).

Apparently, the "investigation grew out of a study of finite differences and first took shape in 1911 as a fantasy which is now relegated to Ch. 11/2" (Richardson, 1922, p. viii). This section describes a forecast factory that, apart from the use of human rather than electronic computers, is remarkably like a modern meteorological office. "A myriad computers are at work upon the weather of the part of the map where each sits, but each computer attends only to one equation or part of an equation." In a large central pulpit

sits the man in charge of the whole theatre; he is surrounded by several assistants and messengers. One of his duties is to maintain a uniform speed of progress in all parts of the globe. In this respect he is like the conductor of an orchestra in which the instruments are slide-rules and calculating machines. But instead of waving a baton he turns a beam of rosy light upon any region which is running ahead of the rest, and a beam of blue light upon those who are behindhand.

Four senior clerks in the central pulpit are collecting the future weather as fast as it is being computed, and dispatching it by pneumatic carrier to a quiet room. There it will be coded and telephoned to the radio transmitting station.

Messengers carry piles of used computer forms down to a storehouse in the cellar.

In a neighbouring building there is a research department, where they invent improvements. But there is much experimenting on a small scale before any change is made in the complex routine of the computing theatre. In a basement an enthusiast is observing eddies in the liquid lining of a huge spinning bowl, but so far the arithmetic proves the better way. In another building are all the usual financial, correspondence and administrative offices. Outside are playing fields, houses, mountains and lakes, for it was thought that those who compute the weather should breathe of it freely (Richardson, 1922, pp. 219–220).

Richardson demonstrated the efficiency of the numerical method by applying it to Laplace's tidal equations, i.e., the shallow-water equations for the globe. Using time steps of $\frac{3}{4}$ hr and a grid of 200 km spacing, he calculated the changes of pressure and "momentum per unit volume" in a limited area, and showed that the agreement with an analytic solution was quite good. However, his attempt to calculate the rate of change of pressure over central Europe using observations for a particular day (May 20, 1910) were not blessed with the same success.

The rate of rise of surface pressure, $\partial p_G/\partial t$, is found on Form P XIII as 145 millibars in 6 hours, whereas observations show that the barometer was nearly steady. This glaring error is examined in detail below in Ch. 9/3, and is traced to errors in the representation of the initial winds (Richardson, 1922, p. 187). ... So we turn to the alternative explanation, which is that stations as far apart as 700 kilometres did not give an adequate representation of the wind in the lower layers. That appears almost certain when one thinks of the irregularities of the surface wind exhibited on the daily weather reports (Richardson, 1922, p. 213).

Thus Chapter 10 of Richardson's book was devoted to the question of "smoothing the initial data." This is necessary, Richardson says, because

> there is a good deal of evidence to show that the wind is full of small "secondary cyclones" or other whirls having the most various diameters. The arithmetical process can only take account individually of such whirls as have diameters greater than the distance between the centers of the red chequers in our co-ordinate chessboard, and this length has been taken provisionally as 400 km (Richardson, 1922, p. 214).

Since "weather prediction by numerical process" became an established method many years later, the problem of initialization or how to appropriately smooth initial data has received a great deal of attention. Many aspects of the problem may be understood in terms of the results of Rossby's adjustment problem. Because of the "gross" or rapid adjustment processes that Rossby studied, the atmosphere is always close to what can be called a "balanced" state. In the simple problem studied by Rossby this balanced state was a geostrophic equilibrium, and indeed the geostrophic balance is nearly achieved over most of the atmosphere. However, in the tropics rotation effects are less important and near the ground friction effects are significant, so it is more general to talk about a "balanced" state rather than to specify that the balance be of the geostrophic variety.

Now when initial data are obtained, these will generally be out of balance due both to instrumental error and the fact that a single measurement will include small-scale and high-frequency components (Richardson's secondary cyclones and other whirls), whereas the model cannot deal with motions that have spatial scales smaller than the grid length or that have time scales smaller than the time interval of the calculation, so such motions are outside the scope of the calculation being attempted. If the unbalanced data are used as an initial condition for the model, and it is capable of handling the rapid adjustment processes such as those studied by Rossby, then the model will undergo a similar adjustment process in an attempt to reach a nearly balanced state. Being a rapid process, the rates of change of pressure, say, will be much larger than the ones observed on weather charts at 6-hr intervals. If this is the case, the rapid changes can be removed by some sort of time averaging to obtain a closer representation of the more subtle changes associated with motion that is always close to a balanced state.

However, models may be designed on the assumption that the changes they are trying to predict have the slower time scales associated with the nearly balanced motion that is observed. In such cases, badly balanced initial data may lead to large errors being generated or to the model calculation becoming unstable. In such cases, appropriate initialization of data is essential. Various methods are used in practice, and references may be found in the works by Wiin-Nielson (1979) and Leith (1980).

Chapter Eight

Gravity Waves
in a Rotating Fluid

8.1 Introduction

The gravity waves that are familiar to us from direct visual observation, whether it be of the sea surface or of Franklin's tumbler (see Section 6.2) of oil and water, are little affected by the rotation of the earth because their scale is too small and hence their frequency is much greater than the frequency f that is associated with rotation effects. It is natural to begin studies of adjustment under gravity with nonrotating systems because of this, but to understand large-scale processes in the atmosphere and ocean it is essential to appreciate how adjustment is affected by rotation, for rotation effects dominate the large-scale behavior. Basic concepts associated with the study of rotating fluids were introduced in Chapter 7. This chapter is specifically about gravity waves in a rotating fluid; in other words, it examines how rotation modifies the waves already studied in Chapters 5 and 6.

In Sections 8.2 and 8.3 we examine the way in which rotation affects a surface wave and/or a given internal wave mode, these effects being important when the horizontal scale is comparable with the Rossby radius of deformation. For internal modes in the ocean, the Rossby radius is only of order 3–30 km, so the horizontal scale does not need to be very large for rotation to be important. A mode affected by rotation is referred to here as a Poincaré wave, and has the distinctive property that the velocity vector continually rotates anticyclonically with time. This property is often observed in the ocean and large lakes. Furthermore, energy is *not* partitioned equally between kinetic and potential forms, the greater part being kinetic.

Sections 8.4–8.6 are about three-dimensional plane waves in an incompressible fluid of constant buoyancy frequency N. The frequency of such waves depends only on the orientation of the wavenumber vector, i.e., depends only on the ratio of horizontal to vertical scale, and lies between the buoyancy frequency and the inertial frequency f. It so happens that in the atmosphere and ocean N is usually much greater than f—an important factor in determining the character of adjustment processes. When $N \gg f$, the high-frequency waves are the ones that are little affected by rotation, and these are the ones with the smallest ratios of horizontal/vertical scale. Rotation becomes important when the horizontal scale exceeds the vertical scale by a factor of N/f, which factor is typically around 100. (Note, however, it can vary a great deal, e.g., N is small in well-mixed regions and f is small in low latitudes.)

The wave properties can be understood by considering the forces on a fluid particle in the plane of motion, which is an inclined plane normal to the wavenumber vector. The restoring force is that due to the component of gravity in the plane and is proportional to the displacement from the equilibrium level. In the absence of rotation, the particle would execute simple harmonic vibrations along the line of greatest slope. However, the component of the Coriolis force in the plane causes the particle to follow an elliptic orbit instead, with movement around the orbit in the anticyclonic sense.

The way the velocity vector rotates with height (at a given time) can be used to distinguish between upward- and downward-propagating waves in the ocean and atmosphere. Near-inertial frequency waves in the ocean appear to propagate mainly downward (i.e., have downward group velocity, which implies upward phase velocity), whereas similar waves observed in the stratosphere have been found to propagate upward.

Sections 8.7–8.10 are devoted to the study of internal waves generated at a horizontal boundary. In particular, this applies to waves generated by topography, whether they be waves in the atmosphere produced by mountains, or waves in the ocean due to abyssal hills. The theory can also be applied to ocean internal waves generated by wind action. Since, however, considerable study has been made of airflow over mountains, the discussion is mainly in terms of this problem.

The character of the response depends very much on the horizontal scale L of the topography in relation to the scales U/N and U/f, where U is the speed of the airflow over the mountains, N the buoyancy frequency, and f the inertial frequency. Waves are produced when L lies between the two scales, with rotation unimportant near the small-scale limit U/N (typically 1 km) and dominant at the large-scale limit U/f (typically 100 km). The magnitude of the effects depends on the vertical structure of the air, i.e., on how U and N vary with height, and on how the height of the topography compares with the scale U/N. When effects are large, considerable damage can be caused by strong winds at ground level on the lee side of the mountains, and the turbulence in the atmosphere above can be a severe hazard to aircraft.

Waves in the atmosphere can travel to considerable heights, but as the atmosphere thins, effects of molecular viscosity and molecular diffusivity become increasingly important. At levels of 100 km or so, this can cause upward-propagating waves to dissipate rapidly, and this and other decay effects are discussed in Section 8.11.

The next topic considered (Section 8.12) is wave propagation in a medium of slowly varying properties, utilizing methods originally developed by Liouville and Green in the 1830s. (Liouville, 1837; Green, 1838). These methods are very useful for dealing with situations commonly encountered in the atmosphere and ocean, and can be used, for instance, to find how energy is propagated from a source region such as a region of large topography at the base of the atmosphere, or to find how internal wave energy varies with depth in the ocean.

The internal wave field observed in practice can be described as a superposition of plane waves, the distribution of amplitudes forming the spectrum of the waves. Properties of the spectrum observed in the ocean are discussed in Section 8.14. The different components of the spectrum do not, in reality, behave independently, as assumed in linear theory, but influence each other through the nonlinear terms in the equations that govern their behavior. Theoretical treatments of wave interactions are discussed in Section 8.13, but without going into details. The interactions cause a wave spectrum to evolve with time, the evolution being such that spectra observed in different parts of the ocean are remarkably similar and thus can be approximately described by a "universal" spectrum.

Another property of waves is that they can cause significant effects a large distance from the place where they were produced. For instance, surface waves on the ocean can propagate from their generation region to the other side of the world and move some sand along the beach where their energy is dissipated. In Section 8.15 the transport of momentum by waves is discussed and illustrated by a simple example that involves *vertical* propagation of internal waves in a nonrotating fluid.

Low-frequency motions have a special character because of a redundancy in the equations, i.e., they are geostrophic to first order, but geostrophic motion happens to satisfy the continuity equation exactly! Hence departures from geostrophy have more significance than might be expected, and the special nature of "quasi-geostrophic" motion in a uniformly rotating system is discussed in the final section.

8.2   Effect of Rotation on Surface Gravity Waves: Poincaré Waves

Gravity waves in a homogeneous fluid of uniform depth H and in the absence of rotation were considered in Section 5.3. It is not very difficult mathematically to generalize this work to the case of a fluid rotating with uniform angular velocity $\frac{1}{2}f$ about a vertical axis. In particular, it may be shown that the dispersion relation (5.3.8) that connects frequency ω with wavenumber κ_H is modified by rotation to become

$$\omega^2 = g\lambda \tanh \lambda H, \tag{8.2.1}$$

where g is the acceleration due to gravity and λ is related to wavenumber κ_H by

$$\kappa_H^2 = \lambda^2 - a^{-2}\lambda H \coth \lambda H, \tag{8.2.2}$$

where a is the Rossby radius of deformation defined by

$$a^2 = gH/f^2. \tag{8.2.3}$$

Although this exact result can be obtained, there is not much point in using it in such a general form because in applications to the ocean and lakes, H is so much smaller (by a factor of at least several hundred) than the Rossby radius that it is a very good approximation to replace (8.2.1) by one of its limiting forms for $\kappa_H^{-1} \ll a$

or $\kappa_H^{-1} \gg H$. For this reason, detailed derivation of (8.2.1) is left as an exercise for the reader and the two limiting forms are discussed instead.

The first limiting form, which applies when the horizontal scale κ_H^{-1} is small compared with the Rossby radius a, corresponds to the case in which rotation effects are weak. Equation (8.2.2) gives $\lambda \simeq \kappa_H$ and therefore (8.2.1) reduces to the non-rotating dispersion relation (5.3.8). This limit has already been discussed in Chapter 5 and will not be considered further.

Rotation has significant effects only on waves with horizontal scale κ_H^{-1} large compared with the depth H. In other words, rotation effects need be studied only in the context of shallow-water waves for which the hydrostatic approximation applies. The free waves in this limit are the "Poincaré waves" found in Chapter 7. Since λH is small in this limit, (8.2.2) is approximated by

$$\lambda^2 \approx \kappa_H^2 + a^{-2}, \tag{8.2.4}$$

and so (8.2.1) reduces to the dispersion relation (7.3.4) for Poincaré waves. These waves were somewhat incidental to the discussion of Chapter 7 and will now be treated more fully.

Before discussing the properties of these "surface" or "barotropic" (see Section 6.2) waves, it should be borne in mind that the discussion will apply equally to "internal" or "baroclinic" modes because of the existence of separable solutions for the stratified-fluid equations when the horizontal scale is large compared with the vertical scale, as found in Chapter 6. Thus when reference is made to surface elevation η and to horizontal velocity components (u, v) for barotropic motion in a fluid of depth H, the same discussion can be applied to internal motions whose properties can be described (see Sections 6.11, 6.14, and 6.17) in terms of equivalent shallow-water variables $\tilde{\eta}(x, y, t)$, $\tilde{u}(x, y, t)$, and $\tilde{v}(x, y, t)$ and an equivalent depth H_e.

Now consider the properties of a progressive shallow-water wave, i.e., one for which η, u, and v are proportional to

$$\exp i(kx + ly - \omega t). \tag{8.2.5}$$

[In this book, all waves with dispersion relation (7.3.4) are referred to as Poincaré waves, although the name is sometimes reserved for the subset that satisfies the boundary conditions for a channel. The plane wave, given by (8.2.5), is sometimes given the special name of Sverdrup wave (see Platzman, 1971).] The polarization relations, i.e., the relations between amplitudes and phases of η, u, and v, are obtained by substituting this form of dependence on (x, y, t) in the governing equations (7.2.1)–(7.2.3). This gives

$$-i\omega u - fv = -ikg\eta,$$

$$-i\omega v + fu = -ilg\eta, \tag{8.2.6}$$

$$-i\omega\eta + iH(ku + lv) = 0.$$

These equations have a nonzero solution only when the dispersion relation (7.3.4) is satisfied, i.e.,

$$\omega^2 = f^2 + \kappa_H^2 gH, \tag{8.2.7}$$

where κ_H is the magnitude of the horizontal wavenumber vector

$$\mathbf{k}_H = (k, l),$$

i.e.,

$$\kappa_H^2 = k^2 + l^2. \tag{8.2.8}$$

Equations (8.2.6) also yield the following relations for u and v in terms of η:

$$u = (k\omega + ilf)\eta/\kappa_H^2 H, \qquad v = (l\omega - ikf)\eta/\kappa_H^2 H. \tag{8.2.9}$$

For internal modes, H is replaced by the equivalent depth H_e. If f is zero, (8.2.9) gives the relation between acceleration and pressure gradient that applies in the nonrotating limit. If ω is zero, (8.2.9) is equivalent to the geostrophic relation (7.2.14).

In addition, the potential vorticity equation (7.2.8) shows that the fluid acquires cyclonic vorticity when the surface is elevated and anticyclonic vorticity when the surface is depressed, i.e.,

$$\zeta/f = i(kv - lu)/f = \eta/H. \tag{8.2.10}$$

The fact that the perturbation potential vorticity is zero for a Poincaré wave follows from (7.2.8) and the assumed form of time dependence, i.e., proportionality with $\exp(-i\omega t)$.

For discussing the properties of a progressive Poincaré wave, it is convenient to choose the x axis in the direction of the wavenumber vector, so that $l = 0$. Then the solution obtained, using the convention that the physical solution is the real part of the complex expression for the wave and using (8.2.9), is

$$\eta = \eta_0 \cos(kx - \omega t), \tag{8.2.11}$$

$$u = (\omega\eta_0/kH) \cos(kx - \omega t), \qquad v = (f\eta_0/kH) \sin(kx - \omega t). \tag{8.2.12}$$

This solution is depicted in Fig. 8.1. The trajectories of fluid particles are, from

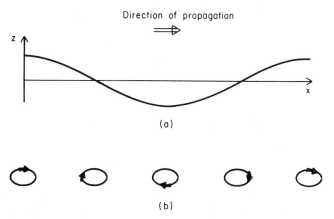

Fig. 8.1. A Poincaré wave progressing from left to right. (a) The surface elevation and (b) the particle trajectories in plan, these being ellipses with major axis in the direction of propagation (i.e., the x axis), and minor axis f/ω times the major axis. Particles move around these trajectories in an anticyclonic sense. The arrows mark the position of the particle when the surface elevation is as shown, and show the direction of motion for the northern hemisphere.

(8.2.12), ellipses with major axes in the direction of the wavenumber vector. The ratio of the lengths of the axes is ω/f.* The velocity vector rotates anticyclonically with time so the particles move around their elliptical orbits in an *anticyclonic* sense.

An example of an internal Poincaré wave observed in Lake Michigan is shown in Fig. 8.2. The thermocline motion, represented by the shaded region, is not a simple sinusoid because more than one wave is present, but a wave with a period of about 17 hr appears to dominate. The vertical variation of velocity is rather like that in a two-layer fluid (see Section 6.2), i.e., the velocity in the thicker layer below the thermocline is smaller than that in the layer above the thermocline, and is in the opposite direction. This indicates that the first baroclinic mode is dominant. The anticyclonic rotation of the velocity vector can be seen both above and below the thermocline.

Now consider the two limiting forms of Poincaré wave. *Short* Poincaré waves (i.e., $\kappa_H^{-1} \ll a$) have high frequency, i.e., ω/f is large and therefore the elliptical trajectories are long and thin. In the limit they become the rectilinear paths associated with ordinary gravity waves in the absence of rotation. *Long* Poincaré waves ($\kappa_H^{-1} \gg a$) have ω only slightly greater than f, so the particle paths are almost circular. In this limit the restoring force of gravity is relatively weak, so the path is the natural one followed by a particle in the absence of forces (or, more precisely, when the sum of the acceleration relative to the rotating frame and the Coriolis acceleration is zero). This is called *inertial motion*, the paths are called *inertial circles*, and the frequency f of rotation around these orbits is called the *inertial frequency*. The speed U is constant by (8.2.12), so the radius R of the circle is U/f. This is, in fact, the same result as that given by (7.10.26) for steady motion in the absence of forces.

In the ocean, energy tends to be concentrated in the lower frequencies and, in particular, Poincaré wave energy tends to be concentrated near the lowest possible frequency for such waves, namely, the inertial frequency. Such waves can be recognized by the nearly circular anticyclonic orbits and the near-inertial frequency. The first clear example of such motion to be observed in the ocean was reported by Gustafson and Kullenberg (1936), and a discussion was presented by Defant (1961, Vol. 1, p. 447). The observations were made in the Baltic (57.8°N, 17.8°E, depth 100 m) and Fig. 8.3 shows a *progressive vector diagram* of the currents measured above the pycnocline at a nominal depth of 14 m. Such a diagram shows the displacement a particle would have if it had the velocity observed at the fixed position of the current meter at all times. (For infinitesimal motion, this coincides with the particle trajectory, but for finite disturbances it need not be the same.) One can see the anticyclonic motion superimposed on a slow drift to the west of north. The period of rotation differed by only about 1% from the inertial period. The Poincaré wave depicted in Fig. 8.2 also has near-inertial frequency.

Other wave forms can be found by superposition of plane progressive waves, e.g., a combination of the form (7.3.10), representing two oppositely directed progressive waves of equal amplitude, gives rise to the standing-wave solution (7.3.12) for which the wave crests remain stationary but the surface moves up and down with frequency ω. At the crests, the motion is purely vertical as for the nonrotating case

---

* When dealing with ratios and inequalities, the expressions for ω and f positive will often be used for convenience.

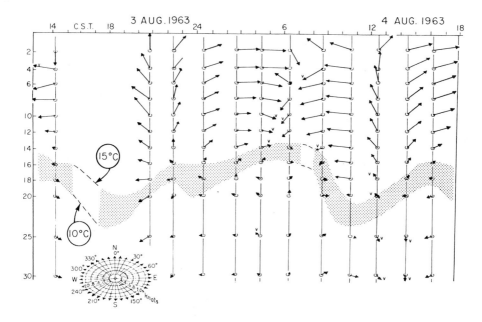

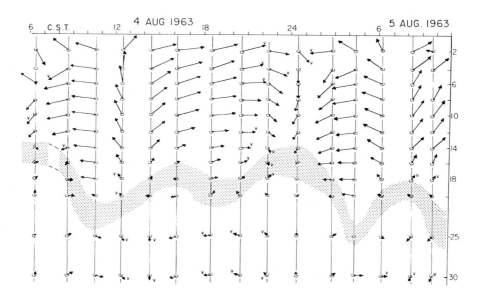

Fig. 8.2.   Observations of currents and thermal structure in the upper 30 m of Lake Michigan, August 3–5, 1963. The arrows indicate direction and speed in accordance with the current rose, and the shaded portion indicates the thermocline as denoted by the 10° and 15° isotherms. The two diagrams, which are overlapping in time, are separated by 17 hr, which is the dominant period. The local inertial period is 17.5 hr. Note the approximate two-layer structure in both temperature and velocity, and the anticyclonic rotation with time of velocity vectors. [From Mortimer (1971, Fig. 85).]

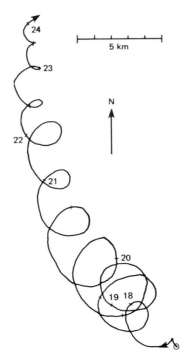

Fig. 8.3.   The historic current measurements in the Baltic by Gustafson and Kullenberg (1936), showing oscillations of near-inertial period. The plot is a progressive vector diagram, showing the displacement a particle would
have, given the velocity observed at the current meter. The inertial period is 14 hr, and marks are shown 1 day
apart. Note the anticyclonic sense of rotation.

depicted in Fig. 5.6b. At the nodes, the motion is purely horizontal, as in the nonrotating case, but now the paths are elliptical. As for the progressive wave, the particles
move around their orbits in an anticyclonic sense, and the ratio of the lengths of the
axes of the ellipse is ω/f.

The fact that the particle trajectories are not linear, as in the nonrotating case,
has an important consequence that was pointed out by Kelvin (Thomson, 1879),
namely, that the boundary condition of no normal motion at a plane boundary
cannot be satisfied by a single plane progressive wave. The ramifications of this fact
will be explored in Chapter 10.

8.3   Dispersion Properties and Energetics of Poincaré Waves

The dispersive properties of Poincaré waves follow from the dispersion relation
(8.2.7), depicted in Fig. 7.2, and have already been the subject of some discussion in
Section 7.3. Poincaré waves have the property that the frequency ω depends only on
the *magnitude* κ_H of the wavenumber vector, and therefore the group velocity $\mathbf{c}_g$ is
in the direction of κ_H. The magnitude c_g of the group velocity is equal to the slope of
the dispersion curve in Fig. 7.2, and its ratio to the phase velocity ω/κ_H is given by

[cf. (7.3.8)]

$$\frac{c_g\kappa_H}{\omega} \equiv \frac{\kappa_H}{\omega}\frac{d\omega}{d\kappa_H} = \frac{\kappa_H^2 gH}{\omega^2} = \frac{\kappa_H^2 a^2}{1 + \kappa_H^2 a^2}, \qquad (8.3.1)$$

which follows from the dispersion relation (8.2.7) and the definition (8.2.3) of the Rossby radius a. As for surface waves in water of finite depth, this ratio is always less than unity, the extreme values being zero in the long-wave limit ($\kappa_H a \ll 1$) and unity in the short-wave limit. For instance, Figs. 5.7 and 6.8a,c could depict a group of Poincaré waves with $\kappa_H a = 1$, for then the group velocity is half the phase velocity just as it is for deep-water surface waves.

The dispersion properties of Poincaré waves are nicely illustrated by the exact solution (7.3.14). In fact, because of this exact solution, Poincaré waves can be taken as the most convenient example of how waves disperse when the group velocity is a maximum for short waves and a minimum (zero) for long waves. This solution is illustrated in Figs. 7.3b and 7.4. The important properties are (i) the front moving out at the maximum group velocity $(gH)^{1/2}$, this front becoming thinner through dispersion as time goes on, and (ii) the near-inertial period waves left behind, these having only very small group velocity. [It should be remembered that the wave properties are based on the shallow-water approximation, and that in practice waves with horizontal scale comparable with depth would satisfy the dispersion equation (5.3.8), so very short waves would have group velocity less than $(gH)^{1/2}$. Thus if one looked at fine detail of the front, one would see fine-scale short-period oscillations similar to those discussed in Section 6.16. These are effectively smoothed or "filtered" out by making the hydrostatic approximation.]

The energy equations for shallow-water motion with rotation are exactly the same as in Section 5.7 since the Coriolis term does zero work, i.e., when the momentum equations (7.2.1) and (7.2.2) are multiplied by $\rho H u$ and $\rho H v$, respectively, and added, the Coriolis terms are eliminated. The total energy equation is (5.7.4), i.e.,

$$\frac{\partial}{\partial t}\left\{\frac{1}{2}\rho H(u^2 + v^2) + \frac{1}{2}\rho g\eta^2\right\} + \frac{\partial}{\partial x}(\rho g H u \eta) + \frac{\partial}{\partial y}(\rho g H v \eta) = 0. \qquad (8.3.2)$$

For a progressive Poincaré wave, it is convenient to average over a wavelength. Denoting this average by an overbar, the mean kinetic energy per unit area is given by

$$\tfrac{1}{2}\rho H\overline{(u^2 + v^2)} = \tfrac{1}{4}\rho(\omega^2 + f^2)\eta_0^2/\kappa_H^2 H = ((\omega^2 + f^2)/(\omega^2 - f^2))\tfrac{1}{4}\rho g\eta_0^2, \qquad (8.3.3)$$

and the mean potential energy per unit area is given by

$$\tfrac{1}{2}\rho g\overline{\eta^2} = \tfrac{1}{4}\rho g\eta_0^2. \qquad (8.3.4)$$

These results follow from (8.2.11), (8.2.12), and the dispersion relation (8.2.7). The total wavenumber κ_H is used rather than k in (8.3.3) to make the resulting expression independent of the choice of axes. Note that the energy is not partitioned equally between the kinetic and potential forms, the kinetic energy always being greater by a factor

$$\frac{\text{KE}}{\text{PE}} = \frac{\omega^2 + f^2}{\omega^2 - f^2} = 1 + \frac{2f^2}{\kappa_H^2 gH} = 1 + \frac{2}{\kappa_H^2 a^2}. \qquad (8.3.5)$$

The second and third expressions for the ratio are obtained from the first by the use of the dispersion relation (8.2.7) and the definition (8.2.3) of the Rossby radius.

The total energy per unit area can be obtained by adding (8.3.3) and (8.3.4) to give

$$\text{energy per unit area} = \frac{1}{2} \frac{\omega^2}{\omega^2 - f^2} \rho g \eta_0^2 = \frac{\omega^2 \rho \eta_0^2}{2 \kappa_H^2 H}, \tag{8.3.6}$$

where the dispersion relation (8.2.7) has been used to obtain the last expression. For short waves $(\kappa_H^{-1} \ll a)$, the ratio (8.3.5) is close to unity and the energy is almost equally divided between the kinetic and potential forms. For long waves $(\kappa_H^{-1} \gg a)$, however, nearly all of the energy is in the kinetic form. This implies that the waves of near-inertial frequency that are associated with motion in nearly circular orbits are more likely to be distinguishable from other waves in *velocity* observations than they are in observations of surface elevation.

The mean energy flux of a progressive Poincaré wave can be calculated from (8.2.11) and (8.2.12), which give

$$\rho g H (\overline{u\eta}, \overline{v\eta}) = \tfrac{1}{2} \rho g \kappa_H^{-2} \omega \eta_0^2 \mathbf{k}_H. \tag{8.3.7}$$

This is equal to the energy density, as given by (8.3.6), times the group velocity, whose magnitude is given by (8.3.1). An important factor in the Rossby adjustment problem discussed in Sections 7.2 and 7.3 was radiation of energy in the form of Poincaré waves—in fact, two-thirds of the potential energy released was lost by radiation. The radiative loss can be calculated from (8.3.7) by summing the contributions from different wavenumbers, the amount lost per unit length being $\rho g \eta_0^2 a$.

Further discussions of Poincaré waves may be found in the works by Proudman (1953), Platzman (1971), and LeBlond and Mysak (1978). The first two authors, however, use the term Poincaré wave in a more restricted sense than in that used here. The waves referred to as plane progressive Poincaré waves in this book are called waves with horizontal crests by Proudman (1953, Section 132) and are called Sverdrup waves by Platzman (1971). The question of terminology is discussed more fully in Chapter 10.

8.4 Vertically Propagating Internal Waves in a Rotating Fluid

Consider the adjustment under gravity of small disturbances to a continuously stratified incompressible fluid, as begun in Section 6.4, but now with rotation effects added. Because of the separation of variables that can be made when the hydrostatic approximation is valid, the previous discussion of Poincaré waves may be applied to stratified fluids as well. However, discussion in those terms has thus far emphasized horizontal propagation characteristics. Now vertical propagation characteristics will also be discussed. It turns out that much of the discussion is only slightly more complicated when the hydrostatic approximation is *not* made, so this restriction will be lifted for the time being.

The basic equations are therefore the same as those in Section 6.4 except for the addition of the Coriolis acceleration. In particular, the incompressibility condition

(6.4.3), Eq. (6.4.6) that expresses conservation of density by a material particle, and the vertical component (6.4.5) of the momentum equation remain unaltered. The equations that are affected are the horizontal momentum equations (6.4.4), which now become (7.12.1) and (7.12.2), i.e.,

$$\partial u/\partial t - fv = -\rho_0^{-1}\,\partial p'/\partial x, \tag{8.4.1}$$

$$\partial v/\partial t + fu = -\rho_0^{-1}\,\partial p'/\partial y. \tag{8.4.2}$$

Sometimes the equations obtained by eliminating u or v from this pair of equations are useful. The variable v can be eliminated by adding the time derivative of (8.4.1) to f times (8.4.2) and u can be eliminated in a similar manner. The resulting equations are

$$\partial^2 u/\partial t^2 + f^2 u = -\rho_0^{-1}\,\partial^2 p'/\partial x\,\partial t - f\rho_0^{-1}\,\partial p'/\partial y, \tag{8.4.3}$$

$$\partial^2 v/\partial t^2 + f^2 v = -\rho_0^{-1}\,\partial^2 p'/\partial y\,\partial t + f\rho_0^{-1}\,\partial p'/\partial x. \tag{8.4.4}$$

Now consider how the number of equations can be reduced to two by eliminating all variables other than w and p'. The procedure adopted in Section 6.4 was to take the divergence of the horizontal components of the momentum equations [which corresponds to adding the x derivative of (8.4.1) to the y derivative of (8.4.2), and the use of the incompressibility condition (6.4.3) to eliminate the horizontal divergence of velocity.] However, with rotation effects included, the result of this procedure [cf. (7.2.4)] is

$$\partial^2 w/\partial z\,\partial t + f\zeta = \rho_0^{-1}(\partial^2 p'/\partial x^2 + \partial^2 p'/\partial y^2), \tag{8.4.5}$$

so a term involving the vertical vorticity component ζ has appeared. Another equation for ζ is obtained by taking the curl of (8.4.1), and (8.4.2), which gives (7.12.4), i.e.,

$$\partial\zeta/\partial t - f\,\partial w/\partial z = 0. \tag{8.4.6}$$

Now either $\partial w/\partial z$ or ζ can be eliminated from (8.4.5) and (8.4.6). If the time derivative of (8.4.6) is added to f times (8.4.5), the result is a relation between vorticity and pressure perturbation, namely,

$$\partial^2\zeta/\partial t^2 + f^2\zeta = f\rho_0^{-1}(\partial^2 p'/\partial x^2 + \partial^2 p'/\partial y^2). \tag{8.4.7}$$

This could also be obtained by taking the curl of (8.4.3) and (8.4.4). Similarly, the equation relating the horizontal divergence $\partial u/\partial x + \partial v/\partial y = -\partial w/\partial z$ with perturbation pressure is

$$\left(\frac{\partial^2}{\partial t^2} + f^2\right)\frac{\partial w}{\partial z} = \frac{1}{\rho_0}\frac{\partial}{\partial t}\left(\frac{\partial^2 p'}{\partial x^2} + \frac{\partial^2 p'}{\partial y^2}\right), \tag{8.4.8}$$

which reduces to (6.4.7) in the absence of rotation. Note that this equation, which can be regarded as one relating the horizontal motion (as expressed by the horizontal divergence) to the perturbation pressure, involves the inertial frequency f but not the buoyancy frequency N.

Another equation relating w and p' is required, and this is provided by (6.4.8), which is unaltered by rotation effects. This equation, namely,

$$\partial^2 w/\partial t^2 + N^2 w = -\rho_0^{-1}\,\partial^2 p'/\partial z\,\partial t, \tag{8.4.9}$$

can be regarded as one that relates the *vertical* motion (as expressed by w) to the perturbation pressure p'. Note that it involves the buoyancy frequency N but not the inertial frequency f.

Elimination of p' between (8.4.8) and (8.4.9) gives an equation for w alone, namely,

$$\frac{\partial^2}{\partial t^2}\left\{\frac{\partial^2 w}{\partial x^2} + \frac{\partial^2 w}{\partial y^2} + \frac{1}{\rho_0}\frac{\partial}{\partial z}\left(\rho_0\frac{\partial w}{\partial z}\right)\right\}$$

$$+ \frac{f^2}{\rho_0}\frac{\partial}{\partial z}\left(\rho_0\frac{\partial w}{\partial z}\right) + N^2\left(\frac{\partial^2 w}{\partial x^2} + \frac{\partial^2 w}{\partial y^2}\right) = 0. \tag{8.4.10}$$

When $f = 0$, this reduces to its nonrotating equivalent (6.4.10). If the vertical scale of the waves is small compared with the scale height, which is necessary if the medium is to be regarded as incompressible, then the Boussinesq approximation (6.4.11) can be made and (8.4.10) becomes

$$\frac{\partial^2}{\partial t^2}\left(\frac{\partial^2 w}{\partial x^2} + \frac{\partial^2 w}{\partial y^2} + \frac{\partial^2 w}{\partial z^2}\right) + f^2\frac{\partial^2 w}{\partial z^2} + N^2\left(\frac{\partial^2 w}{\partial x^2} + \frac{\partial^2 w}{\partial y^2}\right) = 0. \tag{8.4.11}$$

With this approximation, the equation for p' has the same form.

The dispersion relation for internal gravity waves in a rotating fluid with uniform buoyancy frequency N is obtained by substituting the wave form of solution

$$w = w_0 \exp\{i(kx + ly + mz - \omega t)\} \tag{8.4.12}$$

in (8.4.11), giving

$$\omega^2 = (f^2 m^2 + N^2(k^2 + l^2))/(k^2 + l^2 + m^2). \tag{8.4.13}$$

Taking account of the definition (6.5.2) of the wavenumber vector and the fact that rotation is about a vertical axis with angular velocity $\Omega = \tfrac{1}{2}f$, this relation can also be written

$$\kappa^2\omega^2 = (2\mathbf{\Omega}\cdot\mathbf{k})^2 + N^2\kappa_H^2, \tag{8.4.14}$$

a form that is in fact valid whatever the orientation of the rotation vector $\mathbf{\Omega}$ relative to the gravity vector $\mathbf{g}$. Another form of (8.4.13) is obtained when the wavenumber is given in polar coordinates as defined by (6.5.4), this form being a very concise one, namely,

$$\omega^2 = f^2 \sin^2 \varphi' + N^2 \cos^2 \varphi'. \tag{8.4.15}$$

Other forms of this equation which are very useful are

$$N^2 - \omega^2 = (N^2 - f^2)\sin^2 \varphi', \qquad \omega^2 - f^2 = (N^2 - f^2)\cos^2 \varphi'. \tag{8.4.16}$$

Because the frequency ω is a function only of the angle φ' that the wavenumber vector makes with the horizontal and *not* of its magnitude, the dispersion surfaces $\omega = \text{const}$ in wavenumber space are cones, just as in the nonrotating case (see Fig. 6.7). However, the dependence of ω on φ' depends very much on the relative values of the two frequencies f and N, between which ω must lie by (8.4.15). One extreme occurs when the fluid is homogeneous (as in some laboratory experiments, for instance) and therefore

$N = 0$. Then ω is equal to $f \sin \varphi'$ and thus lies between 0 and f. The corresponding waves are called *inertial* waves, which are discussed further by Greenspan (1968).

In the atmosphere and ocean, however, N usually exceeds f by a large factor, so it is more appropriate to think of the waves as internal gravity (or buoyancy) waves whose behavior is influenced by rotation. Typically, N/f is of order 100, so that ω is within 1% of the value (6.5.5) obtained in the absence of rotation until the horizontal scale κ_H^{-1} exceeds the vertical scale m^{-1} by a factor of 14. For such a large-scale ratio, the hydrostatic approximation is a very good one, so there is little need to discuss rotation effects except in the context of the hydrostatic approximation; hence the emphasis on the shallow-water equations that follow from the hydrostatic approximation.

If N/f is 100, the dispersion curves shown in Fig. 6.7 will not look any different because of rotation, except that the vertical axis would have to be labeled $\omega/N = 0.01$ instead of zero. The *group velocity* $\mathbf{c}_g$ is, by the definition (5.4.11), the gradient of ω in wavenumber space and is therefore normal to the conical surfaces of constant ω. From (8.4.13) and (5.4.11), one obtains [cf. (6.6.1)]

$$\mathbf{c}_g = ((N^2 - f^2)/\omega\kappa) \cos \varphi' \sin \varphi'(\sin \varphi' \cos \lambda', \sin \varphi' \sin \lambda', -\cos \varphi'), \quad (8.4.17)$$

i.e., the group velocity has magnitude $(N^2 - f^2) \cos \varphi' \sin \varphi'/\omega\kappa$ and is directed at an angle φ' to the *vertical*.

For some purposes, it is useful to use the aspect ratio α defined by

$$\alpha = \frac{\text{vertical scale}}{\text{horizontal scale}} = \frac{\kappa_H}{m} = \cot \varphi' \quad (8.4.18)$$

rather than the angle φ'. As can be seen from the ratio of the two expressions of (8.4.16), α is uniquely related to frequency ω by

$$\alpha^2 = (\omega^2 - f^2)/(N^2 - \omega^2). \quad (8.4.19)$$

When N/f is large, as is usually the case in the atmosphere and ocean, the wave regime $f \leq \omega \leq N$ can be subdivided into the following three parts for which different approximations apply:

(a) The *nonhydrostatic wave regime* (defined as the range of frequencies for which ω is of order N but $\omega \leq N$).   By (8.4.19) this is equivalent to the range for which the aspect ratio is not small (i.e., is of order unity or larger). In this range, the dispersion relation (8.4.13) is approximated by

$$\omega^2 \approx N^2(k^2 + l^2)/(k^2 + l^2 + m^2) = N^2 \cos^2 \varphi' = N^2\alpha^2/(1 + \alpha^2), \quad (8.4.20)$$

which is the relation obtained when rotation effects are ignored. This regime has been thoroughly explored in Chapter 6.

(b) The *hydrostatic "nonrotating" wave regime* (defined as the range of frequencies for which $f \ll \omega \ll N$).   By (8.4.19) this is equivalent to the range of aspect ratios α given by

$$f/N \ll \alpha \ll 1. \quad (8.4.21)$$

In this range, the dispersion relation (8.4.13) is approximated by

$$\omega^2 \approx N^2(k^2 + l^2)/m^2 = N^2\alpha^2. \tag{8.4.22}$$

Rotation effects do not appear to this order of approximation, which is the reason for calling this a "nonrotating" regime, although it must be remembered that rotation does have an effect at the next order of approximation, and it is sometimes important to consider this. The approximate equation (8.4.22) is merely the long-wave or hydrostatic approximation to the nonrotating dispersion relation, so again this is a regime that has been studied in Chapter 6.

(c) The *rotating wave regime* (defined as the range of frequencies for which ω is of order f but $\omega \geq f$). By (8.4.19) this is equivalent to the range for which the aspect ratio is of order f/N or is small compared with f/N. Since f/N is small, α is small and the hydrostatic approximation applies. The approximate form of the dispersion relation (8.4.13) is

$$\omega^2 \approx f^2 + N^2(k^2 + l^2)/m^2 = f^2 + N^2\alpha^2, \tag{8.4.23}$$

which is effectively the dispersion equation (8.2.7) for Poincaré waves.

The approximations to which these three regimes correspond are used over and over again in this and other chapters, so names for the different regimes are useful. The details of the approximations are listed in Table 8.1, which summarizes many of the most important and basic results of this chapter.

Another topic considered in Section 6.6 is the propagation characteristic of waves when either the horizontal or vertical scale is fixed. Consider how rotation affects the picture. The *horizontal propagation* characteristics of a wave of fixed vertical scale are given by (8.4.13) with m regarded as given. The horizontal part (c_{gx}, c_{gy}) of the group velocity is given by (8.4.17) and has magnitude c_{gH}, which, after use of (6.5.4), (8.4.13), and (8.4.18), can be written

$$
\begin{aligned}
c_{gH} &= \frac{(N^2 - f^2)m^2(k^2 + l^2)^{1/2}}{(k^2 + l^2 + m^2)^{3/2}\{f^2m^2 + N^2(k^2 + l^2)\}^{1/2}} \\
&= \frac{(N^2 - f^2)\alpha}{m(1 + \alpha^2)^{3/2}(f^2 + N^2\alpha^2)^{1/2}}.
\end{aligned} \tag{8.4.24}
$$

At zero horizontal wavenumber κ_H ($\alpha = 0$), the frequency is f and the group velocity is zero. As κ_H increases, the frequency increases and follows the Poincaré wave dispersion curve shown in Fig. 7.2 for as long as the hydrostatic approximation ($\kappa_H \ll m$ or $\alpha \ll 1$) remains valid. The group velocity is given by (8.3.1) or the approximate form of (8.4.24) for α small, namely,

$$c_{gH} \approx N^2\alpha/m(f^2 + N^2\alpha^2)^{1/2}, \tag{8.4.25}$$

and thus increases toward the value N/m as κ_H increases through the "rotating range," where it is of order mf/N. Then follows a large range of wavenumbers ($mf/N \ll \kappa_H \ll m$, i.e., the hydrostatic nonrotating range), where there is very little dispersion and the group velocity remains close to N/m. There is, in fact, a weak maximum (still close to N/m) when $\kappa_H^2 \approx m^2 f/3^{1/2}N$ and $\omega \approx N\kappa_H/m$. For high wavenumbers ($\kappa_H$

TABLE 8.1

Different Regimes of Approximation to the Equations for Small Perturbations to a Rotating Stratified Fluid with $f \ll N$ [a]

Frequency $\hat{\omega}$ (relative to flow): 0.1f — f — 10f = 0.1N — N — 10N

Period (typical value): 1 week — 1 day — 6 hr — 1 hr — 10 min — 1 min

	Quasi-geostrophic flow	Rotating	Hydrostatic nonrotating	Nonhydrostatic	Potential flow
Vertical structure	Evanescent	Evanescent | Wave	Wave	Wave | Evanescent	Evanescent
w equation	$f^2 \dfrac{\partial^2 w}{\partial z^2} + N^2\left(\dfrac{\partial^2 w}{\partial x^2} + \dfrac{\partial^2 w}{\partial y^2}\right) = 0$	$\left(\dfrac{\partial^2}{\partial t^2} + f^2\right)\dfrac{\partial^2 w}{\partial z^2} + N^2\left(\dfrac{\partial^2 w}{\partial x^2} + \dfrac{\partial^2 w}{\partial y^2}\right) = 0$	$\dfrac{\partial^2}{\partial t^2}\dfrac{\partial^2 w}{\partial z^2} + N^2\left(\dfrac{\partial^2 w}{\partial x^2} + \dfrac{\partial^2 w}{\partial y^2}\right) = 0$	$\dfrac{\partial^2}{\partial t^2}\left(\dfrac{\partial^2 w}{\partial x^2} + \dfrac{\partial^2 w}{\partial y^2} + \dfrac{\partial^2 w}{\partial z^2}\right) + N^2\left(\dfrac{\partial^2 w}{\partial x^2} + \dfrac{\partial^2 w}{\partial y^2}\right) = 0$	$\dfrac{\partial}{\partial t}\left(\dfrac{\partial^2 w}{\partial x^2} + \dfrac{\partial^2 w}{\partial y^2} + \dfrac{\partial^2 w}{\partial z^2}\right) = 0$
u equation (v similar)	$-\rho_0 f v = -\dfrac{\partial p'}{\partial x}$	$\rho_0\left(\dfrac{\partial u}{\partial t} - f v\right) = -\dfrac{\partial p'}{\partial x}$	$\rho_0 \dfrac{\partial u}{\partial t} = -\dfrac{\partial p'}{\partial x}$	$\rho_0 \dfrac{\partial u}{\partial t} = -\dfrac{\partial p'}{\partial x}$	$\rho_0 \dfrac{\partial u}{\partial t} = -\dfrac{\partial p'}{\partial x}$
Horizontal divergence equation	$\rho_0 f^2 \dfrac{\partial w}{\partial z} = -\dfrac{\partial}{\partial t}\left(\dfrac{\partial^2 p'}{\partial x^2} + \dfrac{\partial^2 p'}{\partial y^2}\right)$	$\rho_0\left(\dfrac{\partial^2}{\partial t^2} + f^2\right)\dfrac{\partial w}{\partial z} = -\dfrac{\partial}{\partial t}\left(\dfrac{\partial^2 p'}{\partial x^2} + \dfrac{\partial^2 p'}{\partial y^2}\right)$	$\rho_0 \dfrac{\partial^2 w}{\partial t\, \partial z} = -\left(\dfrac{\partial^2 p'}{\partial x^2} + \dfrac{\partial^2 p'}{\partial y^2}\right)$	$\rho_0 \dfrac{\partial^2 w}{\partial t\, \partial z} = -\left(\dfrac{\partial^2 p'}{\partial x^2} + \dfrac{\partial^2 p'}{\partial y^2}\right)$	$\rho_0 \dfrac{\partial^2 w}{\partial t\, \partial z} = -\left(\dfrac{\partial^2 p'}{\partial x^2} + \dfrac{\partial^2 p'}{\partial y^2}\right)$
Vertical motion equation	$\rho_0 N^2 w = -\dfrac{\partial^2 p'}{\partial t\, \partial z}$	$\rho_0 N^2 w = -\dfrac{\partial^2 p'}{\partial t\, \partial z}$	$\rho_0 N^2 w = -\dfrac{\partial^2 p'}{\partial t\, \partial z}$	$\rho_0\left(\dfrac{\partial^2}{\partial t^2} + N^2\right)w = -\dfrac{\partial^2 p'}{\partial t\, \partial z}$	$\rho_0 \dfrac{\partial w}{\partial t} = -\dfrac{\partial p'}{\partial z}$
Dispersion equation	$m^2 f^2 + N^2 \kappa_H^2 = 0$	$(\omega^2 - f^2)m^2 = N^2 \kappa_H^2$	$\omega^2 m^2 = N^2 \kappa_H^2$	$\omega^2(m^2 + \kappa_H^2) = N^2 \kappa_H^2$	$m^2 + \kappa_H^2 = 0$
Horizontal group velocity c_{gH}		$\dfrac{N^2 \kappa_H}{m(f^2 m^2 + N^2 \kappa_H^2)^{1/2}} = \dfrac{N(\omega^2 - f^2)^{1/2}}{\omega m}$	$\dfrac{N}{m}$	$\dfrac{N m^2}{(\kappa_H^2 + m^2)^{3/2}} = \dfrac{(N^2 - \omega^2)^{3/2}}{N^2 m}$	
Vertical group velocity c_{gz}		$\dfrac{N^2 \kappa_H^2}{m^2(f^2 m^2 + N^2 \kappa_H^2)^{1/2}} = \dfrac{(\omega^2 - f^2)^{3/2}}{\omega N \kappa_H}$	$\dfrac{N \kappa_H}{m^2} = \dfrac{\omega^2}{N \kappa_H}$	$\dfrac{N \kappa_H m}{(\kappa_H^2 + m^2)^{3/2}} = -\dfrac{\omega^2(N^2 - \omega^2)^{1/2}}{N^2 \kappa_H}$	
PE (PE + KE)	0.499	0.49 (evan.), 0.4 (wave)	0.499	0.49 (wave), 0.5 (evan.)	0.01
Aspect ratio |z|	0.00999	0.009 (evan.), 0.01 (wave)	0.03	0.1 (wave), 1.1 (evan.)	1.01
Typical scale $U/\hat{\omega}$ for atmosphere	1000 km	100 km	10 km	1 km	100 m
Special phenomena	Wave absorption	Wave absorption	Partial reflection/resonance	Wave reflection	

[a] The top line shows a logarithmic frequency scale for the case $N = 100f$, $\hat{\omega}$ being the frequency sensed by an observer moving with the fluid, and underneath are shown some typical values for the corresponding period. The boundaries between regimes are not, of course, as distinct as shown, each regime merging gradually into the next. The PE (PE + KE) ratio is given by (8.6.6) in the wave regime. The aspect ratio z is the vertical scale divided by the horizontal scale, and is given by (8.4.19). The typical "scale" shown refers to topographically generated waves in the atmosphere. The equations for this problem are obtained from those shown by replacing $\partial/\partial t$ with $U\, \partial/\partial x$. Typical scales for the ocean are about 1/100 of this scale. It will be shown in Chapter 12 that the quasi-geostrophic equations have an additional term (the beta term) for scales L of order $(U/\beta)^{1/2}$ or about 500 km for the atmosphere and 50 km for the ocean. Thus in practice an additional regime (the beta regime) must be added to the left-hand side of the table.

of order m, i.e., the nonhydrostatic range), the dispersion curves are similar to the curves shown in Fig. 6.13a and the frequency approaches N as $\kappa_H \to \infty$. The horizontal group velocity tends to zero in this limit.

The *vertical propagation* characteristics of a wave with fixed horizontal wavenumber κ_H can also be obtained from (8.4.13). The vertical component c_{gz} of the group velocity can be calculated from this equation, or by using (8.4.17), (6.5.4), and (8.4.18) as well, to give

$$
\begin{aligned}
c_{gz} &= \frac{-(N^2 - f^2)(k^2 + l^2)m}{(k^2 + l^2 + m^2)^{3/2}\{f^2m^2 + N^2(k^2 + l^2)\}^{1/2}} \\
&= \frac{-(N^2 - f^2)\alpha^3}{\kappa_H(1 + \alpha^2)^{3/2}(f^2 + N^2\alpha^2)^{1/2}} .
\end{aligned}
\tag{8.4.26}
$$

At zero vertical wavenumber $m = 0$ ($\alpha = \infty$), the frequency has its maximum value of N and c_{gz} is zero. As m increases, the behavior is as described in Section 6.6 for the nonrotating case, with a maximum group velocity of $2N/3^{3/2}\kappa_H$ for propagation at $35°$ to the vertical ($m = \kappa_H/2^{1/2}$). As m increases further, the group velocity decreases, and in the hydrostatic nonrotating range ($\kappa_H \ll m \ll N\kappa_H/f$), it is given approximately by

$$
c_{gz} \approx -N\kappa_H/m^2 = -N\alpha^2/\kappa_H .
\tag{8.4.27}
$$

For m comparable to or large compared with $N\kappa_H/f$ (the rotating wave regime), c_{gz} is given approximately by

$$
c_{gz} \approx \frac{-N^2\kappa_H^2}{m^2(f^2m^2 + N^2\kappa_H^2)^{1/2}} = \frac{-N^2\alpha^3}{\kappa_H(f^2 + N^2\alpha^2)^{1/2}} ,
\tag{8.4.28}
$$

so $c_{gz} \to 0$ as $m \to \infty$ and $\omega \to f$. For example, if N is 10^{-2} s^{-1} and κ_H^{-1} is 30 km, waves with comparable vertical scale would transmit energy vertically at velocities of around 100 m s^{-1} (compressibility effects would need to be considered to obtain an accurate value). However, waves with $m^{-1} = 1$ km would have a group velocity of only 0.3 m s^{-1} (30 km per day), whereas waves with $m^{-1} = 100$ m would have a frequency only 5% above inertial and a vertical group velocity of only 1 mm s^{-1} (100 m per day).

8.5 Polarization Relations

To be able to identify internal waves in the ocean and atmosphere, it is necessary to know how the velocity components and pressure perturbation vary in space and time, and how the different variables are related to each other. For plane progressive waves, these relationships (the polarization relations) are found by substituting the wave form (8.4.12) of solution in the relevant equations.

An important property of plane waves follows directly from the incompressibility condition (6.4.3), which gives

$$
\mathbf{k} \cdot \mathbf{u} = 0,
\tag{8.5.1}
$$

i.e., the motion is confined to the plane perpendicular to the wavenumber vector. The relation between w and p' follows from (8.4.9), which gives

$$w = \frac{-m\omega}{N^2 - \omega^2} \frac{p'}{\rho_0} = \frac{-\kappa\omega}{(N^2 - f^2)\sin\varphi'} \frac{p'}{\rho_0}, \qquad (8.5.2)$$

where use has been made of (8.4.16). The relations between the horizontal velocity components follow from (8.4.3) and (8.4.4), which give [cf. (8.2.9)]

$$u = \frac{k\omega + ilf}{\omega^2 - f^2} \frac{p'}{\rho_0}, \qquad v = \frac{l\omega - ikf}{\omega^2 - f^2} \frac{p'}{\rho_0}. \qquad (8.5.3)$$

If the x axis is chosen to be in the direction of the horizontal component of the wavenumber vector, these equations, after use of (8.4.16), give

$$u = \frac{\kappa\omega}{(N^2 - f^2)\cos\varphi'} \frac{p'}{\rho_0} = -\tan\varphi' w \qquad (8.5.4)$$

and

$$v = \frac{-i\kappa f}{(N^2 - f^2)\cos\varphi'} \frac{p'}{\rho_0} = \frac{-ifu}{\omega} = \frac{if}{\omega}\tan\varphi' w. \qquad (8.5.5)$$

The expressions in terms of w follow from (8.5.2). The perturbation density ρ' is related to w by (6.4.6) and (6.4.9), which give

$$\rho'/\rho_0 = (iN^2/g\omega)w. \qquad (8.5.6)$$

A sketch showing the properties of a plane progressive internal wave in a rotating fluid is shown in Fig. 8.4a, which is similar to the nonrotating case (Fig. 6.6) apart from the motion into and out of the page. It can be seen that the velocity vector rotates anticyclonically with time as already found for Poincaré waves. (Another way of expressing this is to say that the velocity vector moves *cum sole*, i.e., rotates in the same sense as a vector that points toward the sun.) It follows that the velocity vector also rotates anticyclonically in space as one moves in a direction *opposite to* the phase velocity. This property is useful because the distinction can be made between upward- and downward-propagating waves by analyzing a vertical sounding or profile of velocity. Leaman and Sanford (1975) have utilized this idea in examining velocity profiles in the ocean. By comparing drops made an inertial period apart, it was shown that much of the energy was at near-inertial period (Sanford, 1975). A spectral analysis technique was used to compute the contributions of different vertical wavenumbers to both veering (clockwise rotating) and backing components of velocity. The component that rotated anticyclonically with depth was found to have, for a range of wavenumbers (m^{-1} of order 100 m), three times the energy of the component that rotated cyclonically with depth, indicating a predominance of waves with downward group velocity.

A similar result was obtained by Sawyer (1961) from velocity soundings in the stratosphere. Figure 8.5a shows a hodograph of wind velocity for one such sounding, and the veering or anticyclonic rotation with height is evident. This corresponds to downward phase propagation and hence to an upward component of group velocity. Figure 8.5b shows movement of phase lines from soundings made a few days earlier,

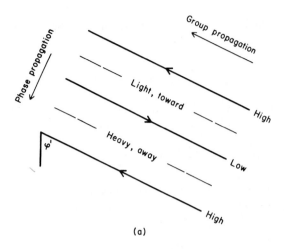

(a)

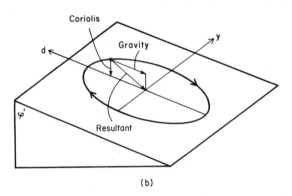

(b)

Fig. 8.4. (a) Phase relationships for a progressive internal wave with downward phase velocity (this implies upward group velocity). The solid lines mark lines of maximum (high) and minimum (low) perturbation pressure, which are also lines of maxima and minima in the component of velocity in the plane of the page. The direction of this velocity component is shown by the arrows. The dashed lines mark the positions of maximum (heavy) and minimum (light) density perturbations, and maxima and minima in the component of velocity normal to the page. The directions of this component for the northern hemisphere are given, and should be reversed for the southern hemisphere. The velocity vector veers (rotates anticyclonically) with time, and hence also in space as one moves in the direction *opposite to* the phase velocity. If the direction of phase propagation is reversed, the only change needed in the diagram is to reverse the directions of all the arrows. (b) Motion of a particle in the plane perpendicular to the wavenumber vector. The particle is subject to a restoring force per unit mass of magnitude $(N \cos \varphi')^2 |d|$ directed toward the line $d = 0$, and the Coriolis acceleration can be considered as equivalent to a Coriolis force per unit mass of magnitude $|f \sin \varphi'|$ times the velocity directed at right angles to the velocity and such that there is always a component toward the inside of the orbit. The resultant of the two forces is always toward the center, $y = d = 0$, and the resulting motion is in an elliptic orbit with major axis along the line of maximum slope and in an anticyclonic sense. The arrows showing the direction of motion are for the northern hemisphere, and should be reversed for the southern hemisphere.

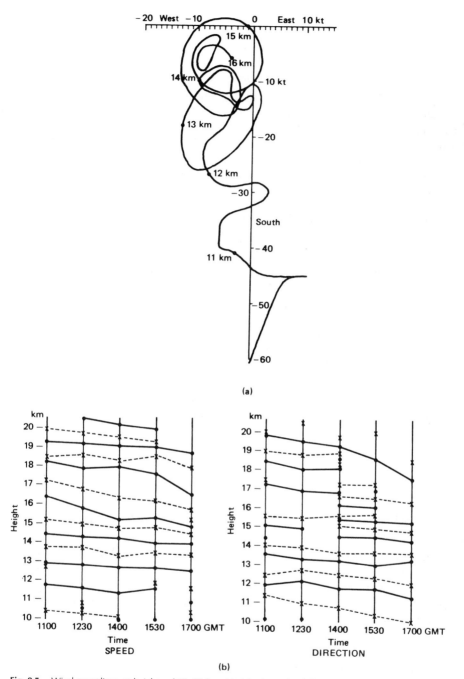

Fig. 8.5. Wind soundings at heights of 10–20 km. (a) A hodograph of the sounding made at 1700 GMT on 24 March 1959, each point representing the velocity observed at the height indicated. The velocity vector, relative to the mean over the neighboring few kilometers, tends to veer (i.e., rotate anticyclonically) with height, indicating a predominance of waves with upward energy propagation. (This sounding can be compared with an oceanic velocity sounding shown in Fig. 8.19.) (b) Heights of maxima and minima of wind velocity and extremes of wind direction on five successive soundings on 19 March 1959. Maxima are indicated by solid circles and joined by continuous lines where appropriate; minima are indicated by crosses and joined by dashed lines. These lines indicate downward phase propagation and hence upward energy propagation. [From Sawyer (1961, Figs. 4 and 5).]

and the downward phase propagation is evident. The vertical wavelength is about 1.5 km ($m^{-1} \sim 250$ m) and the period is about 6 hr ($\omega \sim 2.5f$). Similar soundings have been reported by Newell *et al.* (1966) and Thompson (1978).

By (8.5.1), the motion is entirely in a plane perpendicular to the wavenumber vector, and it is instructive to consider the forces on a fluid particle as it moves in this plane. The gravitational restoring force per unit mass can be found as explained in Section 6.5. The force is due entirely to the component $g \cos \varphi'$ of gravity parallel to the plane, and it is also proportional to the density change $\cos \varphi' \, d\rho/dz$ per unit displacement. Thus if d is the upslope displacement, the upslope force per unit mass is (cf. Section 6.5)

$$-\rho^{-1}g \cos \varphi' \cos \varphi' \, d\rho/dz \, d = -(N \cos \varphi')^2 d.$$

The fluid particle is also subject to a Coriolis acceleration associated with the component $\frac{1}{2}f \sin \varphi'$ of rotation perpendicular to the plane of motion. If y represents along-slope displacement of a fluid particle, then the equations of motion are

$$\ddot{d} + f \sin \varphi' \, \dot{y} = -(N \cos \varphi')^2 d, \qquad \ddot{y} - f \sin \varphi' \, \dot{d} = 0, \qquad (8.5.7)$$

where the overdots denote time derivatives. The solution for the particle orbit is elementary and is given by

$$d = d_0 \cos \omega t, \qquad y = \omega^{-1}f \sin \varphi' \, d_0 \sin \omega t, \qquad (8.5.8)$$

which represents an ellipse with upslope (major) axis $\omega/f \sin \varphi'$ times the along-slope (minor) axis, ω being given by (8.4.15). The elliptical orbit is shown in Fig. 8.4b, the motion around the orbit being in an *anticyclonic* sense because of the action of the Coriolis force.

8.6   Energetics

Since the Coriolis acceleration does not contribute to the energy equation, this has the same form (6.7.1) as that in a nonrotating fluid. Confirmation of this result may be obtained by multiplying (8.4.1) by $\rho_0 u$, (8.4.2) by $\rho_0 v$, (6.4.5) by w, (6.4.6) by $g^2\rho'/\rho_0 N^2$, and adding the results. The mean perturbation energy E per unit volume is still given by (6.7.6), i.e.,

$$E = \tfrac{1}{2}\rho_0\overline{(u^2 + v^2 + w^2)} + \tfrac{1}{2}g^2\overline{\rho'^2}/\rho_0 N^2, \qquad (8.6.1)$$

where each overbar denotes the mean over one wavelength.

The value of E in terms of w_0 can be obtained by using the polarization relations given in the last section. Choosing the x axis to be in the direction of the horizontal component of the wavenumber vector (so that $l = 0$), choosing w_0 to be real, and taking real parts of complex expressions, (8.4.12) gives

$$w = w_0 \cos(kx + mz - \omega t), \qquad (8.6.2)$$

and combined with (8.5.4) and (8.5.5), it gives

$$u = -\tan \varphi' \, w_0 \cos(kx + mz - \omega t),$$
$$v = -(f/\omega)\tan \varphi' \, w_0 \sin(kx + mz - \omega t). \tag{8.6.3}$$

Similarly, (8.5.6) gives

$$\rho' = -(N^2/\omega g)\rho_0 w_0 \sin(kx + mz - \omega t). \tag{8.6.4}$$

Substituting these expressions in (8.6.1) results in the same formula (6.7.7) as that obtained for the nonrotating case, namely,

$$E = \tfrac{1}{2}\rho_0(w_0/\cos \varphi')^2. \tag{8.6.5}$$

However, the energy is not equally partitioned between the kinetic and potential forms as in the nonrotating case. Since ω, which appears in the denominator of (8.6.4), is increased by the addition of the rotation term in (8.4.15), the proportion of potential energy is decreased by rotation. In fact, direct calculation of the individual terms in (8.6.1) gives the ratio

$$\frac{\mathrm{KE}}{\mathrm{PE}} = \frac{\omega^2 + f^2 \sin^2 \varphi'}{\omega^2 - f^2 \sin^2 \varphi'} = 1 + \frac{2f^2}{N^2}\tan^2 \varphi', \tag{8.6.6}$$

use having been made of the dispersion equation (8.4.15). Since N/f is large in the atmosphere and ocean, rotation effects are important only when φ' is near $\pi/2$, and in that case the result (8.6.6) is approximately that obtained already, namely, (8.3.5), using the hydrostatic approximation.

The other quantity that appears in the energy equation is the energy flux density vector $\mathbf{F}'$ defined by (6.7.8), i.e., by

$$\mathbf{F}' = \overline{p'\mathbf{u}}. \tag{8.6.7}$$

This can be evaluated using (8.6.2), (8.6.3), and the expression for p' in terms of w_0, which follows from (8.6.2) and (8.5.4), namely,

$$p'/\rho_0 = -(\kappa\omega)^{-1}(N^2 - f^2)\sin \varphi' \, w_0 \cos(kx + mz - \omega t). \tag{8.6.8}$$

The result can be expressed in the form (6.7.9), namely,

$$\mathbf{F}' = E\mathbf{c}_{\mathrm{g}}, \tag{8.6.9}$$

with E given by (8.6.5), but $\mathbf{c}_{\mathrm{g}}$ now given by (8.4.17). In particular, the vertical component of the flux is given by

$$F'_z = -(2\kappa\omega)^{-1}(N^2 - f^2)\sin \varphi' \, \rho_0 w_0^2, \tag{8.6.10}$$

which follows directly from the definition (8.6.7) and the expressions (8.6.2) and (8.6.8) for w and p'. The horizontal component is given by

$$F'_x = -\tan \varphi' \, F'_z, \tag{8.6.11}$$

which follows from (8.6.3).

Note that, provided that $f < N$, F'_z has the opposite sign from that of the vertical component $\omega/m = \omega/\kappa \sin \varphi'$ of the phase velocity, which (see Section 6.7) is indicative of the property that energy is transferred upward when the phase propagation is downward, and vice versa. This is the result already found in the nonrotating case $f = 0$.

8.7   Waves Generated at a Horizontal Boundary

The subject of waves generated at a horizontal boundary was initiated in Section 6.8. Although the methods have applicability to waves generated by a variety of mechanisms, the discussion was in terms of the particular case of waves generated by flow over small-amplitude topography, and the same procedure will be adopted here. Since any topography can be represented as a superposition of waves each of fixed wavenumber and since the problem is linear, the effect of each wavenumber can be examined in turn. This leads to the study of flow over a sinusoidal range of hills. The x axis is chosen to be perpendicular to the crests of the hills and axes are chosen to be fixed relative to the mean motion of the air, which is assumed to have uniform velocity U in the x direction. In this frame, the hills appear to move with velocity $-U$ in the x direction, i.e., to be given by

$$h = h_0 \sin\{k(x + Ut)\}. \tag{8.7.1}$$

The method of calculating the effects of the topography is the same as that in the nonrotating case, the effects of rotation appearing as modifications of the dispersion relation and of the polarization relations already studied. This section will concentrate on those aspects of the problem that are significantly affected by rotation, and it will be assumed that N/f is large, as is normally found to be the case in the atmosphere and ocean. The fluid is assumed to be incompressible and to have constant buoyancy frequency N.

For hills of small horizontal scale k^{-1}, the behavior was found in Section 6.8 to depend on the nondimensional number

$$|\omega|/N = Uk/N, \tag{8.7.2}$$

which is the ratio of the frequency

$$\omega = -Uk \tag{8.7.3}$$

of encounter of fluid particles with crests to the buoyancy frequency N. When $|\omega|/N$ is of order unity, rotation effects are unimportant and the analysis of Chapter 6 applies. Thus for $k^{-1} < U/N$ (about 1 km for the atmosphere and 300 m for the ocean floor) the waves are evanescent in character, i.e., they decay exponentially with height, and the flow (see Fig. 6.10a) is rather like potential flow (which it is in the limit as $k^{-1} \rightarrow 0$). For $k^{-1} > U/N$, propagating waves are generated (see Fig. 6.10b) and these can transfer wave energy to some remote level where the wave energy is absorbed.

Rotation effects become important when ω is comparable with f, i.e., when the horizontal scale k^{-1} is of order U/f. This is typically some 100 times larger than U/N in the atmosphere, so there is a considerable range of wavenumbers for which

$U/N \ll k^{-1} \ll U/f$, where the hydrostatic approximation applies (because the horizontal scale is large compared with the vertical scale) and rotation effects are small. This will be called the hydrostatic nonrotating range (see Section 8.4 and Table 8.1). The scale $U/|f|$ for which rotation effects become an essential part of the behavior is typically around 100 km for the atmosphere and about 3 km for the ocean floor.

When considering rotation effects, it should be remembered that the wind (or current), which is assumed to be blowing steadily in a fixed direction, must be in geostrophic equilibrium with a pressure gradient, so in the absence of the hills the pressure P at the ground will satisfy

$$\rho_0 f U = -\partial P/\partial y, \tag{8.7.4}$$

i.e., for an observer with his back to the wind, low pressure will be found to the left in the northern hemisphere and to the right in the southern hemisphere.

Now consider the flow over hills of the form (8.7.1) when k^{-1} is in the range $U/N < k^{-1} < U/f$. Then the solution to the wavelike form is given by (6.8.4) and (6.8.5), namely,

$$w = Ukh_0 \cos(kx + mz - \omega t), \tag{8.7.5}$$

where, by the dispersion relation (8.4.13), the vertical wavenumber component m is given by

$$m^2 = k^2(N^2 - \omega^2)/(\omega^2 - f^2) = k^2(N^2 - U^2k^2)/(U^2k^2 - f^2). \tag{8.7.6}$$

The positive root is chosen in order to ensure upward group propagation, which implies downward phase propagation ($\omega/m = -Uk/m < 0$). The perturbation horizontal velocity components are, by (8.5.4) and (8.5.5),

$$\begin{aligned} u &= -Umh_0 \cos(kx + mz - \omega t), \\ v &= k^{-1}fmh_0 \sin(kx + mz - \omega t). \end{aligned} \tag{8.7.7}$$

The particle trajectories relative to an observer moving with the wind at speed U are elliptical as shown in Fig. 8.4, so in the northern hemisphere particles have a velocity component to the left at the crests. In other words, by (8.7.4), they move toward *low* pressure at the crests whatever the sign of f. The maximum displacement of the path toward low pressure is situated at the node that is immediately in the lee of a crest.

The pressure perturbation is given by (8.5.2), i.e., after using (6.8.4) and (6.8.5), by

$$p'/\rho_0 = m^{-1}(N^2 - U^2k^2)h_0 \cos(kx + mz - \omega t). \tag{8.7.8}$$

High pressure is found on the windward side and low pressure on the leeward side, giving a net force on the hills in the direction of the wind. The net drag force τ per unit area can be calculated as follows. Consider a small element of hillside, as shown in Fig. 8.6, with horizontal dimensions δx and δy and height increment δh. If δs is the width of the element as measured along the slope, the force due to the pressure has magnitude $p \, \delta s \, \delta y$ and is directed normal to the surface. The horizontal component of this force has magnitude $p \, \delta s \, \delta y$ times $\delta h/\delta s$, i.e.,

$$p \, \delta h \, \delta y = p(\partial h/\partial x) \, \delta x \, \delta y$$

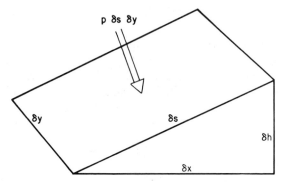

Fig. 8.6. A small element of hillside of area $\delta s\,\delta y$, δy being measured along slope and δs along upslope. δx is the horizontal projection of δs and δh is the increment in height. The force due to the pressure p acting on area $\delta s\,\delta y$ is $p\,\delta s\,\delta y$. The horizontal component is obtained by multiplying by $\delta h/\delta s$, i.e., $p\,\delta h\,\delta y$, or $p\,\partial h/\partial x$ times $\delta x\,\delta y$.

and is directed along the line of maximum slope, i.e., in the direction of ∇h. In other words, the horizontal component of the force is the vector

$$p\,\nabla h\,\delta x\,\delta y.$$

Averaging over a wavelength gives the drag τ per unit area as

$$\tau = \overline{p\,\nabla h} = \overline{p'\,\nabla h}. \tag{8.7.9}$$

The pressure p can be replaced by the perturbation pressure p' (see Section 4.5.4) since the pressure associated with the undisturbed state does not exert a horizontal force.

The magnitude τ of $\boldsymbol{\tau}$ can be calculated from (8.7.1), (8.7.8), and (8.7.6), which give

$$\tau = \tfrac{1}{2}\rho_0\{(N^2 - U^2k^2)(U^2k^2 - f^2)\}^{1/2}h_0^2. \tag{8.7.10}$$

This is related to the vertical component F_z' of the energy flux density vector defined by (6.7.8), the relationship following directly from the definitions and the relation $w = U\,\partial h/\partial x$ between w and h. Thus

$$F_z' = \overline{p'w} = \tau U = \tfrac{1}{2}\rho_0\{(N^2 - U^2k^2)(U^2k^2 - f^2)\}^{1/2}Uh_0^2. \tag{8.7.11}$$

As pointed out in Section 6.8 and further illustrated in Section 8.15, the waves can transmit horizontal stresses from the ground up to considerable heights, depending on the levels at which the waves are absorbed. The rate of transfer vanishes at both ends of the regime in which waves can exist, i.e., when $k^{-1} = U/N$ and when $k^{-1} = U/|f|$. There are, however, some singular features of the solution when k^{-1} is exactly equal to $U/|f|$, for a resonance occurs when the encounter frequency $\omega = -Uk$ is exactly equal to the inertial frequency. For this case (8.7.6) shows that m is infinite (i.e., very short vertical wavelength) and hence the perturbation velocity given by (8.7.7) is also infinite. In practice, the large value of m indicates that friction would tend to remove this wave component and/or nonlinear effects would become important.

The group velocity $\mathbf{c}_g$ relative to the air is given by (8.4.17). This can be expressed in terms of k by using (8.7.3), (8.7.6), and the fact that $m/k = \tan\varphi'$. The components

c_{gx} (horizontal) and c_{gz} (vertical) obtained in this way are

$$c_{gx} = \frac{-(N^2 - U^2 k^2)(U^2 k^2 - f^2)}{(N^2 - f^2)Uk^2} \approx -U + \frac{f^2}{Uk^2}, \tag{8.7.12}$$

$$c_{gz} = \frac{(N^2 - U^2 k^2)^{1/2}(U^2 k^2 - f^2)^{3/2}}{(N^2 - f^2)Uk^2} \approx \frac{(U^2 k^2 - f^2)^{3/2}}{NUk^2}, \tag{8.7.13}$$

the approximate expressions being the values obtained when the hydrostatic approximation is made. The group velocity relative to the air is directed upward into the wind at angle $\pi/2 - \varphi'$ to the horizontal. The group velocity *relative to the ground*, on the other hand, is directed *upward* and *downwind* at an angle with the horizontal whose tangent is equal to

$$\frac{c_{gz}}{U + c_{gx}} = \frac{(N^2 - U^2 k^2)^{1/2}(U^2 k^2 - f^2)^{3/2}}{(U^2 k^2 - f^2)^2 + f^2(N^2 - f^2)} \approx \frac{f}{N}\left(\frac{U^2 k^2}{f^2} - 1\right)^{3/2}, \tag{8.7.14}$$

the approximate expression being the one obtained in the hydrostatic limit. This angle is shown as a function of wavenumber k by the solid line in Fig. 8.7 for the special case $N = 100f$. The maximum value of the angle is approximately

$$\tan^{-1}(3^{3/4}4^{-1}(N/f)^{1/2}) \qquad \text{when} \quad Uk = 3^{1/4}(fN)^{1/2},$$

or about $80°$ for the case $N = 100f$. For comparison, Fig. 8.7 shows the values obtained when various approximations are made that correspond to the different regions in Table 8.1.

An example of the flow patterns found in cases for which rotation is important

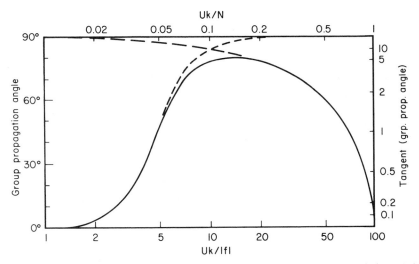

Fig. 8.7.   The angle made with the horizontal by the group velocity relative to the ground, shown (solid line) as a function of the horizontal wavenumber k of the topography for the case $N/f = 100$. If rotation is ignored ($f = 0$), the curve is modified at low wavenumbers to the one shown by large dashes. If the hydrostatic approximation is made, the curve is modified at high wavenumbers to the one shown by small dashes. If rotation is ignored *and* the hydrostatic approximation is made, the angle is $90°$ for all k.

is shown in Fig. 8.8a, which is for the case $kU = 1.25f$, i.e., that for which the horizontal scale k^{-1} is 20% less than the value U/f, where vertical propagation ceases. The vertical wavenumber m in this case is approximately $5N/3U$ and is of the same order as that in the nonrotating case illustrated in Fig. 6.10b (where $m = 3N/5U$). However, the horizontal scale is bigger by the large factor $0.64N/f$, so the diagram is drawn with the vertical scale exaggerated relative to the horizontal by the factor N/f.

If the horizontal scale k^{-1} is larger than $U/|f|$, the waves become *evanescent*, i.e., the amplitude decays exponentially with height. The vertical displacement h of fluid particles now has the form

$$h = h_0 e^{-\gamma z} \sin(kx - \omega t), \tag{8.7.15}$$

where the decay rate γ is given by

$$\gamma^2 = N^2 k^2/(f^2 - U^2 k^2), \tag{8.7.16}$$

assuming that the hydrostatic approximation can be made. γ is the positive root, so the evanescent solutions are obtained by replacing m by $i\gamma$ in the previous formulas. The solution (8.7.15) is given in the frame of reference moving with the air. The solution relative to the ground is obtained by removing the $-\omega t$ term inside the brackets, i.e., by replacing $\sin(kx - \omega t)$ with $\sin kx$.

The formulas for u, v, p', etc., are again derived from Section 8.5, putting $l = 0$, $m = i\gamma$, using the hydrostatic approximation ($k \ll m$ or $Uk \ll N$), and using the expressions (6.8.8) and (6.8.4) for w. In particular, (8.5.2) gives

$$p'/\rho_0 = k^{-1} N h_0 (f^2 - U^2 k^2)^{1/2} e^{-\gamma z} \sin(kx - \omega t), \tag{8.7.17}$$

showing that high perturbation pressure is found on the crests and low perturbation pressure in the valleys (the sign opposite to that found in potential flow!). The perturbation velocity u in the flow direction is, from (8.5.4),

$$u = \frac{NkUh_0}{(f^2 - U^2 k^2)^{1/2}} e^{-\gamma z} \sin(kx - \omega t), \tag{8.7.18}$$

giving, as expected from the continuity equation, faster flow over the crests. Also, because of the constriction of flow over the crests, vortex lines are diminished in length, causing acquisition of anticyclonic relative vorticity ζ. From (8.4.6) or (7.12.5), the value of ζ is given by

$$\zeta = f \frac{\partial h}{\partial z} = -\frac{Nkfh_0}{(f^2 - U^2 k^2)^{1/2}} e^{-\gamma z} \sin(kx - \omega t). \tag{8.7.19}$$

The associated lateral velocity v and lateral displacement y are given by

$$v = \frac{Nfh_0}{(f^2 - U^2 k^2)^{1/2}} e^{-\gamma z} \cos(kx - \omega t) \tag{8.7.20}$$

and

$$y = \frac{Nfh_0}{Uk(f^2 - U^2 k^2)^{1/2}} e^{-\gamma z} \sin(kx - \omega t). \tag{8.7.21}$$

The lateral displacement is toward low pressure over the crests, i.e., to the left in the northern hemisphere and to the right in the southern hemisphere.

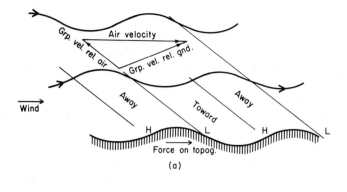

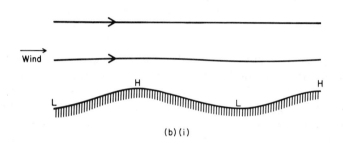

(b) (i)

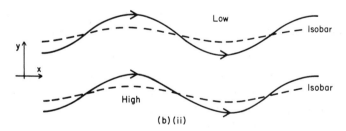

(b) (ii)

Fig. 8.8. The motion produced by uniform flow of a stratified fluid over sinusoidal topography of small amplitude. The format is the same as that in Fig. 6.10, but the horizontal scale is much larger and the vertical scale is exaggerated relative to the horizontal by a factor of N/f. (a) The case of the smaller-wavelength topography, the wavenumber k being given by $k = 1.25f/U$ (a typical value of U/f for the atmosphere is 100 km). The displacement of isopycnals is uniform with height, but the wave crests move upstream with height, i.e., the phase lines are tilted as shown. The group velocity relative to the air is along these phase lines, but the group velocity relative to the ground is directed upward at a shallow angle in the downstream direction as shown. High and low pressures are at the nodes, so there is a net force on the topography in the direction of flow. The directions shown for the component of flow normal to the page are for the northern hemisphere, and should be reversed for the southern hemisphere. (b) The response for larger-wavelength topography, the wavenumber k in this case being equal to $0.8f/U$. (i) A vertical cross section showing how the amplitude decays with height. The phase lines in this case are vertical. (ii) A plan view, the solid lines with the arrows being particle trajectories and the dashed lines isobars. Particles are displaced toward low pressure over the crests and toward high pressure over the valleys. The speed is greater over the hills, so pressure *on a streamline* is lower by Bernoulli's theorem. However, on a line $y = \text{const}$, the pressure is actually higher over the hills as can be seen in (ii), and as is shown in (i).

The fact that the pressure perturbation is high over the crests where the flow speed is also high seems contrary to Bernoulli's theorem. This is not so, however, because streamlines deviate laterally *more* than isobars do, so that on a *streamline* the pressure is actually lower over the crests. This can be seen in Fig. 8.8b, which is drawn for the special case $Uk = 0.8f$ with the same vertical exaggeration N/f and same longitudinal scale as in Fig. 8.8a. The lower part shows the plan view, and it will be noticed that on a straight line $y = $ const, pressure is high over crests, whereas on a streamline, pressure is low over crests.

The limit of very wide topographic features $(k^{-1} \gg U/|f|)$ is also of some interest, especially for the deep ocean, where typical values of $U/|f|$ are a few kilometers. $(U/|f|$ is of order 100 km in the atmosphere, so the effects of the earth's curvature need to be considered for scales large compared with this.) The formula (8.7.16) for the trapping scale shows that as k^{-1} increases, so does γ^{-1}, which becomes proportional to k^{-1} in the limit. It is convenient to call the limiting value

$$\gamma^{-1} = |f|/Nk \tag{8.7.22}$$

the *Rossby height* since the quantity (modified by allowing for compressibility of the atmosphere) was introduced by Rossby in a paper on temperature changes in the stratosphere (Rossby, 1938b). The name is also convenient because of the relationship between Rossby height and Rossby radius. In situations like the one considered in this section, where a horizontal scale k^{-1} is given, there is a natural height scale, the Rossby height, equal to $|f|/N$ times the horizontal scale. On the other hand, when a height (or depth) scale is given, there is a natural length scale, the Rossby radius, proportional to $N/|f|$ times the height scale.

The polarization relations show that in the limit of very wide topography the v component of perturbation velocity is very much greater then the u component, and has amplitude equal to Nh_0, i.e., the limiting form of (8.7.20) is

$$v = Nh_0 e^{-\gamma z} \cos(kx - \omega t). \tag{8.7.23}$$

Another feature of this limit is that streamlines become almost coincidental with isobars, the horizontal amplitude of their deviations [see (8.7.21)] being Nh_0/Uk. This implies that the flow is close to being in geostrophic equilibrium, so the range in which $|Uk| = |\omega| \ll |f|$ will be called the *quasi-geostrophic* range (see Table 8.1).

8.8 Mountain Waves

Localized strong winds in the vicinity of mountains and hills have attracted attention in the past because of damage to crops and buildings etc. An example is the Helm wind [a number of early reports are reprinted in Brunskill (1884)], which occurs on the west side of the Pennines in England. For instance, in a report to the British Association, Watson (1839, pp. 33–34) records:

> Sometimes, when the atmosphere is quite settled, not a breath of wind stirring, and hardly a cloud to be seen, a small but well-known cloud appears on the summit, extends itself to the north and south—the "Helm is on," and in a few

minutes blowing furiously, sufficient to break trees, overthrow stacks of grain, throw a person from his horse, or overturn a horse and cart....When heard or felt for the first time it does not seem to very extraordinary; but when we find it blowing morning, noon and night, for days together, it makes a strong impression on the mind, and we are compelled to acknowledge that it is one of the most singular phenomena of meteorology.

In the 1920s and 1930s, a great deal was learned about mountain waves from observations, using balloons and gliders. In fact, the heights reached (11,400 m) by gliders in the 1930s were viewed with astonishment, particularly as they were achieved on the *leeward* rather than windward side of the mountains [see Alaka (1960)]. A detailed description of the observed phenomenon, as obtained from such information, was given by Kuettner (1939a,b), and Manley (1945) gave similar details for the "Helm wind" mentioned above, his conclusions being based on observations from the ground. (The paper was submitted in 1940 but "withheld from publication during wartime for reasons of national security.") Summaries of these and other studies are given by Alaka (1960), Nicholls (1973), Queney (1977), and R. B. Smith (1979), and a descriptive picture was given by Scorer (1972, Section 5.3).

The first theoretical studies were for flow with uniform velocity U of a fluid with uniform buoyancy frequency N over topography of infinitesimal amplitude. In particular, Lyra (1940, 1943) found solutions for flow over a step profile and over rectangular mountains, whereas Queney (1948) found solutions for flow over "bell-shaped" ridges of the form

$$h = h_m/(1 + (x/L)^2) = h_m L \int_0^\infty e^{-kL} \cos kx \, dk. \tag{8.8.1}$$

The response to flow over hills of any shape can be found by representing the topography as a Fourier synthesis of sinusoidal waves, then taking the appropriate combination of the solutions for sinusoidal topography that were found in the previous section. The bell-shaped ridge (8.8.1) is convenient for study because of the simple form of its Fourier transform. The results obtained by Queney by such a Fourier synthesis will be summarized in what follows.

The character of the response depends on the width L of the mountain. As L increases, the type of response varies in the same way as does that found for sinusoidal topography as the horizontal scale k^{-1} increases. The change in type of response also corresponds to changes of the time L/U taken by an air particle to pass the hill since in a frame of reference moving with the flow the response is forced by the hill, moving relative to the air at velocity $- U$. If N is large compared with the Coriolis parameter f, as is normally the case, five regimes can be distinguished [cf. Queney (1948) and the summary in Table 8.1].

(i) *Potential-Flow Regime* $(L \ll U/N)$. Typical values of U/N are 1 km for the atmosphere and 50 m for the ocean floor, but it should be remembered that considerable departures from these values are possible. This limit applies to such small features that effects associated with the turbulent boundary layer are quite important. [A theory incorporating these effects is given by Jackson and Hunt (1975), also see P. A. Taylor and Gent (1980).] However, the ideal flow solution is given for completeness since it has a simple form for the bell-shaped ridge, namely,

$$h = (1 + z/L)^{-1}h_m/(1 + (x/(L + z))^2). \tag{8.8.2}$$

The amplitude of the vertical displacement of fluid particles falls off with height in a fashion similar to that shown in Fig. 6.10a. By continuity, the flow is faster over the crest, where the flow is more restricted, and so the pressure is lower by Bernoulli's theorem.

(ii) *Nonhydrostatic Wave Regime* $(L \sim U/N)$.   The general solution for the bell-shaped mountain has the form

$$h = h_m L \int_0^\infty \exp(-kL + ikx + imz)\, dk, \tag{8.8.3}$$

the physical solution corresponding, as usual, to taking the real part. The vertical component m of the wavenumber is given by the dispersion equation. When L is comparable with U/N, rotation effects are usually small, so that the value of m is given by (6.8.6), i.e.,

$$m = ((N/U)^2 - k^2)^{1/2}. \tag{8.8.4}$$

Figure 8.9a shows a typical solution that is drawn for the case

$$L = U/N = 1 \quad \text{km} \tag{8.8.5}$$

and is due to Queney (1948). The scale L_s that is shown in the figure is the wavelength associated with the scale U/N, namely,

$$L_s = 2\pi U/N. \tag{8.8.6}$$

Properties of this solution are discussed by Queney (1973). As found in Section 6.8, the group velocity relative to the ground for a single wave component is upward and downwind in the direction of the wavenumber vector and varies between the vertical direction $(k = 0)$ and the horizontal downwind direction $(k = U/N)$. The figure shows that wave energy is confined mainly to this quadrant as expected. The vertical scale of the waves is of the same order as the horizontal scale L. The ground-level wind and pressure variations are also shown in the figure, these being related by Bernoulli's theorem [see (4.8.3) and (7.10.22)], which in its linearized form gives

$$p'/\rho_0 = -Uu \tag{8.8.7}$$

on a streamline. The wind is a maximum at the crest of the hill where there is a minimum pressure.

The pressure is higher on the upwind side of the hill than on the leeward side, indicating a net force on the hill. The horizontal force $\mathscr{F}$ per unit span (i.e., per unit distance in the y direction) is given by [see Section 8.7 and Eq. (8.7.9)]

$$\mathscr{F} = \int p'\, dh = \int_{-\infty}^\infty p'(dh/dx)\, dx. \tag{8.8.8}$$

An alternative formula for the drag force is obtained by using (8.8.7) to substitute for p' in (8.8.8) and using the relation between w and h, i.e., $w = Dh/Dt$. In a frame of reference fixed to the ground, the linearized form of this is

$$w = U \, dh/dx, \qquad (8.8.9)$$

and so (8.8.8) gives

$$\mathcal{F} = \int \rho_0 uw \, dx, \qquad (8.8.10)$$

i.e., the force is equal to the rate at which momentum is transferred vertically by the waves. When the topography is given as a Fourier synthesis of sinusoidal waves,

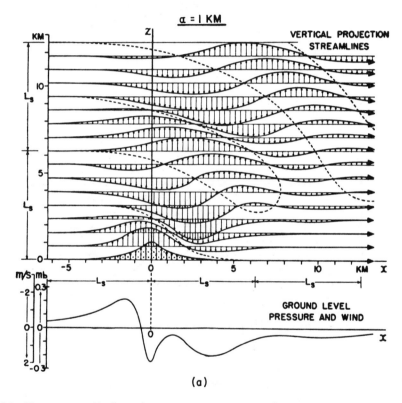

Fig. 8.9.   Waves generated by flow with uniform velocity ($U = 10$ m s^{-1}) of a uniformly stratified ($N = 0.01$ s^{-1}) fluid over bell-shaped ridges of various widths L (from Queney (1948)). The mountain profile is given by (8.8.1) and the solutions are based on linear theory. Case (a) is for $L = U/N = 1$ km and typifies the nonhydrostatic wave regime. Case (b) is for $L = 10$ km and typifies the hydrostatic wave regime in which rotation is not important. Case (c) is for $L = U/f = 100$ km and typifies the wave regime in which rotation effects are important. The upper part of each diagram shows the vertical displacements of air particles, i.e., their trajectories in the vertical plane normal to the ridge. The dashed lines show where the vertical displacement is zero. The scale L_s is defined by $L_s = 2\pi U/N$ and is a good measure of the vertical wavelength found in all three cases. The scale L_f is defined by $L_f = 2\pi U/|f|$, where f is the Coriolis parameter, which is given the value 10^{-4} s^{-1}. The lower panels of (a) and (b) show the ground-level pressure and wind variations associated with the waves. The lower panel of (c) shows a plan view of the particle trajectory and of an isobar at ground level. Amplitudes are based on a maximum height h_m of the ridge of 1 km.

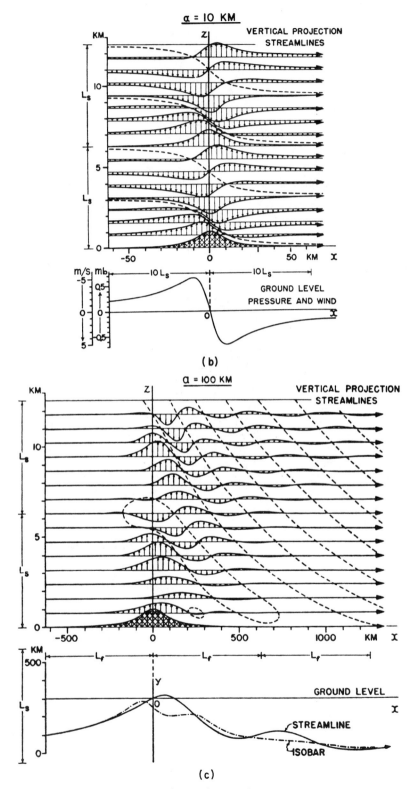

Fig. 8.9. (continued)

namely,

$$h = \int_0^\infty \mathscr{H}(k)e^{ikx}\, dk \tag{8.8.11}$$

(the real part being understood), then the drag force can also be expressed as an integral of the contributions (6.8.11) of each Fourier component. Such contributions come only from wavenumbers $k < N/U$ for which waves are generated, and the result of combining (8.8.10) and (6.8.11) is [see Blumen (1965b) for a detailed derivation]

$$\mathscr{F} = \pi\rho_0 U^2 \int_0^{N/U} |\mathscr{H}(k)|^2 k((N/U)^2 - k^2)^{1/2}\, dk. \tag{8.8.12}$$

The particular case of the bell-shaped mountain (8.8.1) was treated by Sawyer (1959), and the result can be given in terms of special functions (Blumen, 1965a). This result is depicted in the left-hand part of Fig. 8.10, which shows the drag per unit span as a function of width L. The drag increases as L increases up to a limiting value $\pi/4$ that is appropriate to the hydrostatic regime as discussed below.

(iii) *Hydrostatic Nonrotating Wave Regime.* This regime can be defined if N/f is large enough. For the atmosphere, a typical value of N/f is 100, but for the deep ocean a value near 10 is more appropriate. Topographic waves in the atmosphere are generated by the regions in which there is a slope, and for the large mountain chains, which produce the strongest effects, the region in which the slope is maintained at a large value of one sign is typically about 10-km wide. For this scale, the hydrostatic approximation is quite a good one, yet the scale is not large enough for rotation effects to be large. Thus many of the calculations of mountain wave effects have been for this regime.

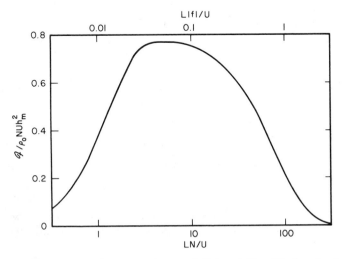

Fig. 8.10. Force $\mathscr{F}$ per unit span due to wave drag on a bell-shaped ridge of the form $h = h_m/(1 + (x/L)^2)$. The buoyancy frequency N and flow speed U are uniform and the Coriolis parameter f is chosen to have a value of $0.01\,N$. The curve for $LN/U < 4$ is based on that given by Blumen (1965a, Fig. 1), whereas that for large width L is based on (8.8.23).

The general equation for lee waves in uniform flow, including nonhydrostatic and rotation effects, is, in a frame of reference moving with the air, (8.4.11). The same equation is satisfied by the vertical particle displacement h since $w = \partial h/\partial t$ and (8.4.11) can be integrated with respect to t. In a frame of reference fixed relative to the ground, the operator $\partial/\partial t$ is replaced by the operator $U \, \partial/\partial x$ so (8.4.11) becomes

$$U^2 \frac{\partial^2}{\partial x^2}\left(\frac{\partial^2 h}{\partial x^2} + \frac{\partial^2 h}{\partial y^2} + \frac{\partial^2 h}{\partial z^2}\right) + f^2 \frac{\partial^2 h}{\partial z^2} + N^2\left(\frac{\partial^2 h}{\partial x^2} + \frac{\partial^2 h}{\partial y^2}\right) = 0. \qquad (8.8.13)$$

When there is no rotation ($f = 0$) and there is no y dependence, the equation can be integrated twice with respect to x to give (assuming no disturbance at $x = \pm\infty$)

$$\partial^2 h/\partial x^2 + \partial^2 h/\partial z^2 + (N/U)^2 h = 0, \qquad (8.8.14)$$

which is the equation for the nonhydrostatic regime. If now the hydrostatic approximation is made (which is equivalent to assuming $\partial/\partial x \ll \partial/\partial z$), this becomes

$$\partial^2 h/\partial z^2 + (N/U)^2 h = 0. \qquad (8.8.15)$$

This very simple equation has the solution

$$h = h_s(x)e^{iNz/U}, \qquad (8.8.16)$$

where h_s is a complex function of x whose real part defines the surface topography. The solution must satisfy the condition of upward group propagation, which in this case means *vertically* upward. For the bell-shaped mountain [cf. Queney (1948)], the solution that satisfies this condition is

$$h = h_m e^{iNz/U}/(1 - ix/L). \qquad (8.8.17)$$

[*Note*: The condition on h_s is that the singularities in the complex x plane are in the negative half-plane. An equivalent formulation in terms of Hilbert transforms is given by Miles and Huppert (1969), Baines (1971), and Drazin and Su (1975). The solution (8.8.17) can be used to develop solutions for other topographies by taking a sum or integral over poles in the complex x plane.]

Figure 8.9b shows the solution (8.8.17). Waves are found only above the mountain because the group propagation is vertical. The associated pressure perturbation p' can be calculated from (6.4.7), which, after putting $w = \partial h/\partial t$, replacing $\partial/\partial t$ by $U \, \partial/\partial x$, and integrating twice with respect to x, yields

$$p' = \rho_0 U^2 \, \partial h/\partial z. \qquad (8.8.18)$$

For the bell-shaped mountain, substitution from (8.8.18) gives at the ground $z = 0$

$$p' = -\rho_0 N U h_m (x/L)/(1 + (x/L)^2), \qquad (8.8.19)$$

and this curve is shown in the lower panel of Fig. 8.9b. Since pressure is high on the windward side and low on the leeward side, there is a net force on the hill whose value per unit span is given by

$$\mathscr{F} = \int_{-\infty}^{\infty} p'(dh/dx) \, dx = \tfrac{1}{4}\pi\rho_0 N U h_m^2, \qquad (8.8.20)$$

which is the value obtained for the nonhydrostatic case in the limit as $NL/U \to \infty$, as can be seen in Fig. 8.10. The integration on the right-hand side of (8.8.20) is readily obtained by elementary means (substitute $x = L \tan \theta$) after (8.8.1) and (8.8.19) have been used to substitute for h and p'.

(iv) *Rotating Wave Regime* $(L \sim U/|f|)$.* In this regime, the solution for the bell-shaped mountain is given by (8.8.3), but the appropriate approximation for m is, from (8.7.6) with $k^{-1} \gg U/N$,

$$m = kN(U^2k^2 - f^2)^{-1/2}. \qquad (8.8.21)$$

As an example, Fig. 8.9c shows the solution for $L = U/|f|$, as obtained by Queney (1948). The wave energy (see Section 8.7, Fig. 8.7) is propagated upward and downwind at an angle that for individual wave components varies between $0°$ (for $k^{-1} = U/|f|$) and $90°$ (for $k \to \infty$). Thus the disturbance shown in the figure is confined largely to this quadrant. The vertical scale is of order U/N just as in cases (ii) and (iii), but the horizontal scale is now much larger. The diagram is drawn with a very much reduced horizontal scale (or, in other words, with a greatly exaggerated vertical scale). Most of the waves appear in the diagram to be in a wedge of angle between $60°$ and $90°$ to the horizontal, but when allowance is made for vertical exaggeration, this corresponds to an actual range of angles between $1°$ and $90°$ to the horizontal.

The lower panel of Fig. 8.9c shows a plan view for the northern hemisphere situation of a streamline and an isobar. The particle is seen to deviate toward low pressure (i.e., to the left), reaching a maximum displacement just past the crest of the hill. The pressure on a section normal to the hill is also given by the "isobar" curve and is a maximum on the windward side, indicating a net drag. The general formula for drag, taking account of both rotation and nonhydrostatic effects, is an integral with respect to wavenumber k of the drag τ for an individual wave, given by (8.7.10), the generalization of (8.8.12) being [cf. Bretherton (1969)]

$$\mathscr{F} = \pi\rho_0 \int_{|f|/U}^{N/U} |\mathscr{H}(k)|^2 \{(N^2 - U^2k^2)(U^2k^2 - f^2)\}^{1/2} \, dk, \qquad (8.8.22)$$

contributions coming only from the range of wavenumbers for which vertically propagating (as opposed to evanescent) wave solutions are obtained. If the hydrostatic approximation is made, i.e., $Uk \ll N$, and the value $\mathscr{H} = h_m L e^{-kL}$ appropriate to the bell-shaped mountain [compare (8.8.1) and (8.8.11)] is substituted, (8.8.22) gives

$$\mathscr{F} = \pi\rho_0 N h_m^2 L^2 \int_{|f|/U}^{\infty} (U^2k^2 - f^2)^{1/2} e^{-2kL} \, dk$$
$$= \tfrac{1}{2}\pi\rho_0 h_m^2 L |f| N K_1(2L|f|/U), \qquad (8.8.23)$$

the value of the integral being taken from Gradshteyn and Ryzhik (1980, formula 3.3873). The drag given by this formula is shown as a function of L in the right-hand

---

* Typical values of $U/|f|$ are 100 km for the atmosphere and 3 km for the ocean floor, but it should be remembered that for light winds or small currents, the value can be considerably smaller, whereas in tropical regions, where f is small, values can be much larger.

part of Fig. 8.10. For small f, the value asymptotically approaches the nonrotating hydrostatic value given by (8.8.20).

(v) *Quasi-Geostrophic Flow Regime* $(L \gg U/|f|)$. In this regime, solutions are again evanescent in character, so waves are not produced. The solutions are of interest for the ocean, for which a typical value of $U/|f|$ is about 3 km, but not for the atmosphere because effects of the earth's curvature must also be considered when $L \gg U/|f|$. The approximation to (8.8.13) for this regime is simply

$$f^2 \frac{\partial^2 h}{\partial z^2} + N^2 \left(\frac{\partial^2 h}{\partial x^2} + \frac{\partial^2 h}{\partial y^2} \right) = 0, \tag{8.8.24}$$

which reduces to the equation for potential flow if a stretched vertical coordinate z_s, defined by

$$z_s = Nz/|f|, \tag{8.8.25}$$

is used in place of z. The solution is therefore of the same form as (8.8.2), namely,

$$h = (1 + z_s/L)^{-1} h_m/(1 + (x/(L + z_s))^2). \tag{8.8.26}$$

Solutions decay with height over distances of the order of the Rossby height $|f|L/N$ [cf. (8.7.22)], which is much *larger* than the typical height scale U/N for propagating wave regimes and increases in proportion with the horizontal scale of the topography. An exact solution, such as (8.8.26), can also be found for a circular hill. (This solution is discussed in Section 8.16 and is illustrated in Fig. 8.20.)

The name "*quasi-geostrophic*" is used for this regime because of a peculiar property of the equations. It is simplest to consider the equations in a frame of reference moving with the fluid, for then they take the form given in Section 8.4. (For a frame at rest relative to the topography, $\partial/\partial t$ is replaced in the equation by $U \partial/\partial x$.) The regime is defined by the condition $\partial/\partial t \equiv U \partial/\partial x \ll f$, so the momentum equations (8.4.1) and (8.4.2) are approximated by the geostrophic balance equations

$$-fv = -\rho_0^{-1} \partial p'/\partial x, \qquad fu = -\rho_0^{-1} \partial p'/\partial y. \tag{8.8.27}$$

However, these equations, with f constant, imply zero horizontal divergence and hence no vertical motion, yet it is the balance between forcing of vertical motion by the topography and gravitational restoring forces that determines the flow! Therefore it is essential to calculate the horizontal divergence, even though it is small, and the appropriate expression comes from the vorticity equation (8.4.6). This shows that the divergence $-\partial w/\partial z$ is *weak*, being smaller than the vorticity ζ by a factor of order ω/f. Note that (8.4.6) can be deduced from the momentum equations only if the (small) acceleration terms are retained. This is the reason for using the name "quasi-geostrophic" rather than geostrophic, i.e., although to a first approximation the flow is geostrophic, the motion cannot be determined without considering the departure from a geostrophic balance.

If now (8.8.27) is used to obtain an expression for the vorticity

$$f\zeta = \rho_0^{-1}(\partial^2 p'/\partial x^2 + \partial^2 p'/\partial y^2), \tag{8.8.28}$$

and this is substituted in (8.4.6), the result is an approximate form of (8.4.8), namely,

$$f^2 \frac{\partial w}{\partial z} = \frac{1}{\rho_0} \frac{\partial}{\partial t} \left(\frac{\partial^2 p'}{\partial x^2} + \frac{\partial^2 p'}{\partial y^2} \right). \tag{8.8.29}$$

An equation for w of the same form as (8.8.24) is now obtained if $\partial p'/\partial t$ is eliminated between (8.8.29) and (6.11.4), which relates vertical motion to buoyancy-restoring forces. A peculiar feature is that although the result is an equation not involving time, it requires elimination of the pressure time derivative (or pressure tendency) $\partial p'/\partial t$!

A summary of the properties of the five regimes is given in Table 8.1. The frequency $\hat{\omega}$ is to be interpreted as the frequency sensed by an observer moving with the medium, and so is equal to ω for a medium at rest (Section 8.4) and to $\omega - Uk$ for flow at velocity U over topography (Section 8.7). In general, if ω is the frequency for an observer at rest and the medium is moving with uniform velocity $\mathbf{U}$, then

$$\hat{\omega} = \omega - \mathbf{U} \cdot \mathbf{k}. \tag{8.8.30}$$

These solutions for the bell-shaped mountain range illustrate the effects that are generally found. Solutions for other topographies have been studied. For example, Blumen (1965b) has discussed the effects of more than one range; Gjevik and Marthinsen (1978) and R. B. Smith (1980) have treated three-dimensional flows, and the ideas have been applied to Martian lee waves by Pickersgill and Hunt (1981); Bretherton (1969) has made detailed calculations of the drag exerted by some real topography (the hills of North Wales—see Section 6.8), and Klemp and Lilly (1980) have reviewed momentum flux effects; Bell (1975) has estimated that the drag on the ocean floor due to the generation of topographic waves is typically of order of 0.05 N m^{-2} (0.5 dyn cm^{-2}); similar calculations for fluctuating flow of fixed frequency can be made, and Bell (1975) has calculated the loss of energy from tidal currents due to internal wave generation at the bottom to be about 0.001 W m^{-2} (1 erg cm^{-2} s^{-1}); spectacular lee waves, generated by tides crossing a sill, have been reported by Farmer and Smith (1980); transient solutions have also been studied, and are reviewed by Alaka (1960) for the case in which vertically propagating waves are generated; Huppert and Bryan (1976) have investigated transient solutions in the quasi-geostrophic regime, showing that isolated topographic features can cause eddies to be generated when the flow is suddenly changed; Baines and Davies (1980) have reviewed laboratory experiments.

8.9 Effects of Variation of Properties with Height

In practice, the fluid velocity U and buoyancy frequency N are not constants, but vary with height, and these variations can have significant effects on the waves as already discussed in Section 6.9 for the case of uniform U but variable N and in the absence of rotation. In particular, waves with certain values of k can be reflected in such a way that they reinforce themselves and so such scales are emphasized. Furthermore, waves can be trapped near the ground if there is a region above in which propagation is not possible, i.e., a region in which m^2 becomes negative. For the range in which rotation is not important, this can only happen when the horizontal scale is of order U/N, i.e., in the nonhydrostatic range.

8.9.1. Total Reflection and Downstream Trains of Waves (Nonhydrostatic Regime)

The models discussed in Section 6.9 have been generalized by Scorer (1949) to include the effect of shear. If density ρ and velocity U are functions only of z in the undisturbed state, then for an incompressible nonrotating fluid, the equations satisfied by a small stationary disturbance are the momentum equations

$$\rho_0\left\{\left(\frac{\partial}{\partial t} + U\frac{\partial}{\partial x}\right)u + \frac{dU}{dz}w\right\} = -\frac{\partial p'}{\partial x}, \tag{8.9.1}$$

$$\rho_0\left(\frac{\partial}{\partial t} + U\frac{\partial}{\partial x}\right)w = -\frac{\partial p'}{\partial z} - \rho'g, \tag{8.9.2}$$

which are the linearized form of (4.5.20), the continuity equation [cf. (6.4.3)]

$$\partial u/\partial x + \partial w/\partial z = 0, \tag{8.9.3}$$

and the linearized form of (6.4.2), namely,

$$\left(\frac{\partial}{\partial t} + U\frac{\partial}{\partial x}\right)\rho' + \frac{d\rho_0}{dz}w = 0. \tag{8.9.4}$$

These equations can be treated in the same way as in Section 6.4. To begin with, the x derivative of (8.9.1), after substitution for $\partial u/\partial x$ from (8.9.3), gives

$$\rho_0\left\{\left(\frac{\partial}{\partial t} + U\frac{\partial}{\partial x}\right)\frac{\partial w}{\partial z} - \frac{dU}{dz}\frac{\partial w}{\partial x}\right\} = \frac{\partial^2 p'}{\partial x^2}. \tag{8.9.5}$$

Elimination of ρ' from (8.9.2) and (8.9.4) gives another equation relating to w and p' namely,

$$\rho_0\left\{\left(\frac{\partial}{\partial t} + U\frac{\partial}{\partial x}\right)^2 w + N^2 w\right\} = -\left(\frac{\partial}{\partial t} + U\frac{\partial}{\partial x}\right)\frac{\partial p'}{\partial z}. \tag{8.9.6}$$

Finally, if the Boussinesq approximation is utilized, elimination of p' from (8.9.5) and (8.9.6) gives

$$\left(\frac{\partial}{\partial t} + U\frac{\partial}{\partial x}\right)^2\left(\frac{\partial^2}{\partial x^2} + \frac{\partial^2}{\partial z^2}\right)w + N^2\frac{\partial^2 w}{\partial x^2} - \frac{d^2 U}{dz^2}\left(\frac{\partial}{\partial t} + U\frac{\partial}{\partial x}\right)\frac{\partial w}{\partial x} = 0. \tag{8.9.7}$$

For steady disturbances, this can be integrated twice with respect to x to give

$$\frac{\partial^2 w}{\partial x^2} + \frac{\partial^2 w}{\partial z^2} + \left(\frac{N^2}{U^2} - \frac{1}{U}\frac{d^2 U}{dz^2}\right)w = 0, \tag{8.9.8}$$

which differs from the equation obtained in the absence of shear only in the replacement of $(N/U)^2$ by a new function of z that is determined by the mean flow, namely,

$$(N/U)^2 - U^{-1} \, d^2U/dz^2. \tag{8.9.9}$$

Scorer (1949) has considered a model for which this parameter is piecewise constant, having a large value of 2.12 km^{-2} in a lower layer of thickness 2.7 km and a small value of 0.33 km^{-2} in the upper layer. This is very similar to the model considered in Section 6.9.4 with $\epsilon = 0.4$ and the resonance condition (6.9.16) is satisfied for $k^{-1} = 0.9$ km, i.e., for a horizontal wavelength $2\pi/k$ of 5.5 km. Figure 8.11 shows that the solution, as modified by Gossard and Hooke (1975), satisfies the correct radiation condition for a bell-shaped mountain of width $L = 1$ km. The difference between this solution and the one for uniform U and N shown in Fig. 8.9a is that the amplitude of the waves found on the lee side does not diminish with distance from the mountain. In other words, the wave energy is channeled into the waveguide instead of propagating upward. These waves have the resonant wavelength. Another difference is the attenuation with height of all but the longest waves in the upper layer. Corby and Wallington (1956) have studied the way in which the amplitude of the resonant wave depends on the width L of the mountain. The maximum amplitude is obtained when L^{-1} is equal to the resonant wavenumber—a condition which was nearly satisfied in Scorer's example $(k^{-1} = 0.9$ km, $L = 1$ km$)$. Other analytic models of this type and more general numerical solutions of (8.9.8) are reviewed by Alaka (1960), Nicholls (1973), and Gossard and Hooke (1975).

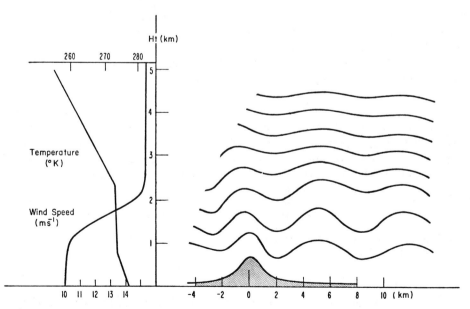

Fig. 8.11. Air flow over a bell-shaped mountain of width $L = 1$ km when the wind and temperature structure is as shown in the left panel. The result is that of Scorer (1949) as corrected by Gossard and Hooke (1975, Fig. 57-1). The vertical structure is such that it strongly emphasizes waves with a horizontal wavelength of 5.5 km, and a regular train of these waves can be seen in the lee of the mountain.

The resonance effect is necessary to produce a regular *series* of waves in the lee of an obstacle. The waves are striking because of their regular spacing, which corresponds to the wavelength of the resonant wave, typically about 10 km, and an example is shown in Fig. 6.12. Such waves are often made visible by clouds and can be seen in satellite pictures, many examples of which have been given by Cruette (1976).

8.9.2   Ray-Tracing Techniques

The effect of variation of properties with height can be studied by ray-tracing techniques when the variations with height are slow enough, i.e., are such that the fractional changes over the vertical scale m^{-1} of a wave are small. Then the so-called WKBJ or Liouville–Green approximation can be made (see Section 8.12). In this approximation, wave properties (such as the vertical component m of the wavenumber) are assumed to depend only on the local properties U and N of the medium, and to depend on U and N just as if the medium were uniform. The dependence on z arises only through the variation of U and N with z. Ray paths along which the wave energy travels are defined as paths such that the tangent at any one point is in the direction of the group velocity (relative to the ground). In the notation of Section 8.7, where c_{gx} and c_{gz} are the horizontal and vertical components of the group velocity relative to the air, the ray path is therefore defined by

$$dz/dx = c_{gz}/(U + c_{gx}), \tag{8.9.10}$$

and the right-hand side is given by (8.7.14). Since the horizontal wavenumber component k must be constant on a ray and U and N vary with z in a prescribed way, (8.9.10) can be integrated to give the ray paths.

In the nonhydrostatic regime (in which kU is of order N, which is large compared with f), (8.9.10), (8.7.14), and (6.8.6) give

$$dz/dx = m/k = [(N^2/U^2k^2) - 1]^{1/2}. \tag{8.9.11}$$

If U/N increases with height (which is usually the case because of the increase of velocity with height), m decreases with height and may eventually become zero. Wave propagation cannot continue above this height, and therefore the waves are reflected and the rays bend downward. If $z = z_c$ is the level where m vanishes, then (8.9.11) can be approximated near this point by

$$dz/dx = [(z - z_c)\, d(N^2/U^2k^2)/dz]^{1/2},$$

which integrates to show that ray paths are approximately parabolic near this point and are of the form

$$z = z_c + \tfrac{1}{4}\{d(N^2/U^2k^2)/dz\}_{z=z_c}(x - x_0)^2, \tag{8.9.12}$$

where x_0 is a constant. Thus ray paths starting upward at an angle to the ground bend over at $z = z_c$ and travel downward again. They then reflect off the ground again and the cycle is repeated, so energy is carried large distances horizontally in a waveguide near the ground. A schematic picture of the ray paths in this case is shown in Fig. 8.12a. Strictly speaking, the slowly varying approximation breaks down when $m = 0$, but still gives a good qualitative picture of the situation.

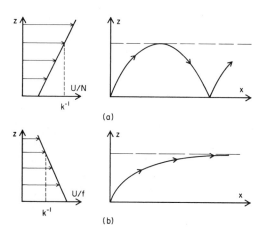

Fig. 8.12. Special types of ray paths for a medium whose properties vary slowly with vertical coordinate z. The paths are those that would be followed by a hypothetical particle traveling at the local group velocity $\mathbf{c_g}$, this being calculated as a function of U and N, using the formula for a uniform medium. The horizontal component k of the wavenumber and frequency ω are fixed for each ray. (a) Case exhibiting wave reflection. This occurs for nonhydrostatic waves when U/N increases with height and exceeds k^{-1}. The waves refract as shown and become horizontal at the level at which $Uk = N$. Thus wave energy can be trapped in a layer near the ground. (b) Case exhibiting absorption. This occurs for hydrostatic waves with wavenumbers such that rotation effects are important when U decreases with height and drops to a value equal to $|f|/k$. Upward-propagating waves approach this level asymptotically. The vertical component of the wavenumber continually decreases, so friction effects cause the wave to dissipate.

8.9.3 Partial Reflection and "Resonance" in the Hydrostatic Nonrotating Range

It was found in Section 6.9 that if the hydrostatic approximation can be made ($k \ll N/U$), then m cannot vanish anywhere (for constant U and N, m is equal to N/U) if the buoyancy frequency is always positive, and therefore total reflection is not possible. However, it was also found that *partial* reflection can occur at a discontinuity in N that can, after a second reflection from the ground, reinforce the original wave and give a marked increase in the amplitude of the response. This does *not* happen in a model with slowly varying U and N, in which the group velocity is always vertically upward if the hydrostatic approximation is made, and therefore rays cannot bend over. It may therefore be inferred that reflection is only possible when N and/or U vary sufficiently rapidly over a scale m^{-1} in some region of the flow for reflection to take place. The effect will give an enhanced response only when the region of rapid change is at the right height above the topography for the reinforcement to take place. For example, in the case shown in Fig. 6.11, the large response at small wavenumbers occurs only when the inverse Froude number $N_1 H/U$ is near an odd multiple of $\pi/2$, N_1 being the buoyancy frequency in the lower layer which has height H.

This wave reinforcement process that is due to partial reflection seems to have been an important factor in producing the enormous mountain waves observed near Boulder, Colorado on January 11, 1972. Klemp and Lilly (1975) analyzed linear models of the situation, utilizing *three* layers as a model of the observed situation.

In this case optimal response was obtained when each of the lower two layers was a quarter of a wavelength deep. The bottom layer had high stability and was about 2-km deep. The second layer had weak stability and was about 6-km deep (so the value of m was smaller by a factor $\epsilon \approx 0.3$). The third layer, representing the stratosphere, had vertical wavenumber only 20% greater than that of the second layer, so the situation was fairly close to that of the model discussed in Section 6.9.4.

The waves observed on this occasion have been described by Lilly (1978), and Fig. 8.13 shows a cross section of the observed fields of potential temperature and velocity. Such waves represent a severe hazard to aircraft, e.g., the research aircraft involved in making the measurements experienced vertical velocities reaching 30 m s^{-1} and the flight recorder of a commercial Boeing 707 showed vertical acceler-ations ranging from $-1.1g$ to $+2.7g$, the normal reading being $+1.0g$. Strong turbu-lence, resulting from the large velocity gradient, was also recorded and analyzed. Another hazard caused by the waves is the strong wind at ground level downstream of the mountain ridge, which can be very destructive. An anemometer in Boulder [see Klemp and Lilly (1975)] recorded gusts up to 50 m s^{-1} (120 mph) and damage from the storm in the Boulder area alone was estimated at $2,000,000. Lilly (1978) also estimated the force per unit span caused by the presence of waves to be about 10^6 N m^{-1}. This is about the value given by (8.8.20), using $U = 25$ m s^{-1}, $N = 10^{-1}$ s^{-1} and $h_m = 2$ km.

The solution in the hydrostatic case is easily obtained because the vertical structure can be determined independently of the horizontal structure. The solution, which has the form (8.8.16) for constant N/U, becomes

$$h = h_s(x)\hat{w}(z), \tag{8.9.13}$$

where $\hat{w}$ satisfies (8.9.8) with the x derivates omitted, i.e.,

$$\frac{d^2\hat{w}}{dz^2} + \left(\frac{N^2}{U^2} - \frac{1}{U}\frac{d^2U}{dz^2}\right)\hat{w} = 0. \tag{8.9.14}$$

For example, for the case considered in Section 6.9.4 with constant velocity U and

$$N = \begin{cases} N_1 & \text{for} \quad z < H_1, \\ \epsilon N_1 & \text{for} \quad z > H_1, \end{cases} \tag{8.9.15}$$

maximum response is obtained when

$$m_1 H_1 = \pi/2, \quad \text{where} \quad m_1 = N_1/U, \tag{8.9.16}$$

in which case the solution of (8.9.14) is

$$\hat{w} = \begin{cases} \cos(m_1 z) + \epsilon^{-1}i\sin(m_1 z) & \text{for} \quad z < H_1, \\ \epsilon^{-1}\exp\{i\epsilon m_1(z - H_1)\} & \text{for} \quad z > H_1. \end{cases} \tag{8.9.17}$$

For the bell-shaped mountain, h_s is given by [cf. (8.8.17)]

$$h_s(x) = h_m/(1 - ix/L), \tag{8.9.18}$$

and the solution in the strongly stable layer near the ground is shown in Fig. 8.14 for

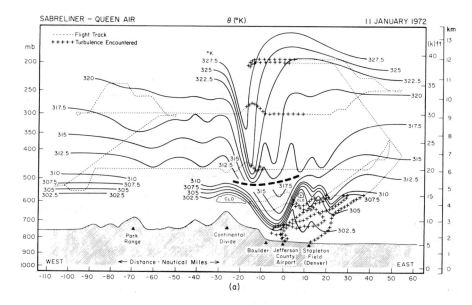

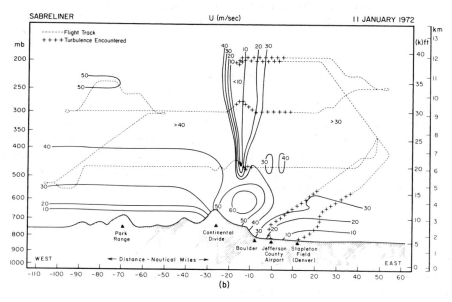

Fig. 8.13.   Observations of (a) potential temperature (in degrees Kelvin) and (b) velocity (in meters per second) in a section across the Rockies in Colorado on January 11, 1972. The dashed lines show aircraft tracks with periods of significant turbulence shown by plus signs. The heavy dashed line separates observations made a few hours apart—it is quite possible that at a given time, the upper dip in the potential temperature contours was directly above the lower one. [From Lilly 1978, Figs. 7 and 9).]

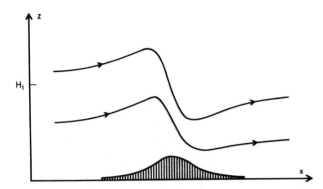

Fig. 8.14. Particle paths for air flow over a bell-shaped mountain, obtained for the linear hydrostatic non-rotating case. The undisturbed flow has uniform velocity U, constant buoyancy frequency N_1 up to height H_1, and constant buoyancy frequency $\frac{1}{3}N_1$ above H_1. The (partial resonance) condition for maximum response is satisfied, namely, $N_1 H_1/U = \pi/2$. Then a quarter of a wavelength is obtained between the ground and the discontinuity in buoyancy frequency, so waves reflected off the discontinuity and the ground reinforce each other. The resonance is only partial in the sense that only partial reflection occurs at the discontinuity. Note that the particle displacements at the discontinuity are significantly greater than at the ground. The pattern at higher levels is the same as that shown in Fig. 8.9b for uniform conditions. Note that the displacements are much larger than the height of the mountain, whereas for constant N (Fig. 8.9b) they are comparable.

$\epsilon^{-1} = 3$. The solution at higher levels is the same as that shown in Fig. 8.9b. The main effect of the enhancement due to partial reflection is that the amplitude of the waves is greater than that of the mountain by a factor ϵ^{-1}.

8.9.4   Wave Absorption: Hydrostatic Rotating Range

When N and U vary slowly enough for ray tracing to be appropriate, the ray path equation (8.9.10), in conjunction with (8.7.14), shows that rays become horizontal not only when $Uk = N$, but also when $Uk = f$. The latter possibility will now be considered. When Uk is comparable with f, which is assumed to be small compared with N, the hydrostatic approximation ($Uk \ll N$) can be made, and then (8.9.10), (8.7.14), and (8.7.6) give approximately

$$\frac{dz}{dx} = \frac{N^2 k^3}{f^2 m^3} = \frac{f}{N}\left(\frac{U^2 k^2}{f^2} - 1\right)^{3/2}. \tag{8.9.19}$$

If U *decreases* with height, the angle of propagation decreases and the ray becomes horizontal at the level $z = z_c$, where $U = f/k$. As this level is approached, m tends to infinity and (8.9.19) is approximated by

$$dz/dx = N^2 k^3/f^2 m^3 = (f/N)[(z - z_c)\, d(U^2 k^2/f^2)/dz]^{3/2}.$$

Integration of this equation shows that near $z = z_c$ the ray path has the form

$$z = z_c - 4A^{-2}(x - x_0)^{-2}, \tag{8.9.20}$$

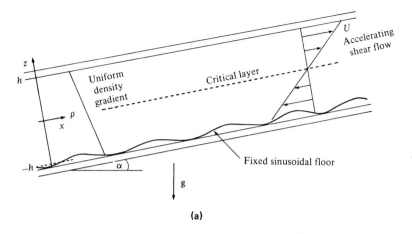

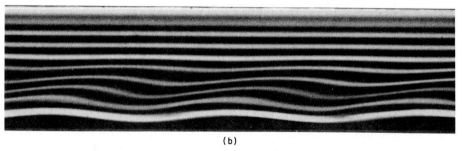

(b)

Fig. 8.15. An experimental demonstration of critical layer absorption of internal waves. (a) The apparatus—an inclined tube with a corrugated floor. (b) Waves observed in an accelerating shear flow over a corrugated floor of amplitude 0.5 cm and wavelength 25 cm. The tube contains stratified fluid with $N = 2.626$ s^{-1} and has been tilted through 5.2°. [From Thorpe (1981, Figs. 1 and 4e).]

where x_0 is a constant and A is given by

$$A^2 = -\frac{f^2}{N^2}\left\{\frac{d}{dz}\left(\frac{U^2 k^2}{f^2}\right)\right\}^3_{z=z_c}. \tag{8.9.21}$$

Details of the waves in the neighborhood of such a level have been studied by W. L. Jones (1967), following earlier work on the nonrotating case by Booker and Bretherton (1967). Dissipative effects become important near the so-called *critical level* $z = z_c$ because the time taken for energy traveling along the ray at the group velocity to reach the critical level is infinite. Thus even effects like Newtonian cooling (see Section 8.11), which are not scale dependent, have time to take effect. Viscosity can be even more effective because its rate of damping increases as the scale decreases. The same effects can occur even *without* a critical level, provided that U falls to a value near enough to f/k for dissipative effects to become significant and hence for wave energy to be absorbed.

Dissipation of waves also implies that the upward momentum flux is decreasing, and so there is an input of momentum at these levels equivalent to a body force acting on the mean flow there. In other words, momentum is transferred from the ground to a remote level through the mean stress set up by waves. The levels to which the momentum is transferred are those at which the wave speed matches the flow speed. In the case of topographic waves, which have zero phase speed, the critical level is that at which the flow speed is zero. A schematic picture of ray paths for a case in which absorption occurs is shown in Fig. 8.12b.

A striking experimental verification of the critical layer absorption process is shown in Fig. 8.15. Initially, a uniformly stratified fluid is at rest in a horizontal tube of rectangular cross section. The tube is then tilted as shown in Fig. 8.15a, leading to an accelerating shear flow that generates internal waves at the sinusoidal floor of the tube. Die lines in the fluid show the wave motion (Fig. 8.15b), and it is clear from the photograph that the waves do not penetrate beyond the critical layer located at the center of the tube.

8.10   Finite-Amplitude Topographic Effects

The studies of topographic effects discussed so far have assumed that the hills producing the effects are small enough for linear theory to be used. In practice, most interest in topographic effects has been in cases for which the effects are large, and therefore linear theory has not been strictly applicable. Because of this attempts to find solutions of the nonlinear equations have been made. One successful approach was indicated by Long (1953), who showed that there is a class of flows for which the equation for the vertical particle displacement h has the same form as that for linear theory, so naturally this class has received considerable attention. The theory applies *only in the nonrotating range* for an incompressible frictionless fluid when the velocity $\bar{u}(z)$ and density $\bar{\rho}(z)$ profiles far upstream satisfy the conditions

$$d\bar{\rho}/dz = \text{const}, \qquad \bar{\rho}\bar{u}^2 = \text{const}. \tag{8.10.1}$$

Then h satisfies the same equation as that for the linear theory, namely [cf. (8.9.8)],

$$\frac{\partial^2 h}{\partial x^2} + \frac{\partial^2 h}{\partial z^2} + \frac{N^2}{\bar{u}^2} h = 0, \tag{8.10.2}$$

where

$$N^2/\bar{u}^2 = -(g/\bar{\rho}\bar{u}^2)\, d\bar{\rho}/dz = \text{const}. \tag{8.10.3}$$

The one point of difference from the linear solution is that h has a prescribed value (the surface elevation) not on $z = 0$, the mean surface level, but on $z = h$, the actual surface level. Solutions have the same form as those discussed in earlier sections, but application of the boundary condition is not so straightforward. In a series of papers, Miles (1968a,b), Miles and Huppert (1969), and Huppert and Miles (1969) found solutions for some special shapes, whereas Lilly and Klemp (1979) have developed an iterative technique that starts with the linear solution, their method being for the case in which the hydrostatic approximation applies.

An interesting feature of the nonlinear solutions is that as the height h_m of the mountain increases the maximum streamline (and hence isopycnal) slope increases and eventually becomes infinite when the inverse

$$F^{-1} = \bar{N}h_m/\bar{u} \tag{8.10.4}$$

of the Froude number F, based on the mountain height, reaches a value of about unity [the appropriate critical value was found by Lilly and Klemp (1979) to be 0.85 for a bell-shaped mountain]. The solution becomes invalid at this point because any further tilt of the isopycnal would cause heavy fluid to overlie light fluid, thus producing rapid mixing. The drag of the mountain at the critical value of the Froude number is usually within a factor of two of the linear value, e.g., for the bell-shaped mountain it is 1.4 times greater. For larger values of h_m, a "rotor" forms behind the mountain, and the drag may change significantly.

For other upstream density and velocity profiles, solutions can be obtained numerically, and considerable success has been achieved in simulating observed conditions. For example, the downslope wind storm of 11 January 1972 has been simulated by Klemp and Lilly (1978), using a hydrostatic model, and by Peltier and Clark (1979), using a nonhydrostatic model. In both models, the strong downdraught over the mountain (as in Fig. 8.14) is followed by a strong updraught in the lee (unlike Fig. 8.14 but found in the observed flow shown in Fig. 8.13), and the maximum ground-level wind is close to that observed. The nonhydrostatic model also exhibited wave breaking just above the tropopause with consequent enhancement of the response (shown in Fig. 8.16). This model also shows a train of lee waves, and the surface drag is some 20 times the value given by linear theory.

A useful insight into the dynamics for finite-amplitude topography comes from the Bernoulli equation (see Section 4.8), which requires that for steady flow

$$p + \tfrac{1}{2}\rho(u^2 + v^2 + w^2) = \bar{p} + \tfrac{1}{2}\bar{\rho}\bar{u}^2 \tag{8.10.5}$$

on a streamline, the values on the right-hand side being those appropriate far upstream. The maximum increase in pressure above the upstream value is therefore

$$p - \bar{p} = \tfrac{1}{2}\bar{\rho}\bar{u}^2. \tag{8.10.6}$$

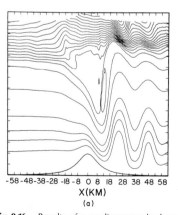

-58 -48 -38 -28 -18 -8 0 8 18 28 38 48 58
X(KM)
(a)

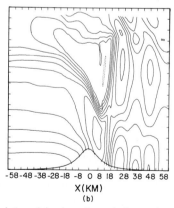

-58 -48 -38 -28 -18 -8 0 8 18 28 38 48 58
X(KM)
(b)

Fig. 8.16. Results of a nonlinear nonhydrostatic numerical simulation of the downslope windstorm shown in Fig. 8.13. (a) Potential temperature contours; (b) horizontal velocity contours (interval 8 m s^{-1}). The maximum in the lee of the peak is in excess of 60 m s^{-1}. The vertical scale is marked in kilometers and goes from 0 to 15 km. [From Peltier and Clark (1979, Figs. 29f and 31f).]

A rough estimate of how far a particle need be raised to have an excess pressure of this magnitude is obtained by using linear theory. A particle with upward displacement h has an excess density of $-(d\bar{\rho}/dz)h$ and this gives an excess pressure of

$$p - \bar{p} \approx -\int_0^h \left(\frac{d\bar{\rho}}{dz}\right) gh' \, dh' = \tfrac{1}{2}\bar{\rho}\bar{N}^2 h^2. \tag{8.10.7}$$

Comparison of (8.10.6) and (8.10.7) gives the maximum particle displacement h to be approximately equal to $\bar{u}/\bar{N}$.

The implication of the above result is that if the mountain height is in excess of $\bar{u}/\bar{N}$, particles more than a distance $\bar{u}/\bar{N}$ below the crest will tend to be "*blocked*" by the mountain, i.e., prevented from flowing over it. In practice these particles may find their way around the mountain instead of going over the top. For instance, Lilly (1978, p. 75), in discussing the event of 11, January 1972, reports that the air reaching Denver (where the standard pressure is about 830 mb) was never below 700 mb on the upstream side. In other words, upstream air below 700 mb was effectively blocked by the Colorado Rockies, but Lilly reports evidence that some of this air, at least, crossed the Divide in Wyoming, to the north, where the Divide drops to about the 760 mb level (compared with about 650 mb in Colorado). An interesting consequence of the blocking is that the air found in the lee of the mountain tends to be very dry, having descended from a high level, thereby making summer conditions near the mountains much more comfortable than those in more distant humid areas.

The same effect occurs on smaller scales and is of importance for pollution control. If effluent from a chimney has no buoyancy of its own and the chimney top is more than a distance $\bar{u}/\bar{N}$ below the crest of a hill on the upstream side, the effluent is not likely to clear the hill. Instead it is likely to flow against the hill at a level within $\bar{u}/\bar{N}$ of the chimney height, and then move along the hill. This effect has been demonstrated in laboratory experiments by Brighton (1978) and by Hunt and Snyder (1980).

Another finite-amplitude effect of importance on small scales ($L < U/N$) is boundary-layer flow separation, which produces a wake flow with a depth scale of the order h_m of the height of the hill and a length scale of the order L of the width of the hill. The drag depends significantly on whether or not flow separation occurs [see the discussion by Scorer (1955), the laboratory experiments of Hunt and Snyder (1980), and the numerical computations of Mason and Sykes (1978)].

8.11   Dissipative Effects in the Upper Atmosphere

Waves can lose energy by a variety of effects. In the troposphere and in the body of the ocean, the main losses are probably due to turbulence and to transfer of energy to other waves or to the mean flow. Waves in the ocean can also lose energy at the bottom by scattering into other waves produced by topographic variations. The rates of loss by these processes are highly variable and are not known with any precision, although empirical formulas are sometimes used to make estimates.

Waves that are traveling upward in the middle atmosphere undergo changes, caused by the thinning of the atmosphere, that are of some interest. A summary of the changes in atmospheric properties with height is given in Section 3.5 and is shown diagrammatically in Fig. 3.3. The pressure and density both fall off at exponential rates with e-folding scales that vary between 5.5 and 8.5 km. At 86 km, the pressure is 1/270,000 of its surface value, whereas the density is 1/180,000 of value it has at the surface. Likewise the mean free path, which measures the average distance between collisions of molecules, is 1 cm at 86 km, which is 180,000 times the surface value. Above 86 km, the composition of the atmosphere can no longer be regarded as constant, and effects of diffusive transport of individual gas species must be taken into account. As a result, the mean molecular weight falls off with height and its value at 300 km is about 60% of its value at 86 km. Also, because of the much higher temperatures, the pressure and density do not fall off as rapidly with height as they do at lower levels, and the corresponding scale heights increase substantially, being about 50 km at an altitude of 300 km.

The kinematic viscosity v, which is equal to μ/ρ by (4.5.15), and the thermal conductivity κ, which is equal to $k/\rho c_p$ by (4.4.8), also undergo huge changes because of the variation of ρ. For instance, the viscosity μ in the lower 86 km does not vary by more than 20% from a constant value of 1.5×10^{-5} N s m^{-2}, so that v attains a value of 1.8 m^2 s^{-1} at 86 km, which is 120,000 times the surface value. The thermal diffusivity varies in a similar manner, and is 1.36 times the kinematic viscosity, according to the kinetic theory of gases.

Now consider how vertically propagating waves will be affected by the variations in v and $\kappa = 1.36v$. If the factor of 1.36 is ignored, the only difference in the linearized equations (see Chapter 4) that is introduced by adding the effects of viscosity and thermal diffusivity is the replacement of the operator $\partial/\partial t$ by the operator

$$\partial/\partial t - v \nabla^2.$$

The effect on waves with vertical scale m^{-1} small compared with the scale height H_ρ, on which ρ and v vary, can be assessed quite easily. For such waves, v can be regarded as constant over a length m^{-1}, and the incompressibility and Boussinesq approximations are applicable. The dispersion equation is therefore obtained from (8.4.13) by replacing the frequency ω by

$$\omega + iv\kappa^2.$$

Consider now a wave of fixed frequency ω and horizontal wavenumber κ_H propagating upward. The rate at which such a wave attenuates with height can be found by solving the dispersion relation for m, which will now be complex. Where the viscosity is small enough, m will be approximately the same as in the inviscid case, and will be changed only by a small amount δm due to the small change $iv\kappa^2$ in ω introduced by viscous effects. The value of δm is given approximately by

$$\delta m = iv\kappa^2/(\partial\omega/\partial m) = iv\kappa^2/c_{gz}, \tag{8.11.1}$$

showing that waves with small vertical group velocity c_{gz} are attenuated more rapidly. The way the attenuation rate varies with frequency ω and vertical component m of the wavenumber can be found by substituting the expression for c_{gz}, given by (8.4.17), putting $\kappa = m/\sin \varphi'$ from (6.5.4), and then using the expressions (8.4.16) for $\cos \varphi'$

and sin φ'. The result is

$$\delta m = \frac{-ivm^3\omega}{\omega^2 - f^2}\left(\frac{N^2 - f^2}{N^2 - \omega^2}\right)^2. \tag{8.11.2}$$

For waves to attenuate as $z \to \infty$, $m\omega$ must be negative, i.e., have downward phase velocity, as already deduced. If N/f is large, the least attenuated waves have $\omega \approx 5^{-1/2}N$, $\kappa_H \approx 0.5m$, and

$$\delta m \approx 3.5ivm^3/N. \tag{8.11.3}$$

The attenuation rate is proportional to $\int \delta m \, dz$, so (8.11.3) implies that short waves (m large) will be dissipated first, and therefore the scale of the predominant waves will increase with height. This is indeed observed (Hines, 1960, 1964; Zimmerman, 1964); the predominant scale m^{-1} seen in distortions of meteor trains being about 1 km at the 90 km level, approaching 10 km at the 140 km level. Observations, using radar techniques, are discussed by Balsley and Gage (1980).

For the above calculations to be valid, m^{-1} must be small compared with the scale height. When this is not so, calculations by Yanowitch (1967) show that if the solution is to decay at infinity, a combination of upward- and downward-propagating waves is required at lower levels, implying that variable viscosity causes partial reflection of waves. This result may seem surprising, but it is fairly obvious for the extreme case of a very viscous fluid overlying an inviscid stratified fluid.

Another effect of the falloff in density is that the velocity amplitude of waves increases with height in order to keep the vertical energy flux constant. This could lead to wave breaking before molecular effects become important, and there is evidence (Holton and Wehrbein, 1980) to suggest that this occurs near the mesopause. The wave drag associated with this has important effects on longer period motions, e.g., Holton and Wehrbein (1980) assume a friction coefficient giving a decay time of 2 days at heights above 80 km as compared with 80 days at heights below 60 km.

Another important effect for longer-period motions is that of infrared radiation, which tends to restore the temperature field to a radiative equilibrium and hence to remove perturbations. A convenient approximation to this process is in terms of a Newtonian cooling coefficient α, the $\partial T'/\partial t$ term in the equation for the perturbation temperature T' being replaced by

$$(\alpha + \partial/\partial t)T'.$$

The temperature relaxation process is important above 25 km, where the rate can be affected significantly by photochemical reactions. Estimates of the rate are given by Dickinson (1973) and Hartmann (1978). α^{-1} is about 20 days at 25 km, decreasing to a value of about 5 days at levels between 45 and 60 km. Then α^{-1} increases again to values of about 30 days between 75 and 85 km.

Detailed calculations of dissipative effect on waves can be made numerically, e.g., Lindzen (1970, 1971) and Lindzen and Blake (1971) have found solutions pertinent to both semidiurnal and diurnal tides and to an internal mode with a period of 3 hr and a vertical scale m^{-1} of 4.5 km. Lindzen and Blake (1972), Francis (1973), and Salby (1980) have calculated effects on Lamb modes, and Francis (1973) has also studied modes that are ducted in the 100–400 km height range and are associated with traveling ionospheric disturbances. Reviews and books on meteorology of the middle atmosphere have been written by Dickinson (1975), Holton (1975, 1980a,b), Venkate-swaran and Sundararaman (1980), and Kato (1980).

8.12   The Liouville–Green or WKBJ Approximation

Much of the discussion of waves thus far has been based on the rather special conditions for which the equations have constant coefficients. For example, when a wavelike form is assumed for variations with respect to the horizontal coordinates and time, the resulting equation may have the form

$$d^2w/dz^2 + m^2w = 0. \tag{8.12.1}$$

If m is constant, the solutions in complex form are

$$w = A \exp(\pm imz). \tag{8.12.2}$$

In practice, conditions in the vertical are rarely uniform, so it is necessary to consider equations like (8.12.1) when m is a function of z. If m varies slowly enough, it is expected that locally the solution will be similar to that of (8.12.2), with the vertical wavenumber m changing slowly with z. This raises the question as to whether a better approximation can be found, i.e., what is the generalization of (8.12.2) that has the widest range of validity when m varies? [More about this problem can be found in Lighthill (1978).]

A problem of this type was tackled by Liouville (1837) and Green (1838) and is discussed in textbooks on asymptotic theory such as those of Erdélyi (1956, Chapter 4) and Olver (1974, Chapter 6). The approximate solution is therefore called the Liouville–Green approximation. It was also (and still is) called the WKB or WKBJ approximation, based on the initials of more recent authors, until it was realized that the method was used much earlier by Liouville and Green.

Basically, Liouville's method was to introduce new coordinates

$$\Phi = \int m \, dz, \qquad W = m^{1/2}w, \tag{8.12.3}$$

thereby transforming (8.12.1) into the equation

$$d^2W/d\Phi^2 + (1 + \delta)W = 0, \tag{8.12.4}$$

where

$$\delta = m^{-3/2} \, d^2(m^{-1/2})/dz^2. \tag{8.12.5}$$

Thus if $\delta \ll 1$, the approximate solution of (8.12.4) is

$$W = e^{\pm i\Phi}, \tag{8.12.6}$$

which is equivalent to giving the approximate solution of (8.12.1) as

$$w = m^{-1/2} \exp\left(\pm i \int m \, dz\right).$$ (8.12.7)

The condition $\delta \ll 1$ means that the vertical scale m^{-1} of the wave is small compared with the scale on which m varies. More precise conditions for the validity of (8.12.7) are given by Olver (1974).

An alternative way to obtain the approximate result is to look for solutions of (8.12.1) in the form

$$w = Ae^{\pm i\Phi}$$ (8.12.8)

with amplitude A and phase Φ real. Substituting in (8.12.1) and taking real and imaginary parts give

$$\left(\frac{d\Phi}{dz}\right)^2 = m^2 + \frac{1}{A}\frac{d^2 A}{dz^2}$$
$$\frac{d^2\Phi}{dz^2}\bigg/\frac{d\Phi}{dz} = -\frac{2}{A}\frac{dA}{dz}.$$ (8.12.9)

If the term involving the second derivative of A is ignored, the solution (8.12.7) is obtained once again.

8.12.1   Vertical Structure of Internal Gravity Waves

As an application of the Liouville–Green method, consider internal gravity waves in an incompressible fluid when the Boussinesq approximation is applied. For small perturbations from a state of rest, Eq. (8.4.11) is satisfied and solutions can be found that are proportional to

$$\exp\{i(kx + ly - \omega t)\}.$$

For such solutions, (8.4.11) reduces to an equation of the form (8.12.1), where

$$m^2 = (k^2 + l^2)(N^2 - \omega^2)/(\omega^2 - f^2),$$ (8.12.10)

i.e., m is given by the dispersion equation (8.4.13), but now N is allowed to depend on z. If N varies sufficiently slowly, the approximate solution (8.12.7) can be used, provided that m^2 is everywhere positive. The same approximation can also be used in a region in which the waves are evanescent, i.e., where m^2 is everywhere negative, but it breaks down in the neighborhood of levels at which m^2 changes sign. The value z at which this occurs is called a turning point, and a different form of approximation is required in this neighborhood [see Erdélyi (1956) and Olver (1974)]. This approximation involves Airy functions that, from Fig. 6.19, can be seen to model a transition from wavelike solutions to evanescent solutions. In Section 8.9.2, the turning point was identified as a place at which wave reflection takes place.

8.12.2 A Stretched Vertical Coordinate for Hydrostatic Waves

For frequencies ω that are everywhere small compared with N, the hydrostatic approximation applies and (8.12.10) gives

$$m = N/c, \qquad (8.12.11)$$

where c is a constant such that

$$c^2 = (\omega^2 - f^2)/(k^2 + l^2), \qquad \text{i.e.,} \qquad \omega^2 = f^2 + (k^2 + l^2)c^2. \qquad (8.12.12)$$

For slowly varying N, the phase Φ is given by (8.12.3), i.e.,

$$\Phi = c^{-1} \int N \, dz. \qquad (8.12.13)$$

An important consequence of this result is that a *modified vertical coordinate* z_m can be introduced such that the phase Φ varies linearly with z_m. This is always possible for a given frequency disturbance, but when the hydrostatic approximation can be made, the required transformation does not depend on frequency. A convenient definition of the stretched coordinate is

$$N_m z_m = \int N \, dz \qquad (8.12.14)$$

(or $N_* dz_*$ in the case of isobaric coordinates—Section 6.17), where N_m is a fixed value, often chosen as a maximum value of the buoyancy frequency. The point of the change of variable is that in the new coordinates the solution (to the extent that the Liouville–Green or WKBJ approximation applies) is the same as that for a *uniform medium*. This is useful in analyzing measurements, e.g., the interpretation of wave-number spectra in the modified coordinate system is the same as that for a uniform medium. The technique has been used, for instance, by Leaman and Sanford (1975) (see Section 8.5) for distinguishing between upward- and downward-propagating waves.

Results, found in previous sections, for hydrostatic solutions in uniform media can readily be modified to give solutions for slowly varying media, using the modified coordinate. For instance, the normal modes (see Section 6.11) between rigid boundaries at $z = 0, -H$ are such that the phase changes by an integral multiple of π over the depth of the fluid, and thus (8.12.13) gives

$$c_n = \int_{-H}^{0} N \, dz/n\pi. \qquad (8.12.15)$$

Then for a given mode $c = c_n$, (8.12.12) gives the dispersion relation for Poincaré waves.

The way wave amplitudes vary with depth can be found from (8.12.7), which shows that in the hydrostatic case

$$w \propto N^{-1/2} e^{\pm i\Phi}, \qquad (8.12.16)$$

and from the polarization relations (8.5.4)–(8.5.6), which for the hydrostatic case give

$$\rho_0^{-1}p', u, v \propto N^{1/2}e^{\pm i\Phi}, \qquad \rho_0^{-1}\rho' \propto N^{3/2}e^{\pm i\Phi}. \tag{8.12.17}$$

Similarly, (8.6.1) shows that all of the terms contributing to the energy vary in like manner with depth, so

$$E \propto N. \tag{8.12.18}$$

Measurements of internal waves in the ocean generally support this relationship in that spectra of E/N from different depths are almost the same, and certainly show much less variation than spectra of E [see, e.g., Briscoe (1975)].

8.12.3   Wave Propagation in a Medium of Variable Properties

The Liouville–Green or WKBJ method, as formulated at the beginning of this section, gives a means of studying waves in a medium that varies in one direction only. It can be generalized, however, to a medium that varies slowly in all spatial dimensions and in time by substituting (8.12.8) in the relevant equations. If all derivatives higher than the first are ignored in the equivalent of the first equation of (8.12.9), an equation relating partial derivatives of the phase is obtained, this being a generalization of the *eikonal equation* of geometrical optics. Now, given any phase function Φ, the frequency ω and wavenumber $\mathbf{k}$ can be *defined* by

$$\omega = -\partial\Phi/\partial t, \quad \mathbf{k} = \nabla\Phi. \tag{8.12.19}$$

With this definition, the generalized eikonal equation becomes a dispersion equation, giving ω as a function of wavenumber $\mathbf{k}$, and also of the properties of the medium, usually through a parameter. For instance, in (8.4.13) the parameter is N, which could be a function of height z, say. In other cases, there is a set λ of parameters such as $N(z)$ and $U(z)$, and the generalized form of the dispersion relation can be written (when λ is a single parameter)

$$\omega = W(\mathbf{k}, \lambda). \tag{8.12.20}$$

Some interesting results follow immediately from these definitions [cf. Lamb (1904) and Whitham (1974)] because there are alternative ways of expressing the mixed derivaties of Φ. For instance,

$$\partial^2\Phi/\partial x\,\partial t = \partial k/\partial t = -\partial\omega/\partial x. \tag{8.12.21}$$

Substituting from (8.12.20) and using

$$\partial^2\Phi/\partial x\,\partial y = \partial k/\partial y = \partial l/\partial x, \qquad \partial^2\Phi/\partial x\,\partial z = \partial k/\partial z = \partial m/\partial x, \tag{8.12.22}$$

there results an equation that can be written (when λ is a single parameter)

$$D_g k/Dt = -(\partial W/\partial\lambda)\,\partial\lambda/\partial x, \tag{8.12.23}$$

where

$$D_g/Dt = \partial/\partial t + \mathbf{c}_g\cdot\nabla, \qquad \mathbf{c}_g = (\partial\omega/\partial k, \partial\omega/\partial l, \partial\omega/\partial m). \tag{8.12.24}$$

Similar results hold for the other components, so in vector form the result is

$$(D_g/Dt)\mathbf{k} = -(\partial W/\partial \lambda)\, \nabla \lambda \qquad (8.12.25)$$

(with summation over all components of λ in the case for which λ is a set of parameters). The operator D_g/Dt is the derivative following a point that moves with the group velocity $\mathbf{c}_g$ of the wave and can be called simply "the derivative following the wave group." The value of this derivative when applied to ω can be found by using (8.12.20) and then substituting from (8.12.25) to give

$$D_g\omega/Dt = (\partial W/\partial \lambda)\, \partial \lambda/\partial t. \qquad (8.12.26)$$

Thus in a medium whose properties do not change with time, the frequency following a wave group (i.e., along a ray path) is constant.

The ray path itself is defined as the path followed by a point moving with the group velocity $\mathbf{c}_g$, i.e., the position $\mathbf{x}_r$ of the point on the ray varies with time according to the equation

$$d\mathbf{x}_r/dt = \mathbf{c}_g. \qquad (8.12.27)$$

It can now be seen that Eqs. (8.12.25)–(8.12.27) specify both the ray path and the way in which the frequency and wavenumber vary along the path. For a given time displacement δt, (8.12.27) gives the change in coordinates $\delta\mathbf{x}_r$ following the wave, whereas (8.12.25) and (8.12.26) give the changes $\delta\omega$ and $\delta\mathbf{k}$ of frequency and wavenumber, respectively. From these, the change in group velocity can be found and hence the integration continued. Note that the ray-tracing technique is often a much more efficient way of obtaining information about waves than is solving the original dynamical equations.

8.12.4   Effect of Weak Shear on the Dispersion Properties of Waves

Suppose that there is a medium of slowly varying properties that is moving in such a way that the undisturbed fluid velocity $\mathbf{U}$ also varies slowly with position and, possibly, time. It is useful to give a special name to the frequency $\hat\omega$ that would be found by an observer moving with the undisturbed fluid, i.e., $\hat\omega$ is defined by

$$\hat\omega = -(\partial\Phi/\partial t + \mathbf{U}\cdot\nabla\Phi). \qquad (8.12.28)$$

$\hat\omega$ is called the *intrinsic frequency* (intrinsic, that is, to the local disturbance dynamics). It is also called the *Doppler-shifted frequency* (Doppler-shifted, that is, to the particular observer in question). It is related to the frequency ω in a fixed frame of reference by

$$\hat\omega = \omega - \mathbf{k}\cdot\mathbf{U}, \qquad (8.12.29)$$

which follows by substitution of (8.12.19) in (8.12.28).

For an observer moving with the undisturbed fluid, the frequency $\hat\omega$ he observes will be the same as if the medium were at rest, i.e.,

$$\hat\omega = W_0(\mathbf{k}, \lambda), \qquad (8.12.30)$$

where $W_0(\mathbf{k}, \lambda)$ is the dispersion relation for a fluid at rest. Equations for ray path, etc., can be found in terms of W_0 by substituting (8.12.30) in (8.12.29) to give

$$\omega = W(\mathbf{k}, \lambda) \equiv \mathbf{k} \cdot \mathbf{U} + W_0(\mathbf{k}, \lambda). \tag{8.12.31}$$

In particular, the group velocity $\mathbf{c}_g$ is related to the value $\mathbf{c}_{g0}$ it has relative to the medium by

$$\mathbf{c}_g = \mathbf{U} + \mathbf{c}_{g0}. \tag{8.12.32}$$

This follows from (8.12.31) and (8.12.24). These relationships can be used, for instance [see, e.g., Whitham (1974) and Lighthill (1967, 1978)], to find wave patterns created by moving obstacles, following the classical work of Kelvin (Thomson, 1891).

8.12.5   Amplitude Variations along a Ray Path

The equations for the variation of frequency and wavenumber along a ray path follow as was seen from the definitions of the quantities involved, and they apply to any sort of wave in a slowly varying medium, whether moving or at rest. The way in which the amplitude varies along a ray path requires quite different information. For general waves in a slowly varying medium, this problem was solved by Whitham (1965), following precedents in quantum mechanics, and was brought into a very useful form by Bretherton and Garrett (1968), who also gave a definitive physical interpretation of their analysis. They found that the *wave-action density* $\mathscr{A}$, defined by

$$\mathscr{A} = E/\hat{\omega}, \tag{8.12.33}$$

where E is the wave-energy density and $\hat{\omega}$ the intrinsic frequency, is conserved following a wave, i.e., the equation satisfied is

$$\partial \mathscr{A}/\partial t + \nabla \cdot (\mathbf{c}_g \mathscr{A}) = 0, \qquad D_g \mathscr{A}/Dt = -\mathscr{A}\, \nabla \cdot \mathbf{c}_g. \tag{8.12.34}$$

A simple derivation of this result has been given by Andrews and McIntyre (1978b) for much more general circumstances in which the medium is *not* required to have slowly varying properties. Note that because of the involvement of spatial derivatives of $\mathbf{c}_g$, (8.12.34) requires consideration of a bundle of neighboring rays [see Hayes (1970) and Lighthill (1978)]. Some pictures of ray bundles for sound rays in shear flows are given by Candel (1977).

8.13   Wave Interactions

In the analysis of waves thus far, the amplitude of the waves has (except in Section 8.10) been assumed to be small enough for nonlinear terms (i.e., terms involving products of perturbation quantities) in the equations to be neglected. A criterion that is often sufficient for this approximation to be valid is that the perturbation velocity of the wave be small compared with the phase velocity of the disturbance. Such a criterion implies that the nonlinear terms do not substantially affect the linear solution over times comparable with the period of the wave. However, the criterion does not rule out the possibility that nonlinear terms produce small systematic changes over a period, and the cumulative effect of these changes over many periods can be large. In practice, systematic changes are the rule rather than the exception, and play an important part in determining how energy is distributed between the various possible wavenumbers and frequencies.

The most elementary example of nonlinear effects is provided by the case of two plane waves with wavenumbers $\mathbf{k}_1$ and $\mathbf{k}_2$ and corresponding frequencies $\omega(\mathbf{k}_1)$ and $\omega(\mathbf{k}_2)$, as given by the dispersion relation. When product terms appear in the equations, they will have the form

$$\exp\{i(\mathbf{k}_1 \cdot \mathbf{x} - \omega(\mathbf{k}_1)t)\} \exp\{i(\mathbf{k}_2 \cdot \mathbf{x} - \omega(\mathbf{k}_2)t)\}$$
$$= \exp\{i(\mathbf{k}_1 + \mathbf{k}_2) \cdot \mathbf{x} - i(\omega(\mathbf{k}_1) + \omega(\mathbf{k}_2))t\}, \tag{8.13.1}$$

where the multiplying coefficient is assumed to be small. Such terms can be regarded as providing a weak forcing for a new wave, with wavenumber $\mathbf{k}_1 + \mathbf{k}_2$ and frequency $\omega(\mathbf{k}_1) + \omega(\mathbf{k}_2)$, and the development of this new wave with time can be worked out from the equations. In most cases, the amplitude of the new wave is small, but if the frequency $\omega(\mathbf{k}_1) + \omega(\mathbf{k}_2)$ of the new wave happens to be equal to the natural frequency $\omega(\mathbf{k}_1 + \mathbf{k}_2)$ for a wave with wavenumber $\mathbf{k}_1 + \mathbf{k}_2$, then resonance occurs and the new wave grows linearly with time, at least while the amplitude of the new wave remains small compared with the amplitudes of the original waves. The time for the new wave to reach a comparable amplitude is called the *interaction time* of the original waves, and is inversely proportional to their amplitudes.

The *condition for resonance* to occur depends only on the geometry of the dispersion surfaces in wavenumber space, and can be written in the form

$$\omega(\mathbf{k}_1 + \mathbf{k}_2) = \omega(\mathbf{k}_1) + \omega(\mathbf{k}_2), \tag{8.13.2}$$

where $\omega(\mathbf{k})$ is the frequency for wavenumber $\mathbf{k}$ as given by the dispersion relation. This condition is satisfied for many pairs of internal waves, and solutions are given by McComas and Bretherton (1977), O. M. Phillips (1977), and Ripa (1981). When (8.13.2) is satisfied, the original pair of waves and the new wave they generate are said to form a *resonant triad*. For some classes of waves (surface gravity waves in deep water are an example), no resonant triads exist, but it is possible to find sets of four waves that are resonant. In fact, resonant wave interactions were first discussed for this case by O. M. Phillips (1960), and wave interactions are discussed by O. M. Phillips (1977) for both surface and internal waves. Equations for the evolution of the amplitudes of waves in a resonant triad can be found, and solutions have been obtained [e.g., Ball (1964), S. A. Thorpe (1966), and W. F. Simmons (1969)]. The total energy of the triad remains constant, but the way in which it is divided among the three waves keeps changing. Laboratory experiments that demonstrate resonant interactions have been reported by Martin *et al.* (1972), McEwan *et al.* (1972), and others.

In naturally occurring situations, there is usually a whole spectrum of waves, i.e., a superposition of waves with wavenumbers varying continuously over some range of values. In such cases, wave interactions occur in the same way as they do when a small number of waves is present, and provided that the wave amplitude is not too large, the transfer of energy is dominated by those waves that are associated with the resonant triads (if such are present, otherwise resonant sets of four waves dominate). The phases of the different wavenumber components in the spectrum are often assumed to be distributed randomly, and this assumption can be utilized to calculate the evolution of the spectrum with time. The theory was developed by Hasselmann (1966, 1967a) following precedents in quantum mechanics, and was first applied to a realistic internal wave spectrum by Olbers (1976). F. P. Bretherton (unpublished results) has shown that the random-phase assumption is justified for propagation in two or three dimensions in an unbounded domain but not for propagation in one dimension.

The contributions to the rate of change of amplitude at a particular wavenumber come from many triads. However, McComas and Bretherton (1977) showed that the behavior can be largely understood by considering three limiting forms of interacting triads. They named the mechanisms associated with those limiting forms as follows:

(a) *Induced Diffusion.*   This occurs when two nearly identical waves interact with another wave of much lower frequency and much smaller wavenumber. The shear of the latter wave (which is like a slowly changing large-scale flow to the small waves) acts to diffuse wave action (wave energy divided by frequency) among vertical wavenumbers.

(b) *Elastic Scattering.*   This occurs when two waves with wavenumbers that are almost mirror images in the horizontal plane interact with a wave of much lower frequency and double the vertical wavenumber. The latter wave acts in a way similar to that of the crystal lattice in Bragg scattering and tends to equalize the energy between upward- and downward-propagating waves. The conditions for elastic scattering to occur are satisfied only for waves with frequency substantially greater that f, so near-inertial frequency waves are little affected. [The predominantly upward- or downward-propagating waves that have been observed (see Section 8.5) had frequencies near f.]

(c) *Parametric Subharmonic Instability.*   This occurs when two waves of nearly opposite wavenumber interact with a wave of much smaller wavenumber and of twice the frequency. The process transfers energy from low-wavenumber energetic waves to high-wavenumber waves of half frequency, and so tends to produce inertial frequency waves (the lowest frequency possible) with high vertical wavenumber.

These processes have a strong influence on the internal wave spectrum in the ocean, and one result is that the spectrum has a shape that varies rather little from one part of the ocean to another. The shape of this spectrum is discussed in the next section.

A question related to wave interactions is that of wave stability, i.e., given a particular form of wave, or of wave spectrum, will that form persist? The most likely answer is "no" because nonlinear interactions will cause the spectrum to evolve. A particular case is that which occurs when a single plane progressive wave is present. Hasselmann (1967b) has shown that this is unstable if two other members of a resonant triad can be found such that the original wave has the highest frequency. This implies, for the case $f \ll N$, that any wave with frequency greater than $2f$ is unstable, for then two other waves with half of the frequency can always be found to make up a resonant triad.

8.14   The Internal Wave Spectrum in the Ocean

Observations of the internal wave spectrum in the ocean indicate the remarkable fact that it has much the same shape wherever it is observed in the deep ocean, unless the observations are made close to a strong source of internal waves. An analytic form that approximates this spectrum was first produced by Garrett and Munk (1972), who later gave an improved version (Garrett and Munk, 1975). Further minor improvements have been put forward in the light of subsequent measurements, e.g., by Cairns and Williams (1976) and by Müller et al. (1978), and reviews are given by Garrett and Munk (1979), Munk (1981), and Olbers (1982).

One type of measurement is of horizontal velocity and of temperature, as sensed by instruments on a mooring. From knowledge of the vertical gradient of temperature, the temperature measurements can be converted into estimates of vertical displacement h and hence of potential energy density [see, e.g., (6.7.3)]. Figure 8.17 shows the spectrum of potential energy density obtained from the IWEX (internal wave experi-

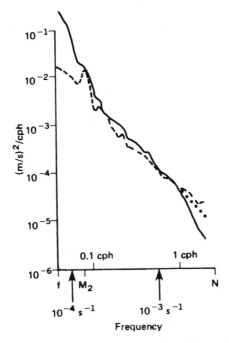

Fig. 8.17.   Spectra of contributions to internal wave energy as obtained from IWEX (internal wave experiment) data by Müller, Olbers, and Willebrand, *J. Geophys. Res.* **81**, 3100 (Fig. 4) (1978); copyright by American Geophysical Union. The solid line shows the spectrum of horizontal kinetic energy divided by ρ, the fluid density. It also shows the spectrum of ρ^{-1} times total kinetic energy, except at high frequencies, where the vertical kinetic energy must be added. If this is estimated from the vertical displacement spectrum, the dotted line is obtained for ρ^{-1} times the total kinetic energy. The dashed line shows ρ^{-1} times the potential energy spectrum. The spectra have been corrected for fine structure contamination and current noise contamination. The observations are for a depth of 600 m, which was near the center of the main thermocline at 27°44'N, 69°51'W. The frequency marked M_2 corresponds to the semidiurnal tide.

ment) data in this way. (A spectrum shows how the quantity is distributed among different frequencies. The integral with respect to frequency over the spectrum gives the quantity.) It also shows the spectrum of kinetic energy density. The ratio between potential and kinetic energy, and between horizontal and vertical velocity components is close to that expected from theory [Eqs. (8.5.4), (8.6.6), and (8.4.15) show that these ratios are functions of frequency only]. Rotational properties of the horizontal vector can also be confirmed. From (8.2.5) and (8.2.9), the horizontal velocity vector at a given place behaves for given frequency and wavenumber and for small perturbations from a state of rest as

$$u = k\omega \cos \omega t + lf \sin \omega t, \qquad v = l\omega \cos \omega t - kf \sin \omega t. \qquad (8.14.1)$$

The velocity vector rotates anticyclonically around an ellipse, but with a random set of waves there is a random orientation of ellipses, and only one item of information about the relationship between u and v can be obtained. This comes from expressing the vector $u + iv$ in the complex plane as the sum of an anticlockwise part [proportional to $\exp(i\omega t)$] and a clockwise-rotating part [proportional to $\exp(-i\omega t)$, ω being positive]. For the particular case (8.14.1),

$$u + iv = \tfrac{1}{2}(k + il)\{(\omega - f)e^{i\omega t} + (\omega + f)e^{-i\omega t}\}. \qquad (8.14.2)$$

In this representation, the energy [cf. Fofonoff (1969)] in the anticyclonically rotating part exceeds that in the cyclonically rotating part by the factor

$$\{(\omega + |f|)/(\omega - |f|)\}^2.$$

This is confirmed by observation (Müller et al., 1978). Also, variation of energy levels with depth was found to be in accordance with (8.12.18) (Briscoe, 1975).

It is more difficult to find the way in which energy is distributed among wavenumbers at each frequency. Information can be obtained from spectra obtained by towing instruments through the water, and from coherences between moored instruments with known separation. This does not give complete detail, but allows fitting to analytic models with only a few disposable parameters. These determine broad features e.g., where the maximum is, how wide the peak is, and how energy falls off at high wavenumbers. An example of a spectrum obtained in this way is shown in Fig. 8.18. The vertical axis gives frequency, but with a distorted scale, which varies as the logarithm of the ratio α of horizontal wavenumber κ_H to vertical wavenumber m [see (8.4.18)]. From (8.4.19), α and ω are related by

$$\omega^2 = (f^2 + \alpha^2 N^2)/(1 + \alpha^2). \qquad (8.14.3)$$

The horizontal axis gives vertical wavenumber m with respect to the *modified* vertical coordinate z_m as given by the Liouville–Green or WKBJ theory (Section 8.12.2), i.e., m is the rate of change of phase Φ with z_m. The quantity plotted is such that the volume under the surface whose contours are shown is equal to the total energy density divided by fluid density. Müller et al. (1978) made refinements by allowing for directionality in the spectrum and for a difference between energy propagating upward as opposed to downward. They found that energy was propagating mainly toward the equator at frequencies less than $5f$, but was isotropic at higher frequencies. Also, about 20% more energy was found to be propagating down-

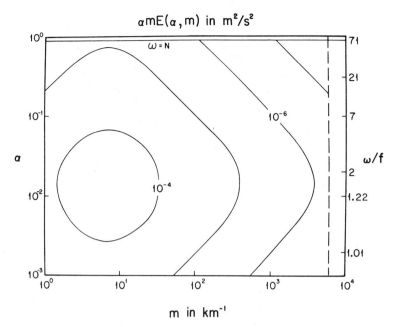

Fig. 8.18. A representation of the spectrum of total energy divided by ρ, the fluid density [from McComas (1977, Fig. 2)] for internal waves, based on the Garrett–Munk spectrum as modified by Cairns and Williams (1976). The quantity plotted times the area $d(\log_{10} \alpha)\, d(\log_{10} m)$ gives the energy of waves within that area. The buoyancy frequency is assumed to be 71 times the inertial frequency f, which is assumed to have the value $7 \times 10^{-5}\ \mathrm{s}^{-1}$ appropriate to a latitude of 30°. The vertical dashed line represents a vertical wavenumber cutoff for the spectrum. α is horizontal wavenumber divided by m, so lines of constant horizontal wavenumber are straight lines with slope -1. The spectrum is assumed to be horizontally isotropic and vertically symmetric, i.e., as much energy is propagating upward as downward.

ward than upward at frequencies less than $2.5f$, but symmetry was found at higher frequencies.

Internal wave spectra from different parts of the ocean are surprisingly similar in shape, and weak nonlinear interactions play an important part in maintaining this shape (McComas and Müller, 1981b). McComas (1977) and McComas and Müller (1981a) have calculated the effect of altering the distribution of energy in the spectrum and find the general tendency is to quickly revert to the equilibrium shape. For instance, if the balance between upward- and downward-propagating waves is upset, elastic scattering restores the balance at frequencies greater than about $3f$ in a time of about $5\omega^{-1}$ for a vertical scale (m^{-1}) of 10 m and $50\omega^{-1}$ for 100 m. The mechanism is not effective for near-inertial waves, and this is consistent with observation (see Section 8.6).

If there is symmetry between upward- and downward-propagating waves, the most important influence on the *shape* of the spectrum at low frequencies (f to $4f$) is parametric subharmonic instability, which transfers energy from waves of large vertical scale (and frequencies over $2f$) to waves of small vertical scale and near-inertial frequency. The dominant mechanism at higher frequencies is induced diffusion, which also results in transfer of energy to smaller vertical scales. The weak

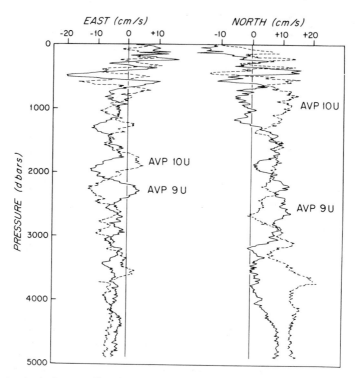

Fig. 8.19.   A pair of velocity profiles taken approximately half an inertial period (i.e., 11 hr) apart in the North Atlantic. The "mirror imaging" in the high-wavenumber components is due to near-inertial period internal wave activity. [From Richardson, Maillard, and Sanford, *J. Geophys. Res.* **84,** 7727 (1979) (Fig. 17a); copyright by the American Geophysical Union.]

interaction theory cannot be applied to vertical scales of 1 m or less, for which mechanisms such as shear instability and overturning will remove the energy that comes from larger scales and dissipate it [see Garrett (1979) and Woods (1982)].

   An important feature of the internal wave spectrum that should always be re-membered is the dominance of *low frequencies* in their contribution to the total energy. In particular, the spectrum given in Fig. 8.17 shows that half the kinetic energy is found in frequencies within about 30% of the inertial frequency. This fact shows up in velocity profiles taken approximately half an inertial period apart, like those shown in Fig. 8.19. The high mode number vertical structure is largely due to internal waves of near-inertial frequency, so an average between the two profiles gives a much smoother profile from which a large part of the internal wave contribu-tions has been removed.

   Not only is most of the energy in low frequencies, but also it is concentrated in the large vertical *scales*. This is presumably where sources [see, e.g., Müller (1977), Bell (1978), and Käse (1979)] supply the energy that is transferred to smaller scales by the interactions. The *shear*, on the other hand, is predominantly in the small vertical scales, and this is important for dissipative mechanisms.

8.15   Wave Transport and Effects on the Mean Flow

Waves have the property that they can produce effects in regions far removed from their source. For instance, surface gravity waves generated in a storm may lose their energy on a beach on the other side of the world and, in so doing, cause significant effects, such as moving sand along the coast. In a similar fashion, waves produced in the lower atmosphere (e.g., by topography or by heating effects) can produce significant effects at higher levels. In particular, waves can cause changes in the winds by virtue of their ability to transport *momentum* vertically.

A simple example [see McIntyre and Weissman (1978) and McIntyre (1980)] of such an effect is provided by that of the flow of a uniformly stratified (i.e., constant N) incompressible Boussinesq fluid over an initially plane surface. The velocity U is initially uniform, but at time $t = 0$, the surface begins to develop sinusoidal undulations with fixed horizontal wavenumber $k < N/U$. The amplitude h_0 of these undulations increases slowly up to some fixed infinitesimal value, i.e.,

$$h_0^2 = F(t), \tag{8.15.1}$$

where F is zero for $t < 0$ and approaches a constant value for t large. The time T for this to happen is assumed to be large compared with the encounter period $2\pi/Uk$ of air particles with crests e.g., F could have the form

$$F(t) = \begin{cases} 0, & t < 0 \\ 1 - e^{-t/T}, & t > 0. \end{cases} \tag{8.15.2}$$

As found in Section 6.8, the corrugations in the boundary will cause internal waves to be generated with vertical wavenumber m given by (6.8.6). After the corrugations have reached their final amplitude, the solution near the ground will be as described in Section 6.8, i.e., there will be waves of the form (6.8.5) that remove momentum from the ground at the rate τ given by (6.8.11) and transport it vertically at the same rate. The region in which waves exist will, however, have only a finite depth of order $c_{gz}t$ since the waves can propagate upward only at a finite rate c_{gz}, where c_{gz} is the vertical component of the group velocity, which is given by (6.6.1). At large heights $z > c_{gz}t$, there will be no waves and therefore no flux of momentum, i.e., no mean stress. It follows that there will be a region in which the momentum flux or stress drops from its surface value τ to zero. In this region, which will have thickness of order $L = c_{gz}T$, a force τ per unit area will be felt by the mean flow, tending to accelerate it.

An explicit solution for the changes in mean flow can be obtained once the equation for the mean flow is derived. Since wave transports are proportional to the square of the wave amplitude, which is assumed to be small, the mean flow equation must be correct to this order. The mean flow velocity will be denoted by $\bar{u}$, where the overbar signifies an average with respect to x. In the present instance, this may be regarded as an average over one wavelength $2\pi/k$, so the overbar has the same meaning as that in Section 6.8. A prime will be used to denote a perturbation from the mean, so the

x component of velocity at any point is given by

$$u = \bar{u} + u'. \tag{8.15.3}$$

Since, in the situation being considered, the only variations are with z and t, the average of the momentum equation (4.10.11) gives

$$\partial\bar{u}/\partial t + \partial(\overline{u'w'})/\partial z = 0. \tag{8.15.4}$$

This equation can be interpreted as saying that the rate of change of momentum of the mean flow is equal to minus the divergence of the mean stress set up by the waves. It depends on the x gradient of mean pressure vanishing [cf. Bretherton (1969)]. Now by (6.8.11), (6.8.6), and the fact that waves propagate upward at speed c_{gz}, the mean stress is given by

$$\rho_0 \overline{u'w'} = -\tfrac{1}{2} kmU^2 \rho_0 F(t - z/c_{gz}). \tag{8.15.5}$$

Hence the solution of (8.15.4) is

$$\bar{u} = U - \tfrac{1}{2} c_{gz}^{-1} kmU^2 F(t - z/c_{gz}) = U - \tfrac{1}{2} U^{-1} N^2 F(t - z/c_{gz}). \tag{8.15.6}$$

The second expression comes from substitution of the value $U \sin \varphi' \cos \varphi'$, given for c_{gz} at the end of Section 6.8, and use of (6.5.4) and (6.8.6). For large times

$$\bar{u} \to U - \tfrac{1}{2}(Nh_0)^2/U. \tag{8.15.7}$$

Thus the effect of "switching on" the corrugations is that the boundary, via the agency of the stress $\overline{u'w'}$ set up by the waves, removes a small fraction $\tfrac{1}{2}(Nh_0/U)^2$ of the momentum from the mean flow. This fraction is half the inverse square of the Froude number [cf. (6.9.18)]

$$U/Nh_0,$$

based on the amplitude of the topography.

The above example shows how waves can transport momentum via the mean stress $\overline{u'w'}$ to remote parts of the flow in a time-developing situation. Similar effects evidently occur when $\overline{u'w'}$ varies with height because of effects of dissipation on the waves. A laboratory experiment that demonstrates such effects has been reported by Plumb and McEwan (1978). General methods for treating such problems have been devised by Andrews and McIntyre (1978a).

It is worth pointing out that the energetics of such problems is dependent on the frame of reference used, a fact that is clearly illustrated by the above example. First, consider a frame of reference fixed relative to the ground. In this frame, since the boundary is stationary, no work is done and thus there is no input of energy from the ground. The loss of energy per unit mass from the mean flow is (to second order)

$$\tfrac{1}{2}U^2 - \tfrac{1}{2}\bar{u}^2 = \tfrac{1}{2}N^2 h_0^2, \tag{8.15.8}$$

using the expression (8.15.7) for $\bar{u}$, and this is precisely equal to the gain of energy per unit mass due to the presence of the wave, as given by (6.8.14). In terms of fluxes, the upward wave energy flux $F_z' = \overline{p'w'}$ due to the wave is balanced by a downward advection of kinetic energy $\rho_0 U \overline{u'w'}$.

Now consider the energy balance in a frame of reference moving with the original undisturbed mean flow. In this frame of reference, the boundary is moving at speed U and so does work at a rate equal to the upward flux $F'_z = \overline{p'w'}$ of energy in the wave. There is no advection (at second order) of kinetic energy in the frame of reference, and the upward flux is used up in the region of the wave front to give particles energy, which to second order is entirely in the form of wave energy.

8.16 Quasi-geostrophic Flow (f Plane): The Isallobaric Wind

The way in which a rotating fluid adjusts to changes when these changes are *slow* (time scale $\gg f^{-1}$) is rather special, and it is of vital importance to understand the nature of such slow adjustment processes because the changes seen on a weather map and the changes in ocean currents from one week to the next have such a character. The key to understanding slow adjustment is an appreciation of the *redundancy* in the equations of geostrophic flow, already discussed in Chapter 7, i.e., the fact that geostrophic flow *exactly* satisfies the continuity equation in shallow-water flow. Thus three equations do not give three independent items of information, so to find the flow, *departures* from geostrophy must be considered, even though these are small. The fact that the departures are small explains the designation of such flows as "quasi-geostrophic."

The redundancy of the equations applies not only to a shallow homogeneous layer, but also to each baroclinic mode and hence applies to a stratified fluid. Another way of deducing this is that geostrophic flow implies zero horizontal divergence and hence vertical motion that is independent of depth. Therefore, if a horizontal boundary is present, no vertical motion is possible for strictly geostrophic flow. However, it is essential to calculate the vertical motion in order to determine gravitational restoring forces, so departures from geostrophy must be considered.

The necessity for considering departures from geostrophy for computing vertical motion in a slowly changing situation was recognized by Brunt and Douglas (1928) in a paper entitled "The modification of the strophic balance for changing pressure distribution, and its effect on rainfall." Their approach was to take the horizontal momentum equations in the form

$$Du/Dt - fv = -\rho^{-1}\,\partial p/\partial x \equiv -fv_g, \tag{8.16.1}$$

$$Dv/Dt + fu = -\rho^{-1}\,\partial p/\partial y \equiv fu_g, \tag{8.16.2}$$

where (u_g, v_g) is the geostrophic wind defined by (7.6.5) and

$$D/Dt = \partial/\partial t + u\,\partial/\partial x + v\,\partial/\partial y \tag{8.16.3}$$

is the derivative following the motion, and to find an approximate expression for (u, v) in terms of (u_g, v_g) based on the requirement that the time derivation D/Dt must be small compared with f. There are several ways of doing this. One method begins by eliminating u and v in turn from (8.16.1) and (8.16.2) to obtain

$$\frac{D^2 u}{Dt^2} + f^2 u = f^2 u_g - f \frac{Dv_g}{Dt} \equiv -\frac{f}{\rho} \frac{\partial p}{\partial y} - \frac{D}{Dt}\left(\frac{1}{\rho} \frac{\partial p}{\partial x}\right), \tag{8.16.4}$$

$$\frac{D^2 v}{Dt^2} + f^2 v = f^2 v_g + f \frac{Du_g}{Dt} \equiv \frac{f}{\rho} \frac{\partial p}{\partial x} - \frac{D}{Dt}\left(\frac{1}{\rho} \frac{\partial p}{\partial y}\right). \tag{8.16.5}$$

Since $D/Dt \ll f$, these expressions can be approximated by

$$u \approx u_g - \frac{1}{f} \frac{Dv_g}{Dt} \equiv -\frac{1}{f\rho} \frac{\partial p}{\partial y} - \frac{1}{f^2} \frac{D}{Dt}\left(\frac{1}{\rho} \frac{\partial p}{\partial x}\right),$$

$$v \approx v_g + \frac{1}{f} \frac{Du_g}{Dt} \equiv \frac{1}{f\rho} \frac{\partial p}{\partial x} - \frac{1}{f^2} \frac{D}{Dt}\left(\frac{1}{\rho} \frac{\partial p}{\partial y}\right). \tag{8.16.6}$$

These equations were obtained by Hesselberg (1915) and by Brunt and Douglas (1928), who also showed that the right-hand sides are the first terms in an infinite expansion for (u, v). In the linear case, the operators D/Dt are replaced by $\partial/\partial t$, the density ρ is replaced by its undisturbed value ρ_0, and p can be replaced by its perturbation value p', i.e.,

$$u \approx u_g - \frac{1}{f} \frac{\partial v_g}{\partial t} \equiv -\frac{1}{f\rho_0} \frac{\partial p'}{\partial y} - \frac{1}{f^2 \rho_0} \frac{\partial^2 p'}{\partial t \, \partial x},$$

$$v \approx v_g + \frac{1}{f} \frac{\partial u_g}{\partial t} \equiv \frac{1}{f\rho_0} \frac{\partial p'}{\partial x} - \frac{1}{f^2 \rho_0} \frac{\partial^2 p'}{\partial t \, \partial y}. \tag{8.16.7}$$

Contours of rate of change of pressure are called *isallobars*, so, as Brunt and Douglas (1928, p. 33) point out, (8.16.7) shows that "the wind is made up of the geostrophic wind blowing around the isobars, with an added component blowing into the isallobaric low of magnitude proportional to the gradient of the isallobars." Thus

> the central regions of low values on the isallobaric charts must be regions of convergence of air, and hence must be the regions in which ascending currents can most readily occur. The regions of high values on the isallobaric charts, being regions of divergence, would appear to be the regions of subsidence of air.... Further, the occurrence of rain in isallobaric lows, and of fine weather in isallobaric highs, is regarded as evidence of convergence of winds in the one case, and of divergence in the other (p. 34).

Brunt and Douglas (p. 37) also recognized that "the effect of friction at the ground and the turbulence which it produces, is to produce a flow of air across the isobars from high pressure to low pressure. This drift can play an important part in the production of rain, though it is not adequate to account for all the facts." This frictional convergence will be considered in the next chapter. (*Note*: Hesselberg (1915) included friction effects, but was concerned with deducing pressure changes from velocity departures rather than the reverse.)

Brunt and Douglas did not write down the expression for the horizontal divergence, which follows from (8.16.7), although they did make extensive use of the idea. They stated (p. 34) that "in a system of purely geostrophic winds, there can be no ascending or descending motion" because "there can be no convergence or divergence and therefore no rainfall except such as may be produced by orography or friction. The occurrence of heavy rain over wider areas appears to be associated with the convergence of the component" $f^{-1}\,\partial(u_g, v_g)/\partial t$ "*i.e.*, to the existence of centres of low values on the isallobaric charts."

For the linear case, (8.16.7) gives for the horizontal convergence $\big[$cf. (8.8.29)$\big]$

$$\frac{\partial u}{\partial x} + \frac{\partial v}{\partial y} = -\frac{1}{f^2\rho_0}\frac{\partial}{\partial t}\left(\frac{\partial^2 p'}{\partial x^2} + \frac{\partial^2 p'}{\partial y^2}\right) \equiv -\frac{1}{f}\frac{\partial}{\partial t}\left(\frac{\partial v_g}{\partial x} - \frac{\partial u_g}{\partial y}\right), \qquad (8.16.8)$$

contributions coming only from the isallobaric terms. On the other hand, the isallobaric terms give no contribution to the vorticity, and (8.16.7) implies

$$\frac{\partial v}{\partial x} - \frac{\partial u}{\partial y} \simeq \frac{1}{f\rho_0}\left(\frac{\partial^2 p'}{\partial x^2} + \frac{\partial^2 p'}{\partial y^2}\right) \equiv \frac{\partial v_g}{\partial x} - \frac{\partial u_g}{\partial y}. \qquad (8.16.9)$$

The fact that the principal terms in (8.16.7) contribute to the *vorticity* but not to the divergence explains why the latter is much smaller than would be expected from normal scaling arguments. In fact, for motion with frequency ω, (8.16.8) and (8.16.9) show that

$$\frac{\text{divergence}}{\text{vorticity}} \quad \text{is of order} \quad \omega/f, \qquad (8.16.10)$$

where ω/f has been assumed to be small.

These equations also show why the vorticity equation plays such an important role in the theory of slow motions of rotating fluids. Elimination of pressure from (8.16.8) and (8.16.9) gives

$$\frac{\partial}{\partial t}\left(\frac{\partial v}{\partial x} - \frac{\partial u}{\partial y}\right) + f\left(\frac{\partial u}{\partial x} + \frac{\partial v}{\partial y}\right) = 0, \qquad (8.16.11)$$

which is the same as the vorticity equation (7.2.7) deduced from the exact form of the linear equations. Therefore, the vorticity equation contains information about *departures* from geostrophy, and so can be used to get around the redundancy problem referred to at the beginning of the section.

The reason why it is so important to calculate the horizontal divergence, though small, is because horizontal divergence is necessary to give vertical motion, and vertical motion must be known if gravitational restoring forces are to be calculated. The equation that combines the vorticity equation with equations for the vertical motion is the potential vorticity equation, perturbation forms of which are discussed in Section 7.12. The potential vorticity equation is of central importance to the theory of rotating fluids, as was discovered by Rossby for the homogeneous case discussed in Section 7.2. In the linear case, this equation can be integrated with respect to time, thereby giving a connection between the initial and final states of a system. This solves the problem of the indeterminancy in the solution of the steady-flow equations by giving an additional condition that the flow must satisfy.

For slow motions, the approximate expression (8.16.9) for the vorticity can be used to obtain approximate forms for the potential vorticity equation. In the incompressible case, for instance, the left-hand side of (8.16.8) is equal to $-\partial w/\partial z$, and w is related to gravitational restoring forces and hence to perturbation pressure through (6.11.4), namely,

$$N^2 w = -\rho_0^{-1} \partial^2 p'/\partial z \, \partial t. \tag{8.16.12}$$

Thus the combination with (8.16.8) yields

$$\frac{\partial}{\partial t}\left\{\frac{1}{f^2}\left(\frac{\partial^2 p'}{\partial x^2} + \frac{\partial^2 p'}{\partial y^2}\right) + \rho_0 \frac{\partial}{\partial z}\left(\frac{1}{\rho_0 N^2}\frac{\partial p'}{\partial z}\right)\right\} = 0, \tag{8.16.13}$$

which is the quasi-geostrophic form of the perturbation potential vorticity equation for an incompressible fluid in uniform rotation (f constant). Equation (8.8.24) is a special case that was discussed in connection with topographic effects.

The quasi-geostrophic approximation is discussed in more detail in Chapter 12, where effects of the variation of the Coriolis parameter with latitude (the beta effect), nonlinear effects, and friction effects are also considered. For uniform flow with speed U over topography with horizontal scale L it turns out that the condition for neglecting beta effects is that $L \ll (U/\beta)^{1/2}$, so that the quasi-geostrophic regime discussed in Section 8.8 covers the range $U/|f|L \ll (U/\beta)^{1/2}$. At 45° latitude, this range extends over a factor of $(700/U)^{1/2}$ when U is in meters per second, and is therefore rather narrow (a factor of 8) for the atmosphere, but fairly wide (a factor of 80) for the ocean, encompassing scales typically between 1 and 80 km.

The illustrative example of f-plane quasi-geostrophic flow considered previously (Section 8) is that of flow over a bell-shaped ridge. The ridge geometry is somewhat special, and so another example will be considered now, namely, that of uniform flow of a uniformly stratified fluid over a circular bump given by

$$h = h_m(1 + (x^2 + y^2)/L^2)^{-3/2}. \tag{8.16.14}$$

If the Boussinesq approximation is made, the potential vorticity equation (8.16.13) can be integrated to give (8.8.24), which reduces to Laplace's equation if the stretched vertical coordinate z_s, defined by (8.8.25), is used. There is an exact solution (R. B. Smith, 1979) corresponding to the topography (8.16.14), namely,

$$h = (1 + z_s/L)h_m[(1 + z_s/L)^2 + (x^2 + y^2)/L^2]^{-3/2}. \tag{8.16.15}$$

The displacement of isopycnals is distributed axisymmetrically, the vertical cross section through the center of the hill being like that shown in Fig. 8.20a. The perturbation pressure p' can be obtained by integrating (8.16.12), which can be written in a form equivalent to (6.11.7). Relative to the field associated with the mean flow, it has an axisymmetric distribution with a high over the hill, but when the part associated with the mean flow [calculated from (8.8.27)] is added, the result is an asymmetric distribution, namely,

$$\rho_0^{-1}p' = -fUy + NfLh_m[(1 + z_s/L)^2 + (x^2 + y^2)/L^2]^{-1/2}. \tag{8.16.16}$$

This result applies even if the gradients associated with the topography are com-

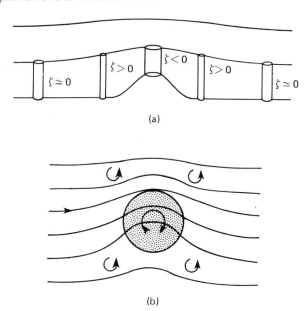

Fig. 8.20.   Quasi-geostrophic flow over a circular hill. (a) Vertical section through the center of the hill, showing displacement of isopycnals and the behavior of vortex tubes. On the flanks of the hill they are stretched slightly, producing cyclonic relative vorticity, whereas the vortex tube is significantly shortened over the hill, producing strong anticyclonic relative vorticity. [From R. B. Smith (1979, Fig. 17), after Buzzi and Tibaldi (1977).] (b) The streamline pattern over the hill, showing the accociated relative vorticity. [From R. B. Smith (1979, Fig. 18).]

parable with those associated with the mean flow. Contours at a fixed level (which are also streamlines) are shown in Fig. 8.20b. Fluid particles are displaced to the left (in the northern hemisphere) over the obstacle, giving accelerated flow over the left flank, where the anticyclonic flow associated with the hill enhances the mean flow. The reason for this behavior can also be explained in terms of vorticity, for Fig. 8.20a shows that vortex lines are stretched slightly on the approaches to the hill and squashed severely over the top of the hill. Thus anticyclonic relative vorticity is generated over the hill and weak cyclonic vorticity away from the hill. Addition of the flow due to this vortex distribution to the mean flow gives the pattern shown in Fig. 8.20b. Observations showing this type of pattern and the decay over a Rossby height are reported by Gould *et al.* (1981) and by Owens and Hogg (1980). Reviews are given by R. B. Smith (1979) for the atmosphere, and by Hogg (1980) for the ocean.

Chapter Nine

Forced Motion

9.1 Introduction

The atmosphere–ocean system is driven by radiation from the sun, which tends to make the tropics warmer than the poles and so induce motion. Thus it is important to understand the response of the atmosphere and ocean to forces of various kinds, e.g., those resulting from heating and cooling, and those due to the action of wind and pressure variations at the ocean surface. In this chapter, different types of forcing are introduced, and the response is studied for a stratified fluid rotating uniformly about a vertical axis. In other words, the response is examined on an f-plane rather than on a sphere.

The first part of the chapter (up to Section 9.7) is about the action of a stress at a horizontal surface. In the first instance, this stress is due the action of the wind on the ocean surface. The wind stress produces a direct response called the Ekman transport, which is principally confined to a thin layer near the ocean surface. In fact, the Ekman transport is thought to be usually found within the upper mixed layer of the ocean, which is mostly between 10 and 100 m deep. A sudden change of wind can cause oscillations in the Ekman transport of inertial period, or can reduce the amplitude of preexisting oscillations. Simple models fit observations of velocity fluctuations in the mixed layer quite well. A similar oscillation can occur in the atmosphere at night owing to a sudden reduction in stress at nightfall. This produces a phenomenon called the nocturnal jet, which is discussed in Section 9.7.

If the wind stress were spatially uniform, the ocean below the mixed layer would be little affected by the wind, which would produce a time-varying Ekman transport that is confined to the near-surface region. However, *spatial* variations in the wind, which, of course, occur, cause spatial variations in Ekman transport and hence convergences and divergences. In other words, the Ekman flow will cause mass to flow horizontally into some regions and out of others. This results in vertical motion.

For instance, if the horizontal flow is converging in a particular region, vertical motion away from the boundary is required in order to conserve mass. The vertical velocity just outside the boundary layer which is so produced is called the Ekman pumping velocity. It is this velocity in the ocean that distorts the density field of the ocean and thereby causes the wind-driven currents.

From the atmospheric point of view, the stress at the underlying ocean (or land) surface is a frictional drag whose magnitude is dependent on the strength of the wind. With the stress is associated an Ekman transport in the atmosphere whose horizontal *mass* flux is equal and opposite to that in the underlying ocean. Consequently, variations in Ekman transport produce Ekman pumping with a *vertical* mass flux that is the *same* in the atmosphere as in the ocean.

The relation between the frictional drag and the velocity near the boundary is discussed in Section 9.5, together with some discussion about the way in which velocities vary with distance from the boundary. In practice, the flow in the frictional layer is turbulent and the drag law is nonlinear. However, the laminar solution is also of some interest, and this is discussed in Section 9.6.

The presence of friction in general tends to reduce motion and make the system tend toward a state of rest. This process is called "spin-down" and is discussed in Section 9.12. Ekman pumping plays an important role in this process.

Another form of forcing for the ocean is the gravitational pull of the sun and moon, which produces tides. The discussion in Section 9.8 concerns the direct effect, which is barotropic, although in practice there are also baroclinic tides produced by interaction of barotropic tides with topography. This effect is particularly important when there are large changes of depth, as along the continental slopes.

The tidal equations were derived at a very early stage in the history of fluid mechanics by Laplace (1778/1779), and these equations, which are a form of forced shallow-water equations, play a central role in geophysical fluid dynamics. In fact, if a shallow layer of homogeneous fluid is forced in any way, whether by wind, by pressure variations at the surface, by evaporation–precipitation differences, or by tide-producing forces, the forced shallow-water equations result. Consequently, the response to any one of these forms of forcing is related to the response to any of the other forms if made an appropriate function of position and time. Thus Laplace's equations apply to a wide range of problems of barotropic flow. They can, however, also be applied to baroclinic flow, as shown in Section 9.10, by using the technique of expansion in terms of normal modes.

A particular application is to the baroclinic response of the ocean to moving disturbances, such as hurricanes. This is dealt with in Section 9.11. The behavior depends very much on whether the storm moves faster or slower than baroclinic waves. If slower, the equations governing the response are elliptic in character, the response decays with distance from the forcing, and therefore effects moving with the storm are confined to the neighborhood of the storm. If the storm moves faster (which is generally the case), the equations are hyperbolic, and a train of lee waves is found moving along behind the storm. Because of the Ekman pumping associated with a cyclone, the storm also leaves in its wake a zone of upwelled water, with the compensating downwelling on the flanks of the storm track.

The prime agent that sets the atmosphere in motion is radiation from the sun, which produces buoyancy forces. The equations for motion so produced are considered in Section 9.13 for the case in which departures from the rest state are small. Particular solutions for cases in which the buoyancy forcing is stationary are considered in the following sections. If the forcing is a function of height z and only one horizontal coordinate y, there is a particular type of solution encountered in a number of problems. The flow in the (y, z) plane is steady in time, being maintained by the steady forcing. The displacement of isopycnals, however, increases linearly with time and this produces motion normal to the plane that is in geostrophic equilibrium with the pressure field so created. This flow therefore increases linearly in time as well. The same is true when the forcing is a function of z and radial distance r from the storm center, as may be the case in a hurricane, for instance. The flow in the (r, z) plane may be steady, but the azimuthal flow increases linearly with time. In fact, rings of fluid moving inward toward the axis conserve their total angular momentum, and thereby acquire cyclonic relative vorticity at a uniform rate. This sort of behavior is found in the developing stages of hurricanes [see, e.g., the review of Gray (1979)] in which flow in the (r, z) plane (relative to the hurricane center) is fairly steady, but the cyclonic motion around the storm center continues to build up with time. Similar flows occur in regions of cooling in the ocean, e.g., in the region where bottom water is formed in the Greenland Sea.

In practice, the buildup will not continue indefinitely because some equilibrating mechanism will come into play. This is usually some form of dissipative mechanism, and examples are considered in the last section of the chapter. One example considered is a model of the Antarctic Circumpolar Current for which the driving force is the wind and the dissipative mechanism is bottom friction. Another example is an atmosphere driven by buoyancy forces with "Newtonian cooling" and "Rayleigh friction" as the dissipative agents. This gives a model of the type of atmospheric circulation envisaged by Halley (1686) and Hadley (1735).

9.2 Forcing Due to Surface Stress: Ekman Transport

When the wind blows over the earth's surface, a stress is exerted on the surface (see Chapter 2) whether the surface be solid earth or sea. This stress represents a retarding force of considerable importance (see, e.g., Section 2.4) for the atmosphere and a driving force of major importance for the ocean. In fact, the main current systems of the world are wind-driven. Some attention will now be given to this type of forcing. Surprisingly, this turns out to have features in common with the topographic forcing already considered!

The horizontal stress (X, Y) at the earth's surface is a horizontal vector representing the force per unit area exerted between the surface (regarded as horizontal) and the neighboring layer of air or water. To incorporate the effect of horizontal stresses into the equations of motion, it is useful to imagine the ocean or atmosphere divided up into a set of thin horizontal layers, rather like a piece of plywood, but with each layer free to move. If a stress (X, Y) is exerted at the top of a layer, it will tend to move and thus exert a stress on the layer underneath. If the layer has thickness

δz, the stress on the layer below will be approximately

$$(X - \delta z \, \partial X/\partial z, \; Y - \delta z \, \partial Y/\partial z).$$

An equal and opposite stress will be exerted on the base of the original layer, so that the net force per unit area on that layer will be the difference between the stress on the top and the bottom, namely,

$$(\partial X/\partial z, \, \partial Y/\partial z) \, \delta z.$$

Multiplying by the area $\delta x \, \delta y$ and dividing by the mass $\rho \, \delta x \, \delta y \, \delta z$ of the layer, it follows that the force per unit mass due to the horizontal stresses is

$$\rho^{-1}(\partial X/\partial z, \, \partial Y/\partial z).$$

This is an additional force per unit mass that tends to accelerate the fluid, and so it must be added to those already included in the horizontal momentum equations. For instance, the linearized equations (8.4.1) and (8.4.2) for a uniformly rotating fluid become, when the stress terms are added,

$$\frac{\partial u}{\partial t} - fv = -\frac{1}{\rho}\frac{\partial p'}{\partial x} + \frac{1}{\rho}\frac{\partial X}{\partial z},$$

$$\frac{\partial v}{\partial t} + fu = -\frac{1}{\rho}\frac{\partial p'}{\partial y} + \frac{1}{\rho}\frac{\partial Y}{\partial z}. \tag{9.2.1}$$

The reason for including only *vertical* derivatives of *horizontal* stresses is due to the fact that the vertical scale of the atmospheric and oceanic boundary layers (i.e., the regions adjacent to the surface where the stresses are comparable with the surface values) is very much smaller than that of the horizontal scale on which the stresses vary. Typically, the atmospheric boundary layer is 1-km thick, whereas the oceanic boundary layer is 10–100 m thick. In constrast, the horizontal scale of stress variations may be 100–1000 km.

Detailed modeling of the boundary layers of the atmosphere and ocean is a complicated business [see, e.g., Lumley and Panofsky (1964), Turner (1973), Monin (1970), Zilitinkevich (1970), Kraus (1977), Zeman (1981)]. Not only is there turbulence caused by shear and wave activity, but there are important mixing processes caused by heating and cooling as well. However, as will be seen, quite simple models can be used to explore some averaged properties of the boundary layer and to calculate effects of the boundary layers on the remainder of the flow.

From (9.2.1) it can be seen that there are two forces tending to accelerate the fluid: that due to the horizontal pressure gradient and that due to the vertical gradient of the stress. The velocities due to the two different forcings can be considered separately. The part (u_p, v_p) of the velocity driven by the pressure gradient satisfies

$$\partial u_p/\partial t - fv_p = -\rho^{-1} \, \partial p'/\partial x, \qquad \partial v_p/\partial t + fu_p = -\rho^{-1} \, \partial p'/\partial y \tag{9.2.2}$$

and in the case of steady flow becomes the geostrophic velocity. The part (u_E, v_E) driven by the stress is confined to the layer in which the stress acts and will be called the *Ekman velocity*, following Ekman's (1905) pioneering work on the boundary-layer problem. The layer in which the stress acts is often referred to as the *Ekman layer* for the same reason. The Ekman velocity satisfies

$$\partial u_E / \partial t - f v_E = \rho^{-1} \partial X / \partial z, \qquad \partial v_E / \partial t + f u_E = \rho^{-1} \partial Y / \partial z. \qquad (9.2.3)$$

Thus the velocity (u, v) that appears in (9.2.1) can be expressed as the sum

$$u = u_p + u_E, \qquad v = v_p + v_E, \qquad (9.2.4)$$

where (u_p, v_p) satisfies (9.2.2) and (u_E, v_E) satisfies (9.2.3).

Now the stress (X, Y) is zero outside the boundary layer (or Ekman layer), so integration of (9.2.3) with respect to z across the layer gives

$$\rho(\partial U_E / \partial t - f V_E) = -X_s, \qquad \rho(\partial V_E / \partial t + f U_E) = -Y_s, \quad \text{when boundary below.}$$

$$(9.2.5)$$

In this equation (X_s, Y_s) is the value of the stress at the boundary and the vector

$$(U_E, V_E) = \int (u_E, v_E) \, dz = \int (u - u_p, v - v_p) \, dz \qquad (9.2.6)$$

is the volume transport, relative to the pressure-driven flow, of the boundary layer. The density is regarded as constant within the boundary layer since it is thin compared with the scale height. The quantity (U_E, V_E) is called the *Ekman volume transport* of the boundary layer, or sometimes just the Ekman transport. The quantity $(\rho U_E, \rho V_E)$ is called the *Ekman mass transport.*

The sign of the stress term in the integral of (9.2.3) depends on whether the boundary surface is above or below the layer. The result (9.2.5) applies to the atmospheric boundary layer or to the ocean's bottom (or benthic) boundary layer. For the ocean surface boundary layer, however, the signs are reversed and the integral of (9.2.3) across the layer gives

$$\rho(\partial U_E / \partial t - f V_E) = X_s, \qquad \rho(\partial V_E / \partial t + f U_E) = Y_s, \quad \text{when boundary above.}$$

$$(9.2.7)$$

Note that since the stress acting at the ocean surface is also the stress acting at the bottom of the atmosphere, adding (9.2.5) and (9.2.7) shows that the sum of the atmospheric and oceanic Ekman *mass* transports is zero (assuming that the sum is zero at some initial time). The same is *not* true of the volume transports because of the large difference between the densities of air and water.

In steady conditions, the Ekman transport is directed *at right angles* to the surface stress. In the atmosphere, the transport direction is backed (i.e., to the left) in the northern hemisphere relative to the surface stress. In the ocean, the transport is veered (i.e., to the right) in the northern hemisphere relative to the surface stress.

The magnitude and direction of the Ekman transport can be obtained quite simply by considering the balance of forces on a horizontal slab of atmosphere or ocean with thickness equal to that of the boundary layer. The balance is simplest when the pressure gradient is zero, for then the force (X_s, Y_s) per unit area exerted on the horizontal boundary of the slab by the surface stress is balanced entirely by the Coriolis force $(\rho f V_E, -\rho f U_E)$ per unit area associated with the mean motion of the slab. Figure 9.1 shows the directions (for the northern hemisphere) of the Ekman

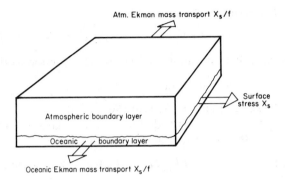

Fig. 9.1.   The directions (for the northern hemisphere) and magnitudes of the steady Ekman mass transports in the atmospheric and oceanic boundary layers when the stress at the surface has the direction shown. In the southern hemisphere, the directions of the Ekman transports are reversed. Note that the sum of the atmospheric and oceanic Ekman mass transports is zero. When there is no pressure gradient, the force per unit area exerted by the surface stress on each boundary layer is equal to the product of mass per unit area and the Coriolis acceleration of the layer. The latter quantity is f times the Ekman mass transport and is directed at right angles to the stress, as shown.

fluxes relative to the surface stress. For a typical value, 0.1 N m^{-2}, of the stress in mid latitudes ($f = 10^{-4}$ s^{-1}), the Ekman mass transport in both layers is 1000 kg m^{-1} s^{-1} or 1 tonne m^{-1} s^{-1}. The theory gives larger values near the equator, where f becomes very small.

Although the concept of Ekman transport is used a great deal, accurate observational confirmation of the relationship (9.2.5) or (9.2.7) is difficult. This is largely because of difficulties in estimating the pressure gradient accurately enough to be able to separate the pressure-driven and Ekman parts of the velocity field. In the atmosphere, for instance, in reasonably steady conditions the stress can be estimated from the properties of the velocity profile near the ground or by direct measurements of stress. The difficulty is in estimating the direction of the pressure gradient because small errors in this lead to significant errors in estimates of the Ekman flux.

9.3   Wind-Generated Inertial Oscillations in the Ocean Mixed Layer

Suppose the sea is at rest and a wind stress in the x direction (i.e., a westerly wind) suddenly rises and is maintained at a constant value X_s. How will the Ekman transport vary? A solution of (9.2.7) can easily be obtained by adding i times the second equation to the first, giving

$$\frac{\partial}{\partial t}(U_E + iV_E) + if(U_E + iV_E) = \frac{X_s}{\rho}, \tag{9.3.1}$$

which has solution [cf. Gold (1908)]

$$U_E + iV_E = -i(X_s/\rho f)(1 - e^{-ift}). \tag{9.3.2}$$

At first the Ekman transport is in the direction of the wind, but as time passes the Coriolis effect causes it to veer (in the northern hemisphere). Ultimately, the transport is given by the sum of a steady Ekman transport at right angles to the wind plus an anticyclonic rotation with the same amplitude around inertial circles. For a particle moving with the average velocity for the layer, the resultant gives a cycloidal path as shown in Fig. 9.2a.

When current meters are placed in the upper mixed layer of the ocean, inertial oscillations are commonly observed, especially in summer or autumn when the mixed layer is relatively thin, and therefore the currents associated with the Ekman

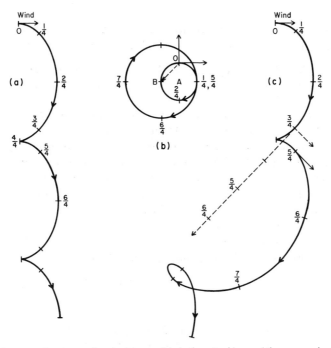

Fig. 9.2. (a) A progressive vector diagram for a particle in the mixed layer of the ocean when a westerly wind (denoted by the arrow at the top of the diagram) suddenly rises at $t = 0$. The particle is assumed to move at the mean velocity for the layer, and marks are shown at intervals of a quarter of an inertial period. The motion is the sum of a steady velocity at right angles to the wind (the southward direction shown in the diagram is the one obtained in the northern hemisphere) and an anticyclonically rotating inertial oscillation. The resultant gives the cycloidal path shown. (b) The variation of Ekman transport with time is shown by the small circle with center A. The value of the transport represented by point A is the steady (southward) Ekman transport appropriate to the westerly wind imposed at $t = 0$. Inertial oscillations about this value are represented by a circle with A as center, and this circle must pass through the origin O in order to satisfy the initial condition. (c) The progressive vector diagram when there is an additional change of wind to the northwest at a later time, the stress after the change being two times the original value. The solid line shows the result when the change occurs at $1\frac{1}{4}$ inertial periods after $t = 0$; the solid arrow shows the new stress at the time the change occurs. The dashed line shows what happens if the change occurs $\frac{3}{4}$ of an inertial period after $t = 0$ instead. The explanation of this result can be seen from the construction shown in (b). When the wind changes, the solution will again consist of a steady part (in this case represented by point B), with direction at right angles to the wind, plus an anticyclonically rotating inertial oscillation. The latter is represented by the circle with center B that passes through the point representing the motion at the time the wind changes. For the case in which the wind changes at $ft = 3\pi/2$, this point coincides with B and thus the circle has zero radius. Consequently, the inertial oscillations are quashed. For the case in which the wind changes at $ft = 5\pi/2$, the point representing the motion at this time is far from B, and the resultant circle has double the original radius. Consequently, the amplitude of the inertial oscillation is doubled.

transport are reasonably large. In fact there is evidence [see, e.g., Halpern (1974, 1980)] to indicate that, especially in conditions of strong wind mixing, the currents associated with the Ekman transport are fairly uniform over the mixed layer. [The same appears to occur in the atmosphere in suitable conditions—see Kaimal *et al.* (1976).] If the mixed-layer depth is H_{mix}, this corresponds to currents being given by [cf. (9.2.4) and (9.2.6)]

$$(u - u_p, v - v_p) \equiv (u_E, v_E) = \begin{cases} (U_E/H_{mix}, V_E/H_{mix}) & \text{for} \quad z > -H_{mix}, \\ 0 & \text{for} \quad z < -H_{mix}. \end{cases} \quad (9.3.3)$$

The stress distribution that gives rise to uniform currents in the mixed layer is one with linear dependence on z within the mixed layer, i.e., is given by

$$\left(\frac{\partial X}{\partial z}, \frac{\partial Y}{\partial z}\right) = \begin{cases} (X_s/H_{mix}, Y_s/H_{mix}) & \text{for} \quad z > -H_{mix}, \\ 0 & \text{for} \quad z < -H_{mix}. \end{cases} \quad (9.3.4)$$

This distribution has the value of being simple: it also appears to be reasonably realistic, and therefore is useful for modeling purposes.

Inertial oscillations in the ocean mixed layer often grow to large amplitude when a storm goes by, and then slowly decay. This is the behavior expected from (9.3.2) when allowance is made for energy loss. However, the opposite is sometimes true, i.e., a storm can kill off preexisting inertial oscillations very rapidly. This seems puzzling at first sight, but can easily be explained from the behavior of solutions of (9.2.7). Suppose that there are inertial oscillations already present at time $t = t_0$ when the wind stress suddenly changes to the value (X_s, Y_s). If (U_0, V_0) is the value of the Ekman transport at $t = t_0$, the solution of (9.2.7), after adding i times the second equation to the first, is

$$U_E + iV_E = -i(X_s + iY_s)/\rho f + \{U_0 + iV_0 + i(X_s + iY_s)/\rho f\} \exp(-if(t - t_0)).$$
$$(9.3.5)$$

Equation (9.3.2) is the special case corresponding to $t_0 = Y_s = U_0 = V_0 = 0$. The amplitude of the inertial oscillations is given by the factor in curly brackets in (9.3.5). If the wind change happens to occur at a phase of the preexisting oscillations when this factor happens to be small, there will be only weak oscillations after the wind change. If, on the other hand, the wind changes occur when the factor is large, strong oscillations will result.

The dependence of the amplitude of oscillation on the time of the wind change is illustrated in Fig. 9.2c. At $t = 0$, a westerly wind rises suddenly to give a progressive vector diagram as in Fig. 9.2a. At a later time, the wind suddenly veers to the northwest and increases in strength slightly. The solid line shows the increased amplitude of oscillations that result if the wind change occurs when the current is in the direction of the wind at the time of change. The dashed line shows the result if the wind change takes place half an inertial period earlier. In this case, the amplitude of the oscillations falls to zero when the wind veers and strengthens.

Confirmation of this general picture has been obtained by Pollard and Millard (1970) [see also Kundu (1976) and Käse and Olbers (1979)], using observations of currents in the mixed layer and simultaneous observations of the wind. Figure 9.3c, d shows the observed currents and the results of a model calculation that uses (9.2.7)

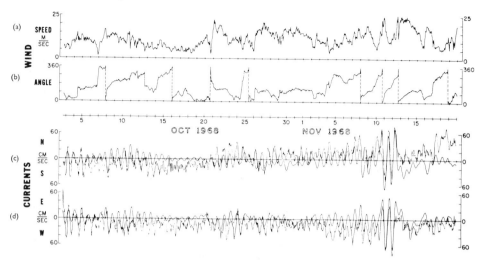

Fig. 9.3. A model simulation of currents in the surface mixed layer. (a) Wind speed and (b) direction toward which the wind blows (measured clockwise from north) as measured from a buoy. (c) North–south and (d) east–west components of velocity, the thin curve being that observed at a depth of 12 m and the thick curve being that computed from the wind by using the model described in the text. [From Pollard and Millard (1970, Fig. 2).]

with one modification, namely, the introduction of an empirical decay term to allow for losses of energy from the surface currents. The modified equation they used is

$$\partial U_E/\partial t - f V_E = -r U_E + X_s/\rho, \qquad \partial V_E/\partial t + f U_E = -r V_E + Y_s/\rho, \quad (9.3.6)$$

where r is a decay constant, sometimes called a "Rayleigh friction" parameter. [This form of friction was used by Airy (1845) in his canal theory of the tides; so it was certainly in use before Rayleigh's time.] The results shown in the figure are for a value of $r^{-1} = 4$ days, which corresponds to 5 inertial periods. Note that the larger amplitudes are found on 11 and 12 October when the wind is veering rapidly. The strongest period of prolonged winds is on October 13 and 14, but in this period the amplitude becomes much smaller, both in fact and in the model calculations.

Bell (1978) has constructed a model to explain the decay of inertial oscillations in the mixed layer. Turbulent motions advected by the mixed-layer currents cause motions of the base of the mixed layer that radiate energy in the form of internal waves at a rate that gives r^{-1} of order 3–4 days. The layer also acts as a significant source of internal waves with frequencies of order N and wavenumbers of order $N H_{mix}/U_E$, U_E/H_{mix} being taken as a typical value of the current at the base of the mixed layer.

The presence of inertial oscillations in the mixed layer of the ocean represents also a possible source of mechanical energy that can be used to entrain water from below the mixed layer into the layer. Since the inertial currents are confined mainly to the mixed layer, they cause a large shear at the base of the mixed layer that can produce turbulent mixing in the form of turbulent Kelvin–Helmholtz billows such as those observed in the laboratory and in a lake by Thorpe (1973, 1977). This idea is utilized in a model for estimating changes in the depth of the mixed layer that has been proposed by Pollard *et al.* (1973).

Entrainment into the mixed layer of dense water from below is one of the important processes that need to be considered when estimating the heat balance of the layer and hence the surface temperature. Radiation balances and turbulent exchange across the surface (see Chapter 2) must of course be calculated, and in wintertime convective overturning due to surface cooling is important (Gill and Turner, 1976). Methods of modeling the heat and salt balance of the mixed layer are discussed by Turner (1973, 1981) and Kraus (1977).

9.4   Ekman Pumping

The stress on the earth's surface varies from place to place and hence so does the Ekman transport. This leads to convergence of mass in some places, and hence to expulsion of fluid from the boundary layer. In other places, the Ekman transport is horizontally divergent, i.e., mass is being lost across the sides of a given area, so fluid must be "sucked" vertically into the boundary layer to replace that which is lost across the sides. This effect is called "Ekman pumping."

For instance in mid-latitudes, where westerly (eastward) winds predominate, the Ekman transport in the ocean boundary layer is *equatorward*. In the trade wind zone, on the other hand, the easterly component of the wind produces a *poleward* Ekman transport, so there is a convergence of mass in the ocean surface layer at latitudes in between. This leads to expulsion of fluid from the boundary layer and hence to *downward* motion just below. The isotherms in the ocean are pushed down in this region, thus creating pressure gradients that produce ocean currents. Details of this process will be considered in Chapter 12.

The magnitude of the vertical velocity w_E just outside the boundary layer that results from convergence or divergence of the Ekman transport can be formed by integrating the continuity equation. Assuming that density variations can be neglected, this takes the form (4.10.12), i.e.,

$$\partial u/\partial x + \partial v/\partial y + \partial w/\partial z = 0,$$

and integration with respect to z across the layer and using the condition $w = 0$ at the boundary give

$$\partial U_E/\partial x + \partial V_E/\partial y - w_E = 0 \quad \text{when boundary above.} \tag{9.4.1}$$

For the steady case, combining this with (9.2.7) gives

$$\rho w_E = \frac{\partial}{\partial x}\left(\frac{Y_s}{f}\right) - \frac{\partial}{\partial y}\left(\frac{X_s}{f}\right). \tag{9.4.2}$$

If the boundary is below, the minus sign in (9.4.1) becomes a plus, and combination with (9.2.5) leads to (9.4.2) again, so this equation applies whether the boundary is above or below. Usually the wind varies much more rapidly than f, so (9.4.2) gives the approximate expression for w_E, called the Ekman pumping velocity,

$$\rho w_E = f^{-1}(\partial Y_s/\partial x - \partial X_s/\partial y). \tag{9.4.3}$$

Thus the Ekman pumping velocity is approximately $(\rho f)^{-1}$ times the *curl* of the wind stress (if the wind is steady or varying slowly relative to the inertial time scale f^{-1}). It has the *same sign* in the atmosphere as in the ocean. For instance (see Fig. 9.4), if a cyclone is situated over the ocean, the Ekman transport in the atmospheric boundary layer is toward the low pressure in the center of the cyclone, i.e., it tends to reduce the low. Consequently, the Ekman pumping velocity outside the boundary layer gives ascending motion (which will tend to produce clouds). In the ocean underneath, the Ekman transport is outward, away from the center of the cyclone, and so produces an upward Ekman pumping velocity below. The vertical mass flux $\rho w_{\rm E}$ associated with this pumping is, by (9.4.2), the same as in the atmosphere. The effect in the ocean is to raise the thermocline and reduce the pressure in the ocean surface layers, thus imprinting a pressure pattern on the ocean that is like the one in the atmosphere above.

Although $w_{\rm E}$ gives only the contribution to the vertical velocity from the Ekman transport terms, it is usually the dominant one, and so the effect of a stress outside the boundary layer can be ascertained by applying the boundary condition that w has a given value $w_{\rm E}$ on a horizontal surface situated just outside the boundary layer. This is just the condition used in Section 8.7 and following sections to study topographically generated waves. It follows that similar methods can be used to study disturbances that are generated in the ocean by storms traveling over the surface. Since these have velocities relative to the ocean of the same magnitude as air velocities relative to topography, criteria about scale that are similar to those that were applied to the atmosphere can be applied to the ocean.

The analogy with lee waves can be taken further in the case of disturbances small

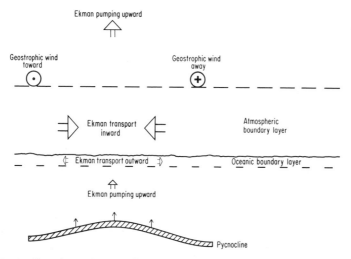

Fig. 9.4. Section through a cyclone over the ocean showing the adjustments due to Ekman transports. The geostrophic wind gives, as shown, a cyclonic rotation around the low-pressure center. Consequently, the Ekman transport in the atmospheric boundary layer is inward, bringing mass in to "fill" the low, and the associated vertical "pumping" velocity is therefore upward. The Ekman mass transport in the oceanic boundary layer is equal and opposite to that in the atmosphere, so there is an outward mass transport and upward pumping velocity in the ocean. This tends to raise the thermocline and create a low-pressure center in the ocean.

enough for linear theory to apply, for then an *Ekman displacement* η_E can be defined by

$$w_E = \partial\eta_E/\partial t. \qquad (9.4.4)$$

The value of η_E can be regarded as equivalent to a topographic elevation. For a storm moving at constant speed U without change of form, η_E has a fixed spatial structure that moves at uniform velocity across the surface of the ocean, and has the same effect as pulling a fixed shape (like an inverted mountain) across the ocean. Relative to the storm, on the other hand, the ocean is moving with fixed speed U past the "Ekman topography," so the effect is just like flow over a mountain, only turned upside down. Solutions for parameters appropriate to the ocean are discussed later in the chapter.

The formula (9.4.3) is valid only for steady conditions, or when the time variations are sufficiently slow. It is not, however, difficult to calculate the equivalent result when time variations are included. If f is regarded as constant, elimination of U_E and V_E in turn from (9.2.7) gives

$$\left(\frac{\partial^2}{\partial t^2} + f^2\right)(U_E, V_E) = \frac{1}{\rho}\left(\frac{\partial X_s}{\partial t} + f Y_s, \frac{\partial Y_s}{\partial t} - f X_s\right) \quad \text{when boundary above.} \quad (9.4.5)$$

Application of (9.4.1) then gives for the Ekman pumping velocity w_E

$$\left(\frac{\partial^2}{\partial t^2} + f^2\right)w_E = \frac{1}{\rho}\frac{\partial}{\partial t}\left(\frac{\partial X_s}{\partial x} + \frac{\partial Y_s}{\partial y}\right) + \frac{f}{\rho}\left(\frac{\partial Y_s}{\partial x} - \frac{\partial X_s}{\partial y}\right), \qquad (9.4.6)$$

and the same formula applies when the boundary is below. In addition to the forcing that is proportional to the curl of the wind stress, there is now also a term proportional to the divergence of the surface stress.

9.5   Bottom Friction: Velocity Structure of the Boundary Layer

When considering wind or ocean currents as a response to driving forces, the stress at the lower boundary is regarded as a drag exerted by the lower surface that tends to reduce the wind or current and therefore is a form of friction—usually called bottom friction. To calculate its magnitude, the stress must be related to the wind or current in some way. The usual formulation is given in terms of a drag law of the form (2.4.1), in which the stress is related to the wind or current at some standard height above the bottom. For the atmosphere, this height is usually 10 m. The wind at this level is usually a considerable fraction of the wind outside the boundary layer (see, e.g., Fig. 2.4), but it is rotated cyclonically relative to the wind outside the boundary layer by an angle α with typical values of about 20°. Thus the stress (X_s, Y_s) at the surface is related to the wind (u_g, v_g) outside the boundary layer by a relationship of the form

$$\begin{aligned}
X_s &= c_{Dg}\rho|\mathbf{u}_g|(u_g \cos\alpha - v_g \sin\alpha), \\
Y_s &= c_{Dg}\rho|\mathbf{u}_g|(u_g \sin\alpha + v_g \cos\alpha),
\end{aligned} \qquad (9.5.1)$$

where α has the same sign as f. Equations (9.5.1) can also be written, in an equivalent complex form,

$$X_s + iY_s = c_{Dg}\rho|\mathbf{u_g}|(u_g + iv_g)\exp(i\alpha). \qquad (9.5.2)$$

Measurements in the atmospheric boundary layer give $c_{Dg} \approx 0.001$ in neutral conditions with $\alpha \approx 20°$. In unstable conditions (heating at the ground) c_{Dg} is nearer 0.002 and α is near 0. In stable conditions (cool surface) c_{Dg} may be 10^{-4} or smaller and α can be near $40°$.

The nonlinear nature of the drag law (9.5.1) needs to be kept in mind when comparing bottom-friction effects at different places or the contributions at a fixed place to bottom friction at different times. Places or times of large wind or current play a much more important part in the dynamics than they would with a linear drag law. Also, the mean stress depends not only on the mean wind or current, but also on the strength and character of the fluctuations. For example, tidal motions often dominate the velocity field at the ocean bottom, mean currents being relatively small. In this situation [cf. Saunders (1977)] the mean stress over a tidal cycle can be calculated by assuming that the stress at any give time is given by (9.5.1) and that the current (u_g, v_g) outside the boundary layer follows a tidal ellipse with center corresponding to the mean current. When the mean current is small compared with the tides, the mean stress over a tidal cycle is found to be related linearly to the mean current, but the coefficient of proportionality and the angle between mean current and mean stress depend on the properties of the tidal ellipse.

The Ekman suction velocity w_E can be found in terms of the current by combining (9.4.6) and (9.5.1). The relationship is complicated and nonlinear. For a steady geostrophic flow (7.6.5) can be used to express w_E in terms of the pressure. The formula is still very complicated, but if variations of c_D, α, and f can be neglected, it simplifies in the special cases for which p depends only on distance y from a line or distance r from a point. Then, assuming $\cos\alpha \approx 1$,

$$w_E \approx \frac{c_{Dg}}{\rho^2 f^3}\left|\frac{dp}{dy}\right|\left(2\frac{d^2p}{dy^2}\right) \qquad \text{when} \quad p = p(y),$$

$$\tag{9.5.3}$$

$$w_E \approx \frac{c_{Dg}}{\rho^2 f^3}\left|\frac{dp}{dr}\right|\left(2\frac{d^2p}{dr^2} + \frac{1}{r}\frac{dp}{dr}\right) \qquad \text{when} \quad p = p(r).$$

For large values, not only must the second-order expression in parentheses be large, but also the modulus of the pressure gradient must be large.

In order to use (9.5.1), knowledge is required of how c_{Dg} and α depend on the nondimensional parameters that characterize the boundary layer. A constraint on the form this dependence can take [see Kazanski and Monin (1961), Csanady (1967) and Gill (1968a)] comes from consideration of the form of the velocity profile in the boundary layer. Near the surface, this has (see Section 2.4 and Fig. 2.4) a logarithmic form:

$$u = (u_*/\kappa)\ln(z/z_0), \qquad (9.5.4)$$

where κ has a value of about 0.4 and is called von Karman's constant, z is height above the bottom, and z_0 is a property of the surface called the roughness length. (For flow over a solid surface, z_0 is related to the size of "roughness elements" on the surface, e.g., a typical value for grass is 1 cm, whereas for forest it is about 1 m.) The quantity u_* is called the *friction* velocity and is defined as the square root of surface stress τ divided by density ρ. The x axis has been chosen to be in the direction of the surface stress. For steady homogeneous flow, the so-called "similarity" form of the drag law is based on the requirement [see, e.g., Gill (1968a)] that the velocity relative to that outside the boundary layer is independent of z_0 for $z \gg z_0$ and matches smoothly with (9.5.4) for $z \ll z_b$, where z_b is a measure of the boundary-layer thickness, which can be taken as

$$z_b = u_*/f = (\tau/\rho)^{1/2}/f. \tag{9.5.5}$$

The "similarity" law can be written, in complex form,

$$u_g + iv_g = (u_*/\kappa)(\ln(z_b/z_0) - A - iB), \tag{9.5.6}$$

where A and B are constants.

The reason for the form (9.5.6) can be seen from the changes produced when z_0 is changed but u_* is held fixed. Equation (9.5.4) shows that the result is a change in velocity of $\kappa^{-1}u_*$ times the change in $\ln z_0$ at all levels where (9.5.4) applies, but this must be true at all higher levels as well because the *relative* velocity there does not change. Hence the geostrophic velocity must change by the same amount, and (9.5.6) expresses this fact.

If the flow is not homogeneous or steady, it is still logarithmic near the surface, so (9.5.6) is still true if A and B are allowed to become functions of the additional parameters introduced by the lack of homogeneity or steadiness. If these parameters can be uniquely identified, the dependence of A and B on them can be found from observations. Results for dependence of $A + iB$ on parameters that measure stability, degree of unsteadiness, etc., are given, for instance, by Csanady (1972a), Clarke and Hess (1974), Arya and Wyngaard (1975), Zilitinkevich (1975), and Yamada (1976). In such cases, it is no longer necessary to use (9.5.5) as a measure of the boundary-layer thickness, and other definitions, such as inversion height, have been found useful.

The boundary layer at the bottom of the ocean (the benthic boundary layer) is much thinner than is the atmospheric boundary layer, which affects the relative importance of topographic effects. The main reason is the factor $\rho^{1/2}$ in (9.5.5), which tends to make the ocean boundary layer about one-thirtieth of the thickness of the atmospheric boundary layer, typically in the range 2–10 m. Observations and theory related to this boundary layer are given by Wimbush and Munk (1970), Weatherly (1972), Armi and Millard (1976), and Weatherly and Martin (1978). The structure of the boundary layer is often dominated by the variations over the tidal cycle. Harvey and Vincent (1977), for instance, found that the tidal ellipse rotates cyclonically as one goes downward, the total angle it turns through being about 14°. Another ocean boundary layer that occurs against a solid surface is that underneath ice. This boundary layer has been studied by McPhee and Smith (1976).

Detailed modeling of the velocity (and possibly temperature) structure of the boundary layer is difficult because of the problem of modeling turbulence. One technique is to develop equations that include representations of turbulent processes [see, e.g., Mellor and Yamada (1974) and Zeman (1981)]. Deardorff (1970) studied a three-dimensional model that was designed to include the turbulent motions explicitly, but such a model is too expensive to be used routinely because of the computer time required.

9.6   The Laminar Ekman Layer

The first model of the frictional boundary layer in a rotating fluid was that obtained by Ekman (1905) for a viscous fluid in laminar motion. In that case, the stress (X, Y) is proportional to the shear, i.e.,

$$\rho^{-1}(X, Y) = v\, \partial(u, v)/\partial z, \tag{9.6.1}$$

where v is the kinematic viscosity of the fluid. Substitution in (9.2.3) in this case gives, after adding i times the second equation to the first,

$$\left(\frac{\partial}{\partial t} + if\right)(u_E + iv_E) = v\frac{\partial^2}{\partial z^2}(u_E + iv_E). \tag{9.6.2}$$

In steady conditions the solution is

$$u_E + iv_E = -(u_g + iv_g)\exp\{-(1 + i)(f/2v)^{1/2}z\} \tag{9.6.3}$$

since $u_E + iv_E$ must vanish as $z \to \infty$ and must equal $-(u_p + iv_p)$ at the ground. The subscript g has been substituted for the subscript p since in steady conditions (9.2.2) shows that the flow is geostrophic. Strictly speaking, $(u_g + iv_g)$ is the value of the geostrophic velocity at the ground. In practice, however, $(u_g + iv_g)$ does not vary much over the boundary layer, so the value just outside the boundary layer will do equally well.

From (9.6.1) and (9.6.3), the surface stress can be obtained in terms of the geostrophic velocity, namely,

$$\rho^{-1}X_s = (fv/2)^{1/2}(u_g - v_g), \qquad \rho^{-1}Y_s = (fv/2)^{1/2}(u_g + v_g). \tag{9.6.4}$$

It follows from (9.4.3) that the Ekman suction velocity w_E is given by

$$w_E = \left(\frac{v}{2f}\right)^{1/2}\left(\frac{\partial v_g}{\partial x} - \frac{\partial u_g}{\partial y}\right) = \frac{1}{\rho f}\left(\frac{v}{2f}\right)^{1/2}\left(\frac{\partial^2 p}{\partial x^2} + \frac{\partial^2 p}{\partial y^2}\right), \tag{9.6.5}$$

the expression in terms of p making use of (9.2.2). Thus for a laminar Ekman layer, the Ekman pumping velocity is proportional to the vorticity outside the layer, which in turn is proportional to the horizontal Laplacian of the pressure.

The fact that the laminar solution does not accord well with observations of the atmospheric boundary layer (because it is turbulent) caused concern from the outset. Taylor (1914), who derived the laminar solution independently of Ekman, suggested that the most appropriate solution of (9.6.2) would perhaps be the one that has the surface stress parallel to the surface wind. Ekman (1906) also made this suggestion, as he noted in a later discussion of his and Taylor's work (Ekman, 1927).

Although laminar Ekman layers are not found in the atmosphere and ocean, they are useful for modeling purposes because many phenomena are not sensitive to details of boundary-layer behavior as long as Ekman pumping takes place at about the right rate. Thus the linear formula (9.6.4) is often substituted for the nonlinear equivalent, but with a value of v chosen to make w_E have, on average, the correct value. An approximate formula for determining the appropriate value of v is obtained by comparison of (9.5.3) and (9.6.5). This gives

$$(\tfrac{1}{2}vf)^{1/2} \approx 2c_{Dg}u_{av},\tag{9.6.6}$$

where u_{av} is an average value of the velocity. For the atmosphere at midlatitudes, the values $u_{av} = 10$ m s^{-1}, $c_{Dg} = 0.002$, and $f = 10^{-4}$ s^{-1} might be considered appropriate, and these give $v \sim 30$ m^2 s^{-1}. For the ocean at a place where the average tidal current at the bottom is 10 cm s^{-1}, the appropriate value of v would be smaller by a factor of 10^4, i.e., $v \sim 30$ cm^2 s^{-1} or 3×10^{-3} m^2 s^{-1}.

9.7   The Nocturnal Jet

There is a feature in the atmosphere called the nocturnal jet that in its dynamics has features in common with the inertial oscillations studied earlier in this chapter. It occurs during the night in inland regions, and the basic reason for its existence was given by Blackadar (1957). The phenomenon tends to occur when, during the heat of the day, the boundary layer is deep and convectively mixed. At nighttime, the ground cools and a stable layer is formed near the ground. Frictional effects are confined to this layer, so the region above, which was in the boundary layer during the day, has effectively been released from frictional influence, i.e., the stress term suddenly drops to zero about sunset. In response, the wind begins an inertial oscillation that proceeds until mixing resumes the next day.

A simple model of the process by Thorpe and Guymer (1977) uses methods already developed in this chapter, and Fig. 9.5 shows a comparison between the model and observation. During the day, there is assumed to be a well-mixed boundary layer of depth $H_{mix} = 800$ m at which the stress varies linearly in accord with (9.3.4). The stress is related to the velocity in the layer by a drag law (a linear relation was used for the model). Consequently, a velocity is obtained in this layer that tends with time toward the equilibrium value, which is rotated cyclonically relative to the geostrophic wind. At sunset, the mixed layer is assumed to suddenly become only 200-m thick. Consequently, the wind in the layer between 200 and 800 m begins to rotate anticyclonically and in the hodograph (i.e., the u, v) plane, is represented by a circle centered on the point that represents the geostrophic wind (see Fig. 9.5). This inevitably leads to a supergeostrophic wind in this layer. At sunrise, mixing over the full 800 m is supposed to resume, so the momentum of the two layers is mixed to give a uniform velocity over 800 m. This wind is not in balance with the pressure gradient and drag, and therefore it adjusts toward equilibrium until the mixed layer suddenly reduces in thickness again at sunset.

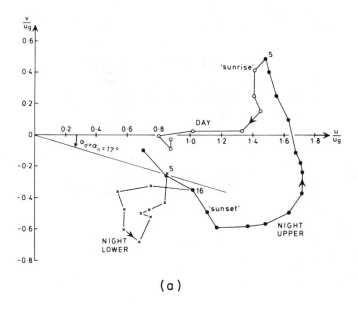

(a)

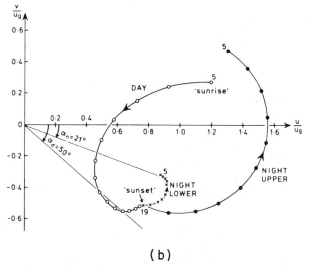

(b)

Fig. 9.5. (a) Averaged velocities at 1-hr intervals on day 13/14 of the Wangara experiment 27/28 (July 1967). The values at night are averaged over 0–200 m (×) and 200–800 m (●). During the day, the values are averaged over 0–800 m (○). Numbers against points denote Eastern Standard Time. [From Thorpe and Guymer (1977, Fig. 6).] (b) Corresponding model results. In the model, the upper layer undergoes half an inertial oscillation, represented by a semicircle centered on the point $u = u_g$, $v = 0$ representing the geostrophic wind. The wind vector rotates anticyclonically (the observations are in the southern hemisphere), giving model nighttime winds in the upper layer that are about 60% greater than the geostrophic wind. (In practice the nocturnal jet was even stronger.) At sunrise, the two layers are assumed to have their momentums mixed together, and the wind adjusts toward a frictional balance. At sunset, the upper layer becomes uncoupled from frictional influence, and so begins another inertial oscillation. [From Thorpe and Guymer (1977, Fig. 7).]

9.8   Tide-Producing Forces

Tides have fascinated men from earliest times, and much has been written about them. [A detailed historical account is given, for instance, by Duhem (1954, Chapter 13).] In fact Aristotle spent his last few months on the island of Evvoia (or Euboea), considering tidal conditions in the narrow straits that separate the island from the Greek mainland. The story has grown [see, e.g., Deacon (1971, p. 10)] that his failure to explain the tidal variations so frustrated him that he flung himself into the offending strait and was drowned!

The modern theory of the tides is founded on Newton's theory of gravitation, which allows the forces of attraction of the moon and sun to be calculated [Newton's contribution to tidal theory is discussed by Proudman (1927)], and on the equations of fluid motion that were developed by Euler. The man who brought these elements together and laid the foundation for the mathematical study of tides was Laplace (1778/1779), whose work is basic not only to tides but also to most of the motions studied in this book. Laplace not only established the equations for fluid motion under gravity on a rotating sphere, but he also calculated the *tide-producing forces* in a form suitable for tidal calculations. The tide-producing force is defined as that part of the attractive force of heavenly bodies that does not affect the motion of the earth as a whole, i.e., it is the remainder after subtracting the value of the force at the center of mass of the earth. The effect is symmetrical about the line joining the earth to the attracting body (the moon or sun), tending to pull the ocean into an ellipsoidal shape with major axis along the line of attraction. This line moves relative to the earth because of the rotation of the earth and the movement of the moon or sun relative to the earth. The ocean does *not* take up the ellipsoidal shape that gravity tries to give it because its response is limited by the speed (~ 200 m s^{-1}) at which gravity waves can propagate. Such waves would take about 2 days to travel around the circumference of the earth, but in practice their course is restricted by the complicated shape of the ocean. Two facts—that (a) the time for waves to travel around the globe is comparable with the rotation period and (b) the ocean has a complicated shape—account for the difficulties and fascination of the subject that has occupied the minds of many great scientists from the time of Laplace to the present day.

Since gravitational forces can be expressed as gradients of a potential, the tide-producing forces can also be represented this way. The distribution of the tide-producing potential Φ_T over the surface of the earth can be expressed as a series expansion in spherical harmonics [these functions form a complete set for describing distributions on the surface of a sphere; the theory is described, for instance, by Jeffreys and Jeffreys (1956, Chapter 24) and by Morse and Feshbach (1953, Chapter 10)]. the coefficients of which can be expanded as Fourier series with frequencies that are linear combinations of the basic frequencies of the solar system. The main periods of interest for the tides are the day ($2\pi/\Omega$), the lunar month ($2\pi/\Omega_m = 27.321$ days), and the tropical year ($2\pi/\Omega_{yr} = 365.242$ days). In practice the frequency

$$\Omega_l = \Omega - \Omega_m + \Omega_{yr} \tag{9.8.1}$$

is used in place of Ω, $2\pi/\Omega_l$ being the lunar day. Then the principal tidal component, the lunar semidiurnal or M_2 tide, has frequency $2\Omega_l$ and period $\pi/\Omega_l = 12.4$ hr. The eight components that contribute more than 10% to the maximum equilibrium tide are listed in Table 9.1. These and many other significant components have been given a letter and subscript designation by Darwin. A very detailed treatment that is due to Doodson, which involves around 390 components, includes not only the three major periods above, but also longer ones ($2\pi/\Omega_{lp} = 8.85$ yr for the lunar perigee, 18.61 yr for the regression of lunar nodes, and 21,000 yr for the solar perigee). The full list is given, for instance, by Godin (1972). The integer coefficient (see Table 9.1) in the expression for the frequency of a given constituent in terms of the fundamental frequencies Ω_l, Ω_m, etc., comprises what is called the Doodson number for that constituent. Further discussion can be found in the works by Defant (1961, Vol. 2, Chapter 7) and by Hendershott and Munk (1970), and in other books on tides.

The equations of motion, when the tide-producing force is included, have an additional force per unit mass $-\nabla \Phi_T$ on the right-hand side of the momentum equation (4.10.11), which becomes

$$D\mathbf{u}/Dt + 2\mathbf{\Omega} \times \mathbf{u} = -\rho^{-1}\nabla p - \mathbf{g} - \nabla\Phi_T. \qquad (9.8.2)$$

However, since the horizontal scale of the forcing is very large compared with the depth, the shallow-water approximation can be used, as it was by Laplace. Now tides are a global phenomenon and of global scale, so it may seem inappropriate to discuss them at this stage when only the "f-plane approximation" (see Section 7.4) to the equations has been introduced, this approximation being appropriate to motions with scales that are small compared with the radius of the earth. However, the semi-

TABLE 9.1

Principal Constituents[a] of the Tides (Those with Amplification Factor > 0.1)[b]

Tidal species	Latitude factor	Amplitude factor	Darwin	Doodson i_1	i_2	i_3	i_4	Period (hr)
Long period	$\frac{1}{4}(1 - 3\sin^2\varphi)$	0.156	M_f	0	2	0	0	327.84
Diurnal	$\sin\varphi\cos\varphi$	0.377	O_1	1	−1	0	0	25.82
		0.176	P_1	1	1	−2	0	24.07
		0.531	K_1	1	1	0	0	23.93
Semidiurnal	$\frac{1}{2}\cos^2\varphi$	0.174	N_2	2	−1	0	1	12.66
		0.908	M_2	2	0	0	0	12.42
		0.424	S_2	2	2	−2	0	12.00
		0.115	K_2	2	2	0	0	11.97

[a] These give the amplitude of contribution of each component to the equilibrium tide at latitude φ as the product of the latitude factor, the amplitude factor, and η_{em}, the maximum amplitude of the equilibrium tide. The latter has a value of 54 cm if no allowance is made for movement of the solid earth, but when this is taken into consideration, its effective value is about 38 cm. The Doodson numbers i_1–i_4 are the coefficients in the expression $\Omega = i_1\Omega_l + i_2\Omega_m + i_3\Omega_{yr} + i_4\Omega_{lp}$ for the frequency.
[b] Adapted from Hendershott and Munk (1970).

diurnal and diurnal tides have frequencies comparable with f (since all these frequencies are related to a rotation of the earth), and at these frequencies differences of scale produce differences of detail rather than differences in fundamental behavior. Thus much of tidal dynamics can be understood in terms of solutions of the "local" shallow-water equations (7.10.1) with a tidal forcing term added, namely,

$$\frac{\partial u}{\partial t} + u\frac{\partial u}{\partial x} + v\frac{\partial u}{\partial y} - fv = -g\frac{\partial}{\partial x}(\eta - \eta_e),$$

$$\frac{\partial v}{\partial t} + u\frac{\partial v}{\partial x} + v\frac{\partial v}{\partial y} + fu = -g\frac{\partial}{\partial y}(\eta - \eta_e),$$

(9.8.3)

where f is the Coriolis parameter defined by (7.4.1), η is the surface elevation, and η_e is given by

$$\eta_e = -\Phi_T/g.$$ (9.8.4)

It is permissible to regard η_e as a function of horizontal position and time only since Φ_T changes very little over the depth of the ocean. η_e is the surface elevation that the ocean would adopt if there were no dynamic effect ($u = v = 0$) and therefore it is called the *equilibrium tide*.

The "local" form of Laplace's tidal equations consists of (9.8.3) and the continuity equation (5.6.7), i.e.,

$$\frac{\partial \eta}{\partial t} + \frac{\partial}{\partial x}\left[(H + \eta)u\right] + \frac{\partial}{\partial y}\left[(H + \eta)v\right] = 0,$$ (9.8.5)

where H is the depth of the ocean. Note that these equations assume that the currents are independent of depth, i.e., that the tide is barotropic. In practice, there is not only a barotropic tide, but a baroclinic tide that is generated by interaction of the currents with bottom topography, so use of (9.8.3) and (9.8.5) assumes that the barotropic tide is not affected significantly by this interaction. Another effect that tends to make currents vary with depth is friction against the bottom, which can be particularly important when strong tidal currents are found in shallow seas. In most calculations, however, depth variations are ignored or else (9.8.5) and (9.8.4) are assumed to be valid for depth-averaged currents. The nonlinear terms in (9.8.3) are usually not important even in shallow seas, where nonlinear effects appear to arise mainly from friction and from the nonlinear terms in the continuity equation.

In calculating the tides, it is important to consider also the solid earth (Hendershott, 1972), which responds elastically to the tide-producing forces. The elastic response is rapid, so the shape that the solid earth assumes is always in equilibrium with the forces concerned. The direct response of the solid earth to the tidal potential produces (Munk and McDonald, 1960) an elevation of $h\Phi_T/g$, where h is a constant with a value of about 0.6. If this were the only effect, the surface elevation of the ocean *relative to the solid earth* would be reduced by this amount. Since η is measured relative to the solid earth, it is necessary to replace η in (9.8.5) by $\eta - h\Phi_T/g$ or, equivalently, to replace $-\eta_e$ by $(1 - h)\Phi_T/g$. But there is another effect that must be included, namely, the additional gravitational potential due to the bulge in the solid earth. This has a value of $k\Phi_T$, where k is a constant with a value of about 0.3

and η_e should be adjusted accordingly. The inclusion of these two effects does not alter the form of (9.8.3) if η is regarded as the elevation relative to the solid earth, and η_e is redefined by

$$\eta_e = -(1 + k - h)\Phi_T/g \approx -0.7\Phi_T/g. \tag{9.8.6}$$

Thus the effect of the solid earth response is to *reduce* the tides by about 30%. The constants k and h in (9.8.6) are called *Love numbers*. (The values quoted apply only to the second spherical harmonic, which dominates the tides.)

Another complication due to elastic yielding of the solid earth is that it is depressed when the sea level is higher. The depression is not as great as that which would be expected on the basis of pure elastic yielding because there is also some gravitational uplift from the larger water mass above. Calculation of the amount of depression, however, is complicated because it depends not only on the local sea level, but on a *weighted mean* of the surface elevation (Farrell, 1973), the weight decreasing with distance from the point in question. The effect is significant for tides (Hendershott, 1977) and therefore will affect details of other barotropic motions as well. The effect is not included in any calculations made in this book but is included in tidal model calculations such as those of Accad and Pekeris (1978) and of Parke and Hendershott (1979). In practice, ocean tides are strongly affected by the disposition of the continents, so boundary effects are very important. Hence the dynamics of ocean tides are not discussed until Chapter 10. The dynamics of atmospheric tides are quite different and are reviewed by Lindzen (1979) and Kato (1980).

9.9   Effect of Atmospheric Pressure Variations and Wind
on Barotropic Motion in the Sea:
The Forced Shallow-Water Equation

Variations in atmospheric pressure p_a at the surface $z = \eta$ can cause the sea to move. The linearized boundary condition on perturbation pressure (5.2.11) becomes

$$p' = p_a' + \rho g\eta \tag{9.9.1}$$

when atmospheric pressure perturbations p_a' are included. This equation may also be written

$$p' = \rho g\eta', \tag{9.9.2}$$

where

$$\eta' = \eta - \eta_a \tag{9.9.3}$$

is called the *adjusted sea level* and η_a is given by

$$\eta_a = -p_a'/\rho g. \tag{9.9.4}$$

η_a is called the surface elevation of an *inverse barometer* since it is equal to the *depression* that would be registered by a water barometer, this being approximately 1 cm of water per millibar of pressure change.

Changes in atmospheric pressure produce in the ocean depth-independent horizontal pressure gradients that, in an ocean of uniform depth, produce depth-independent currents. For a shallow homogeneous ocean, the momentum equation (4.10.2) becomes [cf. (7.10.1)]

$$\frac{\partial u}{\partial t} + u\frac{\partial u}{\partial x} + v\frac{\partial u}{\partial y} - fv = -g\frac{\partial \eta'}{\partial x} \equiv -g\frac{\partial}{\partial x}(\eta - \eta_a),$$

$$\frac{\partial v}{\partial t} + u\frac{\partial v}{\partial x} + v\frac{\partial v}{\partial y} + fu = -g\frac{\partial \eta'}{\partial y} \equiv -g\frac{\partial}{\partial y}(\eta - \eta_a),$$

(9.9.5)

and the continuity equation is (5.6.7) as before. The equations for pressure-induced motion therefore have exactly the same form (9.8.3) and (9.8.5) as for the tides!

Note that by (9.9.5) a state of rest corresponds to the *adjusted* sea level being constant. Thus frictional processes, which tend to reduce velocities to zero, tend to make the topography of the sea surface follow that of an inverse barometer. For this reason, adjustments of pressure in the ocean that keep η' constant are called *isostatic*. Such adjustments do not affect the pressure on the sea floor because the total weight per unit area of the column of air and water above remains unchanged. In practice, the inverse barometer response of sea level to atmospheric pressure changes is found to be a significant part of *seasonal* changes at high lattitudes (Pattullo *et al.*, 1955; Gill and Niiler, 1973; Lisitzin, 1974).

If the motion produced by pressure variations is small enough for linear equations to apply, (9.9.5) and (5.6.7) reduce to

$$\partial u/\partial t - fv = -g\,\partial\eta'/\partial x, \qquad \partial v/\partial t + fu = -g\,\partial\eta'/\partial y, \qquad (9.9.6)$$

and

$$\frac{\partial \eta'}{\partial t} + \frac{\partial}{\partial x}(Hu) + \frac{\partial}{\partial y}(Hv) = -\frac{\partial \eta_a}{\partial t}, \qquad (9.9.7)$$

where H is the ocean depth. (The discussion from this point onward will be restricted to the linear case.) Note that the continuity equation (9.9.7) has been written in terms of the adjusted sea level η', which has the effect of transferring the forcing term from the momentum equation to the continuity equation.

This suggests a useful laboratory analog [cf. Gill *et al.* (1979) and Gill (1979b)] for simulating tidal and pressure forcing, for the arguments of Section 5.6 show that (9.9.7) is the equation for the surface elevation η' of a shallow homogeneous layer when fluid of the same density is added at a volume rate per unit area of $-\partial\eta_a/\partial t$. In fact, the ocean experiences just such a forcing (see Section 2.5) by addition of mass due to precipitation or by loss due to evaporation. If this is also taken into account, (9.9.7) becomes

$$\frac{\partial \eta'}{\partial t} + \frac{\partial}{\partial x}(Hu) + \frac{\partial}{\partial y}(Hv) = -\frac{\partial \eta^F}{\partial t}, \qquad (9.9.8)$$

where η^F is the "forcing displacement" given by

$$\eta^F = \eta_a - \int (P - E)\,dt/\rho. \qquad (9.9.9)$$

This relation makes it easy to compare pressure and precipitation effects. For instance, a depression of 10 mb corresponds to a positive value of η_a equal to 10 cm. To cancel out this effect and make η^F zero would require 10 cm of precipitation!

Wind effects, it turns out, can be incorporated into this formalism as well. The equation for the barotropic response of the ocean to wind forcing is obtained by vertically averaging the momentum equation (9.2.1). In the case of a homogeneous ocean, in which the pressure gradient is proportional to the surface slope and independent of depth, the vertically averaged equation is

$$\frac{\partial u}{\partial t} - fv = -g\,\frac{\partial \eta}{\partial x} + (X_s - X_b)/\rho H,$$

$$\frac{\partial v}{\partial t} + fu = -g\,\frac{\partial \eta}{\partial y} + (Y_s - Y_b)/\rho H,$$

(9.9.10)

where (u, v) now denotes the *vertically averaged* current, (X_s, Y_s) denotes the surface stress imposed by the wind, and (X_b, Y_b) denotes the bottom stress due to friction. Both surface and bottom stress can be estimated from formulas like (9.5.1). For a stratified ocean of constant depth, the same equation is also obtained for the barotropic mode.

Equations (9.9.10) are the formulas used to compute storm surges in shallow seas such as the North Sea [see, e.g., Heaps (1969)]. In fact, the nonlinear terms that appear in (9.9.5) are often added, although this form is not justified for vertically averaged currents when the actual currents are not independent of depth.

The method of converting (9.9.10) to the same form as (9.9.6) is to subdivide the velocity (u, v), as in Section 9.2, into an Ekman velocity, which is locally determined at each horizontal position, and a pressure-driven velocity (u_p, v_p). For a shallow homogeneous layer, the pressure gradient is independent of depth and so (u_p, v_p) is independent of depth. Hence the partition (9.2.4), when expressed in terms of average velocities, becomes

$$u = u_p + U_E/H, \qquad v = v_p + V_E/H,$$

(9.9.11)

where

$$\partial u_p/\partial t - fv_p = -g\,\partial \eta/\partial x, \qquad \partial v_p/\partial t + fu_p = -g\,\partial \eta/\partial y.$$

(9.9.12)

The forcing terms in the momentum equations have now been removed, i.e., (9.9.12) has the same form as (9.9.6). The continuity equation (5.6.8), however, when (9.9.11) is substituted, becomes

$$\frac{\partial \eta}{\partial t} + \frac{\partial}{\partial x}(Hu_p) + \frac{\partial}{\partial y}(Hv_p) = -w_E,$$

(9.9.13)

and so has a forcing term like that of (9.9.7). The forcing in this case is provided by the Ekman pumping velocity w_E, defined by (9.4.1), and which can be calculated from (9.4.6).

Whatever the form of forcing, the above analysis shows that a transformation can be made so that forcing terms appear only in the continuity equations. In particular, with pressure, wind, and evaporation/precipitation forcing terms all included, the equations are

$$\partial u_p/\partial t - f v_p = -g\, \partial\eta'/\partial x, \qquad \partial v_p/\partial t + f u_p = -g\, \partial\eta'/\partial y, \qquad (9.9.14)$$

$$\frac{\partial \eta'}{\partial t} + \frac{\partial}{\partial x}(H u_p) + \frac{\partial}{\partial y}(H v_p) = -\frac{\partial \eta^F}{\partial t}, \qquad (9.9.15)$$

where

$$\eta^F = \eta_a - \int (P - E)\, dt/\rho + \eta_E \qquad (9.9.16)$$

and η_E is the *Ekman vertical displacement* that is defined by (9.4.4).

The incorporation of all forcing effects into one term (9.9.16) allows a comparison to be made among their magnitudes. Values of the Ekman pumping velocity w_E for the North Atlantic Ocean are shown in Fig. 9.6. At mid-latitudes, typical Ekman displacements in a year (Leetma and Bunker, 1978) are around 30 m. This compares with evaporation/precipitation differences of order 1 m and changes in η_a of a few centimeters, showing that wind-driving dominates. It is interesting to note that at one time it was thought that the main currents in the ocean were driven by evaporation/precipitation differences, and Hough (1897) and Goldsbrough (1933) made calculations of currents driven by this means. The above discussion shows that wind-driven currents can be calculated by similar means.

A similar comparison can be made of the relative importance of wind and pressure effects for a storm of wavenumber k moving with speed U. The vertical mass flux ρw_E, which is the same (see Section 9.4) in atmosphere and ocean, can be estimated from (9.5.3), giving

$$\rho_w w_E \approx 2c_D k^3 p_a'^2/\rho_a f^3, \qquad (9.9.17)$$

where p_a' is the atmospheric pressure perturbation, ρ_a the atmospheric density, and c_D the drag coefficient. The water density ρ_w is used on the left-hand side, making w_E the Ekman pumping velocity in the water. By the definition (9.4.4), $\eta_E \approx w_E/Uk$, whereas η_a is given by (9.9.4). Using these relations along with (9.9.17) then gives

$$\eta_E/\eta_a \approx 2c_D \rho_w g^2 k^2 \eta_a/\rho_a U f^3. \qquad (9.9.18)$$

For the values $c_D \rho_w/\rho_a \approx 1$, $k^{-1} \approx 100$ km, $U \approx 10$ m s^{-1}, and $f = 10^{-4}$ s^{-1}, the ratio η_E/η_a is 20 when η_a is 1 cm, and increases in proportion with η_a. Thus wind effects are generally dominant.

Equations (9.9.14) and (9.9.15) can be reduced to a single equation for η' by the same methods as those used in Section 7. First, the potential vorticity perturbation equation is obtained by taking the curl of (9.9.14) and using (9.9.15) to substitute for the divergence. For a *constant-depth* ocean (the discussion from this point onward will be restricted to the case of constant depth), the result is [cf. (7.2.8)]

$$\frac{\partial}{\partial t}\left(\frac{\partial v_p}{\partial x} - \frac{\partial u_p}{\partial y} - \frac{f\eta'}{H} - \frac{f\eta^F}{H}\right) = 0, \qquad (9.9.19)$$

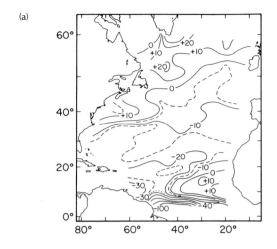

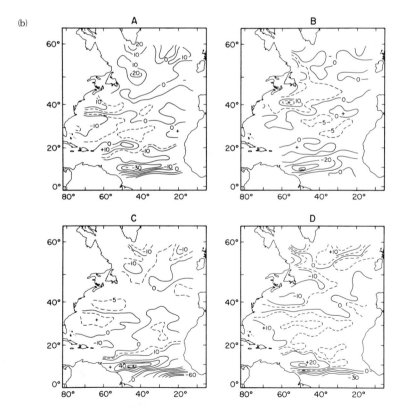

Fig. 9.6.   (a) Annual mean vertical Ekman pumping velocity for the North Atlantic Ocean. Units are 10^{-7} cm s$^{-1} \approx 0.9$ cm day$^{-1} \approx 3.2$ m yr^{-1}. [From Leetmaa and Bunker (1978, Figs. 2 and 5).] (b) Seasonal deviations from the annual mean: (A) winter; (B) spring; (C) summer; (D) autumn.

and this can be integrated to give

$$\frac{\partial v_p}{\partial x} - \frac{\partial u_p}{\partial y} - \frac{f\eta'}{H} - \frac{f\eta^F}{H} = \text{initial value.} \tag{9.9.20}$$

The effect of nonzero initial values was studied in Sections 7.2 and 7.3, so it will not be studied here, and the initial value will be supposed to be zero. The forcing term η^F appears in (9.9.20) in a way similar to that of the initial value, but can vary in *time* as well as in space.

The equation for η' that corresponds only to (7.2.13) is obtained by taking the divergence of (9.9.14), substituting for the divergence of velocity from (9.9.15) and for the vorticity from (9.9.20). The result can be written

$$\frac{\partial^2 \eta'}{\partial x^2} + \frac{\partial^2 \eta'}{\partial y^2} - \frac{1}{gH}\left(\frac{\partial^2}{\partial t^2} + f^2\right)\eta' = \mathscr{F}, \tag{9.9.21}$$

where

$$\mathscr{F} = \frac{1}{gH}\left(\frac{\partial^2}{\partial t^2} + f^2\right)\eta^F, \tag{9.9.22}$$

and hence by (9.4.6) and (9.9.16), $\mathscr{F}$ is given by

$$\frac{\partial \mathscr{F}}{\partial t} = \frac{1}{gH}\left(\frac{\partial^2}{\partial t^2} + f^2\right)\left(\frac{\partial \eta_a}{\partial t} - \frac{P - E}{\rho}\right) + \frac{1}{\rho gH}\frac{\partial}{\partial t}\left(\frac{\partial X_s}{\partial x} + \frac{\partial Y_s}{\partial y}\right) + \frac{f}{\rho gH}\left(\frac{\partial Y_s}{\partial x} - \frac{\partial X_s}{\partial y}\right). \tag{9.9.23}$$

Equation (9.9.21) can be called the *forced shallow-water equation* or the forced Klein–Gordon equation. It governs the behavior of *small* disturbances from an ocean of *uniform* depth when these disturbances are subject to forcing by one of the mechanisms discussed earlier.

9.10   Baroclinic Response of the Ocean to Wind Forcing: Use of Normal Modes

It was shown in Chapter 6 that the free-wave solutions for a shallow homogeneous sea of constant depth could also be applied to a stratified ocean of constant depth. This was first demonstrated for an ocean with two homogeneous layers of different densities. In that case, following Stokes, it was shown that there exist two independent modes, each of which satisfies the shallow-water equations but with different "equivalent depths." Later the concept was extended to a continuously stratified ocean of constant depth for which a countable infinity of normal modes exist. For each of these, the shallow-water equations apply, but with a different equivalent depth for each mode. Now it will be shown that the same technique can be used for *forced* motion, e.g., wind-driven motion. All that is required is to express the forcing function as an expansion in normal modes.

Take, for instance, Eq. (9.2.1), which can be written

$$\partial u/\partial t - fv = -\rho_0^{-1}\,\partial p'/\partial x + \rho_0^{-1}\,\partial X/\partial z,$$
$$\partial v/\partial t + fu = -\rho_0^{-1}\,\partial p'/\partial y + \rho_0^{-1}\,\partial Y/\partial z. \tag{9.10.1}$$

Each of the variables u, v, and p' can be expressed as an expansion in normal modes, as described in Sections 6.11 and 6.13, namely,

$$p' = \sum_{n=0}^{\infty} \tilde{\eta}_n(x, y, t)\hat{p}_n(z),$$

$$(u, v) = \sum_{n=0}^{\infty} (\tilde{u}_n(x, y, t), \tilde{v}_n(x, y, t))\hat{p}_n(z)/\rho_0 g. \tag{9.10.2}$$

This form is chosen to give the equivalent shallow-water equations for each mode. In the same spirit, the forcing term on the right-hand side of (9.10.1) can be written in the form [see, e.g., Gill and Clarke (1974)]

$$\rho^{-1}(\partial X/\partial z, \partial Y/\partial z) = \sum_{n=0}^{\infty} (\tilde{X}_n(x, y, t), \tilde{Y}_n(x, y, t))\hat{p}_n(z)/\rho_0 g. \tag{9.10.3}$$

Then substitution of (9.10.2) and (9.10.3) in (9.10.1) gives for each mode

$$\partial \tilde{u}_n/\partial t - f\,\tilde{v}_n = -g\,\partial \tilde{\eta}_n/\partial x + \tilde{X}_n,$$
$$\partial \tilde{v}_n/\partial t + f\,\tilde{u}_n = -g\,\partial \tilde{\eta}_n/\partial y + \tilde{Y}_n, \tag{9.10.4}$$

i.e., the same as (9.9.10) for a shallow homogeneous ocean, but with the forcing term represented in a more compact form.

The vector quantity $(\tilde{X}_n, \tilde{Y}_n)$ that appears as the forcing term in (9.10.4) can easily be calculated by using the orthogonality property (6.13.4) of the normal modes. Thus if (9.10.3) is multiplied by $\hat{p}_m(z)$ and integrated over the depth, the result for $\tilde{X}_m$ is

$$\tilde{X}_m \int_{-H}^{0} \frac{\hat{p}_m^2}{\rho_0 g}\,dz = \int_{-H}^{0} \frac{1}{\rho_0} \frac{\partial X}{\partial z}\,\hat{p}_m\,dz, \tag{9.10.5}$$

and similarly for $\tilde{Y}_m$. In particular, if the Boussinesq approximation is made so that ρ_0 is regarded as a constant and the approximate solution (6.11.19) that $\hat{p}_0$ is a constant is used for the barotropic mode, then (9.10.5) gives

$$\tilde{X}_0 = g(X_s - X_b)/\hat{p}_0 H. \tag{9.10.6}$$

The quantities $\tilde{\eta}_n$ that appear in (9.10.2) have no particular meaning unless a normalization is chosen for $\hat{p}_n$, and this can be done to suit each particular case. For instance, if the main interest is in the surface elevation, one can normalize the $\hat{p}_n$ so that

$$\hat{p}_n(0) = \rho_0 g \tag{9.10.7}$$

for all n. It then follows from (9.10.2) that $\tilde{\eta}_n$ gives the contribution of mode n to the surface elevation. In particular, the forcing that gives the barotropic contribution is, after using the normalization (9.10.7) in (9.10.6),

$$\tilde{X}_0 = (X_s - X_b)/\rho_0 H. \tag{9.10.8}$$

This makes (9.10.4) the same as the vertically averaged equation (9.9.10).

Estimation of $\tilde{X}_m$ for the baroclinic modes requires knowledge of the way in which $\partial X/\partial z$ varies with depth. If the formula (9.3.4) is deemed appropriate, i.e., if a linear variation of stress over the mixed layer is assumed, then (9.10.5) simplifies to

$$\tilde{X}_m \int_{-H}^{0} \frac{\hat{p}_m^2 \, dz}{\rho_0 g} = \frac{X_s}{H_{\text{mix}}} \int_{-H_{\text{mix}}}^{0} \frac{\hat{p}_m \, dz}{\rho_0}, \tag{9.10.9}$$

and similarly for $\tilde{Y}_m$. The right-hand side involves the average of $\hat{p}_m$ over the mixed layer. For the first mode or two, this average is close to the surface value $\hat{p}_m(0)$.

The expression for $\tilde{X}_m$, $\tilde{Y}_m$ can also be written in a style that mimics (9.10.8) in the absence of bottom friction, namely,

$$(\tilde{X}_m, \tilde{Y}_m) = (X_s, Y_s)/\rho_0 H_m^F, \tag{9.10.10}$$

where H_m^F can be called the *equivalent forcing depth* for mode m. For example, using the normalization (9.10.7), Wunsch and Gill (1976) calculated H_1^F for the equatorial Pacific to be between 270 and 290 m. The forcing depth should not be confused with the ordinary equivalent depth, which is related to the wave speed by (6.11.14) and was found by Wunsch and Gill (1976) to be between 0.7 and 0.8 m for the first baroclinic mode in the equatorial Pacific.

The equations for mode n are (9.10.4) and the continuity equation (6.11.13), namely,

$$\partial \tilde{\eta}_n/\partial t + H_n(\partial \tilde{u}_n/\partial x + \partial \tilde{v}_n/\partial y) = 0, \tag{9.10.11}$$

where H_n is the equivalent depth (not the forcing depth). This set of equations can be manipulated in the same fashion as that used at the end of Section 9.9 to produce the equivalent form of (9.9.21), namely,

$$\frac{\partial^2 \tilde{\eta}_n}{\partial x^2} + \frac{\partial^2 \tilde{\eta}_n}{\partial y^2} - \frac{1}{gH_n}\left(\frac{\partial^2}{\partial t^2} + f^2\right)\tilde{\eta}_n = \mathscr{F}_n, \tag{9.10.12}$$

where $\mathscr{F}_n$ is given by

$$\frac{\partial \mathscr{F}_n}{\partial t} = \frac{1}{g}\frac{\partial}{\partial t}\left(\frac{\partial \tilde{X}_n}{\partial x} + \frac{\partial \tilde{Y}_n}{\partial y}\right) + \frac{f}{g}\left(\frac{\partial \tilde{Y}_n}{\partial x} - \frac{\partial \tilde{X}_n}{\partial y}\right)$$

$$= \frac{1}{\rho_0 g H_n^F}\frac{\partial}{\partial t}\left(\frac{\partial X_s}{\partial x} + \frac{\partial Y_s}{\partial y}\right) + \frac{f}{\rho_0 g H_n^F}\left(\frac{\partial Y_s}{\partial x} - \frac{\partial X_s}{\partial y}\right)$$

$$= \frac{1}{g H_n^F}\left(\frac{\partial^2}{\partial t^2} + f^2\right)\frac{\partial \eta_E}{\partial t}. \tag{9.10.13}$$

The latter alternatives are based on (9.10.10) and (9.4.6).

The above operations have been done for a continuously stratified fluid, but the same equations can be derived for the case of two superposed fluids of different density that was studied in Section 6.2. Making the Boussinesq approximation and adding wind effects as given by (9.3.4) with $H_{\text{mix}} = H_1$, the upper-layer equations

(6.2.3) become, in a rotating fluid,

$$\partial u_1/\partial t - f v_1 = -g\,\partial \eta/\partial x + X_s/H_1, \qquad \partial v_1/\partial t + f u_1 = -g\,\partial\eta/\partial y + Y_s/H_1.$$
$$(9.10.14)$$

The lower-layer equations (6.2.7), with the Boussinesq approximation, are

$$\partial u_2/\partial t - f v_2 = -g\,\partial\eta/\partial x - g'\,\partial h/\partial x, \qquad \partial v_2/\partial t + f u_2 = -g\,\partial\eta/\partial y - g'\,\partial h/\partial y,$$
$$(9.10.15)$$

where g' is reduced gravity (see Section 6.2), h is the *upward* displacement of the interface, H_1 and H_2 are the undisturbed depths of the layers (subscript 1 refers to the upper layer and 2 to the lower layer), and (u_i, v_i) are the horizontal velocity components in layer i.

Subtracting (9.10.15) from (9.10.14) gives

$$\partial\hat{u}/\partial t - f\hat{v} = g'\,\partial h/\partial x + X_s/H_1, \qquad \partial\hat{v}/\partial t + f\hat{u} = g'\,\partial h/\partial y + Y_s/H_1, \quad (9.10.16)$$

where

$$(\hat{u}, \hat{v}) = (u_1 - u_2, v_1 - v_2) \tag{9.10.17}$$

as defined by (6.3.4). These are the momentum equations for the baroclinic mode of the two-layer system, and they have the same form as (9.10.4). The equation equivalent to (9.10.11) is (6.3.5), and the result of eliminating $\hat{u}$ and $\hat{v}$ from (9.10.16) and (6.3.5) is [cf. (9.10.13)]

$$\frac{\partial^2 h}{\partial x^2} + \frac{\partial^2 h}{\partial y^2} - \frac{1}{gH_e}\left(\frac{\partial^2}{\partial t^2} + f^2\right)h = \frac{-1}{\rho g'H_1}\left\{\frac{\partial X_s}{\partial x} + \frac{\partial Y_s}{\partial y} + f\int\left(\frac{\partial Y_s}{\partial x} - \frac{\partial X_s}{\partial y}\right)dt\right\}$$

$$= \frac{-1}{g'H_1}\left(\frac{\partial^2}{\partial t^2} + f^2\right)\eta_E,$$
$$(9.10.18)$$

where the second form of the right-hand side follows from (9.4.4) and (9.4.6) and H_e is given by

$$H_e^{-1} = H_1^{-1} + H_2^{-1}. \tag{9.10.19}$$

Note that both the sign and the factor on the right-hand side of (9.10.18) are different from those in (9.10.12) because (9.10.18) is an equation for interface displacement, whereas (9.10.12) gives the contribution to sea-level change (i.e., to dynamic height) of a given mode.

The discrete modal analysis applied above to the two-layer system can be extended to a system with many homogeneous layers, an additional mode being acquired with the addition of each layer. The modes for a continuously stratified fluid can be obtained by a limiting procedure in which the depth of the layer goes to zero [see, e.g., Lighthill (1967)]. It is also interesting to consider the limit as $H_{mix} \to 0$ [cf. Lighthill (1967)] for the continuously stratified case, for in this limit, the condition

$$w = w_E \tag{9.10.20}$$

that the vertical velocity equals the Ekman pumping velocity becomes applicable at the surface $z = 0$. This is the same condition as the linearized one for topographic forcing, so the methods used in Chapters 6 and 8 are applicable, though they are rarely employed in oceanography problems. Similarly, the methods outlined in Yih's (1980) book for flow over topography in stratified channel flow could be used.

9.11   Response of the Ocean to a Moving Storm or Hurricane

The sea is forced into motion by the atmospheric disturbances that move across it. For barotropic motion, effects of variation of the Coriolis parameter with latitude are important in the deep sea (these will be discussed later), whereas boundaries are important for shallow seas (their effects are discussed in Chapter 10). Hence the main application of the forced shallow-water equations in the absence of boundaries is to internal motion. For instance, Veronis (1956) and Pollard (1970) considered the response to disturbances of limited duration, whereas Leonov and Miropol'skiy (1973) considered resonantly excited waves. Reviews of mechanisms for internal wave generation have been given by Thorpe (1975) and Phillips (1977, Section 5.9).

A particularly interesting case is the response of the ocean to a moving hurricane or depression. If this moves with fixed speed U, then $\mathscr{F}$ has the form

$$\mathscr{F} = \mathscr{F}(\xi, y), \qquad \text{where} \quad \xi = x - Ut. \tag{9.11.1}$$

Solutions for an ocean initially at rest have been obtained by Crease (1956) for the case in which $\mathscr{F}$ is a delta function of ξ and by Geisler (1970) for forcing by a hurricane of realistic planform. The general solution can be expressed as the sum of a particular solution and of transient free waves—just as in the Rossby adjustment problem studied in Chapter 7. The part that is peculiar to the forced problem is the particular solution, which can be taken as the solution for which η' is a function of ξ and y only, and this section will be devoted solely to examining properties of this solution. This gives the form of response when transient effects have died away.

The equation satisfied by $\tilde{\eta}_n$ when there is dependence only on ξ and y is, by (9.10.12) and (9.11.1),

$$\frac{\partial^2 \tilde{\eta}_n}{\partial y^2} - \left(\frac{U^2}{c_n^2} - 1 \right) \frac{\partial^2 \tilde{\eta}_n}{\partial \xi^2} - \frac{f^2}{c_n^2} \tilde{\eta}_n = \mathscr{F}_n, \tag{9.11.2}$$

where $c_n = (gH_n)^{1/2}$ is the wave speed of the mode being studied. There is a different c_n for each mode, and hence a different equation, so evaluation of the full response requires solution of (9.11.2) for each mode and then summation of the contribution from each, as outlined in the previous section. Here, only the contribution from a particular mode will be considered, so the subscript n will be dropped.

The nature of the solutions of (9.11.2) depends very much on whether the storm speed U exceeds the wave speed c or not. For a *slow-moving disturbance* ($U < c$), Eq. (9.11.2) is *elliptic* in character, whereas for a *fast-moving disturbance* ($U > c$), it is *hyperbolic*. The importance of the distinction, and the general nature of the solutions, can be demonstrated conveniently by considering a special case in which the wind is

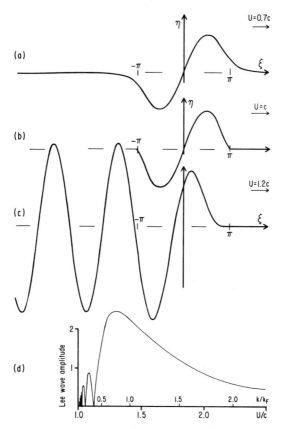

Fig. 9.7. Response of the ocean to a moving two-dimensional storm. The wind stress Y_s is perpendicular to the path of the storm and varies with distance ξ along the storm track as shown in (b) if the vertical axis is taken as Y_s/Y_0, where Y_0 is the maximum value of the stress. The unit of ξ is the Rossby radius c/f, where c is the long gravity wave speed in the absence of rotation and f is the Coriolis parameter. The storm translates to the right at speed U and (a)–(c) show the response for different values of U: (a) $U = (0.5)^{1/2}c$, (b) $U = c$, and (c) $U = (1.5)^{1/2}c$. The unit for η is $cY_0/\rho f g H^F$, where ρ is density, g is the acceleration due to gravity, and H^F is the equivalent forcing depth defined in the text. The response pattern is steady and moves with the storm. In case (a), the governing equation is elliptic and the response is confined to the neighborhood of the storm. In case (c), the governing equation is hyperbolic and lee waves appear behind the storm. Case (b) is the borderline case in which the response is the same as the forcing. (d) How the amplitude (same units as before) of the lee waves varies with the speed of translation of the storm. Corresponding values of the ratio of wavenumber k of the forcing to the wavenumber k_F of the free waves (and hence of the lee waves) are also shown.

unidirectional and has the form (see Fig. 9.7b)

$$X_s = 0, \qquad Y_s = \begin{cases} Y_0 \sin k\xi & \text{if } |k\xi| < \pi, \\ 0 & \text{otherwise.} \end{cases} \tag{9.11.3}$$

This corresponds to the wind blowing one way before the storm center has arrived, and blowing in the opposite direction after the storm center has passed. The forcing function for this case is, by (9.10.13),

$$\mathscr{F} = \begin{cases} -(fY_0/\rho g H^F U) \sin k\xi & \text{if } |k\xi| < \pi, \\ 0 & \text{otherwise.} \end{cases} \tag{9.11.4}$$

For *slow-moving* disturbances, when (9.11.2) is elliptic, the solution decays exponentially with distance from the disturbance, and is given by

$$\eta = \begin{cases} r\{\sin k\xi + ka_F \exp(-\pi/ka_F)\sinh(\xi/a_F)\} & \text{if } |k\xi| < \pi, \\ rka_F \sinh(\pi/ka_F)\exp(-|\xi|/a_F)\operatorname{sgn}(\xi) & \text{otherwise,} \end{cases} \qquad (9.11.5)$$

where

$$r = \frac{c^2 Y_0}{\rho g H^F f U(1 + k^2 a_F^2)} = \frac{c^2 f Y_0}{\rho g H^F U[f^2 + k^2(c^2 - U^2)]} \qquad (9.11.6)$$

is a measure of the response amplitude, and a_F, which is given by

$$a_F^2 = (c^2 - U^2)/f^2, \qquad (9.11.7)$$

is a modified form of the Rossby radius. If ka_F is small, i.e., if the storm is large or its speed is close to that of c, the solution has almost the same shape as that of the forcing and is given approximately by

$$\eta = -c^2 \mathscr{F}/f^2 = c^2 Y_s/\rho g H^F f U. \qquad (9.11.8)$$

This effect may be seen in Fig. 9.7, which shows the response to a storm of fixed size (given by $k = f/c$) for different speeds of the storm. In case (a), $U^2 = 0.5c^2$, which implies $k^2 a_F^2 = 0.5$, and the response looks like a "spread-out" form of the forcing. In case (b), $U = c$, i.e., the storm is moving at the long gravity wave speed (or short Poincaré wave speed) c and the response has the same shape as that of the forcing.

Another interesting limit is the one for which the storm speed U tends to zero. As U gets smaller, the shape of the response does not change much (it looks similar to that shown in Fig. 9.7a), but the amplitude tends to infinity as $1/U$. If the forcing is not moving ($U = 0$), a different approach is needed because Ekman pumping continues at a fixed rate at each place, and therefore a disturbance can build up linearly in time. This type of response is considered in Section 9.14.

For the more usual situation of a *fast-moving* disturbance ($U > c$), Eq. (9.11.2) is *hyperbolic* in character, and therefore the homogeneous equation has sinusoidal rather than exponential solutions. The boundary condition at infinity is now a *radiation condition*, namely, that no waves appear ahead of the disturbance. This is because the maximum group velocity of a Poincaré wave is c, so none can propagate ahead of the disturbance. The solution is therefore given by

$$\eta = \begin{cases} 0 & \text{for } k\xi > \pi, \\ r\{\sin k(k^{-1}\pi - \xi) - (k/k_F)\sin k_F(k^{-1}\pi - \xi)\} & \text{for } |k\xi| < \pi, \\ -2r(k/k_F)\sin(\pi k_F/k)\cos k_F \xi & \text{for } k\xi < -\pi, \end{cases} \qquad (9.11.9)$$

where k_F is the wavenumber of the Poincaré wave that has the same phase speed as the disturbance and is given by

$$(Uk_F)^2 = f^2 + (k_F c)^2, \qquad (9.11.10)$$

i.e., by

$$k_F^2 = f^2/(U^2 - c^2).$$

The amplitude factor r is given by (9.11.6), which can also be written

$$r = c^2 Y_0 / \rho g H^F f U (1 - k^2 / k_F^2). \tag{9.11.11}$$

The solution looks dramatically different from the solution for a slow-moving disturbance because of the train of lee waves behind. For example, for the case $k = f/c$, Fig. 9.7c shows what the response is like when $U^2 = 1.5c^2$. The frequency of the lee waves relative to a fixed observer is given by

$$\omega^2 = (Uk_F)^2 = f^2 U^2 / (U^2 - c^2) \tag{9.11.12}$$

by (9.11.10) and is close to the inertial frequency if $U \gg c$. The amplitude $|2r(k/k_F) \sin(\pi k_F/k)|$ of the lee waves is shown as a function of U/c (or k/k_F) in Fig. 9.7d. The lee waves are insignificant if k/k_F is either small or large, but are prominent when k/k_F is near unity.

If the free waves have the same wavenumber as that of the storm ($k = k_F$), then r is infinite by (9.11.6) and (9.11.10), i.e., there is a type of resonance. The response, however, is not infinite because of the finite duration of the storm, and the solution is

$$\eta = \begin{cases} 0 & \text{for } k\xi > \pi. \\ -(c^2 Y_0 / 2\rho g H^F f U)\{(k\xi - \pi) \cos k\xi - \sin k\xi\} & \text{for } |k\xi| < \pi, \\ (\pi c^2 Y_0 / \rho g H^F f U) \cos k\xi & \text{for } k\xi < -\pi. \end{cases} \tag{9.11.13}$$

This can be obtained either by solving (9.11.2) directly or by taking the limiting form of (9.11.9) as $k_F \to k$. For a more general forcing involving a range of wavenumbers, there is a tendency for the part of the solution that corresponds to $k = k_F$ to be prominent because of the resonance.

The above solutions give a good idea of the baroclinic response characteristics that might be expected to result from the passage of a hurricane across the ocean. Their *speed* of translation is normally much greater (say three times) than that of the first baroclinic mode (for which c is about 2 m s^{-1}, say), so a train of lee waves behind the storm would be expected. The scale k^{-1} of a hurricane is normally much greater (say, three times) than that of the Rossby radius c/f, but the ratio

$$(k/k_F)^2 = k^2 (U^2 - c^2) / f^2 \approx (kU/f)^2 \tag{9.11.14}$$

is usually of order unity. This implies prominent lee waves with wavenumber k_F near f/U by (9.11.10), and frequency near inertial by (9.11.12). Such waves can be seen, for instance, in the numerical solution of Price (1981), shown in Fig. 9.8a. This is a simulation of the vertical velocity pattern associated with Hurricane Eloise (1975), which passed directly over a data buoy in the Gulf of Mexico, and so its structure is relatively well known. Figure 9.8c shows a comparison of the predicted lee-wave pattern (cf. Fig. 9.7c) and that which was observed at the buoy.

There is, however, another important effect due to the component X_s of the stress *parallel* to the storm track. This causes an Ekman transport away from the path of the storm center, resulting in horizontal displacements of particles in the surface layers that can amount to some tens of kilometers in the case of a hurricane. As a consequence, water near the axis of the storm is upwelled, possibly by some tens of meters.

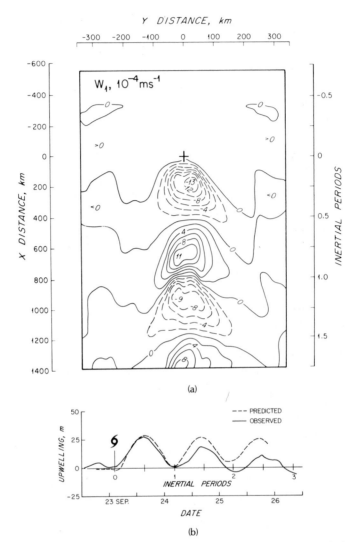

(a)

(b)

Fig. 9.8. (a) Result of a numerical model calculation of the vertical pumping velocity at the base of the mixed layer due to the passage of Hurricane Eloise. The stress pattern was based on wind measurements at a buoy that the hurricane passed over, and uniform translation at a speed of 8.5 m s^{-1} was assumed. The initial density structure in the ocean was based on observations during the week preceding the storm. The contour interval is 2×10^{-4} m s^{-1}, negative values (dashed line contours) indicating upward motion. [From Price (1981, Fig. 18b).] (b) Comparison of the predicted upwelling (dashed line) with that (solid line) observed at the buoy. [From Price (1981, Fig. 22).]

The effect of this upwelling can be calculated by techniques similar to those used in Chapter 7 to calculate the response at large times for the Rossby adjustment problem. In the present case, the task is to calculate the solution for large t, i.e., by (9.11.1), for large negative ξ. For such values (9.10.13) implies that

$$\mathscr{F}(\xi, y) \to \mathscr{F}_{\infty}(y) \equiv \frac{f^2 \eta_{\infty}^{\mathrm{E}}(y)}{gH^{\mathrm{F}}} = \frac{-f}{\rho g H^{\mathrm{F}}} \int_{-\infty}^{\infty} \frac{\partial X_{\mathrm{s}}}{\partial y} \frac{d\xi}{U} \qquad \text{as} \quad \xi \to -\infty, \quad (9.11.15)$$

where η_∞^E is the total vertical Ekman displacement due to passage of the storm. The solution η for ξ large and negative can be expressed as the sum of a particular solution $\eta_\infty(y)$, which is a function of y only, plus lee waves. The particular solution satisfies, by (9.11.2), the equation

$$\partial^2\eta_\infty/\partial y^2 - (f^2/c^2)\eta_\infty = \mathscr{F}_\infty \equiv f^2\eta_\infty^E/gH^F, \qquad (9.11.16)$$

and since, for hurricanes, the scale is normally large compared with the Rossby radius, the approximate solution is

$$\eta_\infty = -c^2\eta_\infty^E/gH^F. \qquad (9.11.17)$$

An example for which a solution is easily calculated is that of a circular storm with tangential stress τ given by

$$\tau = \tau_m(r/L)\exp\{\tfrac{1}{2} - \tfrac{1}{2}(r/L)^2\}, \qquad (9.11.18)$$

r being the radial coordinate, τ_m the maximum stress, and L the radius at which the

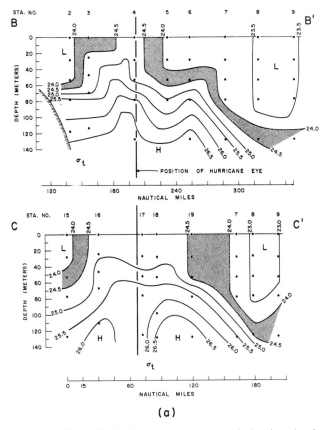

(a)

Fig. 9.9. (a) Two sections BB′ and CC′, showing isopycnals as observed a few days after the passage of a hurricane across the Gulf of Mexico. The sections are approximately normal to the storm track and the position where the eye of the hurricane crossed each section is shown. The hurricane-induced upwelling is clearly shown. (b) The section CC′ as observed the following summer when no hurricane effect was present. [From Leipper (1967, Fig. 14).]

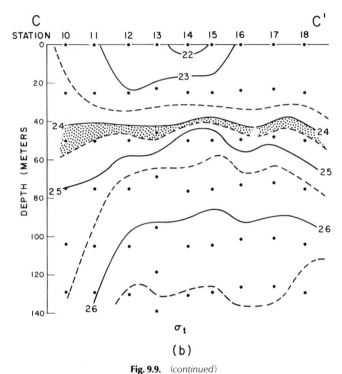

$$\sigma_t$$

(b)

Fig. 9.9. (continued)

maximum stress occurs. In this case, (9.11.15) can be integrated to give

$$\eta_\infty^E = (2\pi)^{1/2}(\tau_m/\rho f U)\{1 - (y/L)^2\} \exp\{\tfrac{1}{2} - \tfrac{1}{2}(y/L)^2\}, \qquad (9.11.19)$$

the maximum value being approximately $4\tau_m/\rho f U$ at $y = 0$. If τ_m is 3 N m^{-2}, f is 5×10^{-5} s^{-1} and U is 6 m s^{-1}, a value of 40 m for the total upwelling during the passage of the hurricane is obtained. The corresponding value of η_∞, the change in dynamic height contributed by the nth mode, is (H_n/H_n^F) times η_∞^E, typically about $\eta_\infty^E/500$ for the first mode. This gives 8 dynamic centimeters when η_∞^E is 40 m.

Such effects were observed by Leipper (1967) soon after the passage of a hurricane. Sections across the path of the hurricane, shown in Fig. 9.9, indicate upwelling due to the hurricane by amounts of 50 m or more. The corresponding dynamic topography had a low near the hurricane path, as expected, with variations normal to the path of several tens of centimeters. This corresponded to currents of around 0.5 m s^{-1}, which were forcibly brought to attention during the cruise because a section across the storm path that was meant to be straight since it was on a fixed compass bearing, was found to be curved considerably because of the strong currents parallel to the storm path! Other examples are discussed by Price (1981).

In practice, the linear theory described above is inadequate for calculating details of the effect of a hurricane for two reasons. One is that hurricane effects are so large that nonlinear terms cannot be ignored in the equations. The second is that thermal effects are important. In fact [see Gray (1979)], hurricanes are maintained by the energy they extract from the heat available in the ocean, so thermal effects are of great importance. Not only is heat removed from the ocean by the hurricane, but it is redistributed within the ocean by the stirring action of the storm in addition to advective effects. All these processes contribute to the isopycnal field seen in Fig. 9.9. Price (1981) and Greatbatch (1982) have developed models that include both mechanical and thermal effects. These show lee waves, upwelling behind the storm, removal of heat from the ocean, and effects of mixed-layer deepening due to stirring by the storm.

9.12 Spin-Down by Bottom Friction

If a strongly rotating fluid is perturbed and the force causing the perturbation is not maintained, the fluid adjusts to a geostrophic equilibrium in which pressure gradients are balanced by Coriolis accelerations associated with steady flow along the isobars. If, however, this flow extends to the bottom, frictional stresses will be exerted on the bottom, Ekman layers (see Section 9.2) will be established, and energy will be removed from the flow. Thus a geostrophic equilibrium, unless it is forcibly maintained, will not be of permanent duration, and the fluid will "spin down" under the action of bottom friction.

The process is most easily considered in a homogeneous laminar shallow fluid whose motion is governed by (9.9.21). In the absence of other forcing, the right-hand side $\mathcal{F}$ is determined by bottom friction, the forcing displacement η^F in (9.9.22) being the Ekman displacement η_E. It will be assumed that the time scale on which bottom friction acts is large compared with f^{-1}, so that (9.9.22) can be approximated by

$$\partial \mathcal{F} / \partial t = -(f^2/gH)\,\partial \eta_E / \partial t = -(f^2/gH)w_E, \tag{9.12.1}$$

use being made of the definition (9.4.4). For laminar flow, w_E is given by (9.6.5), and for a shallow homogeneous fluid, the perturbation pressure is $\rho g \eta$, so (9.12.1) becomes

$$\partial \mathcal{F} / \partial t = -H^{-1}(\tfrac{1}{2}fv)^{1/2}(\partial^2 \eta / \partial x^2 + \partial^2 \eta / \partial y^2). \tag{9.12.2}$$

Substituting in the time derivative of (9.9.21), again using the assumption that $\partial / \partial t \ll f$, the result is

$$\frac{\partial}{\partial t}\left\{\frac{\partial^2 \eta'}{\partial x^2} + \frac{\partial^2 \eta'}{\partial y^2} - \frac{f^2}{gH}\eta'\right\} = -H^{-1}\left(\frac{fv}{2}\right)^{1/2}\left\{\frac{\partial^2 \eta'}{\partial x^2} + \frac{\partial^2 \eta'}{\partial y^2}\right\}. \tag{9.12.3}$$

Note that the equation involves η' rather than η because the rest state corresponds to η', not η, being constant.

For a sinusoidal disturbance of wavenumber κ, this equation becomes

$$(f^2 + \kappa^2 c^2)\,\partial \eta' / \partial t = -(\tfrac{1}{2}fv)^{1/2}g\kappa^2 \eta', \tag{9.12.4}$$

showing that disturbances decay with time on a time scale

$$t_{\text{spin}} = (1 + f^2/\kappa^2 c^2)H(\tfrac{1}{2}fv)^{-1/2} \tag{9.12.5}$$

called the spin-up (or spin-down) time. If the scale κ^{-1} is *small* compared with the Rossby radius $a = c/f$, the spin-down time is independent of wavenumber and is given by

$$t_{\text{spin}} = H(\tfrac{1}{2}fv)^{-1/2}. \tag{9.12.6}$$

This is the value appropriate for a homogeneous fluid with a rigid lid (making $a = \infty$), which was discussed by Greenspan (1968). [*Note:* If the flow is turbulent, a suitable value of eddy viscosity can be estimated from (9.6.6), resulting in the formula

$$t_{\text{spin}} = H/2c_{\text{Dg}}u_{\text{av}}, \tag{9.12.7}$$

where c_{Dg} is the drag coefficient and u_{av} is an average value of the velocity outside the boundary layer. It should be borne in mind, however, that this is an attempt to estimate effects of a nonlinear process using a linear formulation. A more suitable formulation for this case is to use the vertically averaged equations (9.9.10) with the bottom stress estimated from the turbulent drag formula (9.5.1).]

In the opposite extreme, in which the scale is *large* compared with the Rossby radius, (9.12.3) can be approximated by

$$\partial\eta'/\partial t - \kappa_E(\partial^2\eta'/\partial x^2 + \partial^2\eta'/\partial y^2) = 0, \tag{9.12.8}$$

i.e., by a diffusion equation with a diffusivity κ_E, due to Ekman boundary-layer spin-down, given by

$$\kappa_E = g(\tfrac{1}{2}fv)^{1/2}/f^2. \tag{9.12.9}$$

In this case, the free surface moves toward equilibrium as if it were controlled by a diffusion process. [*Note:* A similar result applies to spin-down by viscous effects in the interior, for they act on the density field like a horizontal diffusivity of magnitude vN^2/f^2; see Gill (1981); Garrett and Loder (1981).]

Another way of incorporating friction is through the vertically averaged equations (9.9.10) with the bottom stress given by (9.6.4). Using also the fact that $r \ll f$, these give

$$\partial u/\partial t - fv = -g\,\partial\eta/\partial x - ru, \qquad \partial v/\partial t + fu = -g\,\partial\eta/\partial y - rv, \tag{9.12.10}$$

where

$$r = (\tfrac{1}{2}fv)^{1/2}/H \tag{9.12.11}$$

is the inverse of the spin-down time for scales small compared with the Rossby radius. If v is eliminated from (9.12.10), the result is

$$(r + \partial/\partial t)^2 u + f^2 u = -fg\,\partial\eta/\partial y - (r + \partial/\partial t)g\,\partial\eta/\partial x. \tag{9.12.12}$$

For *slow* adjustment ($\partial/\partial t \ll f$) with *weak* friction ($r \ll f$), the motion is quasi-geostrophic (see Section 8.16), i.e., to the first approximation the flow is geostrophic, but to determine the vertical motion, it is necessary to consider the next approximation to the momentum equations. In other words, (9.12.12) is approximated by [cf. Hesselberg (1915)]

$$\begin{aligned} u &= -f^{-1}g\,\partial\eta/\partial y - f^{-2}(r + \partial/\partial t)g\,\partial\eta/\partial x, \\ v &= f^{-1}g\,\partial\eta/\partial x - f^{-2}(r + \partial/\partial t)g\,\partial\eta/\partial y. \end{aligned} \tag{9.12.13}$$

The geostrophic term does not contribute to the divergence, which comes solely from the isallobaric and the bottom friction terms, i.e.,

$$\partial u/\partial x + \partial v/\partial y = -f^{-2}(r + \partial/\partial t)g(\partial^2 \eta/\partial x^2 + \partial^2 \eta/\partial y^2). \qquad (9.12.14)$$

The time derivative on the right-hand side dominates if the time scale is short compared with the fractional time scale; otherwise the Ekman divergence term dominates. In other words, *isallobaric terms dominate* over frictional convergence if the time scale is *short* compared with the spin-down time. The reverse is true if the time scale is long. The importance of convergence is its connection with surface elevation through the continuity equation (5.6.6). If this is combined with (9.12.14), the governing equation (9.12.3) is recovered.

For motion that is not barotropic, bottom friction can be calculated from bottom currents, as described in Sections 9.5 and 9.6, but these are no longer uniquely related to the currents in the rest of the fluid, except in special circumstances for which, for instance, nearly all of the energy might be in a particular mode. In such a case, the *energy equation* is useful for calculating the spin-down time. For an incompressible fluid, the quasi-geostrophic equations that govern the motion are (8.16.7) and (8.16.12). If the first equation of (8.16.7) is multiplied by $\partial p'/\partial x$, the second is multiplied by $\partial p'/\partial y$, (8.16.12) is multiplied by $\partial p'/\partial z/N^2$, and the results are added, the equation

$$\frac{\partial}{\partial x}(up') + \frac{\partial}{\partial y}(vp') + \frac{\partial}{\partial z}(wp')$$

$$= -\frac{\partial}{\partial t}\left\{ \frac{1}{2}\rho_0^{-1}f^{-2}\left(\left(\frac{\partial p'}{\partial x}\right)^2 + \left(\frac{\partial p'}{\partial y}\right)^2\right) + \frac{1}{2}\rho_0^{-1}N^{-2}\left(\frac{\partial p'}{\partial z}\right)^2\right\} \qquad (9.12.15)$$

is obtained, use being made of the incompressibility condition (4.10.12). The expression on the left-hand side is the divergence of the energy flux density [given by (8.6.7)], whereas the expression in curly brackets is the *quasi-geostrophic* form [cf. (8.6.1)] of the *energy density*.

Integrating (9.12.15) vertically over the depth and averaging horizontally over a wavelength, (9.12.15) becomes

$$\frac{\partial}{\partial t}\int \left\{ \frac{1}{2}\rho_0^{-1}f^{-2}\left(\overline{\left(\frac{\partial p'}{\partial x}\right)^2} + \overline{\left(\frac{\partial p'}{\partial y}\right)^2}\right) + \frac{1}{2}\rho_0^{-1}N^{-2}\overline{\left(\frac{\partial p'}{\partial z}\right)^2}\right\} dz = \overline{[wp']}_b, \qquad (9.12.16)$$

where the subscript b on the right-hand side denotes the value at the bottom of the region considered. This can be taken to be just outside the boundary layer, so that w_b is equal to the Ekman pumping velocity given by (9.6.5). Thus

$$\overline{[wp']}_b = \frac{1}{\rho_0 f}\left(\frac{v}{2f}\right)^{1/2}\overline{\left\{\frac{\partial}{\partial x}\left(p'\frac{\partial p'}{\partial x}\right) + \frac{\partial}{\partial y}\left(p'\frac{\partial p'}{\partial y}\right) - \left(\frac{\partial p'}{\partial x}\right)^2 - \left(\frac{\partial p'}{\partial y}\right)^2\right\}}_b, \qquad (9.12.17)$$

and so (9.12.16) becomes

$$\frac{1}{2}\frac{\partial}{\partial t}\int \left\{ \rho_0^{-1}f^{-2}\left(\overline{\left(\frac{\partial p'}{\partial x}\right)^2} + \overline{\left(\frac{\partial p'}{\partial y}\right)^2}\right) + \rho_0^{-1}N^{-2}\overline{\left(\frac{\partial p'}{\partial z}\right)^2}\right\} dz$$

$$= -\rho_0^{-1}f^{-1}\left(\frac{v}{2f}\right)^{1/2}\overline{\left\{\left(\frac{\partial p'}{\partial x}\right)^2 + \left(\frac{\partial p'}{\partial y}\right)^2\right\}}_b, \qquad (9.12.18)$$

which is the *quasi-geostrophic energy equation* with bottom friction included.

It follows from (9.12.18) that if the motion is periodic in the horizontal and has the separable modal form given by (6.11.5), then the spin-down time $2t_{spin}$ for energy (being quadratic in p', it is twice the value for p' itself) is given by

$$(2t_{spin})^{-1} = f^{-1} \left(\frac{v}{2f}\right)^{1/2} \kappa_H^2 \hat{p}_b^2 \Big/ \int \left\{ (f^{-1}\kappa_H \hat{p})^2 + \left(\frac{1}{N}\frac{\partial \hat{p}}{\partial z}\right)^2 \right\} dz, \quad (9.12.19)$$

where κ_H is the horizontal wavenumber. For the particular case of an ocean with constant buoyancy frequency N, for which $\hat{p}$ is proportional to $\cos(n\pi z/H)$, (9.12.19) gives

$$t_{spin} = \tfrac{1}{2}H(\tfrac{1}{2}fv)^{-1/2}(1 + (\kappa_H a)^{-2}), \quad (9.12.20)$$

where a is the Rossby radius for the first baroclinic mode. Thus if the scale is *small* compared with the baroclinic Rossby radius, the spin-down time is half that of the barotropic mode. If the scale is large, on the other hand, the spin-down time can be very much longer.

9.13   Buoyancy Forcing

The ocean–atmosphere system is driven by radiation from the sun (see Chapter 1), which produces internal sources of heat within the ocean–atmosphere system. The problem of calculating the rate of heating can be quite involved, but a lot can be learned about the *effects* of this heating if the rate of heating is assumed to be known. The equation for changes of internal energy or of temperature produced by such a heat source is derived in Section 4.4 and restated in Section 4.10. In the absence of diffusion, the equation for the effect of heating on *density* is obtained by combining (4.10.3) and (4.10.4) with (4.10.7), the result being

$$D\rho/Dt - c_s^{-2} Dp/Dt = -g^{-1}B_s, \quad (9.13.1)$$

where c_s is the speed of sound, and B_s, the rate of change of buoyancy per unit volume, is given by

$$B_s = \frac{g\alpha'\theta}{Tc_p(p_r, \theta)}(Q - \mathbf{V}\cdot\mathbf{F}^{rad}). \quad (9.13.2)$$

Here Q is the rate of heating per unit volume, which can include effects of latent heat release [see (4.4.9)]; $\mathbf{V}\cdot\mathbf{F}^{rad}$ is the divergence of the radiative heat flux; T is temperature; θ is potential temperature; c_p is the specific heat calculated at the reference pressure p_r used in the definition of potential temperature; and α' (see Section 3.7) is the expansion coefficient with respect to potential temperature. For an ideal atmosphere, c_p is a constant and

$$\alpha' = \theta^{-1}. \quad (9.13.3)$$

The dynamic effects of the heating come through the effect of gravitational forces when there are variations in density, and this is why the equation is given in the above form. The buoyancy of any given mass m of fluid can be defined as the negative of its weight mg. The absolute value is not of importance, but variations are.

As an example, Fig. 9.10 shows an estimate for the atmosphere of the diabatic heating rate $(Q - \mathbf{V} \cdot \mathbf{F}^{\mathrm{rad}})/c_p$, which is proportional to B_s. It should be borne in mind that this heating rate is not the same as it would be in a resting atmosphere, since the motion produced by the heating alters the distribution of the heating, which in turn affects the motion, and so on.

For small perturbations from a state of rest, the procedure outlined in Section 6.14 may be used, the only difference being that (6.14.1) is replaced by (9.13.1), i.e., is modified by the addition of a forcing term. Consequently, the perturbation equation (6.14.3) is also modified by the addition of a forcing term, and it now takes the form

$$g \, \partial(\rho' - c_s^{-2}p')/\partial t - \rho_0 N^2 w = -B_s', \tag{9.13.4}$$

where B_s' is the perturbation value of the rate of change of buoyancy per unit volume.

When the horizontal scale is large compared with the vertical scale, the hydrostatic approximation can be used, and when the upper and lower boundary surfaces (if present) are horizontal, a normal-mode expansion can be used. The forcing term B_s' can also be expanded in normal modes, so again the forced shallow-water equations are obtained. The details depend on whether the fluid is incompressible (as in Section 6.11) or not; if it is compressible, the details then depend on whether height is used as the vertical coordinate (as in Section 6.14) or whether isobaric coordinates are used (as in Section 6.17).

The incompressible case $(c_s = \infty)$ is the most straightforward. If the hydrostatic equation (6.11.2) is used in (9.13.4), it becomes in this case

$$\partial^2 p'/\partial z \, \partial t + \rho_0 N^2 w = B_s'. \tag{9.13.5}$$

If the expansions (6.13.1) are now applied (for a semi-infinite atmosphere, the sums would be replaced by integrals), and the expansion for B_s' is written in the form

$$B_s' = \sum_{n=0}^{\infty} \rho_0 N^2 \hat{h}_n(z) \tilde{b}_n(x, y, t). \tag{9.13.6}$$

The result, which takes the place of (6.11.8), is

$$\tilde{w}_n = \partial \tilde{\eta}/\partial t + \tilde{b}_n, \tag{9.13.7}$$

use having been made of (6.11.7). This equation can be used in conjunction with (6.11.9) to give the forced shallow-water equations. Alternatively, the form (9.9.15) of forced equations is obtained by combining (9.13.7) with the modal form of the incompressibility condition (6.4.3). Substitution of the modal expansions (9.10.2) for (u, v) and (6.13.1) for w then gives

$$\tilde{w}_n + H_n(\partial \tilde{u}_n/\partial x + \partial \tilde{v}_n/\partial y) = 0, \tag{9.13.8}$$

which can be compared with (6.11.13). Now (9.13.7) and (9.13.8) combine to give

$$\partial \hat{\eta}_n/\partial t + H_n(\partial \tilde{u}_n/\partial x + \partial \tilde{v}_n/\partial y) = -\partial \eta_n^{\mathrm{F}}/\partial t, \tag{9.13.9}$$

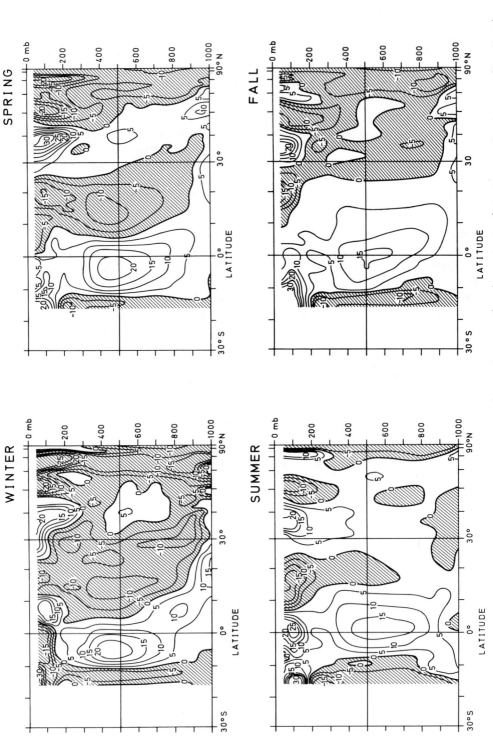

Fig. 9.10. An estimate of the zonally averaged diabatic heating rate $(Q - \nabla \cdot \mathbf{F}^{\mathrm{rad}})/\rho c_p$ (in units of 10^{-6} K s$^{-1} \approx 0.1$ deg day^{-1}), for the atmosphere by season. [From Hantel

i.e., equation (9.9.15) but with the forcing displacement η_n^{F} given by

$$\eta_n^{\text{F}} = \int \tilde{b}_n \, dt. \tag{9.13.10}$$

The compressible case leads to exactly the same equations, and derivation is left to the reader. If isobaric coordinates are used, the appropriate form of expansion, or integral, for B_s', has contributions from each mode of the form

$$B_s' = \rho_0 N N_* \rho_*^{-1/2} \hat{h}_*(z_*) \tilde{b}(x, y, t). \tag{9.13.11}$$

As with the wind forcing, the quantities $\tilde{b}_n$ can readily be calculated by using the orthogonality properties of the modes.

The description in terms of normal modes is not always the most convenient method of dealing with problems, and sometimes it is more useful to consider the full set of equations, which can be reduced to a pair of equations involving only w and p' (or w_* and Φ''). In the incompressible case, the equation for the horizontal divergence, which can be derived from the horizontal momentum equations, is (8.4.8), namely,

$$\left(\frac{\partial^2}{\partial t^2} + f^2 \right) \frac{\partial w}{\partial z} = \frac{1}{\rho_0} \frac{\partial}{\partial t} \left(\frac{\partial^2 p'}{\partial x^2} + \frac{\partial^2 p'}{\partial y^2} \right), \tag{9.13.12}$$

and this can be combined with (9.13.5) to give an equation for w or p' alone. For instance, the equation for the pressure perturbation is

$$\frac{\partial}{\partial t} \left\{ \frac{1}{\rho_0} \left(\frac{\partial^2 p'}{\partial x^2} + \frac{\partial^2 p'}{\partial y^2} \right) + \left(\frac{\partial^2}{\partial t^2} + f^2 \right) \frac{\partial}{\partial z} \left(\frac{1}{\rho_0 N^2} \frac{\partial p'}{\partial z} \right) \right\} = \left(\frac{\partial^2}{\partial t^2} + f^2 \right) \frac{\partial}{\partial z} \left(\frac{B_s'}{\rho_0 w^2} \right). \tag{9.13.13}$$

This equation can be compared with (8.16.13), derived for slow unforced flow.

The equation equivalent to (9.13.12) for the compressible case is [cf. (6.17.19)]

$$\left(\frac{\partial^2}{\partial t^2} + f^2 \right) \left(\frac{\partial w_*}{\partial z_*} - \frac{w_*}{H_s} \right) = \frac{\partial}{\partial t} \left(\frac{\partial^2 \Phi''}{\partial x^2} + \frac{\partial^2 \Phi''}{\partial y^2} \right), \tag{9.13.14}$$

whereas the equivalent of (9.13.5) [cf. (6.17.23)] is

$$N_*^{-2} \partial^2 \Phi v / \partial z_* \, \partial t + w_* = \rho_*^{-1} N^{-2} B_s'. \tag{9.13.15}$$

The combination of the latter two equations results in

$$\frac{\partial}{\partial t} \left\{ \frac{\partial^2 \Phi''}{\partial x^2} + \frac{\partial^2 \Phi''}{\partial y^2} + \left(\frac{\partial^2}{\partial t^2} + f^2 \right) e^{z_*/H_s} \frac{\partial}{\partial z_*} \left(\frac{e^{-z_*/H_s}}{N_*^2} \frac{\partial \Phi''}{\partial z_*} \right) \right\}$$

$$= \left(\frac{\partial^2}{\partial t^2} + f^2 \right) e^{z_*/H_s} \frac{\partial}{\partial z_*} \left(\frac{B_s'}{N^2 \rho_r} \right). \tag{9.13.16}$$

9.14   Response to Stationary Forcing: A Barotropic Example

When the forcing is stationary rather than moving, the response grows with time until friction or some other equilibrating mechanism comes into operation. Particularly simple solutions can be found when there is a suitable symmetry, such as dependence only on distance r from a fixed point or dependence only on distance y from a fixed line. These solutions correspond to a flow that is always close to a state of geostrophic equilibrium, but the equilibrium keeps changing with time because of the forcing. This progression through a series of geostrophic equilibria is typical of the slower adjustment processes in the ocean and atmosphere, and therefore of great importance.

An early example of this type of adjustment was the calculation by Hough (1897) of the barotropic response of the ocean to evaporation/precipitation differences that are a function of latitude only. His calculations were for motion on a sphere, but can be illustrated by the simple case of flow on an f-plane with forcing a function only of y. Then (9.9.14) and (9.9.15) have a simple solution for which the adjusted sea level η' and the zonal velocity u change linearly with time, but the meridional velocity v is independent of time. Thus (9.9.14) and (9.9.15) reduce to

$$\partial u/\partial t = fv, \tag{9.14.1}$$

$$fu = -g\,\partial\eta/\partial y, \tag{9.14.2}$$

$$\partial\eta/\partial t + H\,\partial v/\partial y = -\partial\eta^F/\partial t = (P - E)/\rho. \tag{9.14.3}$$

The subscript p can be dropped because there is no wind effect and the prime can be dropped because there is no pressure forcing. For this solution, (9.14.2) shows that there is an exact geostrophic balance for the zonal-flow component, which is the dominant one because it continually increases with time, but the meridional component is *not* in geostrophic balance with a pressure gradient.

If v and u are eliminated from Eq. (9.14.1)–(9.14.3) the result is

$$\eta - a^2\,\partial^2\eta/\partial y^2 = -\eta^F = \rho^{-1}(P - E)t, \tag{9.14.4}$$

where a is the Rossby radius equal to $(gH)^{1/2}/f$. Suppose, for instance, that evaporation occurs over a strip of ocean given by $|y| < L$, i.e., that

$$-\rho\eta^F/t = P - E = \begin{cases} -E_0 & \text{for } |y| < L, \\ 0 & \text{otherwise.} \end{cases} \tag{9.14.5}$$

The solution is simply given by

$$\eta = \begin{cases} -\rho^{-1}E_0 t(1 - e^{-L/a}\cosh y/a) & \text{for } |y| < L, \\ -\rho^{-1}E_0 t \sinh(L/a)e^{-|y|/a} & \text{for } |y| > L, \end{cases} \tag{9.14.6}$$

and is illustrated in Fig. 9.11. Where there is evaporation, the surface is lowered because of the removal of water, and this effect is spread out by the adjustment process over a distance of the order of the Rossby radius. The surface continues to fall at a uniform rate. A geostrophic zonal current (i.e., with low pressure to the left in the northern hemisphere) results, as given by (9.14.2), but there is also a meridional

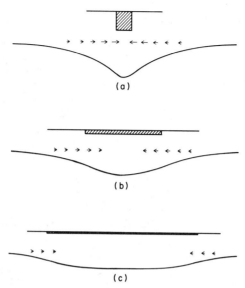

Fig. 9.11. Solution of the type found by Hough (1897) for the effect of evaporation in a homogeneous ocean. The diagram shows the result of uniform evaporation over a strip of ocean of half-width equal to (a) $\frac{1}{5}$, (b) 1, and (c) 2 Rossby radii, the *total* rate of evaporation being the same in each case. The upper part of each section shows the negative of the evaporation rate, whereas the lower part shows the rate of change of surface elevation, which also represents the surface configuration at a given time. There is a current normal to the section that is in geostrophic equilibrium with the surface slope and that increases in proportion with time. In addition there is an isallobaric current whose strength is represented by the arrows. This current does not change with time.

current which, by (9.14.1) and (9.14.2), is given by

$$v = -gf^{-2}\, \partial^2 \eta / \partial y\, \partial t, \tag{9.14.7}$$

i.e., is an isallobaric current (see Section 8.16) flowing toward low pressure (since this is also a low in the rate of change of pressure).

An interesting limiting case is that in which the strip over which the evaporation takes place has half-width L small compared with the Rossby radius a. If $E_1 = 2E_0 L$ is the rate of evaporation per unit length of the strip, then (9.14.4) becomes, in the limit as $L/a \to 0$,

$$\frac{\partial}{\partial t}\left\{\eta - a^2 \frac{\partial^2 \eta}{\partial y^2}\right\} = \frac{-\partial \eta^{\mathrm{F}}}{\partial t} = -\rho^{-1} E_1 \delta(y), \tag{9.14.8}$$

where δ is the Kronecker delta function and the right-hand side has the required integral $-E_1/\rho$ over the whole y domain. If (9.14.8) itself is integrated across the strip, a condition on the discontinuity $[\partial \eta / \partial y]$ in the derivative of η results, namely,

$$\frac{\partial}{\partial t} a^2 \left[\frac{\partial \eta}{\partial y}\right] = \rho^{-1} E_1, \tag{9.14.9}$$

since the integral of η across the strip vanishes in the limit as $L/a \to 0$. Thus η must be such that the left-hand side of (9.14.8) is zero away from the strip, must vanish as

$|y| \to \infty$, and must satisfy (9.14.9). In other words, the solution is

$$\eta = - \tfrac{1}{2}(a\rho)^{-1}E_1 te^{-|y|/a}, \tag{9.14.10}$$

and this is the limiting form of (9.14.6) as required. Note that (9.14.9) implies, by (9.14.2), a discontinuity in velocity across the strip. In other words, the concentrated sink of mass produces a *vortex sheet*. Figure 9.11a shows a case that is close to this limit.

9.15   A Forced Baroclinic Vortex

The solution found in Section 9.14 was an example of motion produced by stationary forcing when only one mode, the barotropic mode, was present. These solutions, however, can easily be generalized to give solutions for baroclinic forced motion, as will be demonstrated shortly.

The problem of *slow* forcing of baroclinic motion was considered by Eliassen (1952) in a paper entitled "Slow thermally or frictionally controlled meridional circulation in a circular vortex." He considered the particular case for which heat sources and sinks (and sources of angular momentum) are arranged symmetrically about an axis. He had particularly in mind the case for which this axis is the axis of the earth, so that the "vortex" is then a zonal flow and the rising motion due to heating generates a meridional circulation. The model can then be used to discuss the general circulation of the atmosphere. The ideas can, however, also be applied to smaller-scale phenomena such as hurricanelike vortices in a uniformly rotating fluid (f-plane).

By "slow" forcing, Eliassen (1952, p. 23) had in mind the following:

> We shall simplify the problem by assuming the sources of heat and angular momentum to be weak, and to change so slowly with time, that resonance phenomena will not occur. The resulting meridional currents may then be considered as being so slow that the accelerations due to these currents are small compared to the centripetal accelerations. The vortex will be very close to the state of balance all the time, so that we may apply the quasi-static approximation: we assume the vortex to be in the state of balance at all times, and determine the meridional motion necessary to maintain this balance. We shall see that the requirement of the maintenance of the balance is sufficient to determine the meridional motion uniquely.

Eliassen (1952) went on to derive the general equation for this sort of motion, and found particular solutions for point sources of heat (or of angular momentum). The theory is further discussed by Charney (1973). In this section, solutions will be constructed for *small* perturbations from a state of rest of a *uniformly rotating* incompressible fluid of constant buoyancy frequency N for circumstances in which the Boussinesq approximation can be made. The equation to be solved is (9.13.13), and the method is to generalize solutions obtained in the last section to baroclinic motion by the device of superposition of modes, as discussed in Section 9.10.

The example chosen is one in which the forcing is by a *buoyancy* force concentrated in a narrow region of the atmosphere. Such forcing occurs, for instance, by latent heat

release in the inner portion of a hurricane. This corresponds to forcing that is concentrated around a vertical line, and consequently solutions with *axial* symmetry can
be obtained. However, for illustrative purposes, it is more convenient to take the case
with forcing concentrated at a *plane* $y = 0$ since simpler functions are involved,
whereas the character of the solution is not very different from the axisymmetric case.

Since N is constant, the normal modes are sinusoids (6.11.21) and their
superposition has the form of a Fourier integral such as (6.12.4). Consequently, it is
appropriate to express the buoyancy forcing as a Fourier integral as well, and the
general form, when this forcing is concentrated along a line $y = 0$, is

$$B'_s = \delta(y) \int_0^\infty B_0(m) \sin mz \, dm. \tag{9.15.1}$$

The corresponding form of the response [cf. (6.11.21) and (6.12.2)] is

$$p'(y, z, t) = \int_0^\infty \tilde{\eta}(y, t; m) \cos mz \, dm, \tag{9.15.2}$$

and substitution of this in the governing equation (9.13.13) leads [cf. (9.10.12)] to the
corresponding equation for a single mode, namely (when $\partial/\partial t \ll f$),

$$\frac{\partial}{\partial t} \left(\frac{\partial^2 \tilde{\eta}}{\partial y^2} - \frac{f^2 m^2}{N^2} \tilde{\eta} \right) = \frac{f^2 m}{N^2} B_0(m) \delta(y). \tag{9.15.3}$$

The form of forcing has been chosen deliberately to give this the same form as (9.14.8)
and thus the solution has the same form as that of (9.14.10)! This can be substituted
in (9.15.2), therefore, to give the general baroclinic solution for flow forced by a source
of buoyancy concentrated along the line $y = 0$, namely,

$$p' = -\frac{1}{2} N^{-1} ft \int_0^\infty B_0(m) \exp(-fm|y|/N) \cos mz \, dm. \tag{9.15.4}$$

The solution is particularly simple for some special choices of $B_0(m)$ such as
[cf. (6.12.5)]

$$B_0 = 2B_1 D \exp(-mD), \tag{9.15.5}$$

which corresponds to the case for which

$$B'_s = \delta(y) 2B_1 Dz/(D^2 + z^2). \tag{9.15.6}$$

This has maximum buoyancy force at height $z = D$, and the solution (9.15.4) in this
case is given by

$$p' = -\frac{ft}{N} \frac{B_1 D(D + f|y|/N)}{(D + f|y|/N)^2 + z^2}. \tag{9.15.7}$$

The solution is illustrated in Fig. 9.12 and can be described as follows. The delta
function in (9.15.6) can be interpreted as a concentration of the buoyancy force in a
region of small horizontal extent, in which the vertical motion will be very large, and

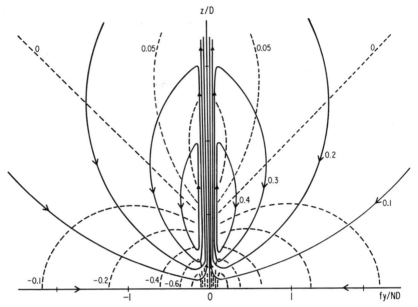

Fig. 9.12. Solution for a buoyancy source concentrated near the y axis. The buoyancy force on the axis varies with height, increasing from zero at the ground to a maximum at $z = D$ (where the dashed line labeled 0 intersects the z axis) and then falls off toward zero as $z \to \infty$. The upward motion along the axis is in proportion with the buoyancy force. The compensating downward flow away from the axis is slow and steady and is a potential flow in a suitably scaled coordinate system. The descending air warms at a steady rate, causing the pressure to fall at a steady rate. The motion toward the axis is an isallobaric wind in balance with the falling pressure. In flowing toward the axis, the fluid acquires momentum normal to the page at a steady rate, contours at a given time being marked by dashed lines. The minus sign corresponds to cyclonic rotation relative to the z axis. In the notation used in the text, the unit used for the stream function is B_1/N^2 and that used for the x component of velocity is $B_1 t/\rho_0 N^2 D$.

therefore (9.13.5) is approximated by

$$\rho_0 N^2 w = B_s'. \tag{9.15.8}$$

In other words, the vertical motion is proportional to the buoyancy force, and therefore the vertical mass flux in the narrow region can be precisely calculated; consequently, the inflow necessary to maintain the given updraft is also known, and this sets a condition on the flow outside the narrow region in which the forcing is concentrated.

   In the outer domain, there is no forcing and p' is proportional to time, so (9.13.13) reduces to the potential flow equation

$$\partial^2 p'/\partial y^2 + \partial^2 p'/\partial z_s^2 = 0 \tag{9.15.9}$$

if the stretched coordinate z_s defined by (8.8.25) is used. The potential flow solution is **uniquely determined by the requirements of given inflow along $y = 0$, no flow** across $z = 0$, and decay as $y \to \infty$. The solution is the one illustrated in Fig. 9.12 and happens to be the same as that for a point source, but with the source located outside the region of physical interest.

The flow in the y, z plane can be represented in terms of a stream function ψ, whose existence follows from the incompressibility condition. The relationships between p' and ψ follow from (9.13.5) and an equation equivalent to (9.14.7), namely,

$$\partial\psi/\partial y = \rho_0 w = N^{-2}B_s' - N^{-2}\,\partial^2 p'/\partial z\,\partial t,$$
$$\partial\psi/\partial z = -\rho_0 v = f^{-2}\,\partial^2 p'/\partial y\,\partial t. \tag{9.15.10}$$

It follows that the solution for ψ that corresponds to (9.15.7) is

$$\psi = \frac{(B_1 Dz/N^2)\ \text{sgn } y}{(D + f|y|/N)^2 + z^2}. \tag{9.15.11}$$

Thus buoyancy drives the fluid upward in the central region and produces a steady circulation in the y, z plane, as given by (9.15.11). This circulation displaces surfaces of constant potential temperature downward outside the central core, i.e., at any given point the air is subsiding and warming up at a constant rate. As a result of this warming the pressure falls at a constant rate, as given by (9.15.7). The wind toward the central region is an isallobaric wind, but is not caused by the dropping of the pressure. Rather, it is part of the circulation driven by the buoyancy force, and it is this circulation that causes the pressure to drop!

Associated with the pressure gradient thus produced are winds normal to the plane of symmetry that, because of the form of the solutions, are always in balance with the pressure gradient. Thus the solution that corresponds to "slow" adjustment, as Eliassen envisaged, happens to be an exact solution in this case. Contours of u are shown by dashed lines in Fig. 9.12 and are quite like those for the swirl velocity in a hurricane, with values increasing toward the center and hence reaching a maximum just outside the core region.

In fact, solutions for buoyancy forcing concentrated along an axis $r = 0$ look to be similar to that shown in Fig. 9.12 if the streamlines are interpreted as those for flow in the vertical plane through the axis and if the dashed lines are interpreted as those of swirl velocity (with negative values corresponding to cyclonic rotation). There is strong cyclonic rotation near the ground, with greatest intensity near the center of the storm, whereas there is a weaker anticyclonic rotation aloft. In the axisymmetric case, this rotation can be explained by the *conservation of angular momentum* principle. Each ring of fluid has angular momentum even when at rest because of the rotation of the whole medium. If the ring contracts through motion toward the center, the absolute angular velocity of the ring increases in order to conserve angular momentum. Thus the ring acquires cyclonic relative rotation. This can be seen in Fig. 9.12, in which cyclonic rotation is found whenever the motion is toward the center and anticyclonic motion is found whenever the motion is outward.

Information that is obtained by combining observations from large numbers of tropical storms and hurricanes shows many features in common with those of the above solution. For instance, the inflow [see Gray (1979)] toward the storm center proceeds at a fairly constant rate throughout the development stages of the hurricane. The average rate of convergence in the region up to a radius of 600 km is found to have a fairly uniform value of about $5 \times 10^{-6}\ \text{s}^{-1}$ up to a level of 400 mbar. A picture of this circulation in the vertical plane through the storm center is given in Fig. 9.13a.

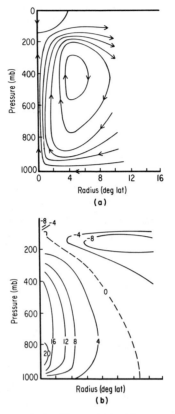

Fig. 9.13. The flow in a hurricane as obtained by combining observations from a large number of them. (a) Streamlines of flow in the vertical plane through the axis of the hurricane. The contour interval is $\frac{1}{3}$ Mt s^{-1}. [Calculated using Fig. 48 of Gray (1979).] (b) Contours of swirl velocity in meters per second. [From Fig. 50 of Gray (1979).]

The total mass flux involved in this circulation is about 2 Mt s^{-1}. The swirl velocity, on the other hand, builds up over a period of days as the linear solution suggests. It does not build up without limit, however, and in the mature stage the structure is like that shown in Fig. 9.13b. In the very center, a complicated substructure (not shown in the figure) develops with a central "eye" where descending air, and hence very little cloud, is found. The eye has a radius of 20–80 km and is bounded by the eye wall, which is marked by very tall clouds. This is the region of strong upward motion, and the maximum swirl velocities are found close to the eye wall. Details are described by Shea and Gray (1973) and Gray and Shea (1973). A general discussion and description of tropical storms can be found in the textbook on tropical meteorology by Riehl (1979). A description is also given by Scorer (1972).

Effects similar to those discussed above can also be found in regions of the ocean where there is concentrated sinking due to cooling. This causes inflow near the ocean surface and hence a cyclonic circulation. Such a circulation is found, for instance, in the Greenland Sea, where a large amount of sinking takes place. A simple model of this situation is discussed by Gill *et al.* (1979).

9.16   Equilibration through Dissipative Effects

When the forcing is stationary and there are no dissipative effects, the solutions of Section 9.15 show that the response grows continually with time. This is an inevitable consequence of the equations; e.g., (9.13.13) shows that if the forcing is fixed, the quantity in curly brackets on the left-hand side must continually grow with time. In practice, effects neglected in this analysis must become important and limit the magnitude of the response. The nature of the response can depend strongly on what equilibrating mechanism is acting.

In this section, some simple solutions will be obtained to illustrate equilibrium situations. They are simple because the mechanisms that limit the response are chosen to be linear, and for that reason may be somewhat artificial.

The simplest examples are for barotropic flow, for which vertically averaged equations can be used. Consider, for instance, the flow in a model ocean confined to a zonal channel and driven by an eastward wind stress X. Such a situation has been studied as a model of the Antarctic Circumpolar Current. In the absence of friction, the behavior would be rather like Hough's solution discussed in Section 9.14, with currents increasing linearly with time. Friction, however, can limit growth, as shown by Munk and Palmén (1951), who used lateral friction, and a similar model can be constructed by using bottom friction as the equilibrating agent. In that case, equilibrium is reached in a time of the order of the spin-up time given by (9.12.6). The steady state consists of an eastward current u of such magnitude that, as the first of Eqs. (9.9.10) requires, the bottom stress balances the surface stress. If the laminar formula (9.6.4) is used to calculate the stress, then the current u is given by

$$(\tfrac{1}{2}fv)^{1/2}u = X/\rho. \tag{9.16.1}$$

This is a very simple solution that illustrates the principle very well, but is not, in practice, applicable to the Antarctic Circumpolar Current, in which other (not clearly identified) equilibrating processes are active.

A baroclinic example may be constructed in a similar way by considering a buoyancy-driven flow in a compressible atmosphere with Newtonian cooling and Rayleigh friction as equilibrating agents. Newtonian cooling (see Section 8.11) is a linear representation of radiation effects that tend to restore equilibrium, and is incorporated in (9.13.15), which represents buoyancy effects by replacing the time derivative $\partial/\partial t$ by the operator $(\partial/\partial t + \alpha)$, α^{-1} being the time constant of the temperature-equilibrating process. A similar device can be used in the momentum equations, in which $\partial/\partial t$ can be replaced by the operator $(\partial/\partial t + r)$, where r^{-1} is the time constant of the decay process, which in this case is called Rayleigh friction. Although somewhat artificial, it allows the arguments of Halley (1686) and Hadley (1735) to be put into a simple analytical form, as shown in the following.

Consider a small perturbation to a compressible atmosphere with constant N_*, the perturbation being forced by a buoyancy term that depends only on y and z_*. The steady-state equivalent of (9.13.15), with Newtonian cooling included, is then

$$\alpha N_*^{-2}\, \partial \Phi''/\partial z_* + w_* = \rho_*^{-1} N^{-2} B_{\rm s}'. \tag{9.16.2}$$

The steady-state form of the momentum equations in the absence of x dependence is

$$ru - fv = 0, \tag{9.16.3}$$

$$rv + fu = -\partial\Phi''/\partial y, \tag{9.16.4}$$

whereas the continuity equation (6.17.11) allows the introduction of a stream function ψ defined by

$$\rho_r \sigma w_* = \partial\psi/\partial y, \qquad H_s\rho_r v = \partial\psi/\partial\sigma \tag{9.16.5}$$

when the σ coordinate is defined [see (6.17.8)] by

$$\sigma = p/p_r = e^{-z_*/H_s}. \tag{9.16.6}$$

If these equations are reduced to a single equation for ψ, the result is

$$\sigma^2 \frac{\partial^2\psi}{\partial\sigma^2} + \frac{N_*^2 H_s^2}{(f^2 + r^2)} \frac{r}{\alpha} \frac{\partial}{\partial y}\left(\frac{\partial\psi}{\partial y} - \frac{B_s'}{N^2}\right) = 0. \tag{9.16.7}$$

The effects of horizontal variations in heating rate can be modeled by assuming that the diabatic heating (which is proportional to B_s'/σ) varies sinusoidally with y, whereas the vertical variations can be modeled by assuming that B_s'/σ varies as $\sigma(\sigma - 1)$. This gives a maximum in the diabatic heating rate near $\sigma = \frac{1}{2}(500 \text{ mbar})$ as observed (see Fig. 9.10). In other words, the forcing is assumed to be of the form

$$B_s' \propto \sigma^2(\sigma - 1) \cos ly. \tag{9.16.8}$$

The solution of (9.16.7) in this case has the form

$$\psi \propto \left\{ \frac{\sigma^3}{6 - \gamma(\gamma + 1)} - \frac{\sigma^2}{2 - \gamma(\gamma + 1)} + \frac{A\sigma^{\gamma+1}}{\gamma + 1} \right\} \sin ly, \tag{9.16.9}$$

where

$$\gamma(\gamma + 1) = N_*^2 H_s^2 l^2 (f^2 + r^2)^{-1} r\alpha^{-1} \tag{9.16.10}$$

and A is determined by the boundary condition at the ground ($\sigma = 1$). The linear bottom friction condition (9.6.5) in this case can be written

$$w_* = f^{-1}(v/2f)^{1/2} \partial^2\Phi''/\partial y^2,$$

which, by (9.16.5), implies that

$$\psi = f^{-1}\rho_r(v/2f)^{1/2} \partial\Phi''/\partial y \qquad \text{at} \quad \sigma = 1. \tag{9.16.11}$$

If this condition is applied, the resulting formula for the coefficient A is

$$A = \frac{6\epsilon + \gamma(\gamma + 1)(4 + \epsilon)}{(\epsilon + \gamma)[6 - \gamma(\gamma + 1)][2 - \gamma(\gamma + 1)]}, \tag{9.16.12}$$

where

$$\epsilon = N_*^2 H_s l^2 \alpha^{-1} f^{-1}(v/2f)^{1/2} \tag{9.16.13}$$

is a parameter that measures the importance of bottom friction.

Solutions are shown in Fig. 9.14 for a small value (0.2) of γ, this being thought reasonable for the large-scale atmospheric circulation. Although the primary balance in (9.16.2) is between Newtonian cooling and the heating term, it is still true, as Halley (1686) stated, that the air that is being heated becomes "less ponderous" and rises. Being a linear system, the reverse is true when the air is being cooled. The "meridional" circulation (i.e., that in the vertical plane along the y axis) that results in the absence of bottom friction is shown in Fig. 9.14a. If bottom friction is added (Fig. 9.14b), part of this circulation is taken up by the boundary layer.

The other aspect of the solution is the tendency to conserve angular momentum, which Hadley (1735) drew attention to. This is reflected in the solution in that wherever the flow is equatorward [see (9.16.3)], easterly winds are produced, and thus easterly "trade" winds (with an equatorward component added) are found in the lower troposphere. Conversely, westerlies are found aloft, where the flow is poleward. In the case for which bottom friction is present, the easterlies near the ground are reduced in magnitude and westerlies tend to dominate the picture.

Although the solution (9.16.9) does not allow for the variation of the Coriolis parameter with latitude, it gives a reasonable simulation of what is now known as the Hadley circulation. If the equilibrating agents in the atmosphere were linear, as in the above solution, this Hadley cell would cover the whole globe, i.e., the left-hand side of

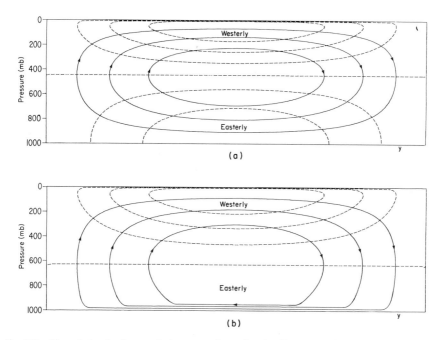

Fig. 9.14. The solution for large-scale buoyancy-driven flow (Hadley cell) in the atmosphere. The diabatic heating rate varies parabolically with pressure, being zero at the ground and at $p = 0$ and a maximum at 500 mb. The variation with y is sinusoidal, and the figures show half a wavelength with heating rate a maximum on the left and cooling rate a maximum on the right. Equilibration is by Newtonian cooling and Rayleigh friction, the parameter γ (see text) being given a value of 0.2. In case (a) there is no bottom friction, whereas in case (b) the bottom friction parameter ϵ (see text) is 0.3. The solid lines are contours of the stream function with contour interval a quarter of the maximum value. The dashed lines are contours of the x component of velocity (normal to the page) with contour interval a quarter of the maximum speed. The horizontal dashed line corresponds to this component being zero.

the figure would represent the equator and the right-hand side the pole. The wind in the lower troposphere at all levels would be a steady easterly wind. [Laboratory simulations of this type of circulation go back to Vettin (1857) and are reviewed by Fultz *et al.* (1959).] In reality, of course, the circulation is not like this because other equilibrating agents are at work. In particular, the above solution has large north–south temperature gradients at mid-latitudes. As shall be seen later, this situation is not a stable one, and therefore small disturbances at mid-latitudes grow and develop into the familiar pattern of cyclones and anticyclones. These disturbances completely alter the meridional circulation in mid-latitudes, and in practice the Hadley cell is found only in the tropics (see Fig. 1.7).

Chapter Ten

Effects
of Side Boundaries

10.1 Introduction

Rotation effects have thus far been studied in the absence of boundaries at the sides. A key feature of rotating-fluid behavior is the geostrophic adjustment process in which pressure and velocity fields adjust to each other in order to reach a geostrophic balance. When the balance is achieved, the flow at any level is along the isobars. If now a boundary is inserted that crosses the isobars, further adjustment would have to take place because no flow is possible across the boundary. This indicates that the adjustment process is strongly affected by the presence of **boundaries**—at least in the neighborhood of those boundaries.

In fact, the presence of a boundary implies that the *longshore* component of the Coriolis acceleration vanishes at the boundary so that the mutual adjustment of the longshore velocity field and the pressure field along the boundary is more like that in a nonrotating fluid than like that in a rotating one. This is certainly true in the extreme case (studied in Section 10.2) in which there are two boundaries close together, as in a narrow gulf or estuary. The rotation effects can be neglected at the first approximation because the motion is mainly along the gulf and the component of Coriolis acceleration in this direction is negligible. At the next order of approximation, rotation modifies the flow in two ways. One is to give a cross-channel *pressure gradient* in order to geostrophically balance the longshore flow, just as concluded by Combes (1859), following the discussion in the French Academy about rotation effects on river flow (see Section 7.6). The other is to produce a *shear* whenever the surface elevation departs from its equilibrium level, this being required in order that potential vorticity be conserved. The narrow-channel approximation can be applied with success to studies of tides and seiches in gulfs, estuaries, and lakes, and even to tides in the Atlantic Ocean.

When the two sides of a channel are *not* close together, the question arises as to how far from the shore the longshore component of the Coriolis force can be neglected. The answer is a distance of the order of the Rossby radius of deformation, so channels must have width small compared with this scale for the narrow-channel approximation to be valid. For wide channels, there is a special form of adjustment near the boundary by means of a wave whose amplitude is only significant within a distance of the order of the Rossby radius from the boundary. This is called a Kelvin wave, and its properties are studied in Section 10.4. A peculiarity of this wave is that it can travel along the coast *in one direction only*, i.e., with the coast on the right in the northern hemisphere and on the left in the southern hemisphere. The reason is that the longshore component of flow is in geostrophic balance with the pressure field, and this can only decay with distance from the shore when the wave propagation is in the appropriate direction.

The presence of boundaries also affects the Poincaré waves, and solutions for a channel are studied in Sections 10.3 and 10.5. If the channel is wide compared with the Rossby radius, effects of the end of the channel can be quite difficult to work out, and these are discussed in Section 10.6. Of particular interest is the behavior of a Kelvin wave at the channel end, and this problem was studied by Taylor (1921) in connection with seiches and tides in the North Sea. A convenient way of describing the solution is in terms of a single Kelvin wave that propagates around the corners at the end of the channel without loss of energy, but with adjustments in phase produced by the process of turning the corner. This idea is used later (Section 10.10) in the study of storm surges.

The adjustment to equilibrium in a channel has some interesting features which are discussed in Section 10.7. In particular, for a wide channel with an initial discontinuity in level, the adjustment *away* from the boundaries is not influenced by the boundaries very much. However, Kelvin waves carry the effects of the discontinuity along the boundaries, but only on one side (the right side, traveling with the wave in the northern hemisphere). Thus when the adjustment is completed, the current that is set up along the line of the initial discontinuity deflects to the left when it reaches the boundary and then continues as a current along the boundary. Upstream of the original discontinuity, the current travels along the opposite bank of the channel.

The dynamics of tides are discussed briefly in Section 10.8. An equilibrium theory is *not* appropriate to the diurnal and semidiurnal constituents because the speed with which the equilibrium tide moves around the earth happens to be similar in magnitude to the speed of long gravity waves. Details are strongly influenced by the complicated shape of the world's ocean, and there is evidence that the semidiurnal tide has period close to that of natural modes of oscillation of the ocean. This means that free waves play an important part in any local description of the tides. In the Pacific Ocean, for instance, Kelvin waves are a major element. The North Atlantic, on the other hand, is thin enough for it to be marginally valid to use the narrow-channel approximation.

The wind produces some very important effects at boundaries. In particular, the longshore component of the wind stress causes an Ekman flow toward or away from the coast. In shallow seas, flow toward the coast causes piling up of water there and hence abnormally high sea levels. This phenomenon is called a storm surge, and severe cases in the past have led to over a million lives being lost, caused great destruction of property, and played a major part in determining the shape of many coastal features. The process by which these surges are formed and sample solutions are discussed in Sections 10.9 and 10.10.

The response to Ekman flow normal to the coast in the deep ocean also has important consequences, especially when the Ekman flow is away from the coast. The water carried seaward is light surface water and this is replaced near the shore by upwelling of dense cold nutrient-rich water from lower levels. When the nutrients are brought into the sunlit zone near the surface, rapid growth of microorganisms occurs, and this leads to such areas being important fisheries. In fact, half of the world's fisheries are located in such regions, which occupy only a tiny fraction of the ocean. The subject of upwelling is introduced in Section 10.11.

Upwelling is not a purely local response to the wind, however, because "coastally trapped waves" that are confined to the near-shore region carry information along the coast. A peculiarity of long waves of this type is that they propagate only in one direction, i.e., poleward on eastern boundaries and equatorward on western boundaries. The response of the coastal region to the wind can be calculated by resolving the driving force into a sum of such waves, then solving a simple equation for the amplitude of each wave and resynthesizing. In practice this has been done in a crude way by assuming that a single wave dominates the response. This matter, and the wave properties, is discussed in Sections 10.12 and 10.13.

The final section is about the boundary currents found on the eastern edges of the Atlantic and Pacific Oceans. These are of some interest, particularly through their relationship with upwelling, which can vary a great deal throughout the year. A comprehensive theory is not given, but various processes that are pertinent to their behavior are discussed.

10.2 Effects of Rotation on Seiches and Tides in Narrow Channels and Gulfs

Seiches and tides in sufficiently narrow channels and gulfs were discussed in Section 5.8, the narrowness being sufficient for rotation effects to be ignored. The solutions were found on the assumption that the motion is everywhere parallel to the axis of the channel. Such solutions have been found to give good approximations to the behavior of seiches and tides in many channels, gulfs, estuaries, and lakes. Suppose that this solution has the form

$$u = u_{nr}(x, t), \qquad \eta = \eta_{nr}(x, t), \tag{10.2.1}$$

where the subscript nr stands for nonrotating.

Now consider what effects rotation will have when these effects are small. First of all, there will be a Coriolis acceleration fu_{nr} directed across the channel, which must be balanced by a small surface slope, i.e., the approximate form of the y

component of the momentum equation is

$$fu = -g \, \partial\eta/\partial y. \tag{10.2.2}$$

Choosing the origin of y suitably (somewhere near the center of the channel—the exact position might have to be determined in retrospect), this integrates to give

$$\eta \approx \eta_{nr}(x, t) - g^{-1}fu_{nr}(x, t)y \tag{10.2.3}$$

for the corrected surface elevation. The condition for the correction term to be small relative to η_{nr} is, using the fact [see, e.g., (5.8.6)] that u_{nr} is of order $(g/H)^{1/2}$ times η_{nr}, that

$$g^{-1}f(g/H)^{1/2}W \ll 1, \qquad \text{i.e.,} \quad W/a \ll 1, \tag{10.2.4}$$

where H is the water depth, W the channel width, and a the Rossby radius defined by (8.2.3). In other words, the condition for rotation effects to be small is that the *width* of the channel be *small compared with the Rossby radius.*

A second effect of rotation comes from potential vorticity equation (7.2.8), which shows that vorticity

$$\zeta = -(f/H)\eta \tag{10.2.5}$$

is created whenever the surface is elevated or depressed. This requires some cross-channel shear, i.e., (10.2.5) becomes

$$\partial u/\partial y = f\eta/H, \tag{10.2.6}$$

which integrates to give

$$u = u_{nr}(x, t) + f\eta_{nr}(x, t)\int dy/H. \tag{10.2.7}$$

Again, the correction term is relatively small if (10.2.4) is satisfied, provided that the channel has vertical sides so that the depth H is not small anywhere, i.e., the depth is everywhere of the same order as that of the value used in (10.2.4) to compute the Rossby radius.

If necessary, further corrections can be found by developing the solution as an expansion in powers of the small ratio W/a. However, the main interest is in the first correction, and particularly in the form (10.2.3) for the corrected surface elevation. This can give a considerable improvement over the nonrotating solution with little increase in effort. The technique was used very successfully by Sterneck (1919) to calculate the tides of the Adriatic, which was divided into 40 sections for the purpose. There is little to distinguish the calculated results from observation. [Sterneck's results are reproduced by Defant (1961, Vol. 2) in his Fig. 169, which shows η_{nr}, and Fig. 170, which shows the fundamental seiche, which he found to be 23 hr.] Calculations for many other special cases are reported in Chapter 12 of Defant (1961, Vol. 2).

The interesting new feature added by rotation is that the crest of the tide, as given by (10.2.3), now moves cyclonically around the basin. This is conveniently illustrated by an analytic solution that gives a fair approximation to the northern end of the

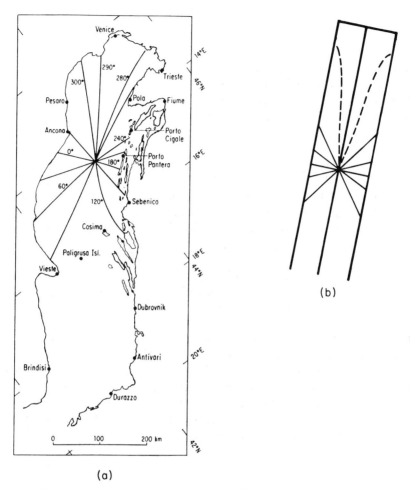

Fig. 10.1.   (a) Co-tidal lines for the northern Adriatic. [After Polli (1960); from Hendershott and Speranza (1971, Fig. 7).] (b) Co-tidal lines for a simple model with depth increasing quadratically with distance from the end. The phase difference between the solid lines is 30°. The phase on the broken lines differs by 10° from that on the axis.

Adriatic, namely, a channel of uniform width $W = 135$ km and depth H that increases with distance x along the channel according to the formula

$$H = \gamma x^2, \tag{10.2.8}$$

where $\gamma = 6.5 \times 10^{-10}$ m^{-1}. The end of the channel (Venice) is situated at $x = x_0 = 150$ km, and the model Adriatic occupies the region $x > x_0$. A median value of the Rossby radius is 250 km, so the ratio W/a is about 0.5. The solutions of (5.8.4) for this geometry and fixed frequency ω have the form of complex powers of x, giving rise to the physical solution

$$\eta_{nr} = (x_0/x)^{1/2}\eta_0\{\beta\,\cos(\beta\,\ln(x/x_0)) + \tfrac{1}{2}\sin(\beta\,\ln(x/x_0))\}\,\cos\,\omega t,$$

$$u_{nr} = (\omega/\gamma x_0)(x_0/x)^{3/2}\eta_0\,\sin(\beta\,\ln(x/x_0))\,\sin\,\omega t, \tag{10.2.9}$$

where

$$\tfrac{1}{4} + \beta^2 = \omega^2/g\gamma.$$

Verification of his result is left as an exercise for the reader. Since $\omega = 1.4 \times 10^{-4}\,\text{s}^{-1}$ and $g = 10\,\text{m s}^{-2}$, $\beta = 1.67$ in this case.

The usual method of displaying tidal variations is in terms of the amplitude A and phase δ, i.e., η is expressed in the form

$$\eta = A \sin(\omega t - \delta). \tag{10.2.10}$$

Contours of A are called *co-range lines*, on which the range $2A$ is given, say, in meters. Contours of δ are called *co-tidal lines* and the phase is given either in degrees or as the time of high water in hours. For instance, Fig. 10.1a shows the observed co-tidal lines for the northern Adriatic. Figure 10.1b shows the corresponding diagram from the model, the equation for the co-tidal lines being, by (10.2.3) and (10.2.9),

$$y \sin \delta = \{\beta \cot(\beta \ln(x/x_0)) + \tfrac{1}{2}\}(x\gamma g/f\omega) \cos \delta. \tag{10.2.11}$$

The more detailed calculation of Sterneck (1919) gives much better agreement, of course, but the simple analytical model illustrates the main features very well. Further discussion of this particular case may be found in Hendershott and Speranza (1971).

The *seiches* in the Adriatic are of importance because they are responsible for the flooding of Venice (Robinson *et al.*, 1973). The fundamental seiche has a period of 22 hr, rather than Sterneck's value of 23 hr, according to more recent calculations, and this seiche is stimulated at times of strong winds. The effect on Venice depends a great deal on whether or not the times of maximum seiche are close to the times of high tide. A model of a wind-produced seiche can be constructed by using the forced shallow-water equations introduced in Chapter 9.

10.3   Poincaré Waves in a Uniform Channel of Arbitrary Width

Although, as Kelvin remarked, the wave referred to in Section 8.2 as a plane progressive Poincaré wave cannot satisfy the boundary conditions at a coast, combinations of such waves can. In particular, this is true of two obliquely directed waves [cf. (5.3.9)] of equal amplitude η_0, which could result, for instance, from reflection of a plane wave at a straight boundary. It is convenient to choose the x axis in the direction of the bisector of the angle between the two wavenumber vectors, and to choose the origin of y as a line of symmetry for v, i.e., a line on which $\partial v/\partial y$ vanishes. The combination of the two waves of the form (8.2.11) can be constructed as in (5.3.9), but with a shift in origin of y so that $\partial v/\partial y$ will vanish at $y = 0$. Then the solution is

$$\eta = (2\eta_0/\kappa\omega_c)(kf \cos ly + \omega l \sin ly) \cos(kx - \omega t),$$

$$u = (2g\eta_0/\kappa\omega_c)(kl \sin ly + (\omega f/gH) \cos ly) \cos(kx - \omega t), \tag{10.3.1}$$

$$v = (2\omega_c\eta_0/\kappa H) \cos ly \sin(kx - \omega t),$$

where ω_c is a frequency defined by

$$\omega_c^2 = f^2 + l^2 gH, \qquad (10.3.2)$$

and hence the dispersion relation (8.2.7) can be written

$$\omega^2 = \omega_c^2 + k^2 gH. \qquad (10.3.3)$$

It follows that ω_c is the minimum frequency a wave with given l can have.

Such a wave (Poincaré, 1910) can satisfy the boundary conditions of no normal flow in a channel of uniform width W provided that l is an integer multiple of π/W, i.e.,

$$l = n\pi/W, \qquad n = 1, 2, 3, \ldots . \qquad (10.3.4)$$

For each n, there is a minimum frequency $\omega_c = \omega_{nc}$ that is necessary for propagation, where, from (10.3.2) and (10.3.4),

$$\omega_{nc}^2 = f^2 + n^2\pi^2 gH/W^2. \qquad (10.3.5)$$

This minimum frequency increases with n, so the smallest value is ω_{1c}. From (10.3.3), the dispersion characteristics for propagation along the channel have the same character as that for a plane Poincaré wave, i.e., ω varies with k as shown in Fig. 7.2, but with ω_c replacing f. Thus if $k^{-1} \ll (gH)^{1/2}/\omega_c$, the waves are only weakly dispersive and propagate along the channel at speed close to $(gH)^{1/2}$ (phase velocity slightly more; group velocity slightly less). If $k^{-1} \gg (gH)^{1/2}/\omega_c$, the waves have relatively small group velocity and the frequency is close to ω_c. When $k = 0$, the wave does not propagate, but is a standing wave spanning the channel.

Figure 10.2 shows an example of a progressive wave of the form (10.3.1). In the northern hemisphere ($f > 0$), the amplitude tends to be higher on the left side of the channel (for an observer moving in the direction of the wave), and the particle trajectories are ellipses. The orbital motion is anticyclonic except for a region near the right bank that occupies a fraction of the channel width that is always less than a half.

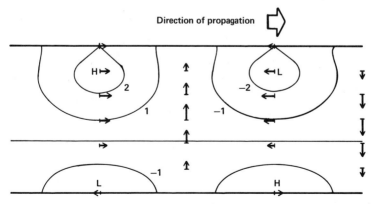

Fig. 10.2.   The progressive Poincaré wave in a channel of width π times the Rossby radius c/f. The cross-channel scale l^{-1} is equal to the Rossby radius, so the minimum frequency ω_c is $\sqrt{2}f$. The scale k^{-1} in the direction of propagation is $\sqrt{\frac{3}{2}}$ Rossby radii, so ω is $\sqrt{\frac{8}{3}}f$ and $\omega i = 2ki$. The signs are those for the northern hemisphere, so the greatest elevations are found on the left side of the channel (facing in the direction of propagation), where particles move anticyclonically. The nodal line is about 65% of the way across the channel. Contours are of surface elevation, and arrows indicate currents.

10.4   Kelvin Waves

In the limit of no rotation ($f = 0$), (10.3.1) does *not* give solutions like those discussed in Section 5.8, but gives what are called "cross-wave" modes, which vary sinusoidally across the channel. Such modes are sometimes found to be of significant amplitudes in lakes and gulfs, but are not usually as important as the "lower modes," i.e., modes that in the nonrotating limit correspond to motion with *no* cross-channel variations. How do these modes change when rotation is introduced? In Section 10.2, it was found that rotation causes u and η to vary with y, but this variation is *not* sinusoidal. Instead, solutions of the form discussed in Section 8.2 must be considered that have l^2 *negative* instead of positive. Of such solutions, only one is compatible with the condition of no normal motion across a straight boundary. This is the wave discovered by Kelvin (Thomson, 1879) and named the "Kelvin wave" in his honor.

Now consider the properties of a Kelvin wave. The variation with y is exponential since if v vanishes on one line $y = $ const (the boundary), it must vanish for other values of y as well. Hence the Kelvin wave solution can be found by putting $v = 0$ in the equations. In particular, (7.2.1) and (7.2.3) give

$$\partial u/\partial t = -g\, \partial \eta/\partial x, \tag{10.4.1}$$

$$\partial \eta/\partial t + H\, \partial u/\partial x = 0. \tag{10.4.2}$$

These determine η and u variations on any line $y = $ const, and contain no Coriolis terms, i.e., are identical with the equations (5.6.4) and (5.6.6) for *nonrotating* shallow-water motion when $v = 0$. Thus in the vertical plane of the bounding wall, and in any parallel vertical plane, the motion is exactly the same as that in a nonrotating system, i.e., a shallow-water gravity wave.

The general solution of (10.4.1) and (10.4.2) consists of the sum of two non-dispersive waves traveling in opposite directions, namely,

$$\begin{aligned}
\eta &= F'(x + ct, y) + G'(x - ct, y), \\
u &= -(g/H)^{1/2}\{F'(x + ct, y) - G'(x - ct, y)\},
\end{aligned} \tag{10.4.3}$$

where F' and G' are functions with properties to be determined and $c = (gH)^{1/2}$ as given by (7.2.5). The way the functions F' and G' vary with y can be found from the one remaining equation, namely, (7.2.2), which for $v = 0$ becomes

$$fu = -g\, \partial \eta/\partial y, \tag{10.4.4}$$

i.e., this component of motion is in geostrophic balance. Substitution from (10.4.3) gives

$$\partial F'/\partial y = f(gH)^{-1/2}F', \qquad \partial G'/\partial y = -f(gH)^{-1/2}G', \tag{10.4.5}$$

showing that one wave (the one represented by G' if $f > 0$) decays exponentially in the positive y direction, whereas the other decays exponentially in the negative y direction. This exponential behavior is very similar to that of the Lamb wave studied in Section 6.14, and so the Kelvin wave is also among the class of waves called boundary waves, edge waves, trapped waves, or surface waves.

If f is positive (northern hemisphere), the solution that decays in the positive y

direction is the one that travels in the positive x direction, i.e., the one that from (10.4.5) is given by

$$F' = 0, \qquad G' = e^{-y/a}G(x - ct), \tag{10.4.6}$$

where G is an arbitrary function. As Kelvin (Thomson, 1879, p. 97) remarked, "the velocity of propagation is the same" as for the non-rotating case, and "the influence of rotation is confined to the factor" $\exp(-y/a)$. "Many interesting results follow from the interpretation of this factor." The important property of this factor is, of course, the scale a, which is now called the Rossby radius of deformation (see Section 7.5), a being so defined by (7.2.23) or (8.2.3). Typical values of a (see Section 7.5) for barotropic Kelvin waves (which are important, for instance, in the description of tides) are 2000 km for the deep sea and about 200 km for coastal waters and shallow seas. Typical values of a for baroclinic Kelvin waves (which are of significance in the description of coastal upwelling) are about 30 km. It has also been suggested (Gill, 1977b) that "coastal lows" observed in the atmosphere are a form of Kelvin wave with a value of a of about 300 km.

The complete Kelvin-wave solution, obtained by substitution of (10.4.6) in (10.4.3), is

$$\eta = e^{-y/a}G(x - ct), \qquad u = (g/H)^{1/2}e^{-y/a}G(x - ct). \tag{10.4.7}$$

In particular, when G is sinusoidal in character, this takes the form

$$\eta = \eta_0 e^{-y/a}\cos(kx - \omega t), \qquad u = (g/H)^{1/2}\eta_0 e^{-y/a}\cos(kx - \omega t), \tag{10.4.8}$$

where the dispersion relation between ω and k is

$$\omega = kc. \tag{10.4.9}$$

This solution is shown in Fig. 10.3.

For an observer traveling with the wave, the coastal boundary (where the wave has maximum amplitude) is always on the right in the northern hemisphere and on the left in the southern hemisphere. Another way of expressing this fact is to say that the wave moves equatorward on a western boundary and moves poleward on an eastern boundary, or that it moves cyclonically around a basin. Also, it is possible to express the Kelvin-wave solution [which has the form (10.4.7) in the northern hemisphere] in a form that is independent of the sign of f, namely,

$$\eta = e^{-y/a}G(x - aft), \qquad u = (g/af)e^{-y/a}G(x - aft). \tag{10.4.10}$$

The wave velocity is af, and therefore it changes sign with f as required (a was defined to be positive), and u follows from (10.4.1).

Now consider the energetics of a Kelvin wave of the form (10.4.8). If the coast is put at $y = 0$, then η_0 is the amplitude of the wave at the coast. Since the solution for a fixed value of y is as in the nonrotating case, the energy of a Kelvin wave is partitioned equally between the kinetic and potential forms. The mean value (denoted by overbar) per unit length of coast is given by

$$\int_0^\infty \frac{1}{2}\rho H\overline{u^2}\,dy = \int_0^\infty \frac{1}{2}\rho g\overline{\eta^2}\,dy = \frac{1}{8}a\rho g\eta_0^2. \tag{10.4.11}$$

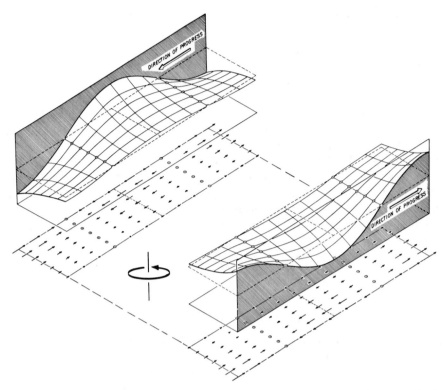

Fig. 10.3. Northern hemisphere Kelvin waves on opposite sides of a channel that is wide compared with the Rossby radius. In each vertical plane parallel to the coast, the currents (shown by arrows) are entirely within the plane and are exactly the same as those for a long gravity wave in a nonrotating channel. However, the surface elevation varies exponentially with distance from the coast in order to give a geostrophic balance. This means Kelvin waves move with the coast on their right in the northern hemisphere and on their left in the southern hemisphere. [From Mortimer (1977).]

The mean value of the energy flux along the coast is equal to

$$\int_0^\infty \rho g H \overline{u\eta}\, dy = \frac{1}{4} \frac{\rho g^2 H \eta_0^2}{|f|}. \tag{10.4.12}$$

This formula is of interest when a Kelvin wave moves through a region in which H or f varies slowly; for the energy flux remains constant, and therefore the amplitude η_0 varies in proportion with $(|f|/H)^{1/2}$. In particular, large amplitudes are produced where the Kelvin wave moves into shallow water, e.g., for the Kelvin-wave component of the tide entering the North Sea and then moving down the west side into even shallower water in the southern bight.

10.5   The Full Set of Modes for an Infinite Channel of Uniform Width

The possible propagating modes for a channel of width W are the Poincaré modes, given by (10.3.1) and (10.3.4), and the two Kelvin waves associated with the two sides of the channel. The dispersion curves for these modes are shown in Fig. 10.4. When the channel is not too wide ($W/a \ll \pi$) as in case (a), the picture is not too different from that for the limit of no rotation. However, as the width of the channel increases, the critical frequencies ω_{nc} for propagation of the Poincaré modes all approach f, and so the picture becomes more like that of case (b).

Kelvin's discussion of waves in a channel was confined to the modes that now bear his name. [His discussion of sinusoidal waves was confined to the plane progressive wave given by (8.2.11) and (8.2.12).] He noted (p. 97), for instance, that

> the more approximately nodal character of the tides on the north coast of the English Channel than on the south or French coast ... is probably to be accounted for on the principle represented by this factor [i.e., $e^{-y/a}$], taken into account along with frictional resistance, in virtue of which the tides of the English Channel may be roughly represented by more powerful waves travelling from west to east, combined with less powerful waves travelling from east to west.

That this gives a reasonable description of the tides in the English Channel can be seen from Fig. 10.5, in which the observed tides are shown. The situation is discussed in more detail by Proudman (1953, Section 131).

Kelvin also noted (p. 97) that "the problem of standing oscillations in an endless rotating canal is solved by the following equations." He then gave the solution for

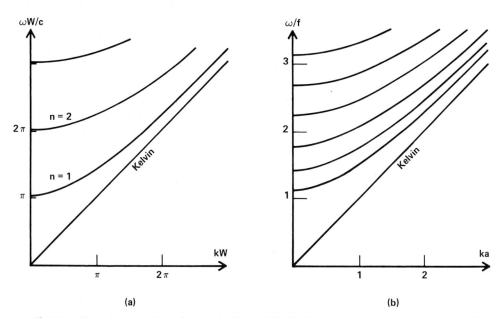

(a)   (b)

Fig. 10.4. Dispersion curves for a channel of uniform width W, k being the wavenumber component parallel to the sides of the channel. In case (a), the width is only a small fraction (0.3) of the Rossby radius and the dispersion relation, given by (10.3.2)–(10.3.4), becomes $(\omega W/c)^2 = 0.09 + (kW)^2 + (n\pi)^2$. Rotation effects which are represented by the constant term 0.09, are barely discernible. In case (b), the width-to-Rossby-radius ratio W/a is large (2π) and the dispersion relation is $(\omega/f)^2 = 1 + (ka)^2 + \frac{1}{4}n^2$, if the width is increased further, the coefficient of n^2 becomes smaller, so the curves move downward on the diagram and become more densely packed. The limiting curve for a Poincaré mode in an infinitely wide channel is the one shown in Fig. 7.2.

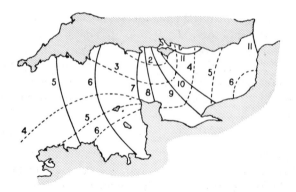

Fig. 10.5.   Co-tidal lines (solid) with time in lunar hours, and co-range lines (dotted with values in meters) for the English Channel. [From Proudman (1953, p. 262); after Doodson and Corkan (1931).]

two Kelvin waves of equal amplitude traveling in opposite directions, i.e.,

$$\eta = H\{e^{-y/a}\cos(kx - \omega t) - e^{y/a}\cos(kx + \omega t)\},$$

$$u = (gH)^{1/2}\{e^{-y/a}\cos(kx - \omega t) + e^{y/a}\cos(kx + \omega t)\}, \qquad (10.5.1)$$

$$v = 0.$$

Kelvin also noted (p. 98) that "if we give ends to the canal we fall upon the unsolved problem referred to above of tesseral oscillations."

When Poincaré (1910) discussed waves in an infinite canal, he added the solutions (10.3.1) to his discussion. For that reason, some authors restrict the name "Poincaré wave" to these solutions. However, that usage has led to many types of waves being distinguished by different scientists' names, where for nonrotating waves the distinction is made merely by a descriptive term like "progressive" or "standing" or "cross" wave. Use of a descriptive terminology is much to be preferred, so in this book only the names of Kelvin and Poincaré are used. Kelvin waves are the boundary waves just introduced, and all classes of nonboundary waves are called Poincaré waves. There seems to be no point to giving a separate name to the plane progressive wave (8.2.11), which is, after all, the limiting form of a wave (10.3.1) in a canal when the canal width goes to infinity.

10.6   End Effects: Seiches and Tides in a Gulf That Is Not Narrow

The addition of the Poincaré wave does not, however, help with the problem of finding the solution in a channel with an end, for as Poincaré (1910, Section 68) said, the wave of the form (10.3.1) "cannot undergo regular reflection" because a combination with the wave with $-k$ replacing k "cannot satisfy formally the conditions expressing that the component of displacement normal to the reflecting wall be zero."

The problem of dealing with the condition at the end of the channel was solved by Taylor (1921), the new feature being to include solutions like (10.3.1), but with imaginary k, i.e., with exponential falloff away from the end of the channel. These terms influence the details of the solution near the end, but do not affect the Kelvin waves except in the determination of the phase of the reflected Kelvin wave relative to that of the incident one. The reflection is subject to a delay that increases as the width of the channel increases. As Taylor suggested, this can be interpreted as the Kelvin wave progressing down one side of the channel, then taking some time to cross the end of the channel before returning along the other side.

The above description assumes that $\omega < \omega_{1c}$, which is true of most natural channels when ω is the tidal frequency, for then no propagating Poincaré waves are possible. From (10.3.5), the condition $\omega < \omega_{1c}$ is automatically satisfied if $\omega < f$, whereas if $\omega > f$, it can also be written

$$W^2 < \pi^2 gH/(\omega^2 - f^2), \tag{10.6.1}$$

i.e., the width of the basin should not be too large relative to the square root of the depth. If $\omega > \omega_{1c}$, then at least one Poincaré mode should be included in the description of the solution [see Brown (1973)].

Figure 10.6 shows the solutions obtained by Taylor for a channel 250 miles (460 km) wide and 40 fathoms (73 m) deep, i.e., a channel that resembles in dimensions and depth the North Sea. This corresponds to $c = 27$ m s^{-1}. Since $\omega = 1.4 \times 10^{-4}$, $f = 1.2 \times 10^{-4}$, the Rossby radius $a = 230$ km and therefore $W = 2a$. The condition (10.6.1) is that the width be less than 1160 km and this is well satisfied. The solution can be compared with that for a narrow channel shown in Fig. 10.1. The configuration of co-tidal lines is similar, but instead of the phase progressing around the head of the channel in a very short time, as in the narrow channel, this rate of progression is now slower and not, in fact, very different from the rate of progression along the sides. As Taylor (1921) states:

> In the lower part of the basin at a distance greater than about 250 miles from the closed end, the cotidal lines and the motion of the particles correspond very nearly to two equal Kelvin waves moving up and down the channel. The tidal streams are very nearly parallel to the sides of the channel and the cotidal lines move in along the right hand shore (*i.e.*, the left-hand side of the figure). The tidal wave then sweeps round the end wall of the basin at a rate rather greater than the velocity of the Kelvin wave, and moves back along the opposite shore to that along which it approached the end. In turning at right angles in order to cross the end of the channel, the wave produces a bigger rise and fall of tide at the two corners than anywhere else in the field. On the scale chosen the range of tide at the corners is represented by the number 1.95, whereas the greatest range in the distant parts of the channel, far from the influence of the end, is represented by 1.61.
>
> In order to show up more conspicuously the nature of the motion, the cotidal lines have been drawn in Fig. 1 [reproduced as Fig. 10.6] with heavy lines in the region where the rise and fall is, on the scale chosen, greater than 1, that is to say in the parts where the range of tide is greater than half the maximum range at the corners. The way in which the strongly marked parts of the cotidal lines move down the left side of the diagram, cross the end and move up the right side, is conspicuous.

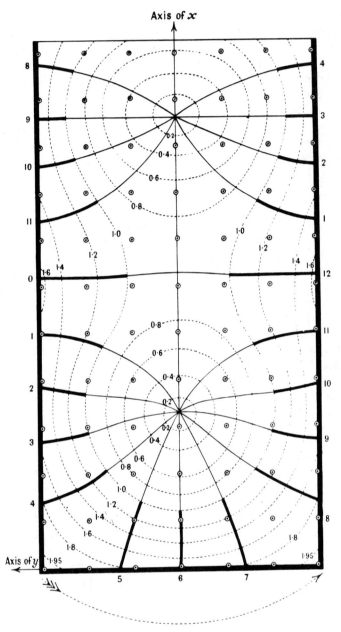

Fig. 10.6.   Solution obtained by Taylor (1921; Fig. 1) for reflection of a Kelvin wave at the end of a channel with dimensions similar to those of the North Sea. Solid lines are co-tidal lines with time given in hours, and dotted lines show co-range lines.

In the distant parts of the channel the tidal streams are parallel to the shores at all states of the tide. At distances from the end less than a distance about equal to the width, however, the particles of water move in ellipses, except, of course, close in shore, where they continue to move parallel to the shore. The direction in which the particles move round all the larger ellipses is the same as that of the rotation of the earth . . . .

The maximum tidal currents occur at certain points close to the parallel shores, the greatest being at a distance from the end about equal to half the breadth of the channel; but the cross-channel current at the mid-point of the end is very nearly as great as this maximum.

The currents near the central part of the basin are considerably smaller than those close to the shore (pp. 150, 153).

With the advent of the computer, much more detailed solutions for particular basins can be calculated, e.g., tides for the North Sea have been calculated by Flather (1976) and seiches in some of the of the Great Lakes have been calculated by Rao and Schwab (1976) and by Rao *et al.* (1976). Further discussion of seiches and tides in channels and bays can be found in Section 28 of LeBlond and Mysak (1978).

10.7 Adjustment to Equilibrium in a Channel

The remarkable experiment by Marsigli (1681), designed to explain basic features of the flow in the Bosphorus, was described in Section 5.1. In picturesque terms, this experiment could be envisaged as putting a barrier across the Bosphorus, with relatively fresh water from the Black Sea on one side and saltier Mediterranean water on the other. On removal of the barrier, Marsigli showed how the lighter water from the Black Sea would flow over the heavier water from the Mediterranean, as it is observed to do. However, the adjustment process would in practice be affected by the rotation of the earth, and the discussion in this section is about such effects.

The problem considered is basically the "Rossby adjustment problem" of Section 7.2, but in a channel of uniform width W and uniform depth H. This problem was considered by Gill (1976). The y axis is chosen to be down the center of the channel and the partition is placed at $x = 0$. The free surface elevation prior to removal of the partition at time $t = 0$ is given by (7.2.11) or (5.6.13), i.e., the surface elevation is $-\eta_0$ at all points $x > 0$ to the right of the partition and $+\eta_0$ at all points $x < 0$ to the left. It is assumed that the hydrostatic approximation can be applied, so that the governing equations are (7.2.1)–(7.2.3). As has been shown in Chapter 6, the same analysis can be applied to any of the normal modes of a stratified fluid, and in particular it can be applied to the baroclinic adjustment of the initial discontinuity at $x = 0$. The results of this section can be translated into results for a two-layer system by exchanging variables as

$$u, v, g\eta, H, c \rightarrow \hat{u}, \hat{v}, g'h, H_e, c_e. \tag{10.7.1}$$

The variables on the left are the ones used in this section, whereas those on the right are the ones appropriate to a two-layer system as defined in Section 6.3.

If the channel is narrow enough, the situation is as described in Section 10.2. The walls force the flow to be everywhere parallel to the walls, and the solution is, to a first approximation, (10.2.1), i.e., the nonrotating solution described in Section 5.6. This solution is the sum of two nondispersive waves moving at speed $c = (gH)^{1/2}$ in opposite directions. For the step-function initial condition, the solution is as illustrated in Fig. 5.9a. After the fronts have passed, the surface is at its equilibrium level

$$\eta_{nr} = 0 \tag{10.7.2}$$

and the flow is uniform,

$$u_{nr} = g\eta_0/c = (g/H)^{1/2}\eta_0, \tag{10.7.3}$$

with the value that corresponds to complete conversion of potential into kinetic energy.

The first corrections for rotation are those discussed in Section 10.2. After the the wave front has passed, therefore, there will be a slope on the interface to balance the flow, i.e., the surface elevation by (10.2.3) is

$$\eta \approx -f\eta_0 y/c. \tag{10.7.4}$$

In addition, the potential vorticity equation (7.2.8) requires a shear given by

$$\partial u/\partial y = (f/H)(\eta - \eta_{init}) \tag{10.7.5}$$

to be established, and the sign of this will be dependent on the sign of x, i.e.,

$$u \approx (f\eta_0/H)y \, \text{sgn}(x). \tag{10.7.6}$$

For a channel of arbitrary width, it is necessary to solve the shallow-water equations (7.2.1)–(7.2.3) with the boundary conditions

$$v = 0 \quad \text{at} \quad y = \pm\tfrac{1}{2}W. \tag{10.7.7}$$

Since the boundary condition is on v, it is useful to obtain an equation in v alone. This equation has a form similar to the one (7.2.13) for η and is obtained by adding $-f$ times (7.2.1), the time derivative of (7.2.2), $-g$ times the y derivative of (7.2.3), and $-gH^2$ times the x derivative of (7.2.10). The result is

$$\frac{\partial^2 v}{\partial t^2} - c^2\left(\frac{\partial^2 v}{\partial x^2} + \frac{\partial^2 v}{\partial y^2}\right) + f^2 v = -gH^2 \frac{\partial Q'}{\partial x}(x, y, 0), \tag{10.7.8}$$

where $Q'(x, y, 0)$ is the initial value of the potential vorticity perturbation defined by (7.2.9).

The relationship of v to the other variables u and η is best expressed by the relationship between v and the sum and difference between u and $(g/H)^{1/2}\eta$, namely,

$$q = (g/H)^{1/2}\eta + u, \tag{10.7.9}$$

$$r = (g/H)^{1/2}\eta - u. \tag{10.7.10}$$

These variables were used by Gill and Clarke (1974) for studying equatorial waves.

The sum and differences between (7.2.1) and $(g/H)^{1/2}$ times (7.2.3) give

$$\frac{\partial q}{\partial t} + c\frac{\partial q}{\partial x} + c\frac{\partial v}{\partial y} - fv = 0, \tag{10.7.11}$$

$$\frac{\partial r}{\partial t} - c\frac{\partial r}{\partial x} + c\frac{\partial v}{\partial y} + fv = 0, \tag{10.7.12}$$

i.e., equations containing only two dependent variables, and thus relating q and r to v. Also, it may be seen that when $v = 0$, each equation corresponds to one Kelvin wave only, so the device of introducing q and r allows the two Kelvin waves to be considered independently.

There are two further equations relating q to v and r to v. These may be obtained as the sum and difference between (7.2.2) and cH times (7.2.10), where Q' is given by (7.2.9). The result is

$$c\frac{\partial q}{\partial y} + fq + \frac{\partial v}{\partial t} - c\frac{\partial v}{\partial x} + cHQ'(x, y, 0) = 0, \tag{10.7.13}$$

$$c\frac{\partial r}{\partial y} - fr + \frac{\partial v}{\partial t} + c\frac{\partial v}{\partial x} - cHQ'(x, y, 0) = 0. \tag{10.7.14}$$

Again, the two Kelvin waves can be considered independently. Note also that (10.7.8) can be obtained either by eliminating q between (10.7.11) and (10.7.13) or by eliminating r between (10.7.12) and (10.7.14).

The solution can now be expressed as a sum of Poincaré modes, as given by (10.3.1), (10.3.4), and (10.3.5), and of the two Kelvin waves that are possible. For simplicity, the initial state will be assumed to have symmetry about the y axis so that the solution for v retains this symmetry, i.e., has the form

$$v = \sum_{n=1}^{\infty} (\omega_{nc}/f)v_n \cos ly, \tag{10.7.15}$$

where l is given by (10.3.4) and the factor ω_{nc}/f, given by (10.3.5), is included for a reason that will soon become apparent. It follows from (10.7.11)–(10.7.14) that q and r may be expanded in the form

$$q = q_0 e^{-fy/c} + \sum_{n=1}^{\infty} q_n(\cos ly + f^{-1}cl \sin ly), \tag{10.7.16}$$

$$r = r_0 e^{fy/c} + \sum_{n=1}^{\infty} r_n(\cos ly - f^{-1}cl \sin ly), \tag{10.7.17}$$

where q_n, r_n, and v_n satisfy

$$\partial q_0/\partial t + c\,\partial q_0/\partial x = 0, \qquad \partial r_0/\partial t - c\,\partial r_0/\partial x = 0 \tag{10.7.18}$$

for $n = 0$, and for positive values of n satisfy

$$\frac{\partial q_n}{\partial t} + c \frac{\partial q_n}{\partial x} - \omega_{nc} v_n = 0, \qquad \omega_{nc} q_n + \frac{\partial v_n}{\partial t} - c \frac{\partial v_n}{\partial x} + cHQ'_n = 0, \quad (10.7.19)$$

$$\frac{\partial r_n}{\partial t} - c \frac{\partial r_n}{\partial x} + \omega_{nc} v_n = 0, \qquad -\omega_{nc} r_n + \frac{\partial v_n}{\partial t} + c \frac{\partial v_n}{\partial x} - cHQ'_n = 0, \quad (10.7.20)$$

Q'_n being the nth term in an expansion of $Q'(x, y, 0)$ in the form (10.7.15). The Kelvin wave parts q_0 and r_0 adjust *as in a nonrotating system*, whereas each of the Poincaré modes adjusts *as in the y-independent case* studied in Sections 7.2 and 7.3, but with f replaced by ω_{nc}. Initial values of q_n and r_n ($n = 1, 2, ...$) can be found as coefficients in the Fourier expansions of

$$c\frac{\partial q}{\partial y} + fq = \sum_{n=1}^{\infty} \frac{\omega_{nc}^2}{f} q_n \cos ly \qquad \text{and} \qquad -c\frac{\partial r}{\partial y} + fr = \sum_{n=1}^{\infty} \frac{\omega_{nc}^2}{f} r_n \cos ly, \quad (10.7.21)$$

these results following from (10.7.16) and (10.7.17). Values of q_0 and r_0 follow from the same two equations and from the property that the average of the values of the Poincaré modes on the walls is zero. Thus, for example, (10.7.16) gives

$$\tfrac{1}{2}\,[q(W/2) + q(-W/2)] = q_0 \cosh (fW/2c),$$

the cosine terms vanishing by (10.3.4). Using the corresponding result for r and the definitions (10.7.9) and (10.7.10), the result in the case of the initial condition $u = v = 0$ and (7.2.11) would be, for instance,

$$q_0 \cosh(f W/2c) = r_0 \cosh(f W/2c) = -(g/H)^{1/2} \eta_0 \, \text{sgn}(x) \qquad \text{at} \quad t = 0. \quad (10.7.22)$$

The solution for the initial step-function distribution (7.2.11) is easily constructed because each Poincaré wave mode behaves as does the solution shown in Fig. 7.3 and the Kelvin wave parts are simply waves traveling at uniform speed. After a long time, each Poincaré mode will be as shown in Fig. 7.1 and therefore will be close to the initial state when $|x|$ is large. However, the Kelvin waves *do* produce a large change at large $|x|$. One Kelvin wave travels along one wall (the one on the right in the northern hemisphere) in the positive x direction, and so changes the solution near that boundary. The other Kelvin wave moves along the opposite wall in the opposite direction, and so changes the solution at large negative values of x. The result is the strongly asymmetric structure shown in Fig. 10.7c and is given by

$$\eta/\eta_0 \sim \begin{cases} -1 + e^{-fy/c} \, \text{sech}(f W/2c) & \text{as} \quad x \to \infty, \\ 1 - e^{fy/c} \, \text{sech}(f W/2c) & \text{as} \quad x \to -\infty; \end{cases} \quad (10.7.23)$$

$$(H/g)^{1/2}(u/\eta_0) \sim \begin{cases} e^{-fy/c} \, \text{sech}(f W/2c) & \text{as} \quad x \to \infty, \\ e^{fy/c} \, \text{sech}(f W/2c) & \text{as} \quad x \to -\infty. \end{cases} \quad (10.7.24)$$

Also illustrated (Fig. 10.7a and b) is the solution at two intermediate times. This shows the wave front moving out at speed c (the wave front for all Poincaré modes and the appropriate Kelvin wave move at this speed) followed by a wake that involves oscillations around the final steady state shown in Fig. 10.7c.

In practice, the flow through straits and channels is much more complicated, but the above solution is a useful beginning for understanding more complex situa-

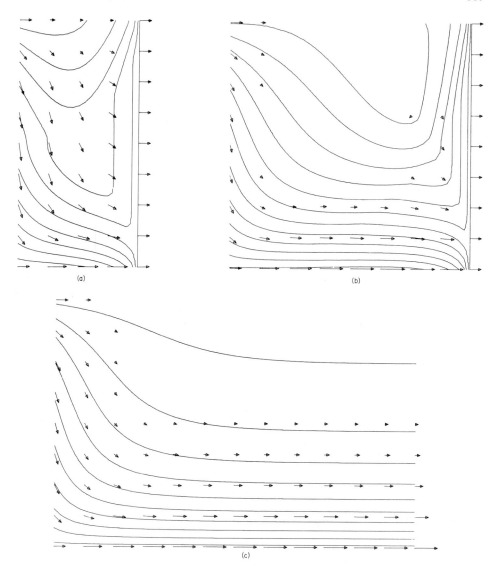

Fig. 10.7. Adjustment under gravity in a rotating channel whose width is 4 Rossby radii. Initially, the flow is at rest and the surface elevation is -1 for $x > 0$ and $+1$ for $x < 0$. The diagrams show contours of surface eleva­tion at odd multiples of 0.1 for the region $x > 0$, and the arrows indicate the direction and magnitude of the velocity. The solution for (a) $ft = 2$, (b) $ft = 4$, and (c) large times. The wave front moving out into the undisturbed fluid at speed c can clearly be seen in (a) and (b), i.e., the front is at $x = 2c/f$ in (a) and at $x = 4c/f$ in (b). The arrows are drawn with their tails on a square grid with a mesh size of $\frac{1}{2}c/f$.

tions. Examples of observed flows are given in Chapter 16 of Defant (1961). If the channel is long, friction effects become important (Anati *et al.*, 1977). If the channel is a relatively shallow link between two ocean basins, it tends to dam up the denser bottom water on one side, and the flow through the channel is hydraulically controlled as is the flow from a reservoir [laboratory experiments and theory are described by Whitehead *et al.* (1974) and Gill (1977a)]. Rotation causes the density surfaces to slope, and the current downstream of the sill tends to follow the right bank in the

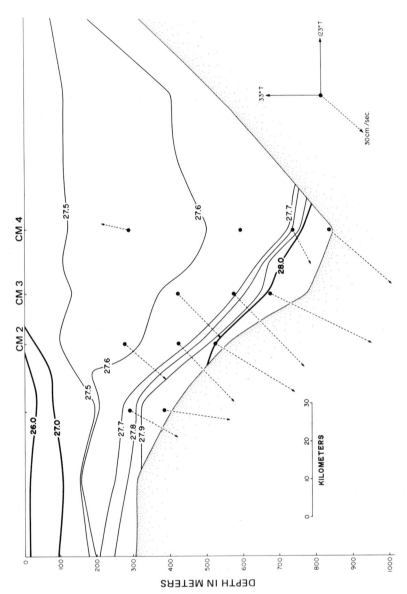

Fig. 10.8 A density (σ_t) section across the Denmark Strait [from P. C. Smith (1976, Fig. 3b)] at about 66°N, 27–29°W. The dense water near the bottom is flowing (out of the page) across the sill into the North Atlantic from the Greenland Sea, and the isopycnals tilt up to the right (for an observer looking downstream) because of the earth's rotation. The dashed lines indicate average current vectors as measured by current meters between 14 August and 15 September 1973. The maximum current is 0.6 m s⁻¹. The section also shows a wedge of light water in the upper left-hand corner of the diagram. This is the East Greenland Current flowing out of the Arctic Ocean. The direction of the flow is the same as for the bottom current, but isopycnals slope the other way because it is a surface current.

northern hemisphere just as seen in Fig. 10.7b. For example, Fig. 10.8 shows a σ_t-section across the outflow of dense water from the Greenland Sea into the North Atlantic. The section is located at the southern end of the Denmark Strait, and the upward slope to the right of the "interface" can be clearly seen. The maximum current was observed to be about 0.6 m s^{-1} toward the southwest.

10.8   Tides

The tide-producing forces were considered in Section 9.8. The problem of finding the response to these forces amounts to the problem of solving forced shallow-water equations on a sphere, with the complicated bottom topography and boundary shape that the ocean has. Basically the problem is that of a *forced linear oscillator*, albeit a complicated one. The response in such problems depends very much on whether the forcing frequency is close to a resonant frequency i.e., a natural frequency of the system. There is evidence from tidal observations to suggest [see, e.g., Garrett and Greenberg (1977) and Heath (1981)] that there are natural modes of oscillation of the ocean with frequencies near that of the semidiurnal tides, implying that the semidiurnal tide is sensitive to details of topography and coastline shape. This means that small differences in the ocean basins in the geological past could make considerable differences to the tides [it also means that a successful numerical simulation of the tides is difficult, which may be the reasons for the differences between different models—see Hendershott (1977)].

Some simple aspects of the problem can be illustrated by considering the semidiurnal tide in a narrow zonal channel of uniform depth H (Airy, 1845). If the channel is narrow enough, rotation may be ignored, and the equations are simply (cf. Sections 5.8 and 9.8)

$$\partial u/\partial t = -g \, \partial(\eta - \eta_e)/\partial x, \qquad \partial \eta/\partial t + H \, \partial u/\partial x = 0, \qquad (10.8.1)$$

where η_e is the equilibrium tide. In a zonal channel, this has the form (cf. Table 9.1)

$$\eta_e = A \sin(2kx - 2\Omega_l t), \qquad (10.8.2)$$

where $2\Omega_l$ is the frequency of the semidiurnal tide (i.e., $2\pi/\Omega_l$ in a lunar day) and

$$k = 2\pi/L \qquad (10.8.3)$$

is the wavenumber of the semidiurnal tide in the channel, i.e., L is half the circumference of the earth at the latitude of the channel (the semidiurnal tide has two wavelengths encircling the globe).

The response has the form

$$\eta = \eta_0 \sin(2kx - 2\Omega_l t), \qquad (10.8.4)$$

where substitution in (10.8.1) gives

$$\eta_0 = A/(1 - \Omega_l^2/c^2 k^2), \qquad (10.8.5)$$

$c = (gH)^{1/2}$ being the speed of long gravity waves. This equation yields some interesting information because it shows the dependence of the response on the ratio of the

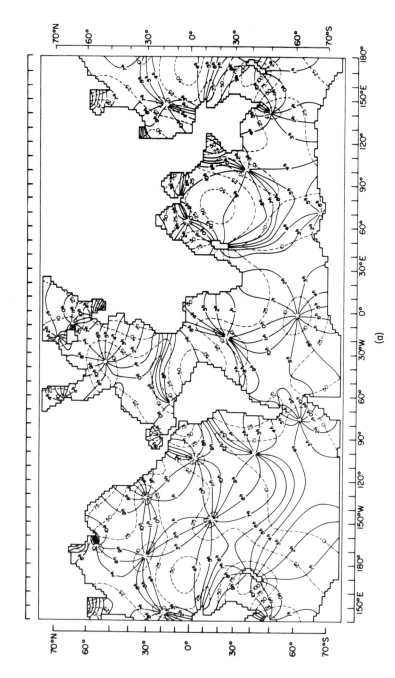

Fig. 10.9. (a) The M_2 tide as computed from a numerical model by Accad and Pekeris (1978, Fig. 8); the phase is shown by solid lines marked in Greenwich hours, and the range is shown by dashed lines, in centimeters. (b) The 12.5-hr free mode of oscillation as computed by Platzman et al. (1981). Phase contours are denoted by solid lines and co-range lines by dashed lines.

(a)

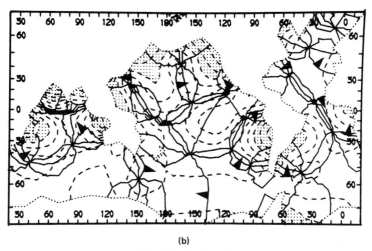

(b)

Fig. 10.9. (continued)

speed Ω_l/k of progress of the equilibrium tide around the earth compared with the long-wave speed c. If the equilibrium tide moved slowly compared with long waves (as it does in the case of the fortnightly tide M_f for instance), then η_0 would be close to A and the response would be close to equilibrium. In fact, however, the semi-diurnal and diurnal tides move around the earth once per day, giving a speed of 330 m s^{-1} at 45° latitude, which is *faster* than the long-wave speed c (about 200 m s^{-1}). This means that the denominator in (10.8.5) is *negative*, so the semidiurnal tide in a zonal channel would have the sign *opposite* to the equilibrium tide!

Another fact shown by (10.8.5) is that if the depth of the canal were increased so that c became closer to Ω_l/k, the amplitude of the response would increase, approaching infinity at the resonant value at which $c = \Omega_l/k$. Further increase in c would result in a change in sign of the response. If the zonal channel is taken as a model of the Southern Ocean, it turns out *not* to be near resonance by itself, but if a side channel, representing, say, the Atlantic Ocean, is added, the whole system can become near-resonant (Gill, 1979a). In fact, near-resonant modes with such a model can give quite realistic looking tides. Calculations of free modes (i.e., resonant modes) of oscillation of the ocean have been made by Platzman *et al.* (1981), and these do, in fact, show modes with natural frequencies not far removed from the semidiurnal frequency. Figure 10.9 shows an example of a free mode along with a numerical solution of the Atlantic tide for comparison, and many similarities will be noted.

In practice, the amplitude near resonance is limited by friction, which is particularly strong in shallow-sea regions (Miller, 1966). This represents a transfer of energy given to the tides from the moon, and so causes very slow changes in the moon's orbit (Munk and McDonald, 1960).

Another interesting question is that of how wide a channel must be before rotation effects become important. The answer is given in Section 10.2 and in particular by formula (10.2.4). This shows that the half-width of the channel must be small compared with the Rossby radius c/f, which is about 2000 km for the deep ocean. The North Atlantic therefore is a borderline case, i.e., the narrow-canal theory gives a reasonable, but not very accurate, model of the tides there. The Pacific, on the other hand, is large compared with the Rossby radius, so the narrow-canal theory is not appropriate. A local description of the tides therefore includes Kelvin waves, and Munk *et al.* (1970) have in fact shown that the Kelvin-wave part of the semidiurnal tide off the California coast is the dominant part.

10.9   Storm Surges on an Open Coastline: The Local Solution

A storm surge refers to an abnormally high sea level, produced by severe meteorological conditions. Certain low-lying areas adjacent to shallow seas are particularly vulnerable to them, the east coast of Britain and the coast of Holland being notable examples. Brooks and Glasspoole (1928, pp. 95, 96) list occurrences of such floods back to A.D. 9, and many features of the North Sea coast were formed during surge conditions. For instance, the Zuyder Zee in Holland was fertile land 1000 years ago, but began to be formed with a surge on 17 November 1218, which overwhelmed a number of parishes and was said to have drowned 100,000 people. The area of the Zuyder Zee was further increased with a surge on 1 October 1250, which first made Wieringen into an island. Another surge on 16 January 1362 drowned 30 parishes in Eastern Friesland and Schleswig, and gave the Friesian islands their present form. The Zuyder Zee reached the form it has maintained for the next five centuries with the great surge of 19 November 1421 in which 100,000 people were drowned. Such events have continued to occur throughout recorded history, a severe example in recent times being the surge of 31 January 1953, which cause widespread devastation in Britain and Holland with coastal defences breached in hundreds of places, flooding 25,000 km^2 of land and resulting in 2000 deaths. This disaster led to considerable work on the problem of surges, culminating in a numerical model of the North Sea that is now used routinely for forecasting surges. The magnitude of the 1953 surge can be seen in Fig. 10.10, where the progress of the surge around the North Sea is charted.

Surges in a narrow sea (i.e., one with width small compared with the Rossby radius) have already been mentioned in Section 10.2. Here rotation can be neglected to first order and longitudinal oscillations of the sea can be stimulated by wind that blows over the sea itself (e.g., winds over the Adriatic set the 22-hr seiche in motion, thereby causing flooding in Venice) or by disturbances that pass the entrance to the sea and thus stimulate oscillations in the same way as do the tides (see Section 5.8); these are called external surges.

However, surges can also be produced on an open coastline, as will now be demonstrated. The situation considered is one of a semi-infinite sea of uniform depth, bounded by a straight boundary $y = 0$. A wind stress parallel to the coast and of magnitude X_s is suddenly applied at time $t = 0$, and the problem is to find the motion and the changes of sea level that result. In the absence of a boundary, inertial oscillations would be set up as described in Section 9.3, with a mean Ekman drift directed normal to the coast. However, there can be no flow across the coastline, so a convergence or divergence is established there with a consequent rise or fall of sea level

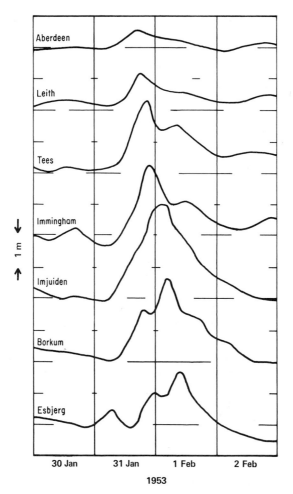

Fig. 10.10.   Progress of the very large storm surges of 1953 around the North Sea. The curves shown are the observed sea level minus the predicted astronomical tide, i.e., the sea level at Ijmuiden was about 3 m above normal in the early hours of February 1. The double peak seen at some parts is often due to surge–tide interaction. [Adapted from Rossiter (1954, Fig. 1).]

in order to conserve mass. The effect is the same as that of having a discontinuity in wind stress with a consequent discontinuity in Ekman current (cf. Section 9.4).

The form of the equations (9.9.10) when there is no friction and no x variation is

$$\partial u/\partial t - fv = X_s/\rho H, \tag{10.9.1}$$

$$\partial v/\partial t + fu = -g\, \partial\eta/\partial y, \tag{10.9.2}$$

and the continuity equation (5.6.6) is

$$\partial\eta/\partial t + H\, \partial v/\partial y = 0. \tag{10.9.3}$$

A single equation for v, the velocity normal to the coast, can be obtained by adding $-f$ times (10.9.1), the time derivative of (10.9.2), and $-g$ times the y

derivative of (10.9.3). This gives

$$\partial^2 v/\partial t^2 + f^2 v - gH\,\partial^2 v/\partial y^2 = -fX_s/\rho H, \tag{10.9.4}$$

which [cf. (9.9.21)] is a form of the forced shallow-water equation. The solution is very similar to that for the Rossby adjustment problem studied in Sections 7.2 and 7.3, and can be expressed as a sum of a steady part and a transient part. The steady part satisfies the condition of no flow across the boundary (i.e., $v = 0$ at $y = 0$) and does not grow at infinity, but tends to the steady Ekman solution. This solution is

$$v = -(X_s/f\rho H)(1 - e^{-y/a}), \tag{10.9.5}$$

where

$$a = c/f = (gH)^{1/2}/f \tag{10.9.6}$$

is the Rossby radius, defined by (7.2.23).

In order to satisfy the initial condition of rest, a transient solution must be added to (10.9.5). This solution has properties similar to one studied in Section 7.3, including a wave front that moves out from the coast at speed c and slowly dispersing near-inertial period waves that are left behind. Details of this solution are given by Crépon (1974) [see also Anderson and Gill (1979)]. The forced shallow-water equations for this problem and some properties of the solution were derived by Nomitsu (1934). However, after a time of order f^{-1}, the solution near the coast is dominated by (10.9.5), and so the transient solution will not be considered further.

Now consider further details of the solution, corresponding to (10.9.5). Because there is a steady flow toward (or away from) the coast, water will continually pile up (or be removed) from that region and therefore the sea level will change linearly with time. In fact, the continuity equation (10.9.3) gives

$$\eta = (X_s/\rho c)e^{-y/a}t. \tag{10.9.7}$$

Because v is independent of time, (10.9.2) shows that there is a current u parallel to the coast that is always in geostrophic equilibrium with the pressure gradient normal to the coast, i.e.,

$$u = -(g/f)\,\partial\eta/\partial y = (X_s/\rho H)e^{-y/a}t. \tag{10.9.8}$$

These properties of the solution are summarized in Fig. 10.11. Another property of the solution comes from the elimination of v between (10.9.1) and (10.9.3) to obtain

$$\frac{\partial}{\partial t}\left(\frac{1}{f}\frac{\partial u}{\partial y} + \frac{\eta}{H}\right) = 0. \tag{10.9.9}$$

In other words, the perturbation potential vorticity remains zero throughout the surge. This is an important property, which will be exploited for more complicated situations later.

The property that the sea-surface slope at the coast increases linearly with time was found by Nomitsu (1934). A complete solution of the form given by (10.9.5)–(10.9.8) was obtained by Charney (1955a), who drew attention to the relationship with the Rossby adjustment problem and pointed out the important features of the solu-

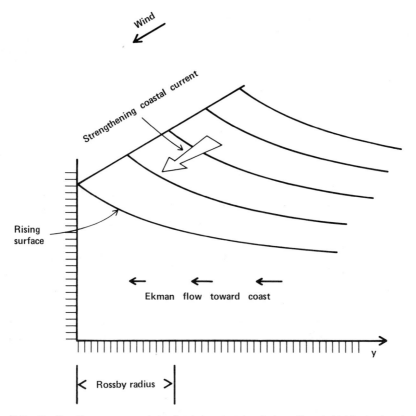

Fig. 10.11.   The "local" storm surge solution for the northern hemisphere. The wind is blowing longshore with the coast on its right, thus producing an Ekman current toward the coast. This causes the surface to rise at a constant rate within a distance of the order of the Rossby radius of the coast. A coastal current in geostrophic equilibrium with the sea-level distribution also increases at a constant rate. This current is in the same direction as that of the wind.

tion, namely, that the sea-level current and the longshore current are affected only within a distance of the coast of the order of the Rossby radius, and there they change linearly with time. The solution is similar in structure to the one obtained by Hough (1897), which was discussed in Section 9.14. In fact, by adding the image system it can be seen that the coastal surge solution is the same as that for an infinite region when the stress is X_s in the region $y > 0$ and $-X_s$ in the region $y < 0$.

Note that (10.9.7) gives the *largest* effect in *shallow* water, where $c = (gH)^{1/2}$ is small. For instance, a large stress of 2 N m^{-2}, which is typical of strong surge conditions, causes the sea level to rise by 1 m every 3 hr in 50 m of water, according to (10.9.7), and this may be compared with the rates of rise seen in the 1953 surge (Fig. 10.10).

The driving force for the surge is the Ekman transport at right angles to the coast, which causes water to pile up (or to be removed) in the coastal region. The question as to what happens to this transport near the coast in the *steady state* was investigated by Jeffries (1923). When the steady state is reached, by continuity, there can be no transport of water normal to the coast anywhere, and therefore the wind stress at the surface [see (9.9.10)] can be balanced only by the bottom stress. In other words, the longshore flow builds up until the bottom stress is big enough to balance the surface stress. If the bottom friction formulation of Section 9.12 [see (9.12.10)] is used, this balance is written

$$0 = (X_s/\rho H) - ru. \tag{10.9.10}$$

The geostrophic balance

$$fu = -g\,\partial\eta/\partial y \tag{10.9.11}$$

is still valid in the steady state, so a uniform slope normal to the coast would eventually be established. Jeffries (1923) calculated the order of magnitude of the time needed to establish this gradient to a distance L from the coast. If this distance is large compared with the Rossby radius, the analysis of Section 9.12 shows that this is the time taken for information to diffuse a distance L if the diffusivity is given by (9.12.9). In practice, of course, surges are essentially a transient phenomenon, and so the steady state is never reached. An example of a calculation that involves observed time variations (storms moving normal to the shore) and bottom friction is given by Heaps (1965).

Note that for the case considered above of water of depth 50 m, a 1-m rise in sea level corresponds to a 0.5-m s^{-1} increase in longshore current. The frictional decay time for phenomena with a scale of the order of the Rossby radius is the spin-down time given by (9.12.7), and this is about 6 hr for a current of 0.5 m s^{-1}.

10.10   Surges Moving along the Coast: Forced Kelvin Waves

In Section 10.9 it was shown that surges are produced by wind-driven Ekman transport toward the coast, this leading to a piling-up of water within a distance of the order of the Rossby radius from the coast. However, the solution was obtained on the assumption that longshore variations could be ignored. Often a surge is found (see, e.g., Fig. 10.10) to move down the coast and therefore it is very important to consider variations with x as well as y. The analysis is simple if the longshore scale L is large compared with the Rossby radius a, which very often is a reasonable assumption. The transient features with time scale f^{-1} or smaller will again be ignored.

The equations are again those of Section 9.9 that lead to the potential vorticity equation (9.9.21), and this has simple solutions when the y variations are rapid compared with x variations. A systematic way of finding the approximate equations that correspond to this assumption is by introducing *nondimensional variables* (which will be marked by an asterisk). Let τ be a scale for the wind stress and L be the length scale on which the stress varies. An additional assumption made is that the time scale is of order L/c, where $c = (gH)^{1/2}$ is the speed of long gravity waves. Then the stress can be written

$$X_s = \tau X_s^*(x/L, y/L, ct/L), \qquad Y_s = \tau Y_s^*(x/L, y/L, ct/L). \tag{10.10.1}$$

The longshore scale is assumed to be L, the scale normal to the shore is the Rossby radius $a = c/f$, and the time scale is L/c, so the nondimensional forms of the independent variables are

$$x^* = x/L, \qquad y^* = y/a, \qquad t^* = ct/L. \tag{10.10.2}$$

The appropriate scales for u, v, and η may be deduced from the solution (10.9.5), (10.9.7), and (10.9.8) of the previous section, which gives

$$u^* = \rho Hcu/\tau L, \qquad v^* = \rho Hfv/\tau, \qquad \eta^* = \rho gH\eta/\tau L. \tag{10.10.3}$$

The corresponding form of the dynamic equations (9.9.10) is

$$\partial u^*/\partial t^* - v^* = -\partial \eta^*/\partial x^* + X_s^*(x^*, \epsilon y^*, t^*), \tag{10.10.4}$$

$$\epsilon^2 \, \partial v^*/\partial t^* + u^* = -\partial \eta^*/\partial y^* + \epsilon Y_s^*(x^*, \epsilon y^*, t^*), \tag{10.10.5}$$

whereas the continuity equation (5.6.6) gives

$$\partial \eta^*/\partial t^* + \partial u^*/\partial x^* + \partial v^*/\partial y^* = 0. \tag{10.10.6}$$

The small parameter ϵ that occurs is the ratio of the two scales involved, i.e.,

$$\epsilon = a/L = c/fL. \tag{10.10.7}$$

In the limit as $\epsilon \to 0$, the dynamic equations (10.10.4) and (10.10.5) become

$$\partial u^*/\partial t^* - v^* = -\partial \eta^*/\partial x^* + X_s^*(x^*, 0, t^*), \tag{10.10.8}$$

$$u^* = -\partial \eta^*/\partial y^*. \tag{10.10.9}$$

Thus only the wind blowing parallel to the shore produces an effect to the leading order of approximation, and variations of this wind with distance from the coast can be ignored. The larger component u^* of the velocity is in geostrophic balance with the gradient of the sea level normal to the shore, but the lesser component v^* is *not* in geostrophic balance.

Now that the nondimensional form of equations has been derived, the asterisks can be dropped and nondimensional units understood. The potential vorticity equation can be obtained by adding (10.10.6) to the y derivative of (10.10.8) and then subtracting the x derivative of (10.10.9), giving

$$\frac{\partial}{\partial t}\left(\frac{\partial u}{\partial y} + \eta\right) = 0. \tag{10.10.10}$$

Integrating and substituting for u from (10.10.9) then give

$$\partial^2 \eta/\partial y^2 - \eta = 0, \tag{10.10.11}$$

and the solution therefore has the form

$$\eta = A(x, t)e^{-y}, \tag{10.10.12}$$

$$u = A(x, t)e^{-y}. \tag{10.10.13}$$

The equation for u is obtained from the geostrophic balance (10.10.9). The equation for the amplitude $A(x, t)$ comes from applying (10.10.8) at the coast $y = 0$, where v

vanishes, and therefore

$$\partial A/\partial t + \partial A/\partial x = X_s(x, 0, t).$$ (10.10.14)

The solution for v turns out to be the same as for that in the x-independent case, and thus is given in dimensional form by (10.9.5).

Equation (10.10.14) is the governing equation for storm surges, A being the quantity of most interest, namely, the sea level at the coast. When there is no dependence on x, A grows linearly with time, reproducing the x-independent solution of 10.9. Another solution, already met with, is the case of no forcing when the solution has the form

$$A = G(x - t),$$ (10.10.15)

where G is an arbitrary function. This corresponds to the Kelvin wave solution studied in Section 10.4.

In general, solutions of (10.10.14) can be thought of as *Kelvin waves* that are *modified by the wind*. In fact, (10.10.14) states that for an observer traveling along the coast at the Kelvin-wave speed (i.e., unit speed in nondimensional terms), the rate of change of coastal sea level A is equal to the wind stress X_s, parallel to the coast as seen by the observer. This is because the Ekman transport toward (or away from) the coast is locally determined and will build up (or reduce) the Kelvin wave depending on signs. For instance, if the observer is at a Kelvin-wave crest ($A > 0$) and X_s is positive (shoreward Ekman transport), the Kelvin-wave amplitude will increase. If the observer is at a trough ($A < 0$) and the Ekman transport is offshore ($X_s < 0$), the trough will deepen. The relationship is obtained mathematically by changing to a coordinate ξ moving with the wave, i.e., introducing new coordinates

$$\xi = x - t, \qquad t' = t$$ (10.10.16)

to replace x and t. Then (10.10.14) becomes

$$\partial A/\partial t' = X_s(\xi + t', 0, t').$$ (10.10.17)

An equation similar to this was obtained by Kajiura (1962).

Because the North Sea has dimensions large compared with the Rossby radius (which is about 200 km), this model can be applied to surges on the east coast of Britain. The wind that produces onshore Ekman transport in this case is in the north to northwest octant, and this has been known as the direction of surge-producing winds for a long time. For instance, Brooks and Glasspoole (1928, pp. 88, 89) quote a contribution to Defoe's history of the great storm of 1703 from the Rev. W. Derham F.R.S.:

> Another unhappy Circumstance with which this Disaster was join'd, was a prodigious Tide, which happen'd the next Day but one, and was occasion'd by the Fury of the Winds; which is also a Demonstration, that the Winds veer'd for Part of the Time to the Northward: and as it is observable, and known by all that understand our Sea Affairs, that a North West Wind makes the Highest Tide, so this blowing to the Northward, and that with such unusual Violence, brought up the Sea raging in such a manner that in some Parts of *England* 'twas

incredible, the Water rising Six or Eight Foot higher than it was ever known to do in the Memory of Man.

Also, appropriate forms of (10.10.17) have been devised on purely empirical grounds by Corkan, and further developed by Rossiter (1959). For instance, a Kelvin wave takes time $T = 8$ hr to travel from Aberdeen to Lowestoft. Thus, integrating (10.10.17) over this time for an observer traveling with the wave gives

$$A_L(t') = A_A(t' - T) + \int_{t'-T}^{t'} X_s(\xi + t, 0, t)\, dt, \qquad (10.10.18)$$

where A_L is the sea level at Lowestoft and A_A the value at Aberdeen. This may be compared with the empirical formula obtained by least squares fitting, namely,

$$A_L(t') = 1.15 A_A(t' - T) + \alpha \overline{X}_s, \qquad (10.10.19)$$

where α is an appropriate coefficient and $\overline{X}_s$ is an average value of X_s at some intermediate time. The factor of 1.15 could be due to shallowing of the North Sea to the south, which [see (10.4.12)] requires A to increase in proportion with $H^{-1/2}$ in order to keep the energy flux constant.

As an example of an analytical solution of (10.10.14), take a case in which there is no surge at $x = 0$, which can be taken to represent the northern end of the east coast of Britain. Suppose a wind blows parallel to the coast, with magnitude independent of space but depending on time, such as

$$X_s = \begin{cases} \pi \sin \pi t & \text{for } 0 < t < 1, \quad 0 < x < 1, \\ 0 & \text{otherwise.} \end{cases} \qquad (10.10.20)$$

The length scale is chosen so that $x = 1$ represents the south end of the east coast. For simplicity's sake, the duration of the storm has been made equal to the time it takes for a wave to traverse the length of the east coast. The solution may be carried beyond $x = 1$, if x is taken as distance along the coast moving in the direction of the wave, and it is assumed that the wave can turn the corner without loss of energy. However, because of the change of orientation of the coast, it is assumed that the forcing is zero from this point onward. Then the solution obtained is

$$A = \begin{cases} 1 - \cos(\pi t) & \text{for } 0 < t < x < 1, \\ 2 \sin(\tfrac{1}{2}\pi x)\cos(\pi(t - \tfrac{1}{2}(1 + x))) & \text{for } x < t < 1, \\ 1 - \cos(\pi(t - 1 - x)) & \text{for } \max(1, x - 1) < t < 1 + x, \\ 0 & \text{otherwise.} \end{cases} \qquad (10.10.21)$$

This solution is shown in Fig. 10.12, and despite the simplicity of the model, has many features in common with the observed surge that is shown in Fig. 10.10. Note that the observed surge decayed along the German and Danish coasts. This is consistent with the estimates of friction that were given at the end of the last section, although other factors can also contribute.

As stated in Section 10.9, surges are produced by an Ekman flux toward the shore, and it is useful to subdivide the flow, as in Section 9.2, into a forced Ekman flow

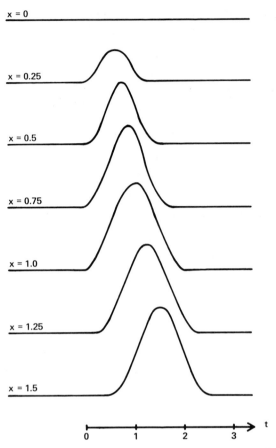

Fig. 10.12. The variation of surface elevation with time t for a model storm surge driven by a wind blowing parallel to the coast in the region $0 \le x \le 1$ and lasting for unit time (half a sine-wave cycle). The interval of one-quarter unit between stations is roughly comparable with that used in Fig. 10.10. The observed storm surge in Fig. 10.10 does not disappear so quickly, suggesting that the wind came up quickly but did not fall off at the same rate. Despite the simplicity of the model, the growth and propagation of the surge is similar to that observed.

and a part associated with the pressure field, i.e.,

$$u = U_E/H + u_p, \qquad v = V_E/H + v_p, \tag{10.10.22}$$

where U_E and V_E satisfy (9.2.7), u_p and v_p satisfy the homogeneous equations (9.2.2). The sea-level changes are associated only with the part (u_p, v_p) that satisfies the homogeneous equations and is forced by the boundary condition

$$v_p = -V_E/H \qquad \text{at} \quad y = 0, \tag{10.10.23}$$

i.e., by having a flow across the boundary that is just sufficient to balance the Ekman flux. Thus a surge could be simulated in the laboratory by having a flexible side wall that could be moved sideways. The horizontal displacement involved is the integral of the Ekman current, which can amount to about 10 km for cases corresponding to severe North Sea surges!

The above discussion shows the general nature of the surge, and equations like (10.10.18) can indeed be useful for surge forecasting. In practice, however, effects of geometrical details, friction, and nonlinear interactions [see, e.g., Prandle and Wolf (1978)] between surge and tide need to be taken into account for accurate prediction, and this can be done by means of numerical models [see Heaps (1969) and Prandle (1975)] that can be linked with atmospheric models in order to predict surges (Flather and Davies, 1976). Review articles on computation of surges are given by Welander (1961) and by Reid *et al.* (1977).

10.11   Coastal Upwelling

The previous two sections were concerned with the barotropic response of the ocean when there is a wind-driven Ekman transport toward or away from the coast, and this response is particularly significant with regard to storm surges in shallow seas. In a stratified ocean, there is a *baroclinic* response of the same form, which is most easily illustrated by the two-layer model discussed in Section 9.10. The main part of the response is confined to a distance from the coast of an internal Rossby radius (typically 30 km in deep-sea regions). The equations for the baroclinic response are (9.10.16) and (6.3.5), which have exactly the same form as those for the barotropic case, and the solutions found in the last two sections can be applied. The region in which the baroclinic response occurs is very small compared with the *barotropic* Rossby radius, so the rigid-lid approximation can be used, i.e., free surface movements can be ignored when discussing the internal motion (see Section 6.3).

The particularly important case of internal motion is the one in which the Ekman transport, which is confined to the upper layer, is *away from* the coast. This requires deep water to *upwell* to replace the surface water being removed. The deep water contains nutrients that, if brought up into the sunlit regions of the upper ocean, can be utilized to produce plant life, which in its turn is necessary for animal life. The result is that the world's most important fisheries are in coastal regions. [Models of the biological factors are discussed, e.g., by Walsh (1977).]

The amount of upwelling in a given region is proportional to the change in Ekman flux across it. In coastal regions, this flux changes from zero at the coast to its deep-sea value in a distance of the order of the internal Rossby radius, i.e., about 30 km. The same change in Ekman flux in the open ocean is only observed over distances of the order of the synoptic scale, i.e., 1000–3000 km, so on this reckoning coastal upwelling is 30–100 times stronger than open ocean upwelling. This explains the importance of coastal fisheries relative to deep-sea fisheries. In fact Ryther (1969, p. 75) put the situation as follows:

> The open sea [90% of the ocean] is essentially a biological desert. It produces a negligible fraction of the world's fish catch at present and has little or no potential for yielding more in the future. Upwelling regions, totaling no more than about one-tenth of 1 percent of the ocean surface (an area roughly the size of California) produce about half the world's fish supply. The other half is produced in coastal waters and the few offshore regions of comparably high fertility.

By coastal waters, Ryther means those within the 100 fathom contour that is not included in the upwelling areas (about 7.5% of the ocean), whereas the offshore regions referred to (about 2.5% of the ocean area) are regions of fronts, divergences, etc.

The local solution of the baroclinic equations (9.10.16) and (6.3.5), i.e., the solution for the case in which there is no variation in x, has the same form as (10.9.5), (10.9.7), and (10.9.8), namely,

$$\hat{v} = -(X_s/\rho f H_1)(1 - e^{-y/a}), \tag{10.11.1}$$

$$h = -(cX_s/\rho g'H_1)e^{-y/a}t, \tag{10.11.2}$$

$$\hat{u} = (X_s/\rho H_1)e^{-y/a}t, \tag{10.11.3}$$

where c is now the speed of long internal waves, given by (6.3.7), i.e.,

$$c^2 = g'H_1 H_2/(H_1 + H_2), \tag{10.11.4}$$

and $a = c/f$ is the Rossby radius. The solution is illustrated in Fig. 10.13, which is the analog of Fig. 10.11 for internal motion.

Solutions of the above form were found by Charney (1955a) and Yoshida (1955). For a stress of 0.1 N m^{-2}, $g' = 0.03$ m s^{-2}, $H_1 = 100$ m, and $H_2 \gg H_1$, the rate of upwelling at the coast is about 5 m day^{-1}. The longshore current $\hat{u}$ for the same conditions increases by 0.1 m s^{-1} day^{-1} at the coast. This current has been called a *coastal jet* by Charney. For a continuously stratified ocean [see Gill and Clarke (1974)], each mode has the above form and the full solution can be found by superposition of modes.

The principal upwelling areas of the world are found on the eastern boundaries of the oceans [see Hart and Currie (1960) and the review by R. L. Smith (1968)], and these areas are also noted for the equatorward currents that are found there [see the review by Wooster and Reid (1963)]. They are also regions in which the wind is equatorward (Wooster and Reid, 1963), but there is by no means a precise coincidence of upwelling with equatorward currents or equatorward winds. Regions particularly noted for upwelling are the coast of Peru, the U.S. West Coast, and the coasts of NW and SW Africa. There is also upwelling off Somalia and Arabia during the Southwest Monsoon. As already mentioned, half the world's fish are caught in these regions.

An example, showing measured currents and temperature structure in an upwelling region, is shown in Fig. 10.14, which may be compared with the transient theoretical solution shown in Fig. 10.13. On 8 March 1974, the wind had been light, but grew during 9 March to reach a strength of 10 m s^{-1} from a little east of north, and this wind was maintained over the period of measurement. The Ekman transport $X_s/\rho f$ (where X_s is the stress, ρ the density, and f the Coriolis parameter), if spread over a depth of 30 m, gives an offshore current of about 0.1 m s^{-1}, and a current of this order can be seen in Fig. 10.14a in the upper 30 m, with an onshore flow at lower levels. The consequent upwelling of the temperature structure can be seen in Fig. 10.14b, and the longshore current structure associated with this is shown in Fig. 10.14c. The Rossby radius (based on a reduced gravity g' of 0.03 m s^{-2} and a depth of 100 m) is about 30 km.

Some good illustrations of the effect come from measurements in the Great Lakes,

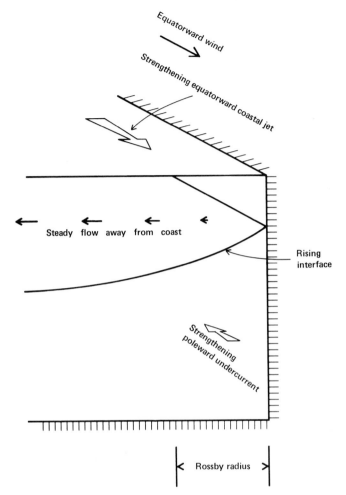

Fig. 10.13. The "local" upwelling solution, which is the baroclinic analog of the surge solution shown in Fig. 10.11. Upwelling at a steady rate is caused by offshore Ekman transport. A coastal jet is generated in the upper layer, this being in the same direction as that of the wind. An undercurrent in the opposite direction is also generated. Directions are for an eastern boundary.

which have widths large compared with the internal Rossby radius. For instance, Fig. 10.15 shows a section across Lake Michigan shortly after a strong N–NE wind impulse that has produced upwelling on the eastern side and downwelling on the western side, as expected from the direction of the Ekman transport. The internal Rossby radius in this case is about 5 km, and the zone affected by the upwelling or downwelling is about twice this distance. Other examples can be found, for instance, in the review of Csanady (1977a) about coastal jets.

Upwelling is very often strong enough for the sea-surface temperature to be affected, and thus low temperatures in a thin strip adjacent to the coasts are a signature of upwelling. It is, in fact, quite common to find coastal waters colder than those offshore. The process is not linear because large displacements are required to make

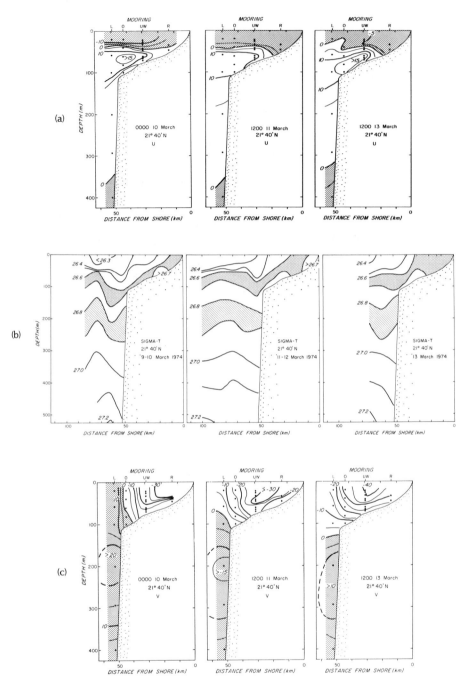

Fig. 10.14. Sections off NW Africa at 21°40′N showing (a) the eastward (onshore) component of velocity (off-shore in stippled regions), (b) density (σ_t), and (c) northward (longshore) component of velocity (poleward in stippled regions). The wind was blowing fairly steadily from a little east of north at 10 m s^{-1}, having risen to this magnitude on 9 March. The Rossby radius is about 30 km. [From Barton *et al.* (1977, Figs. 11, 7, and 10, respectively).]

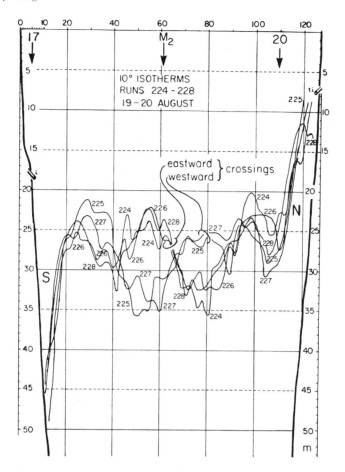

Fig. 10.15. Thermocline depth, as shown by the 10°C isotherm, in some east–west transects of Lake Michigan made on 19–20 August 1963, shortly after a N–NE wind impulse. This produced upwelling on the eastern boundary and downwelling on the western boundary. The Rossby radius is about 5 km. The figure also shows transient oscillations in the middle of the section. [From Mortimer (1968); see also Mortimer (1974, Fig. 26).] The horizontal scale is in km.

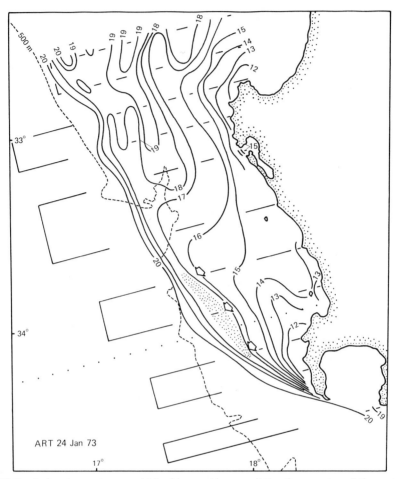

Fig. 10.16.   Surface temperature, as obtained by an airborne radiation thermometer, off the west coast of South Africa [from Bang (1973b)]. Offshore temperatures are above 20°C, but upwelling produces much lower temperatures near the shore. Note the large temperature differences near Cape Point, which is at the southern end of the upwelling zone.

the surface colder and mixing is usually involved. Also, downwelling cannot make the surface any warmer, so a succession of upwelling and downwelling events with mixing still tends to make the coastal zone colder on average. An example of the large surface effects that may be found is shown in Fig. 10.16. The effect is quite obvious at the shore because beaches only a few miles apart can have water temperatures differing by as much as 10°C [see Fig. 1 of Bang (1973a)]. The upwelling can also be seen in the color of the water and in the abundance of sea life.

10.12   Continental Shelf Waves

When spatial variations of the wind are taken into account, the solution for each mode will have the same form as that of the surges studied in Section 10.10, i.e., will consist of an internal Kelvin wave, propagating along the coast with an amplitude that will be growing or decaying according to the sign of the wind effect of the point in question (Gill and Clarke, 1974). Each mode n has a different free speed c_n of propagation (see Section 9.10), and hence has a different Rossby radius $a_n = c_n/f$ and a different coefficient for forcing (i.e., a different equivalent forcing depth—see Section 9.10), so the net effect can be evaluated only when the modes are superposed. In a two-layer system, however, the situation is simple because there is only one baroclinic mode, and therefore the solution given in Section 10.10 can be applied directly.

The theory of forced internal Kelvin waves requires a flat-bottomed ocean, so the coastal boundary needs to be a vertical cliff. If allowance is made for the sloping boundary at the coast, not only is the Kelvin wave modified, but there also exist new forms of wave associated with the sloping bottom. The new forms of wave exist even in the absence of stratification, in which case they are called *continental shelf waves*. The properties of these waves in a homogeneous ocean will now be studied.

Consider a homogeneous ocean whose depth $H(y)$ is a function only of distance y from the coast. It will be supposed that the scale of depth variations (which is also the scale of the response) is small compared with the external Rossby radius, so that the rigid-lid approximation can be made. Then the surface elevation term in the continuity equation (5.6.8) can be ignored, so that it gives

$$\frac{\partial}{\partial x}(Hu) + \frac{\partial}{\partial y}(Hv) = 0 \quad \text{or} \quad \frac{\partial u}{\partial x} + \frac{\partial v}{\partial y} = -\frac{1}{H}\frac{dH}{dy}v. \tag{10.12.1}$$

Substitutions for the divergence in the vorticity equation (7.2.7) then give

$$\frac{\partial}{\partial t}\left(\frac{\partial v}{\partial x} - \frac{\partial u}{\partial y}\right) - \frac{f}{H}\frac{dH}{dy}v = 0. \tag{10.12.2}$$

Now (10.12.1) can also be used to define a stream function ψ such that

$$Hu = -\partial\psi/\partial y, \qquad Hv = \partial\psi/\partial x. \tag{10.12.3}$$

Substituting for u and v in (10.12.2) then gives

$$\frac{\partial}{\partial t}\left\{\frac{\partial}{\partial x}\left(\frac{1}{H}\frac{\partial\psi}{\partial x}\right) + \frac{\partial}{\partial y}\left(\frac{1}{H}\frac{\partial\psi}{\partial y}\right)\right\} - \frac{f}{H^2}\frac{dH}{dy}\frac{\partial\psi}{\partial x} = 0. \tag{10.12.4}$$

Traveling wave solutions of the form

$$\psi = H^{1/2}\phi(y)\exp(ikx - i\omega t) \tag{10.12.5}$$

can be found, where ϕ satisfies

$$\frac{d^2\phi}{dy^2} + \left\{\frac{d}{dy}\left(\frac{1}{2H}\frac{dH}{dy}\right) - \left(\frac{1}{2H}\frac{dH}{dy}\right)^2 - k^2 + \frac{fk}{\omega}\frac{1}{H}\frac{dH}{dy}\right\}\phi = 0. \tag{10.12.6}$$

A particularly simple case is the one in which the depth increases exponentially with distance from the coast, i.e.,

$$H = H_0\exp(2\lambda y). \tag{10.12.7}$$

The solutions have the form

$$\phi \propto \sin ly \qquad (10.12.8)$$

and therefore the dispersion relation is

$$\omega = 2fk\lambda/(k^2 + l^2 + \lambda^2). \qquad (10.12.9)$$

The dispersion curve for a fixed value of l is shown in Fig. 10.17.

Consider some typical values, e.g., when λ^{-1} is 30 km. The value of l depends on the width of the region in which the depth varies exponentially, and also on the properties of the ocean seaward of that region. There is normally an infinite set of values that l may have, the smallest value usually being close to λ. Consider a wave for which $l = \lambda$. Disturbances generated by storms and having similar scales will have k^{-1} large compared with λ^{-1}, so their free wave speed is given by

$$\omega/k = 2f\lambda/(l^2 + \lambda^2). \qquad (10.12.10)$$

In the case $l = \lambda$, this equals f/λ, which is typically about 3 m s^{-1}. The direction of travel depends on the sign of f and is the same as that of a Kelvin wave (assuming depth increases with distance offshore). The *maximum* frequency is given by

$$\omega_{max} = f\lambda(l^2 + \lambda^2)^{-1/2}, \qquad (10.12.11)$$

which is always less than f and is equal to $0.7f$ when $l = \lambda$. At this frequency the longshore group velocity vanishes, so stationary features such as ridges and canyons that run across the shelf could generate lee waves [cf. Martell and Allen (1979)] with the corresponding scale, namely, $(l^2 + \lambda^2)^{-1/2}$—which has typical values of about 20 km. There is evidence for maxima in the coherence spectrum between currents and sea level at this frequency (Cutchin and Smith, 1973). Shorter waves have phase velocity in the same direction as that of the long waves, but group velocity in the opposite direction.

The mechanism [cf. Longuet-Higgins (1965c, Fig. 13)] is illustrated in Fig. 10.18. The solid line represents a line of particles that, in their undisturbed condition, lie along a depth contour parallel to the shore. During the motion, a particle conserves [see (7.10.9)] its potential vorticity Q, which in the case of negligible surface motion is given by

$$Q = (f + \zeta)/H, \qquad (10.12.12)$$

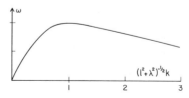

Fig. 10.17. The dispersion curve (10.12.9) for a "continental shelf wave" over topography that increases exponentially with distance from the coast. The phase velocity is in the same direction as for Kelvin waves, i.e., with the shallow water on the right in the northern hemisphere. Long waves have group velocity in the same direction, but short waves have group velocity in the opposite direction.

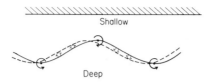

Fig. 10.18. The mechanism for propagation of shelf waves. The solid line represents a line of particles whose undisturbed positions lay along a depth contour. Conservation of potential vorticity implies that particles moving into shallower water acquire anticyclonic relative vorticity and those moving into deeper water acquire cyclonic relative vorticity. The sense of this relative vorticity is shown for the northern hemisphere. The motion induced by this vorticity field is indicated by the broad arrows, so the line of particles moves to the position indicated by the dashed line.

where ζ is the relative vorticity of the particle and H is the depth. If H_{und} was the depth of water when the particle was in its undisturbed state ($\zeta = 0$), $Q = f/H_{\text{und}}$ and thus (10.12.12) gives

$$\zeta/f = (H - H_{\text{und}})/H_{\text{und}}. \tag{10.12.13}$$

Thus particles displaced into deeper water (offshore) acquire cyclonic vorticity and particles displaced into shallower water (onshore) acquire anticyclonic vorticity.

Now consider the motion *induced* by these vortices. This is shown by arrows in the figure for particles near their undisturbed position. The result is that the particles move to the position marked by the dashed line and so, in the northern hemisphere, the wave progresses along the contour with shallow water on the right.

The possible values of l will now be found for a shelf profile given by (10.12.7) for $0 \le y \le B$ and with a flat-bottomed ocean for $y > B$, where B is a measure of the shelf breadth. This profile was used as a model of the topography off the east coast of Australia by Buchwald and Adams (1968). The solution is

$$\psi = \begin{cases} \exp(\lambda(y - B)) \sin ly \exp(ikx - i\omega t) & \text{for} \quad 0 \le y \le B, \\ \exp(k(B - y)) \sin lB \exp(ikx - i\omega t) & \text{for} \quad y > B, \end{cases} \tag{10.12.14}$$

the variation over the exponential shelf being given by (10.12.5), (10.12.7), and (10.12.8), and the variation in the flat-bottomed ocean being potential flow by (10.12.4). The surface elevation η is, by substitution of (10.12.3) in the momentum equation (7.2.1) and assumption of the wavelike form (10.12.5), given by

$$gH\eta = f\psi - (\omega/k)\,\partial\psi/\partial y. \tag{10.12.15}$$

This shows that $\partial\psi/\partial y$ must be continuous at $y = B$, and use of (10.12.14) then gives

$$l^{-1} \tan lB = -(\lambda + k)^{-1}. \tag{10.12.16}$$

There are a countable infinity of possible values of l, and the dispersion curves that correspond to the first few are shown in Fig. 10.19 for the case $\lambda B = 2.7$, as used by Buchwald and Adams. Mode n has the property that

$$(n - \tfrac{1}{2})\pi < lB < n\pi. \tag{10.12.17}$$

In practice, the potential flow solution over the flat-bottomed ocean would not be expected to apply because of stratification effects or free-surface effects. However,

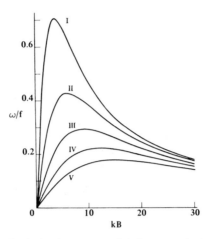

Fig. 10.19. Dispersion curves for the first five continental shelf-wave modes for a shelf whose depth increases exponentially to a value of 5 km at a distance $B = 80$ km offshore and then remains constant. This gives a good model of the shelf topography near Sydney. [From Buchwald and Adams (1968, Fig. 4).]

such effects alter only the magnitude of the right-hand side of (10.12.16) somewhat, and there is still a countable set of modes satisfying (10.12.17). An example in which stratification over the flat-bottomed section alters the structure of the modes is given by Mysak (1967) and Gill and Clarke (1974). Further discussion of shelf waves may be found in the review of LeBlond and Mysak (1977).

The essential property of the ocean that permits continental shelf waves to exist is the variation in depth. A coastal boundary is not necessary—in fact, waves of the same sort can propagate in the deep ocean along topographic features as shown, for instance, by Longuet-Higgins (1968a). However, a coastal boundary *is* important for the *generation* of shelf waves by the wind (Gill and Schumann, 1974). This is so because the Ekman flux produced by the wind is confined to a thin layer near the surface and therefore is unaffected by the underlying topography. The Ekman flux cannot, however, cross a coastal boundary, so a compensating motion below the Ekman layer is required, and the surface must move in such a way as to give the pressure gradients that are associated with the motion below the Ekman layer. Details of the motion below the Ekman layer depend on shelf-wave dynamics. In particular, movement across isobaths (depth contours) causes relative vorticity to be generated, and the field of motion is that induced by the relative vorticity distribution so produced.

The wind that generates shelf waves normally has a scale L that is much larger than the breadth B of the shelf, so a long-wave theory can be utilized. A scaling can be introduced just as in the treatment of storm surges in Section 10.10, except that the Rossby radius a is replaced by the shelf breadth B and the speed $c = af$ is replaced by Bf. The equations, equivalent to (10.10.4)–(10.10.6), that emerge are, for small ϵ and in dimensional form,

$$\partial u/\partial t - fv = -g\,\partial\eta/\partial x + (\rho H)^{-1}X_s(x, t), \qquad (10.12.18)$$

$$fu = -g\,\partial\eta/\partial y, \qquad (10.12.19)$$

$$\partial(Hu)/\partial x + \partial(Hv)/\partial y = 0. \tag{10.12.20}$$

Thus to a first approximation, the longshore flow is in geostrophic balance, only the longshore component of the wind is effective in generating currents and sea-level changes [observational evidence of this has been given by Cragg and Sturges—see Csanady (1981)], and variations in the wind with distance from the coast may be ignored. The potential vorticity equation may be obtained by adding minus the y derivative of (10.12.18), the x derivative of (10.12.19), and $-f/H$ times (10.12.20). When expressed in terms of the stream function ψ defined by (10.12.3), this takes the form

$$\frac{\partial^2}{\partial y\,\partial t}\left(\frac{1}{H}\frac{\partial\psi}{\partial y}\right) - \frac{f}{H^2}\frac{dH}{dy}\frac{\partial\psi}{\partial x} = \frac{X_s}{\rho H^2}\frac{dH}{dy}. \tag{10.12.21}$$

The solution of this equation [see Gill and Schumann (1974)] can be expressed as a sum of shelf modes of the form

$$\psi = \sum_{n=1}^{\infty} A_n(x, t)H^{1/2}\phi_n(y), \tag{10.12.22}$$

where ϕ_n satisfies the long-wave equivalent of (10.12.6), namely,

$$\frac{d^2\phi_n}{dy^2} + \left\{\frac{d}{dy}\left(\frac{1}{2H}\frac{dH}{dy}\right) - \left(\frac{1}{2H}\frac{dH}{dy}\right)^2 + \frac{f}{c_n H}\frac{dH}{dy}\right\}\phi_n = 0. \tag{10.12.23}$$

This equation has solutions only for the eigenvalues c_n, which give the speeds of freely propagating long waves. When substituted in (10.12.21), (10.12.22) gives the following equation for A_n:

$$\frac{1}{c_n}\frac{\partial A_n}{\partial t} + \frac{\partial A_n}{\partial x} = \frac{b_n X_s(x, t)}{\rho f}, \tag{10.12.24}$$

where b_n is a constant that comes from the expansion of the right-hand side of (10.12.21) [see Gill and Schumann (1974)].

This equation (10.12.24) has the same form (10.10.14) as that governing the development of a storm surge, so the amplitude of each shelf-wave mode behaves in the same way. The total effect can be calculated only by superposition of the modes. If, however, one mode dominates, the behavior will be just like that for a surge. If several modes are important, the effect is to spread out the surge—a form of dispersion.

In practice, a shelf wave, in the absence of forcing, would lose energy by friction, interaction with small-scale topographic variations, etc. Gill and Schumann (1974) suggested that allowance be made for such effects by adding a dissipative term to (10.12.24), so that it becomes

$$\frac{1}{c_n}\left(\frac{\partial A_n}{\partial t} + r_n A_n\right) + \frac{\partial A_n}{\partial x} = \frac{b_n X_s(x, t)}{\rho f}. \tag{10.12.25}$$

Brink and Allen (1978) have shown that bottom friction causes the amplitude equation to take precisely this form and estimated the time scale r_1^{-1} for a particular case

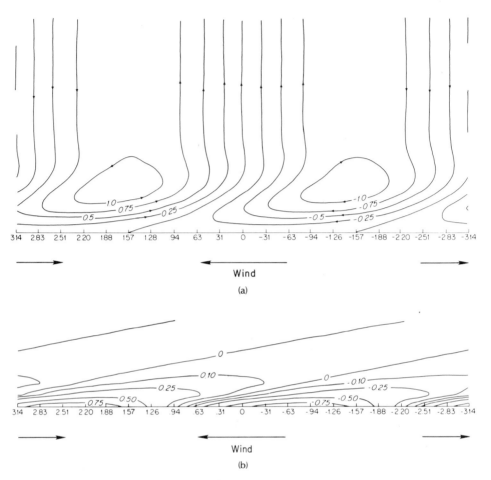

Fig. 10.20.   Contours of (a) stream function and (b) surface elevation for steady flow over a shelf whose depth increases linearly with distance y from the coast. The flow is driven by a periodic wind stress of the form $X_s = -X_{max} \cos kx$ and is in equilibrium with bottom friction of magnitude such that $k(v/2f)^{1/2} = 0.016 \, dH/dy$. The unit of stream function is $X_{max}/fk\rho$ and that of surface elevation is $2X_{max}/\rho g\{k(v/2f)^{1/2} \, dH/dy\}^{1/2}$. The horizontal axis is marked in units of $-kx$. [From Csanady (1978, Figs. 4 and 5).]

to be 5 days. Bottom friction also causes the near-shore flow to lead the offshore flow in phase, and this effect has been observed.

The only modification to the long-wave form of equations when bottom friction is included is the addition of the bottom stress term to (10.12.18). When this has the linear form (9.6.4), (10.12.18) becomes

$$\frac{\partial u}{\partial t} - fv = -g\frac{\partial \eta}{\partial x} + \frac{X_s(x, t)}{\rho H} - \frac{1}{H}\left(\frac{vf}{2}\right)^{1/2} u. \qquad (10.12.26)$$

Solutions of the long-wave equations (10.12.26), (10.12.19), and (10.12.20) in the case of *steady* flow have been discussed by Csanady (1978, 1981). When u and v are eliminated from these, the resulting equation for the surface elevation η is

$$\frac{dH}{dy}\frac{\partial\eta}{\partial x} = \left(\frac{v}{2f}\right)^{1/2}\frac{\partial^2\eta}{\partial y^2}, \tag{10.12.27}$$

which, for a shelf with constant slope, is the heat equation, which can readily be solved. At the shore, where H vanishes, (10.12.26) shows that

$$(vf/2)^{1/2}u = \rho^{-1}X_s, \tag{10.12.28}$$

i.e., the flow is downwind in order for the bottom stress to balance the surface stress. This can be seen in the solution for a wind that varies periodically with longshore distance x, which is shown in Fig. 10.20. Further offshore, the longshore flow pattern is displaced relative to the wind in the direction of propagation of a shelf wave, whereas far offshore the flow is simply the Ekman transport at right angles to the wind. Csanady (1981) has discussed observations relating to shelf circulation cells with this character.

10.13 Coastally Trapped Waves

In a stratified ocean of constant depth, free waves can propagate along a coastal boundary in the form of internal Kelvin waves with a typical scale (the Rossby radius) of order 30 km. In Section 10.12 it was shown that in a homogeneous ocean with shelf topography there also exist free waves (continental shelf waves) that propagate along the boundary and that also have a scale of around 30 km. In practice, stratification *and* topography are both present, so hybrid waves with characteristics of both Kelvin and shelf waves exist. These have been called *coastally trapped waves* by Gill and Clarke (1974), and properties of such waves have been calculated by Allen (1975), Clarke (1977), Wang and Mooers (1976), and Huthnance (1978).

An example of a coastally trapped wave, observed off the Peruvian coast, is shown in Fig. 10.21. Here the structure is similar to that of a Kelvin wave associated with the first baroclinic mode. The offshore length scale is 30–60 km (Huyer, 1980), the poleward propagation speed is about 2 m s^{-1} (R. L. Smith, 1978), and the waves contribute to current fluctuations with periods of days to weeks (Brink *et al.*, 1978). The waves are thought to first propagate eastward along the equator as equatorial Kelvin waves (see Chapter 11) and then propagate poleward along the coastal waveguide in the form shown. They also affect sea level and surface temperature as shown in Fig. 10.21.

Evidence for wave propagation along the eastern boundary of the Pacific Ocean can also be seen in coastal sea-level and temperature records (Enfield and Allen, 1980), the propagation speed being of order 1 m s^{-1}. The existence of coastally trapped waves means that whenever the sea level is high near the equator, it also becomes high further poleward, as can be seen in Fig. 10.22. High sea levels near the equator extend as far as San Francisco, beyond which point wind effects become the predominant determining factor. The high sea levels also correspond to high coastal temperatures.

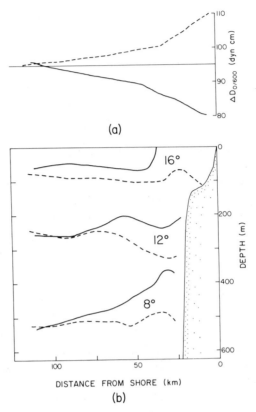

(a)

(b)

DISTANCE FROM SHORE (km)

Fig. 10.21. Offshore profiles of (a) the sea surface (as calculated from the dynamic topography relative to 600 dbar), and (b) selected isotherms at times of very high (31 July 1976, dashed line) and very low (10 August 1976, solid line) sea level at San Juan. The location is at 15°S on the Peruvian coast. The changes are similar to that expected for an internal Kelvin wave associated with the first baroclinic mode, and the poleward-propagation characteristics are consistent with this interpretation [from Huyer (1980, Fig. 5)].

Mortimer (1963) [see also Mortimer (1974)] found evidence from water intake temperature and water-level variations of similar types of waves moving cyclonically around Lake Michigan and Lac Léman. A model for this sort of response in a large lake has been given by Csanady (1972b), and motion of the same type has been observed by Walin (1972a,b). Similar sorts of waves can exist in the atmosphere, and Gill (1977b) has suggested that the "coastal lows" that move around the southern coast of Africa with the land on their left are a form of forced internal wave of the Kelvin type [see also Bannon (1981), Nguyen and Gill (1981)]. The progression of such a low can be seen in Fig. 10.23. A feature such as this can behave like an internal wave because of the strong inversion at about the 1 km level, and an approximate treatment replaces this inversion by the interface in a two-layer model like that used in Section 6.3. This wave can be of the Kelvin type because southern Africa consists largely of an elevated plateau that rises above the inversion and so forms a lateral boundary for the fluid in the lower layer. This boundary blocks the flow that would occur in the absence of a boundary, so a compensating flow must be added in order

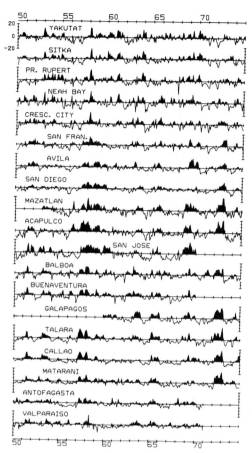

Fig. 10.22. Time series of monthly anomalies of sea level from the American Pacific coast from 1950 to 1974 inclusive. The stations are in order of latitude from Yakutat, Alaska (60°N) to Valparaiso, Chile (33°S). Positive anomalies are shaded black. [From Enfield and Allen (1980, Fig. 4a).]

to satisfy the boundary condition. This additional flow can take the form of a coastally trapped wave. In practice, the amplitude of the coastal lows can become very large, wave steepening can occur, and a sharp front—rather like a tidal wave—can form. When such a front passes, the longshore component of the wind below the inversion level has been observed to increase by 16 m s^{-1} within half an hour (Preston-Whyte, 1975).

Enfield and Allen's (1980) analysis of sea-level variations on the Pacific coast shows that from San Francisco northward the association with tropical anomalies becomes small but that correlations with longshore wind-stress changes become significant. One factor that is involved is the change of structure of coastally trapped waves as latitude (and hence f) increases, so the Rossby radius and its ratio with the shelf width decreases. Sugonihara (1981) found in a numerical experiment that energy that is in Kelvin-like waves at low latitudes is transferred to quasi-barotropic shelf-wave-like modes that lose energy more easily, e.g., by radiation into westward-propagating planetary waves (see Chapter 12).

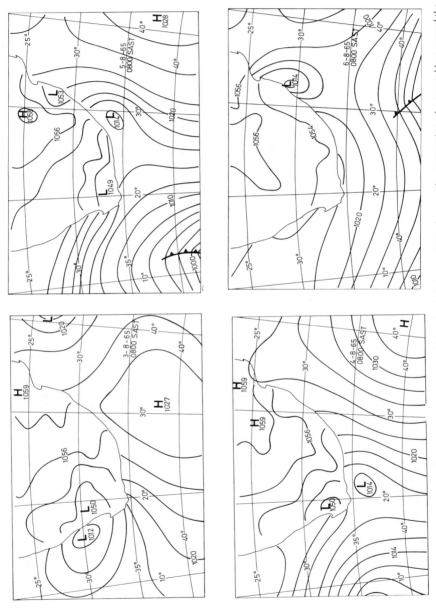

Fig. 10.23. Daily surface pressure maps at 0800 South African Standard Time for 3–6 August 1965, showing the progression of a coastal low around the southern tip of Africa from the Atlantic to the Indian Ocean. The average period between lows is 6 days and the average propagation speed is 6.5 m s^{-1}. The Rossby radius is about 200 km and the free Kelvin-wave speed is about 20 m s^{-1} [from Gill (1977b, Fig. 1)].

The generation of coastally trapped waves by wind blowing over a stratified ocean with a shelf topography can be studied by using a long-wave theory, just as in the discussion of storm surges and shelf waves. The Ekman flux toward the shore needs to be compensated by an equal and opposite flow at the coast, and a sum of coastally trapped waves with this property can be found. Also, it can be shown (Clarke, 1977) that the amplitude A_n of each of the waves satisfies the same equation (10.12.24) as in the particular case of a shelf wave.

Coastally trapped waves have been observed in many situations and reviews are given by LeBlond and Mysak (1977, 1978), Mysak (1980a,b), Allen (1980), and Beardsley and Boicourt (1981). For instance, Hamon's (1962) observation that sea level and air pressure on the east coast of Australia did not satisfy the "inverse barometer" relationship (see Section 9.9) led to the shelf-wave model of Robinson (1964) and the wind-generated shelf-wave theory of Adams and Buchwald (1969). Observed properties of the fluctuating currents off Oregon have been studied by Kundu and Allen (1976), who found mainly barotropic motion along isobaths with a maximum close to the shore. Dynamic balances for the fluctuating currents in three localities have been studied by Allen and Smith (1981).

A method of studying the importance of propagation on the structure of coastal waves has been used by Hamon (1976) and Clarke (1977), who compared the time series of sea level at a certain point with the time series generated by an equation of the form (10.12.25), using winds observed in the appropriate direction along the coast. This can give only a crude representation of the coastally trapped waves since it can include the effect of only one mode, but it does show that taking account of the propagation of one mode improves the "prediction" of sea level relative to the best prediction that can be made by using only local winds. Hamon's results are shown in Fig. 10.24. Curve A is the result of comparing the sea-level record at 29°S with the calculation based on (10.12.25), using the winds over the shelf to the south, a speed c of 4 m s^{-1} (derived from earlier empirical studies), and zero decay rate ($r = 0$). Curve B uses local winds only, i.e., is effectively based on (10.12.25) with infinite decay rate, i.e., the wave has no memory of wind effects experienced as it traveled toward the point of observation. The coherence is much better for case A for periods between 4 days and 2 weeks. A similar analysis by Clarke (1977), using data from the west coast of North America showed that for periods of 4 days and longer the "prediction" made by taking account of wave propagation is better. A similar formulation has also been found useful for Lake Ontario (Bennett and Lindstorm, 1977).

Although the analysis of sea-level data leads to a simple picture of coherent long waves moving along the coast, detailed studies of currents and of temperature/salinity data often show considerable lack of coherence because a great deal of energy is also found in smaller-scale motions. These may be eddies, possibly produced by current instabilities, or wave motions due to small-scale features of bottom topography. Such motions add a random element to upwelling that may lead to an unpredictable patchiness in the biomass distribution. This sort of patchiness is, for instance, a feature of the upwelling region off NW Africa at latitudes near 20°N, where a water-mass boundary is found.

On the other hand, there are certain locations where upwelling is preferred due to local topographic features. For instance, Fig. 10.25 shows the upwelling pattern

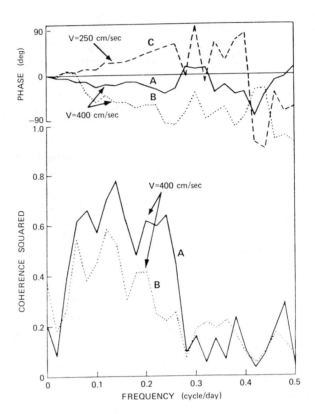

Fig. 10.24. Results of cross-spectral analysis between sea level at Evans Head (on the Australian coast at 29°S) and the solution A_n of (10.12.25), where X_s is the wind-stress component parallel to the coast, b_n is a constant, x is the distance along the coast (positive northward), and r_n is the attenuation rate. The winds were observed over a stretch of coast extending to Gabo Island (about 38°S) and the integration commenced from this point. The record length was 790 days and $v = c_n$ was given the value shown. Curves A and C correspond to the solution for zero attenuation rate, whereas curve B corresponds to a large attenuation rate, i.e., only uses winds at the nearest point. [From Hamon (1976).]

obtained by Peffley and O'Brien (1976) in a model using topography and coastline shape based on that found off Oregon. The strong upwelling found near the model Cape Blanco ($y = -150$ km) was shown to be due to topography rather than coastline shape because a uniform depth model with the same coastline slope gave upwelling rates that were essentially uniform along the coast.

Preferred upwelling locations can also be produced by the topography of the nearby land since strong winds can be directed out over the sea through gaps in the mountains, giving very nonuniform stress distributions. A noted example is found in the Gulf of Tehuantanpec (about 15°N, 95°W), for which Roden (1961) showed that cool surface temperatures are associated with northerly gales that, after being funneled through the mountains, can produce vertical Ekman pumping velocites of the order 10 m day^{-1} (10^{-4} m s^{-1}).

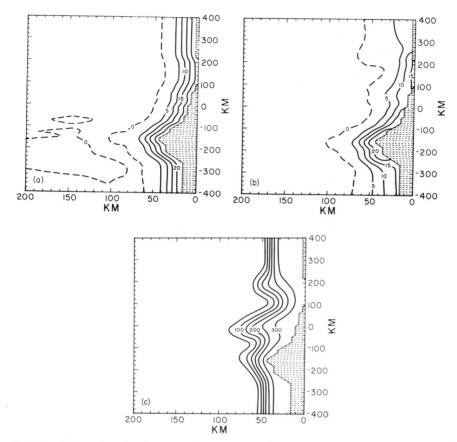

Fig. 10.25.   (a) Pycnocline elevation produced at day 5 in a model ocean of constant depth and with a coastline shape like that of Oregon. The upwelling is produced by an equatorward wind stress and appears to be little affected by the coastline shape. (b) The corresponding picture when the model [shown in (c)] of the shelf topography off Oregon is added. Now the amount of upwelling shows considerable variation along the coast. The results are from a numerical integration by Peffley and O'Brien (1976, Figs. 3b and 5).

10.14   Eastern Boundary Currents

At latitudes below $45°$, the winds along the eastern boundaries of the oceans are predominantly equatorward (Wooster and Reid, 1963), i.e., such as to produce equatorward surface currents, and this is generally what is observed. Reviews about these currents and the associated upwelling regions are given by Wooster and Reid (1963) and by R. L. Smith (1968). However, it should be realized that there is considerable variability on the time scale of a few days (see Fig. 10.14c), that there can be fine spatial structure associated with the fronts that can form (Bang and Andrews, 1974), and that currents in one season can be quite different from those in another season [see, e.g., Wooster *et al.* (1976)].

The best-documented eastern boundary current is the one off the coast of Oregon. Figure 10.26a shows the density field at 44°39′N in two extreme winters, the distribution in other winters lying between these extremes. Figure 10.26b shows two extreme summer distributions in the same format. The differences between the summer and winter patterns are considerably greater than those between extreme examples of

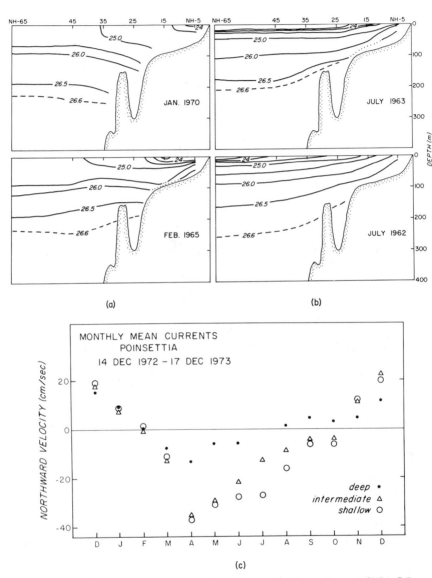

Fig. 10.26. (a) Two extreme winter and (b) two extreme summer density sections at 44°39′N off Oregon. The distance scale is in nautical miles from the coast. [From Huyer (1977).] (c) Monthly mean currents off Oregon at 44°45′N in 100 m of water for (○) a shallow depth (25 m), (△) an intermediate depth (40 m), and (●) a depth of 80 m, i.e., 20 m above the bottom. [From Huyer *et al.* (1975).] (d) Offshore Ekman transport for the eastern boundary of the North Pacific as a function of latitude and time of year. [From Bakun *et al.* (1974).]

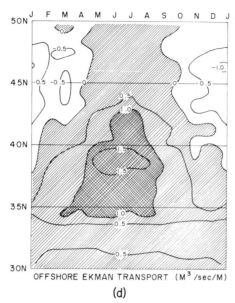

(d)

Fig. 10.26.   *(continued)*

the same season. In winter, light water ($\sigma_t < 25$) is found only in the near-shore zone, i.e., within about 70 km of the coast. In the summer upwelling season, however, water of this density is found only offshore and relatively dense water ($\sigma_t \approx 26.5$) is found near the coast. The associated currents in 100 m of water at 44°45′N are shown in Fig. 10.26c for a particular year. In the winter, the currents were found to be poleward and to vary little with depth. In the summer, the currents were equatorward, but were much stronger at the surface than they were at depth. Note that a seasonal oscillation with a velocity amplitude of 20 cm s^{-1} corresponds to a displacement amplitude of 1000 km, i.e., water found in the coastal current at the end of summer could have originated from 55°N, whereas that found at the end of winter could have originated from 35°N. It is certainly true that there is considerable contrast in water-mass properties between the two seasons (Huyer, 1977). The offshore Ekman transport, which is probably the main driving force for the seasonal oscillations, is shown in Fig. 10.26d as a function of latitude and time of year. Note that at 45°N the mean current is equatorward even though the mean wind is poleward. However, the mean wind does become equatorward a few degrees to the south.

The fact that the seasonal variations in the coastal current are a feature only of the coastal zone, and have little relation to what happens offshore, is illustrated in Fig. 10.27. The first panel (a) shows the variation in dynamic height at various depths as calculated by Reid and Mantyla (1976) for January, whereas (b) shows the equivalent diagram for summer (July). The variations with time of year of the dynamic height at the coast agree well with the observed variations in sea level, as shown in (c). Panels (a) and (b) show that the region of rapid change in dynamic height is confined to a distance of about 200 km from the coast and is possibly related to the width of the continental slope. Comparison of the seasonal variations of dynamic height at different distances from the shore confirms this, showing that the variations

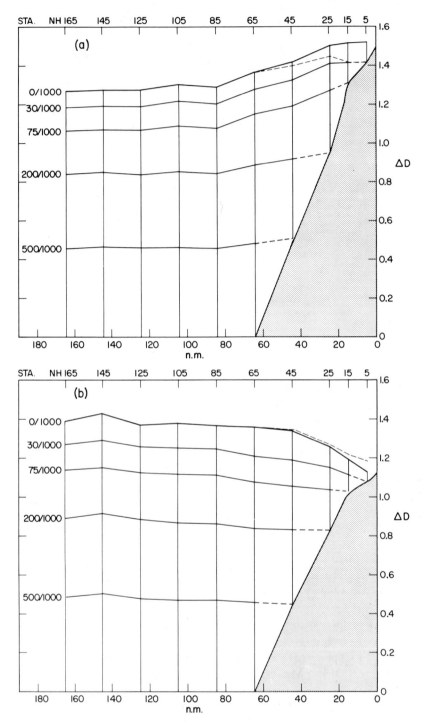

Fig. 10.27. Variation of dynamic height (in dynamic meters) relative to 1000 db for (a) January 1966 and (b) July 1966 for a section off Oregon at 44°38′N. [From Reid and Mantyla (1976); *J. Geophys. Res.* **81**, 3100 (Fig. 4); copyright by the American Geophysical Union.] Values have been extrapolated over the continental slope. The distance scale is in nautical miles (54 NM = 100 km) and station numbers give distance from the coast in nautical miles. (c) The seasonal variations in surface dynamic height calculated by the same method. The dashed line in the first panel shows sea-level variations at a nearby tide gauge. [From Huyer (1977).]

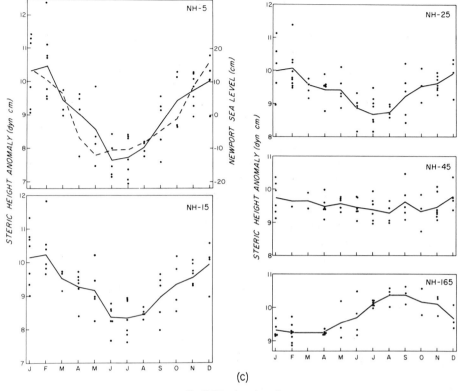

Fig. 10.27.   (*continued*)

are reduced to practically zero at a distance of 80 km and have quite a different phase at a distance of 300 km. Note that since currents are related to the gradient of dynamic height, it is implied that the boundary current has a width of about 200 km and reverses in direction each season.

A complete theory of eastern boundary currents is not available, but various aspects have been studied and a summary is now given.

(a)   The *wind-driven Ekman transport* is a major driving mechanism, the nature of which was illustrated in Section 10.11. For instance, Fig. 10.26d implies that, off the coast of Oregon, megatons (Mton) of water per meter of coast are moved offshore in summer and moved back again in winter. The water moved offshore is light surface water that causes dense water from lower levels to be brought to the surface to replace it, as is seen in Fig. 10.26a,b. The volumes of water involved are of the same order as are those expected from the Ekman transport. The pressure field is therefore altered as shown in Fig. 10.27a,b, and consequently long-shore currents in geostrophic equilibrium with this pressure field are set up.

(b)   The response is *not purely local* because if the thermocline is raised by some mechanism at one latitude, that information is carried poleward by coastally trapped waves, and so the thermocline tends to be raised also at all points poleward of the latitude of forcing. This aspect was studied in Sections 10.11–10.13. This provides a possible explanation of why the mean currents off Oregon are not in the direction of the local mean winds, but are in the direction that the mean winds have further south.

(c)   Another example of nonlocal influence in establishing boundary currents was provided by the example, studied in Section 10.7, of adjustment to equilibrium in a channel. This example emphasized the *one-way* spread of influence, i.e., a boundary current was set up on only one side of the initial discontinuity. The effect is also seen in numerical experiments such as that of Sugonihara (1974).

(d)   The previous example also points to the effect of another form of forcing—by buoyancy. If a river flows out to sea and adds to the volume of light water at a certain point, this will cause the thermocline to be depressed there and also, because of coastal waves, at all points poleward. Since a depression of the thermocline near the coast corresponds to a poleward current, such a current is established from the point at which the river outflow is introduced.

(e)   The description of buoyancy driving in (d) is in terms of linear dynamics, but the same features are found in the nonlinear problem in which a plume of low-density fluid is discharged into a resting fluid of higher density: the plume, if entering on an eastern boundary, tends to be deflected poleward and forms a poleward-flowing boundary current. An example is the Norwegian coastal current that is driven by fresh water inflow from rivers (Monk, 1981). A wedge of light water is thereby formed at the coast, just as was observed in the winter sections off Oregon (Fig. 10.26a), and this corresponds geostrophically to a poleward current. Laboratory and numerical models of this process have been studied by Yoon and Sugonihara (1977), Endoh (1978), and Whitehead (1981). The East Greenland Current, which is on a western boundary, has a similar structure (Wadhams *et al.*, 1979).

(f)   Very strong currents can be subject to hydraulic phenomena caused by changes in shelf topography. The theory is discussed by Gill (1977a) and Gill and Schumann (1979).

(g)   Intense upwelling can lead to strong surface fronts, such as those observed by Bang and Andrews (1974). A model study has been made by Hamilton and Rattray (1978), and Sugonihara (1977) has shown how a double-cell circulation can be set up once the front is established. Observations of upwelling fronts in Lake Ontario have been discussed in relation to a simple conceptual model by Csanady (1977b).

(h)   In practice, because of dissipative effects, there is a limit to the distance to which information is carried poleward by waves. The effect is modeled empirically by the decay term (10.12.25). The decay distance is that traveled by a wave in a decay time, so a decay time of 3 days for waves traveling at 3 m s^{-1} would give a distance of about 1000 km.

(i)   Bottom friction is an important dissipative agent that is much more effective in shallow water than in deep water. An illustration of bottom friction effects is given by the steady solution of (10.12.26), shown in Fig. 10.20. Another example is provided by a source flowing onto a shelf (Beardsley and Hart, 1978): the flow from the source forms a longshore boundary current moving in the same direction as that of the shelf waves.

(j)   Turbulence produced by flow over the bottom can also be an effective stirring agent in shallow water; e.g., stirring by tidal motion can make the whole water column homogeneous [see, e.g., Simpson *et al.* (1978) and Simpson (1981)]. This is one reason for the distinct change in temperature/salinity structure that is often found at the shelf break, i.e., at the point where there is a transition from a gently sloping shelf to a steeply sloping continental slope. Another reason is the difference in the effects of heating and cooling different depths of water. Csanady (1971) has suggested a simple model in which the shelf and offshore zones are forced as if there were a partition at the shelf break. A different structure will emerge in each zone because of the difference in depth. Then the conceptual partition is removed and geostrophic adjustment takes place, leading to distributions similar to those observed.

(k) Steady currents can be produced by oscillating winds even when they have zero mean. The mechanisms involved have been studied by Ou and Bennett (1979) in order to explain currents set up in Lake Kinnaret by the daily sea-breeze cycle.

(l) Lakes have special features because of their size, e.g., upwelling produced on one side of the lake can then propagate as a Kelvin wave around to the other side. Numerical models of lakes are discussed, for example, by Bennett (1977) and Simons (1980).

(m) Although many features of eastern boundary currents are determined entirely by forcing and response in the boundary region, some features must depend on relationships with the flow in the rest of the ocean. For instance, the properties of the water in the boundary current will depend on the properties of the ocean in the source region of the current.

(n) The whole of the discussion thus far makes no distinction between eastern and western boundary currents because the models have been for uniform rotation and for small perturbations from a state of rest. In practice, there are two properties that account for the difference: the curvature of the earth, which causes the Coriolis parameter to vary with local latitude (the so-called "beta" effect), and the fact that the ocean is warmer in the tropics than at the poles, i.e., there are north–south gradients in mean properties. Beta effects, which will be considered in Chapters 11 and 12, have been studied in relation to eastern boundary currents by Anderson and Gill (1975) and by Hurlburt and Thompson (1973). The main reason for the difference between east and west is that the curvature of the earth leads to a form of wave (planetary wave) that propagates information *westward*. This means that eastern boundary layers are little influenced by what happens in the ocean interior but western boundary layers are largely a response to what happens in the ocean interior. In particular, a major driving force for the mean circulation on western boundary shelves is the longshore pressure gradient established as the response (see Chapter 12) to winds over the ocean interior. A discussion of the western shelf circulation has been given by Beardsley and Boicourt (1981).

(o) The existence of planetary waves has important implications for the steady circulation near eastern boundaries, as has been shown by a linear model of McCreary (1981b), which is driven by a north–south wind stress Y_s that depends only on north–south distance y. He chooses special forms of friction and diffusion so that the normal mode approach can be used to study steady flows in a stratified ocean of constant depth. For the first few modes, friction is not very important, but the response does not build up indefinitely as in the model of Section 10.11 because of the radiation of planetary waves, as studied in the model of Anderson and Gill (1975). The inviscid solution of the equation (9.10.4) for the *n*th mode is simply

$$g \, \partial \tilde{\eta}_n / \partial y = \tilde{Y}_n, \tag{10.14.1}$$

i.e., a balance between pressure gradient and wind stress. This relation applies in particular at the coast, and it is of interest to note that Enfield and Allen (1980) find empirically that on the U.S. Pacific coast the relation [see (9.10.10)] '

$$g \, \partial \eta / \partial y = Y_s / \rho_0 H^F \tag{10.14.2}$$

applies both to seasonal changes (best-fit value of $H^F = 157$ m) and to anomalies (best-fit value of $H^F = 383$ m).

The inviscid steady solution has no currents, but currents do occur when friction is included and they become stronger as the mode number increases up to about $n = 5$ or 6. For the higher modes, friction is dominant and the appropriate solution of the second of Eqs. (9.10.4) (with $\partial / \partial t$ replaced by a friction coefficient r_n) is simply

$$r_n \tilde{v}_n = \tilde{Y}_n, \tag{10.14.3}$$

whereas the first of Eqs. (9.10.4) gives a geostrophic balance

$$f \tilde{v}_n = g \, \partial \tilde{\eta}_n / \partial x. \tag{10.14.4}$$

This is the frictional version of the local solution found in Section 10.11 (with the orientation of axes suitably adjusted). As the mode number $n \to \infty$, the currents get weaker because friction gets stronger. Thus the dominant currents are due to intermediate modes. When the modes are summed, the resultant flow shows a surface equatorward coastal current and a poleward undercurrent as observed, with the undercurrent becoming deeper toward the poles. Upwelling is strong in the zone of strong winds, but is quite shallow, being replaced by weak downwelling below 100 m.

Other models of the steady circulation are reviewed by Allen (1980). Some interesting effects occur when shelf topography is included, particularly when there is a sharp change of slope at the shelf break [see Johnson and Killworth (1975) and Johnson (1978)].

(p)   It is of interest to note that eastern boundary phenomena in the ocean affect the climate of the neighboring land because upwelling produces near-shore water temperatures that are considerably lower than those that are normal for a given latitude (R. L. Smith, 1968). For instance, the mean air temperature at San Francisco does not reach its maximum until September, after upwelling has ceased, whereas 100 miles inland the maximum is reached in July. The cold water on the coast cools the air and increases the humidity, producing a shallow stable layer of cool moist air that is often capped by low stratus clouds. Coastal fogs are common, and these can be important for maintaining life in adjacent desert regions. Upwelling can also influence the sea breeze, which carries the cool moist air as much as 50 miles inland. A brief review of these and other effects is given by R. L. Smith (1968).

Chapter Eleven

The Tropics

11.1 Introduction

The rotation of the earth has a dominating influence on the way the atmosphere and ocean respond to imposed changes, as has already been emphasized. The dynamic effect is caused by the Coriolis acceleration, which is equal to the product of the Coriolis parameter and the horizontal velocity, and so the adjustment processes are somewhat special when the Coriolis acceleration vanishes. In Chapter 10, the effects, when one component (the longshore component) of the Coriolis acceleration is zero, were found to be quite substantial, one particularly important effect being the existence of coastally trapped waves that can propagate rapidly along the coastal waveguide. In this chapter, the special class of motions that occur in the vicinity of the equator, where both components of the Coriolis term in the shallow-water equations vanish, will be examined. The equatorial zone is found to be a waveguide just as is a coastal region.

As a preliminary to studying tropical dynamics, the shallow-water equations are derived for a sphere (as Laplace did two centuries ago), so that effects of variations in Coriolis parameter are included in the equations. The concept of conservation of potential vorticity still applies, but the value $f/H = 2\Omega \sin \varphi/H$ for a fluid at rest is not constant (as in the uniformly rotating system studied previously) since it varies with latitude. This has important consequences for the slower adjustment processes and for the nature of possible equilibrium states, and these apply at all latitudes, not just in the tropics.

The calculations of dynamic response in the equatorial region are not done by using spherical polar coordinates, but by using an approximation called the equatorial beta plane. In this approximation, the Coriolis parameter is taken to be the product of a constant (called β) and the distance from the equator. This approximation can usefully be applied over the whole of the tropics and hence over a fair proportion of the earth's surface (half lies between the $30°$ parallels).

The different types of wave that can propagate in the equatorial waveguide are introduced in Sections 11.5 and 11.6. There is an equatorial Kelvin wave, so called because it is similar in structure to the coastal Kelvin wave, and there are equatorially trapped gravity waves, which are the equivalent of the Poincaré waves in a uniformly rotating system. There is also an important new class of waves with much lower frequencies, called planetary waves. These owe their existence to the variations in the undisturbed potential vorticity and thus exist at all latitudes. However, the ray paths along which they propagate bend, as do the paths of gravity waves, because of the variation of Coriolis parameter with latitude, and it is this bending that tends to confine the waves to the equatorial waveguide.

The special nature of slow adjustments (i.e., on a time scale large compared with the inertial period) in a rotating fluid has already been noted in previous chapters. Now variations in Coriolis parameter with latitude are found to be important for these slow adjustment processes, and the nature of the approximations that can be applied to these processes are examined in Section 11.8. The motion is called quasi-geostrophic because the fluid is always close to a state of geostrophic balance, but the character of the adjustment can depend on the departures from geostrophy. This type of motion is obviously important because the day-to-day changes in the pressure and velocity distributions in the atmosphere and ocean are largely in this category.

The properties of vertically propagating, equatorially trapped waves are examined in Section 11.10, and some observations of such waves in the atmosphere and ocean are referred to. These waves can play a significant role in the dynamics of the equatorial zone, for example, by accelerating the mean flow in the stratosphere in regions where the wave energy is absorbed.

Applications of the wave solutions to adjustment problems are then considered in the sections beginning with Section 11.11, where the free-wave aspects are emphasized. For instance, a change of wind in one part of the equatorial ocean (such as the West Pacific) can produce equatorial waves that produce changes in distant locations (such as in the East Pacific and along the coast of Peru), and such processes may be a key element in some of the changes in the Pacific that have a very significant effect on the Peruvian fisheries and on climate. Other solutions can be used to show how such features on the equatorial undercurrent and the Somali Current are established by wind forcing. Similarly, oscillations in heating rate of the troposphere, associated with oscillations in cloud activity, can generate waves that propagate high into the stratosphere, at the same time traveling right around the earth.

Steady forced solutions are discussed in Sections 11.13 and 11.14, beginning with the potential vorticity equation for an ideal fluid. This shows that in the absence of friction vortex lines that are stretched migrate poleward. Stretching in the atmosphere can occur through heating, whereas the main producer of stretching in the ocean is Ekman pumping. When this pumping is related to the wind stress, an important relation known as Sverdrup's equation results. This states that northward transport is produced by a positive wind stress curl.

Solutions with small friction are found in Section 11.14, and these may be used to model many observed features of the tropical atmosphere and ocean. For instance, simple solutions may be found for heating that is confined to a particular region near the equator. In the heating region itself, poleward flow in the lower troposphere tends to be produced, as required by the potential vorticity equation. To the east, there are easterly trade winds set up by Kelvin waves emanating from the heating zone. This flow is parallel to the equator, so the equatorward flow needed to balance the poleward movements in the heating zone is found to the west, and consequently cyclonic circulation is found on the western flank of the heating region. Westerlies are produced on the west side of the heating zone as a planetary wave response. The ocean response to wind can be studied in a similar way.

Finally there are two descriptive sections about the tropical circulations of the atmosphere and ocean, and of their variability. Particular aspects of these circulations have already been covered in previous sections in which relationships with models were discussed, but these do not give an overall view, so the summary sections have been added.

11.2 Effects of Earth's Curvature: Shallow-Water Equations on the Sphere

In previous sections, the effect of rotation has largely been confined to the cases in which the earth's curvature has been neglected, i.e., the horizontal is taken to be a plane surface that has a fixed inclination to the rotation vector. Solutions obtained under these assumptions can be applied to the rotating earth if certain conditions are fulfilled, as discussed in Section 7.4. However, effects of the earth's curvature can be very important at low frequencies. These effects, and the conditions under which they are important, will be considered in this and subsequent chapters.

To begin with, consider as Laplace did, shallow-water motion on a sphere. Vertical motion is neglected so the horizontal momentum equations (4.12.14) and (4.12.15) become, in the absence of friction,

$$\frac{Du}{Dt} - \left(2\Omega + \frac{u}{r \cos \varphi} \right) v \sin \varphi = -\frac{1}{\rho r \cos \varphi} \frac{\partial p}{\partial \lambda}, \tag{11.2.1}$$

$$\frac{Dv}{Dt} + \left(2\Omega + \frac{u}{r \cos \varphi} \right) u \sin \varphi = -\frac{1}{\rho r} \frac{\partial p}{\partial \varphi}, \tag{11.2.2}$$

where λ is the longitude, φ the latitude, and r the radial coordinate. The time derivative D/Dt is one following the *horizontal* motion only (and vertical motion is neglected), and so (4.12.9) gives

$$\frac{D}{Dt} = \frac{D_H}{Dt} \equiv \frac{\partial}{\partial t} + \frac{u}{r \cos \varphi} \frac{\partial}{\partial \lambda} + \frac{v}{r} \frac{\partial}{\partial \varphi}. \tag{11.2.3}$$

Also, since we are considering shallow-water motion, p can be replaced by $\rho g \eta$ on the right-hand sides of (11.2.1) and (11.2.2), and the continuity equation [cf. (5.6.7)] is, taking account of the form (4.12.10) for the divergence operator,

$$\frac{\partial \eta}{\partial t} + \frac{1}{r \cos \varphi} \left\{ \frac{\partial}{\partial \lambda} [(H + \eta)u] + \frac{\partial}{\partial \varphi} [(H + \eta)v \cos \varphi] \right\} = 0. \tag{11.2.4}$$

These sets of equations are the ones derived by Laplace, but with the tide-generating terms omitted. Because a shallow layer is being considered, r can be taken as a constant equal to the radius of the earth.

For small perturbations from a state of rest, the linearized versions of the equations suffice, namely,

$$\partial u/\partial t - 2\Omega \sin \varphi \, v = -r^{-1}g \sec \varphi \, \partial \eta/\partial \lambda, \tag{11.2.5}$$

$$\partial v/\partial t + 2\Omega \sin \varphi \, u = -r^{-1}g \, \partial \eta/\partial \varphi, \tag{11.2.6}$$

$$r \cos \varphi \, \partial \eta/\partial t + \partial(Hu)/\partial \lambda + \partial(Hv \cos \varphi)/\partial \varphi = 0. \tag{11.2.7}$$

If the technique used to derive (7.2.4) for a constant depth layer is repeated, it is found that an additional term occurs. This technique is to take the divergence of the momentum equations (11.2.5) and (11.2.6) and substitute for the velocity divergence from (11.2.7). The result is

$$\partial^2 \eta/\partial t^2 - c^2 \, \Delta \eta + fH\zeta - \beta Hu = 0. \tag{11.2.8}$$

In this formula, $c^2 = gH$, as previously, and

$$f = 2\Omega \sin \varphi, \tag{11.2.9}$$

$$\beta = r^{-1} \, df/d\varphi = 2\Omega \cos \varphi/r, \tag{11.2.10}$$

while ζ is the vertical component of vorticity given by

$$r \cos \varphi \, \zeta = \partial v/\partial \lambda - \partial(u \cos \varphi)/\partial \varphi \tag{11.2.11}$$

and Δ is the Laplacian, which in this case is also the horizontal Laplacian (there being no variation with depth), i.e. [cf. (4.12.20)],

$$\Delta = \Delta_{\rm H} \equiv \frac{1}{r^2 \cos^2 \varphi} \frac{\partial^2}{\partial \lambda^2} + \frac{1}{r^2 \cos \varphi} \frac{\partial}{\partial \varphi} \left(\cos \varphi \, \frac{\partial}{\partial \varphi} \right). \tag{11.2.12}$$

The Coriolis parameter f is defined as in (7.4.1), but now the derivative β of the Coriolis parameter with distance northward is involved in the equation, thus introducing effects of the earth's curvature.

The next step followed in Section 7.2 was to derive the vorticity equation (7.2.7). This is done by taking the curl of (11.2.5) and (11.2.6) to yield

$$\partial \zeta/\partial t + r^{-1}f \sec \varphi(\partial u/\partial \lambda + \partial(v \cos \varphi)/\partial \varphi) + \beta v = 0. \tag{11.2.13}$$

This also involves an effect of the earth's curvature through the beta term, and when (11.2.7) is used to substitute for the divergence of the horizontal velocity, the resulting form of the potential vorticity equation is

$$\partial(\zeta - f\eta/H)/\partial t + \beta v = 0, \tag{11.2.14}$$

and this also involves beta. This is still a perturbation form of (7.10.9) as will be discussed in the next section.

Methods of solving these equations for a sphere or a hemisphere are discussed by Longuet-Higgins (1968b) and Longuet-Higgins and Pond (1969) and, of course, numerical solutions have been found for the tides, allowing for tide-generating forces, variations in depth, and the complicated shape of the ocean basins [see Hendershott

(1981)]. However, approximations can be made for all motions *except* those that have scales comparable with the radius of the earth. For purposes of understanding the nature of the motions that occur, it is advantageous to restrict attention to motions with scale smaller than the radius of the earth, so that full advantage can be taken of the approximations that may thereby be made. Attention will first be given to waves whose energy is mainly confined to the tropics, but before doing that, the potential vorticity equation for a shallow homogeneous ocean will be discussed, using the full spherical polar coordinates.

11.3   Potential Vorticity for a Shallow Homogeneous Layer

The potential vorticity equation for a shallow homogeneous layer was derived in Section 7.10. The form (7.10.7) of the vorticity equation relates vortex stretching to the horizontal divergence, is valid whether or not f varies, and is still valid when applied on a sphere. The same is true of the continuity equation (7.10.8) and hence of the potential vorticity equation (7.10.9), which is, after all, a form of Helmholtz' result for vortex filaments. In polar coordinates the equation is

$$D_H Q/Dt = 0, \tag{11.3.1}$$

with the operator D_H/Dt defined by (11.2.3) and Q defined by (7.10.10), i.e., by

$$Q = (f + \zeta)/(H + \eta) = (2\Omega \sin \varphi + \zeta)/(H + \eta). \tag{11.3.2}$$

Details of the derivation, using polar coordinates, are left as an exercise to the reader.

For discussing small perturbations from a state of rest, Q may be expanded in terms of the small amplitude, the zeroth-order term being

$$\bar{Q} = 2\Omega \sin \varphi/H \equiv f/H \tag{11.3.3}$$

and the first-order perturbation being

$$Q' = \zeta/H - 2\Omega \sin \varphi \, \eta/H^2. \tag{11.3.4}$$

The perturbation equation to this order may be written as

$$\frac{\partial Q'}{\partial t} + \frac{u}{r \cos \varphi} \frac{\partial \bar{Q}}{\partial \lambda} + \frac{v}{r} \frac{\partial \bar{Q}}{\partial \varphi} = 0. \tag{11.3.5}$$

In previous chapters, attention has been confined almost exclusively to a *very special case*, namely that in which $\bar{Q}$ is a constant. Then (11.3.5) reduces to $\partial Q'/\partial t = 0$, so Q' has a fixed value at each point, as given by (7.2.10), and this fact was exploited in Chapter 7. However, when the curvature of the earth is taken into account, i.e., the fact that f, given by (11.2.9), varies with latitude, this result is no longer valid. The same is true of a homogeneous ocean when the depth varies, even if f is constant.

The result is that the properties of the medium are no longer independent of direction, for now there is a *special direction* given by the contours of $\bar{Q}$. In the case of steady flow, (11.3.5) shows that the flow must be along contours of $\bar{Q}$, as found by Hough (1897). For a constant-depth ocean, these contours are circles of latitude.

One case of nonconstant $\bar{Q}$ that has been considered already is the case of continental shelf waves (Section 10.12). This demonstrated how low-frequency waves can propagate along the contours of $\bar{Q}$, and it will be seen that the same is true when H is constant but f varies.

11.4   The Equatorial Beta Plane

For motions near the equator, the approximations

$$\sin \varphi \approx \varphi, \qquad \cos \varphi \approx 1 \tag{11.4.1}$$

may be utilized, giving what is called the equatorial beta-plane approximation. Half of the earth lies at latitudes of less than 30° and the maximum percentage error in the above approximation in that range of latitudes is only 14%. In this approximation, beta is a constant given by

$$\beta = 2\Omega/r = 2.3 \times 10^{-11} \quad \mathrm{m}^{-1}\,\mathrm{s}^{-1} \tag{11.4.2}$$

and f is given by

$$f = \beta y, \tag{11.4.3}$$

where

$$y = r\varphi \tag{11.4.4}$$

is distance northward from the equator. Instead of longitude, eastward distance

$$x = r\lambda$$

is used, so the linear equations (11.2.5)–(11.2.7) take the form

$$\partial u/\partial t - \beta y v = -g \, \partial \eta/\partial x, \tag{11.4.5}$$

$$\partial v/\partial t + \beta y u = -g \, \partial \eta/\partial y, \tag{11.4.6}$$

$$\partial \eta/\partial t + \partial(Hu)/\partial x + \partial(Hv)/\partial y = 0. \tag{11.4.7}$$

The derived equation (11.2.8) for constant depth H is as before, with f given by (11.4.3) and ζ by

$$\zeta = \partial v/\partial x - \partial u/\partial y. \tag{11.4.8}$$

Similarly, the Laplacian has its usual Cartesian form. The derived potential vorticity equation (11.2.14) is also unchanged, except that f and β have their equatorial approximation. The fact that the derived equations as well as the original set can be obtained from the full equations merely by making the approximation (11.4.1) gives confidence that errors are not likely to be any larger than those involved in (11.4.1) itself.

An equation for v alone can be derived by adding $-(\beta y/c^2)\,\partial/\partial t$ of (11.4.5), $(1/c^2)\,\partial^2/\partial t^2$ of (11.4.6), $-(1/H)\,\partial^2/\partial y\,\partial t$ of (11.4.7), and $-\partial/\partial x$ of (11.2.14). The result is, for constant H,

$$\frac{\partial}{\partial t}\left\{\frac{1}{c^2}\left(\frac{\partial^2 v}{\partial t^2} + f^2 v\right) - \left(\frac{\partial^2 v}{\partial x^2} + \frac{\partial^2 v}{\partial y^2}\right)\right\} - \beta\frac{\partial v}{\partial x} = 0. \tag{11.4.9}$$

This differs from the equation (7.2.13) for the f-plane case in that f is now given by (11.4.3) and by the addition of the term involving β.

It will be necessary later on to consider *forced* motion on the equatorial beta plane, so before considering alternative forms of these equations, it is desirable to add forcing terms, so that the way they appear in the derived equations can be seen. The forced versions of (11.4.5)–(11.4.7) are written

$$\frac{\partial u}{\partial t} - \beta y v = -g\frac{\partial \eta}{\partial x} + \frac{X}{\rho H}, \tag{11.4.10}$$

$$\frac{\partial v}{\partial t} + \beta y u = -g\frac{\partial \eta}{\partial y} + \frac{Y}{\rho H}, \tag{11.4.11}$$

$$\frac{\partial \eta}{\partial t} + \frac{\partial}{\partial x}(Hu) + \frac{\partial}{\partial y}(Hv) = -\frac{E}{\rho}. \tag{11.4.12}$$

The symbols used for the forcing terms correspond to those used in (9.9.10) and (9.9.15), i.e., (X, Y) can be interpreted as a surface stress and E as an evaporation rate. However, *any* form of forcing will give the above equations, so X, Y, and E can be given a wider interpretation.

With constant depth H and forcing terms included, the potential vorticity equation, which is obtained by adding $-\partial/\partial y$ of (11.4.10), $\partial/\partial x$ of (11.4.11), and $-\beta y/H$ times (11.4.12), becomes

$$\frac{\partial}{\partial t}\left(\zeta - \frac{f}{H}\eta\right) + \beta v = \frac{1}{\rho H}\left(\frac{\partial Y}{\partial x} - \frac{\partial X}{\partial y} + fE\right). \tag{11.4.13}$$

Using this in place of (11.2.14), the forced version of (11.4.9) takes the form

$$\frac{\partial}{\partial t}\left\{\frac{1}{c^2}\left(\frac{\partial^2 v}{\partial t^2} + f^2 v\right) - \left(\frac{\partial^2 v}{\partial x^2} + \frac{\partial^2 v}{\partial y^2}\right)\right\} - \beta\frac{\partial v}{\partial x}$$
$$= \frac{1}{\rho H}\left\{\frac{\partial}{\partial t}\left(\frac{1}{c^2}\left(\frac{\partial Y}{\partial t} - fX\right) + \frac{\partial E}{\partial y}\right) - \frac{\partial}{\partial x}\left(\frac{\partial Y}{\partial x} - \frac{\partial X}{\partial y} + fE\right)\right\}. \tag{11.4.14}$$

Equation (11.4.14) gives a simple equation for v that can be solved, but to relate v to the other variables, it is better to work with the variables q and r (Gill and Clarke, 1974), defined by (10.7.9) and (10.7.10), rather than with u and v, i.e.,

$$q = g\eta/c + u \equiv (g/H)^{1/2}\eta + u, \tag{11.4.15}$$

$$r = g\eta/c - u \equiv (g/H)^{1/2}\eta - u. \tag{11.4.16}$$

The equations relating v to q and v to r are very similar to those obtained in Section 10.7 for waves in a channel. First, the sum and difference of (11.4.10) and g/c times

(11.4.12) give [cf. (10.7.11) and (10.7.12)]

$$\frac{\partial q}{\partial t} + c\frac{\partial q}{\partial x} + c\frac{\partial v}{\partial y} - \beta yv = \frac{1}{\rho H}(X - cE), \tag{11.4.17}$$

$$\frac{\partial r}{\partial t} - c\frac{\partial r}{\partial x} + c\frac{\partial v}{\partial y} + \beta yv = -\frac{1}{\rho H}(X + cE). \tag{11.4.18}$$

The other two equations relating q to v and r to v are the difference and sum of the time derivative of (11.4.11) and c times (11.4.13), namely,

$$\frac{\partial}{\partial t}\left(c\frac{\partial q}{\partial y} + \beta yq + \frac{\partial v}{\partial t} - c\frac{\partial v}{\partial x}\right) - \beta cv = \frac{1}{\rho H}\left\{\frac{\partial Y}{\partial t} - c\left(\frac{\partial Y}{\partial x} - \frac{\partial X}{\partial y} + fE\right)\right\}, \tag{11.4.19}$$

$$\frac{\partial}{\partial t}\left(c\frac{\partial r}{\partial y} - \beta yr + \frac{\partial v}{\partial t} + c\frac{\partial v}{\partial x}\right) + \beta cv = \frac{1}{\rho H}\left\{\frac{\partial Y}{\partial t} + c\left(\frac{\partial Y}{\partial y} - \frac{\partial X}{\partial x} + fE\right)\right\}. \tag{11.4.20}$$

Although the last two equations relate v in turn to q and r as required, some information is lost in taking the time derivative of (11.4.11), which by itself gives

$$\left(c\frac{\partial q}{\partial y} + \beta yq\right) + \left(c\frac{\partial r}{\partial y} - \beta yr\right) + 2\frac{\partial v}{\partial t} = \frac{2Y}{\rho H}. \tag{11.4.21}$$

The reason why these particular forms of the equation are useful will become apparent later, but it is basically the same reason that applied in the case of the channel. The above equations are, in fact, identical to those for the channel except for the last term involving β in (11.4.19) and (11.4.20), and the fact that f is now equal to βy.

11.5 The Equatorial Kelvin Wave

A very important property of the equatorial zone is that it acts as a *waveguide*, i.e., disturbances are trapped in the vicinity of the equator. [The idea appears to have been first put forward by Yoshida in 1959—see Matsuno (1966).] The simplest wave that illustrates this property is the equatorial Kelvin wave, so named because it is very similar in character to the coastally trapped Kelvin wave studied in Section 10.4. As with the coastal wave, the motion is unidirectional, being everywhere parallel to the equator. Equations (11.4.5) and (11.4.7) therefore give

$$\partial u/\partial t = -g\,\partial\eta/\partial x, \qquad \partial\eta/\partial t + H\,\partial u/\partial x = 0. \tag{11.5.1}$$

These are identical to (10.4.1) and (10.4.2) for the coastal Kelvin wave, and therefore the solution is given by (10.4.3) as in the coastal case. In each plane $y = $ const, the motion is exactly the same as is that in a nonrotating fluid.

Rotation effects do not allow the motion in each plane $y = $ const to be independent because (10.4.4) (now with $f = \beta y$) requires a geostrophic balance between the eastward velocity and the north–south pressure gradient. Substituting (10.4.3) in (10.4.4) again gives (10.4.5) but now with $f = \beta y$. The solution required is the one that decays as $y \to \pm\infty$, i.e., the one represented by G' and satisfying

$$\partial G'/\partial y = -(\beta y/c)G', \tag{11.5.2}$$

where c is the square root of gH. The solution $[\text{cf. (10.4.6)}]$ is

$$G' = \exp(-\tfrac{1}{2}\beta y^2/c)G(x - ct), \tag{11.5.3}$$

showing decay in a distance of order a_e, where a_e is given by

$$a_e = (c/2\beta)^{1/2} \tag{11.5.4}$$

and is called the *equatorial radius of deformation* (Gill and Clarke, 1974) because of its relationship with the decay scale for the f-plane case $[Note:$ The factor of 2 is included because of formula (11.6.4) in Section 11.6.$]$

The complete Kelvin-wave solution is thus, by (10.4.3) and (11.5.3),

$$\begin{aligned}
\eta &= \exp(-\tfrac{1}{2}\beta y^2/c)G(x - ct), \\
u &= (g/c)\exp(-\tfrac{1}{2}\beta y^2/c)G(x - ct), \\
v &= 0.
\end{aligned} \tag{11.5.5}$$

Also, by (11.4.16), $r = 0$ and by (11.4.15) $q = 2u$. (The solution is illustrated in Fig. 11.2a.)

The value of the equatorial Rossby radius for *barotropic* waves in the ocean ($c \approx 200$ m s^{-1}) is about 2000 km, so the idea of a trapped wave is only marginally consistent with use of the equatorial beta plane. However, it will be shown later (Section 11.9) that the same analysis can be applied to *baroclinic* waves in both atmosphere and ocean, with H now being interpreted as the equivalent depth. Typical values of $c = (gH)^{1/2}$ appropriate to the atmosphere are 20–80 m s^{-1}, giving equatorial Rossby radii between 6 and 12 degrees of latitude (650–1300 km). For baroclinic ocean waves, appropriate values of c are typically in the range 0.5–3 m s^{-1}, so the equatorial Rossby radius is 100–250 km.

Equation (11.5.3) shows that equatorial Kelvin waves propagate *eastward* without dispersion at the same speed c as in a nonrotating fluid. The dispersion relation between frequency ω and east–west wavenumber k is simply (10.4.9), i.e.,

$$\omega = kc. \tag{11.5.6}$$

(This curve is shown in the general dispersion diagram for equatorial waves in Fig. 11.1.) For the first baroclinic mode in the ocean, a typical value of c (Wunsch and Gill, 1976) is 2.8 m s^{-1}, so a Kelvin wave would take about 2 months to cross the Pacific from New Guinea to South America. For higher modes in the ocean and for waves that have been observed in the atmosphere, propagation speeds tend to be comparable with flow speeds. In these cases, the wave analysis can be used if the mean flow varies slowly over a wavelength (see Section 8.12), provided that ω in (11.5.6) is interpreted as the intrinsic or Doppler-shifted frequency $\hat{\omega}$ given by (8.12.29).

11.6 Other Equatorially Trapped Waves

In addition to the Kelvin wave, there is an infinite set of other equatorially trapped waves, with trapping scale of the same order as that for Kelvin waves, namely, the equatorial Rossby radius defined by (11.5.4). The properties of these waves were first outlined in detail by Matsuno (1966) and Blandford (1966). These may be found by looking for solutions proportional to

$$\exp(ikx - i\omega t),$$

in which case (11.4.9) reduces to the ordinary differential equation

$$\frac{d^2v}{dy^2} + \left(\frac{\omega^2}{c^2} - k^2 - \frac{\beta k}{\omega} - \frac{\beta^2 y^2}{c^2}\right)v = 0. \tag{11.6.1}$$

The solutions that vanish as $y \to \pm\infty$ are well known and are given (after taking the real part) by

$$v = D_n((2\beta/c)^{1/2}y)\cos(kx - \omega t)$$

$$= 2^{-n/2}H_n((\beta/c)^{1/2}y)\exp(-\beta y^2/2c)\cos(kx - \omega t), \tag{11.6.2}$$

where D_n is the parabolic cylinder function of order n and H_n is a Hermite polynomial of order n (Erdélyi *et al.*, 1953, Chapters 8 and 10). The corresponding dispersion relation is

$$(\omega/c)^2 - k^2 - \beta k/\omega = (2n + 1)\beta/c, \tag{11.6.3}$$

and the corresponding curves are shown in Fig. 11.1 along with the Kelvin wave.

The expressions for the other variables q and r may be obtained using (11.4.17),

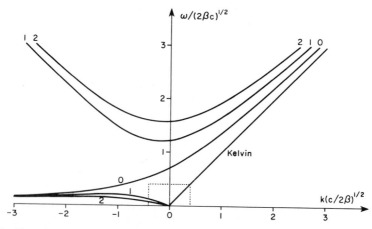

Fig. 11.1. Dispersion curves for equatorial waves. The vertical axis is the frequency in units of $(2\beta c)^{1/2}$ and the horizontal axis is the east–west wavenumber in units of $(2\beta/c)^{1/2}$. The curve labeled 0 corresponds to the mixed planetary–gravity wave. The upper curves labeled 1 and 2 are the first two gravity wave modes and the corresponding lower curves are the first two planetary wave modes. [Reproduced from "Numerical Models of Ocean Circulation," 1975, by permission of the National Academy of Science, Washington, D.C.]

(11.4.18), and the properties

$$(d/d\xi + \tfrac{1}{2}\xi)D_m = mD_{m-1}, \qquad (d/d\xi - \tfrac{1}{2}\xi)D_m = -D_{m+1} \qquad (11.6.4)$$

of parabolic cylindrical functions. The result, using $\xi = (2\beta/c)^{1/2}y$, is

$$q = (ck - \omega)^{-1}(2\beta c)^{1/2}D_{n+1}((2\beta/c)^{1/2}y) \sin(kx - \omega t), \qquad (11.6.5)$$

$$r = (ck + \omega)^{-1}(2\beta c)^{1/2}nD_{n-1}((2\beta/c)^{1/2}y) \sin(kx - \omega t). \qquad (11.6.6)$$

The corresponding expressions for u and η follow from the definitions (11.4.15) and (11.4.16).

For $n \geq 1$, the waves subdivide into two classes. For the upper branches, the term $\beta k/\omega$ in (11.6.3) is small, so the dispersion curves are given approximately by

$$\omega^2 \approx (2n + 1)\beta c + k^2 c^2. \qquad (11.6.7)$$

The fractional error in making this approximation is bounded above by $2 \times 3^{-3/2} \times (2n + 1)^{-1}$, which has a maximum value when $n = 1$ of 13%. The appropriate dispersion relation has the same form as that for Poincaré waves [see, e.g., (10.3.3)], and so these waves are called *equatorially trapped gravity waves* or equatorially trapped Poincaré waves. These waves will be discussed more fully in the next section.

On the lower branches of the curves, the term ω^2/c^2 in (11.6.3) is small, and consequently the dispersion curves are given approximately by

$$\omega = -\beta k/(k^2 + (2n + 1)\beta/c). \qquad (11.6.8)$$

The fractional error is at most $4 \times 3^{-3}/(2n + 1)^2$, which has a maximum value for $n = 1$ of less than 2%. The corresponding waves are called *equatorially trapped planetary waves* or equatorially trapped Rossby waves. The dispersion curve has the same form (10.12.9) as that for continental shelf waves because of similarities in the dynamics, which will be mentioned when planetary waves are discussed further in Section 11.8.

Note that there is a large gap between the minimum gravity wave frequency and the maximum planetary wave frequency, so these waves are easily distinguished. The frequency gap for wave n involves a factor of approximately $2(2n + 1)$, which is equal to 6 for the lowest value $n = 1$. There are, however, two waves with frequencies in this gap. One is the Kelvin wave, sometimes called the $n = -1$ wave because (11.6.3) is satisfied by the Kelvin-wave dispersion relation (11.5.6) when $n = -1$. The other is the wave corresponding to $n = 0$, which will now be considered in detail.

The solution when $n = 0$ is somewhat special because (11.6.6) shows that $r = 0$ in this case, just as for the Kelvin wave. However v is not zero, so (11.4.20) gives the appropriate dispersion relation

$$\omega/c - k - \beta/\omega = 0. \qquad (11.6.9)$$

The equation (11.6.3) gives the same result, but also contains a spurious factor $\omega + ck$. The solution (11.6.2) in this case reduces to the simple form

$$v = \exp(-\beta y^2/2c) \cos(kx - \omega t), \qquad (11.6.10)$$

and (11.6.5) and (11.6.9) give

$$u = g\eta/c = -(\omega y/c) \exp(-\beta y^2/2c) \sin(kx - \omega t). \qquad (11.6.11)$$

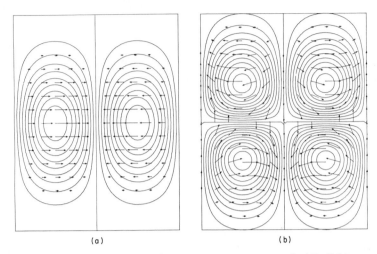

Fig. 11.2. Contours of surface elevation and arrows representing currents for (a) a Kelvin wave and (b) an eastward-propagating mixed planetary–gravity wave. Both waves have eastward phase velocity and eastward group velocity. Fluid particles move parallel to the equator in the case of the Kelvin wave and move anticyclonically around elliptical orbits in the case of the mixed wave. The figures show a range of latitudes corresponding to ±4 equatorial Rossby radii.

The dispersion curve, labeled $n = 0$ in Fig. 11.1, is unique in that for large positive k it behaves like a gravity wave, whereas for large negative k it behaves like a planetary wave. For this reason it is called a *mixed planetary–gravity wave* (or mixed Rossby–gravity wave). The phase velocity can be to the east or west, but the group velocity is always eastward, being a maximum for short waves with eastward group velocity (gravity waves). An example with eastward phase velocity is shown in Fig. 11.2b. For westward phase propagation, the sign of v should be reversed. Particles follow *anticyclonic orbits* everywhere. The case $k = 0$ corresponds to a pure standing wave for which the surface moves sinusoidally up and down, but with opposite sign on opposite sides of the equator. Particles move anticyclonically with eastward motion when the free surface is elevated and with westward motion when it is depressed. The frequency of the standing wave is given by

$$\omega = (\beta c)^{1/2},$$

corresponding to a period of about 9 days for the first baroclinic mode in the ocean, to about 3 weeks for a higher baroclinic mode in the ocean with $c = 0.5$ m s^{-1}, and to 5 days for atmospheric waves with c near 20 m s^{-1}. Evidence for such waves in both atmosphere and ocean has been given [see, e.g., Wallace (1971), Wunsch and Gill (1976), and Weisberg *et al.* (1979a)].

11.7   The Equatorial Waveguide: Gravity Waves

The important property of the solutions of (11.6.1) is the equatorial trapping. In other words, waves are guided along an equatorial waveguide. The waveguide effect is due entirely to the variation of Coriolis parameter with latitude, as can be seen from (11.6.1). For a wave of fixed frequency ω and fixed east–west wavenumber k, the coefficient of v in (11.6.1) may be positive at the equator, giving wavelike behavior, but as $|y|$ increases, $f = \beta y$ increases, and the coefficient of v decreases until it becomes zero at the "turning point" or critical latitude y_c given by

$$f_c^2 = \beta^2 y_c^2 = \omega^2 - k^2 c^2 - \beta k c^2/\omega = (2n + 1)\beta c, \tag{11.7.1}$$

the last equality coming from the dispersion relation (11.6.3). At latitudes higher than the critical one, the coefficient of v becomes negative and solutions of (11.6.1) become exponential in character, thus giving wave trapping.

If the phase of the wave changes rapidly enough with y, the Liouville–Green or WKBJ technique outlined in Section 8.12 can be used, i.e., a north–south wavenumber l can be defined for each value of y and the solution has the approximate form (8.12.7), i.e., in the present notation

$$v = l^{-1/2} \exp\left\{ i \left[\int^{y} l \, dy + kx - \omega t \right] \right\}, \tag{11.7.2}$$

where, from (11.6.1), l is given by

$$l^2 = \frac{\omega^2}{c^2} - k^2 - \frac{\beta k}{\omega} - \frac{\beta^2 y^2}{c^2} = \frac{\beta^2}{c^2}(y_c^2 - y^2), \tag{11.7.3}$$

the last equality making use of (11.7.1). The condition for validity of the approximation is that δ, defined by (8.12.5) (with m replaced by l and z by y), should be small, i.e.,

$$\delta = l^{-3/2} \, d^2(l^{-1/2})/dy^2 \ll 1. \tag{11.7.4}$$

At the equator,

$$\delta = 1/2(2n + 1)^2, \tag{11.7.5}$$

indicating that the theory applies asymptotically in the limit of large n. The asymptotic behavior of parabolic cylinder functions for large n is well known and is given, for instance, by Erdélyi *et al.* (1953, Section 8.4). The approximation breaks down near the turning point at which l vanishes, but a different form of approximation may be used there, as pointed out in Section 8.12.1.

The paths followed by wave groups (ray paths) may be calculated by the methods used in Section 8.9.2 and may be given in general form in Section 8.12.3. The equation (8.12.27) of such a path is

$$dx/dt = c_{gx}, \qquad dy/dt = c_{gy} \qquad \text{or} \qquad dy/dx = c_{gy}/c_{gx} = l/(k + \beta/2\omega), \tag{11.7.6}$$

where c_{gx} and c_{gy} are the components of group velocity whose values are calculated from the dispersion relation (11.6.3) and the definition (8.12.24) of group velocity. When the expression (11.7.3) for l is substituted, the solution for the ray path takes the form

$$y = y_c \sin\left[c^{-1}\beta x/(k + \beta/2\omega) \right], \tag{11.7.7}$$

i.e., a path with sinusoidal oscillations about the equator and reflection at the critical latitudes $y = \pm y_c$.

Effects of the variation of Coriolis parameter with latitude on ray paths are not confined to low latitudes and have been studied, for instance, by Anderson and Gill (1979) for the case of *gravity waves*, for which the term $\beta/2\omega$ can be ignored in (11.7.6) and (11.7.7). In particular, if a uniform wind stress is suddenly applied over a small range of latitudes that are remote from the equator, inertial period oscillations are generated there as described by the f-plane solution of Section 9.3. However, the energy cannot remain at the latitude at which it is generated because it tends to propagate northward or southward in accordance with (11.7.6). Figure 11.3 illustrates the effect for the limiting case of no variation with x [i.e., $k = 0$, so (11.7.7) is not applicable but the first version of (11.7.6) is]. The f-plane solution is valid only for two or three periods, after which the equatorward propagation of energy becomes important. Subsequently, the energy moves as a wave group to and fro across the equator, obeying the ray equation $dy/dt = c_{gy}$ quite well.

Another effect of the waveguide is the separation into a discrete set of modes $n = 1, 2, \ldots$, as occurs also in a channel (Section 10.5). In particular, this means that long Poincaré waves (k small), which are the gravity waves with near-zero group velocity, can have only the discrete set of frequencies [see (11.6.7)] given by

$$\omega^2 \approx (2n + 1)\beta c. \tag{11.7.8}$$

This frequency selection shows up in Pacific sea-level records because variations (of the order of centimeters) that are associated with the first baroclinic mode are large enough to be detected, whereas those associated with higher modes are not, thus providing an effective filter. The value of c for the first baroclinic mode is about 2.8 m s^{-1}, so (11.7.8) gives a period of $5\frac{1}{2}$ days for $n = 1$, 4 days for $n = 2$, and 3 days for $n = 4$. The corresponding peaks can be seen in the sea-level spectrum for Ocean Island (Fig. 11.4a), whereas only the 4-day oscillation shows up prominently at Canton Island (Fig. 11.4b). This is because Canton Island is at a latitude where the amplitude of both 3- and $5\frac{1}{2}$-day waves is small (Wunsch and Gill, 1976). In fact, the amplitude variation in the West Pacific does show changes with latitude very much like those associated with a free mode of oscillation. This is illustrated in Fig. 11.4c for the case of the $5\frac{1}{2}$-day wave. The solid line shows the theoretical variation of the square of the pressure amplitude with latitude for the $n = 1$ mode, whereas the dots show observed energy levels (after the background energy has been removed) at $5\frac{1}{2}$ days for a number of islands in the West Pacific. Wunsch and Gill concluded that these oscillations are due to resonant excitation of the long equatorially trapped modes by the wind.

The waves with the preferred frequencies (11.7.8) are, as pointed out in Section 11.6, the equivalent of inertial period waves at mid latitudes, but the waveguide selects out the waves whose inertial periods correspond to the turning latitudes. Thus the waves seen in the records shown in Fig. 11.4 are the equatorial equivalent of the wind-generated inertial waves discussed in Section 9.3.

Within the ocean, temperature and current measurements show a mix of many modes. The mix appears to be much the same at many tropical locations, and Eriksen (1980) has devised a formula for describing this mix that predicts, in particular, how the shape of the spectrum changes with latitude as the equator is approached.

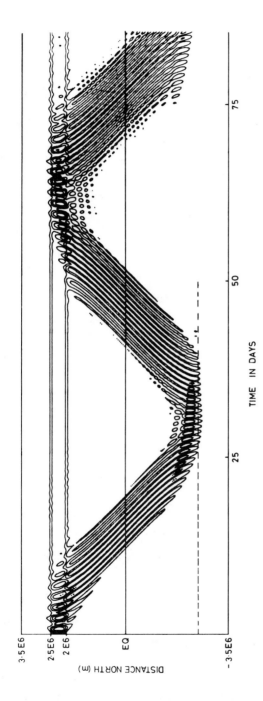

Fig. 11.3 An illustration of beta dispersion of gravity waves. An eastward wind stress is applied in the strip 2000 km < y < 2500 km from t = 0. At first, local inertial waves are generated as on an f-plane, but the variations of f with latitude causes the waves to propagate backward and forward across the equator as shown. Contours are of the meridional velocity, and the ray path, followed by the wave group, is close to that given by (11.7.7). [From Anderson and Gill (1979), *J. Geophys. Res.* **84**, 1836 (Fig. 6); copyrighted by the American Geophysical Union.]

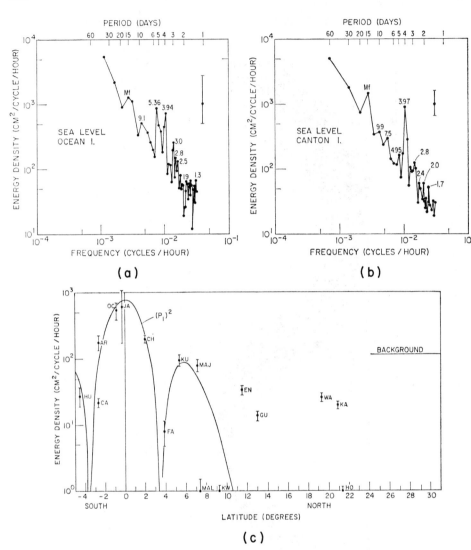

Fig. 11.4.   Sea-level spectra from (a) Ocean I (1°S 170°E) and (b) Canton I (3°S 172°W) showing prominent peaks associated with equatorially trapped baroclinic gravity waves. [From Wunsch and Gill (1976, Figs. 1 and 3).] The $5\frac{1}{2}$-day wave has been identified with $n = 1$, the 4-day wave with $n = 2$, and the 3-day wave with $n = 4$. (c) Energy density at $5\frac{1}{2}$-day period as a function of latitude for various islands in the tropical Pacific Ocean [see Wunsch and Gill (1976, Fig. 10)]. A constant, shown as the horizontal line and taken to represent the background spectrum, has been subtracted from each value. Error bars are one standard deviation of χ^2. The solid line shows the predicted structure for an $n = 1$ wave.

11.8   Planetary Waves and Quasi-geostrophic Motion

The planetary waves, whose dispersion is described by (11.6.8), are a new form of wave whose frequencies are considerably lower than those of gravity waves. They represent an important new class of motion that occurs because the undisturbed potential vorticity on the curved surface of the earth is not a constant, but varies with latitude. A first look at their properties will be made in this section, and further aspects will be considered later.

Consider first the dispersion properties as given by (11.6.8). *Long waves* ($k \to 0$) move approximately nondispersively with wave speed

$$\omega/k = -c/(2n + 1), \tag{11.8.1}$$

i.e., in the direction *opposite* (westward) to the Kelvin wave and with speed reduced by factors of 3, 5, 7, etc. (For example, if $c = 2.8$ m s^{-1}, as for the first baroclinic mode in the Pacific, the $n = 1$ planetary wave speed is 0.9 m s^{-1}, corresponding to a time of 6 months to cross the Pacific from east to west. Other modes are slower.) As $|k|$ increases, the group velocity decreases and eventually becomes zero when

$$k = -[(2n + 1)\beta/c]^{1/2} = -(2n + 1)/y_c = -f_c/c, \tag{11.8.2}$$

the last equalities coming from (11.7.1). At this point, the frequency has a *maximum* value given by

$$\omega = \tfrac{1}{2}[\beta c/(2n + 1)]^{1/2} = \tfrac{1}{2}f_c/(2n + 1) = \tfrac{1}{2}\beta c/f_c. \tag{11.8.3}$$

(For example when $n = 1$, this corresponds to a *minimum* period of 31 days for a first baroclinic ocean mode with $c = 2.8$ m s^{-1}, 74 days for a higher mode with $c = 0.5$ m s^{-1}, and 12 days for an atmospheric mode with $c = 20$ m s^{-1}.)

For *shorter* waves, the group velocity is *eastward*, i.e., in the direction opposite to the phase velocity. The maximum value is

$$c_g = \tfrac{1}{8}c/(2n + 1) \qquad \text{when} \quad k = -[3(2n + 1)\beta/c]^{1/2}. \tag{11.8.4}$$

Thus only short waves can carry information eastward and then at only one-eighth of the speed at which long waves can carry information westward. Some consequences of this will be explored later. For very short waves ($k \to \infty$), the approximate dispersion relation is

$$\omega = -\beta/k, \qquad c_g = \beta/k^2 \qquad \text{for} \quad |k| \to \infty. \tag{11.8.5}$$

Thus the phase and group velocities are equal and opposite in this limit, both tending to zero as $|k| \to \infty$. A summary of the dispersion properties is given in Fig. 11.5.

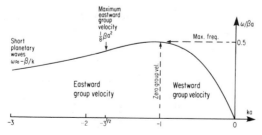

Fig. 11.5. Properties of the planetary wave dispersion relation, which can be written $\omega/\beta a = -ka/(1 + (ka)^2)$, where $a^2 = c/(2n + 1)\beta = c^2/f_c^2$, c is the wave speed in the absence of rotation (i.e., its square is g times the equivalent depth), n the number of the mode, f_c the inertial frequency at the critical latitude, ω the frequency, k the east-west wavenumber, and β the rate of change of the Coriolis parameter with latitude distance.

In the preceding section, considerable insight into the properties of gravity waves in the equatorial waveguide was found by using the Liouville–Green or **WKBJ** technique, which is valid for large n and which approximates the waves at each latitude by a locally plane wave. This was straightforward since the properties of gravity waves in the vicinity of a given latitude ($y = y_0$, say) are already known from earlier chapters. The same technique can be applied to planetary waves, but it is now necessary to *find out* the *local* plane wave properties since these have not already been studied. As a first step, the local dispersion relation may be obtained directly from (11.7.3) by ignoring the term ω^2/c^2, thus giving

$$\omega = -\beta k/(k^2 + l^2 + f^2/c^2). \tag{11.8.6}$$

Dispersion properties associated with this equation will be explored more fully in Chapter 12.

The important task that will be carried out now is to develop the approximations to the equations of motion that are appropriate to these low-frequency waves, so that their dynamics can be appreciated. This is most readily (and systematically) achieved by introducing nondimensional coordinates and using formal asymptotic expansion techniques. Suppose the approximation is required in the neighborhood of $y = y_0$, where the Coriolis parameter is equal to

$$f_0 = \beta y_0. \tag{11.8.7}$$

From the denominator of (11.8.6), the natural scale for the wavenumber is f_0/c, corresponding to the scale c/f_0 (the local Rossby radius) for displacements x and $y - y_0$. Nondimensional displacements x^* and y^* are therefore defined by

$$x^* = f_0 x/c, \qquad y^* = f_0(y - y_0)/c. \tag{11.8.8}$$

Also, from (11.8.6), the frequency scale is $\beta c/f_0$, so a nondimensional time t is defined by

$$t^* = \beta c t/f_0. \tag{11.8.9}$$

Finally, if v_0 is the scale for v, (11.4.17) and (11.4.18) show that the same scale applies to q and r and hence, by (11.4.15) and (11.4.16), to u and $g\eta/c$. Thus the nondimensional dependent variables are defined by

$$v^* = v/v_0, \qquad u^* = u/v_0, \qquad \eta^* = g\eta/cv_0. \tag{11.8.10}$$

With the above definitions, the nondimensional form of the governing equations (11.4.5)–(11.4.7) becomes

$$\epsilon \, \partial u^*/\partial t^* - (1 + \epsilon y^*)v^* = -\partial\eta^*/\partial x^*, \tag{11.8.11}$$

$$\epsilon \, \partial v^*/\partial t^* + (1 + \epsilon y^*)u^* = -\partial\eta^*/\partial y^*, \tag{11.8.12}$$

$$\epsilon \, \partial\eta^*/\partial t^* + \partial u^*/\partial x^* + \partial v^*/\partial y^* = 0, \tag{11.8.13}$$

where

$$\epsilon = \beta c/f_0^2 = 2(a_e/y_0)^2 \tag{11.8.14}$$

is the small parameter of the problem.

If ϵ is put equal to zero in (11.8.11)–(11.8.13) to obtain the leading-order equations, the same situation occurs as for low-frequency motions on an f-plane (see Section 8.16), namely, that the momentum balance is *geostrophic*, so the continuity equation is satisfied exactly. In other words, there is *redundancy* in the zeroth-order equations, so the motion cannot be determined without considering *departures from geostrophy*. That is why the motion is called *quasi-geostrophic*.

Since departures from geostrophy are important, the nature of the motion depends on which of the terms representing such departures is the most consequential. In Section 8.16, the *acceleration terms* gave the most important departure, but the above scaling shows that for planetary waves, the *change in the Coriolis parameter* is equally important.

As found in Section 8.16, it is useful to solve the momentum equations (11.8.11) and (11.8.12) for the velocity components in terms of pressure correct up to the first order in ϵ, so that the dominant departures from geostrophy are included. The result is

$$v^* = (1 - \epsilon y^*) \, \partial \eta^*/\partial x^* - \epsilon \, \partial^2 \eta^*/\partial y^* \, \partial t^*, \tag{11.8.15}$$

$$u^* = -(1 - \epsilon y^*) \, \partial \eta^*/\partial y^* - \epsilon \, \partial^2 \eta^*/\partial x^* \, \partial t^*. \tag{11.8.16}$$

The first term is the geostrophic flow, but now allowing for the leading part of the variation of the Coriolis parameter with latitude, and the second term is the isallobaric flow (see Section 8.16).

The velocity divergence can now be calculated from (11.8.15) and (11.8.16) to give

$$\frac{\partial u^*}{\partial x^*} + \frac{\partial v^*}{\partial y^*} = -\epsilon \frac{\partial \eta^*}{\partial x^*} - \epsilon \frac{\partial}{\partial t^*} \left(\frac{\partial^2 \eta^*}{\partial x^{*2}} + \frac{\partial^2 \eta^*}{\partial y^{*2}} \right), \tag{11.8.17}$$

showing that it is smaller, by a factor of ϵ, than the order one would expect by scaling arguments, i.e., it is smaller than either $\partial u^*/\partial x^*$ or $\partial v^*/\partial y^*$ individually. The vorticity, on the other hand, is to the leading approximation

$$\frac{\partial v^*}{\partial x^*} - \frac{\partial u^*}{\partial y^*} = \frac{\partial^2 \eta^*}{\partial x^{*2}} + \frac{\partial^2 \eta^*}{\partial y^{*2}}, \tag{11.8.18}$$

showing that the *divergence is of order ϵ relative to the vorticity*.

The leading-order equation of motion can now be found by substituting (11.8.17) in (11.8.13) to give

$$\frac{\partial}{\partial t^*} \left(\frac{\partial^2 \eta^*}{\partial x^{*2}} + \frac{\partial^2 \eta^*}{\partial y^{*2}} - \eta^* \right) + \frac{\partial \eta^*}{\partial x^*} = 0 \tag{11.8.19}$$

or, in dimensional form,

$$\frac{\partial}{\partial t} \left(\frac{\partial^2 \eta}{\partial x^2} + \frac{\partial^2 \eta}{\partial y^2} - \frac{f_0^2}{c^2} \eta \right) + \beta \frac{\partial \eta}{\partial x} = 0. \tag{11.8.20}$$

Equation (11.8.19) is none other than the leading-order approximation to the *potential vorticity equation* (11.2.14), showing the *central importance* of that equation for describing quasi-geostrophic flow. It can also be confirmed that (11.8.20) gives the planetary wave dispersion relation (11.8.6), as should be expected for consistency.

The behavior of planetary waves that propagate over a range of latitudes can be studied by using the ray-tracing techniques of the preceding section. Ray paths are curved because of the variation of Coriolis parameter with latitude, and have the sinusoidal shape given by (11.7.7) with total reflection at the critical latitude y_c. Equation (11.7.3) gives the variation of north–south wavenumber l with latitude and shows in particular that

$$l^2 + f^2/c^2 = f_c^2/c^2 \qquad (11.8.21)$$

is a constant on the ray path. Effects of planetary wave propagation along curved ray paths can be seen, e.g., in Fig. 11.6, from Grose and Hoskins (1979), who examined waves generated by a large circular mountain at 30°N. Another effect is that the ray paths that correspond to seasonal oscillations produced at an eastern boundary of the ocean may not be able to reach certain areas, so that shadow zones will exist. Also, focusing effects can lead to large amplitude in places (Schopf *et al.*, 1980). Motions that satisfy the quasi-geostrophic equation (11.8.20) will be further studied in Chapter 12. The approximation is sometimes called the *mid-latitude beta-plane approximation* because it applies for nonzero values of f_0. It can also be derived without first making the equatorial beta-plane approximation, so it is generally valid when the fractional change of the Coriolis parameter f over a distance l^{-1} is small.

The physical mechanism responsible for the propagation of planetary waves is virtually the same as that for continental shelf waves illustrated in Fig. 10.18 and described in Section 10.12. Consider a line of particles that has a fixed value $\bar{Q} = \beta y_0/H$ for potential vorticity, so that in their equilibrium position the particles lie along a circle of latitude $y = y_0$. If a particle is displaced to a position y, the potential vorticity $(\beta y + \zeta)/H$ is conserved, and therefore

$$(\beta y + \zeta)/H = \beta y_0/H, \qquad \text{i.e.,} \quad \zeta = \beta(y_0 - y). \qquad (11.8.22)$$

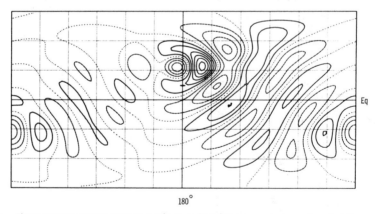

180°

Fig. 11.6. Planetary wave propagation on a sphere, as found in a numerical experiment of Grose and Hoskins (1979). Contours are of perturbation vorticity, and disturbances to a superrotation zonal flow (i.e., an eastward flow with uniform angular velocity about the earth's axis) are produced by a circular mountain centered at 30°N and 180° longitude, and with radius equal to 22.5° of latitude. Waves travel backward and forward across the equator along ray paths that are curved because of variation in the Coriolis parameter f with latitude. The equatorial trapping effect is evident. The amplitude of the wave decays with distance because of dissipative effects included in the model. [From Grose and Hoskins (1979, Fig. 3a).]

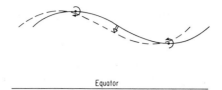

Equator

Fig. 11.7. The mechanism of planetary wave propagation. Because potential vorticity is conserved, a particle displaced equatorward acquires cyclonic vorticity relative to its surroundings (as indicated by the arrow), whereas particles displaced poleward acquire anticyclonic vorticity. The motion induced by this relative vorticity distribution is indicated by the broad arrow, and is such as to produce westward propagation of the wave.

Thus a particle that is displaced equatorward will have cyclonic vorticity relative to its surroundings as illustrated in Fig. 11.7. The motion induced by the vorticity field thus acquired causes the wave to propagate westward as shown in the figure.

11.9 Baroclinic Motion near the Equator

The shallow-water equations on the equatorial beta plane have already been applied to baroclinic modes, but without proper justification. To see what conditions must be satisfied in order to do this, consider the linearized Boussinesq equations for an incompressible stratified fluid. Near the equator, the momentum equations (4.10.11) give

$$\partial u/\partial t + 2\Omega w - \beta y v = -\rho_0^{-1}\,\partial p'/\partial x, \tag{11.9.1}$$

$$\partial v/\partial t + \beta y u = -\rho_0^{-1}\,\partial p'/\partial y, \tag{11.9.2}$$

$$\partial w/\partial t - 2\Omega u = -\rho_0^{-1}(\partial p'/\partial z + \rho' g). \tag{11.9.3}$$

Separation into normal modes, using the method of Section 6.11, is not possible unless the terms $2\Omega w$ in (11.9.1) and $-2\Omega u$ in (11.9.3), which are the components of the Coriolis acceleration that are associated with the *horizontal* component of the rotation vector, can be neglected. It is also necessary, of course, to be able to neglect the vertical acceleration $\partial w/\partial t$, but conditions for this have already been considered in Chapter 6.

The simplest way of dealing with the additional terms is to put the equations in nondimensional form, using the scales appropriate to equatorial waves, namely, the horizontal scale $(c/\beta)^{1/2}$ and the time scale $(\beta c)^{-1/2}$. The scale for p'/ρ_0 is c times the scale for u and v, and the vertical scale (from Section 6.11) is c/N, where N is the buoyancy frequency. The scale for w [from (6.11.4)] is ω/N times the scale for u, with ω of order $(\beta c)^{1/2}$. When these scales are applied to (11.9.1), it turns out that the Coriolis acceleration that is associated with the horizontal component of rotation is of order $2\Omega/N$ compared with the terms retained. Thus the condition for neglect of these terms is

$$2\Omega \ll N, \tag{11.9.4}$$

which is normally satisfied in the ocean and atmosphere. Thus (11.9.1) becomes

$$\partial u/\partial t - \beta y v = -\rho_0^{-1} \partial p'/\partial x \tag{11.9.5}$$

and (11.9.3) becomes the hydrostatic equation (6.11.2). The remaining equations to be satisfied are (11.9.2), the incompressibility condition (6.4.3), and the equation (6.4.6) for the density perturbation.

11.10   Vertically Propagating Equatorial Waves

The equations for small perturbations to an incompressible stratified fluid were found in the preceding section. When the buoyancy frequency N is constant, solutions proportional to

$$\exp(ikx + imz - i\omega t), \tag{11.10.1}$$

representing vertically propagating waves, can be obtained. In fact, as outlined in Chapter 6, problems involving vertically propagating waves in a semi-infinite region can be solved by using the normal-mode technique, the modes consisting of the continuously infinite set (6.11.21). The dispersion relations are the same as those found earlier in this chapter, but now the eigenvalue c [or c_e or $(gH_e)^{1/2}$, where H_e is the equivalent depth], associated with a particular mode, is given by (6.11.21), i.e.,

$$c = N/m. \tag{11.10.2}$$

It should be remembered that c is a property of the mode in question and is *not* equal to the phase speed except in special cases such as that of the Kelvin wave. Note also that for an isothermal *compressible* fluid, c_e is given by (6.17.40) and is approximated by (11.10.2), provided that $(4m^2 H_s^2)^{-1}$ is small, where H_s is the scale height. Even for a vertical wavelength of 20 km, this number is only about 0.03, so the incompressible approximation is reasonable.

Now consider the dispersion relations for the different types of waves. First, the Kelvin-wave relation (11.5.6), after substitution of (11.10.2), may be written

$$m^* = k^*, \tag{11.10.3}$$

where m^* and k^* are nondimensional forms of m and k, defined by

$$m^* = m\omega^2/\beta N, \qquad k^* = k\omega/\beta. \tag{11.10.4}$$

In the same notation, the mixed planetary–gravity wave ($n = 0$) dispersion relation (11.6.9) becomes

$$m^* = k^* + 1, \tag{11.10.5}$$

whereas the remaining waves satisfy (11.6.3), i.e.,

$$m^{*2} - (2n + 1)m^* = k^{*2} + k^* \quad \text{or} \quad m^* = n + \tfrac{1}{2} \pm \{(k^* + \tfrac{1}{2})^2 + n(n + 1)\}^{1/2}, \tag{11.10.6}$$

the positive root corresponding to gravity waves and the negative root to planetary

waves. The full set of dispersion curves is shown in Fig. 11.8. The gravity wave curves
are the hyperbolas in the upper part of the diagram. The planetary waves curves are
also hyperbolas and are shown on an expanded plot in the inset.

The corresponding curves in the m, k plane are, by (11.10.4), the contours of
constant frequency. The group velocity, by definition, is at right angles to these
curves and in the direction of increasing ω. The corresponding direction is shown in
Fig. 11.8 and is always downward when m is positive. For m negative, the curves are
obtained by reflection in the k axis and the group velocity is upward.

The magnitude of the group velocity can be obtained from the definition, i.e.,
the gradient of the frequency in wavenumber space. For $n \geq 1$, the group velocity is
purely vertical at points given by

$$k^* = -\tfrac{1}{2}, \qquad m^* = (n + \tfrac{1}{2}) \pm \{n(n + 1)\}^{1/2}, \qquad (11.10.7)$$

the positive sign corresponding to gravity waves and the negative sign to planetary
waves. The magnitude of the group velocity at these points is given in dimensional
terms by

$$c_g = \tfrac{1}{2}\{n + \tfrac{1}{2} \pm [n(n + 1)]^{1/2}\}(\beta N/m^3)^{1/2}, \qquad (11.10.8)$$

the positive sign again corresponding to gravity waves. For $n = 1$ and $m^{-1} = 1$ km,
this gives the value of about 1 cm s^{-1} (1 km day^{-1}) for gravity waves and 2 mm s^{-1}
(200 m day^{-1}) for planetary waves. These values could apply to both atmosphere
and ocean. As n increases, the gravity wave value increases as $n^{1/2}$, whereas the
planetary wave value decreases as $n^{-1/2}$. This means that away from the equator,

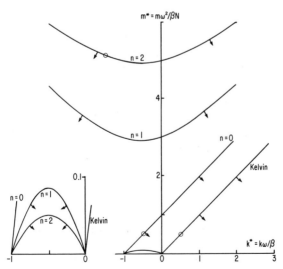

Fig. 11.8.   Dispersion curves for vertically propagating equatorially trapped waves. m is the vertical wavenumber
and k the eastward wavenumber. The curves collapse into a single set when scaled with the frequency ω, buoyancy
frequency N, and beta as indicated. The direction of the group velocity, being the gradient of frequency in wavenumber space, is as indicated. The curves for m negative are obtained by reflection in the k axis and have upward
group velocity. The inset at the left is a blowup of the region near the origin to show the planetary waves $n = 1, 2$.
The upper $n = 1, 2$ curves are the corresponding gravity waves. The circles represent observed waves (see text).

gravity waves have higher vertical group velocities, but planetary waves have lower vertical group velocities.

The polarization relations may be obtained by substituting (11.10.1) in the equations of preceding sections and utilizing the solutions, such as (11.6.2), (11.6.5), and (11.6.6), found previously. For example, the mixed planetary–gravity wave solution [cf. (11.6.10) and (11.6.11)] may be written

$$p'/\rho_0 = y \exp(-\beta y^2 m/2N) \exp(ikx + imz - i\omega t),$$

$$u = N^{-1} my \exp(-\beta y^2 m/2N) \exp(ikx + imz - i\omega t),$$

$$v = -i\omega^{-1} \exp(-\beta y^2 m/2N) \exp(ikx + imz - i\omega t), \qquad (11.10.9)$$

$$w = -N^2 m\omega y \exp(-\beta y^2 m/2N) \exp(ikx + imz - i\omega t).$$

Vertically propagating, equatorially trapped waves have been identified in both the atmosphere and the ocean. First consider examples of Kelvin waves, which can propagate only eastward. Wallace and Kousky (1968) identified such waves in the tropical stratosphere, having a period of about 2 weeks, horizontal wavelengths of about 30,000 km, and vertical wavelengths of about 10 km. The phase velocity is downward, as can be seen in the example shown in Fig. 11.9, so the group velocity is upward. For comparison with theory, properties must be calculated relative to the mean wind. The relative phase speed is 30–50 m s^{-1} to the east, and the corresponding intrinsic (or Doppler-shifted) period is about 8 days. The value of k^* is in the range 0.5–1 and is represented by the circle in Fig. 11.8. Further discussion is presented by Wallace (1971, 1973) and Holton (1975). Another example in the stratosphere has been found by Hirota (1979). The horizontal wavenumber was unity (one wave encircling the earth), the period was 4–9 days, and the vertical wavelength was 17–23 km. The Doppler-shifted phase speed was 60–80 m s^{-1}. The position on Fig. 11.8 is roughly the same, in nondimensional terms, as that for the Kelvin wave previously identified. Zangvil and Yanai (1980) have made a similar identification.

Westward-propagating mixed planetary–gravity waves in the stratosphere have also been identified, following the initial work of Yanai and Maruyama (1966), and are discussed by Wallace (1971, 1973), Holton (1975), and Zangvil and Yanai (1980). The period is 4–5 days, the horizontal wavenumber 4 (four waves encircling the earth, i.e., a wavelength of about 10,000 km), and the vertical wavelength 4–8 km. The Doppler-shifted period is about 3 days. The value of k^* is about -0.5, so the wave occupies the position shown by the circle in Fig. 11.8. These waves also have upward group velocity.

Similar westward-propagating mixed waves have been observed in the Atlantic Ocean by Weisberg et al. (1979a). In this case the period was 31 days, the vertical wavelength about 1 km, and the horizontal wavelength about 1200 km. The phase velocity was upward, corresponding to downward group velocity. The value of k^* for this wave is also about -0.5, so it occupies a similar position in Fig. 11.8 to the stratospheric mixed wave.

Vertically propagating *gravity* waves have also been identified, e.g., Cadet and Teitelbaum (1979) found such waves in the stratosphere, with periods of 35 hr, vertical wavelength 5 km, and horizontal wavelength of about 2400 km. This was

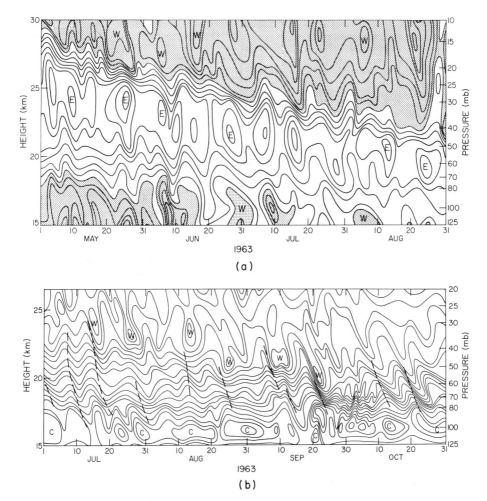

Fig. 11.9. Evidence for Kelvin waves in the equatorial stratosphere. The figure shows contours of (a) zonal wind and (b) temperature at Canton Island (3°S, 172°W) as a function of height and time. Downward-moving phase lines are evident, and these correspond to upward group velocity. The dashed lines in (b) correspond to prominent peaks and troughs in the zonal-wind component. [Courtesy of J. M. Wallace and V. E. Kousky.]

identified with an $n = 1$ gravity wave. In the ocean, Weisberg *et al.* (1979b) observed waves with period of about 9 days that appeared to have properties consistent with $n = 2$ gravity waves. Further analysis by Horigan and Weisberg (1981) showed a good fit to theory for $k^* = -1.5$ (see circle in Fig. 11.8). They also identified gravity waves with $n = 3$, $k^* = 1.3$; $n = 5$, $k^* = -2.1$; and $n = 6$, $k^* = -1.1$.

In the equatorial stratosphere, there are, in addition to the waves reported above, some very interesting longer period variations, as can be seen in Fig. 11.9a, in which the boundary between easterly and westerly winds descends over the 4-month period. This is part of a cycle, known as the quasi-biennial oscillation, which is so-called because the repeat time is usually 2 years, but occasionally $2\frac{1}{2}$ years. It can be seen very clearly in the nearly 30 years of zonal wind data presented in the corrigendum to

Coy (1979). Wave transport effects (see Section 8.15), due to the approach of the shorter period equatorially trapped waves to the critical level at the zero wind line (see Section 8.9.4), play an important role in promoting the descent of this line and are discussed, e.g., by Holton (1975, 1979, 1980a).

11.11 Adjustment under Gravity near the Equator

The way in which a stratified fluid adjusts to equilibrium under gravity has been a continuing theme in this book because it is so fundamental to understanding how the ocean and atmosphere behave. In Chapter 7 this process was examined on an infinite uniformly rotating plane. On a rotating sphere, there are two new effects of particular significance. One is the waveguide effect, so that waves radiated by a disturbance near the equator will be reflected at their critical latitudes and thus will be channeled along the equatorial waveguide. The second effect is due to the existence of a new class of low-frequency waves—the planetary waves. Because their frequencies are distinct from those of the gravity waves, the adjustment tends to take place in two stages. The first stage is the rapid change due to gravity waves, which is much the same as that on an f-plane. This rapid change produces a flow that is close to being in geostrophic equilibrium. The second stage consists of a quasi-geostrophic adjustment by means of planetary waves.

A technique for solving problems of adjustment near the equator is by expansion in terms of the separable wave solutions, just as in the f-plane case. Thus the disturbance is first resolved into vertical modes (see Chapter 6): either the discrete set appropriate to the ocean or the continuous set appropriate to the atmosphere. Each mode then satisfies the shallow-water equations given in Section 11.4, with a different value of the equivalent depth H_e or separation constant $c = (gH_e)^{1/2}$ assigned to each mode.

The shallow-water equations of Section 11.4 are then solved by expanding in terms of the parabolic cylinder functions that appear in the wave solutions (11.6.2), i.e., v, q, and r are expanded in the form

$$(v, q, r) = \sum_{n=0}^{\infty} (v_n, q_n, r_n) D_n((2\beta/c)^{1/2}y). \tag{11.11.1}$$

This is similar to the method used for studying adjustment in a channel in Section 10.7. For forced problems, the forcing terms are expanded in a similar fashion, i.e.,

$$(X, Y, E) = \rho H \sum_{n=0}^{\infty} (X_n, Y_n, E_n) D_n((2\beta/c)^{1/2}y). \tag{11.11.2}$$

Then the equations for the coefficients become

$$\partial q_0/\partial t + c\, \partial q_0/\partial x = X_0 - cE_0, \tag{11.11.3}$$

which corresponds to the Kelvin wave;

$$(\partial/\partial t + c\, \partial/\partial x)q_1 - (2\beta c)^{1/2}v_0 = X_1 - cE_1, \qquad (2\beta c)^{1/2}q_1 + 2\, \partial v_0/\partial t = 2Y_0, \tag{11.11.4}$$

which correspond to the mixed planetary–gravity wave; and for the remaining modes $n \geq 1$,

$$(\partial/\partial t - c\,\partial/\partial x)r_{n-1} + (2\beta c)^{1/2}nv_n = -(X_{n-1} + cE_{n-1}),$$

$$\frac{\partial}{\partial t}\left\{-(2\beta c)^{1/2}r_{n-1} + \left(\frac{\partial}{\partial t} + c\frac{\partial}{\partial x}\right)v_n\right\} + \beta c v_n = \left(\frac{\partial}{\partial t} + c\frac{\partial}{\partial x}\right)Y_n$$

$$-(\tfrac{1}{2}\beta c)^{1/2}[(n+1)X_{n+1} - X_{n-1}] \quad (11.11.5)$$

$$+ c(\tfrac{1}{2}\beta c)^{1/2}[(n+1)E_{n+1} + E_{n-1}],$$

$$(2\beta c)^{1/2}\{-r_{n-1} + (n+1)q_{n+1}\} + 2\,\partial v_n/\partial t = 2Y_n,$$

$$(\partial/\partial t + c\,\partial/\partial x)q_{n+1} - (2\beta c)^{1/2}v_n = X_{n+1} - cE_{n+1}.$$

The last set of equations may be combined into a single equation for r_{n-1}, q_{n+1}, or v_n such as

$$\frac{\partial}{\partial t}\left\{\frac{1}{c^2}\frac{\partial^2 v_n}{\partial t^2} - \frac{\partial^2 v_n}{\partial x^2} + (2n+1)\frac{\beta}{c}v_n\right\} - \beta\frac{\partial v_n}{\partial x} = \text{forcing}, \quad (11.11.6)$$

which can be derived directly by substitution of (11.11.1) in (11.4.9).

The problem of adjustment near the equator has similarities with that of adjustment in a channel because of the waveguide effect. Energy in a given mode, which is concentrated in a certain range of latitudes, will remain in that mode and thus cannot disperse to higher latitudes. This confinement of energy is particularly obvious in cases for which there is no dependence on x. Because energy cannot radiate to $y = \pm\infty$ as it can on the f-plane, the system cannot adjust to a steady state by radiating energy and therefore oscillates instead. A simple example is provided by the exact solution given by (11.6.10) and (11.6.11) when $k = 0$. Initially there is a v field only. The solution does *not* approach an equilibrium but oscillates forever. Other examples of this behavior are provided by Moore and Philander (1977) and Anderson and Gill (1979).

Now consider how energy disperses in the longitudinal direction. The dispersion curves shown in Fig. 11.1 exhibit the frequency gap between gravity waves and planetary waves, and this separation of time scales may be seen in solutions of (11.11.5). For instance, Fig. 11.10 [from Anderson and Rowlands (1976a)] shows the solution for $n = 1$, $x \leq 0$ of (11.11.6), which satisfies $v_1 = 1$ at $x = 0$. (This can also be interpreted as the solution r_0 satisfying $r_0 = 1$ at $x = 0$.) This exhibits a gravity wave front moving at speed c of the same sort as that in the f-plane case drawn in Fig. 7.3. This thins with time and is followed by waves of near-inertial period just as in the classical f-plane solution. However, the solution does *not* tend to a steady state because of the beta term in (11.11.6), and the slow adjustment that follows is determined by planetary wave dynamics. Figure 11.10 also shows [for $t = 20/(2\beta c)^{1/2}$] the corresponding solution of the *planetary-wave approximation* to (11.11.6), namely,

$$\frac{\partial}{\partial t}\left\{-\frac{\partial^2 v_n}{\partial x^2} + (2n+1)\frac{\beta}{c}v_n\right\} - \beta\frac{\partial v_n}{\partial x} = 0, \quad (11.11.7)$$

and this gives a very good fit to the solution after the initial gravity wave front and

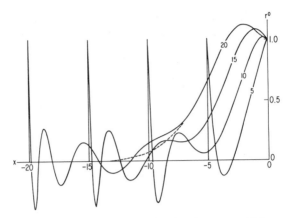

Fig. 11.10.   Solution r_0 for equatorially trapped planetary and gravity waves satisfying $r_0 = 1$ at $x = 0$ and starting from zero. The unit of distance is the equatorial Rossby radius $a_e = (c/2\beta)^{1/2}$ and the unit of time is $(2\beta c)^{-1/2}$. The wave front has the character of a gravity wave front, and may be compared with *f*-plane solutions like those depicted in Fig. 7.3. The slower-moving waves are planetary waves, as can be seen from the dashed line, which depicts the solution of the planetary-wave approximation to the equations when $t = 20$ units. [Adapted from Anderson and Rowlands (1976a, Figs. 4 and 7).]

its wake of gravity waves has passed. In fact, the behavior after $t = 5/(2\beta c)^{1/2}$ is close to that predicted by the *long-wave approximation* to (11.11.6), namely,

$$\partial v_n/\partial t - (c/(2n + 1))\partial v_n/\partial x = 0, \qquad (11.11.8)$$

which gives nondispersive westward propagation at speed $c/(2n + 1)$.

   In the ocean, the adjustment process is strongly influenced by a further factor, namely, the presence of *meridional boundaries* that cross the equator. Waves propagating along the equatorial waveguide eventually reach such a boundary, so it is of some interest to determine what happens. The problem was solved by Moore (1968). Take, for instance, an equatorial Kelvin wave. When this strikes an eastern boundary, part of the energy is reflected in the form of planetary and gravity waves (Fig. 11.10 in fact shows r_0 for this part of the solution when a Kelvin wave of unit amplitude impinges on the eastern boundary $x = 0$ at time $t = 0$). The remainder of the energy is carried poleward along the eastern boundary in the form of coastal Kelvin waves, thereby providing a means for energy to be lost from the equatorial region. Detailed solutions are given by Anderson and Rowlands (1976a). At a western boundary, wave energy can also be reflected into waves with eastward group velocity, but in addition, equatorward propagating coastal Kelvin waves have their energy channeled into the equatorial waveguide.

   An illustration of wave dispersion in the equatorial zone is given in Fig. 11.11. This shows the pycnocline displacement due to a zonal wind stress that is applied from time $t = 0$ in the region marked by the dashed lines in the first panel [from McCreary (1978)]. The stress has a maximum value of 0.05 N m^{-2} (0.5 dyn cm^{-2}) in the center of the region and falls off linearly to zero in the surrounding zones. The

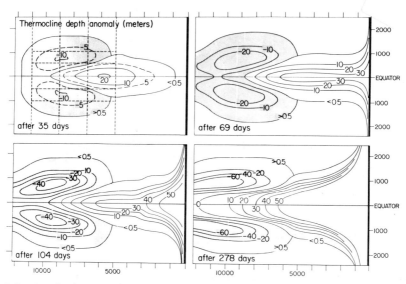

Fig. 11.11.   Time development of a thermocline depth anomaly in response to a wind anomaly that is limited in space to the region shown by the dashed lines (a uniform eastward stress of 0.05 N m^{-2} in the central region that falls off to zero in the outer zones) and that builds up linearly in time over the first 35 days to a constant value. The rapid movement of a Kelvin wave to the east is clearly seen, as is the slower propagation of planetary waves to the west. Poleward-propagating coastal waves on the eastern boundary are in evidence after 69 days. The shape of the 50 m contour after 278 days indicates the presence of long planetary waves reflected from the eastern boundary (these waves are discussed further in Chapter 12). [From McCreary (1978, Fig. 2).]

separation into an eastward-propagating Kelvin wave and westward-propagating planetary wave is clearly seen, and the propagation into the coastal waveguide on the eastern boundary is also evident. Observations of coastal propagation in the Pacific were discussed in Section 10.13. Similar solutions have also been found by Anderson and Rowlands (1976a) and by Hurlburt *et al.* (1976). They illustrate how changes of wind in the central and western Pacific can strongly influence the state of the ocean in the eastern Pacific. In particular, abnormally high temperatures in the surface coastal waters off Peru, which occur in some years, can be produced by a relaxation or reversal of the trade winds far across the Pacific. The abnormally high temperatures have disastrous effects on the Peruvian fisheries, and fishing in these waters has been banned in such years, whereas in good years one-fifth (by weight) of the world's catch has been obtained there. Figure 11.12 shows the sea surface temperature difference between an anomalous year and the preceding one. The similarity in pattern with that of Fig. 11.11 can be seen. The anomalous conditions are associated with the name "El Niño" [the child; reviews may be found, e.g., in Glantz (1980)]. The unusual surface temperatures have profound effects on the atmosphere, which in turn affect the ocean through the winds. In fact, the whole ocean–atmosphere system shows large changes on a time scale of several years, and the El Niño phenomenon is only a part of the system [see, e.g., Julian and Chervin (1978), Horel and Wallace (1981), and Rasmusson and Carpenter (1982)].

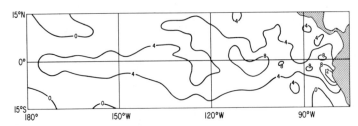

Fig. 11.12.   The sea-surface temperature in the equatorial Pacific in December 1972 (an El Niño year) relative to that in December 1971. Contours are in degrees Farenheit (2°F = 1.1°C). Note the very large values near the equator and the evidence for equatorial confinement of the anomaly. This particular year was disastrous for the Peruvian fishing industry. [Adapted from Fishing Information, December 1972, No. 12, U.S. Dept. of Commerce, National Marine Fisheries Service, La Jolla, California.]

11.12   Transient Forced Motion

The nature of forcing terms and methods of calculating their effect have been discussed in Chapter 9. One method is to expand the forcing functions in terms of vertical normal modes. Then each mode satisfies the forced shallow-water equations. As a next step, the forcing terms can then be expanded in parabolic cylinder functions, i.e., in the form (11.11.2), to give Eqs. (11.11.3)–(11.11.5). The equations in x and t can then be solved by various means, e.g., by expressing solutions as Fourier integrals of the pure wave solutions or by Laplace transform techniques.

For example, the effect of diabatic heating in the troposphere (see Section 9.13) can be calculated by such techniques, and Holton (1972) has worked out examples to show how the waves observed in the stratosphere could be generated. The source was assumed to be stationary but oscillating with a period of 4–5 days. Oscillations in cloud brightness with this period are observed [see, e.g., Murakami and Ho (1972)], being particularly prominent at 5–10°N in the northern hemisphere winter, indicating that they are oscillations of the intertropical convergence zone. Although the forcing is stationary, the response consists of *traveling* waves. This can be understood by considering, as an example, the part of the solution described by (11.11.3) in which the forcing has east–west wavenumber k and oscillates with frequency ω. The equation then has the form

$$\partial q_0/\partial t + c\, \partial q_0/\partial x = 2 \sin kx \cos \omega t$$

$$= \sin(kx + \omega t) + \sin(kx - \omega t), \qquad (11.12.1)$$

and the solution is

$$q_0 = -(\omega + ck)^{-1} \cos(kx + \omega t) + (\omega - ck)^{-1} \cos(kx - \omega t). \quad (11.12.2)$$

In this case, the eastward-moving wave, represented by the second term, dominates because this is most nearly resonant with the free mode, the Kelvin wave. The same is true for other types of waves, and Fig. 11.13 shows a case in which the dominant response consisted of mixed planetary–gravity waves. Further discussion may be found in Holton (1975). Studies by Hayashi (1974) and Hayashi and Golder (1978)

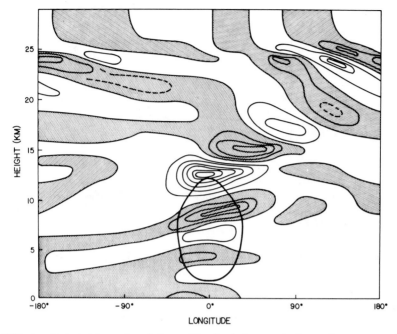

Fig. 11.13. Longitude–height section of the meridional velocity perturbation at the equator as found by Holton (1972, Fig. 9) for an antisymmetric source of diabatic heating that oscillates with an amplitude exceeding $4\,\mathrm{K\,day}^{-1}$ inside the heavy line. Contours are at $2\,\mathrm{m\,s}^{-1}$ intervals. The waves produced are mainly mixed planetary–gravity waves. The mean wind varies with height with a maximum eastward velocity of $8\,\mathrm{m\,s}^{-1}$ at 21 km, zero velocity at 25 km, and westward velocity above that level.

with a general circulation model indicate that diabatic heating is indeed the source for such waves. For a proper understanding, however, the reason for the fluctuations in diabatic heating and their relationship with the field of motion needs to be appreciated.

Another important example of forced response is that of the ocean by the wind. Consider, for instance, the response to a uniform wind blowing parallel to the equator. This should be independent of x, so Eqs. (9.10.4) and (9.10.11) for mode n with $f = \beta y$ become

$$\partial \tilde{u}_n/\partial t - \beta y \tilde{v}_n = \tilde{X}_n,$$

$$\partial \tilde{v}_n/\partial t + \beta y \tilde{u}_n = -g\,\partial \tilde{\eta}_n/\partial y, \qquad (11.12.3)$$

$$\partial \tilde{\eta}_n/\partial t + H_n\,\partial \tilde{v}_n/\partial y = 0.$$

The situation is completely analogous to the storm surge case studied in Section 10.9 and to the corresponding solution for coastal upwelling, there being a solution with $\tilde{v}_n$ independent of time and $\tilde{u}_n$ and $\tilde{\eta}_n$ proportional to time. This solution was first obtained by Yoshida (1959) and is depicted in Fig. 11.14. The equation for $\tilde{v}_n$ obtained from (11.12.3) is

$$(c_n^2/\beta y)\,\partial^2 \tilde{v}_n/\partial y^2 - \beta y \tilde{v}_n = \tilde{X}_n \qquad (11.12.4)$$

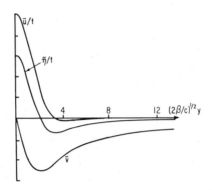

Fig. 11.14.   The analytic solutions obtained by Yoshida (1959) for the longitude-independent equatorial jet produced by a uniform eastward wind. The eastward current $\tilde{u}$ and surface elevation are proportional to time t, whereas the northward velocity $\tilde{v}$ is independent of time.

and can be expressed in terms of a Lommel function of order $\frac{1}{4}$ (Erdélyi *et al.*, 1953, p. 40).

Far from the equator, the first term in (11.12.4) becomes relatively small and the approximate solution is

$$\tilde{v}_n = -\tilde{X}_n/\beta y = -\tilde{X}_n/f, \tag{11.12.5}$$

i.e., the steady-state Ekman current (see Section 9.2). With easterly trade winds, the Ekman transport is away from the equator on both sides, and therefore upwelling is produced at the equator by the same sort of process as that which results in coastal upwelling. The only difference is in the shape of the response function $\tilde{v}_n$ due to the variation of f [(11.12.4) is the same as (10.9.4) in the steady state], and the width of the upwelling region that is obtained by comparing the terms on the left-hand side of (11.12.4) is now the *equatorial Rossby radius* given by (11.5.4). The vertical velocity is proportional to $\tilde{\eta}_n/t$, which is also shown in the figure. Easterlies tend to raise the thermocline at the equator, thus producing a flow along the equator in the same direction as that of the wind. The shape of the jet and a sketch of the situation are shown in Fig. 11.14. The jet is sometimes referred to as the Yoshida jet.

When the wind varies with x and t, solutions can be found by methods similar to those used for coastal upwelling (see Sections 10.11 and 10.13), particularly when the scale of the wind stress is large compared with the equatorial Rossby radius (Gill and Clarke, 1974). An important consideration, even when the wind does not vary, is the *effect of meridional* boundaries, because Yoshida's solution does not satisfy the boundary condition at these places. In order to satisfy this condition, solutions of the homogeneous equations must be added, and these take the form of the equatorially trapped waves already discussed. The most important are the long waves, and these will have no effect at a given point until the wave has had time to reach that point. Thus at midocean the first long wave to arrive will be the Kelvin wave associated with the first baroclinic mode since this travels at speed c_1. The next will be either the second-mode Kelvin wave, which travels at speed c_2, or the first-mode planetary wave, which travels westward at speed $c_1/3$.

The results of a simulation by Gill (1975) of these effects is shown in Fig. 11.15.

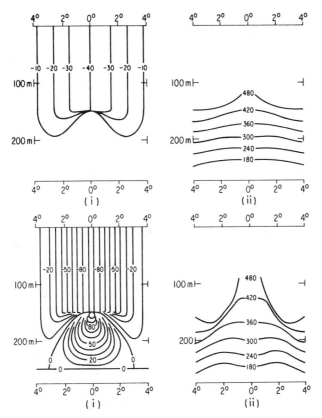

Fig. 11.15. Contours of (i) eastward velocity in centimeters per second, and (ii) thermosteric anomaly in centiliters per tonne as determined analytically in a model calculation by Gill (1975, Fig. 16). Initially, the fluid is at rest; the top 150 m are homogeneous, the thermosteric anomaly changes linearly with depth from 480 to 180 cl t^{-1} over the next 100 m, and the fluid below that level is assumed to be homogeneous over a very large depth. A uniform westward wind stress of 0.35 dyn cm^{-2} is switched on at an initial time, and the upper panels show the current and the displaced position of the isopycnals 20 days later. Up to this stage, the flow near the section shown is independent of longitude and is accelerating uniformly as in Yoshida's solution. At this time, however, a Kelvin wave arrives from the western boundary, bringing with it a pressure gradient and an undercurrent. The lower panels show the solution at 100 days just before other boundary influences are due to affect the further development of the undercurrent. [Reproduced from "Numerical Models of Ocean Circulation," 1975, by permission of the National Academy of Sciences, Washington, D.C.]

Before the boundary effects arrive, the easterly wind causes the surface waters to move away from the equator, thus causing upwelling. As the thermocline rises, a geostrophic current in the direction of the wind develops. Geostrophy can apply even at the equator, where the Coriolis parameter vanishes, since in the limit as $y \to 0$ the steady version of (11.9.2) gives

$$\beta u = -\rho_0^{-1}\, \partial^2 p'/\partial y^2 \qquad \text{as} \quad y \to 0. \tag{11.12.6}$$

Boundary effects cause a fundamental change because of the east–west pressure gradient that comes with the Kelvin wave. Figure 11.15 shows how the situation is altered with only the first-mode Kelvin wave present. A westward (downwind) current is found at the surface, but in the thermocline below this is an eastward under-current, which is driven by the pressure gradient associated with the Kelvin wave. The thermocline has a downward kink under this current, at the same time being upwelled above, thereby "spreading" the isopycnals. This feature is a necessary consequence of the geostrophic balance.

The eastward undercurrent, called the *equatorial undercurrent*, is a major feature of the equatorial ocean circulation, particularly in the Pacific and Atlantic Oceans. It is a strong narrow eastward current found in the region of strong density gradient below the mixed layer and with its core very close (within a degree) to the equator. Its vertical thickness is around 100 m and its half-width is a degree of latitude. The maximum current is typically 1 m s^{-1}. An example of a section through the current is shown in Fig. 11.16.

The undercurrent has an interesting history; for although this unique feature of the circulation was first discovered in the nineteenth century, the fact appears to have been forgotten, and it was rediscovered in the middle of the twentieth century! The original discovery [see Matthäus (1969)] arose as a result of a chemist, J. Y. Buchanan, being invited to join the steamship *Buccaneer*, which was chartered by the Indiarubber, Guttapercha, and Telegraph Works Company of Silvertown to do a survey in 1886 across the Gulf of Guinea in preparation for laying cable. Later that year, Buchanan (1886, p. 761) reported to the Royal Geographical Society a

> very remarkable under-current which is found setting in a south-easterly direc-tion with a velocity of over a mile per hour at three stations almost on the equator ... the surface water was found to have a very slight westerly set. At a depth of 15 fathoms there was a difference, and at 30 fathoms the water was running so strongly to the south-east that it was impossible to make observations of temperature, as the lines, heavily loaded, drifted straight out, and could not be sunk by any weight the strain of which they could bear ....

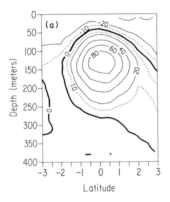 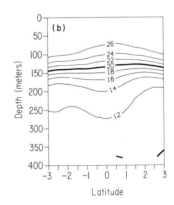

Fig. 11.16. Meridional cross sections through the equatorial undercurrent, showing (a) eastward velocity (in centimeters per second relative to an average over the depth interval 300–500 m and (b) temperature (in degrees Celsius). The values are averages of 34 sections made at three longitudes: 150, 153, and 158°W, during 1979 and 1980. [Courtesy of E. Firing.]

The reason for the existence of the undercurrent is clear and has been brought out in model studies [see, e.g., Gill (1975), Philander (1973), and Philander and Pacanowski (1980)]. The easterly wind stress along the equator is balanced by a pressure gradient that, because of the stratification, is taken up almost entirely in the upper layers. Figure 11.17 [from Lemasson and Piton (1968)] shows a section along the equator in the Pacific Ocean, showing the temperature and relative pressure fields. [The pressure gradient in the Atlantic is discussed by Katz *et al.* (1977).] The pressure falls rapidly toward the east at the surface (the surface slope is about 5×10^{-8}), but this pressure gradient has practically disappeared 250 m below the surface. Away from the equator, this pressure gradient is associated with a geostrophic flow toward the equator. At the equator, however, the Coriolis parameter vanishes, and thus there is a flow directly down the pressure gradient, i.e., toward the east. Near the surface there is, in addition, the flow driven directly by the stress. Away from the equator this is a poleward Ekman transport, but at the equator, where $f = 0$, the stress-driven flow is directly downwind, i.e., toward the west.

The model discussed earlier in this section shows the first stage of the process that sets up the pressure gradient, and the subsequent development has been studied in numerical models, e.g., by Philander and Pacanowski (1980). The model also shows the confinement of the undercurrent to within a Rossby radius of the equator and the upwelling associated with the Ekman divergence at the equator. The east–west slope of the thermocline is also very important since it means that cold water is brought close to the surface on the eastern side. This is easily brought up to cool the surface waters in active periods, but can be covered by a thin warm skin at other times. In the west, the mixed layer is warm and deep, so changes in stirring rates or in rates of input from the atmosphere do not alter the surface temperature a great deal.

Another interesting transient problem is that of the effect of a change of wind on the flow at the western boundary. This is particularly relevant to the establishment of the Somali Current with the onset of the southwest Monsoon over the Indian Ocean. Lighthill (1969) showed how changes that are remote from the boundary can cause changes at the western boundary through the propagation of planetary waves. The mechanism is not specific to the equatorial zone, and therefore it is studied in Chapter 12. A numerical model by Cox (1970) showed that local longshore winds are also important for the Somali Current, the mechanism being the one studied in Section 10.11, i.e., the one associated with a "coastal jet." Further studies by Anderson and Rowlands (1976b), Hurlburt and Thompson (1976), and Cox (1976) have examined the roles of the two mechanisms, showing that local longshore winds dominate the response initially (say for the first month after the winds are turned on), but that remote forcing is important at a later stage. Nonlinear effects can lead not only to downstream intensification of the current, but also to the establishment of an eddylike or meandering structure [Cox (1979); a review is given by Anderson (1979)]. The eddy or eddies move along the coast in the development stages of the numerical computations, and in Cox's (1979) model they become stationary, with the Somali Current leaving the coast at a latitude that depends on the wind field. In practice, the Somali Current does indeed show large meanders and the point of separation from the coast has been observed at different locations.

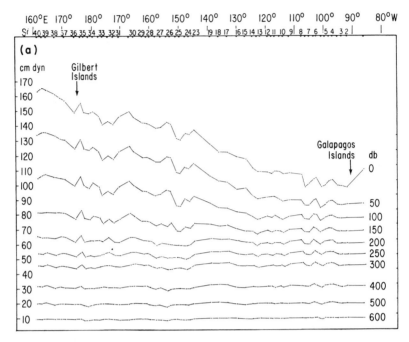

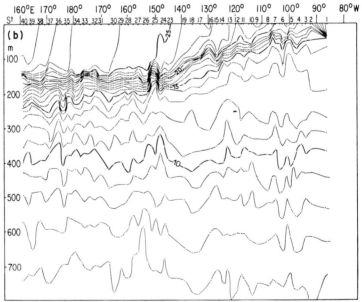

Fig. 11.17.   (a) Variation of dynamic height at various levels relative to 700 db and (b) isotherms along the equator in the Pacific Ocean. The pressure gradient is of the sign required to balance the wind stress and to drive the current below the level at which the stress is acting. In the west, where the mixed layer is deep, the pressure gradient is largely associated with a temperature gradient in the mixed layer. In the east, the pressure gradient is associated with the tilt of the thermocline. This tilt brings cold water close to the surface in the east, thus making the surface temperature here sensitive to small changes in the winds, heating rate, etc. [From Lemasson and Piton (1968, Figs. 1 and 2).]

11.13   Potential Vorticity for Baroclinic Motion: The Steady Limit

For baroclinic motion, the linearized horizontal momentum equations are (for a strongly stratified fluid for which (11.9.4) applies) (11.9.5) and (11.9.2). Taking the curl of these to obtain the vorticity equation gives

$$\partial \zeta / \partial t + f(\partial u / \partial x + \partial v / \partial y) + \beta v = 0, \tag{11.13.1}$$

and combining this with the incompressibility condition (6.4.3) gives

$$\partial \zeta / \partial t + \beta v = f \, \partial w / \partial z. \tag{11.13.2}$$

This is the linearized potential vorticity equation for baroclinic motion. For steady motion, this reduces to

$$\beta v = f \, \partial w / \partial z. \tag{11.13.3}$$

The interpretation of this equation is as follows. Suppose the horizontal velocity field is convergent, so that the right-hand side of (11.13.3) is positive. Vortex tubes, which are material surfaces of fluid whose sides are made up of vortex lines (see Section 7.9), are almost vertical, so the convergent motion will reduce their cross-sectional area and increase their length. In order words, *vortex tubes are being stretched* in the vertical, and so their absolute vorticity (Section 7.9) will increase in magnitude. In steady motion of small amplitude, this can happen only if the *vortex tube moves poleward*, i.e., in a direction in which the background vorticity f increases in magnitude. The rate at which it must move poleward to maintain a vorticity balance is given by (11.13.3). The argument applies equally well in a divergent field of horizontal motion in which vortex tubes shrink vertically and move equatorward.

The balance (11.13.3) can be achieved in *forced* steady motion. Consider, for instance, the case of a shallow homogeneous layer of depth H driven by an Ekman pumping velocity w_E (see Section 9.9) at the surface. Then $\partial w / \partial z$ will have a uniform value of w_E / H over the layer, and (11.13.3) becomes

$$\beta v = f w_E / H \tag{11.13.4}$$

or, using the approximate expression (9.4.3),

$$\beta v = (\rho H)^{-1} (\partial Y / \partial x - \partial X / \partial y), \tag{11.13.5}$$

which is the steady-state version of the shallow-water potential vorticity equation (11.4.13). A more general version, which does not require the layer to be homogeneous, is obtained by taking the vertical integral of (11.13.3) and substituting (9.4.3) for the Ekman velocity. This gives

$$\beta V = \rho^{-1} (\partial Y / \partial x - \partial X / \partial y), \tag{11.13.6}$$

where V [cf. (9.2.6)] is the vertical integral of the northward velocity. Equation (11.13.6), or its equivalent (11.13.5) for a homogeneous layer, is known as Sverdrup's equation, and the northward transport V given by (11.13.6) is called the Sverdrup transport. In words, (11.13.6) states that the northward *mass* transport ρV is equal to the wind-stress curl divided by beta.

Equation (11.13.6) is very important in oceanography because it describes the way in which the steady ocean circulation is driven in linear models. In the subtropical gyres, the wind-stress curl is negative, thus giving convergent motion to the Ekman layer (westerlies at the poleward extreme of the gyre move water equatorward, whereas the easterlies at the equatorward extreme move water poleward). This implies downward motion just below the Ekman layer, so vortex lines shrink and the water moves equatorward as described above. Equation (11.13.6) cannot apply over the whole gyre, however, by continuity of mass, so details of the wind-driven ocean circulation require a more complete analysis (see Chapter 12).

Another way that the steady vorticity balance (11.13.3) can be maintained is by the vortex stretching that results from internal heating. The resultant buoyancy force gives rise to a vertical velocity, given by (9.15.8), and substitution in (11.13.3) gives

$$f^{-1}\beta v = \rho_0^{-1} f^{-2} \beta \, \partial p'/\partial x = \partial(\rho_0^{-1} N^{-2} B_s')/\partial z, \qquad (11.13.7)$$

where B_s' is the rate of change of buoyancy per unit volume and the expression in terms of pressure comes from the geostrophic relation between v and $\partial p'/\partial x$. According to (11.13.7), heating in midatmosphere causes rising motion and hence vortex stretching and poleward motion at lower levels. Conversely, vortex shrinking and equatorward motion is produced at higher levels. Note, however, that (11.13.7) cannot apply for zonally averaged flows for which $\partial p'/\partial x$ must vanish. Examples of buoyancy driven flows will be considered in the next section.

11.14   Steady Forced Motion

The vorticity balance for steady flow without friction was discussed in the preceding section, but generally this solution is applicable only in part of the region of interest and, as found in Section 9.16, complete steady solutions can be obtained only when friction and mixing processes are included in some form. In this section, steady solutions of the forced shallow-water equations will be examined with the simplest form of dissipation, namely, Rayleigh friction with decay rate r and Newtonian cooling, also with decay rate r. Then the equations are the same as those for the transient case, except that $\partial/\partial t$ is replaced everywhere by $r + \partial/\partial t$ or, in the steady case, by r. In particular, when H is constant, (11.4.10)–(11.4.12) become

$$ru - \beta yv = -g \, \partial \eta/\partial x + X/\rho H, \qquad (11.14.1)$$

$$rv + \beta yu = -g \, \partial \eta/\partial y + Y/\rho H, \qquad (11.14.2)$$

$$rg\eta + c^2(\partial u/\partial x + \partial v/\partial y) = -gE/\rho. \qquad (11.14.3)$$

Although formally obtained for a shallow homogeneous layer, these equations also apply for each normal mode, but with a value of c appropriate to that mode and with a magnitude of forcing determined from the expansion of the forcing function in normal modes. In particular, the "stress" terms (X, Y) in (11.14.1) and (11.14.2) may arise from expression of the stress gradient in the ocean mixed layer in terms of modes as shown in Section 9.10, and the "evaporation" term in E may arise from expansion of the buoyancy forcing terms as shown in Sections 9.13 and 9.15. Whatever the form of forcing, equations of the above form result, and these may be condensed into a single equation for v, namely, the steady-state version of (11.4.14):

$$
\frac{r}{c^2}(r^2 + f^2)v - r\left(\frac{\partial^2 v}{\partial x^2} + \frac{\partial^2 v}{\partial y^2}\right) - \beta\frac{\partial v}{\partial x}
$$

$$
= \frac{1}{\rho H}\left\{\frac{r}{c^2}(rY - fX) + r\frac{\partial E}{\partial y} - \frac{\partial}{\partial x}\left(\frac{\partial Y}{\partial x} - \frac{\partial X}{\partial y} + fE\right)\right\}. \qquad (11.14.4)
$$

For small friction ($r \to 0$), the leading terms in (11.14.4) all contain x derivatives. *Zonally independent flows* are thus a rather special case because these "leading" terms all vanish. For such flows (11.14.4) reduces to

$$
\frac{r^2 + f^2}{c^2}v - \frac{\partial^2 v}{\partial y^2} = \frac{1}{\rho H}\left(\frac{\partial E}{\partial y} - \frac{fX}{c^2} + \frac{rY}{c^2}\right). \qquad (11.14.5)
$$

A factor r has been taken from each side, so the v field generated is of order unity when r is small. Equation (11.14.5) is the same as that for an f-plane, except that f now varies with y in accordance with (11.4.3). Thus solutions for v are like distorted versions of those on an f-plane. The free solutions are no longer exponential, but are special functions (parabolic cylinder functions of order $\frac{1}{2}$) that are tabulated by Abramowitz and Stegun (1964, Chapter 19). For example, Fig. 11.18 shows the variation of pressure p (or surface elevation) for evaporation concentrated along the line $y = a_e$, where a_e is the equatorial Rossby radius defined by (11.5.4). The equivalent

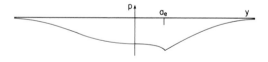

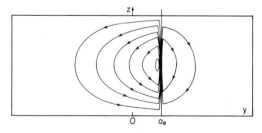

Fig. 11.18.   The linear solution due to a line source of heat (or evaporation) located at $y = a_e$, i.e., one Rossby radius from the equator, as obtained from the forced shallow-water equations. The upper panel shows the pressure distribution (cf. the f-plane solution in Fig. 9.11a), with a trough at the heating latitude. Inflow to the trough comes mainly from the equatorial side. This is depicted in the lower panel, where the solution is interpreted as having a sinusoidal vertical structure associated with a single vertical mode. The picture then has the character of the meridional circulation generated by heating along a particular latitude, as occurs in the intertropical convergence zone.

f-plane solution is the limiting form of that shown in Fig. 9.11a as the width of the forcing region tends to zero. The effect of the variation of f is to give slower falloff in pressure on the equatorial side of the evaporative sink.

Applications of evaporatively forced solutions of the shallow-water equations are much more interesting when applied to baroclinic motions, as was done in Section 9.15. For instance, in an incompressible atmosphere of constant buoyancy frequency N with a "rigid lid" at some height, the baroclinic modes have sinusoidal structure, the "gravest" mode (i.e., the one with largest vertical scale) being a sine of height with a half-wavelength spanning the depth. If *diabatic heating* is applied with this distribution in the vertical, only the gravest mode is stimulated and the equations are the shallow-water equations with heating replacing the evaporation term. Figure 11.18 shows the meridional circulation produced by such heating concentrated on the line $y = a_e$ and is obtained by attaching the appropriate vertical structure to the solutions of (11.14.5). This is a type of Hadley circulation that is stimulated by a line source of heating such as occurs along the intertropical convergence zone (ITCZ). Rising air is found only in the heating zone. Most air is drawn in from the equatorial side, so the most pronounced circulation is on this side. The pressure curve shows how the surface pressure varies with such a solution. The effect of heating in a semi-infinite atmosphere can be found by superposing modes, as was done in Section 9.15, and the same technique can also be used for a compressible atmosphere (see Sections 6.14, 6.15, and 6.17).

As pointed out above, zonally independent flows are rather special examples of forced shallow-water solutions, so it is important also to examine small-friction solutions when the forcing varies with x. Then the leading terms of (11.14.4), when integrated with respect to x, give

$$\beta v = (\rho H)^{-1}(\partial Y/\partial x - \partial X/\partial y + fE), \tag{11.14.6}$$

which is the potential vorticity equation (11.13.5) with evaporative forcing included. To illustrate the behavior of solutions when x variations are important, and to show the part played by equation (11.14.6), consider special cases in which the forcing is limited to a region of finite zonal extent. Particularly simple solutions may be found with evaporative (i.e., buoyancy) forcing given by

$$E/\rho H = F(x) \exp(-\beta y^2/2c) = F(x)D_0((2\beta/c)^{1/2}y), \tag{11.14.7}$$

where F will be chosen to have the form

$$F(x) = \begin{cases} \cos(\pi x/2L) & \text{for} \quad |x| < L, \\ 0 & \text{for} \quad |x| > L. \end{cases} \tag{11.14.8}$$

For this form of forcing, the only nonzero coefficient to occur on the right-hand sides of the equations of Section 11.11 is $E_0 = F(x)$, and so the solution involves only the functions q_0, q_2, v_1, and r_0. If, in addition, the friction coefficient r is small compared with βL, i.e.,

$$r \ll \beta L, \tag{11.14.9}$$

the solutions can be written (Gill, 1980) in the form

$$u = \left[\tfrac{1}{2}q_0(x) + \tfrac{1}{2}q_2(x)(2c^{-1}\beta y^2 - 3)\right] \exp(-\beta y^2/2c),$$

$$g\eta/c = \left[\tfrac{1}{2}q_0(x) + \tfrac{1}{2}q_2(x)(2c^{-1}\beta y^2 + 1)\right] \exp(-\beta y^2/2c), \qquad (11.14.10)$$

$$v = \left[F(x) + 4(r/c)q_2(x)\right]y \exp(-\beta y^2/2c),$$

where q_0 and q_2, the functions defined by (11.4.15) and (11.11.1), satisfy the approximate equations

$$dq_0/dx + (r/c)q_0 = -F, \qquad (11.14.11)$$

$$dq_2/dx - 3(r/c)q_2 = F. \qquad (11.14.12)$$

This form can be deduced most readily from the steady-state equivalents of the equations of Section 11.11. For small r, the solution in the forcing region corresponds to neglecting the terms in r in (11.14.11) and (11.14.12) and this corresponds to the meridional velocity being given by the potential vorticity equation (11.14.6). More comments will be made about this solution later.

Outside the forcing region, the frictionless potential vorticity equation (11.14.6) does *not* apply, because the solutions of (11.14.11) and (11.14.12) decay slowly in the x direction, contrary to the assumptions that give rise to (11.14.6). Equation (11.14.11) is, in fact, the one for a *forced Kelvin wave* that propagates eastward out of the forcing region at speed c and at the same time decays at the rate r. In other words, the wave decays toward the east at the rate r/c per unit distance. It follows from this that the required solution of (11.14.11) is the one that vanishes at $x = -L$, the *western* end of the forcing region, since Kelvin waves can carry information only eastward. Similarly, (11.14.12) corresponds to a *forced long planetary wave* with $n = 1$. This wave propagates westward at speed $c/3$, while decaying at the rate r, so the required solution is the one that vanishes at $x = L$, the *eastern* end of the forcing region.

Clearly, the technique can be applied to forcing with parabolic cylinder functions of any order on the right-hand side of (11.14.7) and the solutions can be superposed to give results for any distribution of forcing. For instance, Fig. 11.19 shows the solution [from Gill (1980)] for

$$E/\rho H = F(x)(1 + y/a_e) \exp(-y^2/4a_e^2) = F(x)\left[D_0(y/a_e) + D_1(y/a_e)\right]. \quad (11.14.13)$$

It is interesting once again to interpret this as a baroclinic solution with heating, for the situation is rather similar to that which obtains in the atmosphere in July. If the equatorial Rossby radius a_e is taken to be about $10°$ of latitude, the solution shown in Fig. 11.19 with $L = 2a_e$ corresponds to maximum heating at about $10°\mathrm{N}$ and covering $40°$ of longitude. According to plate 9.1 of Newell *et al.* (1974), the region of large heating is concentrated as is that of Fig. 11.19, but with maximum at about $15°\mathrm{N}$ and with largest values between $90°$ and $140°\mathrm{E}$ longitude.

In the region of heating, Fig. 11.19a shows vertical motion and mainly northward velocities, as predicted by the potential vorticity equation (11.14.6), although there is some distortion on the west side because of friction. The interpretation is that rising due to heating causes vortex stretching and hence acquisition of cyclonic vorticity.

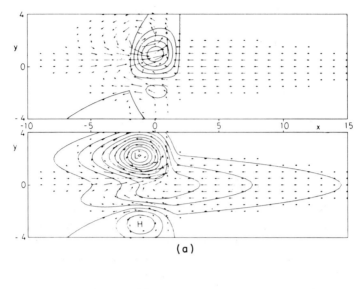

(a)

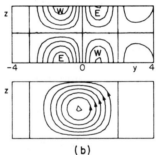

(b)

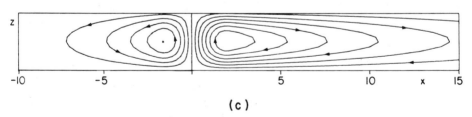

(c)

Fig. 11.19.   The solution of the forced shallow-water equations for heating (or evaporation) that is confined to the range of longitudes $|x| < 2a_e$. The distribution with latitude is given by (11.14.13) and has a maximum north of the equator. The arrows in (a) give the horizontal velocity field and the solid contours in the upper panel give the vertical velocity, which has a distribution close to that of the heating function. The motion is upward within the contours north of the equator with a maximum near $x = 0$, $y = a_e$. The contours in the lower panel are pressure contours, and the axes are labeled in units of a_e. [From Gill (1980, Fig. 3).] (b) The meridional circulation when the response is interpreted as a baroclinic response to heating with sinusoidal distribution in the vertical. The upper panel gives the zonal flow (E for easterlies, W for westerlies) and the lower panel the meridional flow (Hadley circulation). [From Gill (1980, Fig. 3).] (c) The meridionally averaged zonal flow (Walker circulation) with the same interpretation, i.e., as a baroclinic response. [From Gill (1980, Fig. 1c).]

Thus the fluid particles tend to move poleward in order to keep their relative vorticity small.

The only wave that can propagate eastward from the forcing region is a Kelvin wave, so that the region $x > L$ (corresponding to the Pacific Ocean if the model is applied to the effect of forcing over Indonesian longitudes) shows motion with characteristic features of the Kelvin wave, namely, flow parallel to the equator and symmetric about the equator. The winds are easterly toward the heat source and decay eastward at the rate r/c per unit distance. Physically, the decay process represented by r (Geisler, 1981) appears to be "cumulus friction" due to momentum transfer between levels through cumulus activity.

Long planetary waves can propagate westward from the forcing region, but they decay faster than the Kelvin waves and thus cover a smaller area. They include meridional motion, so the equatorward return of the air moving poleward in the heating region is found to the west. The result is a cyclonic center on the west flank of the heating region. (Comparisons can be made with the observed 850-mb flow shown in Fig. 11.21.) Solutions with similar properties were found by Webster (1972), who found numerical solutions with a two-layer model for perturbations on a zonal flow, and Matsuno (1966) found solutions for forcing periodic in x. Solutions of the shallow-water equations with friction on a sphere, as obtained by Margules (1893), also show similar characteristics.

Figure 11.19b shows the zonally averaged flow, exhibiting a strong Hadley cell with rising motion in the latitude of maximum heating. Equatorial motion is associated with easterlies and poleward motion with westerlies. This is due to conservation of angular momentum and the fact that, in a linear formulation, the fluid "remembers" only the angular momentum from the latitude where it has just been. Thus the zonal velocity changes sign on crossing the equator.

Figure 11.19c shows the meridionally averaged circulation, which is due only to the part of the forcing that is symmetric about the equator. Rising motion is found over the longitudes of the heating region, and sinking elsewhere. The corresponding circulation in the atmosphere in the Pacific region is called the Walker circulation.

Forced steady solutions of the above type are also useful for understanding the equatorial ocean circulation, the forcing being due to wind stress distributed over the surface mixed layer. In order to obtain a realistic current structure, many modes are needed, as shown by McCreary (1981a). He uses a model with a vertical eddy viscosity and vertical eddy diffusivity that are equal at each level and vary with depth in such a way that the modal structure is preserved and each mode remains independent of the others. The equations for each mode have the form (11.14.1)–(11.14.3), but now the friction coefficient increases with the mode number (r is in fact proportional to c_n^2). Friction has little effect on the first few modes for which the wind stress is largely balanced by an east–west pressure gradient, and there is consequently little contribution to the currents from these modes. Instead, the main contributions to the current come from the modes for which friction is just large enough to make effects of the eastern and western boundaries of secondary importance to local effects, as given by the x-independent solution. Examples of a solution obtained by this method are shown in Fig. 11.20.

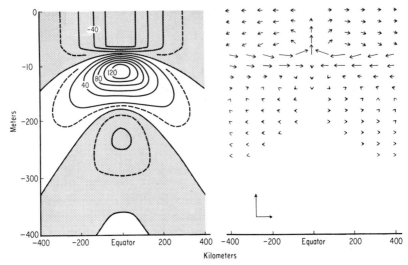

Fig. 11.20.   A meridional cross section of current structure across the equator as obtained from a linear solution by McCreary (1981a, Fig. 4c). The left panel shows zonal currents (positive eastward) and the right panel shows currents in the meridional plane. The model is forced by a surface stress that is westward and applied in a region limited in both latitude and longitude. The region is symmetric about the equator and in the center of a basin with meridional boundaries. The section shown is the middle one. The density profile consists of a mixed layer 75 m deep with a sharp exponential falloff just below and a more gradual decay deeper down. Note the strong eastward undercurrent (contours are in centimeters per second) with westward flows above and below, and the upwelling that draws in water from the thermocline. Calibration vectors in the lower left corner of the right panel have amplitudes of 0.005 cm s^{-1} in the vertical and 10 cm s^{-1} in the horizontal. The horizontal coordinate is in kilometers.

11.15   The Tropical Circulation of the Atmosphere

It was seen in Chapter 1 how, in conditions that are uniform in the horizontal, an equilibrium temperature profile can be established by a balance between radiative and convective processes. A very simple model of convection was used in the examples discussed in Chapter 1, but in practice the convective processes are quite complex because of moisture in the atmosphere. A moist parcel of air that rises adiabatically becomes saturated at some level (cloud base), at which some of the vapor condenses and becomes visible as a cloud. The accompanying release of latent heat of condensation gives extra buoyancy to the air that can help it to rise further. If a parcel remains undiluted by entrainment of air from its environment, all the excess moisture is condensed out as it rises, and all the latent heat released is used to warm the parcel, then the change of temperature with pressure follows a saturation pseudoadiabat (see Section 3.8) and the equivalent potential temperature of the parcel remains constant. The parcel remains buoyant so long as its equivalent potential temperature θ_e exceeds that of the environment, and thus it can rise to a level at which the difference between the environment and the parcel θ_e becomes zero. In practice, most parcels of rising air entrain a lot of environmental air because they are rather small, so small cumulus clouds are formed with a rather short lifetime. Other processes, such as reevaporation of the liquid water droplets in the cloud, take place and further reduce the potential of the air to rise.

On the other hand, large parcels of rising air are not diluted so much and can rise further. For a small number, the entrainment effects are not very important, so they can reach a height at which the equivalent potential temperature difference is zero. These are called "undilute hot towers." The height to which an undilute hot tower can rise varies a lot with surface temperature because warm air can hold so much more moisture than cool air. For instance, at 30°C (303 K) saturated air at surface pressure has an equivalent potential temperature of 386 K, i.e., 83° higher than for dry air, which for typical conditions gives a saturated parcel the potential to rise to around the 100-mb level. For comparison, a saturated parcel at 20°C (292 K) and surface pressure has $\theta_e = 333$ K (42° higher than for dry air), which gives it the potential to rise to about the 300-mb level.

A typical profile of potential temperature and of equivalent potential temperature for the tropical atmosphere is shown in Fig. 3.5. The ability of undilute hot towers to rise to a great height can be seen from the θ_e profile, which shows a decrease with height from the surface to a minimum at about 650 mb, then an increase with height, the surface value not being reached again until above the 200-mb level.

The nature of the convection, as deduced from observations, is outlined, for example, by Riehl (1979). Undilute hot towers are thought to cover only about one-thousandth of the tropical area at any given time, but to contain most of the upward volume flux. A parcel of air may rise to the top of the tower in a matter of hours. Between clouds, air slowly descends, perhaps taking 1 or 2 months to reach the lower levels. If the descent were adiabatic, the air would arrive at the ground some 80° warmer than when it left since its potential temperature would be conserved. However, radiative cooling at a rate of 1 or 2 deg day^{-1} leads to a temperature distribution more like that observed. Thus a radiative–convective equilibrium can be established. Account also needs to be taken of the mixing due to smaller clouds, particularly in the conditionally unstable region (see Section 3.8) in which θ_e decreases with height. In this region, motions with small horizontal scale are the most unstable to small disturbances, but the small clouds entrain a great deal and do not grow very tall. Motions with large horizontal scale can mix over a larger depth because entrainment is less important. Thus account needs to be taken of a whole ensemble of clouds of different sizes. A convection model that includes the whole cloud ensemble and the relevant physical effects has been developed by Arakawa and Schubert (1974) in a form that can be used in large-scale numerical models.

In practice, conditions are not uniform over the tropics, so convection and radiation do not balance everywhere. The difference between the two processes (plus contributions from other processes that are generally less important) gives the diabatic heating rate (see Fig. 9.10 for zonally averaged distributions). Maximum tropospheric values in the tropics tend to be around the 500-mb level, and the greatest heating tends to be concentrated over regions in which the underlying surface is warm and moist (high equivalent potential temperature). The average vertical velocity at 500 mb is approximately given by (9.15.8) [see Table 9.1 from Newell *et al.* (1974)], i.e., is proportional to the heating rate and upward when heating is positive, i.e., when the rate of latent heat release exceeds the radiative cooling rate. Maps of the

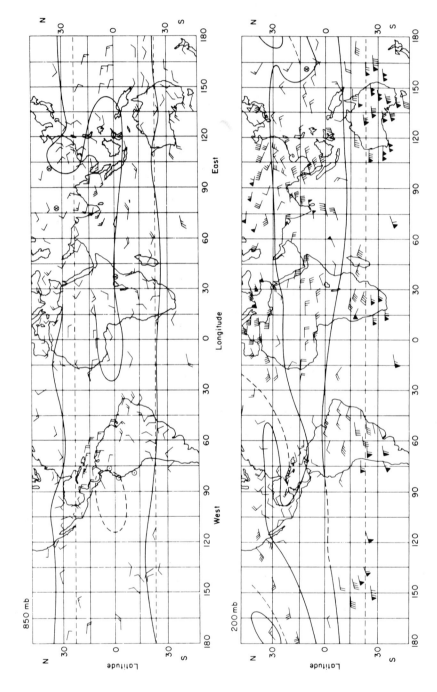

Fig. 11.21. Chart of average 850-mbar (upper panel) and 200-mbar (lower panel) winds for July, after van der Boogaard. [Reproduced with permission from Riehl (1979), "Climate and Weather in the Tropics," Fig. 1.13. Copyright by Academic Press (London) Ltd.] Lines separate areas with east and west wind components. Dashes in lower panel mark positions of semipermanent troughs.

vertical velocity at 500 mb for June–August [Newell *et al.* (1974, plates 9.1 and 9.2)] show a concentration of upward motion in the Indonesian region with maximum values (of order 100 mb day^{-1}) between 90° and 140°E and between 10° and 25°N.

Some idea of the response to such a distribution of heating can be gained from models like that studied in the preceding section (Gill, 1980), although such models do not explain *why* the heating distribution has its observed form. The general pattern of winds for that time of year can be seen in Fig. 11.21, which shows velocity vectors at the 850- and 200-mb levels. Comparison with the simple model result of Fig. 11.19 can be made, with the heating zone centered at about 120°E longitude, and many features of agreement can be seen, namely, easterly trade winds to the east of the heating zone, poleward motion within the heating zone, a cyclonic circulation centered on the western flank of the heating zone, and, the strongest feature of all, the westerly jet into the southern part of the heating zone. The zonally averaged meridional circulation is shown in Fig. 1.7b and shows a pronounced Hadley cell with rising motion in the summer hemisphere. This can also be compared with Fig. 11.19. A representation of the zonal mass flux at 5°N is reproduced in Fig. 11.22. The main rising branch is in Indonesian longitudes, a smaller one being located over South America. Each of the rising branches may be compared with the one found in the model solution depicted in Fig. 11.19c and associated with the heating zone. The Pacific cell, consisting of rising in the West Pacific and sinking in the east Pacific, received particular attention and is usually referred to as the Walker circulation. This name is applied to other cells as well. Another means of representing the motion to highlight the regions of convergence and divergence in the circulation patterns is given by Krishnamurti (1971, 1979).

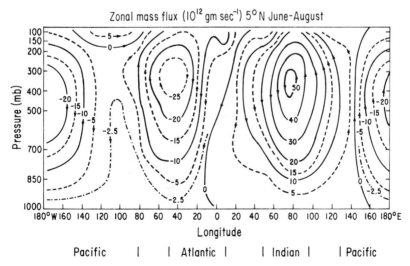

Fig. 11.22.   A representation of the zonal mass flux, averaged over the region 0°–10°N for June–August. Contours do not correspond to streamlines, but give a fairly good representation of the velocity field associated with the Walker circulation. [From Newell *et al.* (1974), "The General Circulation of the Tropical Atmosphere," Vol. 2, p. 151, Fig. 9.3; reproduced by permission of the MIT Press.]

The tropical circulation is by no means constant and is subject to pronounced changes on many time scales. For instance, hurricanes are a violent feature of the variability at time scales of the order of 1 week. A very brief discussion of their character and of their effect on the ocean may be found in Section 9.11, and a review of their properties is given by Gray (1979). They originate in latitudes between 5 and 25° (the equator is excluded because the Coriolis parameter vanishes there) but only in longitudes in which the sea is warm (above 26°C), thus allowing the equivalent potential temperature at the surface to reach values high enough for strong convection to be promoted. Thus hurricanes are mainly a hazard of the summer season. The locations of their points of origin are shown in Fig. 11.23. There are also many less violent features with similar time scales, such as "easterly waves" [see, e.g., Riehl (1979)] that propagate westward at speeds typically around 8 m s^{-1} and are often associated with the *intertropical convergence zones*. The latter are the narrow regions of convergence corresponding to convective activity along lines that have a predominantly east–west orientation and can be seen in satellite pictures (see, e.g., Fig. 1.2) as lines of cloud near the equator. They tend to be found at the latitude at which the sea-surface temperature is maximum and to migrate seasonally with this maximum.

Seasonal variations in the tropics are considerable in some regions, the monsoon winds over the North Indian Ocean (see Fig. 11.24) being an example that is referred to by Pliny and other early Greek authors. Halley (1686) not only described the major seasonal changes but also explained them as a result of seasonal changes in heating. In fact, he stated that the changes in the North Indian Ocean confirm his theory that the trade winds are a result of buoyancy forces.

> That this has no other cause, is clear from the times wherein these Winds set in: *viz.* in *April*, when the Sun begins to warm those Countries to the North, the S.W. *Monsoon* begins, and blows during the Heats until *October*; when the Sun being retired, and all things growing cooler Northward, and the Heat encreasing to the South, the North-East Winds enter and blow all the winter till *April* again (Halley 1686, p. 168).

The onset of the monsoon is usually very rapid, not only with respect to the sudden change to southwesterly winds over the Indian Ocean (Fieux and Stommel, 1977), but also in the spectacular beginning of rains over India and Southeast Asia. Dates for these events vary with location, e.g., monsoon rains near the equator begin in April, whereas there are more northerly locations where the rains do not begin until July. The change in wind over the Indian Ocean is usually in May. An interesting feature of the Southwest Monsoon is the strong low-level jet (Fig. 11.25) that forms as a western boundary jet against the East African escarpment. It is a dynamic feature similar to western boundary currents in the ocean (Anderson, 1976), which are discussed in Chapter 12. Seasonal changes are not only marked over the Indian Ocean, but also in other areas such as the West Pacific and the West African region (see Figs. 11.28 and 11.29).

The tropical regions are noted not only for seasonal changes, but also for large variations from year to year. Many of these changes are coherent over large areas of the globe, and these coherent changes are called the "southern oscillation," following Walker (1924, 1928). (Walker used the term to distinguish this phenomenon from

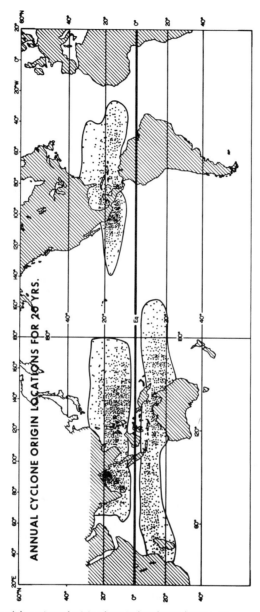

Fig. 11.23. The location of the points of origin of tropical cyclones during a 20-year period. [From Gray (1979).]

two "northern" oscillations, i.e., variability coherent with the changes in strength of the Icelandic low- and of the Aleutian low-pressure centers. The term does *not* refer to the southern hemisphere.) It has been found that time series of such diverse quantities as sea-level pressure, air temperature, sea-surface temperature, precipitation, and sea level from a wide variety of locations are remarkably well correlated. Some examples are shown in Fig. 11.26. Using a sample set of such time series (usually

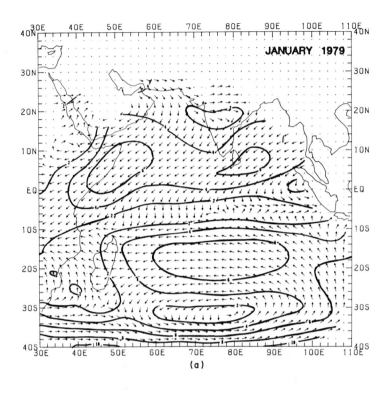

(a)

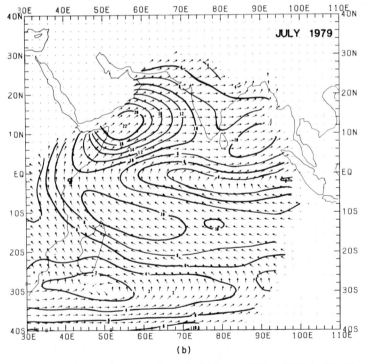

(b)

Fig. 11.24. Low-level winds over the Indian Ocean for (a) January 1979 and (b) July 1979. These are derived from satellite cloud winds and represent the flow at about 1 km above the sea surface. [Courtesy of H. Virji.]

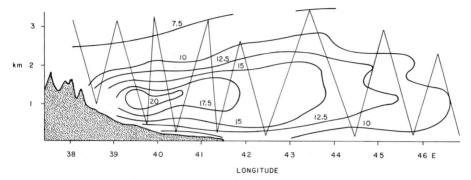

Fig. 11.25.   Southerly wind component (contours in meters per second) in a cross section at the equator on July 4, 1977. The low-level jet is clearly seen. [From Hart *et al.* (1978, Fig. 6a).]

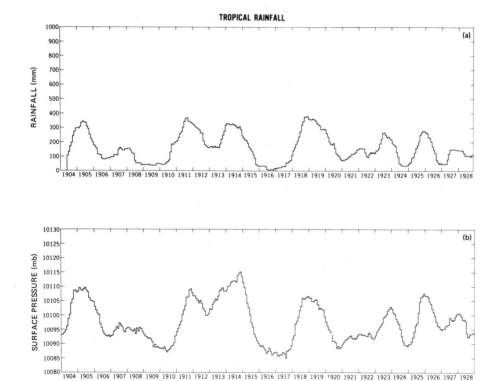

Fig. 11.26.   A comparison between the rainfall at a series of tropical Pacific islands (mostly close to the equator between 160°E and 150°W) and the surface pressure at Darwin. In both cases, 12-month running averages have been taken. [From Allison *et al.* (1972, Fig. 21).]

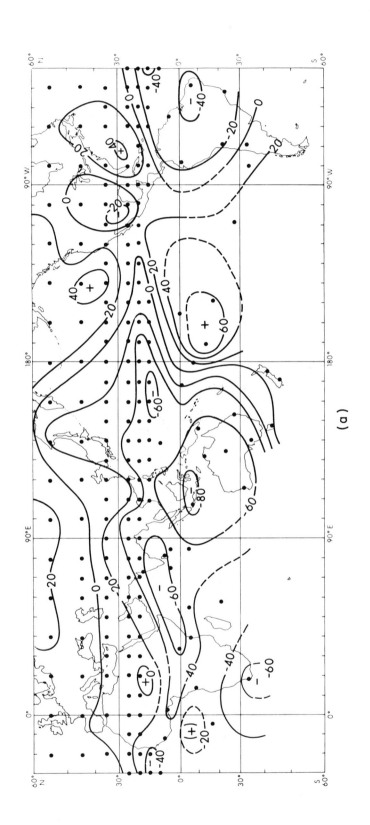

(a)

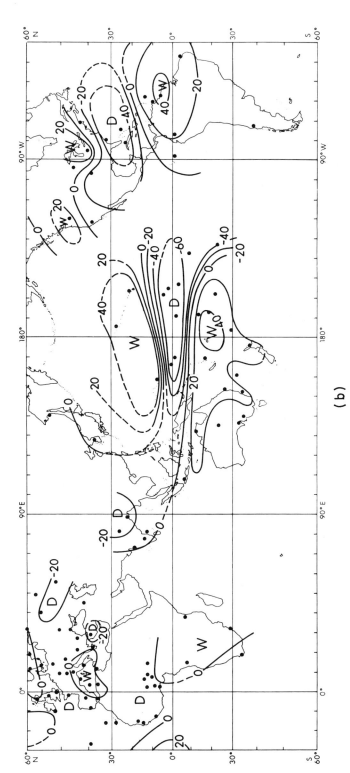

Fig. 11.27. Correlations (%) between (a) sea-level pressure and (b) rainfall with the southern oscillation index for the December–February season. The index was constructed from sea-level pressure time series at eight stations; Cape Town, Bombay, Djakarta, Darwin, Adelaide, Apia, Honolulu, and Santiago, using principal component analysis. The dots show the locations of stations at which the coefficients were calculated. [From Wright (1977).]

(b)

481

ure at Darwin or Djakarta), a southern oscillation index can be
ng with maps of correlations between various quantities and the
ples, namely, surface pressure and rainfall, are shown in Fig. 11.27
...icate the global scale of the phenomenon. Thus when the pressure is higher
than normal over Indonesia, the region of high pressures tends to extend (in northern
hemisphere winters) across the equatorial Indian Ocean and into the tropical West
Pacific on the north side of the equator. Lower than normal pressures tend to be
found in the central South Pacific and in the Northeast Pacific. Rainfall at these times
is high along the equator in the Pacific, whereas air and sea-surface temperatures
(not shown in the figure) are abnormally high along the equator in the East Pacific.
Figure 11.12 shows a striking example of such an anomaly, and the drastic effects on
the Peruvian fishing industry have already been noted in Section 11.11.

Numerical general circulation models of the atmosphere have been used to study
certain features of the variability, e.g., Manabe *et al.* (1974, 1979) have studied the
seasonal cycle by prescribing seasonal changes in insolation and seasonal changes in
sea-surface temperatures. Encouraged by the success of this model, Manabe and Hahn
(1977) applied the same techniques for an ice-age simulation that indicated that
tropical continental regions were much drier then. Models have also been used to
study certain aspects of the southern oscillation. Low-index (as used in Fig. 11.27)
years correspond to high surface temperatures in the tropical East Pacific, with large
anomalies corresponding to El Niño years. For example, Fig. 11.12 shows the huge
temperature differences between an El Niño year and the previous year (which in this
case was a year of abnormal coolness). Bjerknes (1966, 1969) argued that in warm
anomaly years, i.e., in years with small east–west temperature contrast, the Walker
cell over the Pacific would be weaker, and he discussed other consequent effects.
Models [e.g., Rowntree (1972, 1979), Julian and Chervin (1978)] have been used to
look at effects of warm anomalies, and it has been found that not only are there
changes in the tropics, but also significant changes appear to be produced at mid-
latitudes and at high latitudes.

11.16   Tropical Ocean Currents

Currents at the equator have already been discussed in Section 11.12. The surface
current is generally in the direction of the wind, being a direct response to the wind,
but the east–west pressure gradient set up to balance the wind stress drives an eastward undercurrent in the thermocline, i.e., just below the mixed layer where the direct
influence of the wind is felt. In the Atlantic and Pacific Oceans, this normally implies
a westward surface current and an eastward undercurrent, whereas in the Indian
Ocean the currents change direction seasonally [see, e.g., Knox (1976), and Wyrtki
(1973)]. The east–west pressure gradient set up to balance the wind stress in the
Pacific is shown in Fig. 11.17, together with the corresponding thermocline tilt. This
brings the thermocline up against the surface in the east, so a great deal of structure is
"crammed up" near the surface here [see, e.g., Tsuchiya (1975)]. It also means that
surface temperature patterns are very sensitive to small changes in conditions. In the

west, the layer does not become much deeper than 150 m and Fig. 11.17 shows that the surface slope, as calculated from the density field, results from an increase in temperature of the mixed layer toward the west rather than from a thermocline tilt. The Atlantic is not nearly so broad, so these effects are not so great. The east–west pressure gradient in the Atlantic and its seasonal variations are discussed by Katz *et al.* (1977), whereas seasonal variations in the slope of the thermocline in the Pacific are discussed by Meyers (1979b). The core of the undercurrent is not always on the equator and can be displaced up to a degree to the north or south. Such displacements [see, e.g., Gill (1975)] can be produced by the north–south component of the wind, e.g., a northward wind forces the surface layers downwind, thereby depressing the thermocline to the north and raising the surface. The associated pressure gradient corresponds to a reduction of current to the north and to an increase to the south of the equator. In other words, the core of the undercurrent is displaced upwind.

The mean surface elevation pattern for the Pacific Ocean is shown in Fig. 7.8a (part i) along with the associated geostrophic currents. The westward current that spans the equator is associated with a trough at the equator as required by geostrophy. To the south, the pressure continues to rise to about 20°S, so the current remains westward. To the north, a ridge is found at about 4°N. This marks the northern boundary of the South Equatorial Current (i.e., the westward flow extending to about 20°S). The ridge is strongest and extends furthest east in November–December (Wyrtki, 1975). Further north there is another trough at about 10° latitude, north of which a westward current (the North Equatorial Current) is again found, and this extends to about 20° latitude. The trough is deepest in November–December and shallowest in March–June. Variations in the trough and ridge amplitude can be monitored in the West Pacific from island tide gauges (Wyrtki, 1979a).

Between the trough and ridge is found a strong eastward jet known as the Equatorial Countercurrent. Its strength depends on the difference in surface pressure between the ridge and trough. It is strongest (Wyrtki, 1975) in September–October when the pressure difference is 40 dynamic cm or more between 150°W and 130°E. The transport of the countercurrent increases to the west at all times of year, details being given by Kendall (1970). In January–February, however, the pressure difference between ridge and trough decreases rapidly, first in the eastern and central Pacific and later in the western Pacific, reaching minimum values of about 23 dynamic cm in March–April.

The reason for the asymmetry in the currents may be found in the asymmetry of the wind stress pattern, which is shown in Fig. 11.28 for two different times of year. The prominent feature that is responsible for the asymmetry is the intertropical convergence zone (ITCZ), which is found at about 10°N in the East Pacific. This is a region of marked Ekman convergence in the atmospheric boundary layer, and hence one of Ekman divergence in the ocean. The associated Ekman pumping velocity is shown in Fig. 11.28c. The associated mass flux (see Section 9.4) is the same in both atmosphere and ocean. It becomes very large near the equator because the Coriolis parameter f, which appears in the denominator of the formula (9.4.2) for the pumping velocity, vanishes at the equator. Also variations in f can be just as important as variations in stress for determining the value. In fact the Ekman transport is away from the equator over most of the tropical zone because of the easterly trade winds.

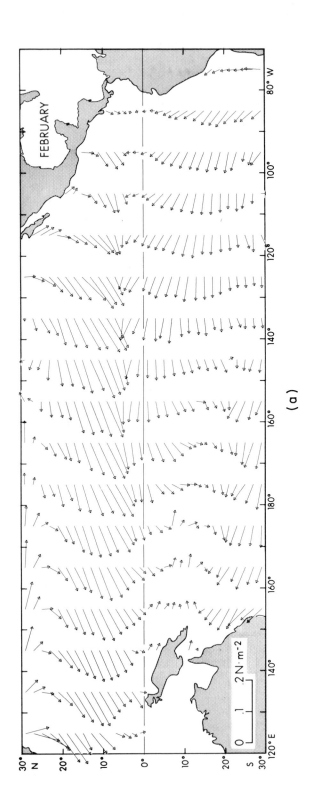

(a)

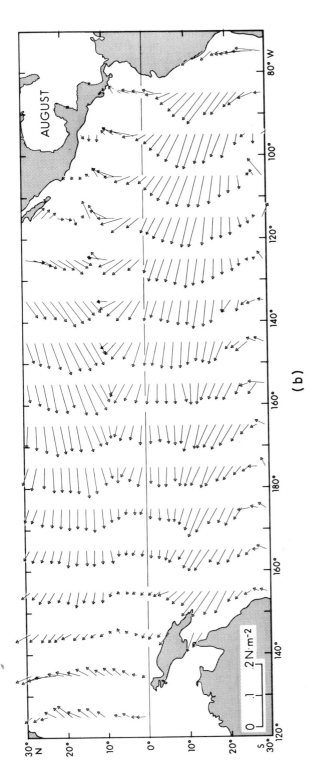

Fig. 11.28a,b. Mean surface wind stress over the tropical Pacific for (a) February and (b) August. The mean wind stress is, to a good approximation, in the same direction as the surface wind with magnitude in newtons per square meter given by 0.002(U^2 + 11), where U is the wind speed in meters per second. [From Wyrtki and Meyers (1975, 1976).]

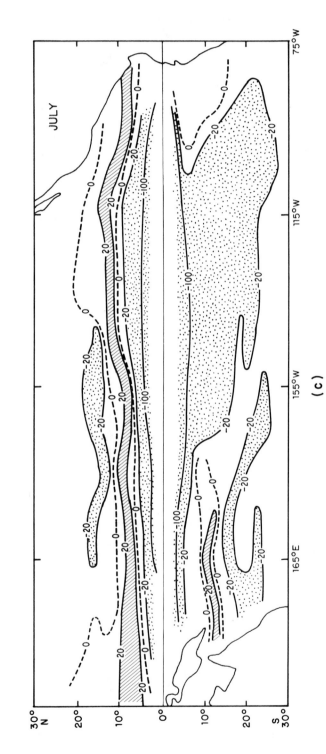

Fig. 11.28c. The corresponding Ekman pumping velocity for the ocean for July in units of 10^{-7} m s^{-1} ≈ 0.9 cm day^{-1} ≈ 3.2 m yr^{-1}, i.e., the same units as those used in Fig. 9.6. Areas with upward values in excess of 20 units (64 m yr^{-1}) are hatched, and areas with downward values in excess of 20 units are stippled. The Ekman pumping velocity in the atmosphere has the same sign but is greater in proportion to the density ratio (i.e., is 800 times bigger; 20 units correspond to 150 m day^{-1}).

The tendency is therefore to upwell water at the equator. However, because f increases with distance from the equator, the Ekman transport tends to fall, and the resulting convergence produces *downwelling*. Within 15° of the equator, it takes very rapid changes of wind with latitude to counteract this tendency, and changes occur only in the intertropical convergence zone, where the downwelling tendency is greatly reduced or even reversed. Similar features occur in the Atlantic (see Fig. 9.6). (*Note*: If the wind field used to calculate the Ekman pumping is not well resolved, the region of upward velocity associated with the ITCZ may not appear, and therefore models driven by such a wind field do not contain a countercurrent.)

To see how the ocean responds to this forcing, consider first the case for which the wind is eastward or westward and the response does not vary with longitude. For any given mode, the equations are (11.14.1)–(11.14.3) and for small friction the first of these gives

$$v = -X/\beta y \rho H, \tag{11.16.1}$$

i.e., a northward or southward Ekman current is driven directly by the wind stress. The corresponding pressure field is found by substitution in (11.14.3), which gives

$$rg\eta = -c^2 \, \partial v/\partial y = c^2 \, \partial(X/\beta y \rho H)/\partial y. \tag{11.16.2}$$

In other words, if upward motion is produced by Ekman convergence, the thermocline stands higher. Consequently, the dynamic height at the sea surface relative to a fixed depth is lower. This is presumably the reason for the trough that is found at about 10°N. Note, however, that the *magnitude* of the response given by (11.16.2) depends on the friction parameter, i.e., mixing processes are important. The remaining feature of the response is the zonal velocity, and this is simply the geostrophic velocity associated with the pressure field (11.16.2), i.e., (11.14.2) gives

$$r\beta yu = -rg \, \partial\eta/\partial y = -c^2 \, \partial^2(X/\beta y \rho H)/\partial y^2. \tag{11.16.3}$$

The above solution is not valid everywhere because of (a) the boundary layer and (b) the effects of meridional boundaries. In the equatorial boundary layer, the friction term in (11.14.1) becomes important, the result being that v becomes the solution of (11.12.4), i.e., the solution found by Yoshida (1959). Nominally the equatorial boundary layer has thickness a_e equal to the equatorial Rossby radius and the asymptotic approximation (11.16.1) is valid at distances large compared with a_e. In practice, the approximation does not give better than 10% accuracy until a latitude of about 6° (for the first baroclinic mode) has been reached [see, e.g., Anderson *et al.* (1979, Fig. 2c)].

The effect of the *western* boundary is felt through Kelvin waves, which are required to satisfy the boundary condition there (Gill, 1975; McCreary, 1981a), whereas the effect of the *eastern* boundary is felt through planetary waves. Because equatorial waves have large phase velocity, they travel a long way before decaying significantly. The decay distance may be expected to drop rapidly with mode number, but despite this, McCreary (1981a) estimates that about 8 modes would carry information from the boundaries to the center of the ocean before decaying. This suggests that a high mode-number structure will be evident near the equator, particularly when the forcing is time dependent, and observations of currents by Luyten and Swallow (1977) in the

Indian Ocean indeed show many reversals of direction with depth. Wunsch (1977) has shown that such reversals can be produced by a seasonally varying wind.

At distances of more than 6° from the equator, the Kelvin wave becomes unimportant, so only the eastern boundary can affect the interior solution. Furthermore, the phase speed [see (11.8.6)] of long planetary waves has a magnitude $\beta c^2/f^2$ that falls off inversely as the *square* of the distance from the equator (cf. Fig. 12.3). Consequently, the distance through which a wave will propagate before decaying falls off rapidly with latitude as well. Thus only the first baroclinic mode is likely to affect the whole width of the ocean at latitudes of 6° or more. The implication is that the *east–west pressure gradients* seen in Fig. 7.8a (part i) are *due to the first baroclinic mode*, except near the equator and the boundaries. The region in which boundary effects are felt is in the east, the width of the affected zone increasing rapidly as the equator is approached.

The planetary wave *correction* for a given mode can be found by using (11.14.1)–(11.14.3) again, but now with x dependence added and forcing removed. For small friction, the motion is quasi-geostrophic, so (11.14.1) and (11.14.2) become

$$\beta y u = -g\, \partial\eta/\partial y, \qquad \beta y v = g\, \partial\eta/\partial x. \tag{11.16.4}$$

However, the weak divergence associated with this field produces vertical motion that advects heat upward or downward. For a steady state, this must be balanced by a heat flux due to mixing processes. With the representation of these processes used in (11.14.3), the resulting equation for a single mode has the form

$$rg\eta + c^2(\partial u/\partial x + \partial v/\partial y) = 0. \tag{11.16.5}$$

Substituting in (11.16.5) from (11.16.4), the potential vorticity equation results, namely,

$$rg\eta = (c^2 g/\beta y^2)\, \partial\eta/\partial x = c^2 v/y. \tag{11.16.6}$$

The solution has the form

$$rg\eta = p_0(y)\, \exp(r\beta y^2 x/c^2), \tag{11.16.7}$$

from which follows, by (11.16.4),

$$r\beta y u = -(2r\beta y x p_0/c^2 + dp_0/dy)\, \exp(r\beta y^2 x/c^2). \tag{11.16.8}$$

If the eastern boundary is the meridian $x = 0$, p_0 is required to have the form

$$p_0 = -c^2\, \partial(X/\beta y \rho H)/\partial y + A, \tag{11.16.9}$$

where A is a constant, in order for the sum of the solutions (11.16.3) and (11.16.8) for u to vanish at $x = 0$. It follows that the complete solution for pressure obtained by adding (11.16.2) and (11.16.7) is

$$rg\eta = c^2(1 - \exp(r\beta y^2 x/c^2))\partial(X/\beta y \rho H)/\partial y + A\, \exp(r\beta y^2 x/c^2). \tag{11.16.10}$$

This has a constant value on the eastern boundary, a condition necessary for ensuring that there will be no geostrophic flow across the boundary. The first term of (11.16.10) gives pressure variations of the same sign as for those in the interior, but which reduce in magnitude as the eastern boundary is approached. The observed surface pressure field shown in Fig. 7.8a (part i) also has this property. The value of A depends on the flow near the equator—this part of the solution can, for instance provide a flow into (and out of) the equatorial zone to feed (or absorb) an equatorial jet.

The tropical Atlantic currents show many features similar to those of the Pacific. The wind forcing in the Atlantic (shown in Fig. 11.29) is also asymmetrical because of an intertropical convergence zone north of the equator. Upward Ekman pumping occurs, particularly in summer (see Fig. 9.6). There is a countercurrent north of the equator as in the Pacific, but it is highly seasonal. In the July–September season, it extends right across the ocean in a belt between 4°N and 10°N, but 6 months later it is found only in a small region near the African coast and a small region in the west. Model studies by Anderson (1982) indicate that a significant part of these variations are associated with the *meridional* component of the wind. The steady-state response to this component of the wind can be calculated from (11.14.1)–(11.14.3). Away from boundaries, the zonal current is an Ekman current that is driven directly by the wind. The planetary wave correction has the effect of setting up a pressure field that, for a given mode, reduces the response as the eastern boundary is approached.

The transient response away from the immediate vicinity of the equator would be expected, on the arguments presented above, to consist of the local response to Ekman pumping plus a planetary wave correction associated with the first mode. This correction would propagate from east to west and carry information about winds near the eastern boundary. When these winds are similar to interior winds, the wave tends to reduce that part of the interior response due to the first mode.

Observations in the Pacific tend to support this view. For instance, Meyers (1975, 1979a) has studied changes with time of the depth of the thermocline and compared them with Ekman pumping displacements. At latitudes above 10°, the two studies indicated that Ekman pumping alone could explain much of the change observed. To model seasonal changes at lower latitudes, however, account must be taken of the planetary wave effect. In particular, the thermocline depth at 6°N shows marked phase propagation of the seasonal wave at a speed of 0.64 m s^{-1}. This agrees with the long planetary wave speed [see (11.8.6)], given by

$$\omega/k = -\beta c^2/f^2,$$

if $c \approx 2.4 \text{ m s}^{-1}$—a reasonable value for the first baroclinic mode.

The seasonal nature of the currents in the Indian Ocean has already been discussed in Section 11.12, the seasonal changes being particularly prominent off the Somali coast. An early reference [see Aleem (1967)] to this phenomenon was made by Ibn Khordazbeh circa A.D. 846, who noted that "the Sea flows during the summer months to the northeast" and "during the winter months to the southwest." The currents found in the summer regime appear to be stronger, and these are shown for a particular year in Fig. 11.30. Observations by Leetmaa (1973) and data studies by Düing and Szekielda (1971) seem to confirm the idea that the reversal of the current at the coast is initiated locally by a mechanism like that discussed in Section 10.11 and is later modified by remote effects propagating to the boundary.

There are significant changes in the ocean on longer-than-seasonal time scales, just as there are in the atmosphere, but these are not very well documented as yet. The failure of upwelling in the East Pacific, and hence failure of the Peruvian fishery, is well recorded and these events are well correlated with the pressure at Darwin,

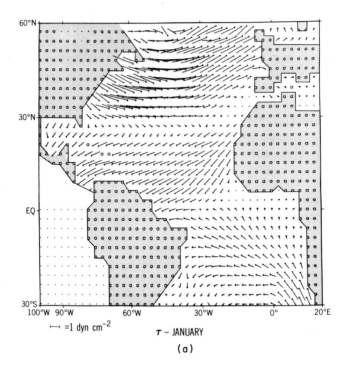

$\longmapsto$ =1 dyn cm^{-2} τ – JANUARY

(a)

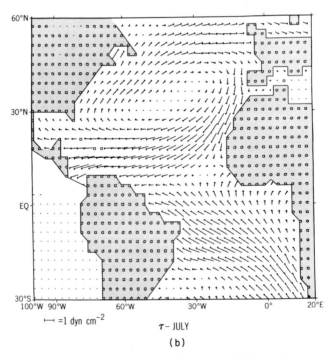

$\longmapsto$ =1 dyn cm^{-2} τ – JULY

(b)

Fig. 11.29. Wind stress over the Atlantic Ocean for (a) January and (b) July, using data compiled by Bunker [see Anderson (1982)].

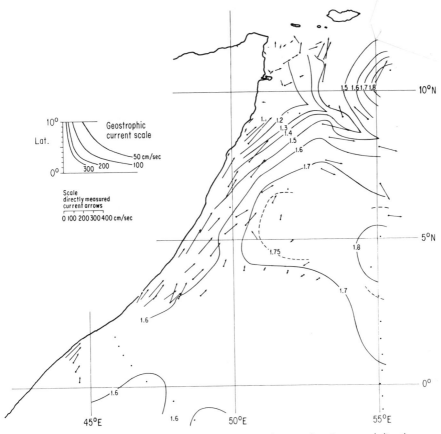

Fig. 11.30. Dynamic topography of the sea surface in the Northwest Indian Ocean and directly measured surface currents at a depth of 10 m during August and September 1964. [From Swallow and Bruce (1966), Fig. 2.]

almost on the other side of the globe, and with other indices of the "southern oscillation" (see Fig. 11.27). Also, it is known that the sea-surface temperature in the central equatorial Pacific is very well correlated with the local rainfall (see Fig. 11.26). More recently, information from hydrographic data (Masuzawa and Nagasaka, 1975) and island sea levels in the West Pacific (Wyrtki, 1977) has shown huge changes in the structure of the ocean surface layers. In particular, subsurface temperature at 137°E has shown changes of up to 8° in a year (at 150 m) with corresponding sea-level changes of about 40 cm. These were so large that they had severe effects on the reef community in Guam (Yamaguchi, 1975). The high sea level in the west coincides with a low sea level in the east, and appears to be built up during a period when the trades in the Central Pacific are stronger than usual in the year preceding an El Niño event (Wyrtki, 1977, 1979b). When these winds reduce, there is a rapid drop of sea level in the west and a corresponding rise in the east, the changes being up to 40 cm over 1 or 2 years. The changes are baroclinic, i.e., there is comparatively little change in pressure at the bottom of the ocean associated with these events. The amount of near-surface water moved in the West Pacific has been estimated by Wyrtki (1979b) to correspond to an average flux of about 27 Mt s^{-1}, so these events are of an impressive magnitude.

Chapter Twelve

Mid-latitudes

12.1 Introduction

The aim of this book is to develop an understanding of the atmosphere and ocean, and in particular, how they adjust under gravity when perturbed from equilibrium. In Chapter 11 we considered these processes in the equatorial region, and this has led to an understanding of the effects of variation of the Coriolis parameter with latitude. That understanding will now be applied to extratropical regions.

A particularly significant fact to emerge from the examination of free waves on the "equatorial beta plane" was the separation of the waves into two classes that become more distinct as the horizontal mode number increases, i.e., as the region affected extends further from the equator. One class has high frequencies, and is little affected by variations of the Coriolis parameter insofar as its local properties are concerned. It consists of the gravity waves studied in earlier chapters, and there is no need to examine these further.

The other class of waves has low frequencies, and its existence depends crucially on the fact that the Coriolis parameter varies with latitude. The *maximum* frequency for this class is given by (11.8.3), i.e., is equal to $\beta c/2f$, corresponding to mid-latitude periods of order 1 year for the ocean, and of order 1 week for the atmosphere. Thus there is a large "spectral gap" at mid-latitudes between internal gravity waves with minimum frequency f and planetary waves with maximum frequency $\beta c/2f$. The ratio of these frequencies is the large factor $2/\epsilon$, where ϵ is given by (11.8.14), i.e., by

$$\epsilon = \beta c/f^2. \tag{12.1.1}$$

For instance, at $45°$ latitude, the frequency ratio is given by

$$2/\epsilon = 2f^2/\beta c = (1300 \text{ m s}^{-1})/c. \tag{12.1.2}$$

Because the new class of motions is so distinct from gravity waves, it can conveniently be studied by approximating the equations in an appropriate way. As

493

first discussed in Section 8.16, low-frequency motions have the property of being in approximate geostrophic balance, but *departures* from geostrophy are very important, as Brunt and Douglas (1928) pointed out, because the vertical motion is associated only with these departures. Brunt and Douglas were interested in vertical motions in connection with rainfall, but it is also necessary to know vertical displacements when considering gravitational restoring forces. The quasi-geostrophic equations were derived in Section 11.8 in a form appropriate to the equatorial region, but simple modifications show that the same form of equations is valid in mid-latitudes. The modifications are outlined in Section 12.2. It is not difficult to include vertical variations and nonlinear effects in the mid-latitude quasi-geostrophic formulation as well, and this is done in Section 12.8. The formulation was developed by Charney (1947, 1948), a review was given by Phillips (1963, 1973), and a detailed discussion was presented by Pedlosky (1979). The concept of potential vorticity, developed by Rossby (1937, 1938a, 1940), is particularly useful for discussing quasi-geostrophic motion.

The freely propagating, quasi-geostrophic waves are called planetary waves, and their horizontal propagation characteristics are studied in Section 12.3 for the case of a single vertical mode. Such waves are generated in the ocean by the wind, and in Section 12.4 we examine the spin-up process that occurs when the wind is suddenly "switched on." This leads to an understanding of the east–west asymmetry of the ocean's response to the wind, in that strong boundary currents are found on the western flanks of ocean basins, these currents completing gyres that involve relatively slow flow in the interior of the ocean. A discussion of steady ocean circulation models is given in Section 12.5.

The vertical propagation of planetary waves is also important, and the structure and dispersive properties of such waves are dealt with in Section 12.7. One way in which these waves can be generated is by flow over topography. Since the frequency of encounter of topographic features must be small enough for planetary waves to exist, such waves are generated only by the major features of the earth's topography with scales of order 1000 km. With such large scales, the concept of perturbations about a *uniform* flow is rather inadequate, so a study is made (in Section 12.9) of small perturbations on a flow that varies with both latitude and height. This leads to a discussion of the stationary waves observed in the atmosphere. These waves can be generated not only by topography, but also by longitudinal variations in heating of the atmosphere, and by effects of smaller-scale waves. In winter, stationary waves can propagate high into the stratosphere because the wind distribution with height is favorable, but in summer this is not so.

The large-scale field of vertical motion in the atmosphere is of great importance because strong upward motion is associated with the development of severe weather conditions. The vertical motion cannot be directly measured, but deductions can be made from properties of the pressure field. The so-called "omega equation" is particularly useful in this regard, and this is derived and discussed in Section 12.10.

12.2 The Mid-latitude Beta Plane

High-frequency motions in mid-latitudes, i.e., motions with frequencies of inertial magnitudes and larger, have been considered already in earlier chapters, so attention will now be focused on mid-latitude motions with lower than inertial frequency. The special "quasi-geostrophic" character of these motions has already been brought to light in Section 8.16 for cases in which the frequency is still not low enough for effects of the earth's curvature ("beta" effects) to be important. The beta effect is necessary to allow freely propagating waves, i.e., planetary waves, and the application of the quasi-geostrophic approximation to those waves was made in Section 11.8 for waves near the equator.

However, the approximation of Section 11.8 can be applied equally well away from the equator, and when used in mid-latitudes, it is called the *mid-latitude beta-plane approximation*. The basic requirement of the approximation was found, in Section 11.8, to be that ϵ, given by (12.1.1), must be small, which means that the locality of interest is more than an equatorial Rossby radius from the equator. Of course, the north–south scale must be small enough for fractional changes in f to be small over such a distance, but this follows automatically if the wave scale is assumed to be of the same order as the local Rossby radius.

The derivation of Section 11.8 was somewhat special in that the equatorial beta-plane approximation was used, but it can be generalized to yield the same equations at mid-latitudes as well. The starting point is the pair of horizontal momentum equations (11.2.1) and (11.2.2) on the sphere, and the object is to obtain a quasi-geostrophic approximation to these equations with the Cartesian form of (11.8.15) and (11.8.16). First, it is necessary to define suitable coordinates and the natural ones are those corresponding to a *local Mercator projection* [see, e.g., Haltiner (1971) and Phillips (1963, 1973)]. Coordinate lines are latitude and longitude as before, but they are relabeled to give an approximately isometric system near the latitude φ_0 about which the expansion is being made. Thus if R is the radius of the earth and λ the longitude, local Mercator coordinates x, y are defined by

$$x = R\lambda \cos \varphi_0, \qquad y = R \int_{\varphi_0}^{\varphi} \frac{d\varphi' \cos \varphi_0}{\cos \varphi'} = R \cos \varphi_0 \ln \left\{ \frac{(1 + \sin \varphi) \cos \varphi_0}{(1 + \sin \varphi_0) \cos \varphi} \right\}.$$

$$(12.2.1)$$

Also, because of the form of the continuity equation, it is useful to use slightly modified velocity components $\tilde{u}$, $\tilde{v}$, defined by

$$\tilde{u} = u/\mu, \qquad \tilde{v} = v/\mu, \tag{12.2.2}$$

where

$$\mu = \cos \varphi_0/\cos \varphi = R^{-1} \, dy/d\varphi \tag{12.2.3}$$

is a factor that is close to unity near the central latitude of the approximation. ($\tilde{v}$ is a useful quantity because it is proportional to the volume flux per unit depth between meridians.) Substituting for $d\lambda$, $d\varphi$, u, and v in (11.2.1) and (11.2.2), there results

$$D\tilde{u}/Dt - f\tilde{v} = -\rho^{-1} \, \partial p'/\partial x, \tag{12.2.4}$$

$$D\tilde{v}/Dt + f\tilde{u} + R^{-1}\mu(\tilde{u}^2 + \tilde{v}^2) \tan \varphi = -\rho^{-1} \, \partial p'/\partial y, \tag{12.2.5}$$

where

$$\frac{D}{Dt} \equiv \frac{\partial}{\partial t} + \mu^2 \left(\tilde{u} \frac{\partial}{\partial x} + \tilde{v} \frac{\partial}{\partial y} \right) + w \frac{\partial}{\partial z} \tag{12.2.6}$$

and

$$f = 2\Omega \sin \varphi. \tag{12.2.7}$$

These are the exact forms of the momentum equations on the sphere, and the corresponding form of the incompressibility relation [see (4.12.11)] is

$$\mu^2(\partial\tilde{u}/\partial x + \partial\tilde{v}/\partial y) + \partial w/\partial z = 0. \tag{12.2.8}$$

Naturally enough, these equations are not precisely the same as the Cartesian ones because of the earth's curvature, but they are approximately so near the chosen latitude since μ is close to unity.

For small-amplitude motion, the quadratic terms are ignored, so the momentum equations (12.2.4) and (12.2.5) in Mercator coordinates become

$$\partial\tilde{u}/\partial t - f\tilde{v} = -\rho^{-1}\,\partial p'/\partial x, \tag{12.2.9}$$

$$\partial\tilde{v}/\partial t + f\tilde{u} = -\rho^{-1}\,\partial p'/\partial y, \tag{12.2.10}$$

which is precisely the form for the f-plane, but now f is a function of y. The beta-plane approximation relies on expanding f about the chosen latitude as a linear function of y, and is obtained just as it was in Section 11.8 for the more special case. The first step is to introduce the scaled variables

$$x^* = x/L, \qquad y^* = y/L, \qquad t^* = \beta Lt,$$

$$u^* = \tilde{u}/v_0, \qquad v^* = \tilde{v}/v_0, \qquad p^* = p'/\rho Lfv_0, \tag{12.2.11}$$

where v_0 is a velocity scale and L is a horizontal length scale. As far as deriving the approximation to the *momentum* equation is concerned, it is not necessary to assume, as in Section 11.8, that L is equal to a, the Rossby radius, nor is it necessary to assume that p' is independent of depth. These two assumptions become relevant only when the continuity equation is being considered.

With the above scalings, the Coriolis parameter f can be expanded in y^*, giving

$$f \approx f_0(1 + \epsilon_L y^*), \tag{12.2.12}$$

where f_0 is the value at the central latitude and ϵ_L is the small parameter defined by

$$\epsilon_L = L \cot \varphi_0/R = \beta L/f_0, \tag{12.2.13}$$

where β is the gradient of the Coriolis parameter, namely,

$$\beta = (2\Omega/R) \cos \varphi_0. \tag{12.2.14}$$

Note that ϵ_L is identical to ϵ as defined by (12.1.1) when L is chosen as the Rossby radius. The expansion procedure relies on ϵ_L being small, i.e., on the horizontal scale being small compared with the earth's radius and the central latitude being not too close to the equator. At first order in ϵ_L, the linearized-momentum equations are (11.8.11) and (11.8.12), just as in the equatorial case, and so the quasi-geostrophic approximation is (11.8.15) and (11.8.16). (Note that if L is not chosen equal to a, then ϵ_L replaces ϵ in the formulas.) In other words, at the first approximation the flow is geostrophic, so $u = u_g$ and $v = v_g$, where in dimensional terms

$$f_0 u_g = -\rho^{-1}\,\partial p'/\partial y, \qquad f_0 v_g = \rho^{-1}\,\partial p'/\partial x. \tag{12.2.15}$$

The tilde over the velocity variables will now be dropped from the approximate equations. (It is used only once more for an exact equation.)

Because of the degeneracy in the equations due to the fact that the geostrophic flow given by (12.2.15) is nondivergent, it is necessary to know the flow to the next order of approximation in order to calculate how the flow changes with time. So it is useful to define an *ageostrophic* part of the flow (u_a, v_a) by

$$u = u_g + u_a, \qquad v = v_g + v_a. \tag{12.2.16}$$

By (11.8.15) and (11.8.16), the ageostrophic motion is of order ϵ_L relative to the geostrophic part, and is given in dimensional terms by

$$f_0 u_a = -\beta y u_g - \partial v_g/\partial t \equiv \rho^{-1} f_0^{-1} \{\beta y\, \partial p'/\partial y - \partial^2 p'/\partial t\, \partial x\}, \tag{12.2.17}$$

$$f_0 v_a = -\beta y v_g + \partial u_g/\partial t \equiv \rho^{-1} f_0^{-1} \{-\beta y\, \partial p'/\partial x - \partial^2 p'/\partial t\, \partial y\}. \tag{12.2.18}$$

These are the *linear quasi-geostrophic momentum equations* that govern the way in which such flows develop.

There are two parts to the ageostrophic velocity field as given above. The beta part (distinguished by the factor β) represents the fact that, for a given pressure gradient, velocities that are in geostrophic balance with that gradient *at each latitude* (not just the central latitude) become larger toward the equator. Consequently the motion is divergent and the

$$\text{beta part of ageostrophic velocity} = -(\beta y/f_0)(u_g, v_g), \tag{12.2.19}$$

which has

$$\text{divergence} = -\beta v_g/f_0 \equiv -(\beta/\rho f_0^2)\, \partial p'/\partial x.$$

The other part is the isallobaric part already encountered in Section 8.16, namely,

$$\text{isallobaric part of ageostrophic velocity} = -\rho^{-1} f_0^{-2}(\partial/\partial x, \partial/\partial y)\, \partial p'/\partial t, \tag{12.2.20}$$

which has

$$\text{divergence} = -f_0^{-1}\, \partial \zeta_g/\partial t \equiv -\rho^{-1} f_0^{-2}(\partial^2/\partial x^2 + \partial^2/\partial y^2)\, \partial p'/\partial t,$$

where ζ_g is the geostrophic part of the vertical component of the relative vorticity. In the formal scaling procedure, the time scale $(\beta L)^{-1}$ was chosen so that the two components of the ageostrophic velocity would formally be of the same order. All that is required for the approximation to be valid, however, is that the ageostrophic motion be small compared with the geostrophic part. The condition $\epsilon_L \ll 1$ ensures that this is true for the beta part of the ageostrophic motion. The condition for the isallobaric part to be small is really an independent one, namely, that the time scale be large compared with f_0^{-1}, as found in Section 8.16.

It is a simple matter to generalize the above equations to their *nonlinear* form. The formal procedure for doing this is merely to choose the velocity scale v_0 in (12.2.11) so that the nonlinear advective terms in (12.2.6) are of the same order as the time derivative, i.e.,

$$v_0 = \beta L^2. \tag{12.2.21}$$

This velocity scale is the product of beta and the square of the horizontal scale. Then all the approximate equations follow as above, but with $\partial/\partial t$ replaced by the first approximation to (12.2.6), namely,

$$D_g/Dt \equiv \partial/\partial t + u_g\,\partial/\partial x + v_g\,\partial/\partial y, \tag{12.2.22}$$

which is the time derivative following the geostrophic part of the motion. The third term on the left-hand side of (12.2.5) is small compared with those retained, provided that

$$L\tan\varphi_0 \ll R, \tag{12.2.23}$$

which is equivalent at mid-latitudes to requiring that ϵ_L be small [see (12.2.13)].

Thus (12.2.17) and (12.2.18) have become

$$f_0 u_a = -\beta y u_g - D_g v_g/Dt \equiv \rho^{-1}f_0^{-1}\left\{\beta y\,\frac{\partial p'}{\partial y} - \frac{D_g}{Dt}\frac{\partial p'}{\partial x}\right\}, \tag{12.2.24}$$

$$f_0 v_a = -\beta y v_g + D_g u_g/Dt \equiv \rho^{-1}f_0^{-1}\left\{-\beta y\,\frac{\partial p'}{\partial x} - \frac{D_g}{Dt}\frac{\partial p'}{\partial y}\right\}, \tag{12.2.25}$$

and there is now a *third* part to the ageostrophic motion, namely,

nonlinear part of ageostrophic velocity

$$= \tfrac{1}{2}f_0^{-1}(-\partial/\partial y, \partial/\partial x)(u_g^2 + v_g^2) - f_0^{-1}\zeta_g(u_g, v_g), \tag{12.2.26}$$

which has

$$\text{divergence} = -f_0^{-1}(u_g\,\partial/\partial x + v_g\,\partial/\partial y)\zeta_g.$$

In (12.2.26), the acceleration term has been written in Lagrange's form (see Section 7.10) as the sum of a Bernoulli term (which gives *no* contribution to the ageostrophic divergence) and a product of vorticity and velocity (which *does* contribute to ageostrophic divergence).

The choice (12.2.21) for velocity scale makes the nonlinear part of the ageostrophic velocity formally of the same order as the other parts. For the geostrophic approximation to be valid, however, the condition on the velocity scale is really independent of the other conditions, and is that the nonlinear part of the ageostrophic velocity be small compared with the geostrophic velocity. The condition is that the parameter Ro, defined by

$$\text{Ro} = v_0/f_0 L, \tag{12.2.27}$$

be small, where L is the horizontal length scale. The parameter is called the Rossby number or Kibel' number, and was first introduced by Kibel' (1940) [see Phillips (1963)]. Thus the quasi-geostrophic approximation requires *three conditions,* associated with the *three components* of the ageostrophic motion: (1) $\epsilon_L \ll 1$ for the *beta* part to be small, (2) the time scale $\gg f_0^{-1}$ for the *isallobaric* part to be small, (3) and Ro $\ll 1$ for the *nonlinear* part to be small.

For the particular case of shallow-water motions, p'/ρ is independent of depth and equal to $g\eta$, where g is the acceleration due to gravity and η is the surface elevation. To complete the set of governing equations, the continuity equation (11.2.4) is required.

In Mercator coordinates for a constant-depth ocean, this becomes $\left[\text{cf. (12.2.8)}\right]$

$$(H + \eta)\mu^2(\partial \tilde{u}/\partial x + \partial \tilde{v}/\partial y) + D\eta/Dt = 0. \tag{12.2.28}$$

The first approximation (a) replaces μ by unity, (b) replaces the divergence of total velocity by the divergence of ageostrophic velocity (this is an identity because the divergence of the geostrophic part is zero), (c) replaces D/Dt by D_g/Dt, implying that terms like $u_a \, \partial \eta/\partial x$ are ignored, and hence (d) terms like $\eta \, \partial u_a/\partial x$ are ignored as well. Thus the approximate equation is

$$H(\partial u_a/\partial x + \partial v_a/\partial y) + D_g\eta/Dt = 0. \tag{12.2.29}$$

The scaled version can be obtained as in Section 11.8, and has its simplest form when the horizontal scale L is chosen to be equal to the Rossby radius a because this makes all the terms in (12.2.29) formally of the same order.

An equation in one variable only can now be obtained by substituting (12.2.24) and (12.2.25) in (12.2.29). The result is

$$\frac{D_g}{Dt}\left(\frac{\partial^2 \eta}{\partial x^2} + \frac{\partial^2 \eta}{\partial y^2} - \frac{f_0^2}{c^2}\eta\right) + \beta\frac{\partial \eta}{\partial x} = 0. \tag{12.2.30}$$

This is the quasi-geostrophic form of the shallow-water potential vorticity equation (11.3.1), which can be seen by expanding (11.3.2) about the central latitude φ_0 and using the approximation $\eta \ll H$. This gives

$$HQ \approx q \equiv f_0 + \beta y + \zeta_g - f_0\eta/H, \tag{12.2.31}$$

where q is called the quasi-geostrophic potential vorticity. This satisfies the equation

$$D_g q/Dt = 0, \tag{12.2.32}$$

which is another way of writing (12.2.30).

The kinetic energy equation for quasi-geostrophic flow can be obtained by multiplying (12.2.24) by v_g and subtracting it from u_g times (12.2.25). The result is

$$\frac{1}{2}\frac{D_g}{Dt}(u_g^2 + v_g^2) = -\frac{u_a}{\rho}\frac{\partial p'}{\partial x} - \frac{v_a}{\rho}\frac{\partial p'}{\partial y}. \tag{12.2.33}$$

For the shallow-water case, H times (12.2.33) can be added to $p'(=\rho g\eta)$ times (12.2.29) to give the total energy equation

$$\frac{D_g}{Dt}\left[\frac{1}{2}H(u_g^2 + v_g^2) + \frac{1}{2}g\eta^2\right] + \frac{\partial}{\partial x}(gH\eta u_a) + \frac{\partial}{\partial y}(gH\eta v_a) = 0. \tag{12.2.34}$$

This may be compared with the nonrotating equivalent (5.7.4). The expression for the potential energy is the same, but only the geostrophic velocities contribute to the kinetic energy in the quasi-geostrophic approximation. Also, the geostrophic flow, being parallel to isobars, does no work and so the flux terms in (12.2.34) involve only the ageostrophic part of the motion.

12.3   Planetary Waves

The dispersion properties of equatorial planetary waves were examined in Chapter 11. Now extratropical planetary waves of the form

$$\eta = \eta_0 \cos(kx + ly - \omega t) \qquad (12.3.1)$$

will be considered. These waves are also known as Rossby waves, following the pioneering work of Rossby *et al.* (1939) [see Platzman (1968)]. Substitution in (11.8.20), which is the linearized form of (12.2.30), gives the dispersion equation

$$\omega = -\beta k/(k^2 + l^2 + f_0^2/c^2), \qquad (12.3.2)$$

showing that all planetary waves have *westward phase velocity*. Contours of constant frequency ω in wavenumber space are the circles

$$(k + \beta/2\omega)^2 + l^2 = (\beta/2\omega)^2 - (f_0/c)^2, \qquad (12.3.3)$$

and these are depicted in Fig. 12.1. By the definition (5.4.11), the group velocity is the gradient of ω in wavenumber space, and is therefore normal to the ω contours, as shown in the figure. The expressions for the components of group velocity obtained by differentiating (12.3.2) are

$$\mathbf{c}_g \equiv (\partial\omega/\partial k, \partial\omega/\partial l) = \beta(k^2 - l^2 - f_0^2/c^2, 2kl)/(k^2 + l^2 + f_0^2/c^2)^2. \quad (12.3.4)$$

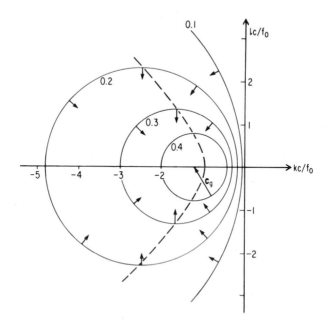

Fig. 12.1.   Dispersion diagram for midlatitude planetary waves. Contours are of frequency in units of $\beta c/f_0$. The group velocity, being the gradient of frequency in wavenumber space, is normal to the contours in the direction shown by the arrows and is inversely proportional to the spacing between contours. The contours are circular in shape and reduce to a single point when $\omega = 0.5\ \beta c/f_0$. No planetary waves exist when the magnitude of the frequency is larger. The dashed line is the hyperbola that separates waves with eastward and westward group velocity.

The line dividing eastward-propagating from westward-propagating waves (in the group-velocity sense) is thus the hyperbola

$$k^2 = l^2 + f_0^2/c^2, \qquad (12.3.5)$$

and this is shown in the figure. The sign of c_{gy} is opposite to that of ω/l, so waves with a northward component of phase velocity have southward group propagation and vice versa. (*Note:* Changing the signs of k, l, and ω in (12.3.1) does not alter the wave, so drawing contours of positive ω in the left half of the wavenumber plane is equivalent to drawing contours of negative ω in the right half of the wavenumber plane.)

The velocity field associated with the planetary wave is, at first order, simply the geostrophic velocity field given by

$$(u_g, v_g) = (l, -k)(g\eta_0/f_0)\sin(kx + ly - \omega t). \qquad (12.3.6)$$

An example is shown in Fig. 12.2. It is interesting to note that this is almost identical to the atmospheric wave sketched by Birt (1847) and reproduced in Fig. 7.7. The only difference is in the direction of the phase velocity, which may be attributed to the fact that Fig. 12.2 shows the direction relative to the medium, whereas Birt's sketch shows the direction relative to the ground.

Although the predominant motion in the planetary wave is the geostrophic motion, an essential feature for the dynamics of the wave is the small ageostrophic

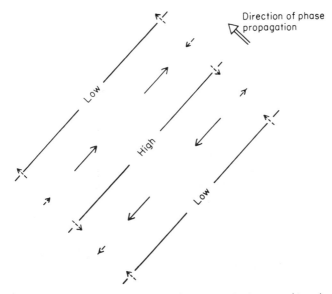

Fig. 12.2. A plane progressive planetary wave. The predominant motion is geostrophic and so is in the direction indicated by the large arrows in the northern hemisphere. There is also a small ageostrophic component shown by the small dashed arrows. The part normal to the isobars is the isallobaric wind, whereas the component parallel to isobars represents the effect of the Coriolis parameter decreasing toward the equator so that, for a given pressure gradient, velocities are larger near the equator and smaller away from the equator. This latter (beta) part is drawn relative to zero on the line of high pressure (or high surface elevation), and is convergent to the west of that line. In the limit of very short (non-divergent or barotropic) waves, this convergence must be balanced by divergence in the isallobaric part, which requires westward phase propagation as shown. For finite wavelengths, the total motion is convergent to the west of the high, consistent with the surface rising there. In the southern hemisphere, the geostrophic part of the motion is reversed, but the ageostrophic part is unchanged.

component given by (12.2.17) and (12.2.18), and shown by small dashed arrows in Fig. 12.2. There are two parts to this field: the isallobaric wind, which is at right angles to pressure surfaces in a plane wave, and the beta part, which is parallel to pressure surfaces. The latter merely represents the fact that for a fixed pressure gradient winds are stronger near the equator because of the smaller Coriolis parameter. Both of these components of the ageostrophic motion lead to convergence to the west of the high-pressure ridge. Consequently, the free surface rises here, thereby raising the pressure and causing the whole pattern to migrate westward.

The energy equation (12.2.34) for quasi-geostrophic flow recognizes only the geostrophic contribution to the kinetic energy, and so for the planetary wave (12.3.1)

$$\text{kinetic energy density} = \tfrac{1}{2}\rho H(\overline{u_g^2 + v_g^2}) = \rho g^2 H(k^2 + l^2)\eta_0^2/4f_0^2. \quad (12.3.7)$$

The overbar represents the average over a wavelength, as in Chapter 6. The potential energy (or available potential energy) density is given by the usual expression for a shallow-water wave, namely,

$$\text{potential energy density} = \tfrac{1}{2}\rho g\overline{\eta^2} = \tfrac{1}{4}\rho g\eta_0^2. \quad (12.3.8)$$

Thus the ratio

$$\frac{\text{kinetic energy density}}{\text{potential energy density}} = (k^2 + l^2)gH/f_0^2 = \kappa_H^2 a^2, \quad (12.3.9)$$

where κ_H is the horizontal wavenumber and a is the Rossby radius. It follows that *long planetary waves*, i.e., those with scale large compared with the Rossby radius, have most of their energy in the *potential* form, whereas *short waves* have most in the *kinetic* form. This statement is generally true of any quasi-geostrophic motion.

Energy fluxes can also be calculated, e.g., (12.2.34) shows that the northward energy flux density is given by

$$\rho Hg\overline{\eta v_a} = -\rho g a^2 \overline{\eta \frac{\partial^2 \eta}{\partial y \, \partial t}} = -\tfrac{1}{2}\rho g a^2 l \omega \eta_0^2 = Ec_{gy}. \quad (12.3.10)$$

[Note that this result is true only if the energy flux density is defined in terms of the ageostrophic velocity. See Longuet-Higgins (1964b).] The first equality is based on the expression (12.2.18) for v_a, the second on substitution from (12.3.1), and the final one uses the expression (12.3.4) for c_{gy}. E is the total energy density given by the sum of (12.3.7) and (12.3.8). It follows that when the group velocity is northward, so is the (ageostrophic) energy flux, and vice versa. Planetary waves also have an associated northward flux of eastward momentum, given by

$$\rho\overline{uv} = -\tfrac{1}{2}\rho kl(g\eta_0/f_0)^2, \quad (12.3.11)$$

which has sign *opposite* to that of the energy flux since ω and k have opposite signs by (12.3.2). In other words, *westward* momentum is carried away from a source in the direction of group propagation. This result is of some interest since when this flux varies because, say, planetary wave energy is dissipated, there must be a corresponding westward acceleration of the mean flow (cf. Section 8.15). The effect has applications to sudden stratospheric warmings (O'Neill and Taylor, 1979; Palmer, 1981), to waves produced by the Gulf Stream (Thompson, 1971), etc., and is reviewed by Dickinson (1978). Further discussion is given by Rhines (1977) and Rhines and Holland (1979).

Additional distinctions can be made between long and short planetary waves. For *short waves* ($\kappa_H a \gg 1$), i.e., those whose energy is mainly kinetic, the dispersion relation (12.3.2) is approximated by

$$\omega = -\beta k/(k^2 + l^2). \tag{12.3.12}$$

This corresponds to ignoring the contribution of depth variations to the potential vorticity (12.2.31). In other words, vertical motion and the corresponding changes in potential energy are unimportant, and the dispersion relation approximates that which would apply if the motion were purely horizontal. [Rossby's analysis (Rossby *et al.*, 1939) was for short waves, and the results for the sphere were found much earlier by Margules (1893) and Hough (1898) as a limiting form of solution of Laplace's equations since the rotation rate goes to zero but ω/Ω remains finite.]

The *long planetary waves* ($\kappa_H a \ll 1$), whose energy is mainly potential, are approximately nondispersive, since (12.3.2) is approximated in this case by

$$\omega/k = -\beta a^2 = -\beta c^2/f_0^2 = -c^2 \cos \varphi_0/2\Omega R \sin^2 \varphi_0, \tag{12.3.13}$$

the equality on the far right making use of (12.2.7) and (12.2.14). This corresponds to ignoring the contribution of the relative vorticity to the potential vorticity (12.2.31) and to considering *only the beta part* of the ageostrophic motion. The formula (12.3.13) for the phase velocity is particularly appropriate for baroclinic waves in the ocean because the Rossby radius is small (typically 30 km), and so (12.3.13) applies to any waves with larger scale. Note, however, the strong inverse dependence of phase (and group) speed on latitude that is implied by (12.3.13) and shown graphically in Fig. 12.3. This graph may also be interpreted as showing the distance that long waves have propagated from a straight (north–south) eastern boundary in a given time. A wave

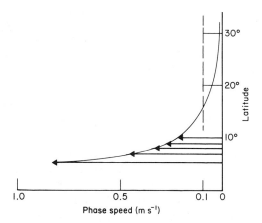

Fig. 12.3. Phase (or group) speed (westward) of long (nondispersive) planetary waves as a function of latitude (solid line). The numerical values shown are for the first baroclinic mode in the ocean with $c = 2.5$ m s^{-1}. A dashed line is drawn at 10 cm s^{-1} to help give the scale for the slower waves. Near the equator the set of possible wave speeds becomes discrete, and the arrows show the first five equatorial modes. They have phase speed given by (11.8.1) and fall on the curve if each mode is associated with its turning latitude, as given by (11.7.1).

front of this shape is often seen when an ocean with a straight eastern boundary is "spun up" from rest by imposing a wind stress at some initial time. Near the equator, (12.3.13) is replaced by (11.8.1), which can be written

$$\omega/k = -c/(2n + 1) = -c^2/2\Omega R\varphi_n^2, \tag{12.3.14}$$

where $\varphi_n = y_c/R$ is the critical latitude defined by (11.7.1). Thus the continuous curve (12.3.13) is replaced by a series of discrete points, given by (12.3.14) (and shown by arrows in Fig. 12.3), which approach one another as the mode number n increases.

Barotropic waves in the ocean can be produced by the wind and therefore have north–south scales l^{-1} that are typical of wind systems, i.e., 1000 km. This is smaller than the barotropic Rossby radius (which is about 2000 km), so they may be regarded as *short* waves whose energy is mainly kinetic, and they propagate westward with phase speeds of order [see (12.3.12)]

$$\beta/l^2 \approx 20 \quad \text{m s}^{-1}. \tag{12.3.15}$$

The nearest equivalent to a barotropic wave in the atmosphere is the "Lamb" mode (see Section 6.14), since this has a velocity that becomes independent of depth in the incompressible limit $H_s \to \infty$. The appropriate value of c for this mode is the speed of sound (about 300 m s^{-1}), so the Rossby radius is about 3000 km. For smaller values of l^{-1} the westward phase progression *relative to the air* of a planetary wave with the Lamb-mode vertical structure is given by (12.3.15). For instance, Pratt and Wallace (1976) identified such a wave with a period of 20 days and with east–west wavenumber 2 (two wavelengths around the globe). The westward phase progression relative to the mean westerly wind of 15 m s^{-1} was 23 m s^{-1}, so that relative to the ground the wave had westward or retrograde (i.e., counter to the direction of rotation of the earth) phase progression at 8 m s^{-1}. Madden (1978) identified a similar wave with a period of between 1 and 3 weeks, which was most easily detectable *north* of 50°N. Pratt and Wallace (1976) also found a wave with *baroclinic* structure with the same period (20 days) and zonal wave number (2) as for the Lamb mode. However, this had eastward or prograde phase progression relative to the ground, implying westward phase progression at $15 - 8 = 7$ m s^{-1} relative to the air.

If the meridional scale l^{-1} becomes comparable with the radius of the earth, the beta-plane approximation breaks down, and wave properties must be examined on the sphere, using spherical polar coordinates. Sinusoidal variations with longitude and time are possible, but the corresponding functions of latitude need to be calculated. They are (for perturbations on a state of rest or solid rotation) called Hough functions, and their properties are given by Longuet-Higgins (1968b). The original equations of Laplace (1778–1779) were, of course, developed for the sphere, and the discovery of planetary waves could be attributed to Margules (1893) and Hough (1897, 1898), who noted that the solution of Laplace's tidal equations could be subdivided into two classes—one giving purely gravitational oscillations in the limit of no rotation, but the other having frequency proportional to the rotation rate in the limit as this goes to zero. This latter class comprises what we now call planetary waves, and their limiting form, obtained by both Margules and Hough, and later rediscovered by Haurwitz (1940), is the nondivergent planetary wave, whose energy is all kinetic. Margules' calculations were for the case of the external or "Lamb" mode (Section 6.14) in an isothermal atmosphere, and include a westward-propagating wave with a period of just over 5 days. Figure 12.4 is a reproduction of his diagram showing this wave, which is now well-known from observational studies (Madden, 1978) and is most easily detected *south* of 50°N.

Hough (1897) showed a remarkable understanding of the nature of adjustment to geostrophic equilibrium for zonally symmetric flows on a sphere.

> Suppose now that a disturbing force ... tending to increase the surface-ellipticity of the ocean, is suddenly applied to our rotating system when in a configuration of relative equilibrium. It will immediately set up oscillations, the initial motion ... [involving] a flow of water directed from the poles towards the equator. The water however coming from higher latitudes into lower will reach these lower latitudes with an amount of rotation less than that which is appropriate for these latitudes if the whole were in a state of steady motion as a rigid body. There are no forces acting which tend to modify the angular momentum about the polar axis of an elementary ring of water, which coincides with a parallel of latitude, and consequently currents will be started, in virtue of which each particle of fluid will move along a parallel of latitude from east to west. The effect of the disturbing force is therefore to modify the state of steady motion about which the free oscillations take place from a uniform rotation of the whole system as a rigid body to a state in which there exist horizontal westerly currents (Hough, 1897, pp. 248–249).

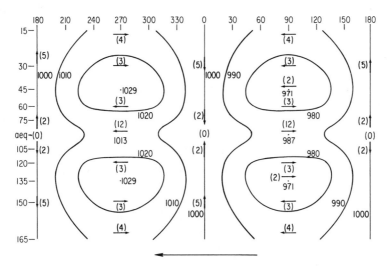

Fig. 12.4. Margules' (1893, Fig. 11) diagram, showing surface pressure pattern (in millibars) for the external ("Lamb") mode of a 273 K isothermal atmosphere. The corresponding surface velocities for this arbitrarily chosen amplitude are shown by arrows and are given in brackets in meters per second. Margules described it as a "Westwärts Wandernde Welle zweiter Art erste Classe Typus I" and calculated the period as 130.7 hr. Here "erste Classe" means wavenumber 1 and "Typus I" refers to the first symmetric meridional mode. The other classification is into two groups or species of waves that behave differently in the limit as the rotation rate tends to zero. The first species become ordinary gravity waves in this limit, whereas the second species are what we now call planetary waves. Their limiting strucure is that of a nondivergent planetary wave with frequency divided by twice the rotation rate, being $2 \times 3 = 6$ for this wave in the limit. The structure calculated by Margules is very close to that of the 5-day wave, whose properties have been identified in long observational records by Madden (1978).

He had already pointed out that if "fluid particles are moving along parallels of latitude, it is possible by a proper adjustment of the free surface to ensure that such a motion should be permanent (p. 248)." Hough's work followed the discussion by Lamb of long-period tides that first appeared in the 1895 edition of his *Hydrodynamics* and referred to the idea of conservation of angular momentum coming from Hadley (1735). Hough (1897) also solved the problem of slow forced adjustment through a succession of geostrophic equilibrium states in connection with ocean circulation driven by evaporation (see Section 9.14). In addition, he proved that for slow motion of a rotating homogeneous fluid, the velocity is the same at all points on lines parallel to the axis of rotation, a result now referred to as the *Taylor–Proudman theorem*. Further, he showed that in steady flow of a homogeneous layer of depth h, the quantity f/h (which is equal to $2\Omega/h$ cosec φ on the sphere) is conserved for fluid particles, and that the same would be true for a stratum of a stratified fluid. In other words, he obtained a special form of the equation for conservation of potential vorticity.

Planetary waves have a maximum frequency (in a frame moving with the medium) that is given [see (12.3.3)] according to the beta-plane approximation by

$$\omega_{\max} = \tfrac{1}{2}\beta a = \beta c/2f_0 = (c/2R)\cot\varphi_0. \qquad (12.3.16)$$

Alternatively, this may be interpreted as giving the maximum latitude at which waves of a certain period can exist. For first-mode baroclinic waves in the ocean, with $c = 2.5$ m s^{-1}, for instance, the maximum latitude for annual period waves is 45°. Kang and Magaard (1980) have studied annual period temperature fluctuations in the North Pacific between 32° and 40° latitude, and found the data to be consistent with dispersive properties close to those of first-mode westward-propagating planetary waves, modified somewhat by the mean flow. Inverse wavenumbers are of order 100 km. White and Saur (1981) have identified annual period waves emanating from a region near 125°W where the amplitude of Ekman pumping is particularly large. The westward phase propagation typical of phenomena for which beta effects are important has also been observed in ocean eddies (Bernstein and White, 1975; Freeland *et al.*, 1975) and McWilliams and Flierl (1976) have found that much of the low-frequency variability observed in the Mid-Ocean Dynamics Experiment can be explained by assuming the presence of barotropic and baroclinic planetary waves with periods of 4–11 months and wavelengths of 170–300 km. Price and Rossby (1981) identified a barotropic planetary wave in the western North Atlantic with a wavelength of 340 km ($\kappa_H^{-1} \approx 50$ km), a period of 2 months ($\omega^{-1} \approx 10$ days), a phase speed $c \approx 0.06$ m s^{-1}, and a velocity amplitude 0.12 m s^{-1}. Such large amplitudes are not uncommon, so nonlinear effects are often likely to be important. Such effects are discussed, for example, by Rhines (1977), Killworth (1979), and Charney and Flierl (1981).

Dispersive properties of planetary waves [given by (12.3.2)] can be illustrated nicely by considering the response of an ocean to a localized forcing. Various idealized examples are discussed by Longuet-Higgins (1965b), Lighthill (1966, 1967), and Rhines (1977). Laboratory experiments are discussed by Platzman (1968), and reviews of planetary wave properties are given by Dickinson (1978, 1980).

12.4 Spin-Up of the Ocean by an Applied Wind Stress

Much of the forcing of the atmosphere by heating or of the ocean by the wind is at time scales large compared with 1 day and consequently produces a response that is at all times close to being in geostrophic equilibrium. The response can be calculated by adding forcing terms to the quasi-geostrophic equations developed in Section 12.3. Here the problem of the response of an ocean of uniform depth to a suddenly imposed wind stress will be considered because this gives useful insight into the nature of the ocean circulation and of its transient behavior.

Before introducing forcing into the quasi-geostrophic equations, it must be remembered that quasi-geostrophic flow is always close to an equilibrium [namely, the geostrophic equilibrium given by (12.2.15)] that is *horizontally nondivergent* at the leading order of approximation. It is the weaker *ageostrophic* flow field that has divergence and causes changes in equilibrium. Forcing effects likewise cause the geostrophic equilibrium to change by the indirect means of creating divergence, and therefore forcing terms are added to the *ageostrophic* equations. (If they were added to the geostrophic equations, an inconsistency would arise if they resulted in nonzero divergence at the leading order). For example, the mechanism by which the wind causes changes in the ocean was discussed in Section 9.4 and is illustrated in Fig. 9.4. The direct effect of the wind is to produce an Ekman transport at right angles in the surface layers. If this flow is convergent or divergent, vertical motion is produced, and it is this vertical motion that alters the pressure field and hence the geostrophic motion balancing the pressure field. In other words, the Ekman pumping is part of the ageostrophic field of motion that changes the pressure field and geostrophic velocities.

For small perturbations in a homogeneous ocean, Eqs. (12.2.17) and (12.2.18) (with p'/ρ replaced by $g\eta$, where η is the surface elevation) for the ageostrophic part of the motion are modified by the addition of the stress term first introduced in Eqs. (9.9.10). Thus if bottom friction is ignored, (12.2.17) and (12.2.18) become

$$f_0 u_a = \frac{g}{f_0}\left(\beta y \frac{\partial \eta}{\partial y} - \frac{\partial^2 \eta}{\partial t\, \partial x}\right) + \frac{Y_s}{\rho H}, \tag{12.4.1}$$

$$f_0 v_a = \frac{g}{f_0}\left(-\beta y \frac{\partial \eta}{\partial x} - \frac{\partial^2 \eta}{\partial t\, \partial y}\right) - \frac{X_s}{\rho H}, \tag{12.4.2}$$

where (X_s, Y_s) is the wind stress at the surface and H is the ocean depth. For a stratified ocean, the equations for each mode have the same form, as shown in Section 9.10, with H replaced by a quantity that can be called the equivalent forcing depth H^F for the mode in question.

The ageostrophic velocity for the linear problem now has three parts: the beta, the isallobaric, and the *Ekman parts*. Each part contributes to the divergence, and when substituted in the linear version of the continuity equation (12.2.29), the result is

$$\frac{\partial}{\partial t}\left(\frac{\partial^2 \eta}{\partial x^2} + \frac{\partial^2 \eta}{\partial y^2} - \frac{f_0^2}{c^2}\eta\right) + \beta \frac{\partial \eta}{\partial x} = \frac{f_0}{\rho g H}\left(\frac{\partial Y_s}{\partial x} - \frac{\partial X_s}{\partial y}\right). \tag{12.4.3}$$

This is the quasi-geostrophic form of the potential vorticity equation [compare with the f-plane from (9.9.19)], which now includes a forcing term equal to the *Ekman divergence* or *curl of the wind stress*.

An example that illustrated the behavior of solutions of this equation is that of an eastward wind stress varying sinusoidally with latitude

$$X_s = X_0 \sin ly. \tag{12.4.4}$$

In this case, the solution has the form

$$\eta = \eta' \cos ly, \tag{12.4.5}$$

where $\eta'(x, t)$ satisfies the equation

$$\frac{\partial}{\partial t}\left(\frac{\partial^2 \eta'}{\partial x^2} - \left(l^2 + \frac{f_0^2}{c^2}\right)\eta'\right) + \beta \frac{\partial \eta'}{\partial x} = -\frac{lf_0 X_0}{\rho g H}. \tag{12.4.6}$$

The boundary conditions of zero normal velocity reduce, at leading order, to

$$\eta' = 0 \qquad \text{at} \quad x = 0, -b \tag{12.4.7}$$

in the case of meridional boundaries at $x = 0, -b$ since η plays the role of a stream function at this order by (12.2.15).

The properties of the solution depend on the value of a single parameter

$$\Lambda = (l^2 + f_0^2/c^2)b^2 \tag{12.4.8}$$

and have been discussed in detail by Anderson and Gill (1975). For the barotropic mode, f_0/c is usually smaller than l, the inverse of the wind stress scale l^{-1} (which is typically 1000 km). Thus Λ is determined by the ratio of the basin width b to the wind scale l^{-1}, and typical values of Λ are in the range 30–200. For the baroclinic modes, on the other hand, f_0/c is much larger than l, so Λ is determined by the ratio of the basin width to the baroclinic Rossby radius. Values of Λ in this case are in excess of 10,000. Because Λ is large, the term involving the second derivative with respect to x is unimportant except near boundaries, so (12.4.6) is approximated by

$$-(l^2 + f_0^2/c^2)\,\partial\eta'/\partial t + \beta\,\partial\eta'/\partial x = -lf_0 X_0/\rho g H. \tag{12.4.9}$$

Away from boundaries, there is no x dependence, so the solution is

$$\eta' = lf_0 X_0 t/\{\rho g H(l^2 + f_0^2/c^2)\}, \tag{12.4.10}$$

and η' grows uniformly with time. This solution represents a local response to *Ekman pumping* of the type illustrated in Fig. 9.4, which in this case can be taken as a meridional section. The wind stress drives an Ekman transport in the surface layers away from the latitude at which the surface air pressure is a minimum, i.e., from the latitude at which the wind stress changes sign. The baroclinic response is therefore a *raising* of the thermocline under the region of low air pressure, as depicted in the figure. The barotropic response, on the other hand, would involve a lowering of sea level, as required by mass conservation. Associated with these pressure changes would be geostrophic zonal currents that would also grow linearly in time.

Now consider *boundary effects*. These will propagate into the interior in the form of free planetary waves, whose behavior is governed by the dispersion equation

(12.3.2), (11.8.6), or (11.6.8). The corresponding graph of ω against k is shown in Fig. 11.5. Effects of the *eastern* boundary are carried by waves with westward group velocity, and the fastest of these are the long nondispersive waves. These will be expected to dominate because of the scale assumption, i.e., the behavior of the solution will be dominated by (12.4.9), except near the western boundary. The solution is (12.4.10) *ahead* of the wave front due to the long waves, i.e., for

$$x < -\beta t/(l^2 + f_0^2/c^2).\qquad(12.4.11)$$

After the wave front has passed, the solution is the steady one, satisfying the boundary condition at $x = 0$, namely,

$$\beta v = g f_0^{-1}\beta\,\partial\eta/\partial x = -\rho^{-1}H^{-1}\,\partial X_s/\partial y \qquad \eta' = (l f_0 X_0/\rho g H)x. \quad(12.4.12)$$

This is Sverdrup's (1947) solution for ocean currents, discussed in Section 11.13, since it is the steady forced solution of the potential vorticity equation. The interpretation is that fluid particles are acquiring potential vorticity at a given rate through Ekman pumping, which forces vortex lines to undergo changes in length and hence in their vorticity. For small slow perturbations, however, the total vertical component of vorticity is constrained to remain close to the local value f. The only way a fluid particle can achieve this in steady motion is to have a meridional velocity v given by (12.4.12). In other words, a material vortex line element that is being stretched will move poleward at a rate given by (11.13.3) and one that is being squashed will move equatorward. The effect has been clearly demonstrated in the laboratory [see Beardsley (1969) and Faller (1981)].

Thus after the long wave from the eastern boundary has passed, the solution becomes virtually steady, with the east–west pressure gradient being given by (12.4.12). The steady solution is sketched out, for the baroclinic case, in Fig. 12.5. The time taken for the long-wave adjustment is very much dependent on the mode concerned. For the barotropic mode, the wave speed of the long wave is approximately given by

$$\beta l^{-2} \approx 20 \quad \text{m s}^{-1},$$

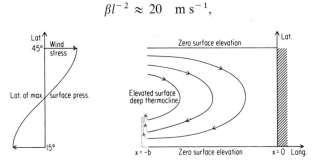

Fig. 12.5. Sverdrup's steady wind-driven circulation for the case of a sinusoidally varying eastward stress. The assumed stress is shown in the panel at the left, with values for the latitudes that approximately apply in practice. Near 30°, where the surface air pressure is a maximum, the Ekman pumping also has a maximum magnitude and is directed downward. The solution is shown by solid contours at the right. The response in practice is mainly baroclinic (in the sense that the largest currents are confined to the upper layers), so the contours may be interpreted as corresponding to either dynamic height or thermocline depth. The direction of the wind-driven currents is shown by the arrows. The results of numerical calculations for real ocean basins look very similar [see, e.g., Anderson *et al.* (1979)]. The dotted lines at the bottom left of the diagram are added to show how the beginnings of the western boundary current (see Section 12.6) relate to the Sverdrup solution.

i.e., this wave would take about 3 days to cross the Atlantic, and this gives the approximate adjustment time for the barotropic part of the flow. For the baroclinic mode, the wave speed is

$$\beta a^2 = \beta c^2 / f_0^2,$$

where a is the Rossby radius. For a typical mid-latitude value of a of 30 km this speed is only 2 cm s^{-1} or 1/1000 of the speed of the barotropic wave. The time to cross the Atlantic therefore is of the order of a decade! At lower latitudes, however, the speed increases rapidly as the equator is approached (see Fig. 12.3), so adjustment is much quicker there.

Figure 11.5 shows that waves with eastward group velocity are very different from those with westward group velocity, and this is the basic reason for the *east–west asymmetry* in the ocean circulation, which is such an obvious feature in any map (like that of Fig. 7.8a) that shows dynamic topography or currents. The largest eastward group velocity is only one-eighth of the speed of the long waves, and the scale of these waves is quite small. For the barotropic mode, it is about

$$(\sqrt{3}\, l)^{-1} \approx 600 \quad \text{km},$$

whereas for the baroclinic mode, it is only $3^{-1/2}$ times the Rossby radius, typically about 20 km. The transient solution in the neighborhood of the western boundary (Lighthill, 1969; Anderson and Gill, 1975) soon becomes dominated by even shorter waves since these have smaller group velocity and remain near the western boundary. The consequence of this dispersive property is a boundary layer that gets thinner with time, its width being inversely proportional to time. The solution for the correction of the long-wave solution has the form

$$(t/(b + x))^{v/2} J_v(2(\beta(b + x)t)^{1/2}), \qquad (12.4.13)$$

where v is a constant and J_v is the Bessel function of order v. When the interior solution is (12.4.10), i.e., is growing with time, the appropriate value of v is unity. After the long-wave has reached the western boundary and the interior solution is Sverdrup's solution, the appropriate value of v is zero.

An example of the complete solution of (12.4.6) is shown in Fig. 12.6 for $\Lambda = 2400$. Graphs of η' are shown at intervals of one-twelfth of the time it takes for a long wave to cross the basin. The sloping solid line near the western boundary marks the wave front corresponding to the wave with greatest eastward group velocity, and behind this the thinning western boundary layer may be observed. Elsewhere it can be seen that the long-wave solution produces a good approximation.

Numerical experiments (Anderson *et al.*, 1979) with a basin that has the shape of the North Atlantic show that the above analysis gives a good description of the response of the ocean to an impulsively applied wind stress when topographic effects are removed, i.e., the ocean has uniform depth. There is one additional feature, however, in the barotropic response, which is that reflections of wave energy from boundaries result in energy being established preferentially in "basin modes" with particular frequencies. The method for calculating the properties of basin modes is given by Longuet-Higgins (1964a, 1965a). When topography is included, the reverberation of basin modes is not so prominent, but a mode with a period of about 2.6 days is still found (Anderson *et al.*, 1979).

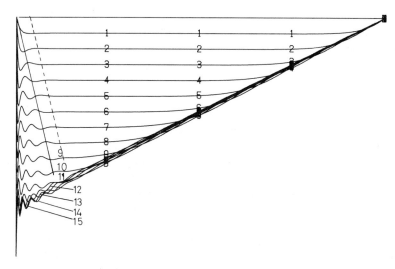

Fig. 12.6.   Displacement of the thermocline as a function of longitude (east to the right) for a baroclinic mode at different times (marked 1–14) after the wind is switched on at time zero. The diagram shows the thermocline moving uniformly downward through Ekman pumping until boundary effects are felt. Long planetary waves, moving rapidly out from the eastern boundary, establish a Sverdrup balance with a uniformly sloping thermocline almost as soon as they arrive. The fastest traveling waves from the western boundary are much slower short waves. Their speed is indicated by the solid diagonal line and its intercepts on the thermocline displacement curves. There is a precursor to these waves, indicated by the dashed line. The diagram may also be interpreted as a graph of the eastward velocity versus longitude. [From Anderson and Gill (1979), *J. Geophys. Res.* **84,** 1836 (Fig. 2); copyrighted by the American Geophysical Union.]

Topography also has a strong influence on the form that the barotropic response takes after the initial transients (basin modes, etc.) have died out. This is so because the potential vorticity Q of the unperturbed state is [see (7.10.10)] given by

$$Q = f/H,$$

i.e., the lines of constant potential vorticity are circles of latitude only when H is constant. When H varies, they are contours of f/H, maps of which are given by Gill and Parker (1970). In the absence of forcing, the only steady motion possible is along contours of f/H (Hough, 1897). With wind forcing, the equivalent of Sverdrup's relation gives the velocity component normal to these contours, and the pressure field can be obtained (Gill and Niiler, 1973) by integrating along contours of f/H from the eastern boundary. The procedure breaks down where there are closed contours of f/H and a larger response might be expected.

Stratification, however, reduces the influence of topography a great deal. In the spin-up calculation, stratification takes some time to remove this influence, however, because of the long time scale of this response. It occurs because vertical motion is induced by flow over topography, and the part of this due to the baroclinic modes acts as a forcing term for the barotropic mode (Anderson and Killworth, 1977). After this mode has come into adjustment, the flow near the bottom of the ocean is very small, and therefore topography has relatively little effect. In fact, the vertically integrated transport in stratified ocean models with topography is very similar to that found in flat-bottom barotropic models.

Although the spin-up problem discussed above was for the special case of a wind that was held steady once imposed, it shows what might be expected of forced transient features of the extratropical ocean circulation. Apart from boundary currents, the main response for that part of the forcing with periods of less than 1 yr is barotropic, the baroclinic part being a direct local response to Ekman pumping (Willebrand et al., 1980). At periods less than 3 days, the forcing falls off rapidly and so does the response, which is mainly direct since planetary waves do not propagate far in $3/2\pi$ days. For periods between a month and several years, the barotropic response is an equilibrium response, i.e., a Sverdrup balance as modified by topography. The forcing does not vary much with frequency and neither does the response. At periods in between, the effects of westward propagation are important and prominent, and resonant responses are possible. Willebrand et al. (1980) calculated the response of the North Atlantic, using observed wind forcing and found the root-mean-square currents were a few centimeters per second with the largest values in the west and near topographic features. There was essentially no coherence between oceanic and atmospheric variables at any frequency, presumably because of the small scale of inhomogeneities associated with topography.

The large-scale response at annual period has been discussed analytically by Gill and Niiler (1973), with contributions to sea-level change from various causes calculated for the *extratropical* North Atlantic and North Pacific. Small-scale features caused by boundary effects are ignored. The *baroclinic* response to the wind is simply a direct local response to Ekman pumping, moving the thermocline up and down (as in Fig. 9.4), if planetary waves cannot propagate far compared with the scale of the forcing in the time scale of the forcing. This is true of large-scale forcing at annual period outside the tropics but is not so in the tropical belt, where planetary waves propagate much faster (see Fig. 12.3). [Seasonal response in the tropics is discussed by Schopf et al. (1981) and Cane and Sarachik (1981).] The barotropic part is quasi-equilibrium and changes the surface elevation by only 1 or 2 cm, but since the associated currents are spread over the whole depth, changes in transport can be 5 Mt s^{-1}. Changes in sea level due to expansion and contraction of the upper 200 m were calculated from observations and were found to have amplitudes of up to 10 cm, whereas changes due to adjustment to atmospheric pressure (see Section 9.9), which have no dynamic effect, were found to have the largest amplitudes of 3–6 cm in higher latitudes. Seasonal changes in currents in the North Atlantic have been studied with the aid of a numerical model by Anderson (1982), who found that many changes, such as those of the Gulf Stream near Florida, are due to changes in the meridional component of the wind.

12.5 Steady Ocean Circulation

It is useful at this stage to consider the picture of the steady ocean circulation that emerges from the calculations of Section 12.4. This shows that a Sverdrup balance is achieved in all the ocean except the western boundary layer, where a steady state cannot be reached until additional effects, such as nonlinearity and dissipation, come into play. Thus it is possible, using Sverdrup's equation, to calculate a steady-state response of the ocean to the wind in all but the western boundary layer, and this has been done many times. Topographic effects are ignored, and this appears to be justified at least partially because of the stratification effect referred to in Section 12.4 (Anderson and Killworth, 1977).

The quantity that is usually calculated is the stream function ψ for the vertically integrated mass transport. Only the barotropic component contributes to this, so ψ is related to v and η by

$$\rho H v = \frac{\partial \psi}{\partial x} = \frac{\rho g H}{f_0} \frac{\partial \eta}{\partial x} = \frac{1}{\beta} \left(\frac{\partial Y_s}{\partial x} - \frac{\partial X_s}{\partial y} \right), \qquad (12.5.1)$$

the last expression coming from the steady-state version of (12.4.3). This form of the equation was also introduced in (11.13.6). The stream function ψ is then calculated by integration westward from the eastern boundary. The result for the North Atlantic is given, e.g., by Leetmaa and Bunker (1978). They also calculated what they called the vertically integrated geostrophic transport, which is shown in Fig. 12.7a. This quantity represents the contribution to the transport from the geostrophic flow, i.e., it ignores the Ekman transport. This flow is not nondivergent, so a stream function cannot be precisely defined. The figure is based on an integration of the meridional geostrophic transport westward from the eastern boundary to either the western boundary or the point at which the curl goes to zero.

The steady geostrophic currents in the ocean are not known precisely since only currents relative to a given level can be calculated from a knowledge of the temperature and salinity fields. Therefore it is not possible to calculate precise transport quantities for comparison with the Sverdrup relation. However, some comparison can be made, e.g., Fig. 12.7b shows the currents in the North Atlantic at 100 m depth relative to those at 1500 m depth (Stommel *et al.*, 1978). Assuming the latter to be relatively small, this gives a picture of near-surface geostrophic currents. There are many striking similarities with Fig. 12.7a, suggesting that wind-driving explains much of the surface current pattern. On the other hand, there are also important differences, suggesting that other processes, such as buoyancy forcing, are also important. Worthington (1970, 1976), for instance, calculated that sinking in the Greenland Sea draws in large volumes of surface water from the North Atlantic, and this must have a significant influence on the circulation pattern.

A feature of Sverdrup's solution is that it allows a calculation to be made of the total transports of the western boundary currents. This is because, for instance, the southward transports in the main ocean parts of the subtropical gyres can be calculated, and these must be balanced by the northward transports of the western boundary currents such as the Gulf Stream and Kuroshio. The calculated Sverdrup transport for the Gulf Stream (Leetmaa and Bunker, 1978) at 31°N is 32 Mt s^{-1}. This compares favorably with the estimated transport of 32 Mt s^{-1} through the Florida Straits (W. S. Richardson *et al.*, 1969) and estimates of the geostrophic transport across the interior of the Atlantic at 32°N (Leetmaa *et al.*, 1977). However, the measured transport of the northward-flowing Gulf Stream increases rapidly northward to values of order 100 Mt s^{-1} [see, e.g., Gill (1971)], contrary to the expectations

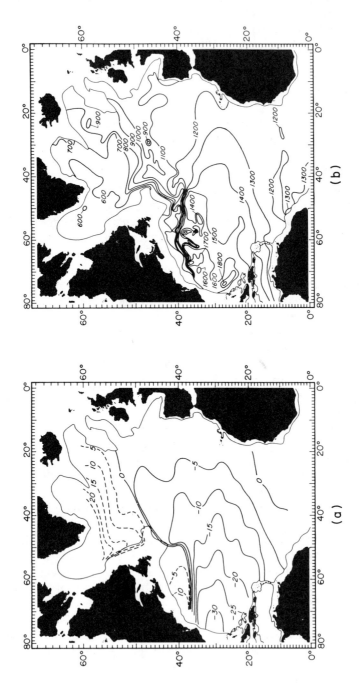

Fig. 12.7. (a) Contours of the annual mean geostrophic transport (Sverdrup minus Ekman) for the North Atlantic. Units are megatons per second (Sverdrups). Values are obtained by integrating the meridional geostrophic transport along circles of latitude from the eastern boundary to either the western boundary or the longitude where the curl of the wind stress goes to zero. (b) Chart of the dynamic topography of the 100-db surface (i.e., at approximately 100 m depth) relative to the 1500-db surface. Units are dynamic millimeters. [Both diagrams are from Leetmaa and Bunker (1978, Fig. 4).]

of Sverdrup theory. There are a number of factors that can lead to enhanced transport, e.g., joint effects of stratification and the continental slope topography (Sarkisyan, 1977; Holland, 1973; Simons, 1979) that tend to displace laterally the southward-flowing deep undercurrent from the northward-flowing current in the upper layers. However, the principal factor is probably the presence of eddies due to instabilities (see Chapter 13) in the currents since models with a fine enough scale for such eddies to appear spontaneously [see Holland (1977)] produce recirculating flow patterns like those observed (Worthington, 1976; Wunsch, 1980), with associated eddy statistics that are also similar to those observed (Schmitz and Holland, 1981). The dynamics of such systems has been discussed by Rhines (1977).

The Antarctic Circumpolar Current is rather unique because it is the only current which can flow entirely around the globe. [It shows up nicely in the tracks of drifting buoys [see J. F. Garrett (1980)]]. This is significant dynamically because, as Hough (1897) pointed out, steady currents can flow freely along circles of latitude (or rather contours of sin φ/H, where φ is latitude and H depth), but cannot cross these contours unless forced. Hough also showed that if there are no meridional barriers and if a zonally symmetric forcing is applied, the response will grow linearly in time, until a frictional balance is achieved. The early models of the circumpolar current [see, e.g., Munk and Palmén (1951)] were of this type, with the current occupying a zonal channel. In practice, the channel does not have walls, but for small friction, a strong current would be expected at the latitudes at which there were no barriers (i.e., the latitudes of the Drake Passage). At other latitudes, the continental barriers (South America and the Antarctic Peninsula) would impose a Sverdrup regime with weak currents, whose amplitude would *not* increase if friction were reduced. Thus there is a problem as to the way in which the two regimes link up, and this question was studied by Gill (1968b) for the case of a homogeneous ocean of constant depth. In practice, however, the Drake Passage is a greater obstacle to the current than that which a uniform-depth model would suggest because it is somewhat *shallower* than the rest of the ocean, and Stommel (1962) proposed a simple model to take account of this fact.

Because of the special character of the dynamics of the circumpolar current, the response of numerical models of the ocean circulation in this region is especially interesting. Numerical models have been reviewed by Gill (1971), Pond and Bryan (1976), Holland (1977), Bryan (1979), and Anderson (1979); the techniques have been outlined by Bryan (1969). It is found, e.g., in the model of Cox (1975), that when the ocean is made homogeneous, there is little flow through the Drake Passage because of the topography, but that a stratified ocean is much less sensitive to topography, and so there is considerable flow when the ocean is stratified. It is interesting to note that Hough (1897) explained the fact that currents do not, in general, follow contours of sin φ/H (where φ is latitude and H is the total depth) by pointing out that in a stratified ocean currents in a particular stratum would follow contours of sin $\varphi/\delta H$ (where δH is the depth of the stratum). He concluded that the only place where they would have to follow contours of sin φ/H would be at the equator, because zonal flow is then implied for each stratum!

A more detailed look at effects of geometry was made in a numerical study by Gill and Bryan (1971). Some interesting results were obtained when a wall was placed like a dam across the bottom part of a gap in the meridional boundary, the gap representing the Drake Passage. In particular, cold water accumulated on the bottom to the east of the barrier (as is observed), leading to zonal pressure gradients below the sill level that actually enhanced the flow in the circumpolar current. More recently, McWilliams *et al.* (1978) have examined models that show the possible role of eddies in the circulation. Wearn and Baker (1980) have studied pressure measurements in the Drake Passage and find that *fluctuations* in transport with periods of longer than 30 days are highly correlated with fluctuations in the wind stress, integrated over the Southern Ocean.

There is little doubt that the major surface currents are predominantly wind-driven, and the mechanisms through which wind stress generates currents have been dealt with in Sections 10.14 (eastern boundary currents), 11.12 and 11.16 (tropical currents), in this section (extratropical gyres and the circumpolar current), and will be continued in the next section (western boundary currents). However, buoyancy driving, as discussed in Section 2.7, is also important, particularly since it helps to establish the thermal structure that is responsible for the wind-driven currents being concentrated near the surface (i.e., above the thermocline). As stated in Section 2.7, bottom-water formation tends to be localized and intermittent, and the associated dynamics is extremely interesting. One major region of bottom-water formation is the Weddell Sea, where water is made dense by brine release (Gill, 1973; Foster and Carmack, 1976; Foster and Middleton, 1980), and a model of the process has been constructed by Killworth (1974). A different form of bottom-water formation, which occurs away from boundaries, has been observed in some detail in the Gulf of Lyons in the Mediterranean Sea (MEDOC Group, 1970; Gascard, 1978), and this may be a prototype of what happens in other locations. Models of the process have been constructed, e.g., by Killworth (1976), and similar effects have been observed in the Labrador Sea (Clarke and Gascard, 1982). The dense water that is formed in such regions flows equatorward in deep western boundary layers (Warren, 1981), whose position is determined by the bottom topography (see Fig. 10.8). The flow may continue into the opposite hemisphere. Little bottom water is formed in the Pacific Ocean, so most of its volume is occupied by dense water that enters from the south and may be traced back to sources in the North Atlantic and in the Atlantic (Reid and Lynn, 1971). Numerical models of the ocean (e.g., Bryan and Lewis, 1979) include effects of buoyancy, but there are many problems involved in simulating these effects accurately.

More detailed discussion of the ocean circulation is given by several authors in Warren and Wunsch (1981), and much useful information about relevant observations can be found in the classical textbooks of Sverdrup *et al.* (1942) and Defant (1961). Dynamical models are discussed by Stommel (1965), Stern (1975), and Pedlosky (1979). Sources of data and regional studies are given in Appendix 5.

12.6   Western Boundary Currents

In Section 12.4, the calculation of the linear response of an inviscid ocean to the wind showed that, apart from some transient basin modes, a steady circulation is eventually reached everywhere except near the western boundary, where a boundary layer is formed that becomes thinner as time goes on; consequently the currents become increasingly strong. At some time t_s, effects not included in the model, such as nonlinearity or friction, will become important, and these may lead to a steady boundary current as well. By (12.4.3), the width W of the boundary layer at this time has order of magnitude given by

$$W = (\beta t_s)^{-1}. \tag{12.6.1}$$

Inspection of the expression (12.4.1) for the ageostrophic velocity component u_a across the boundary layer shows that this is just the width that makes the isallobaric contribution to u_a of the same order as the beta contribution. On the other hand, because x derivatives are large in the boundary layer, the beta term dominates v_a and the other terms are relatively small.

Now consider how friction and nonlinearity affect the behavior of the solution in the boundary layer for the case of a homogeneous ocean that satisfies the "rigid-lid" (see Section 6.3) approximation [c^{-1} equals zero in (12.4.3) or H^{-1} equals zero in (12.2.29)]. Two forms of friction will be considered—one a simple decay process with time scale r^{-1} simulated by replacing $\partial/\partial t$ by $\partial/\partial t + r$ in the equations. The other form of friction that will be considered is lateral friction, with a corresponding lateral eddy viscosity coefficient A. This is introduced by replacing $\partial/\partial t$ by

$$\frac{\partial}{\partial t} - \frac{\partial}{\partial x}\left(A\frac{\partial}{\partial x}\right) - \frac{\partial}{\partial y}\left(A\frac{\partial}{\partial y}\right) \tag{12.6.2}$$

in the equations. In model studies, A is very often taken to be constant, although in practice A varies a great deal and is even negative in places (Webster, 1965; Rhines, 1977). However, it is not yet known how to model the variations of A with any confidence, so constant values are still used. Models with constant A can be useful for showing the way in which friction can affect the motion and where it is important. Inclusion of lateral friction can also be important in numerical simulations to ensure that irregularities on the scale of the mesh spacing do not become too large.

If nonlinearity and the two forms of friction are included, the approximate forms of (12.4.1) and (12.4.2) in the boundary layer become

$$f_0 u_a = \frac{g}{f_0}\left(\beta y\frac{\partial \eta}{\partial y} - \frac{D_g}{Dt}\frac{\partial \eta}{\partial x} - r\frac{\partial \eta}{\partial x} + A\frac{\partial^3 \eta}{\partial x^3}\right) + \frac{Y_s(y)}{\rho H}, \tag{12.6.3}$$

$$f_0 v_a = -\frac{g}{f_0}\beta y\frac{\partial \eta}{\partial x}. \tag{12.6.4}$$

The stress Y_s is regarded as a function of y only since it changes very little across the thin boundary layer. Since the barotropic motion is nondivergent, the divergence of these equations gives

$$\frac{\partial}{\partial x}\left(-\frac{D_g}{Dt}\frac{\partial \eta}{\partial x} - r\frac{\partial \eta}{\partial x} + A\frac{\partial^3 \eta}{\partial x^3}\right) - \beta\frac{\partial \eta}{\partial x} = 0, \tag{12.6.5}$$

which integrates to give

$$\frac{D_g}{Dt}\frac{\partial \eta}{\partial x} + r\frac{\partial \eta}{\partial x} - A\frac{\partial^3 \eta}{\partial x^3} + \beta(\eta - \eta_0(y)) = 0, \tag{12.6.6}$$

where $\eta_0(y)$ is the value of the surface elevation at the outer edge of the boundary layer.

The type of boundary layer to be formed will depend on which of the new terms in (12.6.3) first becomes important as the boundary layer thins. If it is the bottom-friction term, this will happen when $t_s = r^{-1}$, so by (12.6.1) the width $W = W_s$ in this case is given by

$$W_s = r/\beta. \tag{12.6.7}$$

The main attraction of this solution is its simple form, since (12.6.6) in this case gives the steady solution near the western boundary $x = -b$ as

$$\eta = \eta_0(y)(1 - \exp(-\beta(x + b)/r)). \tag{12.6.8}$$

This was used by Stommel (1948) to show how the beta effect causes asymmetry in the ocean circulation. However, if r^{-1} is 100 days, W_s is only 5 km, which is much too thin to match observations.

If, on the other hand, lateral friction becomes important first, this will happen when t_s is of order W^2/A, so (12.6.1) gives

$$W = (\beta t_s)^{-1} = A/W^2\beta \quad \text{or} \quad W = W_M \equiv (A/\beta)^{1/3}. \tag{12.6.9}$$

For typical values of A of 10^2–10^4 m^2 s^{-1}, W_M is 20–80 km. This model was used by Munk (1950), and the corresponding steady solution of (12.6.6) in this case is

$$\eta = \eta_0(y)\left(1 - \frac{2}{\sqrt{3}}\exp\left(-\frac{(x + b)}{2W_M}\right)\sin\left(\frac{\sqrt{3}(x + b)}{2W_M} + \frac{\pi}{3}\right)\right). \tag{12.6.10}$$

The third possibility is that nonlinear terms become important first. In this case, the velocity component *across* the boundary layer will have the same magnitude U as that in the interior (it falls from a value of order U at the outer edge to zero at the boundary), and the time scale t_s will be of order W/U. Consequently, (12.6.1) gives for the width

$$W = (\beta t_s)^{-1} = U/\beta W, \quad \text{i.e.,} \quad W = W_I \equiv (U/\beta)^{1/2}. \tag{12.6.11}$$

Equation (12.6.6), after using the expressions (12.2.22) for D_g/Dt and (12.2.15) for geostrophic velocity, becomes, when the motion is steady,

$$-\frac{\partial \eta}{\partial y}\frac{\partial^2 \eta}{\partial x^2} + \frac{\partial \eta}{\partial x}\frac{\partial^2 \eta}{\partial x \partial y} + \frac{f_0\beta}{g}(\eta - \eta_0(y)) = 0. \tag{12.6.12}$$

An exact solution exists for the special case of a constant *inflow* velocity U at the outer edge of the layer [see, e.g., Morgan (1956)], namely,

$$g\eta = f_0 Uy(1 - \exp(-(x + b)/W_I)). \tag{12.6.13}$$

This solution has the property that

$$f_0 + \beta y + f_0^{-1}g\, \partial^2\eta/\partial x^2 = f_0 + \beta g\eta/f_0 U, \tag{12.6.14}$$

which is the special form of the conservation of potential vorticity relation (7.10.12) that is appropriate to this solution. The left-hand side of (12.6.14) is proportional to the potential vorticity (7.10.10) because the depth of fluid in barotropic flow is approximately constant. The right-hand side of (12.6.14) is a function of the stream function ψ since η is proportional to ψ in quasi-geostrophic flow. The scale assumptions used to derive the quasi-geostrophic equations is Section 12.2 do not strictly apply in the boundary layer because the x scale is now W, not L. However, it has already been shown that u_a/u_g and v_a/v_g are still of order ϵ_L in the boundary layer, so the quasi-geostrophic approximations may still be used.

It is not possible to complete a gyre with a purely inertial boundary layer because fluid leaving the boundary layer would retain the potential vorticity it had acquired in lower latitudes (for the case of a subtropical gyre) before it entered the boundary layer, and so would not match that of the interior flow [see Greenspan (1962)]. Some form of friction or mixing is needed to remove whatever excess relative vorticity exists so that the fluid can reenter the interior. Kamenkovich (1977) discusses barotropic models that involve both inertia and friction. The boundary current is found to go further north than in the noninertial case before returning fluid to the interior, and then the eastward flow out of the boundary layer looks like a damped stationary planetary wave with a structure similar to that given by (12.6.14) when the sign of U is reversed [see Moore (1963)]. Numerical studies show (Bryan, 1963) that the solution may not become steady and that it is sensitive (Blandford, 1971) to boundary conditions. Further discussion of boundary layers is presented by Stern (1975), Kamenkovich (1977), Pedlosky (1979), and Veronis (1981).

The barotropic models are not directly applicable to the ocean circulation because it is highly baroclinic, but it is possible to devise simple baroclinic models that can have a rather similar structure. A particularly simple example consists of an ocean of two homogeneous layers as in Section 6.2, but with the (much deeper) lower layer at rest. Then (6.2.2) and (6.2.6) show that changes of pressure in the upper layer are equal to changes of g' times the depth of the layer, where g' is the reduced gravity given by (6.2.8). This connection between pressure and layer depth is extremely useful and can be exploited to calculate the structure of a *baroclinic* inertial boundary layer, as was done by Morgan (1956) and Charney (1955b). Details are also given by Stommel (1965). In particular, use is made of the fact that the Bernoulli function (7.10.3) is constant on streamlines [see (7.10.13)]. If H is the layer depth, the boundary-layer form of this equation is

$$g'H + \tfrac{1}{2}v^2 = B(\psi), \tag{12.6.15}$$

where B is *known*, as a function of the stream function ψ, from conditions at the outside edge of the layer. When it is combined with the geostrophic equation,

$$f_0 v = g'\, \partial H/\partial x = f_0 H^{-1}\, \partial \psi/\partial x, \tag{12.6.16}$$

a simple first-order equation for the boundary-layer structure results. The equation is usually too complicated for analytic solution, but the procedure for integrating it numerically is straightforward. The behavior near the beginning of the inertial boundary layer is similar to that in the barotropic case, but marked differences between the two models develop as particles progress downstream because of the

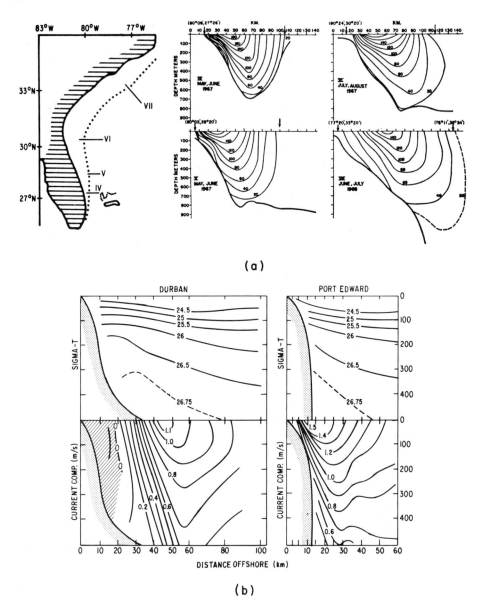

(a)

(b)

Fig. 12.8. (a) The velocity of the Gulf Stream (Florida Current) at four sections (as shown on the map) between the place (27.5°N), where it emerges from the channel between Florida and the Bahamas, and a point (33°N) about 600 km downstream, well before it leaves the coast of Cape Hatteras. Velocity contours are in centimeters per second and are at intervals of 20 cm s^{-1}. Arrows at the top indicate where the mean surface current is. [From W. S. Richardson *et al.* (1969, Fig. 3b).] (b) Mean velocity and density structure of the Agulhas Current off Durban (30°S) and Port Edward, about 140 km downstream to the south (see Fig. 12.9a for the location). The structure can vary considerably from day to day due to eddies and meanders, and examples are given by Pearce (1977). The potential vorticity (Gill, 1977c) is fairly constant offshore from the current maximum, but increases by a factor of 3 in the cyclonic shear zone. The potential vorticity does not change significantly between the sections, but the current structure does because of topographic changes [see Gill and Schumann (1979)]. The 1000-m contour is 60 km offshore at Durban, but only 12 km offshore at Port Edward. [From Pearce (1977, Fig. 9).]

depth changes that occur in the baroclinic case. These can result in the relative vorticity $\partial v/\partial x$ becoming comparable to f. (The structure of boundary layers with vorticity of order f will be examined in Chapter 13 in connection with fronts.) The calculations described above are usually done for a coastline in the form of a vertical wall. If, however, the inertial boundary layer extends over a continental shelf, changes in the structure of the layer can be forced by changes in the geometry of the shelf [see Gill and Schumann (1979)].

Figure 12.8 shows some sections across two poleward-flowing western boundary currents, namely, the Gulf Stream (or Florida Current) and the Agulhas Current. The longshore flow is close to geostrophic balance and consequently the isopycnals slope upward toward the coast. The vorticity seaward of the maximum current is anti-cyclonic and may have values as large as $-0.5f$ (i.e., a velocity difference of 1 m s^{-1} in

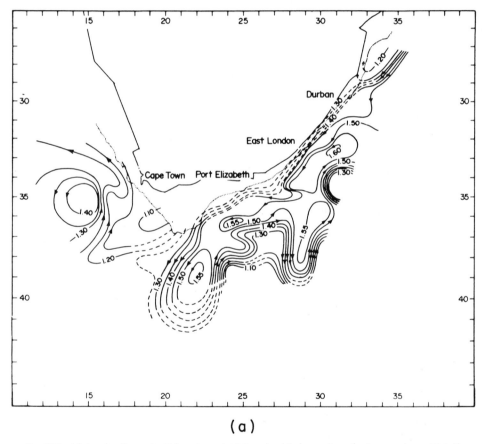

(a)

Fig. 12.9. (a) Acceleration potential contours (at intervals of 5 dyn cm) on the isopycnal $\sigma_t = 26.6$. These contours may be considered as equivalent to dynamic height contours, but at the level of the appropriate density surface rather than at the fixed level (Montgomery and Spilhaus, 1941). (b) Sigma-t (σ_t) section from about 25°E, 35°S (station 183) to about 27°E, 40°S (station 193) across the Agulhas Current, and its retroflection. The 3000-m depth contour on the continental slope is near station 185 and is encountered again near section 190 on the edge of the Agulhas Plateau. [Both parts of the figure are from Harris and van Foreest (1978, Fig. 5 and Appendix, Section J).]

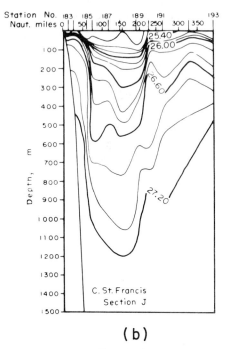

(b)

Fig. 12.9.   (continued)

30 km at 30° latitude). In the shallow water inshore of the stream, velocities are small because of friction, and consequently large cyclonic vorticities, often with values larger than f, are found there. The behavior of western boundary currents, partic- ularly at their downstream ends, can vary a great deal. The Gulf Stream system is discussed by Stommel (1965), Worthington (1976), and Fofonoff (1981). An interesting property of the Kuroshio is that it has two alternative paths south of Japan and may switch from one to the other at intervals of years. A volume on the Kuroshio system has been edited by Stommel and Yoshida (1972). The Agulhas Current (see Fig. 12.9) follows the continental slope to its southernmost point, and then "retroflects" (Bang, 1970) or turns back on itself. Consequently there are two oppositely directed intense currents in the 300-km gap between the mainland continental slope and the Agulhas Plateau. The associated spectacular changes in the density structure are shown in Fig. 12.9b.

Equatorward-flowing western boundary currents have a different character because isopycnals slope *downward* toward the coast. This gives the possibility of a wedge-shaped light water mass being held against the coast and not appearing at all in the ocean interior. An example is the East Greenland Current. This may be con- sidered [see Wadhams *et al.* (1979)] as a drain of light fresh water from the Arctic Ocean that carries pack ice with it as it moves south. It could be driven purely by buoyancy forces in the manner of the boundary currents formed in wide channels and discussed in Section 10.7, but wind effects act in a direction to enhance the current. Other examples of wedge-shaped boundary currents are found in the Mediterranean, and this type of current need not be on a western boundary.

12.7   Vertical Propagation of Planetary Waves in a Medium at Rest

Consider small-amplitude low-frequency perturbations to a uniformly stratified incompressible fluid that have the wavelike form

$$\exp(ikx + ily + imz).$$

As shown in Section 6.11, the dispersion relation is the same as that for a single mode, but with the wave speed c replaced by N/m, where N is the buoyancy frequency. In other words, (12.3.2) becomes

$$\omega = -\beta k/(k^2 + l^2 + f_0^2 m^2/N^2). \tag{12.7.1}$$

The horizontal components of group velocity are given by (12.3.4), and the vertical component is given by

$$c_{gz} = \frac{\partial \omega}{\partial m} = \frac{2f_0^2 \beta k m}{N^2(k^2 + l^2 + f_0^2 m^2/N^2)^2} \tag{12.7.2}$$

and thus has sign opposite to the phase velocity. [For an oceanic wave at mid latitudes with $k^{-1} = 1000$ km and $m^{-1} = 1$ km, $c_{gx} \approx 500$ km yr^{-1} and $c_{gz} \approx 1$ km yr^{-1}. For an atmospheric wave with $k^{-1} = (N/f)m^{-1} = 1000$ km, $c_{gx} = 0$ and $c_{gz} \approx 5$ km day^{-1}.]

It follows that an upward-propagating wave has the structure shown in Fig. 12.10, with phase lines tilting toward the west with height. This tilt is often observed in the atmosphere [see, e.g., van Loon *et al.* (1973) and Lau (1979b)], indicating that upward-propagating waves have larger amplitude than do downward ones. The polarization relations that give the phase relationships between the different fields can be deduced as follows for the wave ($l = 0$) shown in the figure, with pressure given by

$$p = p_0 \cos(kx + mz - \omega t), \tag{12.7.3}$$

where k and m are positive, and hence ω as given by (12.7.1) is negative. The horizontal velocity components are given by (12.2.15)–(12.2.18), and the vertical component is given by (6.11.4), so

$$u = -(k\omega p_0/\rho_0 f_0^2)\cos(kx + mz - \omega t),$$

$$v = -(kp_0/\rho_0 f_0)(1 - \beta y/f_0)\sin(kx + mz - \omega t), \tag{12.7.4}$$

$$w = -(m\omega p_0/\rho_0 N^2)\cos(kx + mz - \omega t).$$

The density perturbation ρ' is given by the hydrostatic equation (6.11.2), i.e.,

$$\rho' = (mp_0/g)\sin(kx + mz - \omega t). \tag{12.7.5}$$

The quantity ρ_0 in (12.7.4) is the undisturbed density. An alternative description of the wave can be obtained by using the log-pressure coordinate z_* in place of z. For small perturbations, the only difference in the formula is that z is replaced by z_*, N is replaced by N_* [see (6.17.25) and (6.17.24)], and that p'/ρ_0 is replaced by Φ'' [see (6.17.17)]. Thus if Φ_0 is the amplitude of geopotential variations on a pressure

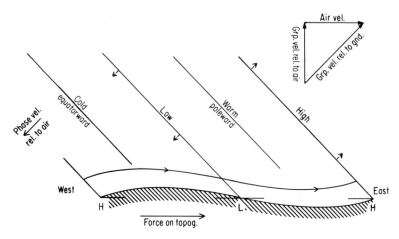

Fig. 12.10.   East–West vertical section through a planetary wave with upward group velocity. Phase lines are drawn a quarter of a wavelength apart. "High" indicates a line of maximum pressure on a level surface or a maximum of geopotential anomaly on a pressure surface. "Low" indicates the opposite phase. "Warm" indicates where the temperature perturbation is a maximum ("Cold," the opposite), and "Poleward" indicates where the meridional velocity is a maximum. The small dashed arrows indicate the ageostrophic motion (which is relative to the medium). Descent along the "Low" corresponds to warming, so the whole pattern moves westward, relative to the medium. The diagram can also be interpreted as a standing planetary wave in which the air is moving from left to right (westerly flow) at a speed that exactly balances the westward phase velocity. The corresponding flow over a sinusoidal topography is shown at the bottom along with a neighboring particle trajectory. The pressure is high (H) on the upstream side and low (L) on the downstream side, so the air exerts a force on the topography in the direction shown. The diagram is drawn with vertical exaggeration N/f for the case in which phase lines then appear to have 45° slope. The group velocity for this case is vertically upward relative to the air, whereas relative to the ground it is at right angles to the phase lines when the exaggerated coordinate system is used. It can be seen from looking at the vertical distance between streamlines that vortex lines are stretched most on the line marked "High" and this stretching gives a cyclonic contribution to their relative vorticity here. However, particles also have their maximum poleward displacement on this line, and the beta effect results in an anticyclonic contribution to stretching. Thus the relative vorticity is anticyclonic on the line marked "High," as can be seen from the meridional velocity field.

surface, p_0 is replaced by

$$p_0 = \rho_0 \Phi_0. \qquad (12.7.6)$$

There is a divergence in the plane of the wave (i.e., $\partial u/\partial x + \partial w/\partial z > 0$) away from the "warm" phase line in Fig. 12.10 that may seem peculiar, but this is in fact balanced by convergence in the meridional velocity ($\partial v/\partial y < 0$) due to the variation of f with latitude, as can be seen from (12.7.4). An important property of the upward-propagating wave is that warm air is carried poleward and cold air equatorward, i.e., there is an apparent net poleward transport of heat. From (12.7.4) and (12.7.5), the corresponding buoyancy flux is given, to first order, by

$$\overline{v\rho'} = -kmp_0^2/2g\rho_0 f_0 = -km\rho_0 \Phi_0^2/2gf_0. \qquad (12.7.7)$$

[For a proper interpretation of this flux, account must be taken of changes in the mean flow to the same order of approximation, i.e., to second order in perturbation amplitude—see McIntyre (1980) and Matsuno (1980).] Such buoyancy fluxes can play an important role in the atmosphere, e.g., in the phenomenon known as a *sudden stratospheric warming*, which occurs in winter and can be driven by upward-propagating planetary waves [see, e.g., Matsuno (1971) and Holton (1980b)]. In a major event of this type, the temperature at the 10-mb (about 31 km) level at the North Pole may increase by 40–60 K in less than 1 week. The near-zonal flow is strongly distorted by the waves that produce the warming, and the associated increase in thickness (see Section 7.7) between 10-mb and lower isobaric surfaces means that the height of the 10-mb surface increases by perhaps 3 km. Reviews of observations of this phenomenon are given by McInturff (1978) and Schoeberl (1978).

The study of topographically generated waves, begun in Chapter 6 for small scales and continued in Chapter 8 for moderate scales, can now be carried to larger topographic scales k^{-1} that correspond to lower encounter frequencies

$$\omega = -Uk \tag{12.7.8}$$

for an observer traveling with the mean flow at speed U. If this flow is uniform, it was found in Chapter 8 that disturbances are trapped (evanescent) at scales k^{-1} larger than that given by

$$-\omega = Uk = f, \tag{12.7.9}$$

i.e., for scales greater than about 100 km. This is because gravity waves do not exist at such frequencies.

If, however, the scale of the topography is further increased, thereby reducing the encounter frequency to levels at which variations with latitude of the Coriolis parameter become important in the dynamics, the situation is changed once again because planetary waves may now be possible. The variation in the vertical is given by the wavenumber m, the expression for which comes from substituting (12.7.8) in (12.7.1) to give

$$m^2 = (N/f)^2 (U^{-1}\beta - k^2 - l^2). \tag{12.7.10}$$

This may be compared with the approximation in Section 8.8 and Table 8.1 for the f-plane quasi-geostrophic regime [regime (v)]. From the discussion in the introductory section of this chapter, it is clear that regime (v) occupies the "spectral gap," defined by

$$|U/f| \ll k^{-1} \ll |U/\beta|^{1/2}, \tag{12.7.11}$$

and there is a new regime

(vi) *beta-plane quasi-geostrophic regime*

for k^{-1} of order $|U/\beta|^{1/2}$. This is about 1000 km for the atmosphere, i.e., the scale of the major topographic features of the earth's surface (these are listed in Table 12.1), so the response to these features falls within regime (vi). The corresponding scale for the ocean is 30–100 km.

In the new regime, there is a major asymmetry [see Charney and Drazin (1961)] between eastward ($U > 0$) and westward directions of the undisturbed flow. Westward currents (or easterly winds) are in the same direction as the phase propagation of planetary waves, so stationary waves are *not* possible. Consequently disturbances

TABLE 12.1

Dominant Topographic Features of the Land Surface[a]

	Height (m)	Horizontal dimensions (km)	Center
Antarctic	3500	5000	90°S
Himalayas	4000	3000	30°N, 90°E
Andes	2500	2700 × 7000	20°S, 70°W
Rockies	2000	2000 × 6000	40°N, 110°W
Greenland	2000	2000	70°N, 40°W
South and East Africa	1000	2000 × 4000	20°S, 30°E

[a] The height and dimensions correspond to those obtained after smoothing on a scale of about 250 km [based on Hoskins (1980)].

remain evanescent no matter how small the wavenumber. The vertical *e*-folding scale has its maximum value [see (12.7.10)] for the longest horizontal waves, this value being

$$|f/N|(|U|/\beta)^{1/2}. \tag{12.7.12}$$

This quantity has values of 5–10 km for the atmosphere in mid-latitudes and 1–3 km for the deep ocean.

For westerly winds or eastward currents ($U > 0$), however, the *e*-folding scale of evanescent waves becomes larger as $k^2 + l^2 \equiv \kappa_H^2$ decreases until κ_H^{-1} reaches the value

$$(U/\beta)^{1/2}. \tag{12.7.13}$$

For larger horizontal scales, propagating waves are possible, and since these must have upward group velocity when the forcing is from below, the structure of the wave is as depicted in Fig. 12.10. The significance of this particular scale can be seen from the *vorticity* balance for a material column. If δy denotes the northward displacement of a column from its mean latitude, the change in *planetary* vorticity is $\beta \, \delta y$. The northward velocity is equal to the rate of change $U \, \partial(\delta y)/\partial x$ of δy, so by geostrophy, it is related to the perturbation geopotential Φ'' by

$$\Phi'' = f_0 U \, \delta y. \tag{12.7.14}$$

Since by (12.2.15) the *relative* vorticity $\partial v/\partial x - \partial u/\partial y$ is equal to the Laplacian of $p'/\rho_0 f_0 = \Phi''/f_0$, the change in *total* vorticity of the wave is given by

$$\beta \, \delta y + (\partial^2/\partial x^2 + \partial^2/\partial y^2)\Phi''/f_0 = (U^{-1}\beta - \kappa_H^2)\Phi''/f_0, \tag{12.7.15}$$

where the last expression makes use of (12.7.14) and the assumed wavelike form (12.7.3). Thus *poleward* displacement gives *cyclonic* total vorticity perturbation [i.e., the right-hand side of (12.7.15) is positive] *only* when the horizontal scale κ_H^{-1} is greater than the value given by (12.7.13).

The conservation of *potential vorticity* [see (7.10.9)] of a material column can be used to relate Φ'' to the field of vertical displacement *h*. For if δz is the height of a column in the undisturbed state, it will be $\delta z + (\partial h/\partial z)\,\delta z$ in the disturbed state, and consequently the fractional change is $\partial h/\partial z$. The conservation principle requires that this be equal to the fractional change in total vorticity, which is given by (12.7.15) over f_0, and hence

$$(U^{-1}\beta + \partial^2/\partial x^2 + \partial^2/\partial y^2)\Phi'' = f_0^2\,\partial h_*/\partial z_*. \tag{12.7.16}$$

(The asterisks are included for consistency with Section 6.17, because the log-pressure coordinate is used.) The streamlines at the bottom of Fig. 12.10 show that because of the westward tilt of phase lines with height, material columns are shortest on the eastern or leeward slopes and consequently [by (12.7.16)] have lowest geopotential or pressure perturbations at those locations and also [by (12.7.14)] have their greatest equatorward displacements there. It follows that the fluid exerts a *force* on the topography in the direction of the wind, as found in other cases (see, e.g., Fig. 8.8) in which propagating waves are generated. In other words, there is wave drag on the topography.

The other equation that can be applied to a material particle is the conservation of potential temperature, so upward displaced particles have low *perturbation* potential temperature. The hydrostatic equation therefore requires that the perturbation geopotential (or perturbation pressure) decrease with height over the mountain, and Fig. 12.10 shows that this indeed occurs when the phase lines tilt westward with height. The equation that expresses the above relationship is (6.17.27), i.e.,

$$N_*^2 h_* + \frac{\partial \Phi''}{\partial z_*} = 0. \tag{12.7.17}$$

If h_* is eliminated from (12.7.16) and (12.7.17) and a wavelike solution is assumed, the dispersion equation (12.7.10) is recovered, thus verifying that conservation of potential temperature and of potential vorticity contains the whole dynamics.

The above analysis for uniform incompressible flow over topography is very useful for a qualitative understanding of what happens and can be extended to the case of an isolated mountain (Queney, 1977), but it is not sufficiently general as it stands. This is because variations of U with height and latitude cannot be ignored at the scales under consideration, so it is necessary to generalize the quasi-geostrophic equations to include advection by nonuniform flows. This is done in Section 12.8 and is applied to small disturbances in Section 12.9.

12.8 Nonlinear Quasi-geostrophic Flow in Three Dimensions

To understand how the atmosphere and ocean respond to imposed changes, it is necessary to appreciate how continuously stratified fluids on a rotating globe adjust under gravitational forces. The process of acquiring this understanding has been spread over many chapters of this book. The first step was to study adjustment in the absence of rotation, and this was done for the continuously stratified case in Chapter 6 after examining the homogeneous case in Chapter 5. The effects of uniform rotation about a vertical axis were introduced in Chapter 7 and applied to a continuously stratified fluid in Chapter 8. This showed that the analysis, neglecting rotation, is applicable for scales small compared with the Rossby radius and for times small compared with the inertial period (typically meaning length scales of kilometers or less and time scales of hours or less). The theory, assuming uniform rotation can in practice be applied on the rotating globe as well if account is taken only of effects of the vertical component of rotation and if the fractional change of this component is small over the length scale of the disturbance. This theory works well for adjustments on time scales comparable with the inertial period, but there is an upper bound on time scales for which it is applicable.

However, many of the most important atmospheric and oceanic phenomena have time scales longer than the inertial period, so it is extremely important to understand these slow adjustment processes. Not only that, it is also important to have a mathematical approximation to the equations that describe only the slow adjustment processes and not the faster ones. For, as Charney (1948, p. 3) states,

> This extreme generality whereby the equations of motion apply to the entire spectrum of possible motions—to sound waves as well as to cyclone waves— constitutes a serious defect of the equations from the meteorological point of view. It means that the investigator must take into account modifications to the large-scale motions of the atmosphere which are of little meteorological importance and which only serve to make the integration of the equations a virtual impossibility.

It was this problem, in fact, which led to the failure of Richardson's early attempts at numerical forecasting (see Section 7.13) since the forecast was dominated by the rapidly adjusting motions associated with errors in the initial field.

The appropriate equations for describing the slow adjustment process were derived in Section 8.16 for the case of small disturbances on an f plane. They were extended to the beta plane in Sections 11.8 and 12.2 for a single mode, and will now be generalized to include nonlinear effects in three-dimensional flow. The basic concepts, as seen already, stem from an appreciation (Shaw, 1908) that the motion is always close to geostrophic balance, but that departures from geostrophy are very important (Brunt and Douglas, 1928) for determining how the motion develops. The appropriate equation for studying such departures, including nonlinear effects, is (8.16.6), first derived by Hesselberg (1915) and used by Brunt and Douglas (1928) to develop the isallobaric method of determining convergence. The connection of this equation to the potential vorticity equation, which Rossby (1937/1938a) showed to be of such fundamental importance, is discussed in Section 8.16. Charney (1947, 1948) was the first to derive systematically the full set of "quasi-geostrophic" equations for baroclinic motion, taking account of the appropriate scales and making direct use of the potential vorticity equation.

In studying the adjustment of a continuously stratified fluid in Chapters 6 and 8, it was found convenient to reduce the equations to two relationships between vertical velocity w and pressure perturbation p' (or if pressure coordinates are used, between w_* and geopotential anomaly Φ''). The first relationship was derived from the horizontal momentum equations, and was really a relationship between horizontal divergence and pressure. The second was derived from the buoyancy equation and the vertical component of the momentum equation.

The quasi-geostrophic approximation to the *momentum equations* and the corresponding expressions for the *horizontal divergence* have already been found in Section 12.2. However, the two-dimensional version (12.2.28) of the continuity equation must now be replaced by the three-dimensional form, which in the case of an *incompressible* fluid is (12.2.8). To obtain the scaled version of this equation, it must be remembered that only the *ageostrophic* velocity components contribute to the divergence, and these have scale $\epsilon_L v_0$, where ϵ_L is given by (12.2.13). If H is the vertical scale, it follows that the appropriate scaled variables are defined $\left[\text{cf. (12.2.11)}\right]$ by

$$u_a^* = f_0 \tilde{u}_a / \beta L v_0, \qquad v_a^* = f_0 \tilde{v}_a / \beta L v_0, \qquad z^* = z/H, \qquad w^* = f_0 w / \beta H v_0. \quad (12.8.1)$$

It follows that the first approximation to (12.2.8), reverting to the unscaled form and dropping the tilde as in the latter part of Section 12.2, is

$$\partial u_a / \partial x + \partial v_a / \partial y + \partial w / \partial z = 0. \quad (12.8.2)$$

The fact that w is related to the *ageostrophic* velocity by (12.8.2) and therefore has the scale given by (12.8.1) is very significant when considering the first approximation to the advection operator (12.2.6). The term $\mu^2 \tilde{v} \, \partial/\partial y$ is of order v_0/L, but by (12.8.1) the term $w \, \partial/\partial z$ is only of order $\epsilon_L v_0/L$ and thus does not contribute to the first approximation. Therefore D/Dt is approximated by D_g/Dt, as defined by (12.2.22), just as in the two-dimensional case, i.e., it is *only the geostrophic velocity that contributes to advection* to the first order, and vertical advection can be ignored. The equation for $\partial w/\partial z$ can now be found by substituting (12.2.24) and (12.2.25) in (12.8.2). The result is

$$\rho_0 f_0^2 \frac{\partial w}{\partial z} = \beta \frac{\partial p'}{\partial x} + \frac{D_g}{Dt}\left(\frac{\partial^2 p'}{\partial x^2} + \frac{\partial^2 p'}{\partial y^2}\right). \quad (12.8.3)$$

This should be compared with the linear f-plane counterpart (8.4.8) (which, of course, does not contain the beta term), the two-dimensional counterpart (12.2.30), and Eq. (12.7.16), derived in a special case from the potential vorticity equation.

The quasi-geostrophic approximation is essentially an approximation to the momentum equations, and the second equation, relating w and p', is just the one used in earlier chapters that was derived by combining the buoyancy equation with the hydrostatic equation. Its linear form is (6.11.4), and the nonlinear quasi-geostrophic form has $\partial/\partial t$ replaced by D_g/Dt, namely,

$$\rho_0 N^2 w = -D_g(\partial p'/\partial z)/Dt. \quad (12.8.4)$$

In the case in which the buoyancy is solely derived from potential temperature perturbations θ' from the value θ_0 at the level concerned, the equations that give rise to (12.8.4) are the temperature equation

$$D_g\theta'/Dt + (N^2/\alpha'g)w = 0, \quad (12.8.5)$$

where α' is a thermal expansion coefficient, defined in Section 3.7.4 (it is equal to θ_0^{-1} for an ideal gas), and the hydrostatic equation

$$\partial p'/\partial z = g\alpha'\rho_0\theta'. \quad (12.8.6)$$

A *single equation* for p' can now be obtained by eliminating w from (12.8.3) and (12.8.4). The result

$$\frac{D_g}{Dt}\left(\frac{\partial^2 p'}{\partial x^2} + \frac{\partial^2 p'}{\partial y^2} + \frac{\partial}{\partial z}\left(\frac{f_0^2}{N^2}\frac{\partial p'}{\partial z}\right)\right) + \beta\frac{\partial p'}{\partial x} = 0 \qquad (12.8.7)$$

is called the equation for *quasi-geostrophic potential vorticity*. It differs from the linear f-plane form (8.16.13) by the inclusion of the beta term and the nonlinear terms. Since the nonlinear advection operator (12.2.22) is defined in terms of pressure by (12.2.15), this equation contains only *one* dependent variable, namely, p'.

The quasi-geostrophic form of the *energy equation* is obtained by adding $\rho_0^{-1}N^{-2}\,\partial p'/\partial z$ times (12.8.4) to ρ_0 times (12.2.33). This gives

$$D_gE/Dt + u_a\,\partial p'/\partial x + v_a\,\partial p'/\partial y + w\,\partial p'/\partial z = 0, \qquad (12.8.8)$$

where E is the energy density (cf. Section 6.7), defined by

$$E = \frac{1}{2}\rho_0(u_g^2 + v_g^2) + \frac{1}{2}\frac{g^2\rho'^2}{\rho_0 N^2} = \frac{1}{2\rho_0 f_0^2}\left[\left(\frac{\partial p'}{\partial x}\right)^2 + \left(\frac{\partial p'}{\partial y}\right)^2\right] + \frac{1}{2\rho_0 N^2}\left(\frac{\partial p'}{\partial z}\right)^2, \quad (12.8.9)$$

where ρ' is the density perturbation at a given level. The energy equation can also be derived by multiplying the potential vorticity equation (12.8.7) by p' and integrating by parts. A divergence form of the energy equation can be obtained by adding (12.8.8) and p' times the incompressibility condition (12.8.2). The result

$$\frac{\partial E}{\partial t} + \frac{\partial}{\partial x}(u_a p' + u_g E) + \frac{\partial}{\partial y}(v_a p' + v_g E) + \frac{\partial}{\partial z}(wp') = 0 \qquad (12.8.10)$$

shows that the energy flux has one part associated with products of the pressure perturbation and the *ageostrophic* part of the flow (which includes the vertical motion) and another part associated with the advection of energy by the *geostrophic* part of the flow.

The equations are easily converted to the form corresponding to use of the log-pressure coordinate z_* in place of the vertical coordinate z. The rules are that z is replaced by z_*, w by w_*, p'/ρ by Φ'' [see (6.17.17)], and N by N_* [see (6.17.25) and (6.17.24)]. Also, effects of *compressibility* are easily included by replacing the incompressibility condition (12.2.8) by the continuity equation (6.17.11). Then (12.8.3) becomes

$$f_0^2\left(\frac{\partial w_*}{\partial z_*} - \frac{w_*}{H_s}\right) = \beta\frac{\partial \Phi''}{\partial x} + \frac{D_g}{Dt}\left(\frac{\partial^2\Phi''}{\partial x^2} + \frac{\partial^2\Phi''}{\partial y^2}\right), \qquad (12.8.11)$$

the buoyancy equation (12.8.4) has the form

$$N_*^2 w_* + D_g(\partial\Phi''/\partial z_*)/Dt = 0, \qquad (12.8.12)$$

and the quasi-geostrophic potential vorticity equation obtained by eliminating w_* from these two equations can be written

$$D_g q/Dt = 0, \qquad (12.8.13)$$

where q is the *quasi-geostrophic potential vorticity*, defined by

$$q = f_0 + \beta y + \frac{1}{f_0}\left[\frac{\partial^2 \Phi''}{\partial x^2} + \frac{\partial^2 \Phi''}{\partial y^2} + \frac{1}{\rho_*}\frac{\partial}{\partial z_*}\left(\frac{f_0^2}{N_*^2}\rho_*\frac{\partial \Phi''}{\partial z_*}\right)\right]. \quad (12.8.14)$$

The relationship with Ertel's (1942a) potential vorticity Q (from which q must be carefully distinguished), can be seen by writing

$$q \approx \rho_0(Q_0 + Q')/(d\theta_0/dz), \quad (12.8.15)$$

where Q_0 and Q' are defined by (7.12.10) and (7.12.11), and the approximation is in accordance with the scalings given above. The incompressible case can be recovered by taking the limit $H/H_s \to 0$.

There are many circumstances in which the quasi-geostrophic equations can themselves be further simplified. For instance, in the ocean, the baroclinic Rossby radius is only about 30 km, and much of the motion has very much larger scale. If the *horizontal scale L is very much larger than* NH/f_0, i.e., if

$$f_0 L \gg NH, \quad (12.8.16)$$

then the horizontal Laplacian in (12.8.7) is small relative to the vertical derivative and therefore (12.8.7) is approximated by

$$\frac{D_g}{Dt}\frac{\partial}{\partial z}\left(\frac{f_0^2}{N^2}\frac{\partial p'}{\partial z}\right) + \beta\frac{\partial p'}{\partial x} = 0, \quad (12.8.17)$$

which is a time-dependent version of what are called the *thermocline equations* in oceanography. In this approximation, only the beta term contributes to the ageostrophic velocity [as defined in (12.2.24) and (12.2.25)]; changes in relative vorticity have negligible effect on change in potential vorticity, and changes in kinetic energy can be ignored relative to changes in potential energy. The vorticity equation is approximated by the form (11.13.3), which applies to steady flow. This equation does not require the horizontal scale to be small compared with the radius of the earth, and so (12.8.17) can be generalized to a form *valid over the whole sphere* (Phillips, 1963).

The approximate equations for the sphere are as follows. First, the horizontal velocity is related to pressure by the general form of the geostrophic relationship [see (11.2.1) and (11.2.2)], namely,

$$2\Omega\rho r u \sin\varphi = -\partial p/\partial\varphi, \quad 2\Omega\rho r v \sin\varphi = \sec\varphi\,\partial p/\partial\lambda. \quad (12.8.18)$$

Secondly, w is determined from the vorticity equation (11.13.3) or

$$r\,\partial w/\partial z = v\cot\varphi = (\operatorname{cosec}^2\varphi/2\Omega\rho r)\,\partial p/\partial\lambda. \quad (12.8.19)$$

The remaining equation is the buoyancy equation (6.4.2), combined with the hydrostatic equation (3.5.8), to give

$$D(\partial p/\partial z)/Dt = 0, \quad (12.8.20)$$

with the possible addition of diffusion terms on the right-hand side. Solutions of this set of equations are discussed by Veronis (1969, 1981), by Welander (1971), and by Anderson and Killworth (1979). Because the balance is valid over the whole sphere, it can be regarded as a type of geostrophic motion that is distinct from the quasi-geostrophic motion first discussed [see Phillips (1963) for a detailed discussion]. The equations may also be used to make deductions about the vertical velocity and the absolute (rather than relative) horizontal velocity field in the ocean (Killworth, 1980b).

In the atmosphere, the horizontal Laplacian can rarely be neglected compared with the vertical derivative, but the beta term can sometimes be neglected compared with the nonlinear terms, so that a nonlinear f- plane approximation suffices. In other words, beta can be put equal to zero in (12.8.14), (12.8.11), or (12.8.7). The condition for this is that v_0 be large compared with the value (12.2.21) at which nonlinear terms are of the same order as beta terms, i.e., that

$$v_0 \gg \beta L^2 \quad \text{or} \quad L \ll (v_0/\beta)^{1/2}. \tag{12.8.21}$$

The scale $(v_0/\beta)^{1/2}$ is about 1000 km for the atmosphere, so beta effects can certainly be ignored on the scale of fronts, and they often play a secondary role even on the scale of developing cyclonic systems.

12.9 Small Disturbances on a Zonal Flow Varying with Latitude and Height

The disturbances generated by flow over topography were studied in Chapters 7 and 8 for scales small enough for beta effects to be ignored, the illustrative examples being for cases of uniform flow. The corresponding solutions for planetary waves were studied in Section 12.7, where it was found that beta effects become important for scales of order $(|U|/\beta)^{1/2}$, which is around 1000 km for the atmosphere. This is the scale (see Table 12.1) of the major topographic features of the earth, and on this scale the atmospheric flow can hardly be treated as uniform. Therefore it is appropriate to consider the quasi-geostrophic equations for small perturbations on a mean-zonal flow $U(y, z)$ that varies with both latitude and altitude. In other words, the geopotential Φ'' is expressed as the sum of a steady longitudinally independent part [i.e., a function of y and z such that $\partial\Phi''/\partial y = -f_0 U(y, z)$] and an infinitesimal perturbation Φ'. (It would perhaps be more logical to use the notation Φ''', but the single prime is more convenient.) The linearized form of (12.8.13) for quasi-geostrophic potential vorticity is then

$$(\partial/\partial t + U \, \partial/\partial x)q' + v \, \partial\bar{q}/\partial y = 0, \tag{12.9.1}$$

where by (12.8.14) and the geostrophic equation, the perturbation potential vorticity q' is given by

$$q' = \frac{1}{f_0}\left(\frac{\partial^2\Phi'}{\partial x^2} + \frac{\partial^2\Phi'}{\partial y^2} + \frac{1}{\rho_*}\frac{\partial}{\partial z_*}\left(\frac{f_0^2}{N_*^2}\rho_*\frac{\partial\Phi'}{\partial z_*}\right)\right), \tag{12.9.2}$$

the perturbation meridional velocity v is given by

$$v = f_0^{-1} \, \partial\Phi'/\partial x, \tag{12.9.3}$$

and $\partial\bar{q}/\partial y$, the latitudinal gradient of the mean quasigeostrophic potential vorticity, is given by

$$\frac{\partial \bar{q}}{\partial y} = \beta - \frac{\partial^2 U}{\partial y^2} - \frac{1}{\rho_*} \frac{\partial}{\partial z_*} \left(\frac{f_0^2}{N_*^2} \rho_* \frac{\partial U}{\partial z_*} \right). \tag{12.9.4}$$

The relationship between Φ' and the associated field of vertical motion is obtained from the perturbation forms of (12.8.11) and (12.8.12), which are

$$f_0^2 \left(\frac{\partial w_*}{\partial z_*} - \frac{w_*}{H_s} \right) = \left(\beta - \frac{\partial^2 U}{\partial y^2} \right) \frac{\partial \Phi'}{\partial x} + \left(\frac{\partial}{\partial t} + U \frac{\partial}{\partial x} \right) \left(\frac{\partial^2 \Phi'}{\partial x^2} + \frac{\partial^2 \Phi'}{\partial y^2} \right) \tag{12.9.5}$$

and

$$N_*^2 w_* + (\partial/\partial t + U \, \partial/\partial x) \, \partial \Phi'/\partial z_* - (\partial U/\partial z_*) \, \partial \Phi'/\partial x = 0. \tag{12.9.6}$$

The second equation is used to give the boundary condition at the surface, which, in the case of small-amplitude topographic forcing, is that w_* is given at the mean-surface pressure. The equation is really one expressing conservation of potential temperature, and the new feature (not present in models with uniform flow) is the term representing *horizontal advection* of mean temperature by the perturbation meridional velocity, which can now occur because the mean temperature varies with latitude. The latitudinal gradient of mean temperature is related to $\partial U/\partial z_*$ by the thermal-wind equation (7.7.10). The new term can give rise to important new effects that do not occur in the uniform-flow case, and such effects will be studied in Chapter 13.

If the disturbance has wavelike form, Φ' can be expressed in the form

$$\Phi' = \psi \exp(z_*/2H_s + ily + ik(x - ct)) \tag{12.9.7}$$

and therefore (12.9.1)–(12.9.3) reduce to

$$\partial^2 \psi / \partial z_*^2 + m^2 \psi = 0, \tag{12.9.8}$$

where

$$m^2 = (N_*/f_0)^2((U - c)^{-1} \, \partial \bar{q}/\partial y - \kappa_H^2) - (2H_s)^{-2} \tag{12.9.9}$$

and

$$\kappa_H^2 = k^2 + l^2 \tag{12.9.10}$$

is the square of the horizontal wavenumber. The sinusoidal variation with latitude can be assumed, of course, only if the fractional change of m^2 (cf. Section 9.12) over a meridional distance l^{-1} is small.

The important information about wave disturbances is contained in the coefficient m^2, which determines the *transmission characteristics* of the atmosphere, and effects that can be produced by variations of m with height have already been studied in Chapters 6 and 8. Application to quasi-geostrophic waves in the atmosphere was first made by Charney and Drazin (1961), who explained why stationary waves (i.e., ones with $c = 0$) propagate into the middle atmosphere in the winter hemisphere but not in the summer hemisphere. The explanation comes from an examination of the zonal wind profiles shown in Fig. 12.11. The temperature pattern in the middle atmosphere is largely determined by a radiational balance, and the corresponding thermal wind gives easterlies in the summer hemisphere and westerlies in the winter

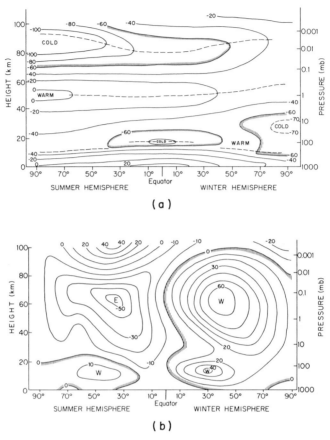

Fig. 12.11. Meridional section showing the longitudinally averaged (a) temperature (°C) and (b) wind speed (m s^{-1}) up to 100 km at the time of the solstices. Dashed lines in (a) indicate the positions of the tropopause, stratopause, and mesopause. (Courtesy of R. K. Reed.) [From Wallace and Hobbs (1977, Figs. 1.10 and 1.12).]

hemisphere. On the other hand, the tropospheric mid-latitude winds tend to be westerly in both hemispheres. This is the direction for which, in uniform winds at least, sufficiently long stationary waves can propagate vertically, and the same is generally true (i.e., $m^2 > 0$) for real conditions in which U is nonuniform. However, Eq. (12.9.9) shows that m^2 changes sign by going from positive infinity to negative infinity at the level where $U = c$, i.e., in the case of stationary disturbances, where $U = 0$. Near such critical levels (see Section 8.9.4) waves tend to be absorbed according to linear theory, although the possibility has been raised that nonlinear effects can cause partial reflection (or overreflection) at such levels instead [see, e.g., Warn and Warn (1978) and Stewartson (1978)]. In any case, since m^2 is mainly negative in the easterly regime, waves cannot propagate and wave energy can be transferred only by "tunneling" (see Section 6.9; the requirement for significant transfer is that the reduction in amplitude across the evanescent region is not too large). A striking verification of the lack of waves in the summer stratosphere is a picture due to Hare (1968) [reproduced by Holton (1975)] of the circulation at 30 mb.

It is so close to zonal that it is hardly worth reproducing! The opposite is true of the winter stratosphere, in which departures from zonal symmetry are often large [see, e.g., van Loon *et al.* (1973)]. Figure 12.12 is another illustration of the effect. This shows variation with latitude and pressure of the amplitude of the wavenumber 1 component of the perturbation of height of pressure surfaces for January and July. In January, the amplitudes above 100 mb are large and increase with altitude (the increase is to be expected because of the reduction in density—see, e.g., Sections 6.14 and 6.17), whereas in July there is a big reduction in amplitude just above 100 mb. This is consistent with the switch from westerly to easterly flow at such levels (see Fig. 12.11).

Near the levels where m^2 goes through infinity, friction is important because of the short vertical wavelength (see Section 8.9). Consequently, waves are dissipated and there are effects on the mean flow like those discussed in Section 8.15. A useful way of examining these meridional fluxes of momentum and heat [see (12.3.11) and (12.7.7)] comes from multiplying (12.9.1) by q' to give

$$(\partial/\partial t + U\,\partial/\partial x)(\tfrac{1}{2}q'^2) + q'v\,\partial\bar{q}/\partial y = 0. \tag{12.9.11}$$

Averaging with respect to x over a wavelength and assuming that a wavelike form of solution (12.9.7) (with steady amplitude) is appropriate, it follows that if $\partial\bar{q}/\partial y$ is nonzero,

$$\overline{q'v} = 0. \tag{12.9.12}$$

In other words, if the wave amplitude is not changing significantly with time and if friction and mixing effects are negligible [as is assumed in the derivation of (12.9.1)], then the meridional flux of quasi-geostrophic potential vorticity is zero. If now the expressions (12.9.2) and (12.9.3) for q' and v are substituted in (12.9.12), and (12.9.12) is integrated by parts, it follows that (12.9.12) can be written in divergence form as

$$\rho_*\overline{q'v} \equiv \frac{\partial}{\partial y}(\mathscr{F}_y) + \frac{\partial}{\partial z_*}(\mathscr{F}_z) = 0, \tag{12.9.13}$$

where $\mathscr{F}_y$ and $\mathscr{F}_z$ are the components of what is known as the quasi-geostrophic approximation to the Eliassen–Palm flux (Eliassen and Palm, 1961; Edmon *et al.*, 1980), given by

$$\mathscr{F}_y = \frac{\rho_*}{f_0^2}\overline{\frac{\partial\Phi'}{\partial x}\frac{\partial\Phi'}{\partial y}} = -\rho_*\overline{u'v},$$

$$\mathscr{F}_z = \frac{\rho_*}{N_*^2}\overline{\frac{\partial\Phi'}{\partial x}\frac{\partial\Phi'}{\partial z_*}} = \alpha_*\frac{f_0 g\rho_*}{N_*^2}\overline{v\theta'}. \tag{12.9.14}$$

The expressions at the extreme right make use of the geostrophic relationship [(12.9.3) and a similar equation for u'] and the expression (6.17.20) for potential temperature. In the neighborhood of any locality, the wave approximately has the form (12.7.3) and (12.7.4), and substitution in (12.9.14) and comparison with (12.3.4) and (12.7.2) show that the Eliassen–Palm flux is in the direction of the wave group velocity projected onto the yz_* plane. If conditions are steady and dissipation-free, as assumed in the

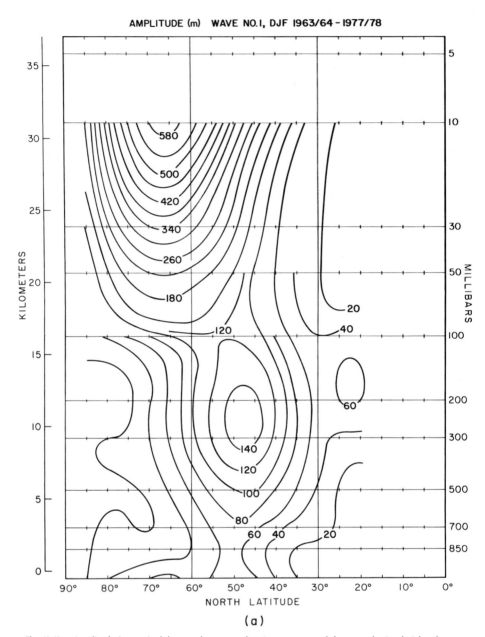

Fig. 12.12.   Amplitude (meters) of the zonal wavenumber 1 component of the perturbation height of pressure surfaces in the northern hemisphere in (a) December–February and (b) July. [From van Loon *et al.* (1973), *J. Geophys. Res* **78** (Figs. 2 and 5); copyrighted by the American Geophysical Union.]

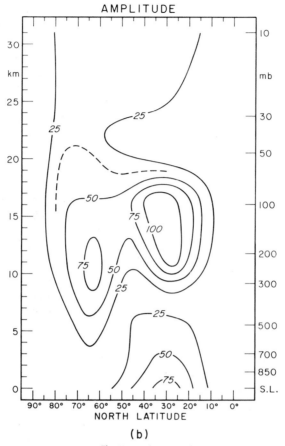

Fig. 12.12. (continued)

derivation of (12.9.13), the flux is divergence-free and there is no effect of the waves on the mean flow [Charney and Drazin (1961): The result is known as the Charney–Drazin nonacceleration theorem and also follows from the exact equations of Andrews and McIntyre (1978a)]. In practice, there is divergence due to nonstationarity and dissipation, so calculations of the flux can be useful in assessing where this divergence and its consequent effects on the mean flow take place. For instance, Dunkerton *et al.* (1981) have used the method to study a stratospheric warming that occurred in a model simulation, and to examine the extent to which critical lines do act as reflectors in the model.

In order to further study stationary waves ($c = 0$), it is useful to write (12.9.9) in the form

$$f_0^2 m^2 / N^2 = \kappa_L^2 - \kappa_H^2, \tag{12.9.15}$$

where $\kappa_L(y, z_*)$ is a known property of the mean flow [cf. Matsuno (1970)], given by

$$\kappa_L^2 = \frac{1}{U}\frac{\partial \bar{q}}{\partial y} - \left(\frac{f_0}{2N_* H_s}\right)^2 = \frac{1}{U}\left(\beta - \frac{\partial^2 U}{\partial y^2} - \frac{f_0^2}{N_*^2}\frac{1}{\rho_*^{1/2}}\frac{\partial^2}{\partial z_*^2}(\rho_*^{1/2}U)\right). \tag{12.9.16}$$

There are two **interpretations** of κ_L that follow from (12.9.15). First, it determines the value of m for very long wave disturbances ($\kappa_H \rightarrow 0$). Second, it gives the limiting value of the horizontal wavenumber above which waves are evanescent. The types of behavior possible when κ_L^2, and hence m^2, is a function of height only were discussed in Chapters 6 and 8, and various investigators [e.g., Tung and Lindzen (1979) and Tung (1979)] have made calculations using profiles of κ_L^2 appropriate to a particular latitude, such as 45°N during mean winter conditions. Then κ_L^{-1} is about 800 km in the troposphere and increases to a value of about 1500 km between altitudes of 15 and 70 km. Consequently, waves with κ_H^{-1} between 800 and 1500 km would be expected to reflect (see Chapters 6 and 8) at about the 15-km level, whereas waves with κ_H^{-1} over 1500 km would be expected to penetrate to 70 km. Above that level, the mean velocity goes through zero, so waves may be absorbed there. For circumstances in which waves reflect (see Section 6.9), resonances are possible, and Tung and Lindzen (1979) suggest these may be important for the *blocking* phenomenon.

In practice, the variations of κ_L with both latitude and height are important, and it is useful to use ray-tracing techniques (see Section 8.12) to explain the nature of the response. **Karoly and Hoskins (1982)** have shown that rays are refracted toward the direction of the gradient of κ_L, and hence a region of maximum κ_L acts as a waveguide, whereas rays tend to avoid regions of low κ_L. For mean wintertime conditions, the effect on waves of zonal wavenumber 1 produced in the troposphere at 60°N is to confine them to two main paths. One set of waves, with more nearly vertical initial propagation, reach levels of 40–50 km before refracting toward the equator and being absorbed near 20° latitude, where the westerlies are starting to give way to easterlies. The other set, with more nearly horizontal initial direction, are trapped in the troposphere and propagate equatorward to a latitude of about 10°.

Stationary waves can be generated not only by topography [the effects of which on planetary-scale motions were first studied by Charney and Eliassen (1949)], but also by diabatic heating (Smagorinsky, 1953), the effects of which can be calculated by adding a buoyancy forcing term to (12.8.12) as in (9.13.5). Stationary waves can also be forced by transient disturbances, tending to occur in preferred locations (storm tracks), which can give rise to a convergence of vorticity [thus adding a forcing term to (12.8.11)] and of heat [thus adding another forcing term to (12.8.12)]. Calculations of these flux convergences have been made by Lau (1979a) and indicate that they are not insignificant. It is generally thought [see reviews by R. B. Smith (1979) and Dickinson (1980)] that heating and topography are about equally important in producing the observed stationary waves, whose structure in the wintertime northern hemisphere is shown in Fig. 12.13. Similar pictures for summer conditions have been calculated by White (1982). The stationary waves make an important contribution to the zonally averaged heat and momentum budgets, values of the fluxes having been computed by Lau (1979b) and White (1982). The westward phase tilt, with height seen in Fig. 12.13, indicates a substantial poleward heat flux (see Section 12.7) in winter; the flux is, in fact, as much as that due to transient eddies above 500 mb, but the transient eddies' contribution becomes rather larger near the surface. In summer, the stationary-wave heat flux is very small. Stationary waves also have a westward momentum flux (see Section 12.3) into the westward-moving jet stream, the flux they carry being about half that of transient eddies. Wintertime fluxes are about twice as big as those in summer.

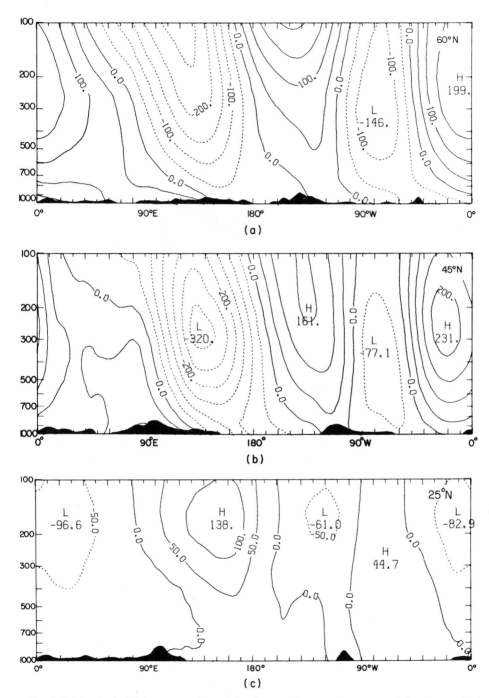

Fig. 12.13.   Longitude–height cross sections of the departure from zonal symmetry of the time-averaged geopotential height field taken along (a) 60°N, (b) 45°N, and (c) 25°N. The contour interval is 50 m, and local orography is depicted at the bottom of the figures. The horizontal pattern can be seen in Fig. 7.8. The associated departure from zonal symmetry of the time-averaged meridional velocity at 300 mb is shown in (d). The contour interval is 2 m s^{-1}. [From Lau (1979b, Figs. 2a–c and 10b).]

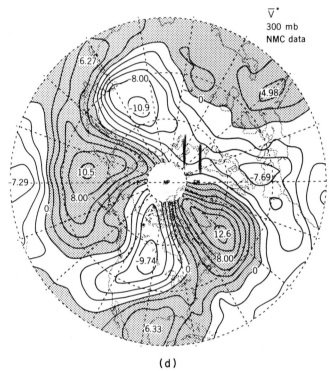

$\overline{V}^{\bullet}$
300 mb
NMC data

(d)

Fig. 12.13. (*continued*)

The vertical structure of the observed stationary wave field is very interesting. In extratropical latitudes (typified by the sections at 45°N and 60°N in Fig. 12.13), the structure has a "barotropic" character, with perturbations having the same sign at all levels shown (up to 100 mb). In the tropics, by contrast, the structure is baroclinic, with the sign of the perturbation at upper levels (say, 200 mb) generally opposite to that at lower levels. The "barotropic" nature of the disturbances in extratropical regions indicates the usefulness of studying the horizontal propagation characteristics of barotropic disturbances. Putting $m = 0$ in (12.9.15) and using the definition (12.9.10), it follows that changes of meridional wavenumber l with latitude are given by

$$l^2 = \kappa_L^2 - k^2, \tag{12.9.17}$$

with the limiting wavenumber κ_L being defined by (12.9.16), where for consistency the variations of U with z should be ignored. Using the 300 mb winds, Hoskins and Karoly (1981) found large values of κ_L, corresponding to zonal wavenumber 7 on the equatorward side of the westerly jet centered at 30°N, with a rapid change to smaller values, corresponding to zonal wavenumbers 3 to 4 on the poleward side of the jet. Consequently, waves propagating through the westerly regime from the tropics tend to split into two groups, with the higher wavenumbers (4–7) confined to the southern side of the jet, whereas the lower wavenumbers (1–3) penetrate to high latitudes, in a time of the order of 1 week. Behavior of perturbations to the winter zonal flow in a five-layer model was found to be similar, and Fig. 12.14a shows the pattern of height perturbation at 300 mb that was produced by a deep elliptical heat source (shaded

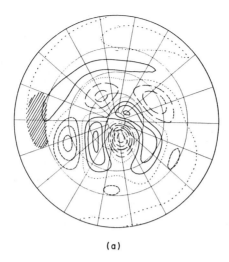

(a)

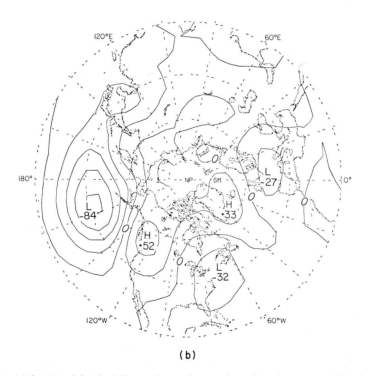

(b)

Fig. 12.14. (a) The 300 mb height field perturbation due to a deep elliptic heat source (shaded area) at 15°N perturbing the northern hemisphere winter zonal flow. The result was obtained by using a linear five-layer numerical model. Negative contours are dashed, the zero contour is dotted, and positive contours are continuous. [From Hoskins and Karoly (1981, Fig. 3c).] (b) Composite anomaly map of 700-mb height for five winters that were cold in the eastern United States and had 700-mb ridges over western Canada, namely, 1960–1961, 1962–1963, 1967–1968, 1969–1970, and 1976–1977. The pattern is known as the Pacific–North American Pattern. The contour interval is 20 m. Note that the outer edge is at 20°N, not at the equator as in (a). [From Wallace and Gutzler (1981).]

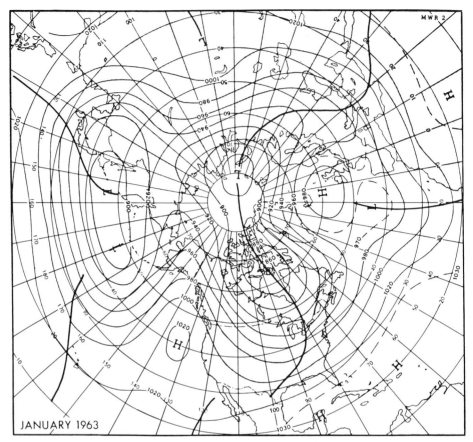

Fig. 12.15. Average contours of the 700-mb surface (tens of feet) for January 1963. Troughs (heavy solid lines) indicate minimum latitudes of contours. Outstanding features were the strong blocking patterns associated with the highs in the eastern Atlantic and eastern Pacific. [From O'Connor (1963).]

area) centered at 15°N. The dominant pattern is that corresponding to the low wave-number paths, and high energies are found near the limiting latitudes (where $l = 0$), as predicted by Liouville–Green or WKBJ theory (Section 8.12). Similar patterns have been identified in the atmosphere by Wallace and Gutzler (1981), and Fig. 12.14b shows an example of such a pattern. There are two patterns of this type—the "Pacific North American" pattern, which appears to originate from the Southeast Asian region of deep convection, and the "West Atlantic" pattern, which appears to originate in the convective zone of northern South America. It appears that the zones produce two types of response: a baroclinic one confined to the tropics (discussed in Chapter 11) and a predominantly barotropic one at higher latitudes.

   Another quasi-stationary pattern of great interest is the so-called "blocking" pattern associated with weather conditions that are unusually persistent [see, e.g., Rex (1950)]. An example of such a pattern is shown in Fig. 12.15, which shows the time-mean height of the 700 mb surface in January 1963, a month of exceptional cold in North America, Europe, and the Far East. Note the huge meridional excursions of the isobars, especially those associated with the strong ridges over the East Pacific and East Atlantic. Studies of forcing of barotropic flow by topography (Charney and DeVore, 1979; Hart, 1979; Davey, 1980) indicate that it is possible to have two stable response patterns for the same forcing. In one, the forcing produces a strong zonal flow, which is so fast that standing waves are not produced by the topography (U/β is larger than the square of the inverse wavenumber). Consequently, there is little wave drag, and so the strong flow can persist. The alternative pattern has the flow slightly subcritical, i.e., just below the value at which standing waves are possible. This produces a very high drag that keeps the flow subcritical. However, changes from one pattern to the other can be produced by changes of forcing and by random internal changes (i.e., effects of other disturbances), and such changes are found to occur in practice. The second (subcritical) pattern is the one similar to observed blocking patterns, and these are important because they can be responsible for persistent, strongly anomalous conditions, e.g., the very cold winter of 1963 (Fig. 12.15) in which the January mean temperature in Warsaw was 10 K below normal (O'Connor, 1963).

12.10   Deductions about Vertical Motion from the Quasi-geostrophic Equations

For weather forecasting, it is vital to be able to make deductions about the field of *vertical motion*, which cannot be measured directly, because upward motion produces precipitation and plays an important role in the development of fronts, etc. One method of doing this has already been discussed in Section 8.16. This is based on the realization by Brunt and Douglas (1928) that the field of horizontal convergence in quasi-geostrophic flow is related to the *isallobaric wind*, i.e., the wind that flows down the gradient of the rate of change of pressure [see (8.16.7)]. This gives rise to a system whereby charts of rate of change of pressure were prepared, contours being called isallobars, and convergence would be associated with flow into an "isallobaric low."

Since taking the convergence of Brunt and Douglas's equation gives rise to the quasi-geostrophic vorticity equation (12.8.3) or (12.8.11), their method is equivalent to using that equation to deduce w from the pressure field. However, there are *two* equations relating the vertical velocity w to the geopotential perturbation Φ'' (or pressure perturbation p') that were derived in Section 12.8, namely, the vorticity equation and the buoyancy equation [(12.8.4) or (12.8.12)]. Either equation, or a combination of the two, could be used to obtain information about w. The combination that is particularly useful is the one that does not involve time derivatives, for then deductions can be made from the field of geopotential at a fixed time. It is convenient to use the compressible form in the following since the incompressible equivalent is easily deduced by letting the scale height H_s tend to infinity.

The form without time derivatives is obtained by adding the derivative of (12.8.11) with respect to z_* to the horizontal Laplacian of (12.8.12). The result is

$$f_0^2 \frac{\partial}{\partial z_*}\left(\frac{\partial w_*}{\partial z_*} - \frac{w_*}{H_s}\right) + N_*^2\left(\frac{\partial^2 w_*}{\partial x^2} + \frac{\partial^2 w_*}{\partial y^2}\right) = \beta \frac{\partial^2 \Phi''}{\partial x\,\partial z_*} + \text{NL}, \quad (12.10.1)$$

where NL represents the contribution of the nonlinear terms. There are several ways of writing the nonlinear terms (Hoskins *et al.*, 1978), and one of these will be deduced in the following by a different method. Equation (12.10.1) is known as the "*omega equation*" because of the use of the symbol omega (cf. Section 6.17) to denote the equivalent of vertical velocity in pressure coordinates, and this is an equation for that quantity. An approximate version of this equation was first developed by Sutcliffe (1947), who wanted a method that took into account differences in rate of divergence at different levels. This was wanted because many highly baroclinic situations were missed through looking at the surface isallobaric field alone, and so Sutcliffe's version of (12.10.1) consisted of a two-level approximation.

A more illuminating approach (Hoskins *et al.*, 1978) is to eliminate the $\partial/\partial t$ terms from the quasi-geostrophic momentum equation and the buoyancy equation directly. Thus if f_0^{-1} times the z_* derivatives of (12.2.24) and (12.2.25) [expressed in terms of Φ'', using (6.17.17)] are subtracted, respectively, from the x and y derivatives of (12.8.12), the result has the form

$$N_*^2 \frac{\partial w_*}{\partial x} - f_0^2 \frac{\partial u_a}{\partial z_*} = 2Q_x - \beta y \frac{\partial^2 \Phi''}{\partial z_* \, \partial y}, \tag{12.10.2}$$

$$N_*^2 \frac{\partial w_*}{\partial y} - f_0^2 \frac{\partial v_a}{\partial z_*} = 2Q_y + \beta y \frac{\partial^2 \Phi''}{\partial z_* \, \partial x}, \tag{12.10.3}$$

where

$$Q_x = \frac{1}{f_0}\left(\frac{\partial^2 \Phi''}{\partial x \, \partial y}\frac{\partial^2 \Phi''}{\partial z_* \, \partial x} - \frac{\partial^2 \Phi''}{\partial x^2}\frac{\partial^2 \Phi''}{\partial z_* \, \partial y}\right) = -\alpha_* g\left(-\frac{\partial v_g}{\partial y}\frac{\partial \theta''}{\partial x} + \frac{\partial v_g}{\partial x}\frac{\partial \theta''}{\partial y}\right), \tag{12.10.4}$$

$$Q_y = \frac{1}{f_0}\left(\frac{\partial^2 \Phi''}{\partial y^2}\frac{\partial^2 \Phi''}{\partial z_* \, \partial x} - \frac{\partial^2 \Phi''}{\partial x \, \partial y}\frac{\partial^2 \Phi''}{\partial z_* \, \partial y}\right) = -\alpha_* g\left(\frac{\partial u_g}{\partial y}\frac{\partial \theta''}{\partial x} - \frac{\partial u_g}{\partial x}\frac{\partial \theta''}{\partial y}\right). \tag{12.10.5}$$

The hydrostatic equation has been used in the form (6.17.20). An interesting aspect of the derivation is that there is a contribution of Q_x to the right-hand side of (12.10.2) from *each* of the original equations. Another interesting property is that the left-hand sides of (12.10.2) and (12.10.3) have the form of horizontal components of vorticity if the scaled vertical coordinate z_s (and a corresponding vertical velocity), defined in (8.8.25), is used. When the x derivative of (12.10.2) is added to the y derivative of (12.10.3), (12.10.1) is obtained with

$$NL = 2(\partial Q_x/\partial x + \partial Q_y/\partial y). \tag{12.10.6}$$

The vector $\mathbf{Q}$ can be calculated if the geopotential and temperature are given on a pressure surface, and the divergence of this vector field gives a clear indication of where rising motion will take place, with the vectors $\mathbf{Q}$ pointing in toward a region of rising motion, as has been demonstrated by Hoskins and Pedder (1980). If s is a coordinate measuring distance along an isotherm (see Section 7.10), and n is the normal coordinate such that s, n, and the vertical form a right-handed system, then $\mathbf{Q}$ is given by the simple formula

$$\mathbf{Q} = (\mathbf{k} \times \partial\mathbf{v_g}/\partial s)g\alpha_* \, \partial\theta''/\partial n, \tag{12.10.7}$$

where **k** is the unit vector pointing vertically upward. Thus **Q** is proportional to the rate of change of velocity along isotherms.

Two model situations (northern hemisphere) are shown in Fig. 12.16. In both, θ'' is decreasing uniformly poleward, i.e.,

$$\theta'' = -Gy. \tag{12.10.8}$$

In the first case, it is the flow *normal* to the isotherms that is varying along the

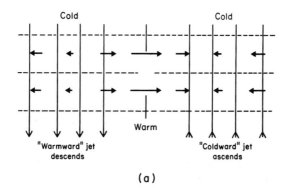

(a)

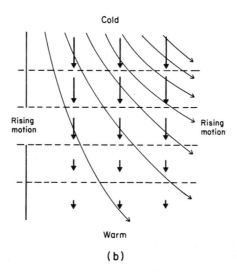

(b)

Fig. 12.16. **Q**-vectors used to deduce vertical motion due to the nonlinear part of the ageostrophic velocity. The vertical motion is associated with the divergence of the **Q**-vector field, with the **Q**-vectors tending to point *toward* regions of rising motion. There are two cases, each with isotherms (dashed line contours) running horizontally across the page, the horizontal temperature gradient being uniform. Solid contours represent geopotential height, and arrows on these contours show the direction of motion (northern hemisphere). In case (a) there is a trough normal to the isotherms, with associated jets running parallel to the trough. The **Q**-vectors (thick arrows) are parallel to the isotherms and point toward the "coldward" jet, which will therefore be ascending. Conversely, the "warmward" jet will be descending. This behavior is typical of a developing cyclone. In case (b) there is a diverging "warmward" jet, with velocity normal to each isotherm independent of distance along the isotherm. The **Q**-vectors (thick arrows) in this case are normal to the isotherms and point toward an ascending zone that runs parallel to the isotherms as shown. This sort of behavior is encountered at cold fronts.

isotherms, thus producing ageostrophic motion. In the second example, it is the flow *along* the isotherms that is varying along them, thus producing the ageostrophic motion.

Model A.   This example represents a developing cyclone with

$$\Phi'' = -\alpha_* gGyz_* - \Phi_0 \exp(-x^2/L^2). \tag{12.10.9}$$

In this case, **Q** is in the x direction and is given by

$$Q_x = \alpha_* gG \, \partial v_g/\partial x = 2\alpha_* g f_0^{-1} L^{-2} G\Phi_0(1 - 2x^2/L^2) \exp(-x^2/L^2). \tag{12.10.10}$$

The vectors point toward the region of maximum "coldward" velocity, and this is where the rising motion takes place, as illustrated in Fig. 12.16. Note that although the horizontal component of motion is "coldward," the vertical component, being upward, is toward higher potential temperature. For the jet to be "coldward" at the actual angle of ascent, and hence to release available potential energy (see Section 7.8), this angle must be less than the slope of the isentropes. Motions that can draw on available potential energy will be discussed in Chapter 13.

Model B.   This example represents a frontogenetic situation with the geostrophic flow tending to cause isotherms to come closer together. Suppose

$$\Phi'' = -\alpha_* gGyz_* - xF(y),$$

so that

$$f_0 u_g = xF'(y), \qquad f_0 v_g = -F(y). \tag{12.10.11}$$

The **Q**-vector is now directed normal to the isotherm and is given by

$$Q_y = -\alpha_* gG \, \partial u_g/\partial x = -\alpha_* g f_0^{-1} GF'(y). \tag{12.10.12}$$

Figure 12.16b illustrates this, with the case

$$F(y) = F_0(y/L + (1 + (y/L)^2)^{1/2}) \tag{12.10.13}$$

corresponding to a decelerating and diverging "warmward" jet. The **Q**-vectors are uniform far upstream, where the motion is one of plain strain, and zero far downstream, where the flows tend to zero. Consequently, the divergence in the **Q**-vector field and the associated rising motion are centered on $y = 0$ (and marked "rising motion" in the figure).

In a real situation, both types of situation arise, e.g., Fig. 12.17a shows the analysis of the 700 mb chart for 0000 GMT on 10 November 1975, at a time when a major storm was developing over the midwest of North America. The position of the surface fronts is indicated and the **Q**-vectors are shown in Fig. 12.17b. A region of rising motion is indicated northeastward of the low, where a strong "coldward" jet is found as in model A. Conversely, there is a region of descent in the "warmward" jet southwestward of the low. This jet diverges at the cold front, as in model B, so rising motion is indicated along the front.

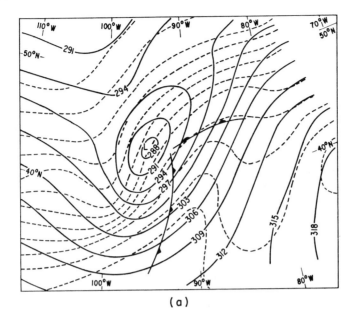

(a)

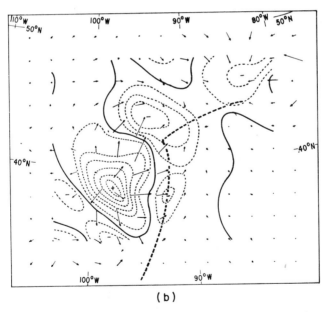

(b)

Fig. 12.17.   (a) Analysis of the 700 mb chart for 0000 GMT on November 10, 1975. Height contours are drawn every 30 dynamic meters and temperature contours every 2°. The surface frontal analysis is indicated. (b) **Q**-vectors (arrows) and contours of their divergence (zero lines marked solid) for the situation shown in (a). [From Hoskins and Pedder (1980).]

Chapter Thirteen

Instabilities,
Fronts,
and the General Circulation

13.1 Introduction

Halley's (1686) idea of the atmospheric circulation, with hot air rising in the tropics and cooler air descending at higher latitudes, was presumably based on experience with nonrotating fluids. In a nonrotating system, however, a zonally symmetric distribution of heating and cooling would give us no eastward or westward motion, and it was in this respect that Halley's scheme was deficient. The importance of rotation was later recognized by Hadley (1735), who showed that the tendency to conserve angular momentum can explain the eastward component of the trade winds (although he wrongly used conservation of angular velocity rather than angular momentum). Subsequent developments of models of the circulation are discussed by Lorenz (1967). In the nineteenth century, these were largely attempts to construct models that were qualitatively consistent with the observed surface distributions and with principles such as those outlined by Hadley. A rather different approach was the attempt by Vettin (1857) [see Fultz *et al.* (1959)] to model the circulation by using a rotating vessel that contained air as the working fluid and sources and sinks (such as ice) of heat to drive the motion. This approach was promising, but appears not to have been followed up until nearly a century later.

A major obstacle to progress was the lack of appreciation of the role of instabilities and of the transient and nonaxisymmetric motions that develop thereby. Helmholtz (1888) saw that instabilities could be important, but he placed most emphasis on the sort that gives rise to billow clouds and thereby aids vertical mixing. The instability idea was later utilized by V. Bjerknes (1937) [see Lorenz (1967)] to develop a picture in close accord with present ideas. He concluded that the circulation would be rather different if it were forced to be zonally symmetric, and that such a zonally symmetric flow would be unstable to small longitude-dependent disturbances. Hence the observed circulation contains fully developed disturbances that take the form of cyclones and anticyclones.

Mathematical models of the instability that leads to cyclone development were developed by Charney (1947) and Eady (1949), and these are discussed in Sections 13.4 and 13.3, respectively. The process that they studied is called baroclinic instability, and the source of energy for the disturbances is the available potential energy (see Section 7.8) of the original zonally symmetric flow. The mere presence of available energy does *not*, however, imply instability, as the counterexample of Section 13.2 shows. In fact, certain conditions are necessary for instability to be possible, and these are considered in Section 13.5.

Another form of instability of geophysical interest is called barotropic instability. In this case, the source of energy is associated with horizontal variations in the velocity of the mean flow. The example that is chosen (in Section 13.6) to illustrate this process is based on Rayleigh's (1880) study of parallel-flow instability. As well as being directly applicable to the barotropic instability problem, the mathematics of this example is very similar to that of the Eady problem studied in Section 13.3.

The instability theories deal only with the initial development of small disturbances, whereas the role of eddies in the general circulation depends on their mean effect over a life cycle. The life cycle of a baroclinic disturbance is discussed in Section 13.9 for a model pertinent to the atmospheric circulation. The eddies (i.e., cyclones and anticyclones) transport heat poleward, as expected from the fact that they take available potential energy from the mean flow. However, they also transport zonal momentum poleward, apparently because planetary waves propagate upward and equatorward from the seat of the instability and tend to be absorbed in the equatorward side of the jet stream. The eddy momentum transports have direct consequences for the surface wind distribution because of the angular momentum balance requirement. This is discussed in Section 13.10 along with other aspects of the circulation problem. An important feature is that the zonal flow is close to being in hydrostatic and geostrophic balance, as realized by Ferrel (1859/1860) (see Section 7.6).

Baroclinic eddies are a prominent feature of the ocean as well as of the atmosphere, and these are discussed in Section 13.7. Although they are dynamically similar to their atmospheric counterparts, their horizontal scale is about a tenth of that for the atmosphere (100 km instead of 1000 km) and their time scale is much longer. Another phenomenon of great interest is that of fronts. In the atmosphere, they are usually associated with developing baroclinic disturbances, and an example of a front forming through nonlinear development of an Eady wave is considered in Section 13.8.

13.2   Free Waves in the Presence of a Horizontal Temperature Gradient

In Chapter 7 it was found that a rotating fluid adjusts to a geostrophic equilibrium rather than to a state of rest, and this equilibrium state is characterized by having potential energy that is available for conversion into other forms. It is of great interest to examine the behavior of small disturbances to such equilibria to see whether the dynamic constraints allow the disturbances to draw on this supply of available potential energy. If they can, such disturbances will grow spontaneously and become an important feature of the flow. If not, the behavior of the disturbances is still of interest, and it is useful to consider why potential energy is *not* released.

To begin with, the problem will be studied in a uniformly rotating system (f plane) and complications due to the beta effect will be ignored. The fluid will be assumed to have reached an equilibrium state in which the temperature Θ has a uniform gradient in both the y (horizontal) and z_* (vertical) directions. It will be convenient to refer to the y direction as northward, although the significance of this direction is due to the temperature gradient rather than to the beta effect. Because of the horizontal temperature gradient, the system has available potential energy, as discussed in Section 7.8, and this energy could be released if the isotherms could be made horizontal.

By the thermal wind equation (7.7.10), the x component of velocity U has uniform shear in the vertical that is related to the horizontal temperature gradient by

$$f\, dU/dz_* = -\alpha_* g\, \partial\Theta/\partial y, \tag{13.2.1}$$

where f is the Coriolis parameter, g the acceleration due to gravity, and α_* the effective "expansion" coefficient defined by (6.17.21). The fluid is assumed to be incompressible (so the scale height H_s is infinite), log-pressure coordinates will be used, and only quasi-geostrophic processes will be considered.

The equations satisfied by small perturbations to this basic state are given in Section 12.9 and have an especially simple form in the present case in which $\beta = 0$, dU/dz_* is uniform, and the frequency N_* is constant. In fact (12.9.1) and (12.9.2) reduce to

$$\frac{\partial^2 \Phi'}{\partial x^2} + \frac{\partial^2 \Phi'}{\partial y^2} + \frac{f^2}{N_*^2}\frac{\partial^2 \Phi'}{\partial z_*^2} = 0, \tag{13.2.2}$$

which becomes Laplace's equation if the stretched vertical coordinate $z_s = N_* z_*/f$ [see (8.8.25)] is used in place of z_*. Solutions exist that are wavelike in the horizontal and in time, and have a form such as

$$\Phi' = \Phi_0 \sin ly \, \sin(k(x - ct)) \exp(-z_*/H_R), \tag{13.2.3}$$

where (k, l) is the horizontal wavenumber and c is the phase speed of the disturbance in the x direction (i.e., in the direction of surface level isotherms). The solution decays with altitude on the scale [see (8.7.22)] of the Rossby height H_R, which is given by

$$H_R = f/N_*\kappa_H, \tag{13.2.4}$$

where $\kappa_H = (k^2 + l^2)^{1/2}$ is the horizontal wavenumber. The associated potential temperature perturbation θ is given by the hydrostatic equation (6.17.20), i.e., by

$$\alpha_* g\theta = \partial\Phi'/\partial z_* = -\Phi'/H_R. \tag{13.2.5}$$

Thus for perturbations that decay upward, *cold* is associated with *high* geopotential (i.e., with high pressure on level surfaces) and *warm* is associated with *low* pressure. Lines of constant phase are vertical.

Laplace's equation does not allow solutions that are wavelike in all directions, so the only possible form of wave is a "surface" wave trapped against a boundary such as a deep-water surface gravity wave. For such a wave to exist, the appropriate surface condition must be satisfied. In the present case, the condition to be applied at the horizontal boundary $z_* = 0$ is $w_* = 0$, i.e., by the temperature equation [cf. (12.9.6)]

$$(\partial/\partial t + U\,\partial/\partial x)\theta + v\,\partial\Theta/\partial y = 0 \qquad \text{at}\quad z_* = 0, \tag{13.2.6}$$

where v is the perturbation velocity component in the y direction. Using (12.9.3) for v and (13.2.5) for Φ' in terms of θ, this becomes

$$\partial\theta/\partial t + (U(0) + H_R\,dU/dz_*)\,\partial\theta/\partial x = 0 \qquad \text{at}\quad z_* = 0, \tag{13.2.7}$$

showing that the wave translates at a speed c given by

$$c = U(0) + H_R\,dU/dz_*, \tag{13.2.8}$$

i.e., at the wind speed one Rossby height (or one *e*-folding scale) above the surface. This level, where phase speed equals wind speed, is called the "*steering level.*"

The structure of the boundary wave is shown in Fig. 13.1 in a frame of reference that is stationary relative to the wave. At ground level, the streamlines are sinusoidal and air is warm for its latitude when displaced furthest poleward. Consequently, by (13.2.5) the surface pressure is low, and this is consistent with the streamline pattern only if the flow is easterly as shown in Fig. 13.1d. At the steering level there is no mean flow relative to the wave, so the motion is purely "coldward" east of the low and purely "warmward" (i.e., in the direction of the mean horizontal temperature gradient) west of the low. This is just the situation found in model A of Section 12.10, and consequently rising motion is expected in the "coldward" jet and descent in the "warmward" jet. Such is indeed the case. The value of w_* is given by (12.9.6), using (13.2.3) and (13.2.8), with the result (for $c = 0$)

$$w_* = \Phi_0 N_*^{-2} k\,(dU/dz_*)\sin ly \cos kx\,(z_*/H_R)\exp(-z_*/H_R). \tag{13.2.9}$$

Streamlines of the ageostrophic motion in the east–west vertical plane are shown in Fig. 13.1b and are obtained by integrating (13.2.9) with respect to x. Figure 13.1c shows isotherms (dashed lines) and contours of meridional velocity in the east–west vertical plane. Where isotherms are depressed the most, the air at that level is warmest, but the meridional velocity is zero. Figure 13.1e shows particle trajectories in the meridional plane. These can be calculated from (13.2.9), (12.9.3), and (13.2.3), which give

$$\frac{w_*}{v} = \frac{fz_*}{N_*^2 H_R}\frac{dU}{dz_*} = -\frac{z_*}{H_R}\frac{\partial\Theta/\partial y}{\partial\Theta/\partial z_*}. \tag{13.2.10}$$

The last equality makes use of the definition (6.17.24) of N_*^2. The formula shows that trajectories are less steep than are isotherms below the steering level and steeper

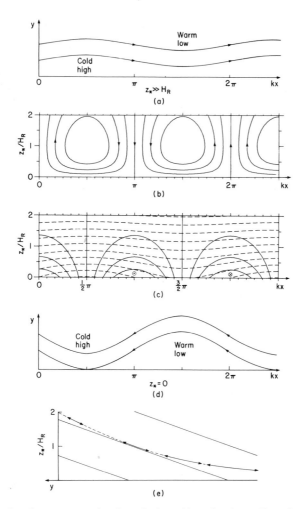

Fig. 13.1.   Properties of a wave trapped against a horizontal boundary in a uniform shear flow in a uniformly rotating environment. Isotherms are uniformly sloping in the y–z plane as shown by the solid lines in (e). (a) Streamlines (which coincide with isobars and isotherms) in the horizontal for the flow relative to the wave at a high level where the disturbance is weak. As at all levels high pressure (or high geopotential) is associated with cold air. The coldness is due to the air being displaced upward. (b) Streamlines of the ageostrophic flow (i.e., of the disturbance in the y–z_* plane. Ascent is associated with a "coldward" flow, coldward meaning that there is a horizontal component in the y-direction, i.e., toward the direction where the mean temperature at a given level is colder (see Section 12.10). (c) Contours of v, the y component of velocity (solid) and of potential temperature in the x–z_* plane. Where the air is warmest (isentropes most depressed), there is no poleward flow, and where the poleward flow is strongest, the temperature perturbation is zero. Thus there is no poleward heat transfer by the wave. (d) Surface streamlines relative to the wave. The flow is easterly, and high pressure (where streamlines are displaced furthest equatorward) is associated with cold air, the coldness being due to the equatorward displacement. (e) Particle trajectories (arrows) in the y–z_* plane relative to the isentropes (solid sloping lines). Near the ground, where the amplitude is large, the slope of these trajectories is more nearly horizontal than are those of the isentropes, so equatorward-displaced air is cold. At high levels where the amplitude is small, the slope of the trajectories is greater than that of the isentropes, so equatorward-displaced air is warm because of its relatively large downward displacement.

above. Because trajectories are steeper than are isotherms above the steering level, poleward-displaced particles are *cold* relative to their surroundings, as seen in Fig. 13.1a and c.

A similar analysis can be done for an *interfacial wave* on a horizontal boundary at which there is a discontinuity in N_*^2. This provides a model of the *tropopause* if the high value of N_* is above the interface. The geopotential anomaly Φ' is continuous at the interface and decreases exponentially away from the interface. It follows that the temperature perturbation changes sign at the interface and therefore is discontinuous! This is possible because the isotherm slope changes discontinuously at the interface and the particle trajectories in the meridional plane have a slope in between the two isotherm slopes. If the value of N_* above the interface is very large compared with that below, the tropopause behaves as a solid boundary and the solution is the same as that in Fig. 13.1, but with the signs of w_*, z_*, v, U, and Φ' reversed, those of θ and x being unchanged.

Another variant of the solution arises when the boundary slopes in the y direction. Waves propagate relative to the flow at the boundary, provided there is a temperature gradient there. The solution has a form similar to that found above [it is more appropriate to give the solution in terms of $(\partial\Theta/\partial y)_b$ rather than of dU/dz_*, where $(\partial\Theta/\partial y)_b$ is the temperature gradient along the boundary] even if the isothermal surfaces are horizontal and the boundary slopes. In that case the waves are those that were studied by Rhines (1970) not only for the quasi-geostrophic case but also for frequencies not small compared with f.

The above solution (i.e., that depicted in Fig. 13.1) is interesting because disturbances do *not* grow despite the availability of potential energy in the mean flow. For some reason, the dynamic constraints do not allow the disturbances to tap this energy source, and it is worth investigating why. Figure 13.2 is a reminder of the concept of available potential energy (see Section 7.8) in the form of an example in which isopycnals are sloping (Fig. 13.2a). For simplicity, it is assumed there are six homogeneous layers as illustrated, the large dots indicating the center of gravity of each layer. Figure 13.2b shows the minimum potential energy configuration of the layers with the new positions of the centers of gravity and arrows indicating the change in position of these centers. This illustrates the principle that release of available potential energy is associated with heavy (cold) fluid moving equatorward and light (warm) fluid moving poleward.

In the wave solution above, the dynamic constraints do not allow this to happen. The mean poleward heat flux over a wave is proportional to

$$\overline{v\theta} = \frac{1}{f}\overline{\frac{\partial\Phi'}{\partial x}\theta} = \frac{1}{\alpha_*gf}\overline{\frac{\partial\Phi'}{\partial x}\frac{\partial\Phi'}{\partial z_*}}, \tag{13.2.11}$$

where the overbar denotes an average over a wavelength in the x direction. The relation (13.2.5), however, requires this to be identically zero at all levels, because θ is proportional to Φ', and so

$$\overline{v\theta} \propto \overline{\Phi'\,\partial\Phi'/\partial x} = \overline{\partial(\tfrac{1}{2}\Phi'^2)/\partial x} = 0. \tag{13.2.12}$$

Thus no heat is carried poleward and no energy is released.

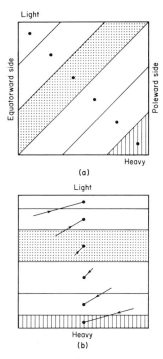

Fig. 13.2. Illustration of displacements associated with rearranging a fluid with sloping isentropes (a) to a minimum energy condition with horizontal isentropes (b). The fluid has been divided into six layers, each treated as being homogeneous. The solid circles denote the center of gravity of each layer and the arrows in (b) show the displacement of these centers that is required to achieve the state of minimum potential energy. Heavy (cold) fluid moves downward and equatorward, whereas light (warm) fluid moves upward and poleward. Consequently, there is a net poleward transfer of heat. One layer is shown hatched and another stippled for ease of identification.

To escape from the consequences of (13.2.5), it must be possible for the v and θ fields to be phase-shifted relative to each other. This can happen in various ways. One is confining the disturbance between *two* horizontal boundaries or between the surface and tropopause as shown in Eady's (1949) celebrated paper, for then there are two solutions in the vertical, one decaying away from the upper surface and one from the lower surface. If the two surfaces are many Rossby heights apart, the waves on the two boundaries affect each other only slightly, so no new effects are found. For the waves to have a large effect on each other, the Rossby height must be comparable with the height of the model tropopause, i.e., the inverse wave number must be comparable with the Rossby radius based on the tropopause height (i.e., N/f times the tropopause height, or about 1000 km). The phase shift can have different signs, depending on the relative positioning of the waves on the upper and lower boundaries. The sign that gives release of potential energy can be seen from the last expression in (13.2.11) because if *phase lines tilt westward with height* (as in Fig. 12.10, for instance), Φ' increases with x when it increases with z_*, so the poleward heat flux is positive. When the two waves are able to coexist with this phase shift, it will be shown in Section 13.3 that they grow spontaneously. Conversely, if the phase shift results in eastward tilt with height, the disturbances decay.

13.3   Baroclinic Instability: The Eady Problem

Now consider the same problem as that in Section 13.2, but with an upper boundary provided by the tropopause. To simplify the mathematics, the limiting case is taken for which the value of N_*^2 above the tropopause is large, so the upper boundary behaves as a solid boundary. Then the symmetry of the problem can be exploited by taking a frame of reference fixed in the flow at the level $z_* = 0$, midway between the two boundaries that are located at $z_* = \pm H$. The flow is thus given by

$$U = z_* \, dU/dz_*, \tag{13.3.1}$$

where dU/dz_* is a constant satisfying (13.2.1). The solution of (13.2.2) for a disturbance of fixed horizontal wavenumber κ_H can now be written in the form

$$\Phi' = A(x, y, t) \sinh(z_*/H_R) + B(x, y, t) \cosh(z_*/H_R), \tag{13.3.2}$$

where H_R is given by (13.2.4). The boundary condition of no vertical motion, which gives rise to the temperature condition (13.2.6), now applies on the two boundaries at $z_* = \pm H$. The alternative form in terms of Φ' is (12.9.6) with $w_* = 0$. Substituting (13.3.2) and taking odd and even parts give

$$\frac{\partial A}{\partial t} + \frac{dU}{dz_*}\left(H \tanh \frac{H}{H_R} - H_R\right)\frac{\partial B}{\partial x} = 0,$$

$$\frac{\partial B}{\partial t} + \frac{dU}{dz_*}\left(H \coth \frac{H}{H_R} - H_R\right)\frac{\partial A}{\partial x} = 0. \tag{13.3.3}$$

Solutions exist in which A and B are proportional to

$$\cos ly \, \exp(ik(x - ct)), \tag{13.3.4}$$

where the wave speed c is given by

$$c^2 = (dU/dz_*)^2(H \tanh(H/H_R) - H_R)(H \coth(H/H_R) - H_R). \tag{13.3.5}$$

This is one of the results obtained by Eady (1949).

In the limit, in which the two boundaries are many Rossby heights apart, i.e., $H \gg H_R$ [by (13.2.4), this corresponds to the *shortwave limit*], the tanh and coth functions both tend to unity and (13.3.5) gives

$$c \approx \pm(dU/dz_*)(H - H_R). \tag{13.3.6}$$

In other words, c is equal to the wind speed one Rossby height from the boundary, i.e., these are the boundary waves found in Section 13.2.

In the *long-wave limit*, where the boundaries are a small fraction of a Rossby height apart, i.e., $H \ll H_R$, (13.3.5) gives

$$c^2 \approx -\tfrac{1}{3}H^2(dU/dz_*)^2, \tag{13.3.7}$$

i.e., c has become purely imaginary. In general, if c is expressed in terms of its real part c_r and imaginary part c_i, i.e.,

$$c = c_r + ic_i, \qquad (13.3.8)$$

the amplitudes A and B vary as does (13.4.4), i.e., as

$$\cos ly \, \exp(ik(x - c_r t)) \, \exp(kc_i t). \qquad (13.3.9)$$

If c_i is positive, disturbances grow spontaneously, whereas $c_i < 0$ corresponds to a decaying disturbance. Formula (13.3.7) shows that both types of disturbance exist, but the growing one will soon dominate because of its exponentially growing amplitude. In fact, there will be a wavenumber selection in favor of the fastest-growing disturbance. Figure 13.3 shows the growth rate $\sigma \equiv kc_i$ as a function of wavenumber (k, l). It is small for long waves because (13.3.7) gives

$$\sigma \equiv kc_i \approx 3^{-1/2}kH \, dU/dz_*, \qquad (13.3.10)$$

whereas (13.3.5) shows that growth occurs only when

$$H \tanh(H/H_R) < H_R, \quad \text{i.e.,} \quad H < 1.1997 H_R \quad \text{or} \quad N_* \kappa_H H < 1.1997 f \quad (13.3.11)$$

by (13.2.4). Maximum growth is achieved when

$$l = 0 \quad \text{and} \quad H = 0.8031 H_R, \quad \text{i.e.,} \quad N_* kH = 0.8031 f, \quad (13.3.12)$$

the maximum value being given by

$$\sigma_{max} \equiv 0.3098(f/N_*) \, dU/dz_*. \qquad (13.3.13)$$

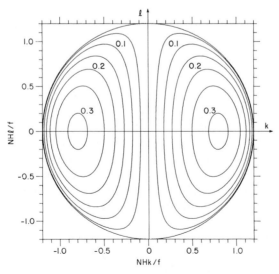

Fig. 13.3. Growth rate σ of an Eady wave as a function of wavenumber (k, l). Contours are shown in units of $(f/N_* \, dU/dz_*$. Values are zero on $k = 0$ and when the magnitude κ_H of the wavenumber equals 1.1997. The maximum value $0.3098(f/N_*) \, dU/dz_*$ is achieved when $l = 0$ and $N_* H\kappa_H/f = 0.8031$. For fixed ratio k/l, the maximum is at the same value of κ_H. The maximum for a fixed l (corresponding to a baroclinic zone of fixed width) is at a value of k that decreases as l increases (longer unstable waves for narrower baroclinic zones).

For this value, (13.3.3) shows that

$$\frac{A}{B} = \left(\frac{H_R - H \tanh(H/H_R)}{H \coth(H/H_R) - H_R}\right)^{1/2} = 1.502. \qquad (13.3.14)$$

This gives the general formula for A/B. For the fastest-growing mode it is also equal to $\coth(H/H_R)$, and thus the solution (13.3.2), taking the real part in (13.3.9), has the form

$$\Phi' = \left[\cos kx \frac{\sinh(z_*/H_R)}{\sinh(H/H_R)} + \sin kx \frac{\cosh(z_*/H_R)}{\cosh(H/H_R)}\right] \exp(\sigma t),$$

$$\alpha_* g H_R \theta = \left[\cos kx \frac{\cosh(z_*/H_R)}{\sinh(H/H_R)} + \sin kx \frac{\sinh(z_*/H_R)}{\cosh(H/H_R)}\right] \exp(\sigma t). \qquad (13.3.15)$$

Figure 13.4 shows the structure of this fastest-growing mode solution in a format similar to that in Fig. 13.1. Near the lower boundary, the Eady wave is very much like the boundary wave except for the very important phase shift of 21° between the positions of isotherms and isobars, i.e., the warmest air is just ahead (eastward) of the surface trough. Conversely, the Eady wave is similar to a trapped wave on an upper boundary at that boundary. For the growing wave, the highs and lows are *displaced 90° to the west* on the upper boundary relative to the lower boundary. Thus the phase lines for the Φ' field tilt westward with height and the same applied to v, which is proportional to the x derivative of Φ'. The same is also true for the ageostrophic field because u_a is proportional to $U \partial v/\partial x$ by (12.2.24). The phase lines for temperature, however, do *not* tilt westward, but are displaced about 48° *eastward* between the lower and upper boundaries. At the middle level, the warmest air is actually the air that is moving most rapidly poleward, so v and θ are perfectly correlated at this level. The heat flux $v\theta$ at any level can be found by substituting the solution (13.3.2) in (13.2.11). This gives

$$\alpha_* f g H_R \overline{v\theta} = \overline{A \, \partial B/\partial x} = -\overline{B \, \partial A/\partial x}, \qquad (13.3.16)$$

which is independent of altitude for all waves and can be shown to be positive for all growing waves by use of (13.3.3). Particle paths in the y–z_* plane are ellipses that expand with time. At the middle level $z_* = 0$, and (13.3.2), (12.9.3), and (12.9.6) show that particles follow rectilinear paths along a line of constant slope equal to 0.5347 times the slope of the isentropes. (It is of interest to note, before discussing cases with $l = 0$, that the $l = 0$ solution exactly satisfies the *nonlinear* quasi-geostrophic equations.)

Figure 13.5 shows the structure in the horizontal of a square $(k = l)$ Eady wave at the steering level. Warm air is carried poleward as shown and ascends at 0.5347 times the slope of the isentropes. The characteristics are very similar to those of observed cyclonic and anticyclonic disturbances at the 700- or 600-mb level (see, e.g., Fig. 12.17). The zonal wavelength of this wave is $2^{1/2}$ times the value given by (13.3.12), namely, $11.1NH/f \approx 4000$ km or about zonal wavenumber 6 at the latitudes where disturbances are generally found. The shortest *e*-folding time for growth of a

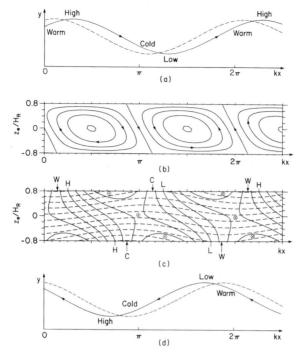

Fig. 13.4. Properties of the most unstable Eady wave, i.e., the most unstable wave in a uniform shear flow between two horizontal boundaries in a uniformly rotating environment. The solution is independent of y. (a) The pattern on the upper surface and (d) the pattern on the lower surface, the solid line being an isobar and the dashed line an isotherm. The pattern on the lower surface is very similar to that of the boundary wave shown in Fig. 13.1d, except that isotherms are now phase-shifted 21° to the east relative to the isobars. (The phase shift is exaggerated in (a) and (d) for clarity, but (b) and (c) are accurate representations.) The poleward flow is now on the warm side and the equatorward flow is on the cold side so there is net poleward heat flux. At the furthest poleward point of the isotherm, the flow is still poleward because displacements are increasing with time. The pattern at the upper surface can be obtained from that at the lower surface by symmetry. (b) The stream function for the ageostrophic flow in the x–z, plane. Ascent is associated with "coldward" flow and descent with "warmward" flow as was found in Section 12.10. The most unstable wave has wavenumber such that there are 1.6 Rossby heights H_R between the two horizontal boundaries. (c) Contours of normal velocity v (solid) and isentropes (dashed) in the x–z, plane. The lines marked H (high geopotential) and L (low geopotential) are zero lines for v. The points marked W (warm) and C (cold) on the boundary show where the air is warmest and coldest. At all levels, air going poleward is generally warmer than air going equatorward, so there is a net poleward heat flux. The phase lines of the v field (as for Φ') tilt westward with height, the total change in phase between the two boundaries being 90°.

square Eady wave is $2^{1/2}$ times the value given by (13.3.13), and a typical value for the atmosphere is 2 or 3 days.

Thus the Eady problem provides a qualitatively plausible (and mathematically straightforward) model for developing disturbances. It explains how they can form spontaneously through instability of the mean flow and draw on the energy available there. It also explains how a particular structure will tend to emerge by selection of the disturbance that grows most quickly. The properties of this disturbance also agree well with observation, i.e., (a) there are growing waves of wavenumber 6, (b) there is usually a warm tongue a little ahead of the surface trough, (c) at the steering level (located roughly midway between the surface and the tropopause) warm air

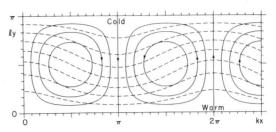

Fig. 13.5. Geopotential anomaly contours (solid) and temperature contours (dashed) for a growing square $(k = l)$ Eady wave at the steering level. The warm poleward jet is descending at about half the slope of the isentropes and the cold equatorward jet ascends at the same angle, so there is a net poleward heat transfer and release of potential energy. The relationship between the two fields is similar to that in the synoptic situation shown in Fig. 12.17 although the fields are considerably distorted in the real situation.

flows poleward and descends at about half the slope of the isentropes, etc. (*Note*: The observed steering level tends to be a bit lower than the Eady model predicts. Models that include beta effects can explain this lowering—see Section 13.4.)

Laboratory models have been found to be very useful for studying the properties of baroclinic systems and, in particular, the variety of behavior of disturbances that can be found in such systems. A review of this work has been given by Hide and Mason (1975). Numerical models [e.g., Williams (1971)] show that the initial structure of freshly growing disturbances is very similar to that predicted by Eady's analysis, especially if this is generalized to take account of nonuniform gradients (but still with uniformly sloping isentropes (G. P. Williams, 1974)), and of Ekman friction at the horizontal boundaries (Williams and Robinson, 1974). Some very interesting phenomena occur when the disturbances reach finite amplitude, and these are discussed in Hide and Mason's review.

The Eady problem also points to the possibility of eddylike disturbances that grow spontaneously in the ocean, although the model of the mean flow is not very realistic. However, since the preferred scale of the disturbances is of the order of the Rossby radius, it might be expected that the preferred wavelength in the ocean would be of the order of $2\pi \times 30 \approx 200$ km and this is in agreement with observed scales. Also, since the shear is so much smaller in the ocean, the growth rate given by (13.3.13) is small and therefore e-folding times of order 100 days are to be expected (Gill *et al.*, 1974).

13.4 Baroclinic Instability: The Charney Problem

The previous two sections ignored the beta effect, and it is important to find out how this affects the stability problem. Consider first the situation of Section 13.2, i.e., that with uniformly sloping isentropes over a single horizontal boundary, but with the beta effect included. This is the problem (with compressibility effects also included) considered in Charney's (1947) pioneering paper on baroclinic instability.

The equation for a small wavelike disturbance on any zonal mean flow $U(y, z_*)$ with the y variations sufficiently slow is (12.9.8) with the potential vorticity gradient given by (12.9.4). The boundary condition $w_* = 0$ on a horizontal boundary gives, after substitution of the wavelike form (12.9.7) in (12.9.6),

$$(U - c)(\partial \psi / \partial z_* + \tfrac{1}{2}\psi / H_s) = (\partial U / \partial z_*)\psi. \qquad (13.4.1)$$

For "purely baroclinic" instability problems, U is a function only of z_*, so the partial derivatives in (12.9.8) and (13.4.1) become ordinary derivatives. In the Eady problem, $\partial \bar{q}/\partial y$ was zero, so m^2 given by (12.9.9) was constant, and analytic solutions of (12.9.8) were thus easy to obtain. If $\partial \bar{q}/\partial y$ is nonzero, m^2 is not constant, but solutions are easily obtained numerically when the imaginary part c_i of c is nonzero. For "neutral" disturbances, i.e., freely propagating waves with $c_i = 0$, the calculation is complicated by the singularity that exists at the steering level, but this does not occur for growing disturbances.

In the Charney problem

$$d\bar{q}/dy = \beta \qquad (13.4.2)$$

is a constant and there is no upper boundary. Solutions are discussed, e.g., by Kuo (1952, 1973), Charney (1973), and Pedlosky (1979). The solution for the fastest-growing mode (which has $l = 0$) is shown in Fig. 13.6 for the incompressible limit $H_s \rightarrow \infty$, so comparison can be made with the f-plane solution of Fig. 13.1 and the Eady solution of Fig. 13.4. The maximum growth rate is achieved for $l = 0$ and is independent of β, being given by

$$\sigma_{\max} = 0.286(f_0/N_*) \, dU/dz_*. \qquad (13.4.3)$$

Surprisingly, this is almost the same result as that (13.3.13) found for the Eady

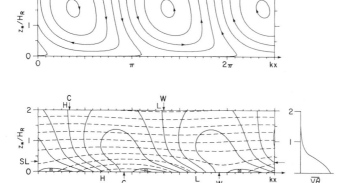

Fig. 13.6. Properties of the most unstable Charney wave, i.e., of the most unstable perturbation to a uniform shear flow on a beta plane, in the incompressible limit. The solution is independent of y. (a) The stream function for the ageostrophic flow in the x–z_* plane. Near the ground, ascent is associated with "coldward" flow and descent with "warmward" flow as found in Section 12.10. (b) Contours of normal velocity v (solid) and isentropes (dashed) in the x–z_* plane. The lines marked H (high geopotential) and L (low geopotential) are zero lines for v. The points marked W (warm) and C (cold) show where the air is warmest and coldest on the lower boundary and also at $z_* = 2H_R$, which is the edge of the picture but not of the flow. The small panel at the side shows how the poleward heat flux varies with height. The steering level (SL) is also marked. The Rossby height H_R for this wave is given by (13.4.4). (Solution courtesy of P. D. Killworth.)

problem. The functional dependence is exactly the same, with no dependence on β, and the value is only about 8% less than that in the Eady case Thus an e-folding time of about 2 days for the mid-latitude atmosphere is also predicted by this model.

Qualitatively, the solution looks like the Eady solution (Fig. 13.4) near the ground. Perturbation geopotential contours tilt westward with height, warm air is ahead of the trough line, poleward-moving air is warmer than average and descends at an angle less than that of the isentropes. Quantitatively, of course, there are differences. For example, the surface warm tongue is now $41°$ ahead of the surface trough. The phase difference between temperature and pressure increases to a maximum of $92°$ at $z_* = 0.5H_R$ and then falls off toward zero. The steering level is at $z_* = 0.33H_R$ (this is lower than for the Eady problem, in which it was $0.40H_R$ above the surface). The poleward heat flux falls off with height and is less than 7% of its surface value at $z_* = H_R$. At higher levels, the solution approaches asymptotically the f-plane solution shown in Fig. 13.1 and the visual similarity can be seen in the figures. In particular, the phase differences between Φ' and θ become small. This is also a feature of observed transient disturbances in the atmosphere (Lau and Wallace, 1979).

The important feature of the incompressible Charney solution that is not in the Eady solution is that the horizontal and vertical scales are set by the value of β and not by the tropopause height. Scales for the fastest-growing wave (which has $l = 0$) are given by

$$k^{-1} = 1.26(f_0/\beta N_*) \, dU/dz_*,$$

$$H_R \equiv (f_0/kN_*) = 1.26(f_0^2/\beta N_*^2) \, dU/dz_* \qquad (13.4.4)$$

$$= 1.26f_0/\beta \times \text{isentrope slope}.$$

The last expression comes from using the thermal wind equation (13.2.1) and the definition (6.17.24) of N_*^2. It so happens that the value of H_R, at midlatitudes, that is given by this formula is of order 10 km, so these scales are very similar to those found by Eady.

Green (1960) studies the problem when both the beta effect *and* a lid are present, thereby combining the Eady and Charney problems. The parameter that determines the solutions is the ratio of the height scale given in (13.4.4) to the height of the lid. The maximum growth rate varies little, as expected from the fact that it is very similar in the two limiting cases. However, the beta effect does strongly influence the properties of the very short and very long waves. There are now short wave modes that are rendered unstable by the beta effect, whereas the longest waves are unstable through a new mode with a more complicated vertical structure and a weaker growth rate.

The height scale given by (13.4.4) also is comparable with the scale height H_s, so compressibility effects should also be included. These produce differences in detail but not in the basic structure of the fastest-growing mode [see Lindzen *et al.* (1980) for the compressible case].

The Charney and Eady problems provide good illustrations of the baroclinic instability process in a continuously stratified fluid. The instability can also be obtained in the single situation of two superposed homogeneous layers of different densities and differing mean velocities. This was studied by Phillips (1951) for the *f*-plane case [see also Pedlosky (1979)], with layers of equal depth, and was extended, e.g., by Gill *et al.* (1974), to include effects of beta, bottom slope, and unequal layer depth. The model has been applied by P. C. Smith (1976) to explain the energetic fluctuations observed in the very strong (0.6 m s^{-1}) bottom current (Fig. 10.8) that flows through Denmark Strait between Iceland and Greenland. The fluctuations have a period of 1.8 days and the *e*-folding time given by the model is about 4 days.

13.5 Necessary Conditions for Instability

In Sections 13.3 and 13.4 it was shown by studies of particular cases how disturbances can grow spontaneously by drawing on the supply of potential energy available in the mean flow. This is a very important demonstration because it shows a means by which atmospheric depressions and oceanic eddies can be formed. However, the example of Section 13.2 shows that the mere presence of available energy is *not* sufficient to ensure instability since dynamical constraints may not allow the energy to be released. This raises the question about what conditions are required for instability to occur. It turns out that it is possible to find conditions that are *necessary* in order for instability to occur. These are very useful because if they are *not* satisfied, it can be concluded that the dynamical constraints will *not* allow the energy to be released. If they *are* satisfied, the possibility of instability is indicated, but this cannot be verified without making detailed calculations.

The necessary conditions for instability are most easily derived for perturbations with no *y* dependence and for flow between two horizontal boundaries. Suppose that the flow is unstable, so that the imaginary part c_i of c is nonzero. Then multiplying (12.9.8) by the complex conjugate of ψ, integrating by parts, and using the boundary condition (13.4.1), an equation results whose imaginary (divided by c_i) and real parts yield, respectively,

$$\int \frac{N_*^2 |\psi|^2}{f_0^2 |U - c|^2} \frac{\partial \bar{q}}{\partial y} dz_* - \left[\frac{\alpha_* g |\psi|^2}{f_0 |U - c|^2} \frac{\partial \Theta}{\partial y} \right] = 0, \qquad (13.5.1a)$$

$$\int (U - c_r) \frac{N_*^2 |\psi|^2}{f_0^2 |U - c|^2} \frac{\partial \bar{q}}{\partial y} dz_* - \left[(U - c_r) \frac{\alpha_* g |\psi|^2}{f_0 |U - c|^2} \frac{\partial \Theta}{\partial y} \right] = P. \qquad (13.5.1b)$$

where P is a positive definite expression.

In each equation, the first term is an integral from the lower to the upper boundary and the square brackets enclosing the second term denote the value at the upper boundary minus the value at the lower boundary. The thermal-wind equation (13.2.1) has been used to express the boundary contributions in terms of the temperature gradient on the boundary. The result is easily generalized to the case in which *y* dependence is allowed, in which case (13.5.1) is integrated over the *y* domain, at the boundaries of which either periodicity or a condition of no normal flow is assumed. Also, the upper and lower boundaries can have a small slope, provided that $\partial \Theta / \partial y$ in (13.5.1) is interpreted as the temperature gradient along the boundary rather than that at a fixed level. It is also possible to remove one or both of the boundaries to infinity by taking the appropriate limit.

The application of (13.5.1a) to the problems already studied can now be made by considering the *signs* of the various terms on the left-hand side. In the example of Section 13.2, the only contribution is a negative one from the lower boundary, which cannot equal zero, and therefore the flow must be stable, as is indeed the case. In the Eady problem, the negative contribution from the lower boundary is balanced by a positive contribution from the upper boundary, thereby allowing the possibility of instability. In the Charney problem, there is no upper boundary contribution, but there is instead a positive contribution from the interior, so instability is again possible. In general, the *necessary condition for instability* that is required for (13.5.1) to hold is (Green, 1960; Charney and Stern, 1962) that the set of functions

$$(\partial \bar{q}/\partial y)_{\text{interior}}, \qquad (\partial \Theta/\partial y)_{\text{lower}}, \qquad -(\partial \Theta/\partial y)_{\text{upper}} \tag{13.5.2}$$

must *not* have the same sign throughout, but must include both positive and negative values. Conversely, a sufficient condition for stability is that the set of functions (13.5.2) have the same sign everywhere. An alternative derivation and interpretation of this result is given by Bretherton (1966).

The condition can be strengthened by taking into account (13.5.1b), in which c_r can be replaced by an arbitrary constant U_r because an arbitrary multiple of (13.5.1a) can be added to (13.5.1b). It follows that the *flow is stable if* (Pedlosky, 1964) a number U_r can be found such that the functions

$$((U - U_r) \, \partial \bar{q}/\partial y)_{\text{interior}}, \qquad ((U - U_r) \, \partial \Theta/\partial y)_{\text{lower}}, \qquad -((U - U_r) \, \partial \Theta/\partial y)_{\text{upper}} \tag{13.5.3}$$

are nowhere positive. The result can also be obtained [see Pedlosky (1979)] without assuming a wavelike dependence on x and t by balancing the rate of change of disturbance energy with rates of change of other integrals that are negative definite when the quantities (13.5.3) are everywhere negative or zero. The technique can be generalized to nonparallel, quasi-geostrophic flows. Then the geopotential perturbation Φ'' acts as a stream function for the steady flow whose stability is being investigated, since the horizontal flow is geostrophic and thus horizontally nondivergent at the leading order of approximation. Consequently (12.8.13) shows that the potential vorticity $\bar{q}$ is constant on streamlines $\Phi'' = \text{const.}$, i.e.,

$$\bar{q} = \bar{q}(\Phi''). \tag{13.5.4}$$

Similarly, the temperature Θ of a fluid particle on the boundary is conserved, and so

$$\Theta = \Theta(\Phi'') \tag{13.5.5}$$

on the upper and lower boundaries. A sufficient condition for stability that was found by Blumen (1968), using a method due to Arnold (1965), is that

$$(\partial \bar{q}/\partial \Phi'')_{\text{interior}}, \qquad (\partial \Theta/\partial \Phi'')_{\text{lower}}, \qquad -(\partial \Theta/\partial \Phi'')_{\text{upper}} \tag{13.5.6}$$

be positive everywhere. The relationship with (13.5.3) can be seen by noting that for parallel flow

$$\frac{\partial \bar{q}}{\partial \Phi''} = \frac{\partial \bar{q}/\partial y}{\partial \Phi''/\partial y} = -\frac{\partial \bar{q}/\partial y}{f_0 U} \tag{13.5.7}$$

The above conditions are useful, for example, in discussing baroclinic instability in the ocean. Gradients are generally weak below the surface layers, so $\partial \bar{q}/\partial y$ is close to beta except near the surface. If $\partial \bar{q}/\partial y$ does not change sign and the bottom temperature gradient is negligible, (13.5.2) implies that instability is possible only if the surface temperature increases toward the poles! This rarely happens, but in regions of westward surface flow, the temperature increases poleward at the thermocline level, as can be seen in meridional temperature sections (see also Fig. 12.5). This can produce a change in sign of $\partial \bar{q}/\partial y$ with quite modest currents. For instance, Gill *et al.* (1974) found instabilities with *e*-folding times of order 100 days for westward surface currents of 0.05 m s^{-1}. A quite different situation exists in the cold waters of Drake Passage, where the observed eddies appear to be due to an instability associated with the equatorward increase of temperature on the bottom (Wright, 1981).

13.6 Barotropic Instability

The stability problems examined so far have been for the case in which U is a function only of z_*, i.e., the "pure baroclinic" case. In general U is a function of both y and z_*, and the potential vorticity gradient (12.9.4), which can be responsible for instability, as it is in the Charney problem, involves y derivatives as well as z_* derivatives. The condition for the terms involving y derivatives to be small relative to those involving z_* derivatives is that the y scale L should satisfy

$$L \gg N_* H/f_0, \tag{13.6.1}$$

where H is the z_* scale of the U profile, assumed not to be larger than the scale height H_s. In other words, the pure baroclinic problem is generally applicable only when the undisturbed flow has horizontal scales large compared with the Rossby radius. The condition (see Section 12.3) is also the one for the energy of the undisturbed flow to be principally available potential energy rather than kinetic energy.

The opposite limit occurs when U is a function only of y, and is called the "pure barotropic" case. This applied when variations with height can be neglected. When both types of instability are potentially present, a quantitative assessment of their relative importance can be made by using the disturbance energy equation. For the incompressible case, this equation can be deduced by starting from the disturbance form of the quasi-geostrophic momentum equations (12.2.24) and (12.2.25), which give

$$f_0 u_a = -\beta y u_g - (\partial/\partial t + U \partial/\partial x) v_g, \tag{13.6.2}$$

$$f_0 v_a = -\beta y v_g + (\partial/\partial t + U \partial/\partial x) u_g + (\partial U/\partial y) v_g, \tag{13.6.3}$$

where $U(y, z_*)$ is the undisturbed flow,

$$u_g = -f_0^{-1} \partial \Phi'/\partial y, \qquad v_g = f_0^{-1} \partial \Phi'/\partial x \qquad (13.6.4)$$

is the disturbance geostrophic velocity, and (u_a, v_a) is the disturbance ageostrophic velocity. If u_g times (13.6.3) is subtracted from v_g times (13.6.2), and (13.6.4) is used, the result [cf. (12.2.33)] can be written

$$\tfrac{1}{2}(\partial/\partial t + U \partial/\partial x)(u_g^2 + v_g^2) + (\partial U/\partial y)u_g v_g + u_a \partial \Phi'/\partial x + v_a \partial \Phi'/\partial y = 0. \qquad (13.6.5)$$

Adding $N_*^{-2} \partial \Phi'/\partial z_*$ times (12.9.6) and using (12.8.2), (13.6.4), (13.2.5), and (13.2.1) gives

$$\frac{1}{2}\left(\frac{\partial}{\partial t} + U \frac{\partial}{\partial x}\right)\left(u_g^2 + v_g^2 + \left(\frac{\alpha_* g \theta}{N_*}\right)^2\right) + \frac{\partial U}{\partial y} u_g v_g + \left(\frac{\alpha_* g}{N_*}\right)^2 \frac{\partial \Theta}{\partial y} v_g \theta$$

$$+ \frac{\partial}{\partial x}(u_a \Phi') + \frac{\partial}{\partial y}(v_a \Phi') + \frac{\partial}{\partial z_*}(w_* \Phi') = 0. \qquad (13.6.6)$$

This can be averaged with respect to x over a wavelength, the average being denoted by an overbar and periodicity in x being assumed, to give on integration with respect to y and z_*

$$\frac{1}{2}\frac{\partial}{\partial t} \iint \overline{\left(u_g^2 + v_g^2 + \left(\frac{\alpha_* g \theta}{N_*}\right)^2\right)} \, dy \, dz_* = -\iint \frac{\partial U}{\partial y} \overline{u_g v_g} \, dy \, dz_*$$

$$-\iint \left(\frac{\alpha_* g}{N_*}\right)^2 \frac{\partial \Theta}{\partial y} \overline{v_g \theta} \, dy \, dz_*. \qquad (13.6.7)$$

The normal velocity is assumed to vanish on the boundary of the domain of integration. The right-hand side contains two terms representing sources or sinks of disturbance energy. Only the second arises in a "pure baroclinic" problem, in which this represents conversion of mean available potential energy to disturbance energy. Only the first arises in "pure barotropic" problems, in which it represents conversion of mean kinetic energy to disturbance energy. When both processes are active, the ratio of the terms may be taken to define the relative importance of the two effects.

The "purely barotropic" problem will now be illustrated by a simple f-plane example. When beta is zero, the problem has exactly the same form as in the non-rotating case, so the classical theory of the stability of undirectional flows can be applied (Lin, 1955; Drazin and Howard, 1966; Drazin and Reid, 1981). The example chosen is that of a uniform shear flow

$$U = y \, dU/dy \qquad \text{for} \quad |y| < L \qquad (13.6.8)$$

(with dU/dy constant) sandwiched between two regions of uniform flow as shown in Fig. 13.7. This problem was first studied by Rayleigh (1880) and is very similar mathematically to the Eady problem studied in Section 13.3. This is because the potential vorticity gradient $\partial \bar{q}/\partial y$, given by (12.9.4), is zero (except at $y = \pm L$), so (12.9.1) again reduces to (13.2.2), although this time there is no z_* variation. The solution has structure similar to that of (13.3.2), and for a growing mode it takes the

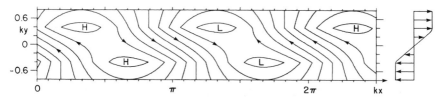

Fig. 13.7.   Perturbation geopotential (or perturbation pressure) for the most unstable disturbance to the split-line velocity profile shown at the right. The tilt of the phase lines is such that it is correlated with $-v$, i.e., if y points northward, eastward momentum is carried southward and westward momentum is carried northward.

form

$$\Phi' = (a \cos kx \sinh ky + b \sin kx \cosh ky) \exp(\sigma t) \qquad \text{for} \quad |y| < L. \quad (13.6.9)$$

For $|y| > L$, Φ' decays exponentially away from the value given by continuity of Φ' at $y = L$. The relationships between a, b, and σ follow from the requirement of continuity of v_a, as given by (13.6.3), at $y = \pm L$. They give

$$\sigma a = b(dU/dy)(\tfrac{1}{2} - kL + \tfrac{1}{2}\exp(-2kL)),$$
$$\sigma b = a(dU/dy)(kL - \tfrac{1}{2} + \tfrac{1}{2}\exp(-2kL)), \qquad (13.6.10)$$

and hence

$$\sigma^2 = (dU/dy)^2(\tfrac{1}{4}\exp(-4kL) - (\tfrac{1}{2} - kL)^2). \qquad (13.6.11)$$

The maximum value for the growth rate σ is given by

$$\sigma_{\max} = 0.2012\, dU/dy \qquad \text{when} \quad kL = 0.3984, \qquad (13.6.12)$$

and the corresponding solution is shown in Fig. 13.7. As for the Eady problem, instability occurs only for wavenumbers that are less than a cutoff value, which in this case is given by $kL = 0.6392$.

The geopotential perturbation Φ' shown in Fig. 13.7 has phase lines tilting in the direction opposite to the shear profile shown in the panel at the side, i.e., if eastward mean velocity increases to the north, phase lines tilt westward with increasing latitude. This is characteristic of a growing wave because (13.6.7) shows $\overline{u_g v_g}$ must have the sign opposite to dU/dy for growth. Equations (13.6.4) then show that where Φ' increases with x, the y gradients of Φ' and U must have the same sign, and this implies that phase lines tilt the way opposite to that of the shear flow. In the present example $\overline{u_g v_g}$ has a constant value for $|y| < L$ and is zero for $|y| > L$. The sign is that associated with bringing momentum from outside the shear layer in toward the center of the layer. This has the effect of reducing the mean flow energy by transfer to the disturbance.

The above example is useful for illustrating the characteristics of barotropic instability, but for geophysical applications the beta effect is often of prime importance. The required modification is discussed by Kuo (1949, 1973). The first condition (13.5.2), found in Section 13.5, shows that barotropic zonal flow is stable if

$$\beta - d^2U/dy^2$$

does not change sign, i.e., if the maximum value of the vorticity gradient d^2U/dy^2

is less than β. To give an indication of the implications of this, define a length scale L and velocity scale ΔU such that $\Delta U/L^2$ equals the maximum vorticity gradient. The instability is possible only when

$$L < (\Delta U/\beta)^{1/2}, \tag{13.6.13}$$

i.e., for atmospheric examples with $\Delta U = 20$ m s^{-1}, L must be less than 1000 km for instability, whereas for an oceanic example with $\Delta U = 20$ cm s^{-1}, L must be less than 100 km for barotropic instability to be possible.

Stability calculations have also been made for barotropic planetary waves (Lorenz, 1972; Gill, 1974; Coaker, 1977; Ripa, 1981). Except for the special case of zonal flow, planetary waves on an infinite beta plane are always unstable (Gill, 1974). If their inverse wavenumber is well below the value given by (13.6.13) (or, looked at from another viewpoint, if their amplitude ΔU is sufficiently large), the instability is just like that of parallel flow in the absence of beta. For large scales (or small amplitude), on the other hand, the unstable disturbance consists of two waves that form a resonant triad (see Section 8.13) with the primary wave. The results have been generalized to a two-layer system by Jones (1979). On a sphere, the geometric constraints reduce the possibilities of resonant interactions, so not all waves are unstable (Hoskins, 1973; Baines, 1976).

In practical examples, there is often a mix of baroclinic and barotropic effects. Parameters that affect the situation are the ratios of the length scale L on which the flow varies to the Rossby radius and to the beta scale (13.6.13). In oceanographic examples the ratio of the depth scale, on which the mass flow varies, to the ocean depth is also important, and Killworth (1980a) has discussed the stability properties for the variety of limiting cases that are possible. The presence of side boundaries is also important in the case of boundary currents like the Gulf Stream, and bottom topography also has an effect. Laboratory experiments on the stability of boundary currents are discussed, e.g., by Griffiths and Linden (1981).

The remaining sections show how disturbances can grow spontaneously in certain circumstances, and there are many other examples that occur in nature. In fact, instabilities are very common and give rise to such natural features as turbulence in wind and water, both tropical and extratropical storms, a great variety of cloud forms, rain bands, thermals, etc. The instabilities are not studied here, but the following authors have written pertinent books: Betchov and Criminale (1967), Chandrasekhar (1961), Charney (1973), Drazin and Reid (1981), Gossard and Hooke (1975), Lilly (1979), Turner (1973), Wallace and Hobbs (1977), Woods (1982), and Yih (1980).

13.7   Eddies in the Ocean

Many aspects of the dynamics of the ocean circulation were discussed earlier in Sections 10.14 (eastern boundary currents), Sections 11.12, 11.14, and 11.16 (tropical currents), and Sections 12.5 and 12.6 (extratropical currents). In particular, it was seen that in mid latitudes, Ekman pumping (introduced in Section 9.4) causes vertical displacements of the thermocline and the establishment of a "Sverdrup balance," as depicted in Fig. 12.5. The associated density field contains a large amount of available potential energy that is generated at an average rate of order 10^{-3} W m^{-2} (Gill *et al.*, 1974) by Ekman pumping. The scale of the gyre is set by the meridional scale $l^{-1} \approx 1000$ km of the wind stress, which is about 30 times the baroclinic Rossby radius (see Section 12.5). It follows [see (7.5.2)] that the available potential energy is about 30^2, i.e., about 1000 times bigger than the kinetic energy associated with the Sverdrup mean circulation. If this were the whole story, the ocean currents away from the boundaries would be very weak (of order 1 cm s^{-1}), and this was thought by many to be the case until current measurements in such regions began to be made in the late 1950s.

In practice, however, currents observed in the ocean interior are typically of order 10 cm s^{-1} rather than 1 cm s^{-1}, so how do such strong currents arise? It has been seen in the earlier sections of this chapter that eddies can arise through instability of the mean flow and can draw on the potential energy available in the mean field. Such eddies tend to have a scale of the order of the Rossby radius, i.e., a scale for which [see (7.5.2)] the kinetic energy is comparable with the available potential energy. It follows that if the available potential energy of the ocean gyre were suddenly used to create eddies on the scale of the Rossby radius, their available potential energy would be about half that of the original gyre (i.e., the eddy available potential energy would be the same order as that of the original gyre), and the eddy kinetic energy would be about the same, i.e., very much *larger* than that of the original gyre. This argument gives a possible source of energy for the eddies, but it does not explain why and how that energy source might be utilized. Numerical experiments [see, e.g., Schmitz and Holland (1981); Robinson *et al.* (1977)] with sufficient resolution to simulate eddies demonstrate how these can be generated and also contain many features that can be compared with observation. In particular, the geographical distribution of eddies indicates that the major production zones for eddies are in the regions of strong currents such as the Gulf Stream. Also, the models indicate that the deep recirculating flow found in the western basin of the North Atlantic is eddy-driven.

The observed properties of eddies vary considerably, but the word eddy is generally used to describe features with length scales (inverse wavenumbers) of order 10–100 km, and with time scales (inverse frequency) of 10–30 days. They tend to be in approximate geostrophic balance and are generally found to move westward at a few centimeters per second. Their amplitude, in terms of vertical displacement of isopycnals, can be 100 m or more, and the associated currents can be 1 m s^{-1} or more, although magnitudes of order 10 cm s^{-1} are more typical. In the North Atlantic between 0° and 50°N, the distribution of eddy potential energy density, calculated by Dantzler (1977), indicates that the largest values (500–2000 cm^2 s^{-2}) are confined to the general area of the Gulf Stream. Although there is little doubt that eddies in this region are due mostly to instability (Schmitz and Holland, 1982), other generation mechanisms such as wind forcing (see Section 9.11) or flow over topography (Sections 8.7–8.10) could be significant elsewhere (Müller and Frankignoul, 1981).

Instabilities may take a variety of forms, and the mechanisms are often difficult to identify observationally. However, one form clearly seen in observations is due

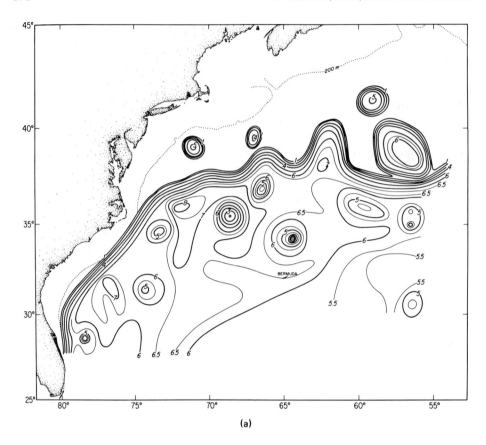

(a)

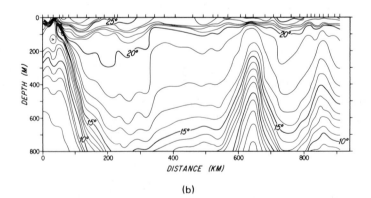

(b)

Fig. 13.8. (a) Chart of the depth, in hundreds of meters, of the 15° isothermal surface, showing the Gulf Stream, nine cyclonic rings, and three anticyclonic rings. Contours are based on data obtained between 16 March and 9 July 1975. (b) A temperature section through the Gulf Stream and two cyclonic (cold-core) rings south of the stream. The section is a "dog-leg" from 36°N, 75°W (left-hand end) to 35°N, 70°W (middle) and then to 37°N, 65°W (right-hand end). [From Richardson *et al.* (1978, Figs. 1a and 4a, Section 3).]

to Gulf Stream meanders forming large loops that pinch off and form Gulf Stream "rings." If they break off on the equatorward side, they have cold water in the middle (cold-core eddies), and the Labrador Sea water they contain may be trapped in the eddy for a year or so, in which time the ring may move a considerable distance to the west and south. Similarly, warm-core eddies form on the poleward side and drift westward also, sometimes being reabsorbed into the Gulf Stream. For example, Fig. 13.8 shows the Gulf Stream, nine cold-core (cyclonic) rings and three warm-core (anticyclonic) rings observed in the spring of 1975, together with a cross section through the Gulf Stream and two of the rings. A summary of the properties of rings is given by Richardson (1982).

Further information about eddies may be found in the survey article by Wunsch (1981) and in the comprehensive volume on eddies edited by Robinson (1982). Reviews of the observational base are given by Richman *et al.* (1977) and the MODE Group (1978). Modeling of eddies and their interactions with the mean flow is discussed by Rhines (1977, 1979) and by Rhines and Holland (1979). Analogies between ocean eddies and baroclinic disturbances in the atmosphere are considered by Charney and Flierl (1981).

13.8 Fronts

The solutions found in Section 13.3 nicely illustrate the *initial* development of baroclinic disturbances, but sooner or later effects that have been neglected in the model will come into play and give rise to new features. One is the development of the sharp fronts that are such a familiar feature of surface weather charts. These have a variety of forms and develop in a variety of ways. Descriptions are given e.g., by Wallace and Hobbs (1977) and by Palmén and Newton (1969); many ideas of structure go back to Bjerknes (1919). Here attention will be concentrated on one illustrative example, namely, the fronts formed by the fastest-growing Eady wave (see Fig. 13.4). This is somewhat special because the disturbance is independent of y and in fact satisfies the nonlinear quasi-geostrophic equations. However, the quasi-geostrophic approximation itself breaks down when the ageostrophic velocity component u_a becomes comparable with its geostrophic counterpart, which in this case is U. Then the derivative D/Dt, following the motion, can no longer be approximated by D_g/Dt.

A formal procedure for finding the equations that apply in the neighborhood of a front is to introduce appropriate nondimensional variables (in a frame of reference moving with the front), allowing for the fact that the cross-front scale L_x is small compared with the along-front scale L_y. Suitable scales for t, z_*, u, v, w, Φ', and θ are $(L_y/L_x)f^{-1}$, $(f/N_*)L_x$, $(L_x/L_y)fL_x$, fL_x, $(fL_x)^2/N_*L_y$, $(fL_x)^2$, and fN_*L_x/α_*g. The result of introducing these scalings is that the only term that can be neglected in the equations is the acceleration term Du/Dt in the x component of the momentum equation, this term being of order $(L_x/L_y)^2$ relative to the Coriolis term. Thus flow along the front is in geostrophic balance with the cross-front pressure gradient, i.e.,

$$v = v_g = f^{-1} \, \partial \Phi' / \partial x. \tag{13.8.1}$$

The cross-front flow is *not* in geostrophic balance, but it is still useful to express it as the sum of a geostrophic part and an ageostrophic part u_a, where

$$f u_a = -Dv/Dt = -Dv_g/Dt, \tag{13.8.2}$$

the equality on the far right using (13.8.1). The operator D/Dt has its full three-dimensional form (4.1.7), i.e., it includes the term $w_* \, \partial / \partial z_*$. The remaining equations to be satisfied are the hydrostatic equation (6.17.20), the continuity equation (12.8.2), and the potential temperture equation (4.10.8). These equations were used by Sawyer (1956) and Eliassen (1962) to discuss the vertical circulation at fronts.

It so happens that this set of equations can be transformed into the quasi-geostrophic equations by a change of variables (Eliassen, 1962; Hoskins and Bretherton, 1972; Hoskins and Draghici, 1977), i.e., by a replacement of the variables on the left of the following tabulation with those on the right (denoted by changing lowercase into capitals and vice versa)

$$
\begin{aligned}
\Phi' \qquad & \phi' = \Phi' + \tfrac{1}{2} v^2, \\[4pt]
x \qquad & X = x + v_g / f, \\[4pt]
z_* \qquad & Z_* = z_*, \\[4pt]
t \qquad & T = t, \\[4pt]
u_a \qquad & U_a = u_a + (fJ)^{-1} w_* \, \partial v / \partial z_*, \\[4pt]
w_* \qquad & W_* = w_* / J \\[4pt]
N_*^2 \qquad & n_*^2 = \frac{\alpha_* g}{f} \left[\left(f + \frac{\partial v}{\partial x} \right) \frac{\partial \theta}{\partial z_*} - \frac{\partial v}{\partial z_*} \frac{\partial \theta}{\partial x} \right],
\end{aligned}
\tag{13.8.3}
$$

where

$$J = 1 + \frac{1}{f} \frac{\partial v}{\partial x} = \frac{\partial X}{\partial x} = \frac{\partial(X, Z_*)}{\partial(x, z_*)} \tag{13.8.4}$$

is the Jacobian of the coordinate transformation. The dependent variables θ, u_g, and v are unchanged, so there is no need to introduce new symbols for them. The independent variables z_* and t are not changed either, but capital letters are used to indicate that partial derivatives such as $\partial / \partial Z_*$ that involve capital letters, are for fixed X, whereas $\partial / \partial z_*$ is for fixed x.

The proof of the result about the transformation is a matter of manipulation of partial derivatives. For instance, by the definition (13.8.3) of X, it follows that its derivative, following the motion, is

$$DX/Dt = Dx/Dt + f^{-1} \, Dv/Dt = u + f^{-1} \, Dv/Dt = u_g, \tag{13.8.5}$$

the last equality using (13.8.2) and the definition of u_a. It follows that the expression for the derivative following the motion is, in the new coordinates,

$$\frac{D}{Dt} \equiv \frac{\partial}{\partial T} + \frac{DX}{Dt}\frac{\partial}{\partial X} + v\frac{\partial}{\partial y} + \frac{DZ_*}{Dt}\frac{\partial}{\partial Z_*} = \frac{D_g}{Dt} + w_*\frac{\partial}{\partial Z_*}, \quad (13.8.6)$$

where D_g/Dt is defined by

$$D_g/Dt = \partial/\partial T + u_g\,\partial/\partial X + v\,\partial/\partial y \quad (13.8.7)$$

and $\partial/\partial T$ represents the time derivative with X (not x) kept constant. The transformed versions of (13.8.2) and (4.10.8) are now

$$fU_a = -D_gv/Dt, \qquad D_g\theta/Dt + (n_*^2/\alpha_*g)W_* + v\,\partial\Theta/\partial y = 0, \quad (13.8.8)$$

and the remaining equations are

$$fv = \partial\phi'/\partial X, \qquad \alpha_*g\theta = \partial\phi'/\partial Z_*, \qquad \partial U_a/\partial X + \partial W_*/\partial Z_* = 0. \quad (13.8.9)$$

These follow by expanding derivatives with respect to X and Z_* in terms of x and z_* derivatives, using for any function F

$$\frac{\partial F}{\partial Z_*} = \frac{\partial(F, X)}{\partial(Z_*, X)} = \frac{1}{J}\frac{\partial(F, X)}{\partial(z_*, x)} = \frac{1}{J}\left[\frac{\partial F}{\partial z_*}\left(1 + \frac{1}{f}\frac{\partial v}{\partial x}\right) - \frac{\partial F}{\partial x}\frac{1}{f}\frac{\partial v}{\partial z_*}\right] \quad (13.8.10)$$

and a similar result for $\partial F/\partial X$.

Now the quantity n_*^2 is proportional to Ertel's potential vorticity Q, defined by (7.11.17), and therefore is conserved following the motion by (7.11.13). The semi-geostrophic equivalent can be derived from the above equations. For the Eady problem, n_*^2 is uniform for the undisturbed flow and thus remains uniform throughout the motion. It follows [as was first shown by Hoskins and Bretherton (1972)] that the solution in the transformed coordinate has the same form (13.3.15) as that found in section 13.3. In particular θ and v are given on a section $y = 0$ by

$$\alpha_*g\theta = n_*^2Z_* + \frac{\exp(\sigma T)}{H_R}\left[\cos kX\,\frac{\cosh(Z_*/H_R)}{\sinh(H/H_R)} + \sin kX\,\frac{\sinh(Z_*/H_R)}{\cosh(H/H_R)}\right],$$
$$(13.8.11)$$

$$f^2(X - x) \equiv fv = k\exp(\sigma T)\left[-\sin kX\,\frac{\sinh(Z_*/H_R)}{\sinh(H/H_R)} + \cos kX\,\frac{\cosh(Z_*/H_R)}{\cosh(H/H_R)}\right].$$
$$(13.8.12)$$

This solution is shown in Fig. 13.9, which was constructed by first calculating contours in (X, Z_*) space and then using (13.8.12) to transform to (x, z_*) space. It has the property that a singularity develops on the boundaries (at the points marked L) after a finite time, the singularities corresponding to the formation of fronts with infinite gradients. The singularity on the ground is at the point where $J = \infty$, i.e., $\partial x/\partial X = 0$, i.e., by (13.8.12) it occurs at $kX = \frac{7}{4}\pi$ when $k^2\exp(\sigma T) = 2^{-1/2}f^2$. Figure 13.9 shows the solution when the disturbance amplitude is 90% of the value at this time. For comparison, Fig. 13.10 shows an observed section through a front, and the

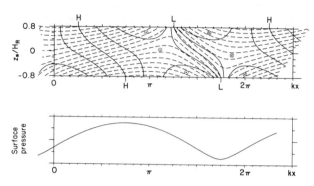

Fig. 13.9. Large-amplitude solution for the fastest-growing (two-dimensional) Eady wave. Contours are of normal velocity (solid) and potential temperature (dashed) in a vertical section. The amplitude of the perturbation is 90% of the value at which singularities develop at the low pressure (L) points on the boundary. At these points, which corresponds to the fronts, the normal velocity is zero. The contour interval in potential temperature is the same as in the small-amplitude solution shown in Fig. 13.4c. The solution was obtained by coordinate transformation from the latter, so the boundaries of the contoured region do not coincide with the frame. The vertical exaggeration is N_*/f. Note that the velocity and potential temperature contours become nearly parallel at the front. The lower panel shows the corresponding variation of surface pressure with distance x. Note the sharpness of the trough as compared with the broadness of the ridge.

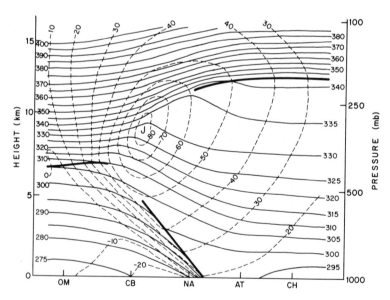

Fig. 13.10. A vertical cross section through a frontal zone from Omaha, Nebraska (OM), to Charleston, South Carolina (CH), showing contours of velocity in meters per second (dashed) and potential temperature in degrees Kelvin (solid). The horizontal distance across the section is about 2000 km. The time was 00 GCT on 20 November 1964. The figure is from Wallace and Hobbs (1977, Fig. 3.20), where further details are described. The front may be compared with the one obtained for an ideal fluid in Fig. 13.9.

similarities in the lower troposphere are apparent. (The structure near the tropopause can be modeled in a similar way, as discussed below.)

There are several features of the front that are independent of the details of the formation process. By definition, the front is a region of large gradients, so the dominant terms in the potential vorticity balance [the last equation of (13.8.3)] give

$$\frac{\partial v}{\partial x}\frac{\partial \theta}{\partial z_*} - \frac{\partial v}{\partial z_*}\frac{\partial \theta}{\partial x} \equiv \frac{\partial(v, \theta)}{\partial(x, z_*)} = 0, \tag{13.8.13}$$

i.e., *v and θ contours tend to coincide* in the front. This feature can be seen in both the theoretical and the observed fronts. Also, potential vorticity conservation requires $f + \partial v/\partial x$ to remain positive, so it can be large only when the relative vorticity $\partial v/\partial x$ is large and *cyclonic*. To give an idea of possible magnitudes, detailed observations by Sanders (1955) of a front in the United States showed $\partial v/\partial x$ with a value of $10f$ at a height of 300 m, and $\partial \theta/\partial x$ at the same height was 0.5 deg km^{-1}.

Another general statement concerns the cross-frontal flow, for (13.8.2) gives on the surface

$$\partial v/\partial t + u(f + \partial v/\partial x) = u_\mathrm{g} = -f^{-1}\,\partial\Phi'/\partial y. \tag{13.8.14}$$

Assuming that the along-front pressure gradient and along-front acceleration do not change much, the cross-front flow u must become small when $f + \partial v/\partial x$ becomes large. This statement can be made more precise for the Eady case since v (and hence $\partial v/\partial t$) is zero at the point L of minimum pressure and maximum cyclonic vorticity at which the front forms. From Fig. 13.4b it can be seen that the ageostrophic flow opposes the geostrophic flow at this point. The latter remains constant, but u_a increases exponentially with time so (13.8.14) implies that $(f + \partial v/\partial x)$ will become infinite in a finite time. Because there is little cross-front flow, the poleward flow ahead of the front continues to bring in warm air, whereas the equatorward flow behind brings in cold air, and so builds up the temperature contrast even further. Other features of the front shown in Fig. 13.9 follow from the properties of baroclinic disturbances, e.g., the slope to the west with height, with a value for the slope of order f/N. Typical observed values are of this order [the value is about f/N in Fig. 13.10 and was about 0.3 f/N in the case studied by Sanders (1955)].

In reality, infinite gradients are not observed, so clearly other effects such as y variations (see below), friction, mixing, and latent heat release modify the structure of fronts, which can become quite complicated when viewed on a small scale [see, e.g., Bennetts and Hoskins (1979)]. The maximum velocity in the solution of Fig. 13.9 is $0.90k^{-1}f = 1.1N_*H$ or about 100 m s^{-1}. Sanders found changes in *geostrophic* velocity v_g of this magnitude across the front, but the change in the *measured* velocity v was only 20 m s^{-1} at 300 m. At 1200 m the changes in measured and geostrophic velocities were both about 35 m s^{-1}. One obvious discrepancy in the theory is the neglect of the surface boundary layer. The large shear (see Section 9.5) leads to a large Ekman flux into the surface low (Sanders measured convergence at a rate equal to $5f$ at 300 m) that enhances the temperature gradient and leads to increased upward velocity at the front. Descriptions of the vertical motion observed at fronts have been given, e.g., by Browning *et al.* (1975). Blumen (1980) has added surface friction effects to the solution shown in Fig. 13.9 and has made comparisons with Sanders' observations. Friction effects outside the boundary layer have been considered by Gill (1981) in another context and by Garrett and Loder (1981) in connection with fronts in the ocean. Shapiro (1981) has discussed the effects of turbulent fluxes in the free atmosphere and R. T. Williams (1974) obtained a steady-state front numerically. Latent heat release may also be important and has been modeled, e.g., by Orlanski and Ross (1978).

The development of fronts in *three-dimensional flows* can be handled by the same methods as those used above. The y component of the momentum equation is approximated in the same way as is (13.8.2), i.e.,

$$f v_a = D u_g / D t, \tag{13.8.15}$$

and will provide a good approximation in frontal regions, whatever their orientation, provided that the along-front scale is large compared with the cross-front scale. The approximation is known as the *geostrophic momentum* approximation, first introduced by Eliassen (1949), and the equations are known as the semigeostrophic equations. The equations can be transformed into the quasi-geostrophic equations (Hoskins, 1975; Hoskins and Draghici, 1977) by using the variables of (13.8.3) and also by using

$$Y = y - u_g / f. \tag{13.8.16}$$

The coordinates X, Y, called *geostrophic coordinates*, were used by Yudin (1955) and the transformation has been further discussed by Blumen (1981). The quantity n_*^2 is proportional to Ertel's potential vorticity as before, and J is the ratio of total vorticity to f. The approximation has been applied to the development of the square Eady wave (shown in Fig. 13.5) by Hoskins (1976). Hoskins and West (1979) added the effect of horizontal shear in the mean flow, and the types of fronts that develop have been discussed by Hoskins and Heckley (1981). The mathematical theory of frontogenesis has been reviewed by Hoskins (1982).

As Figs. 13.9 and 13.10 indicate, fronts can also form on the tropopause. The rigid lid model is not very good for describing their development, but representation of the tropopause as a discontinuity in N_*^2 (or rather n_*^2) gives remarkably realistic results, as found, e.g., by Hoskins (1972). A significant feature of these fronts is the descent of a tongue of stratospheric air well below normal tropopause heights, and such tongues can be traced, e.g., by measuring ozone concentration. Descriptions may be found, e.g., in Reed and Danielsen (1959) and Shapiro (1974).

In the ocean, fronts are produced by a variety of mechanisms. Sometimes they are very distinct in the temperature and salinity fields but *not* in the density field, and this distinction is important so far as the dynamics is concerned. A survey of temperature fronts as seen by satellites is given by Legeckis (1978). Figure 13.11 shows the main climatological frontal regions (i.e., regions where fronts are most commonly observed) in the North Pacific, and these are discussed by Roden (1975). One important type of front is that produced by *Ekman convergence* in the surface layer, the *subtropical fronts* found at about 30°N and 40°S being important examples. Roden and Paskausky (1978) have studied changes in such fronts due to changing Ekman

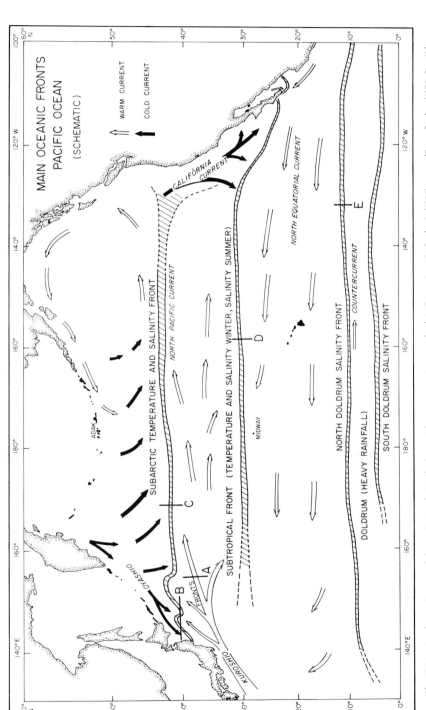

Fig. 13.11. A schematic map of the main North Pacific fronts. The cross sections marked A–E are given in Roden's (1975) paper. [From Roden (1975, Fig. 1).]

convergence. A second type of front is that formed at a *water mass boundary* (see Welander, 1981) such as that which separates the subarctic and subtropical gyres. In the North Pacific (Fig. 13.11) this front is located at about 42°N and is formed where the cold equatorward flowing Oyashio meets the warm poleward flowing Kuroshio. The surface front is a prominent feature in temperature and salinity sections but not in density sections. A factor that may be important in determining the properties of the front is that it is close to the boundary between upward Ekman pumping in the subarctic gyre and downward Ekman pumping in the subtropical gyre. The *Antarctic Convergence* (Mackintosh, 1946) has properties similar to those of the North Pacific subarctic front (Taylor *et al.*, 1978). A detailed study of the Antarctic Convergence, where it passes through Drake Passage, has been made by Joyce *et al.* (1978). Fronts can also be formed by *coastal upwelling* of cold water (Mooers *et al.*, 1976; Foo, 1981), an example being shown in Fig. 10.16. McVean and Woods (1980) have studied the formation of fronts by *convergent mean flow*, using the method of Hoskins and Bretherton (1972). A further type of front is formed by *gradients in rates of stirring*, e.g., by tides (Simpson, 1981), and such fronts have been modeled by James (1978). They are distinctive features of shallow seas and form a boundary between well-mixed water on the shallow side of the front and stratified water on the deep side. Large-amplitude waves are commonly observed on fronts (Woods *et al.*, 1977) and models have been constructed by James (1981). Yet another way of producing temperature contrasts in models of well-mixed shallow seas is by applying the same amount of cooling to different depths of water (Gill and Turner, 1969).

13.9   The Life Cycle of a Baroclinic Disturbance

The baroclinic instability theory indicates how depressions form in the atmosphere and how their initial structure is determined. Similarly, eddies can be generated in the ocean through instabilities of the mean flow. Eddies (the term is taken to include atmospheric disturbances) cannot, however, grow indefinitely, and their mean effect cannot be assessed without some knowledge of the way in which they mature, decay, interact with other disturbances, etc. For the atmosphere, a useful picture is provided by considering the behavior of disturbances to a realistic zonal flow such as that shown in Fig. 13.12a. The horizontal temperature gradient is mainly confined to the 30–60° latitude band, giving a jet stream centered at 45° and 200 mb. Simmons and Hoskins (1978, 1980) studied the evolution of a disturbance to this flow that initially had the structure of the fastest growing mode of zonal wavenumber 6 with small amplitude (such that the maximum disturbance pressure was 1 mb). The disturbance grows rapidly by drawing on the potential energy available in the mean flow, and develops a realistic surface structure with fronts, as shown in Fig. 13.12c. The occlusion process eventually chokes off the disturbance near the surface, and the disturbance energy reaches a peak during the seventh day. Then the disturbance energy falls off as rapidly as it grew and by the end of the tenth day is reduced to about a tenth of its peak value. Thereafter the decay is much slower. About a quarter of the available potential energy is released during this process, the zonal mean flow and zonal mean temperature distribution at the end of the life cycle being as shown in Fig. 13.12b. This distribution is only weakly unstable, so no further significant release of available potential energy takes place.

A useful means of depicting the changes in the disturbance during its life cycle (Edmon *et al.*, 1980) is by means of "Eliassen–Palm cross sections," which show the quasi-geostrophic Eliassen–Palm flux, defined in Section 12.9, and its divergence. This flux has the direction of the group velocity (when that concept applies); its horizontal component [see (12.9.14)] is proportional to the horizontal momentum flux, its vertical component is proportional to the horizontal heat flux, and its divergence is proportional to the quasi-geostrophic potential vorticity flux. Figure 13.13

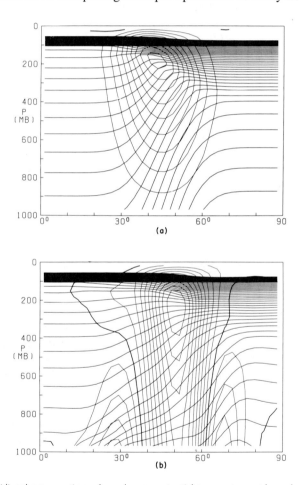

Fig. 13.12.   Meridional cross sections of zonal-mean potential temperature and zonal wind for (a) The basic zonal-mean state at the beginning of the life cycle experiment. [From Simmons and Hoskins (1980, Fig. 1a).] (b) The zonal-mean state at the end of the life cycle. Contour intervals are 5 K and 5 m s^{-1}. The zero velocity contour is drawn relatively dark. [Courtesy of B. J. Hoskins.] (c) North polar stereographic plot showing surface pressure (solid contours) and near-surface temperature (dashed contours) after 5 days of integration for a disturbance of zonal wavenumber 6 to the flow shown in (a). The initial surface pressure amplitude was 1 mbar and the initial perturbation was the fastest-growing normal mode. Contour intervals are 8 mb and 8 K.

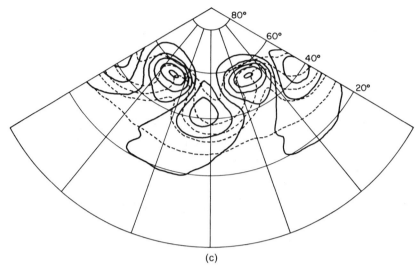

(c)

Fig. 13.12. (continued)

shows this flux at two stages in the life cycle, and the average values over the cycle. The arrows give the direction of the flux and the contours its divergence, i.e., they give the potential vorticity flux. The initial (fastest-growing) disturbance, shown in Fig. 13.13a, has a structure remarkably similar to that of the Charney mode, studied in Section 13.4, but confined to a channel, so that the horizontal planform is like that shown in Fig. 13.5. The channel walls would be placed at about 35 and 55° to match the structure shown in Fig. 13.13a, with half a sine wave spanning the channel. The arrows for a Charney mode are exactly vertical because it has no momentum flux, and their magnitude varies with height in a manner like that shown in the inset at the right of the lower panel in Fig. 13.6. The pattern seen in Fig. 13.13a is very close to this.

The structure at day 8 (Fig. 13.13b), when the disturbance has begun to decay rapidly, is very different. The fluxes near the surface have become relatively small and spread over a wider range of latitude. The baroclinic instability process is now ceasing to operate (see above), and it appears that planetary waves have been radiated upward and equatorward from the seat of the instability. The strongest fluxes are now in the jet stream within the 10 m s^{-1} contour, the significance of this contour being that this is where the disturbance wave speed equals the flow speed. At such places planetary waves tend to be absorbed and perhaps partially reflected (see Section 8.9), so that disturbance energy can be trapped in the jet stream by the "critical line." The arrows show the preferred direction of wave propagation for these disturbances (which presumably could be found approximately by ray-tracing techniques), namely, equatorward and upward. This determines where the waves are absorbed, namely, on the equatorial flank of the jet stream just below the tropopause. The energy lost from the disturbance as it propagates to regions where the velocity relative to the wave is weaker is largely converted to mean flow kinetic energy. A calculation of the rates of transfer of disturbance energy to zonal kinetic energy in fact shows that it begins to rise rapidly at the end of the sixth day, reaches a maximum 2 days later (the time corresponding to Fig. 13.13b), then falls to near zero during the next 2 days. The resultant zonal flow at the end of the cycle is shown in Fig. 13.12b. The westerlies are too strong at the surface, but this can be rectified by including surface friction (Simmons and Hoskins, 1980).

The mean Eliassen–Palm cross section for the cycle is shown in Fig. 13.12c. The upward arrows represent a poleward heat flux, and this is large over the region of large initial horizontal gradient. The equatorward pointing component of the flux represents a *poleward* transfer of westerly momentum, corresponding to equatorward planetary-wave propagation [see (12.9.14)]. The flux is particularly large between 150 and 400 mb within the jet stream. This transfer is extremely important in

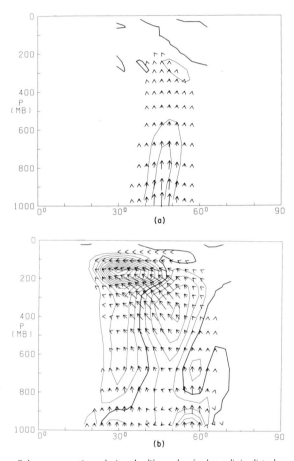

Fig. 13.13. Eliassen–Palm cross sections during the life cycle of a baroclinic disturbance. The vectors, shown by arrows, give the Eliassen–Palm flux (see Section 12.9), which for linear waves is in the direction of the group velocity. Because of the coordinates used, the horizontal and vertical components of the vector are defined, respectively, by $-2\pi r^2 g^{-1}\overline{u'v'}\cos^2\varphi$ and $2\pi f r^3 g^{-1}\overline{v'T'}\cos^2\varphi/(\partial\Theta/\partial p)$, where r is the earth's radius, φ the latitude, g the acceleration due to gravity, and f the Coriolis parameter. [This is $2\pi r^2\cos^2\varphi/\rho_* g$ times the Eliassen–Palm flux as defined in (12.9.14).] With this choice, the "divergence," defined as the sum of the φ derivative of the horizontal component and the p derivative of the vertical component, is equal to $2\pi r^3 g^{-1}\overline{v'q'}\cos^2\varphi$ and therefore is proportional to the quasi-geostrophic potential vorticity flux. The "divergence," so defined, is shown by contours [contour interval 1.5×10^{15} m^3 for (a) and (b) and 4×10^{15} m^3 for (c)]. The zero contour is drawn dark. (a) At day zero for the life cycle, i.e., represents the fastest-growing mode; (b) at day 8; and (c) the average over the cycle. [From Edmon *et al.* (1980, Fig. 3).]

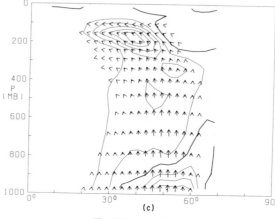

Fig. 13.13. (continued)

the angular momentum budget of the earth, and is responsible for the midlatitude surface westerlies that are such an important feature of the circulation. Further discussion may be found in Section 13.10.

13.10 General Circulation of the Atmosphere

13.10.1 Introduction

An introductory discussion about the way in which the atmosphere responds to the radiative driving from the sun was given at the beginning of the book, and it seems fitting to conclude with a further discussion in light of the concepts that have been developed in the intervening chapters. In Chapter 1 it was shown that a purely radiative equilibrium is unstable in the sense that air at the surface would be lighter than that above the surface. Thus convection takes place over the depth of the troposphere and radiative–convective equilibrium models can be constructed to simulate a local balance between radiative and convective effects. One could imagine an atmosphere in which such a balance applied on average at each latitude independently. This would give much higher temperatures at the equator and much lower ones at the poles, with the equator-to-pole temperature difference being perhaps four times the 30 K that appears in the zonal average distribution shown in Fig. 7.9. Such a temperature distribution could be in thermal-wind balance with the zonal flow, giving a very much stronger jet stream than that observed.

Such a state is not observed, however, because of the action of baroclinic disturbances as described in Section 13.9. These derive their energy from the potential energy available in the meridional temperature gradient, and act to reduce this gradient as illustrated in Fig. 13.12. Radiation, on the other hand, tends to restore the gradient, so the mean temperature field observed is largely due to a balance between the competing effects of the instability and of radiation.

To gain further insight into the zonal mean circulation, consider the equations for $[\bar{u}]$, $[\bar{v}]$, and $[\bar{w}_*]$, the temporally and zonally averaged velocity components associated with the coordinates λ, φ, z_*, where λ is longitude, φ is latitude, and z_* is the log-pressure coordinate introduced in Section 6.17, and for the time and zonally averaged temperature $[\overline{T}]$. The square brackets denote a zonal mean (and the departure from this mean will be denoted by a superscript asterisk) and the overbar denotes a mean with respect to time (the depature from the time mean will be denoted by a prime). The two components $[\overline{v}]$ and $[\overline{w}_*]$ can be expressed [cf. (9.15.10)] in terms of a stream function ψ associated with the meridional circulation. From the continuity equation (6.17.11) and the definition (6.17.29) of ρ_*, it follows that the stream function can be defined by

$$\rho_*[\overline{v}]\cos\varphi = -\partial\psi/\partial z_*, \qquad \rho_*[\overline{w}_*]\cos\varphi = r^{-1}\,\partial\psi/\partial\varphi, \qquad (13.10.1)$$

where r is the radius of the earth (the spherical polar form of the equations is discussed in Section 4.12). Similarly, the two components $[\overline{u}]$ and $[\overline{T}]$ can be expressed in terms of a single variable $[\overline{\Phi}]$, the time and zonally averaged geopotential, because the "rapid" adjustment processes studied in earlier chapters will keep them in approximate thermal-wind balance. Taking from (4.12.15), the meridional momentum equation in spherical polar coordinates, the terms that contribute to the geostrophic balance (7.6.7) (but also including a nonlinear term that can be significant in low latitudes), and appending the hydrostatic equation (6.17.20), there results

$$(2\Omega + r^{-1}[\overline{u}]\sec\varphi)[\overline{u}]\sin\varphi = -r^{-1}\,\partial[\overline{\Phi}]/\partial\varphi,$$
$$H_s^{-1}R[\overline{T}] = \partial[\overline{\Phi}]/\partial z_*. \qquad (13.10.2)$$

For simplicity, the distinction between virtual temperature T_v and temperature T will be dropped, i.e., moisture effects on buoyancy will be ignored.

Thus the zonal mean circulation can be described in terms of two variables ψ and $[\overline{\Phi}]$. The observed field of ψ for the two extreme seasons is shown in Fig. 1.7, whereas the fields of $[\overline{u}]$ and $[\overline{T}]$ that are associated with $[\overline{\Phi}]$ are shown in Fig. 7.9. The equations that determine the distributions of $[\overline{\Phi}]$ and ψ are the remaining two equations, namely, the averaged forms of the zonal momentum equation (4.12.14) and of the temperature equation [see (4.4.6) and (6.17.13)]. These are coupled equations and can be solved only when taken together, but for discussion purposes it is useful to consider the two equations separately.

13.10.2 The Zonal Momentum and Angular Momentum Balances

A useful form of the zonal momentum equations (4.12.14) is in terms of the angular momentum

$$M = (\Omega r\cos\varphi + [\overline{u}])r\cos\varphi. \qquad (13.10.3)$$

The averaged form of this equation can then be written as

$$r^{-1}[\overline{v}]\,\partial M/\partial\varphi + [\overline{w}_*]\,\partial M/\partial z_* = r\cos\varphi(\mathscr{V}_{\text{eddy}} + \mathscr{V}_{\text{friction}}), \qquad (13.10.4)$$

where

$$\mathscr{V}_{\text{eddy}} = -r^{-1}\sec^2\varphi\,\partial\{(uv)_{\text{eddy}}\cos^2\varphi\}/\partial\varphi - \rho_*^{-1}\,\partial\{\rho_*(uw)_{\text{eddy}}\}/\partial z_* \qquad (13.10.5)$$

is the mean rate of convergence of zonal eddy momentum per unit volume and $\mathscr{V}_{\text{friction}}$ represents the "friction" term due to subsynoptic scale processes (i.e., processes with a scale smaller than that of the baroclinic eddies). In the above expression, the brackets and subscript "eddy" denote

$$(uv)_{\text{eddy}} = [\bar{u}^*\bar{v}^*] + [\overline{u'v'}], \tag{13.10.6}$$

i.e., the sum of contributions from both time variations (transient eddies) and longitudinal variations (stationary eddies). Values of the eddy momentum fluxes, as estimated directly from observations, are given by Oort and Rasmusson (1971) and by Newell *et al.* (1972). Values may also be calculated indirectly from data, using the analyses of geopotential, temperature, and wind fields that are made routinely for use in numerical forecast models, and values for the northern hemisphere (latitudes above 20°) so calculated are given by Lau (1978, 1979b) for the winter and by White (1981) for the summer. Values for the southern hemisphere at 500 mb are given by Trenberth (1982).

   In practice, $\mathscr{V}_{\text{eddy}}$ depends on the mean Φ field, in a way very similar to that studied in Section 13.9. To solve the two equations for $[\bar{\Phi}]$ and ψ, this relationship must be known. Examples of methods that attempt to approximate this relationship for use in simplified models of the circulation are given by Green (1970) and by Held and Suarez (1978). For this discussion, however, the eddy fluxes of both heat and momentum will be taken as *given functions* that are known from observations. The principal components of these fluxes, as assumed in quasi-geostrophic theory, are those that contribute to the quasi-geostrophic Eliassen–Palm flux [see (12.9.14)], so that data on observed fluxes can be displayed in an Eliassen–Palm cross section. Figure 13.14 shows the observations (including both transient and stationary eddy contributions) for the northern hemisphere winter as calculated from two different data sets. The similarity with the model result of Fig. 13.13 is remarkable and indicates that much of the observed distribution is achieved through the same processes as those discussed in the previous section.

   Consider now the flow that would be produced if initially there were some specified state, such as the hypothetical radiative–convective equilibrium mentioned earlier, and then the eddy fluxes and any other known forcing terms were suddenly switched on. The response would be determined by the two coupled equations for $[\bar{\Phi}]$ and ψ. In the initial stages, restoring effects would not be important, so the behavior would be as discussed for the frictionless, time-dependent case by Eliassen (1952) [see also Charney (1973)]. However, equilibrium can be reached only when frictional and thermal restoring effects have come into play, the dominant restoring effect depending on the relative time constants. Examples of equilibrium solutions that were determined in this way are given by Schneider and Lindzen (1977) and by Held and Hou (1980), who calculated Hadley circulations driven by thermal forcing alone. Their results showed that the equilibrium reached depends strongly on the nature of the restoring terms. Unfortunately, those terms are not known in practice with any degree of accuracy.

   In the momentum equation (13.10.4), however, $\mathscr{V}_{\text{friction}}$ is thought to be very small except in the surface boundary layer and in the tropics, where momentum transfer in cumulus clouds may be significant [the point is discussed by Thompson and Hartmann (1979)]. Consider, therefore, the solution of (13.10.4), for which $\mathcal{V}_{friction}$ is small, or alternatively, for which $\mathcal{V}_{friction}$ is assumed to be known. The left-hand side of (13.10.4) may be written, after use of (13.10.1), as the Jacobian of ψ and M, so that if M were known, ψ could be determined by integration downward along isolines of M. (The equations are closely related to those that apply to fronts, and this term is simplified if M is used as a horizontal coordinate in place of φ.) In practice, the isolines of M do not differ too much ($\pm 2°$ of latitude) from isolines of φ (i.e., vertical lines). Thus the integration does not differ very much from integration along the vertical, which is equivalent to saying that the term involving $[\bar{w}_*]$ on the left-hand side of (13.10.4) is small compared with the first. Thus $[\bar{v}]$ can be calculated from $\mathcal{V}_{eddy}$ and $\mathcal{V}_{friction}$ if the coefficient $r^{-2} \sec \varphi\, \partial M/\partial \varphi$ is known. In practice, the value of this coefficient is close to f in mid latitudes and at low levels. At high levels in the tropics it may be much smaller than f, but it still has the same sign as f for positions more than a few degrees from the equator.

Since the contribution from vertical eddy fluxes is small in (13.10.4), it may be approximated, when $\mathcal{V}_{friction}$ is small, by

$$[\bar{v}]\, \partial M/\partial \varphi = -r \sec \varphi\, \partial((uv)_{eddy} \cos^2 \varphi)/\partial \varphi. \tag{13.10.7}$$

Alternatively, (13.10.7) may be taken to define the part of the meridional flow associated with eddy fluxes alone. In practice, $(uv)_{eddy}$ (see Figs. 13.14 and 13.16) is mainly positive in the northern hemisphere, with maximum values in the upper troposphere near 30°N, i.e., near the latitude of the jet stream. Thus in the upper troposphere (13.10.7) gives equatorward flow to the north of 30°N, and poleward flow to the south, as seen in Fig. 1.7 [a meridional cross section of $[\bar{v}]$ is given by Lau (1978, Fig. 17)]. Typical values of $[\bar{v}]$ are 0.1–0.3 m s^{-1}. The mid-latitude cell associated with the upper level equatorward flow is called the *Ferrel cell*, and this appears to be entirely a response to the eddy fluxes.

Between the equator and 30°N, (13.10.7) gives northward flow, so at least *part* of the observed meridional flow in the Hadley cell seen in Fig. 1.7a is associated with eddy fluxes. There may well be additional flow associated, say, with heating effects, but any enhancement above the boundary-layer requires other terms in (13.10.7) to be important there. If M is assumed to have its observed distribution, the only possibility left is for $\mathcal{V}_{friction}$ to be important, i.e., transfers of momentum by sub-synoptic scale motions outside the boundary layer must have significant effects. Models showing the possible role of "cumulus friction" have been studied by Schneider and Lindzen (1977) and by Schneider (1977). These models show how a strong Hadley cell could be driven even in the absence of the $\mathcal{V}_{eddy}$ term. In these models, the dominant restoring mechanism can be determined only by considering the coupled equations for $[\bar{\Phi}]$ and ψ. For friction to be important, and thus make a Hadley cell possible even in the absence of $\mathcal{V}_{eddy}$, the time scale for this process must be small compared with other "restoring" effects like Newtonian cooling. The connection between friction and the thermal field can be seen from the fact that friction tends to reduce the vertical shear and hence, by the thermal-wind equation, $\partial[\bar{T}]/\partial \varphi$. Thus friction can serve to flatten isotherms (relative, e.g., to their radiative–convective equilibrium slopes) and reduce available potential energy. (*Note*: In practice the distribution of M is *not* fixed, but is determined as part of the response. In the limiting case considered by Held and Hou (1980), for instance, the M and ψ fields adjust to make the left-hand side of (13.10.4) zero above the boundary layer, i.e., so as to make angular momentum conserved along streamlines.)

Now comes the question of what determines the *surface winds*. In mid-latitudes, the answer is a simple consequence of the fact that (13.10.7) applies except in the surface boundary layer. Calculation of $[\bar{v}]$ from this equation gives a net equatorward mass flux above the boundary layer, so a *poleward Ekman flux* is required within the boundary layer. It follows from (9.2.5) that an eastward stress is required at the surface, and hence from (9.5.1) that the surface wind must be westerly in mid-latitudes. It also follows, since the oceanic Ekman transport is directed oppositely to the atmospheric Ekman transport (see Section 9.2 and Fig. 9.1), that the oceanic Ekman flux beneath the Ferrel cell is equatorward. [An extension of this idea has been used by Holopainen (1967, 1978) to deduce the ocean circulation from upper-wind statistics!]

A more general argument is obtained by integrating ρ_* times (13.10.4) with respect to z_* to obtain the angular momentum balance for a zonal ring of the atmosphere. If the friction term $\mathscr{V}_{\text{friction}}$ is assumed to be dominated by the vertical gradient of the horizontal stress between layers, its effects integrate out to give only a surface contribution and the balance obtained after integration may be written

$$d \left(\int_0^\infty ([\bar{u}][\bar{v}] + (uv)_{\text{eddy}})\rho_* \cos^2 \varphi \, dz_* \right) \Big/ r \, d\varphi = -\tau^x(\varphi) \cos^2 \varphi, \quad (13.10.8)$$

where $\tau^x(\varphi)$ is the eastward surface stress. In other words, the surface torque discussed in Section 2.3 (which includes the mountain torque) is balanced by the gradient of the angular momentum flux across latitude circles that are associated with the motion of the air. (The flux of "planetary angular momentum" associated with the Coriolis term is proportional to the meridional mass flux, and hence is zero.) The variation of τ^x with latitude is shown in Fig. 2.2 and, with the exception of low latitudes, is balanced almost entirely by the eddy term, i.e., the term involving $(uv)_{\text{eddy}}$ (which can be written as the sum of a transient wave contribution and a stationary wave contribution). The contribution from the mean motion is important only at low latitudes in the Hadley cell. The peak value of this contribution in the middle of the cell is about equal to the eddy contribution at this latitude (Newell *et al.*, 1972, Fig. 4.10).

The angular momentum balance is easy to understand once eddy fluxes of momentum, and the reason for these fluxes, are appreciated. However, the information now available about eddy fluxes is relatively recent, and the reason for the mid-latitude surface westerlies was a puzzle for a long time. Lorenz (1967) provides a very interesting discussion of the history and development of ideas about the general circulation. The suggestion that eddies are important came from Jeffreys (1926), who first presented a shallow-water model of the general circulation that gave surface easterlies everywhere as in the model presented in Section 9.16 and Fig. 9.14. He then pointed out that the only way to reconcile (13.10.8) with the existence of surface westerlies is to assume that the eddy contribution is important. Further discussion of the angular momentum budget is given by Newton (1972b) and Newell *et al.* (1972, Chapter 4).

To summarize, the following features of the zonal mean wind field shown in Fig. 7.9 are worth noting:

(i)   The surface wind is generally much smaller than the thermal wind, as represented, say, by the difference between the surface wind and the wind at 200 mb. Thus it is perhaps appropriate to think of the surface winds as being zero to a first level of approximation, and the upper winds as those following from a thermal-wind balance.

(ii)   The surface winds are largely determined by the need for (13.10.8) to be satisfied. In mid latitudes at least, this means that they result from the eddy momentum fluxes since the $[\bar{u}][\bar{v}]$ term is relatively small. In Section 13.9 it was found that eddy momentum fluxes arise mainly during the later stages of the life cycle of eddies, and the direction of the fluxes appears to be the result of wave refraction by the mean field. The structure of the jet stream also influences where and how strongly the waves are absorbed, and thereby is instrumental in determining flux divergence. Since the main ocean currents are driven by the surface winds, they may be said to be largely a consequence of such processes in the upper troposphere!

(iii)   Since the *surface pressure* (shown in Fig. 2.3) is largely in geostrophic balance with the wind near the surface, the surface pressure distribution follows from the surface-wind distribution and is therefore determined by the same processes.

13.10.3   The Heat Balance

Consider first the heat balance in a vertical column. In Chapter 1 it was seen how a radiative–convective equilibrium could be established with heat being absorbed at the ground and transferred upward by convection through the depth of the troposphere. This model therefore gives a balance in the troposphere between radiative cooling and upward transfer by convection. The observed situation at most latitudes is qualitatively similar, with radiative cooling rates [see, e.g., Dopplick (1972)] over the depth of the troposphere of the order of 1 K day^{-1}. The main heating term, which can be identified with a convective effect, is that due to latent heat release. This reaches values of over 2 K day^{-1} (Newell *et al.*, 1974). In the tropics this form of heating extends over the whole depth of the troposphere, but in mid latitudes it is confined to much shallower depths. The difference between the radiative and "convective" terms is called the *diabatic heating rate*, and values calculated by Hantel and Baader (1978) are shown in Fig. 9.10 (these were actually estimated from the advective terms that balance them). If there were no meridional transfer of heat, there would be a radiative–convective equilibrium at each latitude, and the diabatic heating term would be zero. This would give a very much greater equator-to-pole temperature difference than that observed. Because meridional transfers take place, however, the diabatic heating term would be expected to be positive in the tropics and negative at high latitudes. In practice, Fig. 9.10 shows a somewhat more complicated pattern with heating in the tropics (0–10°), but also between 30 and 40°N. Cooling is found both in the subtropics and in high latitudes. This structure is a result of the latent-heating distribution whose geographical variations follows the precipitation pattern shown in Fig. 2.6. The secondary maximum at 30–40°N is associated with the rain belt at these latitudes. In other words, because the "convective" term involves latent heat release that depends on moisture transport, it cannot be regarded as a purely local response to heating at the surface. [The contributions of different terms to the moisture balance are discussed by Rasmusson in Chapter 6 of Newell *et al.* (1972).]

The diabatic heating term can be regarded as a forcing term that is balanced through two types of advective flux: one associated with eddies and the other with the mean meridional circulation. The zonally and time averaged form of the heat equation [cf. (4.4.6) and (6.17.13)] is

$$r^{-1}[\bar{v}]\,\partial[\bar{T}]/\partial\varphi + [\bar{w}_*](\kappa H_s^{-1}[\bar{T}] + \partial[\bar{T}]/\partial z_*) = (Q_{\text{eddy}} + Q_{\text{diab}})/\rho c_p, \quad (13.10.9)$$

where

$$Q_{\text{eddy}}/\rho c_p = -r^{-1}\sec\varphi\,\partial((vT)_{\text{eddy}}\cos\varphi)/\partial\varphi$$

$$-\exp(z_*/\gamma H_s)\,\partial((w_*T)_{\text{eddy}}\exp(-z_*/\gamma H_s))/\partial z_* \quad (13.10.10)$$

is the eddy flux convergence, the eddy terms being the sum of the transient and stationary parts as in (13.10.6), and $Q_{\text{diab}}/\rho c_p$ is the diabatic heating rate in degrees per unit time.

Before considering the terms in this equation, it is useful to look at the vertically integrated energy balance shown in Fig. 1.8. Here, the poleward flux of energy (the upper curve) is required to balance the radiation surplus at the equator and deficit at the poles. The shaded area represents the energy flux in the ocean and the unshaded area the atmospheric flux. The diagram shows that transient eddies carry most of the flux at mid-latitudes with peak values at 40–50°N. [Note that the latent heat flux contributes about 35% to the peak. Details of the various contributions are tabulated by Oort (1971).] The flux associated with the mean meridional circulation is important near the equator.

Information about $(vT)_{\text{eddy}}$ and $(w_*T)_{\text{eddy}}$ is given by Oort and Rassmusson (1971), Newell *et al.* (1974), Lau (1978, 1979a,b), van Loon (1980), and White (1982). According to quasi-geostrophic theory, the main contribution to Q_{eddy} comes from the term in (13.10.10) that involves the *horizontal* flux $(vT)_{\text{eddy}}$, which is proportional to the *vertical* component of the Eliassen–Palm flux displayed in Fig. 13.14. [The flux $[\overline{w_*'T'}]$ may, however, play a significant role in redistributing heat in the vertical (Edmon *et al.*, 1980).] The main features of the $(vT)_{\text{eddy}}$ field (see Figs. 13.14 and 13.15) are that this flux is mainly poleward outside the tropics with largest values around 40–50° latitude, maxima being at the surface and at 200 mb. Thus the eddy heat flux is divergent (giving cooling) between 20° and 40°N but convergent (warming) north of 50°N. The other advective flux is that due to the mean meridional circulation, but outside the tropics this circulation is driven by the eddies, as seen earlier in the section. The heat carried by this circulation therefore can *also* be regarded as being due to the eddies. Thus the extratropical temperature pattern can be regarded as a balance between diabatic heating and the advective fluxes resulting from the presence of eddies.

The dominant term on the left-hand side of (13.10.9) is the one involving $[\bar{w}_*]$, which represents heating due to subsidence or cooling due to upward motion. The zonal mean field of $[\bar{w}_*]$ is given by Lau (1979b, Fig. 16a) for the northern hemisphere

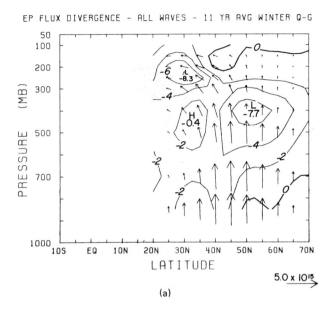

(a)

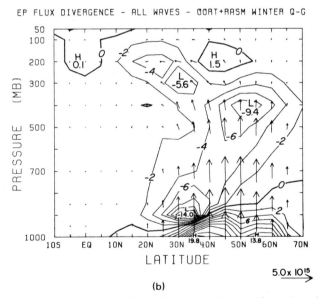

(b)

Fig. 13.14.   Eliassen–Palm cross sections for the total wintertime (transient plus stationary) fluxes in the same format as for that of Fig. 13.13. The contour interval is 2×10^{15} m^3. (a) Based on an 11-year average of NMC data and (b) based on a 5-year average from Oort and Rasmusson (1971). [From Edmon *et al.* (1980, Fig. 5).] Calibration arrows for the vectors are shown at the bottom.

$\mathbb{F}(T), \overline{T}$
850 mb

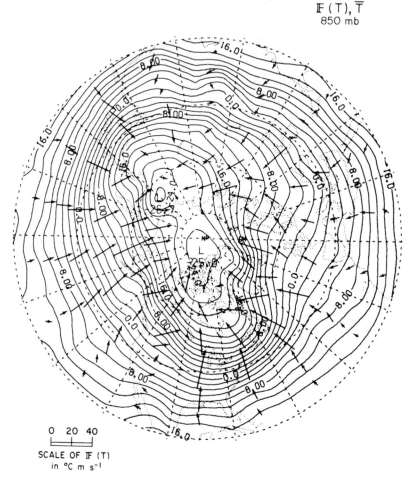

Fig. 13.15.   Vectors showing the wintertime heat flux by transient eddies at 850 mb. The contours show the corresponding mean 850-mb temperature (contour interval 2 K). [From Lau and Wallace (1979, Fig. 3).]

winter, and can also be estimated from Fig. 1.7. Values are mostly under 1 mm s^{-1}. At midlevels near the equator, where Q_{diab} is particularly large and positive, it is balanced largely by cooling due to mean upward motion (Newell *et al.*, 1974, Fig. 7.21). Note, however, that if the $[\bar{v}]$ term dominates the left-hand side of (13.10.4), as seems to be the case, there can be no meridional circulation and hence no $[\overline{w}_{*}]$ term without eddy momentum transfer of some kind, i.e., by synoptic scale eddies or by subsynoptic processes like cumulus friction. This is indicative of the coupled nature of the zonal momentum and temperature equations, and both equations need to be considered to determine the response to a specified forcing.

13.10.4 Longitudinal Structure

Stationary wave patterns in the atmosphere were discussed in Section 12.9 and the structure in the northern hemisphere winter is depicted in Figs. 12.13 and 7.8. Numerical models such as those of Simmons (1982) indicate the importance of low-latitude forcing on these patterns, and forcing by topography and heating variations were discussed in Section 12.9. These patterns and longitudinal variations in surface properties strongly influence the way eddies behave. In particular, eddies tend to grow in favored locations and dissipate further downstream, so different stages in the life cycle tend to occur at different longitudes. This influences the longitudinal distribution of eddy fluxes, which can therefore play a role in maintaining longitudinal variations in the mean flow.

Meridional sections of the mean flow and various eddy statistics are given for the northern hemisphere winter and for different longitudinal sectors by Lau (1978). The jet stream varies a great deal in strength as can be seen in Fig. 7.8b (iii), which shows geopotential topography at the level of the jet stream core. Figure 13.16 shows the zonal velocity field at 250 mb explicitly. The jet stream is strongest over East Asia, where it is located at 30°N and has a maximum of 61 m s^{-1}. It weakens over the Pacific to a minimum of 27 m s^{-1} over western North America, then accelerates to 40 m s^{-1} over eastern North America. It decelerates again over the Atlantic.

The heat flux due to eddies tends to be strongest near places where the jet stream is strongest, horizontal temperature gradients being strongest near these locations as well, e.g., Fig. 13.15 shows vectors of the transient eddy flow at 850 mb super-imposed on the mean temperature distribution [cf. Fig. 7.8b (iii)]. The flux is nearly everywhere down gradient, although Lau and Wallace (1979) found no systematic relation between the magnitude of the flux and the temperature gradient. (Their scatter diagram, however, gives an upper bound of 6×10^6 m^2 s^{-1} on eddy diffusivity, with a median value of 2×10^6 m^2 s^{-1}.) The eddy momentum flux $[\overline{u'v'}]$ (see Fig. 13.16) has its largest value of 100 m^2 s^{-2} at 250 mb over western North America, i.e., at a longitude where the jet stream is weak. This is consistent with the idea that eddies develop (as in the early stages of a life cycle) where the jet stream is strong and decay (as in the later changes of the life cycle) further downstream where the jet stream is relatively weak.

The equations corresponding to (13.10.4) and (13.10.9) for the longitudinally varying flow may be taken as the one for the meridional component of the ageo-strophic flow, which is discussed by Lau (1978), and the heat equation, which is dis-cussed by Lau (1979b). The stationary wave pattern is inevitably associated with accelerations and decelerations of the zonal flow, and these may be seen in Fig. 13.16. The accelerations must be balanced by a meridional ageostrophic flow, and this turns out to be bigger than the ageostrophic flow forced by the eddies by a factor of 3–4. When zonally averaged, however, the acceleration terms for the zonal flow disappear, and the zonal mean meridional flow is in balance with eddy terms as has been seen already.

In the heat balance, the horizontal advection terms associated with the mean motion are *not* small, as in the zonal average, but are as large as or larger than the

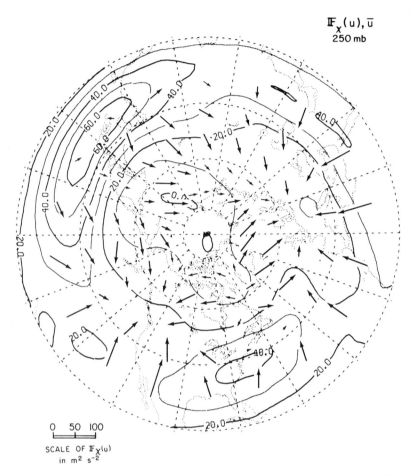

Fig. 13.16. Vectors showing the "irrotational" transient eddy flux of westerly momentum at 250 mb in the winter season. Contours are for the corresponding mean 250-mb zonal wind field (contour interval 10 m s^{-1}). The vector field shown is obtained by expressing the flux as the sum of a nondivergent part (for which a stream function can be defined) and an irrotational part, which can be expressed as the gradient of a scalar. [From Lau and Wallace (1979, Fig. 9).]

other terms. However, (Lau, 1979b, Fig. 20) the vertical advection (by the time-mean flow) term tends to be of sign opposite to that of the horizontal advection (by the time-mean flow) term (it would cancel the horizontal advection term if flow were along isentropes). The diabatic heating, deduced by computing the balancing advective term, is shown in Fig. 13.17 at 700 mb. Note the large positive values off the east coasts of the continents. The region of large heating at 700 mb over the western North Atlantic corresponds to the area of large heat flux from the ocean to the atmosphere seen in Fig. 2.7. This draws attention once more to the connection with the ocean, and the coupled nature of the atmosphere–ocean system. Some insight into the way in which this system operates can be gleaned from the above discussion, but nature is complex and there is still much to be learned!

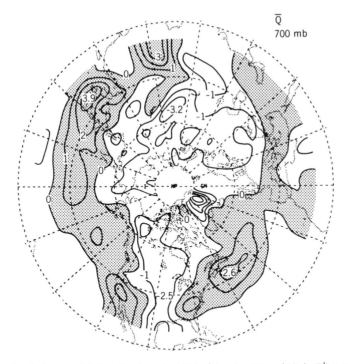

Fig. 13.17.   The distribution of diabatic heating rate at 700 mb (contour interval 1 K day^{-1}) as estimated by Lau (1979b, Fig. 21).

Units
and Their SI Equivalents*

Physical quantity	Unit	Equivalent	Reciprocal
Length	nautical mile (NM)	1.85318 km	km = 0.5396 NM
Mass	tonne (t)	10^3 kg = 1 Mg	
Time	min	60 s	
	hr	3600 s	
	day	86,400 s	s = 1.1574×10^{-5} day
	yr	3.1558×10^7 s	s = 3.1688×10^{-8} yr
Temperature	°C	°C = K − 273.15	
Velocity	knot	0.51477 m s^{-1}	m s^{-1} = 1.9426 knot
Density	gm cm^{-3}	tonne m^{-3} = 10^3 kg m^{-3}	
Force	dyn	10^{-5} N	
Pressure	dyn cm^{-2}	10^{-1} N m^{-2} = 10^{-1} Pa	
	bar	10^5 N m^{-2} = 10^5 Pa	
Energy	erg	10^{-7} J	
	cal (I.T.)	4.1868 J	
	cal (15°C)	4.1855 J	
	cal (thermochemical)	4.184 J	J = 0.239 cal
(Note: The last value is the one used for subsequent conversions involving calories.)			
Energy flux	langley (ly) min^{-1} ⎫ cal cm^{-2} min^{-1} ⎭	697 W m^{-2}	W m^{-2} = 1.434×10^{-3} ly min^{-1}
	ly hr^{-1}	11.6 W m^{-2}	W m^{-2} = 0.0860 ly hr^{-1}
	ly day^{-1}	0.484 W m^{-2}	W m^{-2} = 2.065 ly day^{-1}
	kcal cm^{-2} yr^{-1}	1.326 W m^{-2}	W m^{-2} = 0.754 kly yr^{-1}
Volume flux	Sverdrup	10^6 m^3 s^{-1}	
Latent heat	cal gm^{-1}	4184 J kg^{-1}	J kg^{-1} = 2.39×10^{-4} cal gm^{-1}

* Values are taken from or derived from "The Royal Society Conference of Editors. Metrication in Scientific Journals," 1968, The Royal Society, London.

595

Appendix Two

Useful Values

Molecular mass of dry air, $m_a = 28.966$
Molecular mass of water, $m_w = 18.016$
Universal gas constant, $R_* = 8.31436$ J mole^{-1} K^{-1}
Gas constant for dry air, $R = R_*/m_a = 287.04$ J kg^{-1} K^{-1}
Gas constant for water vapor, $R_v = R_*/m_w = 461.50$ J kg^{-1} K^{-1}
Molecular weight ratio $\epsilon \equiv m_w/m_a = R_a/R_v = 0.62197$
Stefan's constant $\sigma = 5.67 \times 10^{-8}$ W m^{-2} K^{-4}
Acceleration due to gravity, g (in ms^{-2}) as a function of latitude φ and height z (in m)

$$g = (9.78032 + 0.005172 \sin^2 \varphi - 0.00006 \sin^2 2\varphi)(1 + z/a)^{-2}$$

Mean surface value, $\bar{g} = \int_0^{\pi/2} g \cos \varphi \, d\varphi = 9.7976$
Radius of sphere having the same volume as the earth, $a = 6371$ km (equatorial radius = 6378 km, polar radius = 6357 km)
Rotation rate of earth, $\Omega = 7.292 \times 10^{-5}$ s^{-1}
Mass of earth = 5.977×10^{24} kg
Mass of atmosphere = 5.3×10^{18} kg
Mass of ocean = 1.4×10^{21} kg
Mass of water in sediments and rocks = 2×10^{20} kg
Mass of ice on earth = 2.2×10^{19} kg
Mass of water in lakes and rivers = 5×10^{17} kg
Mass of water vapor in atmosphere = 1.3×10^{16} kg
Area of earth = 5.10×10^{14} m^2
Area of ocean = 3.61×10^{14} m^2
Area of land = 1.49×10^{14} m^2
Area of ice sheets and glaciers = 1.62×10^{13} m^2
Area of sea ice = 1.75×10^{13} m^2 in March and 2.84×10^{13} m^2 in September

Appendix Three

Properties of Seawater

A3.1 The Equation of State

It is necessary to know the equation of state for the ocean very accurately to determine stability properties, particularly in the deep ocean. The equation of state defined by the Joint Panel on Oceanographic Tables and Standards (UNESCO, 1981) fits available measurements with a standard error of 3.5 ppm for pressure up to 1000 bars, for temperatures between freezing and 40°C, and for salinities between 0 and 42 (Millero *et al.*, 1980; Millero and Poisson, 1981). The density ρ (in kilograms per cubic meter) is expressed in terms of pressure p (in bars), temperature t (in °C), and practical salinity S. The last quantity is defined in such a way (Dauphinee, 1980) that its value (in practical salinity units or psu) is very close to the old value expressed in parts per thousand (ppt). Its relation to previously defined measures of salinity is given by Lewis and Perkin (1981).

The equation for ρ is obtained in a sequence of steps. First, the density ρ_w of pure water ($S = 0$) is given by

$$\rho_w = 999.842594 + 6.793952 \times 10^{-2}t - 9.095290 \times 10^{-3}t^2 + 1.001685$$
$$\times 10^{-4}t^3 - 1.120083 \times 10^{-6}t^4 + 6.536332 \times 10^{-9}t^5. \tag{A3.1}$$

Second, the density at one standard atmosphere (effectively $p = 0$) is given by

$$\rho(S, t, 0) = \rho_w + S(0.824493 - 4.0899 \times 10^{-3}t + 7.6438 \times 10^{-5}t^2$$
$$- 8.2467 \times 10^{-7}t^3 + 5.3875 \times 10^{-9}t^4)$$
$$+ S^{3/2}(-5.72466 \times 10^{-3} + 1.0227 \times 10^{-4}t$$
$$- 1.6546 \times 10^{-6}t^2) + 4.8314 \times 10^{-4}S^2. \tag{A3.2}$$

Finally, the density at pressure p is given by

$$\rho(S, t, p) = \rho(S, t, 0)/(1 - p/K(S, t, p)), \tag{A3.3}$$

where K is the secant bulk modulus. The pure water value K_w is given by

$$K_w = 19652.21 + 148.4206t - 2.327105t^2 + 1.360477 \times 10^{-2}t^3$$
$$- 5.155288 \times 10^{-5}t^4. \tag{A3.4}$$

The value at one standard atmosphere ($p = 0$) is given by

$$K(S, t, 0) = K_w + S(54.6746 - 0.603459t + 1.09987 \times 10^{-2}t^2$$
$$- 6.1670 \times 10^{-5}t^3) + S^{3/2}(7.944 \times 10^{-2} + 1.6483 \times 10^{-2}t$$
$$- 5.3009 \times 10^{-4}t^2) \tag{A3.5}$$

and the value at pressure p by

$$K(S, t, p) = K(S, t, 0) + p(3.239908 + 1.43713 \times 10^{-3}t$$
$$+ 1.16092 \times 10^{-4}t^2 - 5.77905 \times 10^{-7}t^3) + pS(2.2838 \times 10^{-3}$$
$$- 1.0981 \times 10^{-5}t - 1.6078 \times 10^{-6}t^2) + 1.91075 \times 10^{-4}pS^{3/2}$$
$$+ p^2(8.50935 \times 10^{-5} - 6.12293 \times 10^{-6}t + 5.2787 \times 10^{-8}t^2)$$
$$+ p^2S(-9.9348 \times 10^{-7} + 2.0816 \times 10^{-8}t + 9.1697 \times 10^{-10}t^2). \tag{A3.6}$$

Values for checking the formula are $\rho(0, 5, 0) = 999.96675$, $\rho(35, 5, 0) = 1027.67547$, and $\rho(35, 25, 1000) = 1062.53817$.

Since ρ is always close to 1000 kg m^{-3}, values quoted are usually those of the difference $(\rho - 1000)$ in kilograms per cubic meters as is done in Table A3.1. The table is constructed so that values can be calculated for 98% of the ocean (see Fig. 3.2). The maximum errors in density on straight linear interpolation are 0.013 kg m^{-3} for both temperature and pressure interpolation and only 0.006 for salinity interpolation in the range of salinities between 30 and 40. The error when combining all types of interpolation for the 98% range of values is less than 0.03 kg m^{-3}.

A3.2   Other Quantities Related to Density

Older versions of the equation of state usually gave formulas not for calculating the absolute density ρ, but for the *specific gravity* ρ/ρ_m, where ρ_m is the maximum density of pure water. Since this is always close to unity, a quantity called σ was defined by

$$\sigma = 1000((\rho/\rho_m) - 1) = (1000/\rho_m)(\rho - \rho_m). \tag{A3.7}$$

Since the value of ρ_m is

$$\rho_m = 999.975 \quad \text{kg m}^{-3}, \tag{A3.8}$$

it follows that σ, as defined above, is related to the $(\rho - 1000)$ values by

$$\sigma = (\rho - 1000) + 0.025, \tag{A3.9}$$

i.e., 0.025 must be added to the values of $(\rho - 1000)$ in the table to obtain the old σ value. The notation σ_t (sigma-t) was used for the value of σ calculated at zero pressure, and σ_θ (sigma-theta) for the quantity corresponding to potential density. Another quantity commonly used in oceanography is the specific volume (or steric) *anomaly* δ defined by

$$\delta = v_s(S, t, p) - v_s(35, 0, p) \tag{A3.10}$$

and usually reported in units of 10^{-8} m^3 kg^{-1}.

A3.3 Expansion Coefficients

The thermal expansion coefficient α is given in Table A3.1 in units of 10^{-7} K^{-1} along with its S derivative. The maximum error from pressure interpolation is two units, that from temperature interpolation is three units, and that for salinity interpolation $(30 < S < 40)$ is two units plus a possible round-off error of two units. The salinity expansion coefficient β can be calculated by using the given values of $\partial\rho/\partial S$.

A3.4 Specific Heat

The specific heat at surface pressure is given by Millero *et al.* (1973) and can be calculated in two stages. First, the value in joules per kilogram per degree Kelvin for fresh water is given by

$$c_p(0, t, 0) = 4217.4 - 3.720283t + 0.1412855t^2 - 2.654387 \times 10^{-3}t^3$$
$$+ 2.093236 \times 10^{-5}t^4. \tag{A3.11}$$

Second,

$$c_p(S, t, 0) = c_p(0, t, 0) + S(-7.6444 + 0.107276t - 1.3839 \times 10^{-3}t^2)$$
$$+ S^{3/2}(0.17709 - 4.0772 \times 10^{-3}t + 5.3539 \times 10^{-5}t^2). \tag{A3.12}$$

The formula can be checked against the result $c_p(40, 40, 0) = 3981.050$. The standard deviation of the algorithm fit is 0.074. Values at nonzero pressures can be calculated by using (3.3.1) and the equation of state. The values in Table A3.1 are based on the above formula and a polynomial fit for higher pressures derived from the equation of state by Dr. N. P. Fofonoff. The intrinsic interpolation errors in the table are 0.4, 0.1, and 0.3 J kg^{-1} K^{-1} for pressure, temperature, and salinity interpolation, respectively, and there are additional obvious roundoff errors.

A3.5   Potential Temperature

The *adiabatic lapse rate* Γ is given by (3.6.5) and therefore can be calculated from the above formulas. The definition in Section 3.7.2 can then be used to obtain θ. The following algorithm, however, was derived by Bryden (1973), using experimental compressibility data, to give θ (in °C) as a function of salinity S, temperature t (in °C), and pressure p (in bars) for $30 < S < 40$, $2 < t < 30$, and $0 < p < 1000$:

$$\begin{aligned}
\theta(S, t, p) = {} & t - p(3.6504 \times 10^{-4} + 8.3198 \times 10^{-5}t - 5.4065 \times 10^{-7}t^2 \\
& + 4.0274 \times 10^{-9}t^3) - p(S - 35)(1.7439 \times 10^{-5} \\
& - 2.9778 \times 10^{-7}t) - p^2(8.9309 \times 10^{-7} - 3.1628 \times 10^{-8}t \\
& + 2.1987 \times 10^{-10}t^2) + 4.1057 \times 10^{-9}(S - 35)p^2 \\
& - p^3(-1.6056 \times 10^{-10} + 5.0484 \times 10^{-12}t).
\end{aligned} \tag{A3.13}$$

A check value is $\theta(25, 10, 1000) = 8.4678516$, and the standard deviation of Bryden's polynomial fit was 0.001 K. Values in Table A3.1 are given in millidegrees, the intrinsic interpolation errors being 2, 0.3, and 0 millidegrees for pressure, temperature, and salinity interpolation, respectively.

A3.6   Speed of Sound

The speed of sound c_s can be calculated from the equation of state, using (3.7.16). Values given in Table A3.1 use algorithms derived by Chen and Millero (1977) on the basis of direct measurements. The formula applies for $0 < S < 40$, $0 < t < 40$, $0 < p < 1000$ with a standard deviation of 0.19 m s^{-1}. Values in the table are given in meters per second, the intrinsic interpolation errors being 0.05, 0.10, and 0.04 m s^{-1} for pressure, temperature, and salinity interpolation, respectively.

A3.7   Freezing Point of Seawater

The freezing point t_f of seawater (in °C) is given (Millero, 1978) by

$$\begin{aligned}
t_f(S, p) = {} & -0.0575S + 1.710523 \times 10^{-3}S^{3/2} - 2.154996 \times 10^{-4}S^2 \\
& - 7.53 \times 10^{-3}p.
\end{aligned} \tag{A3.14}$$

The formula fits measurements to an accuracy of ± 0.004 K.

p (bar)	S	t (°C)	$\rho - 1000$ (kg m⁻³)	$\frac{\partial \rho}{\partial S}$	α (10⁻⁷ K⁻¹)	$\frac{\partial \alpha}{\partial S}$	c_p (J kg⁻¹ K⁻¹)	$\frac{\partial c_p}{\partial S}$	θ (10⁻³ °C)	$\frac{\partial \theta}{\partial S}$	c_s (m s⁻¹)	$\frac{\partial c_s}{\partial S}$
0	35	-2	28.187	0.814	254	33	3989	-6.2	-2000	0	1439.7	1.37
0	35	0	28.106	0.808	526	31	3987	-6.1	0	0	1449.1	1.34
0	35	2	27.972	0.801	781	28	3985	-5.9	2000	0	1458.1	1.31
0	35	4	27.786	0.796	1021	26	3985	-5.8	4000	0	1466.6	1.29
0	35	7	27.419	0.788	1357	23	3985	-5.6	7000	0	1478.7	1.25
0	35	10	26.952	0.781	1668	20	3986	-5.5	10000	0	1489.8	1.22
0	35	13	26.394	0.775	1958	17	3988	-5.3	13000	0	1500.2	1.19
0	35	16	25.748	0.769	2230	15	3991	-5.2	16000	0	1509.8	1.16
0	35	19	25.022	0.764	2489	14	3993	-5.1	19000	0	1518.7	1.13
0	35	22	24.219	0.760	2734	12	3996	-4.9	22000	0	1526.8	1.10
0	35	25	23.343	0.756	2970	11	3998	-4.9	25000	0	1534.4	1.08
0	35	28	22.397	0.752	3196	9	4000	-4.8	28000	0	1541.3	1.06
0	35	31	21.384	0.749	3413	8	4002	-4.7	31000	0	1547.6	1.03
100	35	-2	32.958	0.805	552	31	3953	-5.8	-2029	-2	1456.1	1.38
100	35	0	32.818	0.799	799	28	3953	-5.7	-45	-2	1465.5	1.35
100	35	2	32.629	0.793	1031	26	3954	-5.6	1939	-2	1474.5	1.33
100	35	4	32.393	0.788	1251	24	3955	-5.5	3923	-2	1483.1	1.30
100	35	7	31.958	0.781	1559	21	3957	-5.3	6901	-1	1495.1	1.26
100	35	10	31.431	0.774	1844	18	3960	-5.2	9879	-1	1506.3	1.22
100	35	13	30.818	0.769	2111	16	3963	-5.1	12858	-1	1516.7	1.19
100	35	16	30.126	0.763	2363	14	3967	-5.0	15838	-1	1526.4	1.16
100	35	19	29.359	0.759	2603	13	3970	-4.9	18819	-1	1535.3	1.13
200	35	-2	37.626	0.797	834	28	3922	-5.5	-2076	-3	1472.8	1.39
200	35	0	37.429	0.791	1058	26	3923	-5.4	-107	-3	1482.3	1.36
200	35	2	37.187	0.786	1269	24	3925	-5.3	1862	-3	1491.2	1.33
200	35	4	36.903	0.781	1469	22	3927	-5.2	3832	-3	1499.8	1.30
200	35	7	36.402	0.774	1750	19	3931	-5.1	6789	-3	1511.8	1.26
300	35	-2	42.191	0.789	1101	26	3893	-5.2	-2140	-5	1489.9	1.39
300	35	0	41.941	0.783	1303	24	3896	-5.1	-186	-5	1499.3	1.36
300	35	2	41.649	0.778	1494	22	3899	-5.0	1771	-5	1508.2	1.33
300	35	4	41.319	0.774	1676	20	3903	-5.0	3728	-5	1516.6	1.30
400	35	-2	46.658	0.781	1351	24	3867	-4.9	-2221	-7	1507.2	1.39
400	35	0	46.356	0.776	1534	22	3871	-4.8	-279	-6	1516.5	1.36
400	35	2	46.017	0.771	1707	20	3876	-4.8	1665	-6	1525.3	1.33
400	35	4	45.643	0.767	1872	19	3880	-4.7	3610	-6	1533.7	1.30
500	35	-2	51.029	0.773	1587	22	3844	-4.7	-2316	-8	1524.8	1.38
500	35	0	50.678	0.769	1751	20	3849	-4.6	-386	-8	1534.0	1.35
500	35	2	50.293	0.764	1907	19	3854	-4.6	1546	-7	1542.7	1.32
600	35	-2	55.305	0.766	1807	20	3824	-4.4	-2426	-9	1542.6	1.37
600	35	0	54.908	0.762	1954	18	3829	-4.4	-506	-9	1551.6	1.34
600	35	2	54.481	0.758	2094	17	3835	-4.4	1416	-9	1560.2	1.31

Properties of Moist Air

A4.1 Methods of Specifying Moisture Content

(a) *The vapor concentration*, or absolute humidity ρ_v is the mass of vapor per unit volume of moist air.

(b) *The specific humidity q* is the mass of vapor per unit mass of moist air:

$$q = \rho_v/\rho. \tag{A4.1}$$

(c) *The mixing ratio r* is the ratio of the mass of vapor to the mass of dry air:

$$r = q/(1 - q). \tag{A4.2}$$

(d) *The vapor pressure e'* of water vapor in moist air is defined as the function of p and q given by (3.1.12)

$$e'/p = q/(\epsilon + (1 - \epsilon)q) = r/(\epsilon + r) = r/(0.62197 + r). \tag{A4.3}$$

If air were an ideal gas mixture, e' would be exactly equal to the partial pressure e of water vapor. In practice it will be slightly different.

(e) *The relative humidity U* is the ratio of r to the saturation mixing ratio r_w of moist air relative to a plane water surface:

$$U = r/r_w = q(1 - q_w)/(q_w(1 - q)) \tag{A4.4}$$

It is usually expressed as a percentage.

A4.2   Saturation Vapor Pressure

(a)   The *saturation vapor pressure* $e_w(T)$ of pure water vapor over a plane water surface is discussed in Section 3.4. Values are given in Table 94 of the Smithsonian Meteorological Tables. For temperatures t between $\pm 40°C$, the value in millibars is given correct to 1 part in 500 by

$$\log_{10} e_w(t) = (0.7859 + 0.03477t)/(1 + 0.00412t). \tag{A4.5}$$

(b)   In air, the partial pressure e'_w of water vapor at saturation is not exactly e_w but is given by

$$e'_w = f_w e_w(T), \tag{A4.6}$$

where f_w lies between 1 and 1.006 for observed atmospheric conditions, values being given in Table 89 of the Smithsonian Meteorological Tables. The value of f_w is given correct to 2 parts in 10^4 by

$$f_w = 1 + 10^{-6}p(4.5 + 0.0006t^2), \tag{A4.7}$$

where p is the pressure in millibars. Values of r_w and q_w follow from (A4.3) with $e' = e'_w$, and are shown as functions of p and T in Fig. 3.6a.

(c)   The saturation vapor pressure $e_i(T)$ of pure water vapor over ice is given in Table 96 of the Smithsonian Meteorological Tables and satisfies (correct to 3 parts in 1000 for $0 > t > -40$)

$$\log_{10} e_i(t) = \log_{10} e_w(t) + 0.00422t. \tag{A4.8}$$

(d)   The saturation partial pressure e'_i in moist air is f_i times e_i. Values of f_i are given in Table 90 of the Smithsonian Meteorological Tables, and f_i is given correct to 1 part in 10^4 by (A4.7).

(e)   The saturation vapor pressure over a salt solution is less than over fresh water. For seawater, the reduction is about 2% (Kraus, 1972, p. 46).

A4.3   Further Quantities Related to Moisture Content

(a)   *Dew point* T_d is the temperature to which air must be cooled at constant pressure and constant mixing ratio to reach saturation with respect to a plane water surface. The *frost point* is the equivalent with respect to a plane ice surface.

(b)   *Lifting condensation level* is the level at which a parcel of moist air lifted adiabatically becomes saturated.

(c)   *Wet-bulb temperature* T_w is the temperature to which a parcel of air is cooled by evaporating water into it gradually, adiabatically, and at constant pressure until it is saturated. It is measured directly by a thermometer whose bulb is covered by a moist cloth over which air is drawn.

(d)   From the above definitions, it follows that for a parcel with pressure p the θ contour through (p, T), the θ_e^* or θ_w^* contour through (p, T_w), and the r_w or q_w contour through (p, T_d) all intersect at the lifting condensation level (Normand's rule).

A4.4   Latent Heats

The latent heat of vaporization L_v is given by (3.4.6), i.e.,

$$L_v(t) = 2.5008 \times 10^6 - 2.3 \times 10^3 t \quad \text{J kg}^{-1}, \tag{A4.9}$$

where t is the temperature (in degrees Celsius). The latent heat of sublimation L_s is given by

$$L_s(t) = 2.839 \times 10^6 - 3.6(t + 35)^2 \quad \text{J kg}^{-1}. \tag{A4.10}$$

A4.5   Lapse Rates

The dry adiabatic lapse rate is given to within 0.3% by

$$\Gamma =_, g/c_p. \tag{A4.11}$$

The saturation adiabatic lapse rate (for liquid water) is given approximately by

$$\Gamma_s = 6.4 - 0.12t + 2.5 \times 10^{-5} t^3$$
$$+ [-2.4 + 10^{-3}(t - 5)^2](1 - p/p_r) \quad \text{K km}^{-1}, \tag{A4.12}$$

where $p_r = 1000$ mb and t is the temperature (in °C). The maximum error in the range $|t| < 40$ and $500 < p < 1000$ is 0.2 K km^{-1}. Accurate values of Γ_s are given in Table 79 of the Smithsonian Meterological Tables, whereas Table 80 gives values for the ice stage.

A List of Atlases
and Data Sources

Note that (i) many references could be placed in more than one subdivision but are given only once; (ii) if a work is listed in the main reference section, only the author and year are quoted; (iii) the list can be extended by consulting the bibliographies given in the references below.

A. Catalogs and Bibliographies

Jenne, R. L. (1975). "Data Sets for Meteorological Research," NCAR Tech. Note 1A-111. Natl. Center Atmos. Res., Boulder, Colorado.
Keehn, P. A., and Rigby, M., eds. (1968). "Bibliography on Marine Atlases." Am. Meteorol. Soc., Washington, D.C.
Ropelewski, C. F., Predoehl, M. C., and Platto, M. (1980). "A Quick Reference to Selected Climate Data." Center for Environmental Assessment Services, Washington, D.C.
Stommel, H., and Fieux, M. (1978). "Oceanographic Atlases: A Guide to Their Geographic Coverage and Contents." Woods Hole Press, Woods Hole, Massachusetts.

B. Atmospheric Circulation

Atkinson, G. D., and Sadler, J. C. (1970). "Mean Cloudiness and Gradient Level Wind Charts over the Tropics." Tech. Rep. 215, Vols. 1 and 2. Air Weather Service, U.S. Air Force, Washington, D.C.
Lau, N.C., White, G. H., and Jenne, R. L. (1981). Circulation statistics for the extratropical northern hemisphere based on NMC analyses. NCAR Tech. Note TN-171 + STR. Natl. Center Atmos. Res., Boulder, Colorado.
Oort, A. H. (1982). Global atmospheric circulation statistics, 1958–1973. *NOAA Prof. Pap.*
Ramage, C. S., and Raman, R. V. (1972). "Meteorological Atlas of the Indian Ocean Expedition," Vol. 2. Natl. Sci. Found., Washington, D.C.'
Sadler, J. C. (1975). The upper tropospheric circulation over the global tropics. *Univ. Hawaii* **UH MET 75-05**, 1–35. Dept. of Meteorol., Univ. of Hawaii, Honolulu.
Sadler, J. C., Oda, L., and Kilonsky, B. J. (1976). Pacific Ocean cloudiness from satellite observations. *Univ. Hawaii* **UH MET 76-01**, 1–137. Dept. of Meteorol., Univ. of Hawaii, Honolulu.
Shideler, D. H., and Sadler, J. C. (1979). Pacific Ocean cloudiness from satellite observations. Part II. *Univ. Hawaii* **UH MET 79-13**, 1–56.

U.S. Air Force and U.S. Dept. of Commerce (1971). "Global Atlas of Relative Cloud Cover 1967–1970." USAF/USDC, Washington, D.C.

van de Boogard, H. (1977). The mean circulation of the tropical and subtropical atmosphere—July. *NCAR* **TN-118 STR**, 1–41. Natl. Center Atmos. Res., Boulder, Colorado.

Wahl, E. W., and Lahey, J. F. (1969). "A 700mb Atlas for the Northern Hemisphere." Univ. of Wisconsin Press, Madison.

Corby (1969), Crutcher and Meserve (1970), Newell *et al.* (1972–1974), Newton (1972a), Oort and Rasmusson (1971), Palmén and Newton (1969), Schutz and Gates (1971, 1972), Taljaard *et al.* (1969), Winston *et al.* (1979), and other references in Section 13.10.

C. Precipitation

Dorman, C. E., and Bourke, R. H. (1979). Precipitation over the Pacific Ocean 30°N to 60°N. *Mon. Weather Rev.* **107,** 896–910.

Dorman, C. E., and Bourke, R. H. (1981). Precipitation over the Atlantic Ocean, 30°N to 70°N. *Mon. Weather Rev.* **109,** 554–563.

Jaeger, L. (1976). Monatskarten des Niederschlage für die ganze Erde, "Berichte des Deutschen Wetterdienstes," No. 139, Offenbach.

Rao, M. S. V., Abbott, W. V., and Theon, J. S. (1976). "Satellite Derived Oceanic Rainfall Atlas." NASA, Washington, D.C.

Taylor, R. C. (1973). "An Atlas of Pacific Islands Rainfall." University of Hawaii, Inst. Geophys., Honolulu.

Baumgartner and Reichel (1975).

D. Surface Winds†

Fernandez-Partagas, J., Samuels, G., and Schott, F. (1980). "Surface Wind Maps for the Western Indian Ocean Jan–Sep 1979," Tech. Rep. TR80-4. Rosenthiel School Mar. Atmos. Sci., University of Miami, Miami, Florida.

Goldenberg, S. B., and O'Brien, J. J. (1981). Time and space variability of tropical Pacific wind stress. *Mon. Weather Rev.* **109,** 1190–1207.

Han, Y.-J., and Lee, S.-W. (1981). A new analysis of monthly mean wind stress over the global ocean. Climatic Res. Inst. Rep. No. 26, Oregon State Univ., Corvallis.

Hellerman, S. (1980). Charts of the variability of the wind stress over the tropical Atlantic. *Deep-Sea Res.* **26,** Suppl. 2, 63–75.

Wearn, R. B., and Baker, D. J. (1982). Wind stress variability over the Southern Ocean 1976–78. *J. Phys. Oceanogr.* **12.**

Anderson (1982). Bunker (1976), Fieux and Stommel (1977), Hellerman (1967), Leetmaa and Bunker (1978), White and Saur (1981), Wyrtki and Meyers (1975, 1976).

E. Surface Temperature, Pressure, Currents, Winds, Sea Ice, etc.

Deutsches Hydrographisches Institut, Hamburg. Monatskarten für den Indischen Ozean (Publ. No. 2422, 1960), Nordatlantischen Ozean (Publ. No. 2420, 1967), Südatlantischen Ozean (Publ. No. 2421, 1971).

Hastenrath, S., and Lamb, P. (1977). "Climatic Atlas of the Tropical Atlantic and Eastern Pacific Ocean." Univ. of Wisconsin Press, Madison.

Japan Oceanographic Data Center (1979). "Marine Environmental Atlas. Currents; Adjacent Seas of Japan." Japan Hydrographic Association, Tokyo.

Koninklijk Nederlands Meteorologisch Instituut (1952). "Oceanografische en Meteorologische Gegevens," Publ. 135 (Indische Oceaan).

Landsberg, H. E., ed. (1969). "World Survey of Climatology" (A series of many vols). Elsevier, Amsterdam.

† See also Section E.

Ministerstvo Oborony SSSR (1980). "Atlas okeanov Severnyy Ledovityy Okean" [Atlas of the Oceans. The Arctic Ocean]. Glavnoye Upravleniye Navigatsii i Okeanografii Ministerstvo Oborony, Moscow, SSSR.

Stidd, C. K. (1974). Ship drift components: Means and standard deviations. *Scripps Inst. Oceanogr. Ref. Ser.* **74-33,** 1–57.

USSR Academy of Science (1964). "Physical Geographical Atlas of the World." Moscow.

Zwally, H. J., Comiso, J. C., Parkinson, C. L., Campbell, W. J., Carsey, F., and Gloerson, P. (1982). "Antarctic Sea Ice Cover 1973–76 from Satellite Passive Microwave Observations." NASA, Greenbelt, Maryland.

J. F. Garrett (1980), Hastenrath and Lamb (1979), Perlroth (1969), U.S. Department of Commerce (1963, 1965, 1974, 1976, 1977, 1978, 1979).

F. Air Sea Heat Exchanges, etc.

Clark, N. E., Eber, L., Laurs, R. M., Renner, J. A., and Saur, J. F. T. (1974). Heat exchange between ocean and atmosphere in the eastern North Pacific for 1961–71. *NOAA Tech. Rep., NMFS SSRF* **NMFS SSRF-682,** 1–108.

Hastenrath, S., and Lamb, P. (1978). "Heat Budget Atlas of the Tropical Atlantic and Eastern Pacific Oceans." Univ. of Wisconsin Press, Madison.

Hastenrath, S., and Lamb, P. (1979). "Climatic Atlas of the Indian Ocean. Part II. The Oceanic Heat Budget." Univ. of Wisconsin Press, Madison.

Privett, D. W. (1960). The exchange of energy between the atmosphere and the oceans of the southern hemisphere. *Geophys. Mem.* (U.K. Met. Off.) No. 104, 1–61.

Ramage, C. S., Miller, F. R., and Jefferies, C. (1972). "Meteorological Atlas of the Indian Ocean Expedition," Vol. I. The Surface Climate of 1963 and 1964. Natl. Sci. Found., Washington, D.C.

Ramage, C. S., Adams, C. W., Hori, A. H., Kilonsky, B. H., and Sadler, J. C. (1980). Meteorological atlas of the 1972–73 El Niño. *Univ. Hawaii UH Met* **90-03,** 1–101. Dept. of Meteorol., Univ. of Hawaii, Honolulu.

Weare, B. C., Strub, P. T., and Samuel, M. D. (1980). "Marine Climate Atlas of the Tropical Pacific Ocean." University of California, Davis.

Budyko (1963, 1974), Bunker (1976, 1980), Hastenrath (1980), Malkus (1962), Stommel (1980), Wyrtki (1965).

G. Sea Level

Permanent Service for Mean Sea Level (1978). "Monthly and Annual Mean Heights of Sea Level," 3 vols. Inst. Ocean Sci., Bidston, Merseyside, United Kingdom.

H. Time Series and Variability

Fofonoff, N. P., and Tabata, S. (1966). Variability of oceanographic conditions between ocean station P and Swiftsure Bank off the Pacific coast of Canada. *J. Fish. Res. Board Can.* **23,** 825–868.

Helland, P. (1963). Temperature and salinity variations in the upper layers at Ocean Weather Station M (66°N 2°E). *Arbok Univ. Bergen, Mat.-Naturvitensk. Ser.* No. 6.

Husby, D. M., and Seckel, G. R. (1975). Large-scale air-sea interactions at Ocean Weather Station V, 1951–71. *NOAA Tech. Rep. NMFS SSRF* **NMFS SSRF-696,** 1–44.

Lazier, J. R. N. (1980). Oceanographic conditions at O.W.S.BRAVO 1964–1974. *Atmos.-Ocean* **18,** 227–238.

Lee, V., and Wunsch, C., eds. (1977). Atlas of the Mid-Ocean Dynamics Experiment (MODE-1). MIT, Cambridge, Massachusetts.

Schroeder, E., and Stommel, H. (1969). How representative is the series of Panulirus stations of monthly mean conditions off Bermuda? *Prog. Oceanogr.* **5,** 31–40.

Smed, J. (1947). Monthly anomalies of surface temperature in the sea around South Greenland during the years 1876–1939. *Anne. Biol.* **2,** 16–22 (updated in later issues).

Tabata, S. (1965). Variability of oceanographic conditions at Ocean Station P in the northeast Pacific Ocean. *Trans. R. Soc. Can.* **3,** 367–418.

White, W. B., and Walker, E. A. (1974). Time and depth scales of anomalous sub-surface temperature at ocean weather stations P, N and V in the North Pacific. *J. Geophys. Res.* **79**, 4517–4522.

J.   Surface and Near-Surface Ocean Data

Robinson, M. K. (1976). "Atlas of North Pacific Ocean Monthly Mean Temperatures and Mean Salinities of the Surface Layer." Dept. of the Navy, Naval Oceanogr. Office, Washington, D.C.

Robinson, M. K., Bauer, R. A., and Schroeder, E. H. (1979). "Atlas of North Atlantic-Indian Ocean Monthly Mean Temperatures and Mean Salinities of the Surface Layer." Naval Oceanogr. Office, Bay St. Louis, Mississippi.†

K.   Oceanographic Atlases

Dietrich, G. (1969). "Atlas of the Hydrography of the Northern North Atlantic." Cons. Int. Explor. Mer., Charlottenlund Slot, Denmark.

Düing, W., Ostapoff, F., and Merle, J., eds. (1980). "Physical Oceanography of the Tropical Atlantic during GATE." University of Miami, Coral Gables, Florida.

Fuglister, F. C. (1960). "Atlantic Ocean Atlas." WHOI, Woods Hole, Massachusetts.

GEOSECS (1981) "Atlantic Expedition" (2 vols.), "Pacific Atlas." Natl. Sci. Foundation, Washington, D.C.

Gordon, A. L., Molinelli, E., and Baker, T. (1982). "Oceanographic Atlas of the Southern Ocean." Columbia Univ. Press, New York.

Japan Oceanographic Data Center (1975). "Marine Environmental Atlas: North-western Pacific Ocean," Vol. 1. Japan Hydrographic Association, Tokyo (Vol. 2, 1978).

Kolesnikov, A. G. (1973). "Equalant I and II." UNESCO, Paris.

Levitus, S. (1982). Climatological Atlas of the World Ocean. *NOAA Prof. Pap.*

Love, C. M., ed. (1972). "EASTROPAC Atlas," Vol. 1. U.S. Dept of Commerce, NOAA, Washington, D.C. (Vol. 3, 1971; Vol. 5, 1972; Vol. 7, 1973; Vol. 9, 1975; Vol. 11, 1977).

Miller, A. R., Tchernia, P., Charnock, H., and McGill, D. A. (1970). "Mediterranean Sea Atlas." WHOI, Woods Hole, Massachusetts.

Reid, J. L. (1965). "Intermediate Waters of the Pacific Ocean." Johns Hopkins Press, Baltimore, Maryland.

Sayles, M. A., Aagaard, K., and Coachman, L. K. (1979). "Oceanographic Atlas of the Bering Sea Basin." Univ. of Washington Press, Seattle.

Tsuchiya, M. (1968). "Upper Waters of the Intertropical Pacific Ocean." Johns Hopkins Press, Baltimore, Maryland

Worthington, L. V., and Wright, W. R. (1970). "North Atlantic Ocean Atlas." WHOI, Woods Hole, Massachusetts.

Wyrtki, K. (1971). "Oceanographic Atlas of the International Indian Ocean Expedition." Natl. Sci. Found., Washington, D.C.

L.   Data Centers

Meteorological and oceanographic data can be obtained from Data Centers. Many countries have their own National Data Centers, and there are also World Data Centers, such as

World Data Center A for Meteorology
National Climatic Center
Federal Building, Asheville, North Carolina 28801

World Data Center A for Oceanography
National Oceanographic Data Center/NOAA
Washington, D.C. 20235

† Also available on magnetic tape as "Bauer–Robinson Numerical Atlas" from Fleet Numerical Oceanography Center, Monterey, California.

References

Abbe, C. (1877). "The Mechanics of the Earth's Atmosphere. A Collection of Translations," First collection in Smithsonian Institute Report, pp. 376–478. Smithsonian Institution, Washington, D.C.

Abbe, C. (1893). "The Mechanics of the Earth's Atmosphere. A Collection of Translations," Second collection in Smithson. Misc. Collect. No. 34. Smithsonian Institution, Washington, D.C.

Abbe, C. (1910). "The Mechanics of the Earth's Atmosphere. A Collection of Translations," Third collection in Smithson. Misc. Collect. No. 51. Smithsonian Institution, Washington, D.C.

Abramowitz, M., and Stegun, I. A. (1964). "Handbook of Mathematical Functions." National Bureau of Standards, Washington, D.C.

Accad, Y., and Pekeris, C. P. (1978). Solution of the tidal equations for the M_2 and S_2 tides in the world oceans from a knowledge of the tidal potential alone. *Philos. Trans. R. Soc. London, Ser. A* **290**, 235–266.

Adams, J. K., and Buchwald, V. T. (1969). The generation of continental shelf waves. *J. Fluid Mech.* **35**, 815–826.

Airy, G. B. (1845). "Tides and Waves," Vol. 5. Encyclopaedia Metropolitana, London.

Alaka, M. A., ed. (1960). "The Airflow over Mountains," Tech. Note 34. World Meteorol. Organ., Geneva.

Aleem, A. A. (1967). Concepts of currents, tides and winds among medieval Arab geographers in the Indian Ocean. *Deep-Sea Res.* **14**, 459–465.

Allen, J. S. (1975). Coastal trapped waves in a stratified ocean. *J. Phys. Oceanogr.* **5**, 300–325.

Allen, J. S. (1980). Models of wind-driven currents on the continental shelf. *Annu. Rev. Fluid Mech.* **12**, 389–433.

Allen, J. S., and Smith, R. L. (1981). On the dynamics of wind-driven shelf currents. *Philos. Trans. R. Soc. London, Ser. A* **302**, 617–634.

Allison, L. J., Steranka, J., Holub, R. J., Hansen, J., Godshall, F. A., and Prabhakara, C. (1972). "Air–Sea Interaction in the Tropical Pacific Ocean," Tech. Note D-6684. Natl. Aeronaut. Space Admin., Washington, D.C.

Anati, D. A., Assaf, G., and Thompson, R. O. R. Y. (1977). Laboratory models of sea straits. *J. Fluid Mech.* **81**, 341–351.

Anderson, D. L. T. (1976). The low-level jet as a western boundary current. *Mon. Weather Rev.* **104**, 907–921.

Anderson, D. L. T. (1979). Basin models: The general circulation of the world ocean. *Dyn. Atmos. Oceans* **3**, 345–371.

Anderson, D. L. T. (1982). Low latitude seasonal adjustment in the Atlantic. *Prog. Oceanogr.* (in press).

Anderson, D. L. T., and Gill, A. E. (1975). Spin-up of a stratified ocean, with application to upwelling. *Deep-Sea Res.* **22**, 583–596.

Anderson, D. L. T., and Gill, A. E. (1979). Beta-dispersion of inertial waves. *J. Geophys. Res.* **84**, 1836–1842.

Anderson, D. L. T., and Killworth, P. D. (1977). Spin-up of a stratified ocean with topography. *Deep-Sea Res,* **24**, 709–732.

Anderson, D. L. T., and Killworth, P. D. (1979). Non-linear propagation of long Rossby waves. *Deep-Sea Res.* **26**, 1033–1050.

Anderson, D. L. T., and Rowlands, P. B. (1976a). The role of inertia-gravity and planetary waves in the response of a tropical ocean to the incidence of an equatorial Kelvin wave on a meridional boundary. *J. Mar. Res.* **34**, 295–312.

Anderson, D. L. T., and Rowlands, P. B. (1976b). The Somali current response to the S.W. Monsoon: The relative importance of local and remote forcing. *J. Mar. Res.* **34**, 395–417.

Anderson, D. L. T., Bryan, K., Gill, A. E., and Pacanowski, R. C. (1979). The transient response of the North Atlantic—some model studies. *J. Geophys. Res.* **84**, 4795–4815.

Andrews, D. G., and McIntyre, M. E. (1978a). An exact theory of nonlinear waves on a Lagrangian-mean flow. *J. Fluid Mech.* **89**, 609–646.

Andrews, D. G., and McIntyre, M. E. (1978b). On wave-action and its relatives. *J. Fluid Mech.* **89**, 647–664.

Arakawa, A. (1966). Computational design for long term numerical integration of the equations of fluid motion: Two-dimensional incompressible flow. Part I. *J. Comput. Phys.* **1**, 119–143.

Arakawa, A., and Schubert, W. (1974). Interaction of cumulus cloud ensemble with the large scale environment. *J. Atmos. Sci.* **31**, 674–701.

Armi, L., and Millard, R. C. (1976). The bottom boundary layer of the deep ocean. *J. Geophys. Res.* **81**, 4983–4990.

Arnold, V. I. (1965). Conditions for nonlinear stability of stationary plane curvilinear flows of an ideal fluid. *Dokl. Akad. Nauk. SSSR* **162**, 975–978; *Sov. Math. (Engl. Transl.)* **6**, 331–334.

Arya, S. P. S., and Wyngaard, J. C. (1975). Effect of baroclinicity on wind profiles and the geostrophic drag law for the convective planetary boundary layer. *J. Atmos. Sci.* **32**, 767–778.

Babinet, M. (1859). Influence du mouvement de rotation de la terre sur le cours des rivières. *C. R. Hebd. Seances Acad. Sci.* **49**, 638–641.

Baines, P. G. (1971). The reflection of internal/inertial waves from bumpy surfaces. *J. Fluid Mech.* **46**, 273–291.

Baines, P. G. (1976). The stability of planetary waves on a sphere. *J. Fluid Mech.* **73**, 193–213.

Baines, P. G., and Davies, P. A. (1980). Laboratory studies of topographic effects in rotating and/or stratified fluids. *In* "Orographic Effects in Planetary Flows," Chapter 8, pp. 233–299. GARP Publ. Ser. No. 23, WMO-ICSU, Geneva.

Baines, W. D., and Turner, J. S. (1969). Turbulent buoyant convection from a source in a confined region. *J. Fluid Mech.* **37**, 51–80.

Bakun, A., McLain, D. R., and Mayo, F. V. (1974). The mean annual cycle of coastal upwelling off western North America as observed from surface measurements. *Fish. Bull.* **72**, 843–844.

Ball, F. K. (1964). Energy transfer between external and internal gravity waves. *J. Fluid Mech.* **19**, 465–478.

Balsley, B. B., and Gage, K. S. (1980). The MST radar technique: Potential for middle atmosphere studies. *Pure Appl. Geophys.* **118**, 452–493 (reprinted in Venkateswaran and Sundararaman, 1980).

Bang, N. D. (1970). Dynamic interpretations of a detailed surface temperature chart of the Agulhas Current retroflexion and fragmentation area. *S. Afr. Geogr. J.* **52**, 67–76.

Bang, N. D. (1973a). Characteristics of an intense ocean frontal system in the upwell régime west of Capetown. *Tellus* **25**, 256–265.

Bang, N. D. (1973b). The southern Benguela system: Finer oceanic structure and atmospheric determinants, Ph.D. Thesis, University of Capetown.

Bang, N. D., and Andrews, W. R. H. (1974). Direct current measurements of a shelf-edge frontal jet in the southern Benguela system. *J. Mar. Res.* **32**, 405–417.

Bannon, P. R. (1981). Synoptic-scale forcing of coastal lows: Forced double Kelvin waves in the atmosphere. *Q. J. R. Meteorol. Soc.* **107**, 313–327.

Barber, B. F., and Ursell, F. (1948). The generation and propagation of ocean waves and swell. *Philos. Trans. R. Soc. London, Ser. A* **240**, 527–560.

Barton, E. D., Huyer, A., and Smith, R. L. (1977). Temporal variation observed in the hydrographic régime near Cabo Corveiro in the northwest African region, February to April, 1974. *Deep-Sea Res.* **24**, 7–23.

Batchelor, G. K. (1967). "An Introduction to Fluid Dynamics." Cambridge Univ. Press, London and New York.

Baumgartner, A., and Reichel, E. (1975). "The World Water Balance." Elsevier/North-Holland, Amsterdam.

Beardsley, R. C. (1969). A laboratory model of the wind-driven circulation. *J. Fluid Mech.* **38**, 255–272.

Beardsley, R. C., and Boicourt, W. C. (1981). On estuarine and continental-shelf circulation in the Middle Atlantic Bight. *In* "Evolution of Physical Oceanography" (B. A. Warren and C. Wunsch, eds.), Chapter 7. MIT Press, Cambridge, Massachusetts.

Beardsley, R. C., and Hart, J. (1978). A simple theoretical model for the flow of an estuary onto a continental shelf. *J. Geophys. Res.* **83**, 873–883.

Bell, T. H. (1975). Topographically generated internal waves in the open ocean. *J. Geophys. Res.* **80**, 320–327.

Bell, T. H. (1978). Radiation damping of inertial oscillations in the upper ocean. *J. Fluid Mech.* **88**, 289–308.

Bennett, J. R. (1977). A three-dimensional model of Lake Ontario's summer circulation. I. Comparison with observations. *J. Phys. Oceanogr.* **7**, 591–601.

Bennett, J. R., and Lindstrom, E. J. (1977). A simple model of Lake Ontario's coastal boundary layer. *J. Phys. Oceanogr.* **7**, 620–625.

Bennetts, D. A., and Hoskins, B. J. (1979). Conditional symmetric instability—a possible explanation for frontal rainbands. *Q. J. R. Meteorol. Soc.* **105**, 945–962.

Berger, A. L. (1979). Insolation signatures of quaternary climatic changes. *Nuovo Cimento Soc. Ital. Fis., C* [1] **2C**, 63–87.

Bernstein, R. L., and White, W. B. (1975). Time and length scales of baroclinic eddies in the central North Pacific Ocean. *J. Phys. Oceanogr.* **4**, 613–624.

Betchov, R., and Criminale, W. O. Jr. (1967). "Stability of Parallel Flows." Academic Press, New York.

Birkhoff, G., and Rota, G.-C. (1962). "Ordinary Differential Equations." Ginn, Boston, Massachusetts.

Birt, W. R. (1847). Third report on atmospheric waves. *Br. Assoc. Rep.* **16**, 119–168.

Bjerknes, J. (1919). On the structure of moving cyclones. *Geofys. Publ.* **1**(1), 1–8.

Bjerknes, J. (1966). A possible response of the atmospheric Hadley circulation to equatorial anomalies of ocean temperature. *Tellus* **18**, 820–829.

Bjerknes, J. (1969). Atmospheric teleconnections from the equatorial Pacific. *Mon. Weather Rev.* **97**, 163–172.

Bjerknes, V. (1898). Über die Bildung von Circulationsbewegungen und Wirbeln in reibunglosen Flüssigkeiten. *Skr. Nor. Vidensk.-Akad. [Kl.] 1: Mat.-Naturvidensk. Kl.* **5**, pp. 1–29.

Bjerknes, V. (1901). Zirkulation relativ zu der Erde. *Öevers. Fin. Vetensk.-Soc. Foerh.* pp. 739–775.

Bjerknes, V. (1937). Application of line integral theorems to the hydrodynamics of terrestrial and cosmic vortices. *Astrophys. Norv.* **2**, 263–339.

Blackadar, A. K. (1957). Boundary layer wind maxima and their significance for the growth of the nocturnal inversion. *Bull. Am. Meteorol. Soc.* **38**, 283–290.

Blandford, R. R. (1966). Mixed gravity-Rossby waves in the ocean. *Deep-Sea Res.* **13**, 941–961.

Blandford, R. R. (1971). Boundary conditions in homogeneous ocean models. *Deep-Sea Res.* **18**, 739–751.

Blumen, W. (1965a). On drag and wave resistance. *Pure Appl. Geophys.* **60**, 137–140.

Blumen, W. (1965b). A random model of momentum flux by mountain waves. *Geofys. Publ.* **26**, (2), 1–33.

Blumen, W. (1968). On the stability of quasi-geostrophic flow. *J. Atmos. Sci.* **25**, 929–931.

Blumen, W. (1972). Geostrophic adjustment. *Rev. Geophys. Space Phys.* **10**, 485–528.

Blumen, W. (1980). A comparison between the Hoskins–Bretherton model of frontogenesis and the analysis of an intense surface frontal zone. *J. Atmos. Sci.* **37**, 64–77.

Blumen, W. (1981). The geostrophic coordinate transformation. *J. Atmos. Sci.* **38**, 1100–1105.

Booker, J. R., and Bretherton, F. P. (1967). The critical layer for internal gravity waves in a shear flow. *J. Fluid Mech.* **27**, 513–539.

Boussinesq, J. (1903). "Théorie analytique de la chaleur," Vol. 2. Gauthier-Villars, Paris.

Brandes, H. W. (1820). "Beitrage zur Witterungskunde." Leipzig.

Brass, G. W., Southam, J. R., and Peterson, W. H. (1982). Warm saline waters in the ancient oceans. *Nature (London)* **296**, 620–623.

Brekhovskikh, L. M. (1980). "Waves in Layered Media," 2nd ed. Academic Press, New York.

Bretherton, F. P. (1966). Critical layer instability in baroclinic flows. *Q. J. R. Meteorol. Soc.* **92**, 325–334.

Bretherton, F. P. (1967). The time-dependent motion due to a cylinder moving in an unbounded rotating or stratified fluid. *J. Fluid Mech.* **28**, 545–570.

Bretherton, F. P. (1969). Momentum transport by gravity waves. *Q. J. R. Meteorol. Soc.* **95**, 213–243.

Bretherton, F. P., and Garrett, C. J. R. (1968). Wave trains in inhomogeneous moving media. *Proc. R. Soc. London, Ser. A* **302**, 529–554.

Brighton, P. W. M. (1978). Strongly stratified flow past three dimensional objects. *Q. J. R. Meteorol. Soc.* **104**, 284–308.

Brink, K. H., and Allen, J. S. (1978). On the effect of bottom friction on barotropic motion over the continental shelf. *J. Phys. Oceanogr.* **8**, 919–922.

Brink, K. H., Allen, J. S., and Smith, R. L. (1978). A study of low-frequency fluctuations near the Peru coast. *J. Phys. Oceanogr.* **8**, 1025–1041.

Briscoe, M. G. (1975). Preliminary results from the trimoored internal wave experiment. (IWEX). *J. Geophys. Res.* **80**, 3872–3884.

Brooks, C. E. P., and Glasspoole, J. (1928). "British Floods and Droughts." Benn Ltd., London.

Brown, P. J. (1973). Kelvin-wave reflection in a semi-infinite canal. *J. Mar. Res.* **31**, 1–10.

Browning, K. A., Pardoe, C. W., and Hill, F. F. (1975). The nature of oceanographic rain at wintertime cold fronts. *Q. J. R. Meteorol. Soc.* **101**, 333–352.

Brunskill, J. (1884). The Helm wind. *Q. J. R. Meteorol. Soc.* **10**, 267–275.

Brunt, D. (1927). The period of simple vertical oscillations in the atmosphere. *Q. J. R. Meteorol. Soc.* **53**, 30–32.

Brunt, D., and Douglas, C. K. M. (1928). The modification of the strophic balance for changing pressure distribution, and its effect on rainfall. *Mem. R. Meteorol. Soc.* **3**, 29–51.

Bryan, K. (1963). A numerical investigation of a nonlinear model of a wind-driven ocean. *J. Atmos. Sci.* **20**, 594–606.

Bryan, K. (1969). A numerical method for the study of the circulation of the world ocean. *J. Comp. Phys.* **3**, 347–376.

Bryan, K. (1979). Models of the world ocean. *Dyn. Atmos. Oceans* **3**, 327–338.

Bryan, K., and Cox, M. D. (1972). An approximate equation of state for numerical models of ocean circulation. *J. Phys. Oceanogr.* **2**, 510–514.

Bryan, K., and Lewis, L. J. (1979). A water mass model of the world ocean. *J. Geophys. Res.* **84**, 2503–2517.

Bryden, H. L. (1973). New polynomials for thermal expansion, adiabatic temperature gradient and potential temperature gradient of sea water. *Deep-Sea Res.* **20**, 401–408.

Bryden, H. L., and Hall, M. M. (1980). Heat transport by currents across 25°N latitude in the Atlantic Ocean. *Science* **207**, 884–886.

Buchanan, J. Y. (1886). On similarities in the physical geography of the great oceans. *Proc. R. Geogr. Soc. London* **8**, 753–770.

Buchwald, V. T., and Adams, J. K. (1968). The propagation of continental shelf waves. *Proc. R. Soc. London, Ser. A* **305**, 235–250.

Budden, K. G. (1961). "The Wave-Guide Mode Theory of Wave Propagation." Academic Press, New York.

Budyko, M. I., ed. (1963). "Atlas teplovogo balansa zemnogo shara" (Atlas of the heat balance of the earth). Gidrometeoizdat, Moscow.

Budyko, M. I. (1974). "Climate and Life." Academic Press, New York.

Bunker, A. F. (1976). Computations of surface energy flux and annual air-sea interaction cycles of the North Atlantic. *Mon. Weather Rev.* **104**, 1122–1140.

Bunker, A. F. (1980). Trends of variables and energy fluxes over the Atlantic Ocean from 1948 to 1972. *Mon. Weather Rev.* **108**, 720–732.

Buzzi, A., and Tibaldi, S. (1977). Inertial and frictional effects on rotating and stratified flow over topography. *Q. J. R. Meteorol. Soc.* **103**, 135–150.

Cadet, D., and Teitelbaum, H. (1979). Observational evidence of internal inertia-gravity waves in the tropical stratosphere. *J. Atmos. Sci.* **36**, 892–907.

Cahn, A. (1945). An investigation of the free oscillations of a simple current system. *J. Meteorol.* **2**, 113–119.

Cairns, J. L., and Williams, G. O. (1976). Internal wave observations from a mid-water float. Part 2. *J. Geophys. Res.* **81**, 1943–1950.

Candel, S. M. (1977). Numerical solution of conservation equations arising in linear wave theory: Application to aeroacoustics. *J. Fluid Mech.* **83**, 465–493.

Cane, M., and Sarachik, E. (1981). The response of a linear baroclinic equatorial ocean to periodic forcing. *J. Mar. Res.* **39**, 651–693.

Chamberlain, J. W. (1978). "Theory of Planetary Atmospheres: An introduction to their Physics and Chemistry." Academic Press, New York.

Chandrasekhar, S. (1961). "Hydrodynamic and Hydromagnetic Stability." Oxford Univ. Press (Clarendon), London and New York.

Charney, J. G. (1947). The dynamics of long waves in a baroclinic westerly current. *J. Meteorol.* **4**, 135–163.

Charney, J. G. (1948). On the scale of atmospheric motions. *Geofys. Publ. Oslo* **17**(2), 1–17.

Charney, J. G. (1955a). The generation of ocean currents by wind. *J. Mar. Res.* **14**, 477–498.

Charney, J. G. (1955b). The Gulf Stream as an inertial boundary layer. *Proc. Natl. Acad. Sci. U.S.A.* **41**, 731–740.

Charney, J. G. (1973). Planetary fluid mechanics. *In* "Dynamical Meteorology" (P. Morel, ed.), pp. 97–351. Reidel Publ., Dordrecht, Netherlands.

Charney, J. G., and DeVore, J. G. (1979). Multiple flow equilibria in the atmosphere and blocking. *J. Atmos. Sci.* **36**, 1205–1216.

Charney, J. G., and Drazin, P. G. (1961). Propagation of planetary scale disturbances from the lower into the upper atmosphere. *J. Geophys. Res.* **66**, 83–109.

Charney, J. G., and Eliassen, A. (1949). A numerical method for predicting the perturbation of the middle-latitude westerlies. *Tellus* **1**, 38–54.

Charney, J. G., and Flierl, G. R. (1981). Oceanic analogues of large-scale atmospheric motions. *In* "Evolution of Physical Oceanography" (B. A. Warren and C. Wunsch, eds.), Chapter 18. MIT Press, Cambridge, Massachusetts.

Charney, J. G., and Stern, M. E. (1962). On the stability of internal baroclinic jets in a rotating atmosphere. *J. Atmos. Sci.* **19**, 159–172.

Charnock, H. (1981). Air–sea interaction. *In* "Evolution of Physical Oceanography" (B. A. Warren and C. Wunsch, eds.), Chapter 17. MIT Press, Cambridge, Massachusetts.

Chen, C.-T., and Millero, F. J. (1977). Speed of sound in sea-water at high pressures. *J. Acoust. Soc. Am.* **62**, 1129–1135.

Clarke, A. J. (1977). Observational and numerical evidence for wind-forced coastal trapped long waves. *J. Phys. Oceanogr.* **7**, 231–247.

Clarke, R. A., and Gascard, J. C. (1982). The formation of Labrador sea water: Part 1, Large scale processes (unpublished).

Clarke, R. H., and Hess, G. D. (1974). Geostrophic departure and the functions A and B of Rossby-number similarity theory. *Boundary-Layer Meteorol.* **7**, 267–287.

Clarke, R. H., Dyer, A. J., Brook, R. R., Reid, D. G., and Troup, A. J. (1971). "The Wangara Experiment: Boundary Layer Data." Div. Meteorol. Phys. Tech. Pap. No. 19. CSIRO, Australia.

Coaker, S. A. (1977). The stability of a Rossby Wave. *Geophys. Astrophys. Fluid Dyn.* **9**, 1–17.

Combes (1859). Observations au sujet de la communication de M. Perrot et de la note de M. Babinet. *C. R. Hebd. Seances Acad. Sci.* **49**, 775–780.

Corby, G. A., ed. (1969). "The Global Circulation of the Atmosphere." Royal Meteorological Society, London.

Corby, G. A., and Wallington, C. E. (1956). Air flow over mountains: The lee-wave amplitude. *Q. J. R. Meteorol. Soc.* **82**, 266–274.

Coriolis, G. (1835). Mémoire sur les équations du mouvement rélatifs des systèmes de corps. *J. Ec. Polytech. (Paris)* **15**, 142.

Cox, M. D. (1970). A mathematical model of the Indian Ocean. *Deep-Sea Res.* **17**, 47–75.

Cox, M. D. (1975). A baroclinic model of the world ocean: Preliminary results. *In* "Numerical Models of Ocean Circulation," pp. 107–120. Natl. Acad. Sci., Washington, D.C.

Cox, M. D. (1976). Equatorially trapped waves and the generation of the Somali Current. *Deep-Sea Res.* **23**, 1139–1152.

Cox, M. D. (1979). A numerical study of Somali current eddies. *J. Phys. Oceanogr.* **9**, 311–326.

Coy, L. (1979). An unusually large westerly amplitude of the quasi-biennial oscillation. *J. Atmos. Sci.* **36**, 174–176; corrigendum in **37**, No. 4.

Crease, J. (1956). Propagation of long waves due to atmospheric disturbances in a rotating sea. *Proc. R. Soc. London, Ser. A* **233**, 556–569.

Crépon, M. (1974). Genèse d'ondes internes dans un milieu à deux couches. *Houille Blanche* **8**, 631–636.

Cruette, D. (1976). Experimental study of mountain lee-waves by means of satellite photographs and aircraft measurements. *Tellus* **28**, 499–523.

Crutcher, H. K., and Meserve, J. M. (1970). Selected level heights, temperatures and dew points for the north hemisphere, NAVAIR 50-1C-52. Naval Weather Service Command, Washington, D.C.

Csanady, G. T. (1967). On the resistance law of a turbulent Ekman layer. *J. Atmos. Sci.* **24**, 467–471.

Csanady, G. T. (1971). On the equilibrium shape of the thermocline in a shore zone. *J. Phys. Oceanogr.* **1**, 263–270.

Csanady, G. T. (1972a). Geostrophic drag, heat and mass transfer coefficients for the diabatic Ekman layer. *J. Atmos. Sci.* **29**, 488–496.

Csanady, G. T. (1972b). Response of large stratified lakes to wind. *J. Phys. Oceanogr.* **2**, 3–13.

Csanady, G. T. (1977a). The coastal jet conceptual model in the dynamics of shallow seas. *In* "The Sea" (E. D. Goldberg, I. N. McCave, J. J. O'Brien, and J. H. Steele, eds.) Vol. 6, pp. 117–144. Wiley, New York.

Csanady, G. T. (1977b). Intermittent 'full' upwelling in Lake Ontario. *J. Geophys. Res.* **82**, 397–419.

Csanady, G. T. (1978). The arrested topographic wave. *J. Phys. Oceanogr.* **8**, 47–62.

Csanady, G. T. (1981). Shelf-circulation cells. *Philos. Trans. R. Soc. London, Ser. A* **302**, 515–530.

Cutchin, D. L., and Smith, R. L. (1973). Continental shelf waves: Low-frequency variations in sea level and currents over the Oregon continental shelf. *J. Phys. Oceanogr.* **3**, 73–82.

d'Alembert, J. le Rond (1761). Remarques surs les loix du mouvement des fluides. *Opusc. Math.* **1**, 157–168.

Dantzler, H. L. (1977). Potential energy maxima in the tropical and sub-tropical North Atlantic. *J. Phys. Oceanogr.* **7**, 512–519.

Dauphinee, T. M. (1980). Introduction to the special issue on the Practical Salinity Scale 1978. *IEEE, J. Oceanic Eng.* **OE-5**, 1–2.

Davey, M. K. (1980). A quasi-linear theory for rotating flow over topography. Part I. Steady β-plane channel. *J. Fluid Mech.* **99**, 267–292.

Davies, J. T., and Rideal, E. K. (1963). "Interfacial Phenomena." Academic Press, New York.

Deacon, M. B. (1971). "Scientists and the Sea 1650–1900: A Study of Marine Science." Academic Press, New York.

Deardorff, J. W. (1970). A three-dimensional numerical investigation of the idealized planetary boundary layer. *Geophys. Fluid Dyn.* **1**, 377–410.

Defant, A. (1961). "Physical Oceanography," Vols. 1 and 2. Pergamon, Oxford.

Delauney (1859). Remarques concernant le question de l'influence de la rotation de la terre sur la direction des courants d'eau. *C. R. Hebd. Seances Acad. Sci.* **49**, 688–692.

Dickinson, R. E. (1973). Method of parameterization of infrared cooling between altitudes of 30 and 70 kilometers. *J. Geophys. Res.* **78**, 4451–4457.

Dickinson, R. E. (1975). Meteorology of the upper atmosphere. *Rev. Geophys. Space Phys.* **13**, 771–862.

Dickinson, R. E. (1978). Rossby waves—long period oscillations of oceans and atmospheres. *Annu. Rev. Fluid Mech.* **10**, 159–195.

Dickinson, R. E. (1980). Planetary waves: theory and observation. *In* "Orographic Effects in Planetary Flows," Chapter 2, pp. 53–84. GARP Publ. Ser. No. 23, WMO-ICSU, Geneva.

Donn, W. L., and Shaw, D. M. (1967). Exploring the atmosphere with nuclear explosions. *Rev. Geophys.* **5**, 53–82.

Doodson, A. T., and Corkan, R. H. (1931). The principal constituent of the tides in the English and Irish Channels. *Philos. Trans. R. Soc. London, Ser. A* **231**, 29–53.

Dopplick, T. G. (1972). Radiative heating of the global atmosphere. *J. Atmos. Sci.* **29**, 1278–1294.

Drazin, P. G., and Howard, L. N. (1966). Hydrodynamic stability of parallel flow of inviscid fluid. *Adv. Appl. Mech.* **9**, 1–89.

Drazin, P. G., and Reid, W. H. (1981). "Hydrodynamic Stability." Cambridge Univ. Press, Cambridge, England.

Drazin, P. G., and Su, C. H. (1975). A note on long-wave theory of airflow over a mountain. *J. Atmos. Sci.* **32**, 437–439.

Drummond, A. J. (1970). Precision radiometry and its significance in atmospheric and space physics. *Adv. Geophys.* **14**, 1–52.

Dugas, R. (1957). "A History of Mechanics." Routledge & Kegan Paul, London.

Duhem, P. (1954). "Le système du monde: Histoire des doctrines cosmologiques de Platon à Copernic," Vol. 2. Hermann, Paris.

Düing, W., and Szekielda, K.-H. (1971). Monsoonal response in the western Indian Ocean. *J. Geophys. Res.* **76**, 4181–4187.

Dunkerton, T., Hsu, C. P. F., and McIntyre, M. E. (1981). Some Eulerian and Lagrangian diagnostics for a model stratospheric warming. *J. Atmos. Sci.* **38**, 819–843.

Dutton, J. A., and Johnson, D. R. (1967). The theory of available potential energy and a variational approach to atmospheric energetics. *Adv. Geophys.* **12**, 333–435.

Eady, E. T. (1949). Long waves and cyclone waves. *Tellus* **1** (3), 33–52.

Eckart, C. (1960). "Hydrodynamics of Oceans and Atmospheres." Pergamon, Oxford.

Edmon, H. J., Hoskins, B. J., and McIntyre, M. E. (1980). Eliassen-Palm cross-sections for the troposphere. *J. Atmos. Sci.* **37**, 2600–2616; corrigendum in **38**, 1115).

Ekman, V. W. (1904). On dead water. *Sci. Results Norw. North Polar Expedi. 1893–96* **5** (15).

Ekman, V. W. (1905). On the influence of the earth's rotation on ocean currents. *Arch. Math. Astron. Phys.* **2**, No. 11.

Ekman, V. W. (1906). Beiträge zur Theorie der Meeresströmungen. *Ann. Hydrogr. (Berlin)* **34**, 484.

Ekman, V. W. (1927). Eddy-viscosity and skin-friction in the dynamics of winds and ocean-currents. *Mem. R. Meteorol. Soc.* **2**, 161–172.

Eliassen, A. (1949). The quasi-static equations of motion with pressure as independent variable. *Geofys. Publ., Oslo* **17**(3), 1–44.

Eliassen, A. (1952). Slow thermally or frictionally controlled meridional circulation in a circular vortex. *Astrophys. Norv.* **5**, 19–60.

Eliassen, A. (1962). On the vertical circulation in frontal zones. *Geophys. Publ.* **24**, No. 4, 147–160.

Eliassen, A., and Kleinschmidt, E. (1957). Dynamic meteorology. *In* "Handbuch der Physik," Vol. 48 (J. Bartels, ed.), pp. 1–154. Springer-Verlag, Berlin and New York.

Eliassen, A., and Palm, E. (1961). On the transfer of energy in stationary mountain waves. *Geofys. Publ. Oslo* **22**(3), 1–23.

Endoh, M. (1978). Three-dimensional structures of gravity currents in a rotating basin. Part I. A local discharge of buoyancy. *J. Oceanogr. Soc. Jpn.* **34**, 303–306.

Enfield, D. B., and Allen, J. S. (1980). On the structure and dynamics of monthly mean sea level anomalies along the Pacific coast of North and South America. *J. Phys. Oceanogr.* **10**, 557–578.

Erdélyi, A., ed. (1953). "Higher Transcendental Functions," Vol. II. McGraw-Hill, New York.

Erdélyi, A. (1956). "Asymptotic Expansions." Dover, New York.

Erdélyi, A., Magnus, W., Oberhettinger, R., and Tricomi, F. G. (1954). "Tables of Integral Transforms," Vol. I. McGraw-Hill, New York.

Eriksen, C. (1980). Evidence for a continuous spectrum of equatorial waves in the Indian Ocean. *J. Geophys. Res.* **85**, 3285–3303.

Ertel, H. (1942a). Ein neuer hydrodynamischer Wirbelsatz. *Meteorol. Z.* **59**, 271–281.

Ertel, H. (1942b). Über des Verhältnis des neuen hydrodynamischen Wirbelsatzes zum Zirkulationssatz von V. Bjerknes. *Meteorol. Z.* **59**, 385–387.

Euler, L. (1755). Principes généraux du mouvement des fluides. *Mém. Acad. Berlin* **11**, 274–315, 316–361; see also Principia motus fluodorum. *Novi commentarii academiae scientiarum Petropolitanae* **6**, 271–311. ("Opera Omnia," Ser. Secunda, Vol. XII, pp. 54–91, 92–132, 133–168. Societatis Scientiarum Naturalium Helveticae; Orell Füssli Turici, Lausanne, 1954.)

Eyre, W. S. (1973). "The Spherical Harmonic Analysis of Global Wind Stress Field and Atmospheric Angular Momentum," Res. Rep. 6. Flinders Institute for Atmospheric and Marine Sciences. Flinders Univ. of S. Australia.

Faller, A. J. (1981). The origin and development of laboratory models and analogues of the ocean circulation. *In* "Evolution of Physical Oceanography" (B. A. Warren and C. Wunsch, eds.), Chapter 16. MIT Press, Cambridge, Massachusetts.

Farmer, D. M., and Smith, J. D. (1980). Tidal interaction of stratified flow with a sill in Knight Inlet. *Deep-Sea Res.* **27**, 239–254.

Farrell, W. E. (1973). Earth tides, ocean tides and tidal loading. *Philos. Trans. R. Soc. London, Ser. A* **274**, 253–259.

Ferrel, W. (1859/1860). The motions of fluids and solids relative to the earth's surface. *Math. Mon.* **1**, 140–148, 210–216, 300–307, 366–373, 397–406; **2**, 89–97, 339–346, 374–391.

Fieux, M., and Stommel, H. (1977). Onset of the Southwest Monsoon over the Arabian Sea from marine reports of surface winds: Structure and variability. *Mon. Weather Rev.* **105**, 231–236.

Flather, R. A. (1976). A tidal model of the north-west European continental shelf. *Mém. Soc. R. Sci. Liège* **10**, 141–174.

Flather, R. A., and Davies, A. M. (1976). Note on a preliminary scheme for storm surge prediction using numerical models. *Q. J. R. Meteorol. Soc.* **102**, 123–132.

Fofonoff, N. P. (1962). Physical properties of sea water. *In* "The Sea," vol. 1 (M. N. Hill, ed.), pp. 3–30. Wiley, New York.

Fofonoff, N. P. (1969). Spectral characteristics of internal waves in the ocean. *Deep-Sea Res.* **16**, Suppl., 58–71.

Fofonoff, N. P. (1981). The Gulf Stream system. *In* "Evolution of Physical Oceanography" (B. A. Warren and C. Wunsch, eds.), Chapter 4. MIT Press, Cambridge, Massachusetts.

Foo, E.-C. (1981). A two-dimensional diabatic isopycnal model—simulating the coastal upwelling front. *J. Phys. Oceanogr.* **11**, 604–626.

Foster, T. D., and Carmack, E. C. (1976). Frontal zone mixing and Antarctic bottom water formation in the southern Weddell Sea. *Deep-Sea Res.* **23**, 301–317.

Foster, T. D., and Middleton, J. H. (1980). Bottom water formation in the western Weddell Sea. *Deep-Sea Res.* **27**, 367–381.

Foucault, L. (1851). Démonstration physique de mouvement de rotation de la Terre au moyen de pendule. *C- R. Hebd. Seances Acad. Sci.* **32**, 135–138.

Francis, S. H. (1973). Acoustic-gravity modes and large-scale traveling ionospheric disturbances of a realistic, dissipative atmosphere. *J. Geophys. Res.* **78**, 2278–2301.

Franklin, B. (1762). Letter published in "Experiments and Observations on Electricity," pp. 438–40, London, 1769 [*Also in* "The Papers of Benjamin Franklin" (L. W. Labarce, ed.), Vol. 10, pp. 158–160. Yale Univ. Press, New Haven, 1966].

Freeland, H., Rhines, P., and Rossby, T. (1975). Statistical observations of the trajectories of neutrally buoyant floats in the North Atlantic. *J. Mar. Res.* **34**, 69–92.

Fultz, D., Long, R. R., Owens, G. V., Bohan, W., Kaylor, R., and Weil, J. (1959). Studies of thermal convection in a rotating cylinder with some application for large-scale atmospheric motion. *Meteorol. Monogr.* **4**, No. 21, 1–104.

Garratt, J. R. (1977). Review of drag coefficients over oceans and continents. *Mon. Weather Rev.* **105**, 915–929.

Garrett, C. J. R. (1969). Atmospheric edge waves. *Q. J. R. Meteorol. Soc.* **95**, 731–753.

Garrett, C. J. R. (1979). Mixing in the ocean interior. *Dyn. Atmos. Oceans* **3**, 239–265.

Garrett, C. J. R., and Greenberg, D. A. (1977). Predicting changes in tidal régime: The open boundary problem. *J. Phys. Oceanogr.* **7**, 171–181.

Garrett, C. J. R., and Loder, J. (1981). Dynamical aspects of shallow sea fronts. *Philos. Trans. R. Soc. London, Ser. A* **302**, 563–581.

Garrett, C. J. R., and Munk, W. H. (1972). Space-time scales of internal waves. *Geophys. Fluid Dyn.* **3**, 225–264.

Garrett, C. J. R., and Munk, W. H. (1975). Space-time scales of internal waves: A progress report. *J. Geophys. Res.* **80**, 281–297.

Garrett, C., and Munk, W. H. (1979). Internal waves in the ocean. *Annu. Rev. Fluid Mech.* **11**, 339–369.

Garrett, J. F. (1980). Availability of the FGGE drifting buoy system data set. *Deep-Sea Res.* **27A**, 1083–1086.

Garrett, W. D. (1967). The organic chemical composition of the ocean surface. *Deep-Sea Res.* **14**, 221–227.

Gascard, J. C. (1978). Mediterranean deep-water formation, baroclinic instability and oceanic eddies. *Oceanol. Acta* **1**, 315–330.

Gascard, J. C., and Clarke, R. A. (1982). The formation of Labrador sea water: Part 2, Mesoscale and smaller scale processes (unpublished).

Geisler, J. E. (1970). Linear theory of the response of a two-layer ocean to a moving hurricane. *Geophys. Fluid Dyn.* **1**, 249–272.

Geisler, J. E. (1981). A linear model of the Walker Circulation. *J. Atmos. Sci.* **38**, 1390–1400.

Gill, A. E. (1968a). Similarity theory and geostrophic adjustment. *Q. J. R. Meteorol. Soc.* **94**, 581–585.

Gill, A. E. (1968b). A linear model of the Antarctic circumpolar current. *J. Fluid Mech.* **32**, 465–488.

Gill, A. E. (1971). Ocean models. *Philos. Trans. R. Soc. London, Ser. A* **270**, 391–413.

Gill, A. E. (1973). Circulation and bottom water formation in the Weddell Sea. *Deep-Sea Res.* **20**, 111–140.

Gill, A. E. (1974). The stability of planetary waves on an infinite beta-plane. *Geophys. Fluid Dyn.* **6**, 29–47.

Gill, A. E. (1975). Models of equatorial currents. *In* "Numerical Models of Ocean Circulation," pp. 181–203. Natl. Acad. Sci., Washington, D.C.

Gill, A. E. (1976). Adjustment under gravity in a rotating channel. *J. Fluid Mech.* **77**, 603–621.

Gill, A. E. (1977a). The hydraulics of rotating channel flow. *J. Fluid Mech.* **80**, 641–671.

Gill, A. E. (1977b). Coastally-trapped waves in the atmosphere. *Q. J. R. Meteorol. Soc.* **103**, 431–440.

Gill, A. E. (1977c). Potential vorticity as a tracer. Appendix to Pearce (1977).

Gill, A. E. (1979a). A simple model for showing effects of geometry on the world's tides. *Proc. R. Soc. London, Ser. A* **367**, 549–571.

Gill, A. E. (1979b). Constructing a two-layer model for studying the effects of deep convection. *In* "Séminaire de dynamique des fluides géophysiques et de modélisation numérique appliqué à l'océanographie," CNEXO, Paris.

Gill, A. E. (1980). Some simple solutions for heat-induced tropical circulation. *Q. J. R. Meteorol. Soc.* **106**, 447–462.

Gill, A. E. (1981). Homogeneous intrusions in a rotating stratified fluid. *J. Fluid Mech.* **103**, 275–295.

Gill, A. E., and Bryan, K. (1971). Effects of geometry on the circulation of a three-dimensional southern-hemisphere ocean model. *Deep-Sea Res.* **18**, 685–721.

Gill, A. E., and Clarke, A. J. (1974). Wind-induced upwelling, coastal currents, and sea-level changes. *Deep-Sea Res.* **21**, 325–345.

Gill, A. E., and Niiler, P. (1973). The theory of seasonal variability in the ocean. *Deep-Sea Res.* **20**, 141–177.

Gill, A. E., and Parker, R. L. (1970). Contours of 'h cosec θ' for the world's oceans. *Deep-Sea Res.* **17**, 823–824.

Gill, A. E., and Schumann, E. H. (1974). The generation of long shelf waves by the wind. *J. Phys. Oceanogr.* **4**, 83–90.

Gill, A. E., and Schumann, E. H. (1979). Topographically induced changes in the structure of an inertial coastal jet: application to the Agulhas Current. *J. Phys. Oceanogr.* **9**, 975–991.

Gill, A. E., and Turner, J. S. (1969). Some new ideas about the formation of Antarctic bottom water. *Nature (London)* **224**, 1287–1288.

Gill, A. E., and Turner, J. S. (1976). A comparison of seasonal thermocline models with observation. *Deep-Sea Res.* **23**, 391–401.

Gill, A. E., Green, J. S. A., and Simmons, A. J. (1974). Energy partition in the large-scale ocean circulation and the production of mid-ocean eddies. *Deep-Sea Res.* **21**, 499–528.

Gill, A. E., Smith, J. M., Cleaver, R. P., Hide, R., and Jonas, P. R. (1979). The vortex created by mass transfer between layers of rotating fluid. *Geophys. Astrophys. Fluid Dyn.* **12**, 195–220.

Gjevik, B. (1980). Orographic effects revealed by satellite pictures: Mesoscale flow phenomena. *In* "Orographic Effects in Planetary Flows," Chapter 9, pp. 301–316. GARP Publ. Ser. No. 23, WMO-ICSU, Geneva.

Gjevik, B., and Marthinsen, T. (1978). Three-dimensional lee-wave patterns. *Q. J. R. Meteorol. Soc.* **104**, 947–957.

Glantz, M. H., ed. (1980). "Resource Management and Environmental Uncertainty." Wiley, New York.

Godin, G. (1972). "The Analysis of Tides." Liverpool Univ. Press, Liverpool.

Godske, C. L., Bergeron, T., Bjerknes, J., and Bundgaard, R. C. (1957). "Dynamic Meteorology and Weather Forecasting." Am. Meteorol. Soc., Boston, Massachusetts.

Gold, E. (1908). "Barometric Gradient and Wind Force." Meteorol. Off., M.O. 190, HM Stationery Office, London.

Goldsbrough, G. R. (1933). Ocean currents produced by evaporation and precipitation. *Proc. R. Soc. London, Ser. A* **141**, 512–517.

Goody, R. M. (1964). "Atmospheric Radiation." Oxford Univ. Press (Clarendon), London and New York.

Gossard, E. E., and Hooke, W. H. (1975). "Waves in the Atmosphere." Elsevier, Amsterdam.

Gossard, E. E., and Munk, W. H. (1954). On gravity waves in the atmosphere. *J. Meteorol.* **11**, 259–269.

Gould, W. S., Hendry, R., and Huppert, H. E. (1981). An abyssal topographic experiment. *Deep-Sea Res.* **28**, 409–440.

Gradshteyn, I. S., and Ryzhik, I. M. (1980). "Table of Integrals, Series, and Products," corrected and enlarged ed. (Translation from Russian), 4th ed. Academic Press, New York.

Gray, W. M. (1979). Hurricanes: Their formation, structure and likely role in the tropical circulation. *In* "Meteorology Over the Tropical Oceans" (D. B. Shaw, ed.), pp. 151–218. Royal Meteorological Society, London.

Gray, W. M., and Shea, D. J. (1973). The hurricane's inner core region. II. Thermal stability and dynamic characteristics. *J. Atmos. Sci.* **30**, 1565–1576.

Greatbatch, R. J. (1982). On the response of the ocean to a moving storm *J. Phys. Oceanogr.* (to be published).

Green, G. (1838). On the motion of waves in a variable canal of small depth and width. *Trans. Cambridge Philos. Soc.* [See Mathematical Papers, pp. 223–230. Macmillan, London, 1871.]

Green, J. S. A. (1960). A problem in baroclinic stability. *Q. J. R. Meteorol. Soc.* **86**, 237–251.

Green, J. S. A. (1970). Transfer properties of the large-scale eddies and the general circulation of the atmosphere. *Q. J. R. Meteorol. Soc.* **96**, 157–185.

Greenspan, H. P. (1962). A criterion for the existence of inertial boundary currents in ocean circulation. *Proc. Natl. Acad. Sci. U.S.A.* **48**, 2034–2039.

Greenspan, H. P. (1968). "The Theory of Rotating Fluids." Cambridge Univ. Press, London and New York.

Griffiths, R. W., and Linden, P. F. (1981). The stability of buoyancy-driven coastal currents. *Dyn. Atmos. Oceans* **5**, 281–306.

Grose, W. L., and Hoskins, B. J. (1979). On the influence of orography on large-scale atmospheric flow. *J. Atmos. Sci.* **36**, 223–234.

Gustafson, T., and Kullenberg, B. (1936). Untersuchungen von Trägheitssttrömungen in der Ostsee. *Sven. Hydrogr.-Biol. Komm. Skr., Hydrogr.* No. 13.

Hadley, G. (1735). Concerning the cause of the general trade-winds. *Philos. Trans. R. Soc. London* **39**, 58–62.

Halley, E. (1686). An Historical Account of the Trade Winds, and Monsoons, observable in the Seas between and near the Tropicks, with an attempt to assign the Phisical cause of the said Winds. *Philos. Trans. R. Soc. London* **16**, 153–168.

Halpern, D. (1974). Observations of the deepening of the wind-mixed layer in the North-East Pacific Ocean. *J. Phys. Oceanogr.* **4**, 454–466.

Halpern, D. (1980). Variability of near-surface currents in the Atlantic North Equatorial Counter-current during GATE. *J. Phys. Oceanogr.* **10**, 1213–1220.

Haltiner, G. J. (1971). "Numerical Weather Prediction." Wiley, New York.

Hamilton, P., and Rattray, M. (1978). A numerical model of the depth-dependent, wind-driven upwelling circulation on a continental shelf. *J. Phys. Oceanogr.* **8**, 437–457.

Hamon, B. V. (1962). The spectrums of mean sea level at Sydney, Coff's Harbour, and Lord Howe Island. *J. Geophys. Res.* **67**, 5147–5155; correction in **68**, 4635.

Hamon, B. V. (1976). Generation of shelf waves on the east Australian coast by wind stress. *Mém. Soc. R. Sci. Liège* **10**, 359–367.

Haney, R. L. (1971). Surface thermal boundary condition for ocean circulation models. *J. Phys. Oceanogr.* **1**, 241–248.

Hantel, M., and Baader, H. (1978). Diabatic heating climatology of the zonal atmosphere. *J. Atmos. Sci.* **35**, 1180–1189.

Hare, F. K. (1968). The Arctic. *Quart. J. R. Meteorol. Soc.* **94**, 439–459.

Harris, T. F. W., and van Foreest, D. (1978). The Agulhas Current in March 1969. *Deep-Sea Res.* **25**, 549–561.

Hart, J. E. (1979). Barotropic geostrophic flow over anisotropic mountains. *J. Atmos. Sci.* **36**, 1736–1746.

Hart, J. E., Rao, G. V., van de Boogaard, H., Young, J. A., and Findlater, J. (1978). Aerial observations of the East African low-level jet stream. *Mon. Weather Rev.* **106**, 1714–1724.

Hart, T. J., and Currie, R. I. (1960). The Benguela Current. *Discovery Rep.* **31**, 123–298.

Hartmann, D. L. (1978). A note concerning the effect of varying extinction on radiative-photochemical relaxation. *J. Atmos. Sci.* **35**, 1125–1130.

Harvey, J. G., and Vincent, C. E. (1977). Observations of shear in near-bed currents in the southern North Sea. *Estuarine Coastal Mar. Sci.* **5**, 715–731.

Hasselmann, K. (1966). Feynman diagrams and interaction rules of wave-wave scattering processes. *Rev. Geophys.* **4**, 1–32.

Hasselmann, K. (1967a). Non-linear interactions treated by the methods of theoretical physics. *Proc. R. Soc. London, Ser. A* **299**, 77–100.

Hasselmann, K. (1967b). A criterion for non-linear wave instability. *J. Fluid Mech.* **30**, 737–739.

Hastenrath, S., and Lamb, P. J. (1979). "Climatic Atlas of the Indian Ocean. Part I. Surface Climate and Atmospheric Circulation." Univ. of Wisconsin Press, Madison.

Hastenrath, S. (1980). Heat budget of tropical ocean and atmosphere. *J. Phys. Oceanogr.* **10**, 159–170.

Haurwitz, B. (1940). The motion of atmospheric disturbances on the spherical earth. *J. Mar. Res.* **3**, 254–267.

Hayashi, Y. (1974). Spectral analysis of tropical disturbances appearing in a GFDL general circulation model. *J. Atmos. Sci.* **31**, 180–218.

Hayashi, Y., and Golder, D. G. (1978). The generation of equatorial transient planetary waves: Control experiments with a GFDL general circulation model. *J. Atmos. Sci.* **35**, 2068–2082.

Hayes, W. D. (1970). Kinematic wave theory. *Proc. R. Soc. London, Ser. A* **320**, 209–226.

Hays, J. D., Imbrie, J., and Shackleton, N. J. (1976). Variations in the earth's orbit: Pacemaker of the ice ages. *Science* **194**, 1121–1132.

Heaps, N. S. (1965). Storm surges on a continental shelf. *Philos. Trans. R. Soc. London, Ser. A* **257**, 351–383.

Heaps, N. S. (1969). A two-dimensional numerical sea model. *Philos. Trans. R. Soc. London, Ser. A* **265**, 93–137.

Heath, R. A. (1981). Estimates of the resonant period and Q in the semi-diurnal tidal band in the North Atlantic and Pacific Oceans. *Deep-Sea Res.* **28**, 481–493.

Held, I. M., and Hou, A. Y. (1980). Nonlinear axially symmetric circulations in a nearly-inviscid atmosphere. *J. Atmos. Sci.* **37**, 515–533.

Held, I. M., and Suarez, M. J. (1978). A two-level primitive equation atmospheric model designed for climate sensitivity experiments. *J. Atmos. Sci.* **35**, 206–229.

Hellerman, S. (1967). An updated estimate of the wind stress on the world ocean. *Mon. Weather Rev.* **95**, 607–626 (with corrected tables in **96**, 62–74).

Helmholtz, H. von (1858). Über Integrale der hydrodynamischen Gleichungen welche den Wirbelbewegungen entsprechen. *J. Reine Angew. Math.* **55**, 25–55. [Also in *Wiss. Abh.* **1**, 101–134 (1882).] [English transl. in *Phil. Mag.* **23**, 485–510 (1867) and in Abbe (1893, pp. 31–57).]

Helmholtz, H. von (1888). Über atmospharische Bewegungun I. *Sitzungsberichte Akad. Wissenschaften Berlin* **3**, 647–663 [English transl. in Abbe (1893, pp. 78–93).] [Also in *Wiss. Abh.* **3**, 289–308, 1895.]

Hendershott, M. C. (1972). The effects of solid earth deformation on global ocean tides. *Geophys. J. R. Astron. Soc.* **29**, 389–402.

Hendershott, M. C. (1977). Numerical models of ocean tides. *In* "The Sea" (E. D. Goldberg *et al.*, eds.), Vol. 6, pp. 47–95. Wiley, New York.

Hendershott, M. C. (1981). Long waves and ocean tides. *In* "Evolution of Physical Oceanography" (B. A. Warren and C. Wunsch, eds.), Chapter 10. MIT Press, Cambridge, Massachusetts.

Hendershott, M. C., and Munk, W. H. (1970). Tides. *Annu. Rev. Fluid Mech.* **2**, 205–224.

Hendershott, M. C., and Speranza, A. (1971). Co-oscillating tides in long narrow bays: The Taylor problem revisited. *Deep-Sea Res.* **18**, 959–980.

A. M. Soward, eds.), pp. 205–238. Academic Press, New York.

Hess, S. L. (1959). "Introduction to Theoretical Meteorology." Holt, New York.

Hesselberg, T. (1915). Über eine Beziehung zwischen Druckgradient, Wind und Gradientenänderungen. *Veroeff. Geophys. Inst. Univ. Leipzig* **1**, 207–216.

Hide, R., and Mason, P. J. (1975). Sloping convection in a rotating fluid. *Adv. Phys.* **24**, 47–100.

Hildebrandsson, H. H., and Teisserenc de Bort, L. (1898). "Les bases de météorologie dynamique," Part 1. Gauthier-Villars, Paris.

Hines, C. O. (1960). Internal atmospheric waves at ionospheric heights. *Can. J. Phys.* **38**, 1441–1481.

Hines, C. O. (1964). Minimum vertical scale sizes in the wind structure above 100 km. *J. Geophys. Res.* **69**, 2847–2848.

Hirota, I. (1979). Kelvin waves in the equatorial middle atmosphere observed by the Nimbus 5 SCR. *J. Atmos. Sci.* **36**, 217–222.

Hogg, N. G. (1980). Effects of bottom topography on ocean currents. *GARP Publ. Ser.* **23**, 167–205.

Holland, W. R. (1973). Baroclinic and topographic influences on the transport in western boundary currents. *Geophys. Fluid Dyn.* **4**, 187–210.

Holland, W. R. (1977). Ocean general circulation models. *In* "The Sea" (E. D. Goldberg, I. N. McCave, J. J. O'Brien, and J. H. Steele, eds.) Vol. 6, pp. 3–45. Wiley (Interscience), New York.

Holopainen, E. O. (1967). A determination of the wind-driven ocean circulation from the vorticity budget of the atmosphere. *Pure Appl. Geophys.* **67**, 156–165.

Holopainen, E. O. (1978). On the dynamic forcing of the long-term mean flow by the large-scale Reynold's stresses in the atmosphere. *J. Atmos. Sci.* **35**, 1597–1604.

Holton, J. R. (1972). Waves in the equatorial stratosphere generated by tropospheric heat sources. *J. Atmos. Sci.* **29**, 368–375.

Holton, J. R. (1975). "The Dynamic Meteorology of the Stratosphere and Mesosphere." Am. Meteorol. Soc., Boston, Massachusetts.

Holton, J. R. (1979). "An Introduction to Dynamic Meteorology," 2nd ed. Academic Press, New York.

Holton, J. R. (1980a). Wave propagation and transport in the middle atmosphere. *Philos. Trans. R. Soc. London, Ser. A* **296**, 73–85.

Holton, J. R. (1980b). The dynamics of sudden stratospheric warmings. *Annu. Rev. Earth Planet. Sci.* **8**, 169–180.

Holton, J. R., and Wehrbein, W. M. (1980). A numerical model of the zonal mean circulation of the middle atmosphere. *Pure Appl. Geophys.* **118**, 284–306 (reprinted in Venkateswaran and Sundararaman, 1980).

Horel, J. D., and Wallace, J. M. (1981). Planetary scale atmospheric phenomena associated with the inter-annual variability of sea-surface temperatures in the equatorial Pacific. *Mon. Weather Rev.* **109**, 813–829.

Horigan, A. M., and Weisberg, R. H. (1981). A systematic search for trapped equatorial waves in the GATE velocity data. *J. Phys. Oceanogr.* **11**, 497–509.

Hoskins, B. J. (1972). Non-Bousinessq effects and further development in a model of upper tropospheric frontogenesis. *Q. J. R. Meteorol. Soc.* **98**, 532–541.

Hoskins, B. J. (1973). Stability of Rossby–Haurwitz wave. *Q. J. R. Meteorol. Soc.* **99**, 723–745.

Hoskins, B. J. (1975). The geostrophic momentum approximation and the semi-geostrophic equations. *J. Atmos. Sci.* **32**, 233–242.

Hoskins, B. J. (1976). Baroclinic waves and frontogenesis. Part I. Introduction and Eady waves. *Q. J. R. Meteorol. Soc.* **102**, 103–122.

Hoskins, B. J. (1980). Representation of the Earth topography using spherical harmonics. *Mon. Weather Rev.* **108**, 111–115.

Hoskins, B. J. (1982). The mathematical theory of frontogenesis. *Ann. Rev. Fl. Mech.* **14**, 131–151.

Hoskins, B. J., and Bretherton, F. P. (1972). Atmospheric frontogenesis models: mathematical formulation and solution. *J. Atmos. Sci.* **29**, 11–37.

Hoskins, B. J., and Draghici, J. (1977). The forcing of ageostrophic motion according to the semi-geostrophic equations and in an isentropic coordinate model. *J. Atmos. Sci.* **34**, 1859–1867.

Hoskins, B. J., and Heckley, W. A. (1981). Cold and warm fronts in baroclinic waves. *Q. J. R. Meteorol. Soc.* **107,** 79–90.

Hoskins, B. J., and Karoly, D. (1981). The steady linear response of a spherical atmosphere to thermal and orographic forcing. *J. Atmos. Sci.* **38,** 1179–1196.

Hoskins, B. J., and Pedder, M. A. (1980). The diagnosis of middle latitude synoptic development. *Q. J. R. Meteorol. Soc.* **106,** 707–719.

Hoskins, B. J., and West, N. V. (1979). Baroclinic waves and frontogenesis. Part II. Uniform potential vorticity jet flows in cold and warm fronts. *J. Atmos. Sci.* **36,** 1663–1680.

Hoskins, B. J., Draghici, I., and Davies, H. C. (1978). A new look at the ω-equation. *Q. J. R. Meteorol. Soc.* **104,** 31–38.

Hough, S. S. (1897). On the application of harmonic analysis to the dynamical theory of the tides. Part I. On Laplace's 'oscillations of the first species', and on the dynamics of ocean currents. *Philos. Trans. R. Soc. London, Ser. A* **189,** 201–257.

Hough, S. S. (1898). On the application of harmonic analysis to the dynamical theory of the tides. Part II. On the general integration of Laplace's dynamical equations. *Philos. Trans. R. Soc. London, Ser. A* **191,** 139–185.

Hunt, J. C. R., and Snyder, W. H. (1980). Experiments on stably and neutrally stratified flow over a model three-dimensional hill. *J. Fluid Mech.* **96,** 671–704.

Huppert, H. E., and Bryan, K. (1976). Topographically generated eddies. *Deep-Sea Res.* **23,** 655–679.

Huppert, H. E., and Miles, J. W. (1969). Lee waves in a stratified flow. Part 3. Semi-elliptical obstacle. *J. Fluid Mech.* **35,** 481–496.

Hurlburt, H. E., and Thompson, J. D. (1973). Coastal upwelling on a beta-plane. *J. Phys. Oceanogr.* **3,** 16–32.

Hurlburt, H. E., and Thompson, J. D. (1976). A numerical model of the Somali Current. *J. Phys. Oceanogr.* **6,** 646–664.

Hurlburt, H. E., Kindle, J. C., and O'Brien, J. J. (1976). A numerical study of the onset of El Niño. *J. Phys. Oceanogr.* **6,** 621–631.

Hutchins, L. M., ed-in-chief (1952). "Great Books of the Western World," Vol. 2, pp. 538–542. Encyclopaedia Brittanica, Chicago, Illinois.

Huthnance, J. M. (1978). On coastal trapped waves: Analysis and numerical calculation by inverse iteration. *J. Phys. Oceanogr.* **8,** 74–92.

Huyer, A. (1977). Seasonal variation in temperature, salinity and density over the continental shelf off Oregon. *Limnol. Oceanogr.* **22,** 442–453.

Huyer, A. (1980). The offshore structure and subsurface expression of sea level variations off Peru 1976–1977. *J. Phys. Oceanogr.* **10,** 1755–1768.

Huyer, A., Pillsbury, R. D., and Smith, R. L. (1975). Seasonal variation in the alongshore velocity field over the continental shelf off Oregon. *Limnol. Oceanogr.* **20,** 90–95.

Imbrie, J., and Imbrie, J. Z. (1980). Modelling the climatic response to orbital variations. *Science* **207,** 943–954.

Imbrie, J., and Imbrie, K. P. (1979). "Ice Ages: Solving the Mystery." Enslow, Short Hills, New Jersey.

Jackson, P. S., and Hunt, J. C. R. (1975). Turbulent wind flow over a low hill. *Q. J. R. Meteorol. Soc.* **101,** 929–955.

James, I. D. (1978). A note on the circulation induced by a shallow sea front. *Estuarine Coastal Mar. Sci.* **7,** 197–202.

James, I. D. (1981). Fronts and shelf circulation models. *Philos. Trans. R. Soc. London, Ser. A* **302,** 597–604.

Jeffreys, H. (1926). On the dynamics of geostrophic winds. *Q. J. R. Meteorol. Soc.* **52,** 85–104.

Jeffreys, H., and Jeffreys, B. S. (1956). "Methods of Mathematical Physics." 3rd ed. Cambridge Univ. Press, London and New York.

Jeffreys, H. (1923). The effect of a steady wind on the sea-level near a straight shore. *Philos. Mag.* **46,** 114–125.

Jerlov, N. G. (1968). "Optical Oceanography." Elsevier, Amsterdam.

Johnson, J. A. (1978). Topics in oceanography. *In* "Rotating Fluids in Geophysics" (P. H. Roberts and A. M. Soward, eds.), pp. 205–238. Academic Press, New York.

Johnson, J. A., and Killworth, P. D. (1975). A bottom current along the shelf break. *J. Phys. Oceanogr.* **5**, 185–188.

Jones, S. (1979). Rossby wave interactions and instabilities in a rotating, two-layer fluid on a beta-plane. Part 2. Stability. *Geophys. Astrophys. Fluid Dyn.* **12**, 1–33.

Jones, W. L. (1967). Propagation of internal gravity waves in fluids with shear flow and rotation. *J. Fluid Mech.* **30**, 439–448.

Jordan, C. L. (1958). Mean soundings for the West Indies area. *J. Meteorol.* **15**, 92–93.

Joyce, T. M., Zenk, W., and Toole, J. M. (1978). The anatomy of the Antarctic Polar Front in the Drake Passage. *J. Geophys. Res.* **83**, 6093–6113.

Julian, P. R., and Chervin, R. M. (1978). A study of the southern oscillation and Walker circulation phenomenon. *Mon. Weather Rev.* **105**, 1433–1451.

Käse, R. H. (1979). Calculations of the energy transfer by the wind to near-inertial waves. *Deep-Sea Res.* **26**, 227–232.

Käse, R. H., and Olbers, D. J. (1979). Wind-driven inertial waves observed during phase III of GATE. *Deep-Sea Res.* **26**, Suppl. 1, 191–216.

Kaimal, J. C., Wyngaard, J. C., Haugen, D. A., Coté, O. R., Izumi, Y., Caughey, S. J., and Readings, C. J. (1976). Turbulence structure in the convective boundary layer. *J. Atmos. Sci.* **33**, 2152–2169.

Kajiura, K. (1962). A note on the generation of boundary waves of Kelvin type. *J. Oceanogr. Soc. Jpn.* **18**, 49–58.

Kamenkovich, V. M. (1977). "Fundamentals of Ocean Dynamics." Elsevier, Amsterdam.

Kang, Y. Q., and Magaard, L. (1980). Annual baroclinic Rossby waves in the central North Pacific. *J. Phys. Oceanogr.* **10**, 1159–1167.

Karoly, D., and Hoskins, B. J. (1982). The three-dimensional propagation of planetary waves. *J. Meteorol. Soc. Jpn.* **60**, 109–123.

Kasahara, A. (ed.) (1979). "Numerical Values Used in Atmospheric Models," GARP Publ. Ser. No. 17, Vol. 2. World Meteorol. Organ.—Int. Counc. Sci. Unions, Geneva.

Kato, S. (1980). "Dynamics of the Upper Atmosphere." Center for Academic Publications, Tokyo.

Katz, E. J. *et al.* (1977). Zonal pressure gradient along the equatorial Atlantic. *J. Mar. Res.* **35**, 293–307.

Kazanski, A. B., and Monin, A. S. (1961). On the dynamic interaction between the atmosphere and the earth's surface. *Bull. Acad. Sci. USSR, Geophys. Ser.* **5**, 514–515.

Kendall, T. R. (1970). "The Pacific Equatorial Countercurrent." International Center for Environmental Research, Laguna Beach, California.

Khrgian, A. Kh. (1970). "Meteorology. A Historical Survey." Israel Program for Scientific Translations, Jerusalem (translation of "Ocherki razvitiya meteorologii." Gidrometeorol., Leningrad, 1959).

Kibel', I. A. (1940). Prilozhenie k meteorologii uravnenii mekhaniki baroklinnoi zhidkosti. *Izv. Akad. Nauk SSSR, Ser. Geogr. Geofiz.* No. 5.

Killworth, P. D. (1974). A baroclinic model of motions on Antarctic continental shelves. *Deep-Sea Res.* **21**, 815–837.

Killworth, P. D. (1976). The mixing and spreading phase of MEDOC.1. *Prog. Oceanogr.* **7**, 59–90.

Killworth, P. D. (1979). On the propagation of stable baroclinic Rossby waves through a mean shear flow. *Deep-Sea Res.* **26A**, 997–1031.

Killworth, P. D. (1980a). Barotropic and baroclinic instability in rotating stratified fluids. *Dyn. Atmos. Oceans* **4**, 143–184.

Killworth, P. D. (1980b). On determination of absolute velocities and density gradients in the ocean from a single hydrographic section. *Deep-Sea Res.* **27A**, 901–929.

Kinsman, B. (1965). "Wind Waves." Prentice-Hall, Englewood Cliffs, New Jersey.

Klemp, J. B., and Lilly, D. K. (1975). The dynamics of wave-induced downslope winds. *J. Atmos. Sci.* **32**, 320–339.

Klemp, J. B., and Lilly, D. K. (1978). Numerical simulation of hydrostatic mountain waves. *J. Atmos. Sci.* **35**, 78–107.

Klemp, J. B., and Lilly, D. K. (1980). Mountain waves and momentum flux. *In* "Orographic Effects in Planetary Flows," Chapter 4, pp. 116–141. GARP Publ. Ser. No. 23, WMO-ICSU, Geneva.

Klostermeyer, J. (1977). Lamb waves originating in nongeostrophic disturbances: A case study. *J. Geophys. Res.* **82**, 1441–1448.

Knox, R. A. (1976). On a long series of measurements of Indian Ocean equatorial currents near Addu Atoll. *Deep-Sea Res.* **23**, 211–221.

Kondratyev, K. Ya. (1969). "Radiation in the Atmosphere." Academic Press, New York.

Kraus, E. B. (1972). "Atmosphere-ocean Interaction." Oxford Univ. Press (Clarendon), London and New York.

Kraus, E. B., ed. (1977). "Modelling and Prediction of the Upper Layers of the Ocean." Pergamon, Oxford.

Krauss, W. (1966). "Interne Wellen." Borntraeger, Berlin.

Krishnamurti, T. N. (1971). Tropical east-west circulations during the Northern summer. *J. Atmos. Sci.* **28**, 1342–1347.

Krishnamurti, T. N. (1979). Large-scale features of the tropical atmosphere. *In* "Meteorology Over the Tropical Oceans" (D. B. Shaw, ed.), pp. 31–56. Royal Meteorological Society, London.

Kuettner, J. (1939a). Moazagotl and Föhnwelle. *Beitr. Phys. Freien Atmos.* **25**, 79–114.

Kuettner, J. (1939b). Zur Entstehung der Föhnwelle. *Beitr. Phys. Freien Atmos.* **25**, 251–299.

Kundu, P. K. (1976). An analysis of inertial oscillations observed near Oregon coast. *J. Phys. Oceanogr.* **6**, 879–893.

Kundu, P. K., and Allen, J. S. (1976). Some three-dimensional characteristics of low-frequency current fluctuations near the Oregon coast. *J. Phys. Oceanogr.* **6**, 181–199.

Kuo, H. L. (1949). Dynamic instability of two-dimensional non-divergent flow in a barotropic atmosphere. *J. Meteorol.* **6**, 105–122.

Kuo, H. L. (1952). Three-dimensional disturbances in a baroclinic zonal current. *J. Meteorol.* **9**, 260–278.

Kuo, H. L. (1973). Quasi-geostrophic flows and instability theory. *Adv. Appl. Mech.* **13**, 247–330.

La Fond, E. C. (1962). Internal waves. *In* "The Sea" (M. N. Hill, ed.) Vol. 1, pp. 731–751. Wiley (Interscience), New York.

Lagrange, J. L. (1781). Mémoire sur le théorie du mouvements des fluides. *Nouv. Mem. Acad. R. Sci. Bellelett. Berlin.* [Reprinted in "Oeuvres," Vol. IV, pp. 695–750. Gauthier-Villars, Paris, 1869.]

Lamb, H. (1904). On group velocity. *Proc. London Math. Soc.* **1**, 473–479.

Lamb, H. (1910). On atmospheric oscillations. *Proc. R. Soc. London* **84**, 551–572.

Lamb, H. (1916). "Hydrodynamics," 4th ed. Cambridge Univ. Press, London and New York.

Lamb, H. (1932). "Hydrodynamics," 6th ed. Cambridge Univ. Press, London and New York.

Landau, L. D., and Lifshitz, E. M. (1959). "Fluid Mechanics" (Engl. transl.). Pergamon, Oxford.

Laplace, P. S. (1778/1779). Recherches sur plusieurs points du système du monde. *Mem. Acad. R. Sci. Paris* (published with Histoire de l'Académie), 1775; 75–182 (publ. 1778), 1776; 117–267, 525–552 (publ. 1779). [Reprinted in "Oeuvres," Vol. 9, pp. 71–183, 187–280, 283–310. Gauthier-Villars, Paris, 1893.]

Lau, N.-C. (1978). On the three-dimensional structure of the observed transient eddy statistics of northern hemisphere wintertime circulation. *J. Atmos. Sci.* **35**, 1900–1923.

Lau, N.-C. (1979a). The structure and energetics of transient disturbances in the northern hemisphere wintertime circulation. *J. Atmos. Sci.* **36**, 982–995.

Lau, N.-C. (1979b). The observed structure of tropospheric stationary waves and the local balance of vorticity and heat. *J. Atmos. Sci.* **36**, 996–1016.

Lau, N.-C., and Wallace, J. M. (1979). On the distribution of horizontal transports by transient eddies in the northern hemisphere wintertime circulation. *J. Atmos. Sci.* **36**, 1844–1861.

Leaman, K. D., and Sanford, T. B. (1975). Vertical energy propagation of inertial waves: A vector spectral analysis of velocity profiles. *J. Geophys. Res.* **80**, 1975–1978.

LeBlond, P. H., and Mysak, L. A. (1977). Trapped coastal waves and their role in shelf dynamics. *In* "The Sea" (E. D. Goldberg *et al.*, eds.), Vol. 6, pp. 459–495. Wiley (Interscience), New York.

LeBlond, P. H., and Mysak, L. A. (1978). "Waves in the Ocean." Elsevier, Amsterdam.

Lee, O. S. (1961). Observations of internal waves in shallow water. *Limnol. Oceanogr.* **6**, 312–321.

Leetmaa, A. (1973). The response of the Somali Current at 2°S to the southwest monsoon of 1971. *Deep-Sea Res.* **20**, 397–400.

Leetmaa, A., and Bunker, A. F. (1978). Updated charts of the mean annual wind stress, convergence in the Ekman layers, and Sverdrup transports in the North Atlantic. *J. Mar. Res.* **36**, 311–322.

Leetmaa, A., Niiler, P., and Stommel, H. (1977). Does the Sverdrup relation account for the mid-Atlantic circulation? *J. Mar. Res.* **35**, 1–10.

Legeckis, R. (1978). A survey of worldwide sea surface temperature fronts detected by environmental satellites. *J. Geophys. Res.* **83,** 4501–4522.

Leipper, D. F. (1967). Observed ocean conditions and Hurricane Hilda 1964. *J. Atmos. Sci.* **24,** 182–196.

Leith, C. E. (1980). Nonlinear normal mode initialization and quasigeostrophic theory. *J. Atmos. Sci.* **37,** 958–968.

Lemasson, L., and Piton, B. (1968). Anomalie dynamique de la surface de la mer le long de l'équateur dans l'Océan Pacifique. *Cah. ORSTOM, Sér. Océanogr.* **6,** 39–45.

Leonov, A. I., and Miropol'skiy, Yu. Z. (1973). Resonant excitation of internal gravity waves in the ocean by atmospheric pressure fluctuations. *Izv. Atmos. Ocean Phys.* **9,** 851–862 (Engl. Transl. pp. 480–485).

Lerch, F. H., Klosko, S. M., Laubscher, R. E., and Wagner, C. A. (1979). Gravity model improvements using Geos 3 (GEM 9 and 10). *J. Geophys. Res.* **84,** 3897–3916.

Lewis, E. L. (1980). The Practical Salinity Scale 1978 and its antecedents. *IEEE J. Oceanic Eng.* **OE5,** 3–8.

Lewis, E. L., and Perkin, R. G. (1981). The Practical Salinity Scale 1978: conversion of existing data. *Deep-Sea Res.* **28A,** 307–328.

Lighthill, M. J. (1958). "Fourier Analysis and Generalised Functions." Cambridge Univ. Press, London and New York.

Lighthill, M. J. (1965). Group velocity. *J. Inst. Math. Appl.* **1,** 1–28.

Lighthill, M. J. (1966). Dynamics of rotating fluids. A survey. *J. Fluid Mech.* **26,** 411–431.

Lighthill, M. J. (1967). On waves generated in dispersive systems by travelling forcing effects, with applications to the dynamics of rotating fluids. *J. Fluid Mech.* **27,** 725–752.

Lighthill, M. J. (1969). Dynamic response of the Indian Ocean to onset of the Southwest Monsoon. *Philos. Trans. R. Soc. London, Ser. A* **265,** 45–92.

Lighthill, M. J. (1978). "Waves in Fluids." Cambridge Univ. Press, London and New York.

Lilly, D. K. (1971). Observations of mountain induced turbulence. *J. Geophys. Res.* **76,** 6585–6588.

Lilly, D. K. (1972). Wave momentum flux—a GARP problem. *Bull. Am. Meteorol. Soc.* **53,** 17–23.

Lilly, D. K. (1978). A severe downslope windstorm and aircraft turbulence event induced by a mountain wave. *J. Atmos. Sci.* **35,** 59–77.

Lilly, D. K. (1979). The dynamic structure and evolution of thunderstorm and squall lines. *Annu. Rev. Earth Planet. Sci.* **7,** 117–162.

Lilly, D. K., and Klemp, J. B. (1979). The effects of terrain shape on mountain waves. *J. Fluid Mech.* **95,** 241–262.

Lin, C. C. (1955). "The Theory of Hydrodynamic Stability." Cambridge Univ. Press, London and New York.

Lindzen, R. S. (1970). Internal gravity waves in atmospheres with realistic dissipation and temperature. Part 1. Mathematical development and propagation of waves into the thermosphere. *Geophys. Fluid Dyn.* **1,** 303–355.

Lindzen, R. S. (1971). Internal gravity waves in atmospheres with realistic dissipation and temperature. Part 3. Daily variations in the thermosphere. *Geophys. Fluid Dyn.* **2,** 89–121.

Lindzen, R. S. (1979). Atmospheric tides. *Annu. Rev. Earth Planet. Sci.* **7,** 199–225.

Lindzen, R. S., and Blake, D. (1971). Internal gravity waves in atmospheres with realistic dissipation and temperature. Part II. Thermal tides excited below the mesopause. *Geophys. Fluid Dyn.* **2,** 31–61.

Lindzen, R. S., and Blake, D. (1972). Lamb waves in the presence of realistic distributions of temperature and dissipation. *J. Geophys. Res.* **77,** 2166–2176.

Lindzen, R. S., Farrell, B., and Tung, K.-K. (1980). The concept of wave over-reflection and its application to baroclinic instability. *J. Atmos. Sci.* **37,** 44–63.

Liouville, J. (1837). Sur le développement des fonctions ou parties de fonctions en séries . . . . *J. Math. Pure Appl.* **2,** 16–35.

Lisitzin, E. (1974). Sea-level Changes. Elsevier, Amsterdam.

List, R. J. (1951). "Smithsonian Meteorological Tables," 6th ed. Smithson. Misc. Collect. No. 114. Smithson. Inst. Washington, D.C.

Liu, W. T., Katsaros, K. B., and Businger, J. A. (1979). Bulk parameterisation of air-sea exchanges of heat and water vapour including the molecular constraints at the interface. *J. Atmos. Sci.* **36,** 1722–1735.

London, J., and Sasamori, T. (1971). Radiative energy budget of the atmosphere. *Space Res.* **11,** 639–649.

Long, R. R. (1953). Some aspects of the flow of stratified fluids. I. A theoretical investigation. *Tellus* **5**, 42–58.

Longuet-Higgins, M. S. (1964a). Planetary waves on a rotating sphere. *Proc. R. Soc. London, Ser. A* **279**, 446–473.

Longuet-Higgins, M. S. (1964b). On group velocity and energy flux in planetary wave motions. *Deep-Sea Res.* **11**, 35–43.

Longuet-Higgins, M. S. (1965a). Planetary waves on a rotating sphere. II. *Proc. R. Soc. London, Ser. A* **294**, 40–54.

Longuet-Higgins, M. S. (1965b). The response of a stratified ocean to stationary and moving wind systems. *Deep-Sea Res.* **12**, 923–973.

Longuet-Higgins, M. S. (1965c). Some dynamical aspects of ocean currents. *Q. J. R. Meteorol. Soc.* **91**, 425–457.

Longuet-Higgins, M. S. (1968a). Double Kelvin waves with continuous depth profiles. *J. Fluid Mech.* **34**, 49–80.

Longuet-Higgins, M. S. (1968b). The eigenfunctions of Laplace's tidal equations over a sphere. *Proc. R. Soc. London, Ser. A* **262**, 511–607.

Longuet-Higgins, M. S., and Pond, G. S. (1969). The free oscillations of fluid on a hemisphere bounded by meridians of longitude. *Philos. Trans. R. Soc. London, Ser. A* **266**, 193–223.

Lorenz, E. N. (1955). Available potential energy and the maintenance of the general circulation. *Tellus* **7**, 157–167.

Lorenz, E. N. (1960). Energy and numerical weather prediction. *Tellus* **12**, 364–373.

Lorenz, E. N. (1967). "The Nature and Theory of the General Circulation of the Atmosphere." World Meteorol. Organ., Geneva.

Lorenz, E. N. (1969). The nature of the global circulation of the atmosphere: A present view. *In* "The Global Circulation of the Atmosphere" (G. A. Corby, ed.), pp. 3–23. Royal Meteorological Society, London.

Lorenz, E. N. (1972). Barotropic instability of Rossby wave motion. *J. Atmos. Sci.* **29**, 258–269.

Lucassen-Reynders, E. H., and Lucassen, J. (1969). Properties of capillary waves. *Adv. Colloid Interface Sci.* **2**, 347–395.

Lumb, F. E. (1964). The influence of cloud on hourly amount of total solar radiation at the sea surface. *Q. J. R. Meteorol. Soc.* **90**, 43–56.

Lumley, J. L., and Panofsky, H. A. (1964). "The Structure of Atmospheric Turbulence." Wiley (Interscience), New York.

Luyten, J., and Swallow, J. (1977). Equatorial undercurrents. *Deep-Sea. Res.* **23**, 999–1001.

Lynn, R. J., and Reid, J. L. (1968). Characteristics and circulation of deep and abyssal waters. *Deep-Sea Res.* **15**, 577–598.

Lyra, G. (1940). Über den Einfluss von Bodener hebungen auf die Strömung einer stabil geschichteten Atmosphäre. *Beitr. Phys. Freien Atmos.* **26**, 197–206.

Lyra, G. (1943). Theorie der stationären Leewellenströmung in freier Atmosphäre. *Z. Angew. Math. Mech.* **23**, 1–28.

McAlister, E. D., and McLeish, W. (1969). Heat transfer in the top millimeter of the ocean. *J. Geophys. Res.* **74**, 3408–3414.

McComas, C. H. (1977). Equilibrium mechanisms within the internal wave field. *J. Phys. Oceanogr.* **7**, 836–845.

McComas, C. H., and Bretherton, F. P. (1977). Resonant interaction of oceanic internal waves. *J. Geophys. Res.* **82**, 1397–1412.

McComas, C. H., and Müller, P. (1981a). Time scales of resonant interactions among oceanic internal waves. *J. Phys. Oceanogr.* **11**, 139–147.

McComas, C. H., and Müller, P. (1981b). The dynamic balance of internal waves. *J. Phys. Oceanogr.* **11**, 970–986.

McCreary, J. P. (1978). Eastern tropical ocean response to changing wind systems. *In* "Review Papers of Equatorial Oceanography–FINE Workshop Proceedings," Chapter 7. Nova N.Y.I.T. Univ. Press, Fort Lauderdale, Florida.

McCreary, J. P. (1981a). A linear stratified ocean model of the equatorial undercurrent. *Philos. Trans. R. Soc. London, Ser. A* **298**, 603–635.

McCreary, J. P. (1981b). A linear stratified model of the coastal undercurrent. *Philos. Trans. R. Soc. London, Ser. A* **302**, 385–413.

McEwan, A. D., Mander, D. W., and Smith, R. K. (1972). Forced resonant second-order interaction between damped internal waves. *J. Fluid Mech.* **35**, 589–608.

McInturff, R. M., ed. (1978). Stratospheric warmings: Synoptic dynamic and general circulation aspects. *NASA Ref. Publ.* **1017**, 1–174.

McIntyre, M. E. (1980). An introduction to the generalized Lagrangian-mean description of wave, mean-flow interaction. *Pure Appl. Geophys.* **118**, 152–176. [Reprinted in "The Middle Atmosphere" (S. V. Venkatswaran and N. Sundararaman, eds.), pp. 152–176. Birkhäuser, Basel, 1980).]

McIntyre, M. E., and Weissman, M. A. (1978). On radiating instabilities and resonant over-reflection. *J. Atmos. Sci.* **35**, 1190–1196.

Mackintosh, N. A. (1946). The Antarctic convergence and the distribution of surface temperature in Antarctic waters. *Discovery Rep.* **23**, 177–212.

McPhee, M. G., and Smith, J. D. (1976). Measurements of the turbulent boundary layer under pack ice. *J. Phys. Oceanogr.* **6**, 696–711.

McVean, M. K., and Woods, J. D. (1980). Redistribution of scalars during upper ocean frontogenesis: A numerical model. *Q. J. R. Meteorol. Soc.* **106**, 293–311.

McWilliams, J. C., and Flierl, G. R. (1976). Optimal quasi-geostrophic wave analysis of MODE array data. *Deep-Sea Res.* **23**, 285–300.

McWilliams, J. C., Holland, W. R., and Chow, H. S. (1978). A description of numerical Antarctic circumpolar currents. *Dyn. Atmos. Oceans* **2**, 213–291.

Madden, R. A. (1978). Further evidence of traveling planetary waves. *J. Atmos. Sci.* **35**, 1605–1618.

Malkus, J. S. (1962). Interchange of properties between sea and air: Large-scale interactions. *In* "The Sea" (M. N. Hill, ed.), Vol. 1, Chapter 4, pp. 88–294. Wiley (Interscience), New York.

Manabe, S., and Hahn, D. G. (1977). Simulation of the tropical climate of an Ice Age. *J. Geophys. Res.* **82**, 3889–3911.

Manabe, S., and Strickler, R. F. (1964). Thermal equilibrium of the atmosphere with a convective adjustment. *J. Atmos. Sci.* **21**, 361–385.

Manabe, S., and Wetherald, R. T. (1967). Thermal equilibrium in the atmosphere with a given distribution of relative humidity. *J. Atmos. Sci.* **24**, 241–259.

Manabe, S., Hahn, D. G., and Holloway, J. L. (1974). The seasonal variation of the tropical circulation as simulated by a global model of the atmosphere. *J. Atmos. Sci.* **31**, 43–83.

Manabe, S., Bryan, K., and Spelman, M. J. (1979). A global ocean–atmospheric climate model with seasonal variation for future studies of climate sensitivity. *Dyn. Atmos. Oceans* **3**, 393–426.

Manley, G. (1945). The Helm Wind of Crossfell, 1937–1939. *Q. J. R. Meteorol. Soc.* **71**, 197–220.

Manton, M. J. (1972). On the wave field generated by a variable wind. *Geophys. Fluid Dyn.* **3**, 91–104.

Margules, M. (1893). Luftbewegungen in einer rotierenden Sphäroidschale (II. Teil). *Sitzungsber. Kais. Akad. Wiss. Wien, Math.-Nat. Cl.* **102**, Abt. IIA, 11–56. [English transl.: "Air motion in a rotating spherical shell, by Max Margules" (B. Haurwitz, transl.). Natl. Cent. Atmos. Res. Tech. Note NCAR/TN-156 + STR.]

Margules, M. (1903). Über die Energie der Stürme. *Jahrb. Zentralanst. Meteorol. Wien* **40**, 1–26; translation in Abbe (1910, pp. 533–595).

Margules, M. (1906). Über Temperaturschichtung in stationär bewegter und ruhender Luft. *Meteorol. Z.* **23**, 243–254.

Marsigli, L. M. (1681). Osservazioni intorno al Bosforo Tracio o vero Canale di Constantinopli, rappresentate in lettera alla Sacra Real Maesta Cristina Regina di Svezia, Roma. [Reprinted in *Boll. Pesca, Piscic. Idrobiol.* **11**, 734–758 (1935).]

Martell, C. M., and Allen, J. S. (1979). The generation of continental shelf waves by alongshore variations in bottom topography. *J. Phys. Oceanogr.* **9**, 696–711.

Martin, S., Simmons, W., and Wunsch, C. (1972). The excitation of resonant triads by single internal waves. *J. Fluid. Mech.* **53**, 17–44.

Mason, P. J., and Sykes, R. I. (1978). On the interaction of topography and Ekman boundary layer pumping in a stratified atmosphere. *Q. J. R. Meteorol. Soc.* **104**, 475–490.

Masuzawa, J., and Nagasaka, K. (1975). The 137°E oceanographic section. *J. Mar. Res.* **33**, 109–116.

Matijević, E. (ed.) (1969ff). "Surface and Colloid Science." Wiley (Interscience), New York.

Matsuno, T. (1966). Quasi-geostrophic motions in the equatorial area. *J. Meteorol. Soc. Jpn.* **44**, 25–43.

Matsuno, T. (1970). Vertical propagation of stationary planetary waves in the winter northern hemisphere. *J. Atmos. Sci.* **27**, 871–884.

Matsuno, T. (1971). A dynamic model of the stratospheric sudden warming. *J. Atmos. Sci.* **28**, 1479–1494.

Matsuno, T. (1980). Lagrangian motion of air parcels in the stratosphere in the presence of planetary waves. *Pure Appl. Geophys.* **118**, 189–216. [Reprinted in "The Middle Atmosphere" (S. V. Venkateswaran and N. Sundararaman, eds.), pp. 189–216. Birkhäuser, Basel, 1980.]

Matthäus, W. (1969). Zur entdeckungsgeschichte des Aquatorialen Unterstroms im Atlantischen Ozean. *Beitr. Meereskd.* **23**, 37–70.

Maykut, G. A., and Untersteiner, N. (1971). Some results of a time-dependent thermodynamic model of sea ice. *J. Geophys. Res.* **76**, 1550–1575.

MEDOC Group (1970). Observations of formation of deep water in the Mediterranean Sea. *Nature (London)* **227**, 1037–1040.

Mellor, G. L., and Yamada, T. (1974). A hierarchy of turbulence closure models for planetary boundary-layers. *J. Atmos. Sci.* **31**, 1791–1806.

Mesinger, F., and Arakawa, A. (1976). "Numerical Methods Used in Atmospheric Models," GARP Publ. Ser. No. 17, Vol. 1. World Meteorol. Organ.—Int. Counc. Sci. Unions, Geneva.

Meyers, G. (1975). Seasonal variations in transport of the Pacific North Equatorial Current relative to the wind field. *J. Phys. Oceanogr.* **5**, 442–449.

Meyers, G. (1979a). On the annual Rossby wave in the tropical North Pacific Ocean. *J. Phys. Oceanogr.* **6**, 663–674.

Meyers, G. (1979b). Annual variation in the slope of the 14°C isotherm along the equator in the Pacific Ocean. *J. Phys. Oceanogr.* **9**, 885–891.

Milankovich, M. (1930). Mathematische Klimalehre und astronomische Theorie der Klimaschwankungen. *In* "Handbuch der Klimatologie," Vol. I, Part A, pp. 1–176. Koppen & Geiger, Berlin.

Milankovich, M. (1941). Kanon der Erdbestrahlung und seine Anwendung auf das Eiszeitenproblem. Königlich Serbische Akademie, Special Publ. Vol. 132. [Transl. by Israel Program for Scientific Translations, 1969.]

Miles, J. W. (1968a). Lee waves in a stratified flow. Part I. Thin barrier. *J. Fluid Mech.* **32**, 549–567.

Miles, J. W. (1968b). Lee waves in a stratified flow. Part II. Semi-circular obstacle. *J. Fluid Mech.* **33**, 803–814.

Miles, J. W., and Huppert, H. E. (1969). Lee waves in a stratified flow. Part IV. Perturbation approximations. *J. Fluid Mech.* **35**, 497–525.

Miller, G. R. (1966). The flux of tidal energy out of the deep oceans. *J. Geophys. Res.* **71**, 2485–2489.

Millero, F. J. (1978). Freezing point of seawater. *In* "Eighth Report of the Joint Panel on Oceanographic Tables and Standards," UNESCO Tech. Pap. Mar. Sci. No. 28, Annex 6. UNESCO, Paris.

Millero, F. J., and Poisson, A. (1981). International one-atmosphere equation of state for seawater. *Deep-Sea Res.* **28A**, 625–629.

Millero, F. J., Perron, G., and Desnoyers, J. E. (1973). Heat capacity of seawater solutions from 5 to 35°C and 0.5 to 22% chlorinity. *J. Geophys. Res.* **78**, 4499–4507.

Millero, F. J., Chen, C.-T., Bradshaw, A., and Schleicher, K. (1980). A new high pressure equation of state for seawater. *Deep-Sea Res.* **27A**, 255–264.

Minzner, R. A. (1977). The 1976 standard atmosphere and its relationship to earlier standards. *Rev. Geophys. Space Phys.* **15**, 375–384.

MODE Group (1978). The mid-ocean dynamics experiment. *Deep-Sea Res.* **25**, 859–910.

Möller, F., and Manabe, S. (1961). Über des Strahlungsgleichgewicht der Atmosphäre. *Z. Meteorol.* **15**, 3–20.

Monin, A. S. (1970). The atmospheric boundary layer. *Annu. Rev. Fluid Mech.* **2**, 225–250.

Monin, A. S. (1972). "Earth's Rotation and Climate." Gidrometeoizdat, Leningrad (English translation published by Radhakrishna Prakashan, Delhi, 1974).

Monin, A. S. (1975). The role of the oceans in climate models. *In* "The physical basis of climate and climate modelling," GARP Publ. Ser. No. 16, Appendix 6, pp. 201–205. World Meteorol. Organ./Int. Counc. Sci. Unions, Geneva.

Montgomery, R. B., and Spilhaus, A. F. (1941). Examples and outlines of certain modifications in upper-air analysis. *J. Atmos. Sci.* **8**, 276–283.

Mooers, C. N. K., Collins, C. A., and Smith, R. L. (1976). The dynamic structure of the frontal zone in the coastal upwelling region off Oregon. *J. Phys. Oceanogr.* **6**, 3–21.

Moore, D. W. (1963). Rossby waves in ocean circulation. *Deep-Sea Res.* **10**, 735–748.

Moore, D. W. (1968). Planetary-gravity waves in an equatorial ocean. Ph.D. Thesis, Harvard University, Cambridge, Massachusetts.

Moore, D. W., and Philander, S. G. H. (1977). Modeling of the tropical ocean circulation. *In* "The Sea" (E. D. Goldberg, I. N. McCave, J. J. O'Brien, and J. H. Steele, eds.) Vol. 6, pp. 319–362. Wiley (Interscience), New York.

Morgan, G. W. (1956). On the wind-driven ocean circulation. *Tellus* **8**, 301–320.

Mork, M. (1981). Circulation phenomena and frontal dynamics of the Norwegian coastal current. *Philos. Trans. R. Soc. London, Ser. A* **302**, 635–647.

Morse, P. M. (1964). "Thermal Physics." Benjamin, New York.

Morse, P. M., and Feshbach, H. (1953). "Methods of Theoretical Physics," 2 parts. McGraw-Hill, New York.

Mortimer, C. H. (1963). Frontiers in physical limnology with particular reference to long waves in rotating basins. *Publ.—Great Lakes Res. Div., Univ. Mich.* **10**, 9–42.

Mortimer, C. H. (1968). Internal waves and associated currents observed in Lake Michigan during the summer of 1963. *Spec. Rep.—Univ. Wis.-Milwaukee, Cent. Great Lakes Stud.* **1**, 1–24.

Mortimer, C. H. (1971). Large scale oscillatory motions and seasonal temperature changes in Lake Michigan and Lake Ontario. *Spec. Rep.—Univ. Wis.-Milwaukee. Cent. Great Lakes Stud.* **12**, Parts I and II.

Mortimer, C. H. (1974). Lake hydrodynamics. *Mitt. Int. Ver. Theor. Angew. Limnol.* **20**, 124–197.

Mortimer, C. H. (1977). Internal waves observed in Lake Ontario during the International Field Year for the Great Lakes (IFYGL), 1972. *Spec. Rep.—Univ. Wis-Milwaukee, Cent. Great Lakes Stud.* **32**, 1–122.

Mowbray, D. E., and Rarity, B. S. H. (1967). A theoretical and experimental investigation of the phase configuration of internal waves of small amplitude in a density stratified fluid. *J. Fluid Mech.* **28**, 1–16.

Müller, P. (1977). Spectral features of the energy transfer between internal waves and a large scale shear flow. *Dyn. Atmos. Oceans* **2**, 49–72.

Müller, P. and Frankignoul, C. (1981). Direct atmospheric forcing of geostrophic eddies. *J. Phys. Oceanogr.* **11**, 287–308.

Müller, P., Olbers, D. J., and Willebrand, J. (1978). The IWEX spectrum. *J. Geophys. Res.* **83**, 479–500.

Munk, W. H. (1950). On the wind-driven ocean circulation. *J. Meteorol.* **7**, 79–93.

Munk, W. H. (1981). Internal waves and small-scale processes. *In* "Evolution of Physical Oceanography" (B. A. Warren and C. Wunsch, eds.), Chapter 9. MIT Press, Cambridge, Massachusetts.

Munk, W. H., and McDonald G. J. F. (1960). "The rotation of the Earth." Cambridge Univ. Press, London and New York.

Munk, W. H., and Palmén, E. (1951). Note on the dynamics of the Antarctic Circumpolar Current. *Tellus* **3**, 53–56.

Munk, W. H., Snodgrass, F., and Wimbush, M. (1970). Tides off-shore: Transition from California coastal to deep-sea waters. *Geophys. Fluid Dyn.* **1**, 161–235.

Murakami, T., and Ho, F. P. (1972). Spectrum analysis of cloudiness over the North Pacific. *J. Meteorol. Soc. Jpn.* **50**, 285–300.

Mysak, L. A. (1967). On the theory of continental shelf waves. *J. Mar. Res.* **25**, 205–226.

Mysak, L. A. (1980a). Topographically trapped waves. *Annu. Rev. Fluid Mech.* **12**, 45–76.

Mysak, L. A. (1980b). Recent advances in shelf wave dynamics. *Rev. Geophys. Space Phys.* **18**, 211–241.

National Committee for Fluid Mechanics Films (1972). "Illustrated Experiments in Fluid Mechanics." MIT Press, Cambridge, Massachusetts.

Newell, R. E., Mahoney, J. R., and Lenhard, R. W. (1966). A pilot study of small-scale wind variations in the stratosphere and mesosphere. *Q. J. R. Meteorol. Soc.* **92**, 41–54.

Newell, R. E., Vincent, D. G., Dopplick, T. G., Ferruzza, D., and Kidson, J. W. (1969). The energy balance of the atmosphere. *In* "The Global Circulation of the Atmosphere" (G. A. Corby, ed.), pp. 42–90. Royal Meteorological Society, London.

Newell, R. E., Kidson, J. W., Vincent, D. G., and Boer, G. J. (1972). "The General Circulation of the Tropical Atmosphere," Vol. 1. MIT Press, Cambridge, Massachusetts.

Newell, R. E., Kidson, J. W., Vincent, D. G., and Boer, G. J. (1974). "The General Circulation of the Tropical Atmosphere," Vol. 2. MIT Press, Cambridge, Massachusetts.

Newton, C. W. (1971). Global angular momentum balance: Earth torques and atmospheric fluxes. *J. Atmos. Sci.* **28,** 1329–1341.

Newton, C. W., ed. (1972a). "Meteorology of the Southern Hemisphere," Meteorol. Monogr., Vol. 13. Am. Meteorol. Soc., Boston, Massachusetts.

Newton, C. W. (1972b). Southern hemisphere general circulation in relation to global energy and momentum balance requirements. *In* "Meteorology of the Southern Hemisphere" (C. W. Newton, ed.), Chapter 9, pp. 215–246. Am. Meteorol. Soc., Boston, Massachusetts.

Nguyen, N. A., and Gill, A. E. (1981). Generation of coastal lows by synoptic-scale waves. *Q. J. R. Meteorol. Soc.* **107,** 521–530.

Nicholls, J. M., ed. (1973). "The Airflow Over Mountains: Research 1958–76," Tech Note 127. World Meteorol. Organ., Geneva.

NOAA/NASA/USAF (1976). "US Standard Atmosphere 1976." Washington, D.C.

Nomitsu, T. (1934). Coast effect upon the ocean current and the sea level. II. Changing state. *Mem. Coll. Sci., Univ. Kyoto, Ser. A* **17,** 249–280.

O'Connor, J. F. (1963). The weather and circulation of January 1963. *Mon. Weather Rev.* **91,** 209–218.

Olbers, D. J. (1976). Non-linear energy transfer and the energy balance of the internal wave field in the deep ocean. *J. Fluid Mech.* **74,** 375–399.

Olbers, D. J. (1982). Internal waves. *In* "Turbulence in the Ocean" (J. D. Woods, ed.), Chapter 6. Springer-Verlag, Berlin and New York.

Olver, F. W. J. (1974). "Asymptotics and Special Functions." Academic Press, New York.

O'Neill, A., and Taylor, B. F. (1979). A study of the major stratospheric warming of 1976/77. *Q. J. R. Meteorol. Soc.* **105,** 71–92.

Oort, A. H. (1971). The observed annual cycle in the meridional transport of atmospheric energy. *J. Atmos. Sci.* **28,** 325–339.

Oort, A. H., and Rasmusson, E. M. (1970). On the annual variation of the mean meridional circulation. *Mon. Weather Rev.* **98,** 423–442.

Oort, A. H., and Rasmusson, E. M. (1971). "Atmospheric Circulation Statistics," NOAA Prof. Pap. 5. U.S. Dept. of Commerce, Washington, D.C.

Orlanski, I., and Ross, B. B. (1978). The circulation associated with a cold front. Part II. Moist case. *J. Atmos. Sci.* **35,** 445–465.

Ou, H. W., and Bennett, J. R. (1979). A theory of the mean flow driven by long internal waves in a rotating basin, with application to Lake Kinneret. *J. Phys. Oceanogr.* **9,** 1112–1125.

Owens, W. B., and Hogg, N. G. (1980). Oceanic observations of stratified Taylor columns near a bump. *Deep-Sea Res.* **27,** 1029–1045.

Palmén, E., and Newton, C. W. (1969). "Atmospheric Circulation Systems." Academic Press, New York.

Palmer, T. N. (1981). Diagnostic study of a wavenumber-2 stratospheric sudden warming in a transformed Eulerian-mean formalism. *J. Atmos. Sci.* **38,** 844–855.

Paltridge, G. W., and Platt, C. M. R. (1976). "Radiative Processes in Meteorology and Climatology." Elsevier, Amsterdam.

Parke, M. E., and Hendershott, M. C. (1979). M_2, S_2 and K_1 models of the global ocean tide on an elastic earth. *Mar. Geodesy* **3,** 379–408.

Pattullo, J., Munk, W. H., Revelle, R., and Strong, E. (1955). The seasonal oscillation in sea level. *J. Mar. Res.* **14,** 88–155.

Pearce, A. F. (1977). Some features of the upper 500m of the Agulhas current. *J. Mar. Res.* **35,** 731–753.

Pedlosky, J. (1964). The stability of currents in the atmosphere and the oceans. Part I. *J. Atmos. Sci.* **27,** 201–219.

Pedlosky, J. (1979). "Geophysical Fluid Dynamics." Springer-Verlag, Berlin and New York.

Peffley, M. B., and O'Brien, J. J. (1976). A three-dimensional simulation of coastal upwelling off Oregon. *J. Phys. Oceanogr.* **6,** 164–180.

Peltier, W. R., and Clark, T. L. (1979). The evolution and stability of finite-amplitude mountain waves. Part II. Surface wave drag and severe downslope windstorms. *J. Atmos. Sci.* **36,** 1498–1529 (see also **37,** 2119–2125).

Perlroth, I. (1969). Effects of oceanographic media on equatorial Atlantic hurricanes. *Tellus* **21,** 230–244.

Perrot, A. (1859). Nouvelle expérience pour rendre manifeste le mouvement de rotation de la terre. *C. R. Hebd. Seances Acad. Sci.* **49,** 637–638.

Peterson, W. H. (1982). On the interaction of multiple buoyancy sources in a simple steady-state convection model. *J. Fluid Mech.* (to be published).

Philander, S. G. H. (1973). Equatorial undercurrent: Measurements and theories. *Rev. Geophys. Space Phys.* **11,** 513–570.

Philander, S. G. H., and Pacanowski, R. C. (1980). The generation and decay of equatorial currents. *J. Geophys. Rev.* **85,** 1123–1136.

Phillips, N. A. (1951). A simple three-dimensional model for the study of large-scale extratropical flow patterns. *J. Meteorol.* **8,** 381–394.

Phillips, N. A. (1963). Geostrophic motion. *Rev. Geophys.* **1,** 123–126.

Phillips, N. A. (1973). Principles of large scale numerical weather prediction. *In* "Dynamic Meteorology" (P. Morel, ed.), pp. 1–96. Reidel Publ., Dordrecht, Netherlands.

Phillips, O. M. (1960). On the dynamics of unsteady gravity waves of finite amplitude. *J. Fluid Mech.* **9,** 193–217.

Phillips, O. M. (1977). "The Dynamics of the Upper Ocean," 2nd ed. Cambridge Univ. Press, London and New York.

Pickersgill, A. O., and Hunt, G. E. (1981). An examination of the formation of linear lee waves generated by giant Martian volcanoes. *J. Atmos. Sci.* **38,** 40–51.

Platzman, G. W. (1968). The Rossby wave. *Q. J. R. Meteorol. Soc.* **94,** 225–246.

Platzman, G. W. (1971). Ocean tides and related waves. *In* "Mathematical Problems in the Geophysical Sciences" (W. H. Reid, ed.), Vol. 2, pp. 239–291. Am. Math. Soc., Providence, Rhode Island.

Platzman, G. W., Curtis, G. A., Hansen, K. S., and Slater, R. D. (1981). Normal modes of the world ocean. Part II. Description of modes in the period range 8–80 hours. *J. Phys. Oceanogr.* **11,** 579–603.

Plumb, R. A., and McEwan, A. D. (1978). The instability of a forced standing wave in a viscous, stratified fluid: A laboratory analogue of the quasi-biennial oscillation. *J. Atmos. Sci.* **35,** 1827–1839.

Poincaré, S. (1910). "Théorie des Marées." Leçons de Mécanique Celeste, Vol. 3. Gauthiers-Villars, Paris.

Pollard, R. T. (1970). On the generation by winds of inertial waves in the ocean. *Deep-Sea Res.* **17,** 795–812.

Pollard, R. T., and Millard, R. C. (1970). Comparison between observed and simulated wind-generated inertial oscillations. *Deep-Sea Res.* **17,** 813–821.

Pollard, R. T., Rhines, P. B., and Thompson, R. O. R. Y. (1973). The deepening of the wind-mixed layer. *Geophys. Fluid Dyn.* **3,** 381–404.

Polli, S. (1960). La propagazione delle maree nell'Adriatico. *Atti IX Convegno Ass. Geofis. Ital., Roma, 1959.* [Also: Publ. Inst. Talassogr. 370.]

Pond, S., and Bryan, K. (1976). Numerical models of the ocean circulation. *Rev. Geophys. Space Phys.* **14,** 243–263.

Prandle, D. (1975). Storm surges in the southern North Sea and River Thames. *Proc. R. Soc. London, Ser. A* **344,** 509–539.

Prandle, D., and Wolf, J. (1978). The interaction of surge and tide in the North Sea and River Thames. *Geophys. J. R. Astron. Soc.* **55,** 203–216.

Pratt, R. W., and Wallace, J. M. (1976). Zonal propagation characteristics of large-scale fluctuations in the mid-latitude troposphere. *J. Atmos. Sci.* **33,** 1184–1194.

Preston-Whyte, R. A. (1975). A note on some bioclimatic consequences of coastal lows. *S. Afr. Geogr. J.* **57,** 17–25.

Price, J. F. (1981). Upper ocean response to a hurricane. *J. Phys. Oceanogr.* **11,** 153–175.

Price, J. F., and Rossby, T. H. (1982). Observations of a barotropic planetary wave in the western North Atlantic. *J. Mar. Res. Suppl.* **40.**

Price, P. G. (1975). A comparison between available potential and kinetic energy estimates for the southern and northern hemispheres. *Tellus* **27,** 443–452.

Proudman, J. (1927). Newton's work on the theory of the tides. *In* "Isaac Newton 1642–1727" (W. J. Greenstreet, ed.), pp. 87–95. Bell, London.

Proudman, J. (1953). "Dynamical Oceanography." Methuen, London, and Wiley, New York.

Queney, P. (1948). The problem of air flow over mountains: A summary of theoretical studies. *Bull. Am. Meteorol. Soc.* **29,** 16–26.

Queney, P. (1973). Transfer and dissipation of energy by mountain waves. *In* "Dynamical meteorology" (P. Morel, ed.), pp. 97–351. Reidel Publ., Dordrecht, Netherlands.

Queney, P. (1977). Synthèse de travaux theoriques sur les perturbations de relief. Seconde partie. *Meteorologie* **6**(9) 111–163.

Rao, D. B., and Schwab, D. J. (1976). Two-dimensional normal modes in arbitrary enclosed basins on a rotating earth: Application to Lakes Ontario and Superior. *Philos. Trans. R. Soc. London, Ser. A* **281,** 63–96.

Rao, D. B., Mortimer, C. H., and Schwab, D. J. (1976). Surface normal modes of Lake Michigan: Calculations compared with spectra of observed water level fluctuations. *J. Phys. Oceanogr.* **6,** 575–588.

Raschke, E., Vonder Haar, T. H., Bandeen, W. R., and Pasternak, M. (1973). The annual radiation balance of the earth-atmosphere system during 1969–70 from Nimbus 3 measurements. *J. Atmos. Sci.* **30,** 341–364.

Rasmusson, E. M., and Carpenter, T. H. (1982). Variations in tropical sea surface temperature and surface wind fields associated with the Southern Oscillation/El Niño. *Mon. Weather Rev.* (to be published).

Rayleigh, Lord (1880). On the stability, or instability, of certain fluid motions. *Proc. London Math. Soc.* **9,** 57–70. [Also in "Scientific Papers," Vol. 1, pp. 474–487. Cambridge Univ. Press.]

Rayleigh, Lord (1883). Investigation of the character of the equilibrium of an incompressible heavy fluid of variable density. *Proc. London Math. Soc.* **14,** 170–177. [Also in "Scientific Papers," Vol. 2, pp. 200–207. Cambridge Univ. Press, 1900.]

Reed, R. S., and Danielsen, E. F. (1959). Fronts in the vicinity of the tropopause. *Arch. Meteorol. Geophys. Bioklimatol., Ser. A* **11,** 1–17.

Reid, J. L., and Arthur, R. S. (1975). Interpretation of maps of geopotential anomaly for the deep Pacific Ocean. *J. Mar. Res.* **33,** Suppl., 37–52.

Reid, J. L., and Lynn, R. S. (1971). On the influence of the Norwegian–Greenland and Weddell Seas upon the bottom waters of the Indian and Pacific Oceans. *Deep-Sea Res.* **18,** 1063–1088.

Reid, J. L., and Mantyla, A. W. (1976). The effect of the geostrophic flow upon coastal sea elevations in the northern North Pacific Ocean. *J. Geophys. Res.* **81,** 3100–3110.

Reid, R. O., Vastano, A. C., Whitaker, R. E., and Wanstrath, J. J. (1977). Experiments in storm surge simulation. *In* "The Sea" (E. D. Goldberg *et al.,* eds.), Vol. 6, Chapter 5. Wiley (Interscience), New York.

Rex, D. F. (1950). Blocking actions in the middle troposphere and its effect upon regional climate. 1. An aerological study of blocking action. *Tellus* **2,** 196–211.

Rhines, P. B. (1970). Edge-, bottom-, and Rossby waves in a rotating stratified fluid. *Geophys. Fluid Dyn.* **1,** 273–302.

Rhines, P. B. (1977). The dynamics of unsteady currents. *In* "The Sea" (E. D. Goldberg *et al.,* eds.), Vol. 6, Chapter 7. Wiley (Interscience), New York.

Rhines, P. B. (1979). Geostrophic turbulence. *Annu. Rev. Fluid Mech.* **11,** 401–444.

Rhines, P. B., and Holland, W. R. (1979). A theoretical discussion of eddy-driven mean flows. *Dyn. Atmos. Oceans* **3,** 289–325.

Richardson, L. F. (1922). "Weather Prediction by Numerical Process." Cambridge Univ. Press, London and New York.

Richardson, P. L. (1982). Gulf Stream rings. *In* "Eddies in Marine Science" (A. R. Robinson, ed.), Chapter 3. Springer-Verlag, Berlin and New York.

Richardson, P. L., Cheney, R. E., and Worthington, L. V. (1978). A census of Gulf Stream rings, Spring 1975. *J. Geophys. Res.* **83,** 6136–6144.

Richardson, P. L., Maillard, C., and Sanford, T. B. (1979). The physical structure and life history of cyclonic Gulf Stream ring Allen. *J. Geophys. Res.* **84,** 7727–7741.

Richardson, W. S., Schmitz, W. S., and Niiler, P. (1969). The velocity structure of the Florida Current from the Florida Straits to Cape Fear. *Deep-Sea Res.* **16,** Suppl., 225–231.

Richman, J. G., Wunsch, C., and Hogg, N. G. (1977). Space and time scales of mesoscale motion in the sea. *Rev. Geophys. Space Phys.* **15,** 385–420.

Riehl, H. (1979). "Climate and Weather in the Tropics." Academic Press, New York.

Ripa, P. (1981). On the theory of non-linear wave-wave interactions among geophysical waves. *J. Fluid Mech.* **103**, 87–115.

Roberts, J. (1975). "Internal Gravity Waves in the Ocean." Dekker, New York.

Robinson, A. R. (1964). Continental shelf waves and the response of sea level to weather systems. *J. Geophys. Res.* **69**, 367–368.

Robinson, A. R., ed. (1982). "Eddies in Marine Science." Springer-Verlag, Berlin and New York.

Robinson, A. R., Tomasin, A., and Artegiani, A. (1973). Flooding of Venice: Phenomenology and prediction of the Adriatic storm surge. *Q. J. R. Meteorol. Soc.* **99**, 688–692.

Robinson, A. R., Harrison, D. E., Mintz, Y., and Semtner, A. J. (1977). Eddies and the general circulation of an idealised oceanic gyre: A wind and thermally-driven primitive equation numerical experiment. *J. Phys. Oceanogr.* **7**, 182–207.

Roden, G. I. (1961). On the wind-driven circulation in the Gulf of Tehuantepec and its effects upon sea surface temperature. *Geofis. Int.* **1**(3), 55–76.

Roden, G. I. (1975). On North Pacific temperature salinity, sound velocity, and density fronts and their relation to the wind and energy flux fields. *J. Phys. Oceanogr.* **5**, 557–571.

Roden, G. I., and Paskausky, D. F. (1978). Estimation of rates of frontogenesis and frontolysis in the North Pacific Ocean using satellite and surface meteorological data from January 1977. *J. Geophys. Res.* **88**, 4545–4550.

Rossby, C. G. (1936). Dynamics of steady ocean currents in the light of experimental fluid mechanics. *Pap. Phys. Oceanogr. Met.* **5**, No. 1, 1–43.

Rossby, C. G. (1937). On the mutual adjustment of pressure and velocity distributions in certain simple current systems. I. *J. Mar. Res.* **1**, 15–28.

Rossby, C. G. (1938a). On the mutual adjustment of pressure and velocity distributions in certain simple current systems. II. *J. Mar. Res.* **2**, 239–263.

Rossby, C. G. (1938b). On temperature changes in the stratosphere resulting from shrinking and stretching. *Beitr. Phys. Freien Atmos.* **24**, 53–59.

Rossby, C. G. (1940). Planetary flow patterns in the atmosphere. *Q. J. R. Meteorol. Soc.* **66**, Suppl., 68–97.

Rossby, C. G., *et al.* (1939). Relation between variations in the intensity of the zonal circulation of the atmosphere and the displacements of the semi-permanent centers of action. *J. Mar. Res.* **2**, 38–55.

Rossiter, J. R. (1954). The North Sea surge of 31 January and 1 February 1953. *Philos. Trans. R. Soc. London, Ser. A* **246**, 371–400.

Rossiter, J. R. (1959). Results on methods of forecasting storm surges on the east and south coasts of Great Britain. *Q. J. R. Meteorol. Soc.* **85**, 262–277.

Rowntree, P. R. (1972). The influence of tropical east Pacific Ocean temperatures on the atmosphere. *Q. J. R. Meteorol. Soc.* **98**, 290–321.

Rowntree, P. R. (1979). The effects of changes in ocean temperature on the atmosphere. *Dyn. Atmos. Oceans* **3**, 373–390.

Rumford, B., Count (1800). Essay VII. The propagation of heat in fluids. *In* "Essays, Political, Economical and Philosophical, A New Edition, 2" (T. Cadell and W. Davies, eds.), pp. 197–386. London. [Also *in* "Collected Works" (S. C. Brown, ed.), Vol. 1, pp. 117–285. Harvard Univ. Press, Cambridge, Massachusetts].

Russell, J. Scott (1844). Report on waves. *Br. Assoc. Rep.* **13**, 311–390.

Ryther, J. H. (1969). Photosynthesis and fish production in the sea. *Science* **166**, 72–76.

Salby, M. L. (1979). On the solution of the homogeneous vertical structure problem for long period oscillations. *J. Atmos. Sci.* **36**, 2350–2359.

Salby, M. L. (1980). The influence of realistic dissipation on planetary normal mode structures. *J. Atmos. Sci.* **37**, 2186–2199.

Sanders, F. (1955). An investigation of the structure and dynamics of an intense surface frontal zone. *J. Meteorol.* **12**, 542–552.

Sanford, T. B. (1975). Observations of the vertical structure of internal waves. *J. Geophys. Res.* **80**, 3861–3871.

Sarkisyan, A. S. (1977). The diagnostic calculations of a large-scale ocean circulation. *In* "The Sea" (E. D. Goldberg *et al.*, eds.), Vol. 6. Wiley (Interscience), New York.

Saunders, P. M. (1973). The instability of a baroclinic vortex. *J. Phys. Oceanogr.* **3**, 61–65.

Saunders, P. M. (1977). Average drag in an oscillatory flow. *Deep-Sea Res.* **24**, 381–384.

Sawyer, J. S. (1956). The vertical circulation at meteorological fronts and its relation to frontogenesis. *Proc. R. Soc. London, Ser A.* **234**, 346–362.

Sawyer, J. S. (1959). The introduction of the effects of topography into methods of numerical forecasting. *Q. J. R. Meteorol. Soc.* **85**, 31–43.

Sawyer, J. S. (1961). Quasi-periodic wind variations with height in the lower stratosphere. *Q. J. R. Meteorol. Soc.* **87**, 24–33.

Schmitz, W. J., and Holland, W. R. (1982). A preliminary comparison of selected numerical eddy-resolving general circulation experiments with observations. *J. Mar. Res.* **40**, 75–117.

Schneider, E. K. (1977). Axially symmetric steady-state models of the basic state for instability and climate studies II. Nonlinear circulations. *J. Atmos. Sci.* **34**, 280–296.

Schneider, E. K., and Lindzen, R. S. (1977). Axially symmetric steady-state models of the basic state for instability and climate studies. Part I. Linearised calculations. *J. Atmos. Sci.* **34**, 263–279.

Schoeberl, M. R. (1978). Stratospheric warming: Observation and theory. *Rev. Geophys. Space Phys.* **16**, 521–538.

Schopf, P., Anderson, D. L. T., and Smith, R. (1981). Beta-dispersion of low frequency Rossby waves. *Dyn. Atmos. Oceans* **5**, 187–214.

Schutz, C., and Gates, W. L. (1971). "Global Climate Data for Surface, 800mb, 400mb: January," R-915-ARPA. Rand Corporation, Santa Monica, California.

Schutz, C., and Gates, W. L. (1972). "Global Climate Data for Surface, 800mb, 400mb: July," R-1029-ARPA. Rand Corporation, Santa Monica, California.

Scorer, R. S. (1949). Theory of waves in the lee of mountains. *Q. J. R. Meteorol. Soc.* **75**, 41–56.

Scorer, R. S. (1955). Theory of air flow over mountains. IV. Separation of flow from the surface. *Q. J. R. Meteorol. Soc.* **81**, 340–350.

Scorer, R. S. (1972). "Clouds of the World." Lothian, Melbourne.

Shapiro, M. A. (1974). A multiple-structured frontal zone-jet stream as revealed by meteorologically instrumented aircraft. *Mon. Weather Rev.* **102**, 244–253.

Shapiro, M. A. (1981). Frontogenesis and geostrophically forced secondary circulations in the vicinity of jet stream-frontal zone systems. *J. Atmos. Sci.* **38**, 954–973.

Shaw, N. (1908). Barometric gradient and wind force. Preface to Gold (1908). [Reprinted in selected meteorological papers of Sir Napier Shaw FRS. McDonald, London, 1955.]

Shaw, N. (1916). "Gradient Wind in Meteorological Glossary." Meteorol. Off., HM Stationery Office, London.

Shea, D. J., and Gray, W. M. (1973). The hurricane's inner core region. I. Symmetric and asymmetric structure. *J. Atmos. Sci.* **30**, 1544–1564.

Silberstein, L. (1896). Über die Enstehung von Wirbelbewegungen in ein reibunglosen Flüssigkeit. *C. R. Acad. Sci. Cracovie.* 280–290.

Simmons, A. J. (1982). The forcing of stationary wave motion by tropical diabatic heating. *Q. J. R. Meteorol. Soc.* **108**, 503–534.

Simmons, A. J., and Hoskins, B. J. (1978). The life cycles of some non-linear baroclinic waves. *J. Atmos. Sci.* **35**, 414–432.

Simmons, A. J., and Hoskins, B. J. (1980). Barotropic influences on the growth and decay of non-linear baroclinic waves. *J. Atmos. Sci.* **37**, 1679–1684.

Simmons, W. F. (1969). A variational method for weak resonant wave interaction. *Proc. R. Soc. London, Ser. A* **309**, 551–575.

Simons, T. J. (1978). Generation and propagation of downwelling fronts. *J. Phys. Oceanogr.* **8**, 571–581.

Simons, T. J. (1979). On the joint effect of baroclinicity and topography. *J. Phys. Oceanogr.* **9**, 1283–1287.

Simons, T. J. (1980). Circulation models of lakes and inland seas. *Can. Bull. Fish. Aquat. Sci.* **203**, 1–146.

Simpson, J. H. (1981). Shelf sea fronts, implications of their existence and behaviour. *Philos. Trans. R. Soc. London, Ser. A* **302**, 531–546.

Simpson, J. H., Allen, C. M., and Morris, N. C. G. (1978). Fronts on the continental shelf. *J. Geophys. Res.* **83**, 4607–4614.

Smagorinsky, J. (1953). The dynamical influence of large-scale heat sources and sinks on the quasi-stationary mean motions of the atmosphere. *Q. J. R. Meteorol. Soc.* **79**, 342–366.

Smith, P. C. (1976). Baroclinic instability in the Denmark Strait overflow. *J. Phys. Oceanogr.* **6**, 355–371.

Smith, R. B. (1979). The influence of mountains on the atmosphere. *Adv. Geophys.* **21**, 87–230.

Smith, R. B. (1980). Linear theory of stratified hydrostatic flow past an isolated mountain. *Tellus* **32**, 348–364.

Smith, R. L. (1968). Upwelling. *Oceanogr. Mar. Biol. Ann. Rev.* **6**, 11–46.

Smith, R. L. (1978). Poleward propagating disturbances in currents and sea-levels along the Peru coast. *J. Geophys. Res.* **83**, 6083–6092.

Smith, S. D. (1980). Wind stress and heat flux over the ocean in gale force winds. *J. Phys. Oceanogr.* **10**, 709–726.

Snodgrass, F. E., Groves, G. W., Hasselmann, K., Miller, G. R., Munk, W. H., and Powers, W. H. (1966). Propagation of ocean swell across the Pacific. *Philos. Trans. R. Soc. London, Ser. A* **259**, 431–497.

Spiegel, E. A., and Veronis, G. (1960). On the Boussinesq approximation for a compressible fluid. *Astrophys. J.* **131**, 442–447.

Spiers, H. B., and Spiers, A. G. H. (1937). "Collection of Pascal's Works." Columbia Univ. Press, New York.

Stephens, G. K., Campbell, G. G., and Vonder Haar, T. H. (1981). Earth radiation budgets. *J. Geophys. Res.* **86**, 9739–9760.

Stern, M. E. (1968). T-S gradients on the micro-scale. *Deep-Sea Res.* **15**, 245–250.

Stern, M. E. (1975). "Ocean Circulation Physics." Academic Press, New York.

Sterneck, R. V. (1919). Die Gezeitenerscheinungen in der Adria. II Teil. Die theoretische Erklärung der Beobachtungstatsachen. *Denkschr. Akad. Wiss. Wien* **96**, 277–324.

Stewartson, K. (1978). The evolution of the critical layer of a Rossby wave. *Geophys. Astrophys. Fluid Dyn.* **9**, 185–200.

Stilke, G. (1973). Occurrence and features of ducted modes of internal gravity waves over western Europe and their influence on microwave propagation. *Boundary-Layer Meteorol.* **4**, 493–509.

Stokes, G. G. (1847). On the theory of oscillatory waves. *Trans. Cambridge Philos. Soc.* **8**, 441–455.

Stokes, G. G. (1876). Smith's Prize examination paper for Feb. 2, 1876, question 11. *Math. Phys. Pap.* **5**, 362 (1905). [See also "Memoirs and Scientific Correspondence of Sir George Gabriel Stokes," Vol. 2, p. 146. Cambridge Univ. Press, London and New York, 1907.]

Stoker, J. J. (1957). "Water Waves." Wiley (Interscience), New York.

Stommel, H. (1948). The westward intensification of wind-driven ocean currents. *Trans. Am. Geophys. Union* **99**, 202–206.

Stommel, H. (1957). A survey of ocean current theory. *Deep-Sea Res.* **4**, 149–184.

Stommel, H. (1962). An analogy to the Antarctic circumpolar current. *J. Mar. Res.* **20**, 92–96.

Stommel, H. (1965). "The Gulf Stream: A Physical and Dynamical Description," 2nd ed. Univ. of California Press, Berkeley.

Stommel, H. (1980). Asymmetry of interoceanic fresh-water and heat fluxes. *Proc. Natl. Acad. Sci. U.S.A.* **77**, 2377–2381.

Stommel, H., and Schott, F. (1977). The beta spiral and the determination of the absolute velocity field from hydrographic station data. *Deep-Sea Res.* **24**, 325–329.

Stommel, H., and Yoshida, K., eds. (1972). "Kuroshio, Physical Aspects of the Japan Current." Univ. of Washington Press, Seattle.

Stommel, H., Niiler, P., and Anati, D. (1978). Dynamic topography and recirculation of the North Atlantic. *J. Mar. Res.* **36**, 449–468.

Stone, P. H. (1972). A simplified radiative-dynamical model for the static stability of rotating atmospheres. *J. Atmos. Sci.* **29**, 405–418.

Suginohara, N. (1974). Onset of coastal upwelling in a two-layer ocean by wind stress with longshore variation. *J. Oceanogr. Soc. Jpn.* **30**, 23–33.

Suginohara, N. (1977). Upwelling front and two-cell circulation. *J. Oceanogr. Soc. Jpn.* **33**, 115–130.

Suginohara, N. (1981). Propagation of coastal trapped waves at low latitudes in a stratified ocean with continental shelf slope. *J. Phys. Oceanogr.* **11**, 1113–1122.

Sutcliffe, R. C. (1947). A contribution to the problem of development. *Q. J. R. Meteorol. Soc.* **73**, 370–383.

Sutcliffe, R. C., and Godart, O. (1942). "Isobaric Analysis," S.D.T.M. No. 50. Meteorol. Off., London.

Sverdrup, H. U. (1947). Wind-driven currents in a baroclinic ocean, with application to the equatorial currents of the eastern Pacific. *Proc. Natl. Acad. Sci. U.S.A.* **33**, 318–326.

Sverdrup, H. U., Johnson, M. W., and Fleming, R. H. (1942). "The Oceans: Their Physics, Chemistry and General Biology." Prentice-Hall, Englewood Cliffs, New Jersey.

Swallow, J. C., and J. G. Bruce, J. G. (1966). Current measurements off the Somali coast during the southwest monsoon of 1964. *Deep-Sea Res.* **13**, 861–888.

Symons, G. (ed.) (1888). "The Eruption of Krakatoa and Subsequent Phenomena." Trubner, London.

Taljaard, J. S., Van Loon, H., Crutcher, H. L., and Jenne, R. L. (1969). "Climate of the Upper Air: Southern Hemisphere," Vol. 1, NAVAIR 50-1C-55. Naval Weather Service Command, Washington, D.C.

Taylor, G. I. (1914). Eddy motion in the atmosphere. *Philos. Trans. R. Soc. London, Ser. A* **215**, 1–26.

Taylor, G. I. (1921). Tidal oscillations in gulfs and rectangular basins. *Proc. London Math. Soc.* **20**, 148–181. [Also in "Scientific Papers," Vol. 2, pp. 144–171. Cambridge Univ. Press, London and New York, 1960.]

Taylor, G. I. (1936). The oscillations of the atmosphere. *Proc. R. Soc. London, Ser. A* **156**, 318–326. [Also in "Scientific Papers," Vol. 2, pp. 365–371. Cambridge Univ. Press, London and New York, 1960.]

Taylor, H. W., Gordon, A. L., and Molinelli, E. (1978). Climatic characteristics of the Antarctic Polar Front zone. *J. Geophys. Res.* **83**, 4572–4578.

Taylor, P. A., and Gent, P. R. (1980). Modification of the boundary layer by orography. *In* "Orographic Effects in Planetary Flows," Chapter 5, pp. 145–165. GARP Publ. Ser. No. 23, WMO-ICSU, Geneva.

Thompson, R. (1971). Topographic Rossby waves at a site north of the Gulf Stream. *Deep-Sea Res.* **18**, 1–19.

Thompson, R. O. R. Y. (1978). Observation of inertial waves in the stratosphere. *Q. J. R. Meteorol. Soc.* **104**, 691–698.

Thompson, S. L., and Hartmann, D. L. (1979). "Cumulus friction": estimated influence on the tropical mean meridional circulation. *J. Atmos. Sci.* **36**, 2022–2026.

Thompson, W. (Lord Kelvin) (1869). On vortex motion. *Trans. R. Soc. Edinburgh* **25**, 217–260. [Also in *Math. Phys. Pap.* **4**, 13–66 (1910).]

Thomson, W. (Lord Kelvin) (1879). On gravitational oscillations of rotating water. *Proc. Roy. Soc. Edinburgh* **10**, 92–100. [Reprinted in *Philos. Mag.* **10**, 109–116 (1880); *Math. Phys. Pap.* **4**, 141–148 (1910).]

Thomson, W. (Lord Kelvin) (1891). "Popular Lectures," Vol. 3. Macmillan, London.

Thorpe, S. A. (1966). On wave interactions in a stratified fluid. *J. Fluid Mech.* **24**, 737–751.

Thorpe, S. A. (1973). Experiments on instability and turbulence in a stratified shear flow. *J. Fluid Mech.* **32**, 693–704.

Thorpe, S. A. (1975). The excitation, dissipation and interaction of internal waves in the deep ocean. *J. Geophys. Res.* **80**, 328–338.

Thorpe, S. A. (1977). Billows in Loch Ness. *Deep-Sea Res.* **24**, 371–379.

Thorpe, S. A. (1981). An experimental study of critical layers. *J. Fluid Mech.* **103**, 321–344.

Thorpe, A. J., and Guymer, T. H. (1977). The nocturnal jet. *Q. J. R. Meterol. Soc.* **103**, 633–653.

Thoulet, J. (1894). Contribution à l'étude des lacs des Vosges. *Bull. Soc. Géogr. Paris* **15**, 557–604.

Tolstoy, I. (1973). "Wave Propagation." McGraw-Hill, New York.

Tracy, C. (1843). On the rotary action of storms. *Am. J. Sci. Arts* **45**, 65–72 (also in Abbe, 1910, pp. 16–22).

Trenberth, K. E. (1981). Observed southern hemisphere eddy statistics at 500mb: Frequency and spatial dependence. *J. Atmos. Sci.* **38**, 2585–2605.

Truesdell, C. (1954a). "Kinematics of Vorticity," Indiana Univ. Sci. Ser. No. 19. Indiana University, Bloomington.

Truesdell, C. A. (1954b). "Rational Fluid Mechanics 1687–1765," Editor's introduction to Euleri Opera Omnia, Ser. II, Vol. 12. Orell Füssli Turici, Lausanne.

Tsuchiya, M. (1975). Subsurface countercurrents in the eastern equatorial Pacific Ocean. *J. Mar. Res.* **33**, Suppl., 145–175.

Tung, K. K. (1979). A theory of stationary long waves. Part III. Quasi-normal modes in a singular waveguide. *Mon. Weather Rev.* **107**, 751–774.

Tung, K. K., and Lindzen, R. S. (1979). A theory of stationary long waves. Part I. A simple theory of blocking. Part II. Resonant Rossby waves in the presence of realistic vertical shears. *Mon. Weather Rev.* **107**, 714–734, 735–750.

Turner, J. S. (1973). "Buoyancy Effects in Fluids." Cambridge Univ. Press, London and New York.

Turner, J. S. (1981). Small-scale mixing processes. *In* "Evolution of Physical Oceanography" (B. A. Warren and C. Wunsch, eds.), Chapter 8. MIT Press, Cambridge, Massachusetts.

UNESCO (1981). Tenth report of the joint panel on oceanographic tables and standards. UNESCO Technical Papers in Marine Sci. No. 36. UNESCO, Paris.

U.S. Department of Commerce (1963). "U.S. Navy Marine Climatic Atlas of the World." Arctic Ocean. National Climate Center, Washington, D.C.

U.S. Department of Commerce (1965). "U.S. Navy Marine Climatic Atlas of the World," Antarctic. National Climate Center, Washington, D.C.

U.S. Department of Commerce (1974). "U.S. Navy Marine Climatic Atlas of the World," North Atlantic Ocean. National Climate Center, Washington, D.C. **NAVAIR 50-IC-528.**

U.S. Department of Commerce (1976). "U.S. Navy Marine Climatic Atlas of the World," Indian Ocean. National Climate Center, Washington, D.C. **NAVAIR 50-IC-530.**

U.S. Department of Commerce (1977). "U.S. Navy Marine Climatic Atlas of the World," North Pacific Ocean. National Climate Center, Washington, D.C. **NAVAIR 50-IC-529.**

U.S. Department of Commerce (1978). "U.S. Navy Marine Climatic Atlas of the World," South Atlantic Ocean. National Climate Center, Washington, D.C. **NAVAIR 50-IC-531.**

U.S. Department of Commerce (1979). "U.S. Navy Marine Climatic Atlas of the World," South Pacific Ocean. National Climate Center, Washington, D.C. **NAVAIR 50-IC-532.**

Väisälä, V. (1925). Über die Wirkung der Windschwankungen auf die Pilotbeobachtungen. *Soc. Sci. Fenn. Commentat. Phys.-Math.* **2** (19), 19–37.

van Loon, H. (1980). Transfer of sensible heat by transient eddies in the atmosphere on the southern hemisphere: an appraisal of data before and during FGGE. *Mon. Weather Rev.* **108**, 1774–1781.

van Loon, H., Jenne, R. L., and Labitzke, K. (1973). Zonal harmonic standing waves. *J. Geophys. Res.* **78**, 4463–4471.

Van Mieghem, J. (1956). The energy available in the atmosphere for conversion into kinetic energy. *Beitr. Phys. Atmos.* **29**, 129–142.

Van Mieghem, J. (1957). Energies potentielle et interne convertibles en énergies cinétique dans l'atmosphère. *Beitr. Phys. Atmos.* **30**, 5–17.

Venkateswaran, S. V., and Sundararaman, N. (1980). "The Middle Atmosphere." Birkhaeuser, Basel.

Veronis, G. (1956). Partition of energy between geostrophic and nongeostrophic oceanic motions. *Deep-Sea Res.* **3**, 157–177.

Veronis, G. (1969). On theoretical models of the thermocline circulation. *Deep-Sea Res.* **16**, Suppl., 301–323.

Veronis, G. (1981). Dynamics of large-scale ocean circulation. *In* "Evolution of Physical Oceanography" (B. A. Warren and C. Wunsch, eds.), Chapter 6. MIT Press, Cambridge, Massachusetts.

Vettin, F. (1857). Ueber den aufsteigenden Luftstrom, die Entstehung des Hagels und der Wirbel-Stürme. *Ann. Phys. (Leipzig)* **102**, 246–255.

Vonder Haar, T. H., and Oort, A. H. (1973). New estimate of annual poleward energy transport by northern hemisphere oceans. *J. Phys. Oceanogr.* **3**, 169–172.

Wadhams, P., Gill, A. E., and Linden, P. F. (1979). Transects by submarine of the East Greenland Polar Front. *Deep-Sea Res.* **269**, 1311–1327.

Walin, G. (1972a). On the hydrographic response to transient meteorological disturbances. *Tellus* **24**, 1–18.

Walin, G. (1972b). Some observations of temperature fluctuations in the coastal region of the Baltic. *Tellus* **24**, 187–198.

Walker, G. T. (1924). Correlation in seasonal variations of weather. IX. A further study of world weather. *Mem. Indian Meteorol. Dep.* **24**, 275–332.

Walker, G. T. (1928). World weather. *Q. J. R. Meteorol. Soc.* **54**, 79–87.

Wallace, J. M. (1971). Spectral studies of tropospheric wave disturbances in the tropical Western Pacific. *Rev. Geophys. Space Phys.* **9**, 557–612.

Wallace, J. M. (1973). General circulation of the tropical lower stratosphere. *Rev. Geophys. Space Phys.* **11**, 191–222.

Wallace, J. M., and Gutzler, D. S. (1981). Teleconnections in the geopotential height field during the Northern Hemisphere winter. *Mon. Weather Rev.* **109**, 784–812.

Wallace, J. M., and Hobbs, P. V. (1977). "Atmospheric Science: An Introductory Survey." Academic Press, New York.

Wallace, J. M., and Kousky, V. E. (1968). Observational evidence of Kelvin waves in the tropical stratosphere. *J. Atmos. Sci.* **25,** 900–907.

Wallace, W. J. (1974). "The Development of the Chlorinity/Salinity Concept in Oceanography." Elsevier, Amsterdam.

Walsh, J. J. (1977). A biological sketchbook for an eastern boundary current. *In* "The Sea" (E. D. Goldberg, I. N. McCave, J. J. O'Brien, and J. H. Steele, eds.), Vol. 6, pp. 923–968. Wiley (Interscience), New York.

Wang, D.-P., and Mooers, C. N. K. (1976). Coastal-trapped waves in a continuously stratified ocean. *J. Phys. Oceanogr.* **6,** 853–863.

Warn, T., and Warn, J. (1978). The evolution of a nonlinear Rossby wave critical level. *Stud. Appl. Math.* **59,** 37–71.

Warren, B. A. (1981). Deep-circulation of the world ocean. *In* "Evolution of Physical Oceanography" (B. A. Warren and C. Wunsch, eds.), Chapter 1. MIT Press, Cambridge, Massachusetts.

Warren, B. A., and Wunsch, C., eds. (1981). "Evolution of Physical Oceanography." MIT Press, Cambridge, Massachusetts.

Watson, E. R. (1903). Internal oscillation in the waters of Loch Ness. *Nature (London)* **69,** 174.

Watson, E. R. (1904). Movements of Loch Ness as observed by temperature observations. *Geog. J.* **24,** 430–437.

Watson, J. (1839). On the helm wind of Crossfell. *Br. Assoc. Rep.* **7,** Sec. II, 33–34. [reprinted in *Q. J. Meteorol. Soc.* **10,** 270–271 (1884).]

Wearn, R. B., and Baker, D. J. (1980). Bottom pressure measurements across the Antarctic Circumpolar Current and their relation to the wind. *Deep-Sea Res.* **27A,** 875–888.

Weast, R. C., ed. (1971–1972). "Handbook of Physics and Chemistry." 52nd ed. Chem. Rubber Publ. Co., Cleveland, Ohio.

Weatherly, G. L. (1972). A study of the bottom boundary layer of the Florida current. *J. Phys. Oceanogr.* **2,** 54–72.

Weatherly, G. L., and Martin, P. J. (1978). On the structure and dynamics of the ocean bottom boundary layer. *J. Phys. Oceanogr.* **8,** 557–570.

Webster, F. (1965). Measurement of eddy fluxes of momentum in the surface layer of the Gulf Stream. *Tellus* **17,** 239–245.

Webster, P. J. (1972). Response of the tropical atmosphere to local steady forcing. *Mon. Weather Rev.* **100,** 518–541.

Weisberg, R. H., Horigan, A., and Colin, C. (1979a). Equatorially trapped Rossby-gravity wave propagation in the Gulf of Guinea. *J. Mar. Res.* **37,** 67–86.

Weisberg, R. H., Miller, L., Horigan, A., and Knauss, J. A. (1979b). Velocity observations in the equatorial thermocline during GATE. *Deep-Sea Res.* **26,** Suppl. 2, 217–242.

Welander, P. (1961). Numerical prediction of storm surges. *Adv. Geophys.* **8,** 315–379.

Welander, P. (1971). Some exact solutions to the equations describing an ideal fluid thermocline. *J. Mar. Res.* **29,** 60–68.

Welander, P. (1981). Mixed layer and fronts in ocean circulation models. *J. Phys. Oceanogr.* **11,** 148–152.

Wetherald, R. T., and Manabe, S. (1972). Response of the joint ocean-atmosphere model to the seasonal variation of the solar radiation. *Mon. Weather Rev.* **100,** 42–59.

Wetherald, R. T., and Manabe, S. (1981). Influence of seasonal variations upon the sensitivity of a model climate. *J. Geophys. Res.* **86,** 1194–1204.

Whipple, F. J. W. (1930). The great Siberian meteor, and the waves, seismic and aerial, which it produced. *Q. J. R. Meteorol. Soc.* **56,** 287–298.

White, G. H. (1982). An observational study of the northern hemisphere extratropical summertime general circulation. *J. Atmos. Sci.* **39,** 24–40.

White, W. B., and Saur, T. (1981). The source of annual baroclinic waves in the eastern subtropical North Pacific. *J. Phys. Oceanogr.* **11,** 1452–1462.

Whitehead, J. A. (1981). Laboratory models of circulation in shallow seas. *Philos. Trans. R. Soc. London, Ser. A* **302,** 583–595.

Whitehead, J. A., Leetmaa, A., and Knox, R. A. (1974). Rotating hydraulics of strait and sill flows. *Geophys. Fluid Dyn.* **6,** 101–125.

Whitham, G. B. (1965). A general approach to linear and non-linear dispersive waves using a Lagrangian. *J. Fluid Mech.* **22,** 273–283.

Whitham, G. B. (1974). "Linear and Nonlinear Waves." Wiley (Interscience), New York.

Wiin-Nielsen, A. (1979). On normal mode linear initialization on the sphere. *J. Atmos. Sci.* **36,** 2040–2048.

Willebrand, J., Philander, S. G. H., and Pacanowski, R. C. (1980). The oceanic response to large-scale atmospheric disturbances. *J. Phys. Oceanogr.* **10,** 411–429.

Williams, G. P. (1971). Baroclinic annulus waves. *J. Fluid Mech.* **49,** 417–449.

Williams, G. P. (1974). Generalized Eady waves. *J. Fluid Mech.* **62,** 643–655.

Williams, G. P., and Robinson, J. B. (1974). Generalized Eady waves and Ekman pumping. *J. Atmos. Sci.* **31,** 1768–1776.

Williams, R. T. (1974). Numerical simulation of steady-state fronts. *J. Atmos. Sci.* **31,** 1286–1296.

Willson, R. C. (1984). Measurement of solar total irradiance and its variability. *Space Sci. Rev.* **38,** 203–242.

Wimbush, M., and Munk, W. H. (1970). The benthic boundary layer. *In* "The Sea" (A. Maxwell, ed.), Vol. 4, pp. 731–758. Wiley (Interscience), New York.

Winston, J. S., Gruber, A., Gray, T. I., Varnadore, M. S., Earnest, C. L., and Mannello, L. P. (1979). "Earth-atmosphere Radiation Budget Analyses Derived from NOAA Satellite Data June 1974–February 1978," Vols. 1 and 2,. U.S. Dept. of Commerce, Washington, D.C.

Woods, J. D., ed. (1982). "Turbulence in the Ocean." Springer-Verlag, Berlin and New York.

Woods, J. D., Wiley, R. L., and Briscoe, M. G. (1977). Vertical circulation at fronts in the upper ocean. *In* "A Voyage of Discovery" (M. Argel, ed.) [*Deep-Sea Res. Suppl.* **24**], pp. 253–275. Pergamon, Oxford.

Woods, J. D., Wiley, R. L., and Briscoe, M. G. (1977). Vertical circulation at fronts in the upper ocean. *Deep-Sea Res.* **24,** Suppl. 253–275.

Wooster, W. S., and Reid, J. L. (1963). Eastern boundary currents. *In* "The Sea" (M. N. Hill, ed.), Vol. 2, pp. 253–280. Wiley (Interscience), New York.

Wooster, W. S., Bakun, A., and McLain, D. R. (1976). The seasonal upwelling cycle along the eastern boundary of the North Atlantic. *J. Mar. Res.* **34,** 131–141.

Worthington, L. V. (1970). The Norwegian Sea as a Mediterranean basin. *Deep-Sea Res.* **17,** 17–84.

Worthington, L. V. (1976). "On the North Atlantic Circulation." Johns Hopkins Univ. Press, Baltimore, Maryland.

Wright, D. G. (1981). Baroclinic instability in Drake Passage. *J. Phys. Oceanogr.* **11,** 231–246.

Wright, P. B. (1977). The Southern Oscillation-patterns and mechanisms of the teleconnections and the persistence. *Hawaii Inst. Geophys.* [*Rep.*] **HIG-77-13.**

Wu, J. (1980). Wind-stress coefficients over sea surface near neutral conditions—a revisit. *J. Phys. Oceanogr.* **10,** 727–740.

Wunsch, C. (1977). Response of an equatorial ocean to a periodic monsoon. *J. Phys. Oceanogr.* **7,** 497–511.

Wunsch, C. (1980). Meridional heat flux of the North Atlantic Ocean. *Proc. Natl. Acad. Sci. U.S.A.* **77,** 5043–5047.

Wunsch, C. (1981). Low frequency variability of the sea. *In* "Evolution of Physical Oceanography" (B. A. Warren and C. Wunsch, eds.), Chapter 11. MIT Press, Cambridge, Massachusetts.

Wunsch, C., and Gill, A. E. (1976). Observations of equatorially trapped waves in Pacific sea level variations. *Deep-Sea Res.* **23,** 371–390.

Wyrtki, K. (1965). The average annual heat balance of the North Pacific Ocean and its relation to ocean circulation. *J. Geophys. Res.* **70,** 4547–4559.

Wyrtki, K. (1973). An equatorial jet in the Indian Ocean. *Science* **181,** 262–264.

Wyrtki, K. (1975). Fluctuations of the dynamic topography in the Pacific Ocean. *J. Phys. Oceanogr.* **5,** 450–459.

Wyrtki, K. (1977). Sea level during the 1972 El Niño. *J. Phys. Oceanogr.* **7,** 779–787.

Wyrtki, K. (1979a). Sea-level variations: Monitoring the breath of the Pacific. *Trans. Am. Geophys. Union* **60,** 25–27.

Wyrtki, K. (1979b). The response of the sea-surface topography to the 1976 El Niño. *J. Phys. Oceanogr.* **9**, 1223–1231.

Wyrtki, K., and Meyers, G. (1975). The trade wind field over the Pacific Ocean. *Hawaii Inst. Geophys.* [*Rep.*] **HIG-75-1.**

Wyrtki, K., and Meyers, G. (1976). The trade wind field over the Pacific Ocean. *J. Appl. Meteorol.* **15**, 698–704.

Yamada, T. (1976). On the similarity functions A, B and C of the planetary boundary layer. *J. Atmos. Sci.* **33**, 781–793.

Yamaguchi, M. (1975). Sea-level fluctuations and mass mortalities of reef animals in Guam, Mariana Islands. *Micronesica* **11**, 227–243.

Yanai, M., and Maruyama, T. (1966). Stratospheric wave disturbance propagating over the equatorial Pacific. *J. Meteorol. Soc. Jpn.* **44**, 291–294.

Yanowitch, M. (1967). Effect of viscosity on gravity waves and the upper boundary condition. *J. Fluid Mech.* **29**, 209–231.

Yih, C. S. (1980). "Stratified Flows." Academic Press, New York.

Yoon, J.-H., and Suginohara, N. (1977). Behaviour of warm water flowing into a cold ocean. *J. Oceanogr. Soc. Jpn.* **33**, 272–282.

Yoshida, K. (1955). Coastal upwelling off the California coast. *Rec. Oceanogr. Works Jpn.* **2** (2), 1–13.

Yoshida, K. (1959). A theory of the Cromwell current (the equatorial undercurrent) and of the equatorial upwelling—an interpretation in a similarity to a coastal circulation. *J. Oceanogr. Soc. Jpn.* **15**, 159–170.

Yudin, M. I. (1955). Invariant quantities in large-scale atmospheric processes. (In Russian.) *Tr. Gl. Geofiz. Obs.* **55**, 1–12.

Zangvil, A., and Yanai, M. (1980). Upper tropospheric waves in the tropics. Part I. Dynamical analysis in the wavenumber–frequency domain. *J. Atmos. Sci.* **37**, 283–298.

Zeman, O. (1981). Progress in the modelling of planetary boundary layers. *Annu. Rev. Fluid Mech.* **13**, 253–272.

Zilitinkevich, S. S. (1970). "Dynamics of the Atmospheric Boundary Layer" Hydrometeorol. Press, Leningrad (in Russian).

Zilitinkevich, S. S. (1975). Resistance laws and prediction equations for the depth of the planetary boundary-layer. *J. Atmos. Sci.* **32**, 741–752.

Zimmerman, S. P. (1964). Small scale wind structure above 100km. *J. Geophys. Res.* **69**, 784–785.

Index